Undergraduate Texts in Mathematics

Undergraduate Texts in Mathematics

Undergraduate Texts in Mathematics are generally aimed at third- and fourth-year undergraduate mathematics students at North American universities. These texts strive to provide students and teachers with new perspectives and novel approaches. The books include motivation that guides the reader to an appreciation of interrelations among different aspects of the subject. They feature examples that illustrate key concepts as well as exercises that strengthen understanding.

More information about this series at http://www.springer.com/series/666

David A. Cox • John Little • Donal O'Shea

Ideals, Varieties, and Algorithms

An Introduction to Computational Algebraic Geometry and Commutative Algebra

Fourth Edition

 Springer

David A. Cox
Department of Mathematics
Amherst College
Amherst, MA, USA

John Little
Department of Mathematics
and Computer Science
College of the Holy Cross
Worcester, MA, USA

Donal O'Shea
President's Office
New College of Florida
Sarasota, FL, USA

ISSN 0172-6056   ISSN 2197-5604   (electronic)
Undergraduate Texts in Mathematics
ISBN 978-3-319-16720-6   ISBN 978-3-319-16721-3   (eBook)
DOI 10.1007/978-3-319-16721-3

Library of Congress Control Number: 2015934444

Mathematics Subject Classification (2010): 14-01, 13-01, 13Pxx

Printed on acid-free paper

Springer International Publishing AG Switzerland is part of Springer Science+Business Media (www.
springer.com)

To Elaine,
for her love and support.
D.A.C.

To the memory of my parents.
J.B.L.

To Mary and my children.
D.O'S.

Preface

We wrote this book to introduce undergraduates to some interesting ideas in algebraic geometry and commutative algebra. For a long time, these topics involved a lot of abstract mathematics and were only taught at the graduate level. Their computational aspects, dormant since the nineteenth century, re-emerged in the 1960s with Buchberger's work on algorithms for manipulating systems of polynomial equations. The development of computers fast enough to run these algorithms has made it possible to investigate complicated examples that would be impossible to do by hand, and has changed the practice of much research in algebraic geometry and commutative algebra. This has also enhanced the importance of the subject for computer scientists and engineers, who now regularly use these techniques in a whole range of problems.

It is our belief that the growing importance of these computational techniques warrants their introduction into the undergraduate (and graduate) mathematics curriculum. Many undergraduates enjoy the concrete, almost nineteenth century, flavor that a computational emphasis brings to the subject. At the same time, one can do some substantial mathematics, including the Hilbert Basis Theorem, Elimination Theory, and the Nullstellensatz.

Prerequisites

The mathematical prerequisites of the book are modest: students should have had a course in linear algebra and a course where they learned how to do proofs. Examples of the latter sort of course include discrete math and abstract algebra. It is important to note that abstract algebra is *not* a prerequisite. On the other hand, if all of the students have had abstract algebra, then certain parts of the course will go much more quickly.

The book assumes that the students will have access to a computer algebra system. Appendix C describes the features of Maple™, *Mathematica*®, Sage, and other computer algebra systems that are most relevant to the text. We do not assume any prior experience with computer science. However, many of the algorithms in the book are described in pseudocode, which may be unfamiliar to students with no background in programming. Appendix B contains a careful description of the pseudocode that we use in the text.

How to Use the Book

In writing the book, we tried to structure the material so that the book could be used in a variety of courses, and at a variety of different levels. For instance, the book could serve as a basis of a second course in undergraduate abstract algebra, but we think that it just as easily could provide a credible alternative to the first course. Although the book is aimed primarily at undergraduates, it could also be used in various graduate courses, with some supplements. In particular, beginning graduate courses in algebraic geometry or computational algebra may find the text useful. We hope, of course, that mathematicians and colleagues in other disciplines will enjoy reading the book as much as we enjoyed writing it.

The first four chapters form the core of the book. It should be possible to cover them in a 14-week semester, and there may be some time left over at the end to explore other parts of the text. The following chart explains the logical dependence of the chapters:

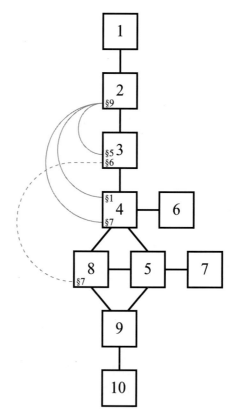

The table of contents describes what is covered in each chapter. As the chart indicates, there are a variety of ways to proceed after covering the first four chapters. The three solid arcs and one dashed arc in the chart correspond to special dependencies that will be explained below. Also, a two-semester course could be designed that covers the entire book. For instructors interested in having their students do an independent project, we have included a list of possible topics in Appendix D.

Features of the New Edition

This fourth edition incorporates several substantial changes. In some cases, topics have been reorganized and/or augmented using results of recent work. Here is a summary of the major changes to the original nine chapters of the book:

- Chapter 2: We now define *standard representations* (implicit in earlier editions) and *lcm representations* (new to this edition). Theorem 6 from §9 plays an important role in the book, as indicated by the solid arcs in the dependence chart on the previous page.
- Chapter 3: We now give two proofs of the Extension Theorem (Theorem 3 in §1). The resultant proof from earlier editions now appears in §6, and a new Gröbner basis proof inspired by SCHAUENBURG (2007) is presented in §5. This makes it possible for instructors to omit resultants entirely if they choose. However, resultants are used in the proof of Bezout's Theorem in Chapter 8, §7, as indicated by the dashed arc in the dependence chart.
- Chapter 4: There are several important changes:
    - In §1 we present a Gröbner basis proof of the Weak Nullstellensatz using ideas from GLEBSKY (2012).
    - In §4 we now cover saturations $I : J^\infty$ in addition to ideal quotients $I : J$.
    - In §7 we use Gröbner bases to prove the Closure Theorem (Theorem 3 in Chapter 3, §2) following SCHAUENBURG (2007).
- Chapter 5: We have added a new §6 on Noether normalization and relative finiteness. Unlike the previous topics, the proofs involved in this case are quite classical. But having this material to draw on provides another illuminating viewpoint in the study of the dimension of a variety in Chapter 9.
- Chapter 6: The discussion of the behavior of Gröbner bases under specialization in §3 has been supplemented by a brief presentation of the recently developed concept of a *Gröbner cover* from MONTES and WIBMER (2010). We would like to thank Antonio Montes for the Gröbner cover calculation reported in §3.

In the biggest single change, we have added a new Chapter 10 presenting some of the progress of the past 25 years in methods for computing Gröbner bases (i.e., since the improved Buchberger algorithm discussed in Chapter 2, §10). We present Traverso's Hilbert driven Buchberger algorithm for homogeneous ideals, Faugère's F_4 algorithm, and a brief introduction to the signature-based family of algorithms including Faugère's F_5. These new algorithmic approaches make use of several interesting ideas from previous chapters and lead the reader toward some of the next steps in commutative algebra (modules, syzygies, etc.). We chose to include this topic in part because it illustrates so clearly the close marriage between theory and practice in this part of mathematics.

Since software for the computations discussed in our text has also undergone major changes since 1992, Appendix C has been completely rewritten. We now discuss Maple, *Mathematica*, Sage, CoCoA, Macaulay2, and Singular in some detail and list several other systems that can be used in courses based on our text. Appendix D has also been substantially updated with new ideas for student projects. Some of the wide range of applications developed since our first edition can be seen from the new topics there. Finally, the bibliography has been updated and expanded to reflect some of the continuous and rapid development of our subjects.

Acknowledgments

When we began writing the first edition of *Ideals, Varieties, and Algorithms* in 1989, major funding was provided by the New England Consortium for Undergraduate Science Education (and its parent organization, the Pew Charitable Trusts). This project would have been impossible without their support. Various aspects of our work were also aided by grants from IBM and the Sloan Foundation, the Alexander von Humboldt Foundation, the Department of Education's FIPSE program, the Howard Hughes Foundation, and the National Science Foundation. We are grateful to all of these organizations for their help.

We also wish to thank colleagues and students at Amherst College, George Mason University, College of the Holy Cross, Massachusetts Institute of Technology, Mount Holyoke College, Smith College, and the University of Massachusetts who participated in courses based on early versions of the manuscript. Their feedback improved the book considerably.

We want to give special thanks to David Bayer and Monique Lejeune-Jalabert, whose thesis BAYER (1982) and notes LEJEUNE-JALABERT (1985) first acquainted us with this wonderful subject. We are also grateful to Bernd Sturmfels, whose book STURMFELS (2008) was the inspiration for Chapter 7, and Frances Kirwan, whose book KIRWAN (1992) convinced us to include Bezout's Theorem in Chapter 8. We would also like to thank Steven Kleiman, Michael Singer, and A. H. M. Levelt for important contributions to the second and third editions of the book.

We are extremely grateful to the many, many individuals (too numerous to list here) who reported typographical errors and gave us feedback on the earlier editions. Thank you all!

Corrections, comments, and suggestions for improvement are welcome!

Amherst, MA, USA   David Cox
Worcester, MA, USA   John Little
Sarasota, FL, USA   Donal O'Shea
January 2015

Contents

Preface vii

Notation for Sets and Functions xv

Chapter 1. Geometry, Algebra, and Algorithms 1
   §1. Polynomials and Affine Space 1
   §2. Affine Varieties 5
   §3. Parametrizations of Affine Varieties 14
   §4. Ideals 29
   §5. Polynomials of One Variable 37

Chapter 2. Gröbner Bases 49
   §1. Introduction 49
   §2. Orderings on the Monomials in $k[x_1, \ldots, x_n]$ 54
   §3. A Division Algorithm in $k[x_1, \ldots, x_n]$ 61
   §4. Monomial Ideals and Dickson's Lemma 70
   §5. The Hilbert Basis Theorem and Gröbner Bases 76
   §6. Properties of Gröbner Bases 83
   §7. Buchberger's Algorithm 90
   §8. First Applications of Gröbner Bases 97
   §9. Refinements of the Buchberger Criterion 104
   §10. Improvements on Buchberger's Algorithm 109

Chapter 3. Elimination Theory 121
   §1. The Elimination and Extension Theorems 121
   §2. The Geometry of Elimination 129
   §3. Implicitization 133
   §4. Singular Points and Envelopes 143
   §5. Gröbner Bases and the Extension Theorem 155
   §6. Resultants and the Extension Theorem 161

Chapter 4.   The Algebra–Geometry Dictionary 175
  §1.   Hilbert's Nullstellensatz 175
  §2.   Radical Ideals and the Ideal–Variety Correspondence 181
  §3.   Sums, Products, and Intersections of Ideals 189
  §4.   Zariski Closures, Ideal Quotients, and Saturations 199
  §5.   Irreducible Varieties and Prime Ideals 206
  §6.   Decomposition of a Variety into Irreducibles 212
  §7.   Proof of the Closure Theorem 219
  §8.   Primary Decomposition of Ideals 228
  §9.   Summary 232

Chapter 5.   Polynomial and Rational Functions on a Variety 233
  §1.   Polynomial Mappings 233
  §2.   Quotients of Polynomial Rings 240
  §3.   Algorithmic Computations in $k[x_1, \ldots, x_n]/I$ 248
  §4.   The Coordinate Ring of an Affine Variety 257
  §5.   Rational Functions on a Variety 268
  §6.   Relative Finiteness and Noether Normalization 277

Chapter 6.   Robotics and Automatic Geometric Theorem Proving 291
  §1.   Geometric Description of Robots 291
  §2.   The Forward Kinematic Problem 297
  §3.   The Inverse Kinematic Problem and Motion Planning 304
  §4.   Automatic Geometric Theorem Proving 319
  §5.   Wu's Method 335

Chapter 7.   Invariant Theory of Finite Groups 345
  §1.   Symmetric Polynomials 345
  §2.   Finite Matrix Groups and Rings of Invariants 355
  §3.   Generators for the Ring of Invariants 364
  §4.   Relations Among Generators and the Geometry of Orbits 373

Chapter 8.   Projective Algebraic Geometry 385
  §1.   The Projective Plane 385
  §2.   Projective Space and Projective Varieties 396
  §3.   The Projective Algebra–Geometry Dictionary 406
  §4.   The Projective Closure of an Affine Variety 415
  §5.   Projective Elimination Theory 422
  §6.   The Geometry of Quadric Hypersurfaces 436
  §7.   Bezout's Theorem 451

Chapter 9.   The Dimension of a Variety 469
  §1.   The Variety of a Monomial Ideal 469
  §2.   The Complement of a Monomial Ideal 473
  §3.   The Hilbert Function and the Dimension of a Variety 486
  §4.   Elementary Properties of Dimension 498

§5.   Dimension and Algebraic Independence 506
§6.   Dimension and Nonsingularity 515
§7.   The Tangent Cone 525

Chapter 10.   Additional Gröbner Basis Algorithms 539
§1.   Preliminaries 539
§2.   Hilbert Driven Buchberger Algorithms 550
§3.   The F_4 Algorithm 567
§4.   Signature-based Algorithms and F_5 576

Appendix A.   Some Concepts from Algebra 593
§1.   Fields and Rings 593
§2.   Unique Factorization 594
§3.   Groups 595
§4.   Determinants 596

Appendix B.   Pseudocode 599
§1.   Inputs, Outputs, Variables, and Constants 600
§2.   Assignment Statements 600
§3.   Looping Structures 600
§4.   Branching Structures 602
§5.   Output Statements 602

Appendix C.   Computer Algebra Systems 603
§1.   General Purpose Systems: Maple, *Mathematica*, Sage 604
§2.   Special Purpose Programs: CoCoA, Macaulay2, Singular 611
§3.   Other Systems and Packages 617

Appendix D.   Independent Projects 619
§1.   General Comments 619
§2.   Suggested Projects 620

References 627

Index 635

Notation for Sets and Functions

In this book, a *set* is a collection of mathematical objects. We say that x is an *element* of A, written $x \in A$, when x is one of objects in A. Then $x \notin A$ means that x is not an element of A. Commonly used sets are:

$$\mathbb{Z} = \{\ldots, -2, -1, 0, 1, 2, \ldots\}, \text{ the set of integers,}$$
$$\mathbb{Z}_{\geq 0} = \{0, 1, 2, \ldots\}, \text{ the set of nonnegative integers,}$$
$$\mathbb{Q} = \text{the set of rational numbers (fractions),}$$
$$\mathbb{R} = \text{the set of real numbers,}$$
$$\mathbb{C} = \text{the set of complex numbers.}$$

Sets will often be specified by listing the elements in the set, such as $A = \{0, 1, 2\}$, or by *set-builder notation*, such as

$$[0, 1] = \{x \in \mathbb{R} \mid 0 \leq x \leq 1\}.$$

The *empty set* is the set with no elements, denoted $\emptyset$. We write

$$A \subseteq B$$

when every element of A is also an element of B; we say that A *is contained* in B and A is a *subset* of B. When in addition $A \neq B$, we write

$$A \subsetneq B$$

and say that A is a *proper subset* of B. Basic operations on sets are

$$A \cup B = \{x \mid x \in A \text{ or } x \in B\}, \text{ the } union \text{ of } A \text{ and } B,$$
$$A \cap B = \{x \mid x \in A \text{ and } x \in B\}, \text{ the } intersection \text{ of } A \text{ and } B,$$
$$A \setminus B = \{x \mid x \in A \text{ and } x \notin B\}, \text{ the } difference \text{ of } A \text{ and } B,$$
$$A \times B = \{(x, y) \mid x \in A \text{ and } y \in B\}, \text{ the } cartesian\ product \text{ of } A \text{ and } B.$$

We say that f is a *function* from A to B, written

$$f : A \longrightarrow B,$$

when for every $x \in A$, there is a unique $f(x) \in B$. We sometimes write the function f as

$$x \longmapsto f(x).$$

Given any set A, an important function is the *identity function*

$$\mathrm{id}_A : A \longrightarrow A$$

defined by $x \mapsto x$ for all $x \in A$. Given functions $f : A \to B$ and $g : B \to C$, their *composition*

$$g \circ f : A \longrightarrow C$$

is defined by $x \mapsto g(f(x))$ for $x \in A$.

A function $f : A \to B$ is *one-to-one* if $f(x) = f(y)$ implies $x = y$ whenever $x, y \in A$. A function $f : A \to B$ is *onto* if for all $y \in B$, there is $x \in A$ with $f(x) = y$. If f is one-to-one and onto, then f has an *inverse function*

$$f^{-1} : B \longrightarrow A$$

defined by $f^{-1}(y) = x$ when $f(x) = y$. The inverse function satisfies

$$f^{-1} \circ f = \mathrm{id}_A \quad \text{and} \quad f \circ f^{-1} = \mathrm{id}_B.$$

Chapter 1
Geometry, Algebra, and Algorithms

This chapter will introduce some of the basic themes of the book. The geometry we are interested in concerns *affine varieties*, which are curves and surfaces (and higher dimensional objects) defined by polynomial equations. To understand affine varieties, we will need some algebra, and in particular, we will need to study *ideals* in the polynomial ring $k[x_1, \ldots, x_n]$. Finally, we will discuss polynomials in one variable to illustrate the role played by *algorithms*.

§1 Polynomials and Affine Space

To link algebra and geometry, we will study polynomials over a field. We all know what polynomials are, but the term *field* may be unfamiliar. The basic intuition is that a field is a set where one can define addition, subtraction, multiplication, and division with the usual properties. Standard examples are the real numbers $\mathbb{R}$ and the complex numbers $\mathbb{C}$, whereas the integers $\mathbb{Z}$ are not a field since division fails (3 and 2 are integers, but their quotient $3/2$ is not). A formal definition of field may be found in Appendix A.

One reason that fields are important is that linear algebra works over *any* field. Thus, even if your linear algebra course restricted the scalars to lie in $\mathbb{R}$ or $\mathbb{C}$, most of the theorems and techniques you learned apply to an arbitrary field k. In this book, we will employ different fields for different purposes. The most commonly used fields will be:

- The rational numbers $\mathbb{Q}$: the field for most of our computer examples.
- The real numbers $\mathbb{R}$: the field for drawing pictures of curves and surfaces.
- The complex numbers $\mathbb{C}$: the field for proving many of our theorems.

On occasion, we will encounter other fields, such as fields of rational functions (which will be defined later). There is also a very interesting theory of finite fields— see the exercises for one of the simpler examples.

© Springer International Publishing Switzerland 2015
D.A. Cox et al., *Ideals, Varieties, and Algorithms*, Undergraduate Texts in Mathematics, DOI 10.1007/978-3-319-16721-3_1

We can now define polynomials. The reader certainly is familiar with polynomials in one and two variables, but we will need to discuss polynomials in n variables $x_1, \ldots, x_n$ with coefficients in an arbitrary field k. We start by defining monomials.

Definition 1. A **monomial** in $x_1, \ldots, x_n$ is a product of the form

$$x_1^{\alpha_1} \cdot x_2^{\alpha_2} \cdots x_n^{\alpha_n},$$

where all of the exponents $\alpha_1, \ldots, \alpha_n$ are nonnegative integers. The **total degree** of this monomial is the sum $\alpha_1 + \cdots + \alpha_n$.

We can simplify the notation for monomials as follows: let $\alpha = (\alpha_1, \ldots, \alpha_n)$ be an n-tuple of nonnegative integers. Then we set

$$x^\alpha = x_1^{\alpha_1} \cdot x_2^{\alpha_2} \cdots x_n^{\alpha_n}.$$

When $\alpha = (0, \ldots, 0)$, note that $x^\alpha = 1$. We also let $|\alpha| = \alpha_1 + \cdots + \alpha_n$ denote the total degree of the monomial x^α.

Definition 2. A **polynomial** f in $x_1, \ldots, x_n$ with coefficients in a field k is a finite linear combination (with coefficients in k) of monomials. We will write a polynomial f in the form

$$f = \sum_\alpha a_\alpha x^\alpha, \quad a_\alpha \in k,$$

where the sum is over a finite number of n-tuples $\alpha = (\alpha_1, \ldots, \alpha_n)$. The set of all polynomials in $x_1, \ldots, x_n$ with coefficients in k is denoted $k[x_1, \ldots, x_n]$.

When dealing with polynomials in a small number of variables, we will usually dispense with subscripts. Thus, polynomials in one, two, and three variables lie in $k[x]$, $k[x, y]$, and $k[x, y, z]$, respectively. For example,

$$f = 2x^3y^2z + \frac{3}{2}y^3z^3 - 3xyz + y^2$$

is a polynomial in $\mathbb{Q}[x, y, z]$. We will usually use the letters f, g, h, p, q, r to refer to polynomials.

We will use the following terminology in dealing with polynomials.

Definition 3. Let $f = \sum_\alpha a_\alpha x^\alpha$ be a polynomial in $k[x_1, \ldots, x_n]$.
   (i) We call a_α the **coefficient** of the monomial x^α.
   (ii) If $a_\alpha \neq 0$, then we call $a_\alpha x^\alpha$ a **term** of f.
   (iii) The **total degree** of $f \neq 0$, denoted $\deg(f)$, is the maximum $|\alpha|$ such that the coefficient a_α is nonzero. The total degree of the zero polynomial is undefined.

As an example, the polynomial $f = 2x^3y^2z + \frac{3}{2}y^3z^3 - 3xyz + y^2$ given above has four terms and total degree six. Note that there are two terms of maximal total degree, which is something that cannot happen for polynomials of one variable. In Chapter 2, we will study how to *order* the terms of a polynomial.

The sum and product of two polynomials is again a polynomial. We say that a polynomial f *divides* a polynomial g provided that $g = fh$ for some polynomial $h \in k[x_1, \ldots, x_n]$.

One can show that, under addition and multiplication, $k[x_1, \ldots, x_n]$ satisfies all of the field axioms except for the existence of multiplicative inverses (because, for example, $1/x_1$ is not a polynomial). Such a mathematical structure is called a commutative ring (see Appendix A for the full definition), and for this reason we will refer to $k[x_1, \ldots, x_n]$ as a *polynomial ring*.

The next topic to consider is affine space.

Definition 4. Given a field k and a positive integer n, we define the n-dimensional **affine space** over k to be the set

$$k^n = \{(a_1, \ldots, a_n) \mid a_1, \ldots, a_n \in k\}.$$

For an example of affine space, consider the case $k = \mathbb{R}$. Here we get the familiar space $\mathbb{R}^n$ from calculus and linear algebra. In general, we call $k^1 = k$ the *affine line* and k^2 the *affine plane*.

Let us next see how polynomials relate to affine space. The key idea is that a polynomial $f = \sum_\alpha a_\alpha x^\alpha \in k[x_1, \ldots, x_n]$ gives a function

$$f : k^n \longrightarrow k$$

defined as follows: given $(a_1, \ldots, a_n) \in k^n$, replace every x_i by a_i in the expression for f. Since all of the coefficients also lie in k, this operation gives an element $f(a_1, \ldots, a_n) \in k$. The ability to regard a polynomial as a function is what makes it possible to link algebra and geometry.

This dual nature of polynomials has some unexpected consequences. For example, the question "is $f = 0$?" now has two potential meanings: is f the zero polynomial?, which means that all of its coefficients a_α are zero, or is f the zero function?, which means that $f(a_1, \ldots, a_n) = 0$ for all $(a_1, \ldots, a_n) \in k^n$. The surprising fact is that these two statements are not equivalent in general. For an example of how they can differ, consider the set consisting of the two elements 0 and 1. In the exercises, we will see that this can be made into a field where $1 + 1 = 0$. This field is usually called $\mathbb{F}_2$. Now consider the polynomial $x^2 - x = x(x - 1) \in \mathbb{F}_2[x]$. Since this polynomial vanishes at 0 and 1, we have found a nonzero polynomial which gives the zero function on the affine space $\mathbb{F}_2^1$. Other examples will be discussed in the exercises.

However, as long as k is infinite, there is no problem.

Proposition 5. *Let k be an infinite field and let $f \in k[x_1, \ldots, x_n]$. Then $f = 0$ in $k[x_1, \ldots, x_n]$ if and only if $f : k^n \to k$ is the zero function.*

Proof. One direction of the proof is obvious since the zero polynomial clearly gives the zero function. To prove the converse, we need to show that if $f(a_1, \ldots, a_n) = 0$ for all $(a_1, \ldots, a_n) \in k^n$, then f is the zero polynomial. We will use induction on the number of variables n.

When $n = 1$, it is well known that a nonzero polynomial in $k[x]$ of degree m has at most m distinct roots (we will prove this fact in Corollary 3 of §5). For our particular $f \in k[x]$, we are assuming $f(a) = 0$ for all $a \in k$. Since k is infinite, this means that f has infinitely many roots, and, hence, f must be the zero polynomial.

Now assume that the converse is true for $n - 1$, and let $f \in k[x_1, \ldots, x_n]$ be a polynomial that vanishes at all points of k^n. By collecting the various powers of x_n, we can write f in the form

$$f = \sum_{i=0}^{N} g_i(x_1, \ldots, x_{n-1})x_n^i,$$

where $g_i \in k[x_1, \ldots, x_{n-1}]$. We will show that each g_i is the zero polynomial in $n-1$ variables, which will force f to be the zero polynomial in $k[x_1, \ldots, x_n]$.

If we fix $(a_1, \ldots, a_{n-1}) \in k^{n-1}$, we get the polynomial $f(a_1, \ldots, a_{n-1}, x_n) \in k[x_n]$. By our hypothesis on f, this vanishes for every $a_n \in k$. It follows from the case $n = 1$ that $f(a_1, \ldots, a_{n-1}, x_n)$ is the zero polynomial in $k[x_n]$. Using the above formula for f, we see that the coefficients of $f(a_1, \ldots, a_{n-1}, x_n)$ are $g_i(a_1, \ldots, a_{n-1})$, and thus, $g_i(a_1, \ldots, a_{n-1}) = 0$ for all i. Since $(a_1, \ldots, a_{n-1})$ was arbitrarily chosen in k^{n-1}, it follows that each $g_i \in k[x_1, \ldots, x_{n-1}]$ gives the zero function on k^{n-1}. Our inductive assumption then implies that each g_i is the zero polynomial in $k[x_1, \ldots, x_{n-1}]$. This forces f to be the zero polynomial in $k[x_1, \ldots, x_n]$ and completes the proof of the proposition. □

Note that in the statement of Proposition 5, the assertion "$f = 0$ in $k[x_1, \ldots, x_n]$" means that f is the zero polynomial, i.e., that every coefficient of f is zero. Thus, we use the same symbol "0" to stand for the zero element of k and the zero polynomial in $k[x_1, \ldots, x_n]$. The context will make clear which one we mean.

As a corollary, we see that two polynomials over an infinite field are equal precisely when they give the same function on affine space.

Corollary 6. *Let k be an infinite field, and let $f, g \in k[x_1, \ldots, x_n]$. Then $f = g$ in $k[x_1, \ldots, x_n]$ if and only if $f : k^n \to k$ and $g : k^n \to k$ are the same function.*

Proof. To prove the nontrivial direction, suppose that $f, g \in k[x_1, \ldots, x_n]$ give the same function on k^n. By hypothesis, the polynomial $f - g$ vanishes at all points of k^n. Proposition 5 then implies that $f - g$ is the zero polynomial. This proves that $f = g$ in $k[x_1, \ldots, x_n]$. □

Finally, we need to record a special property of polynomials over the field of complex numbers $\mathbb{C}$.

Theorem 7. *Every nonconstant polynomial $f \in \mathbb{C}[x]$ has a root in $\mathbb{C}$.*

Proof. This is the Fundamental Theorem of Algebra, and proofs can be found in most introductory texts on complex analysis (although many other proofs are known). □

We say that a field k is *algebraically closed* if every nonconstant polynomial in $k[x]$ has a root in k. Thus $\mathbb{R}$ is not algebraically closed (what are the roots of $x^2 + 1$?), whereas the above theorem asserts that $\mathbb{C}$ is algebraically closed. In Chapter 4 we will prove a powerful generalization of Theorem 7 called the Hilbert Nullstellensatz.

EXERCISES FOR §1

1. Let $\mathbb{F}_2 = \{0, 1\}$, and define addition and multiplication by $0 + 0 = 1 + 1 = 0, 0 + 1 = 1 + 0 = 1, 0 \cdot 0 = 0 \cdot 1 = 1 \cdot 0 = 0$ and $1 \cdot 1 = 1$. Explain why $\mathbb{F}_2$ is a field. (You need not check the associative and distributive properties, but you should verify the existence of identities and inverses, both additive and multiplicative.)
2. Let $\mathbb{F}_2$ be the field from Exercise 1.
   a. Consider the polynomial $g(x, y) = x^2y + y^2x \in \mathbb{F}_2[x, y]$. Show that $g(x, y) = 0$ for every $(x, y) \in \mathbb{F}_2^2$, and explain why this does not contradict Proposition 5.
   b. Find a nonzero polynomial in $\mathbb{F}_2[x, y, z]$ which vanishes at every point of $\mathbb{F}_2^3$. Try to find one involving all three variables.
   c. Find a nonzero polynomial in $\mathbb{F}_2[x_1, \ldots, x_n]$ which vanishes at every point of $\mathbb{F}_2^n$. Can you find one in which all of $x_1, \ldots, x_n$ appear?
3. (Requires abstract algebra). Let p be a prime number. The ring of integers modulo p is a field with p elements, which we will denote $\mathbb{F}_p$.
   a. Explain why $\mathbb{F}_p \setminus \{0\}$ is a group under multiplication.
   b. Use Lagrange's Theorem to show that $a^{p-1} = 1$ for all $a \in \mathbb{F}_p \setminus \{0\}$.
   c. Prove that $a^p = a$ for all $a \in \mathbb{F}_p$. Hint: Treat the cases $a = 0$ and $a \neq 0$ separately.
   d. Find a nonzero polynomial in $\mathbb{F}_p[x]$ that vanishes at all points of $\mathbb{F}_p$. Hint: Use part (c).
4. (Requires abstract algebra.) Let F be a finite field with q elements. Adapt the argument of Exercise 3 to prove that $x^q - x$ is a nonzero polynomial in $F[x]$ which vanishes at every point of F. This shows that Proposition 5 fails for *all* finite fields.
5. In the proof of Proposition 5, we took $f \in k[x_1, \ldots, x_n]$ and wrote it as a polynomial in x_n with coefficients in $k[x_1, \ldots, x_{n-1}]$. To see what this looks like in a specific case, consider the polynomial

$$f(x, y, z) = x^5y^2z - x^4y^3 + y^5 + x^2z - y^3z + xy + 2x - 5z + 3.$$

   a. Write f as a polynomial in x with coefficients in $k[y, z]$.
   b. Write f as a polynomial in y with coefficients in $k[x, z]$.
   c. Write f as a polynomial in z with coefficients in $k[x, y]$.
6. Inside of $\mathbb{C}^n$, we have the subset $\mathbb{Z}^n$, which consists of all points with integer coordinates.
   a. Prove that if $f \in \mathbb{C}[x_1, \ldots, x_n]$ vanishes at every point of $\mathbb{Z}^n$, then f is the zero polynomial. Hint: Adapt the proof of Proposition 5.
   b. Let $f \in \mathbb{C}[x_1, \ldots, x_n]$, and let M be the largest power of any variable that appears in f. Let $\mathbb{Z}_{M+1}^n$ be the set of points of $\mathbb{Z}^n$, all coordinates of which lie between 1 and $M + 1$, inclusive. Prove that if f vanishes at all points of $\mathbb{Z}_{M+1}^n$, then f is the zero polynomial.

§2 Affine Varieties

We can now define the basic geometric objects studied in this book.

Definition 1. Let k be a field, and let $f_1, \ldots, f_s$ be polynomials in $k[x_1, \ldots, x_n]$. Then we set

$$\mathbf{V}(f_1,\ldots,f_s) = \{(a_1,\ldots,a_n) \in k^n \mid f_i(a_1,\ldots,a_n) = 0 \text{ for all } 1 \leq i \leq s\}.$$

We call $\mathbf{V}(f_1,\ldots,f_s)$ the **affine variety** defined by $f_1,\ldots,f_s$.

Thus, an affine variety $\mathbf{V}(f_1,\ldots,f_s) \subseteq k^n$ is the set of all solutions of the system of equations $f_1(x_1,\ldots,x_n) = \cdots = f_s(x_1,\ldots,x_n) = 0$. We will use the letters V, W, etc. to denote affine varieties. The main purpose of this section is to introduce the reader to *lots* of examples, some new and some familiar. We will use $k = \mathbb{R}$ so that we can draw pictures.

We begin in the plane $\mathbb{R}^2$ with the variety $\mathbf{V}(x^2 + y^2 - 1)$, which is the circle of radius 1 centered at the origin:

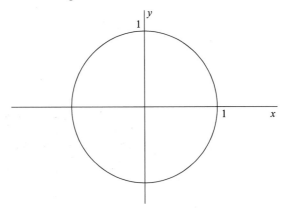

The conic sections studied in school (circles, ellipses, parabolas, and hyperbolas) are affine varieties. Likewise, graphs of polynomial functions are affine varieties [the graph of $y = f(x)$ is $\mathbf{V}(y - f(x))$]. Although not as obvious, graphs of rational functions are also affine varieties. For example, consider the graph of $y = \frac{x^3-1}{x}$:

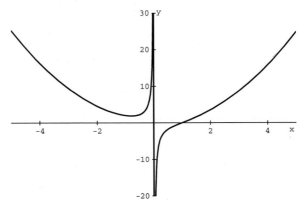

It is easy to check that this is the affine variety $\mathbf{V}(xy - x^3 + 1)$.

Next, let us look in the 3-dimensional space $\mathbb{R}^3$. A nice affine variety is given by paraboloid of revolution $\mathbf{V}(z - x^2 - y^2)$, which is obtained by rotating the parabola

$z = x^2$ about the z-axis (you can check this using polar coordinates). This gives us the picture:

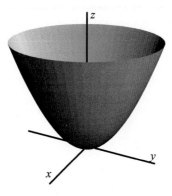

You may also be familiar with the cone $\mathbf{V}(z^2 - x^2 - y^2)$:

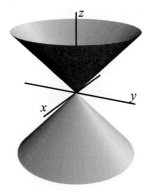

A much more complicated surface is given by $\mathbf{V}(x^2 - y^2 z^2 + z^3)$:

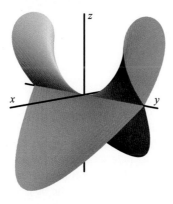

In these last two examples, the surfaces are not smooth everywhere: the cone has a sharp point at the origin, and the last example intersects itself along the whole *y*-axis. These are examples of *singular points*, which will be studied later in the book.

An interesting example of a curve in $\mathbb{R}^3$ is the *twisted cubic*, which is the variety $\mathbf{V}(y - x^2, z - x^3)$. For simplicity, we will confine ourselves to the portion that lies in the first octant. To begin, we draw the surfaces $y = x^2$ and $z = x^3$ separately:

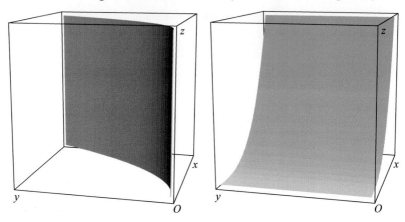

Then their intersection gives the twisted cubic:

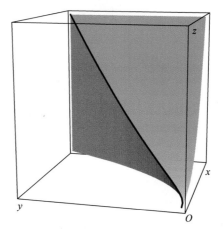

Notice that when we had one equation in $\mathbb{R}^2$, we got a curve, which is a 1-dimensional object. A similar situation happens in $\mathbb{R}^3$: one equation in $\mathbb{R}^3$ usually gives a surface, which has dimension 2. Again, dimension drops by one. But now consider the twisted cubic: here, two equations in $\mathbb{R}^3$ give a curve, so that dimension drops by two. Since each equation imposes an extra constraint, intuition suggests that each equation drops the dimension by one. Thus, if we started in $\mathbb{R}^4$, one would hope that an affine variety defined by two equations would be a surface. Unfortunately, the notion of dimension is more subtle than indicated by the above examples. To illustrate this, consider the variety $\mathbf{V}(xz, yz)$. One can easily check that the equations $xz = yz = 0$ define the union of the (x, y)-plane and the z-axis:

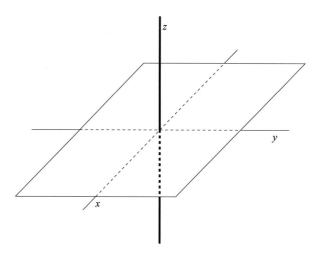

Hence, this variety consists of two pieces which have different dimensions, and one of the pieces (the plane) has the "wrong" dimension according to the above intuition.

We next give some examples of varieties in higher dimensions. A familiar case comes from linear algebra. Namely, fix a field k, and consider a system of m linear equations in n unknowns $x_1, \ldots, x_n$ with coefficients in k:

$$
\begin{aligned}
a_{11}x_1 + \cdots + a_{1n}x_n &= b_1, \\
&\vdots \\
a_{m1}x_1 + \cdots + a_{mn}x_n &= b_m.
\end{aligned}
\tag{1}
$$

The solutions of these equations form an affine variety in k^n, which we will call a *linear variety*. Thus, lines and planes are linear varieties, and there are examples of arbitrarily large dimension. In linear algebra, you learned the method of row reduction (also called Gaussian elimination), which gives an algorithm for finding all solutions of such a system of equations. In Chapter 2, we will study a generalization of this algorithm which applies to systems of polynomial equations.

Linear varieties relate nicely to our discussion of dimension. Namely, if $V \subseteq k^n$ is the linear variety defined by (1), then V need not have dimension $n - m$ even though V is defined by m equations. In fact, when V is nonempty, linear algebra tells us that V has dimension $n - r$, where r is the rank of the matrix (a_{ij}). So for linear varieties, the dimension is determined by the number of *independent* equations. This intuition applies to more general affine varieties, except that the notion of "independent" is more subtle.

Some complicated examples in higher dimensions come from calculus. Suppose, for example, that we wanted to find the minimum and maximum values of $f(x, y, z) = x^3 + 2xyz - z^2$ subject to the constraint $g(x, y, z) = x^2 + y^2 + z^2 = 1$. The method of Lagrange multipliers states that $\nabla f = \lambda \nabla g$ at a local minimum or maximum [recall that the gradient of f is the vector of partial derivatives $\nabla f = (f_x, f_y, f_z)$]. This gives us the following system of four equations in four unknowns, x, y, z, λ, to solve:

(2)
$$3x^2 + 2yz = 2x\lambda,$$
$$2xz = 2y\lambda,$$
$$2xy - 2z = 2z\lambda,$$
$$x^2 + y^2 + z^2 = 1.$$

These equations define an affine variety in $\mathbb{R}^4$, and our intuition concerning dimension leads us to hope it consists of finitely many points (which have dimension 0) since it is defined by four equations. Students often find Lagrange multipliers difficult because the equations are so hard to solve. The algorithms of Chapter 2 will provide a powerful tool for attacking such problems. In particular, we will find all solutions of the above equations.

We should also mention that affine varieties can be the empty set. For example, when $k = \mathbb{R}$, it is obvious that $\mathbf{V}(x^2 + y^2 + 1) = \emptyset$ since $x^2 + y^2 = -1$ has no real solutions (although there are solutions when $k = \mathbb{C}$). Another example is $\mathbf{V}(xy, xy - 1)$, which is empty no matter what the field is, for a given x and y cannot satisfy both $xy = 0$ and $xy = 1$. In Chapter 4 we will study a method for determining when an affine variety over $\mathbb{C}$ is nonempty.

To give an idea of some of the applications of affine varieties, let us consider a simple example from robotics. Suppose we have a robot arm in the plane consisting of two linked rods of lengths 1 and 2, with the longer rod anchored at the origin:

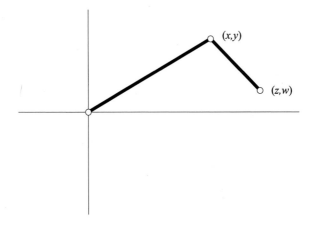

The "state" of the arm is completely described by the coordinates (x, y) and (z, w) indicated in the figure. Thus the state can be regarded as a 4-tuple $(x, y, z, w) \in \mathbb{R}^4$. However, not all 4-tuples can occur as states of the arm. In fact, it is easy to see that the subset of possible states is the affine variety in $\mathbb{R}^4$ defined by the equations

$$x^2 + y^2 = 4,$$
$$(x - z)^2 + (y - w)^2 = 1.$$

Notice how even larger dimensions enter quite easily: if we were to consider the same arm in 3-dimensional space, then the variety of states would be defined by two equations in $\mathbb{R}^6$. The techniques to be developed in this book have some important applications to the theory of robotics.

So far, all of our drawings have been over $\mathbb{R}$. Later in the book, we will consider varieties over $\mathbb{C}$. Here, it is more difficult (but not impossible) to get a geometric idea of what such a variety looks like.

Finally, let us record some basic properties of affine varieties.

Lemma 2. *If $V, W \subseteq k^n$ are affine varieties, then so are $V \cup W$ and $V \cap W$.*

Proof. Suppose that $V = \mathbf{V}(f_1, \ldots, f_s)$ and $W = \mathbf{V}(g_1, \ldots, g_t)$. Then we claim that

$$V \cap W = \mathbf{V}(f_1, \ldots, f_s, g_1, \ldots, g_t),$$
$$V \cup W = \mathbf{V}(f_i g_j \mid 1 \leq i \leq s, 1 \leq j \leq t).$$

The first equality is trivial to prove: being in $V \cap W$ means that both $f_1, \ldots, f_s$ and $g_1, \ldots, g_t$ vanish, which is the same as $f_1, \ldots, f_s, g_1, \ldots, g_t$ vanishing.

The second equality takes a little more work. If $(a_1, \ldots, a_n) \in V$, then all of the f_i's vanish at this point, which implies that all of the $f_i g_j$'s also vanish at $(a_1, \ldots, a_n)$. Thus, $V \subseteq \mathbf{V}(f_i g_j)$, and $W \subseteq \mathbf{V}(f_i g_j)$ follows similarly. This proves that $V \cup W \subseteq \mathbf{V}(f_i g_j)$. Going the other way, choose $(a_1, \ldots, a_n) \in \mathbf{V}(f_i g_j)$. If this lies in V, then we are done, and if not, then $f_{i_0}(a_1, \ldots, a_n) \neq 0$ for some i_0. Since $f_{i_0} g_j$ vanishes at $(a_1, \ldots, a_n)$ for all j, the g_j's must vanish at this point, proving that $(a_1, \ldots, a_n) \in W$. This shows that $\mathbf{V}(f_i g_j) \subseteq V \cup W$. $\qquad\square$

This lemma implies that finite intersections and unions of affine varieties are again affine varieties. It turns out that we have already seen examples of unions and intersections. Concerning unions, consider the union of the (x, y)-plane and the z-axis in affine 3-space. By the above formula, we have

$$\mathbf{V}(z) \cup \mathbf{V}(x, y) = \mathbf{V}(zx, zy).$$

This, of course, is one of the examples discussed earlier in the section. As for intersections, notice that the twisted cubic was given as the intersection of two surfaces.

The examples given in this section lead to some interesting questions concerning affine varieties. Suppose that we have $f_1, \ldots, f_s \in k[x_1, \ldots, x_n]$. Then:

- (Consistency) Can we determine if $\mathbf{V}(f_1, \ldots, f_s) \neq \emptyset$, i.e., do the equations $f_1 = \cdots = f_s = 0$ have a common solution?
- (Finiteness) Can we determine if $\mathbf{V}(f_1, \ldots, f_s)$ is finite, and if so, can we find all of the solutions explicitly?
- (Dimension) Can we determine the "dimension" of $\mathbf{V}(f_1, \ldots, f_s)$?

The answer to these questions is yes, although care must be taken in choosing the field k that we work over. The hardest is the one concerning dimension, for it involves some sophisticated concepts. Nevertheless, we will give complete solutions to all three problems.

EXERCISES FOR §2

1. Sketch the following affine varieties in $\mathbb{R}^2$:
   a. $\mathbf{V}(x^2 + 4y^2 + 2x - 16y + 1)$.
   b. $\mathbf{V}(x^2 - y^2)$.
   c. $\mathbf{V}(2x + y - 1, 3x - y + 2)$.
   In each case, does the variety have the dimension you would intuitively expect it to have?
2. In $\mathbb{R}^2$, sketch $\mathbf{V}(y^2 - x(x-1)(x-2))$. Hint: For which x's is it possible to solve for y? How many y's correspond to each x? What symmetry does the curve have?
3. In the plane $\mathbb{R}^2$, draw a picture to illustrate

$$\mathbf{V}(x^2 + y^2 - 4) \cap \mathbf{V}(xy - 1) = \mathbf{V}(x^2 + y^2 - 4, xy - 1),$$

   and determine the points of intersection. Note that this is a special case of Lemma 2.
4. Sketch the following affine varieties in $\mathbb{R}^3$:
   a. $\mathbf{V}(x^2 + y^2 + z^2 - 1)$.
   b. $\mathbf{V}(x^2 + y^2 - 1)$.
   c. $\mathbf{V}(x + 2, y - 1.5, z)$.
   d. $\mathbf{V}(xz^2 - xy)$. Hint: Factor $xz^2 - xy$.
   e. $\mathbf{V}(x^4 - zx, x^3 - yx)$.
   f. $\mathbf{V}(x^2 + y^2 + z^2 - 1, x^2 + y^2 + (z-1)^2 - 1)$.
   In each case, does the variety have the dimension you would intuitively expect it to have?
5. Use the proof of Lemma 2 to sketch $\mathbf{V}((x-2)(x^2 - y), y(x^2 - y), (z+1)(x^2 - y))$ in $\mathbb{R}^3$. Hint: This is the union of which two varieties?
6. Let us show that all finite subsets of k^n are affine varieties.
   a. Prove that a single point $(a_1, \ldots, a_n) \in k^n$ is an affine variety.
   b. Prove that every finite subset of k^n is an affine variety. Hint: Lemma 2 will be useful.
7. One of the prettiest examples from polar coordinates is the four-leaved rose

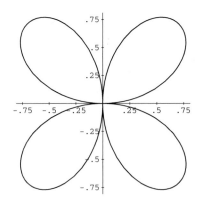

This curve is defined by the polar equation $r = \sin(2\theta)$. We will show that this curve is an affine variety.

a. Using $r^2 = x^2 + y^2$, $x = r\cos(\theta)$ and $y = r\sin(\theta)$, show that the four-leaved rose is contained in the affine variety $\mathbf{V}((x^2+y^2)^3 - 4x^2y^2)$. Hint: Use an identity for $\sin(2\theta)$.

b. Now argue carefully that $\mathbf{V}((x^2 + y^2)^3 - 4x^2y^2)$ is contained in the four-leaved rose. This is trickier than it seems since r can be negative in $r = \sin(2\theta)$.

Combining parts (a) and (b), we have proved that the four-leaved rose is the affine variety $\mathbf{V}((x^2 + y^2)^3 - 4x^2y^2)$.

8. It can take some work to show that something is *not* an affine variety. For example, consider the set

$$X = \{(x,x) \mid x \in \mathbb{R},\ x \neq 1\} \subseteq \mathbb{R}^2,$$

which is the straight line $x = y$ with the point $(1,1)$ removed. To show that X is not an affine variety, suppose that $X = \mathbf{V}(f_1, \ldots, f_s)$. Then each f_i vanishes on X, and if we can show that f_i also vanishes at $(1,1)$, we will get the desired contradiction. Thus, here is what you are to prove: if $f \in \mathbb{R}[x, y]$ vanishes on X, then $f(1,1) = 0$. Hint: Let $g(t) = f(t,t)$, which is a polynomial $\mathbb{R}[t]$. Now apply the proof of Proposition 5 of §1.

9. Let $R = \{(x,y) \in \mathbb{R}^2 \mid y > 0\}$ be the upper half plane. Prove that R is not an affine variety.

10. Let $\mathbb{Z}^n \subseteq \mathbb{C}^n$ consist of those points with integer coordinates. Prove that $\mathbb{Z}^n$ is not an affine variety. Hint: See Exercise 6 from §1.

11. So far, we have discussed varieties over $\mathbb{R}$ or $\mathbb{C}$. It is also possible to consider varieties over the field $\mathbb{Q}$, although the questions here tend to be *much* harder. For example, let n be a positive integer, and consider the variety $F_n \subseteq \mathbb{Q}^2$ defined by

$$x^n + y^n = 1.$$

Notice that there are some obvious solutions when x or y is zero. We call these *trivial solutions*. An interesting question is whether or not there are any nontrivial solutions.

a. Show that F_n has two trivial solutions if n is odd and four trivial solutions if n is even.

b. Show that F_n has a nontrivial solution for some $n \geq 3$ if and only if Fermat's Last Theorem were false.

Fermat's Last Theorem states that, for $n \geq 3$, the equation

$$x^n + y^n = z^n$$

has no solutions where x, y, and z are nonzero integers. The general case of this conjecture was proved by Andrew Wiles in 1994 using some very sophisticated number theory. The proof is *extremely* difficult.

12. Find a Lagrange multipliers problem in a calculus book and write down the corresponding system of equations. Be sure to use an example where one wants to find the minimum or maximum of a polynomial function subject to a polynomial constraint. This way the equations define an affine variety, and try to find a problem that leads to complicated equations. Later we will use Gröbner basis methods to solve these equations.

13. Consider a robot arm in $\mathbb{R}^2$ that consists of three arms of lengths 3, 2, and 1, respectively. The arm of length 3 is anchored at the origin, the arm of length 2 is attached to the free end of the arm of length 3, and the arm of length 1 is attached to the free end of the arm of length 2. The "hand" of the robot arm is attached to the end of the arm of length 1.

a. Draw a picture of the robot arm.

b. How many variables does it take to determine the "state" of the robot arm?

c. Give the equations for the variety of possible states.

d. Using the intuitive notion of dimension discussed in this section, guess what the dimension of the variety of states should be.

14. This exercise will study the possible "hand" positions of the robot arm described in Exercise 13.

a. If (u, v) is the position of the hand, explain why $u^2 + v^2 \leq 36$.

   b. Suppose we "lock" the joint between the length 3 and length 2 arms to form a straight angle, but allow the other joint to move freely. Draw a picture to show that in these configurations, (u, v) can be *any* point of the annulus $16 \leq u^2 + v^2 \leq 36$.
   c. Draw a picture to show that (u, v) can be any point in the disk $u^2 + v^2 \leq 36$. Hint: Consider $16 \leq u^2 + v^2 \leq 36$, $4 \leq u^2 + v^2 \leq 16$, and $u^2 + v^2 \leq 4$ separately.
15. In Lemma 2, we showed that if V and W are affine varieties, then so are their union $V \cup W$ and intersection $V \cap W$. In this exercise we will study how other set-theoretic operations affect affine varieties.
   a. Prove that finite unions and intersections of affine varieties are again affine varieties. Hint: Induction.
   b. Give an example to show that an infinite union of affine varieties need not be an affine variety. Hint: By Exercises 8–10, we know some subsets of k^n that are not affine varieties. Surprisingly, an infinite intersection of affine varieties is still an affine variety. This is a consequence of the Hilbert Basis Theorem, which will be discussed in Chapters 2 and 4.
   c. Give an example to show that the set-theoretic difference $V \setminus W$ of two affine varieties need not be an affine variety.
   d. Let $V \subseteq k^n$ and $W \subseteq k^m$ be two affine varieties, and let

$$V \times W = \{(x_1, \ldots, x_n, y_1, \ldots, y_m) \in k^{n+m} \mid (x_1, \ldots, x_n) \in V, (y_1, \ldots, y_m) \in W\}$$

be their Cartesian product. Prove that $V \times W$ is an affine variety in k^{n+m}. Hint: If V is defined by $f_1, \ldots, f_s \in k[x_1, \ldots, x_n]$, then we can regard $f_1, \ldots, f_s$ as polynomials in $k[x_1, \ldots, x_n, y_1, \ldots, y_m]$, and similarly for W. Show that this gives defining equations for the Cartesian product.

§3 Parametrizations of Affine Varieties

In this section, we will discuss the problem of describing the points of an affine variety $\mathbf{V}(f_1, \ldots, f_s)$. This reduces to asking whether there is a way to "write down" the solutions of the system of polynomial equations $f_1 = \cdots = f_s = 0$. When there are finitely many solutions, the goal is simply to list them all. But what do we do when there are infinitely many? As we will see, this question leads to the notion of parametrizing an affine variety.

To get started, let us look at an example from linear algebra. Let the field be $\mathbb{R}$, and consider the system of equations

(1)
$$x + y + z = 1,$$
$$x + 2y - z = 3.$$

Geometrically, this represents the line in $\mathbb{R}^3$ which is the intersection of the planes $x + y + z = 1$ and $x + 2y - z = 3$. It follows that there are infinitely many solutions. To describe the solutions, we use row operations on equations (1) to obtain the equivalent equations

$$x + 3z = -1,$$
$$y - 2z = 2.$$

Letting $z = t$, where t is arbitrary, this implies that all solutions of (1) are given by

(2)
$$\begin{aligned} x &= -1 - 3t, \\ y &= 2 + 2t, \\ z &= t \end{aligned}$$

as t varies over $\mathbb{R}$. We call t a *parameter*, and (2) is, thus, a *parametrization* of the solutions of (1).

To see if the idea of parametrizing solutions can be applied to other affine varieties, let us look at the example of the unit circle

(3)
$$x^2 + y^2 = 1.$$

A common way to parametrize the circle is using trigonometric functions:

$$\begin{aligned} x &= \cos(t), \\ y &= \sin(t). \end{aligned}$$

There is also a more algebraic way to parametrize this circle:

(4)
$$\begin{aligned} x &= \frac{1 - t^2}{1 + t^2}, \\ y &= \frac{2t}{1 + t^2}. \end{aligned}$$

You should check that the points defined by these equations lie on the circle (3). It is also interesting to note that this parametrization does not describe the whole circle: since $x = \frac{1-t^2}{1+t^2}$ can never equal -1, the point $(-1,0)$ is not covered. At the end of the section, we will explain how this parametrization was obtained.

Notice that equations (4) involve quotients of polynomials. These are examples of *rational functions*, and before we can say what it means to parametrize a variety, we need to define the general notion of rational function.

Definition 1. Let k be a field. A **rational function** in $t_1, \ldots, t_m$ with coefficients in k is a quotient f/g of two polynomials $f, g \in k[t_1, \ldots, t_m]$, where g is not the zero polynomial. Furthermore, two rational functions f/g and f'/g' are equal provided that $g'f = gf'$ in $k[t_1, \ldots, t_m]$. Finally, the set of all rational functions in $t_1, \ldots, t_m$ with coefficients in k is denoted $k(t_1, \ldots, t_m)$.

It is not difficult to show that addition and multiplication of rational functions are well defined and that $k(t_1, \ldots, t_m)$ is a field. We will assume these facts without proof.

Now suppose that we are given a variety $V = \mathbf{V}(f_1, \ldots, f_s) \subseteq k^n$. Then a *rational parametric representation* of V consists of rational functions $r_1, \ldots, r_n \in k(t_1, \ldots, t_m)$ such that the points given by

$$x_1 = r_1(t_1, \ldots, t_m),$$
$$x_2 = r_2(t_1, \ldots, t_m),$$
$$\vdots$$
$$x_n = r_n(t_1, \ldots, t_m)$$

lie in V. We also require that V be the "smallest" variety containing these points. As the example of the circle shows, a parametrization may not cover all points of V. In Chapter 3, we will give a more precise definition of what we mean by "smallest."

In many situations, we have a parametrization of a variety V, where $r_1, \ldots, r_n$ are polynomials rather than rational functions. This is what we call a *polynomial parametric representation* of V.

By contrast, the original defining equations $f_1 = \cdots = f_s = 0$ of V are called an *implicit representation* of V. In our previous examples, note that equations (1) and (3) are implicit representations of varieties, whereas (2) and (4) are parametric.

One of the main virtues of a parametric representation of a curve or surface is that it is easy to draw on a computer. Given the formulas for the parametrization, the computer evaluates them for various values of the parameters and then plots the resulting points. For example, in §2 we viewed the surface $\mathbf{V}(x^2 - y^2z^2 + z^3)$:

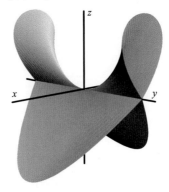

This picture was not plotted using the implicit representation $x^2 - y^2z^2 + z^3 = 0$. Rather, we used the parametric representation given by

(5)
$$x = t(u^2 - t^2),$$
$$y = u,$$
$$z = u^2 - t^2.$$

There are two parameters t and u since we are describing a surface, and the above picture was drawn using the range $-1 \leq t, u \leq 1$. In the exercises, we will derive this parametrization and check that it covers the entire surface $\mathbf{V}(x^2 - y^2z^2 + z^3)$.

At the same time, it is often useful to have an implicit representation of a variety. For example, suppose we want to know whether or not the point $(1, 2, -1)$ is on the above surface. If all we had was the parametrization (5), then, to decide this question, we would need to solve the equations

$$1 = t(u^2 - t^2),$$
(6)
$$2 = u,$$
$$-1 = u^2 - t^2$$

for t and u. On the other hand, if we have the implicit representation $x^2 - y^2 z^2 + z^3 = 0$, then it is simply a matter of plugging into this equation. Since

$$1^2 - 2^2(-1)^2 + (-1)^3 = 1 - 4 - 1 = -4 \neq 0,$$

it follows that $(1, 2, -1)$ is not on the surface [and, consequently, equations (6) have no solution].

The desirability of having both types of representations leads to the following two questions:

- (Parametrization) Does every affine variety have a rational parametric representation?
- (Implicitization) Given a parametric representation of an affine variety, can we find the defining equations (i.e., can we find an implicit representation)?

The answer to the first question is no. In fact, most affine varieties cannot be parametrized in the sense described here. Those that can are called *unirational*. In general, it is difficult to tell whether a given variety is unirational or not. The situation for the second question is much nicer. In Chapter 3, we will see that the answer is always yes: given a parametric representation, we can always find the defining equations.

Let us look at an example of how implicitization works. Consider the parametric representation

(7)
$$x = 1 + t,$$
$$y = 1 + t^2.$$

This describes a curve in the plane, but at this point, we cannot be sure that it lies on an affine variety. To find the equation we are looking for, notice that we can solve the first equation for t to obtain

$$t = x - 1.$$

Substituting this into the second equation yields

$$y = 1 + (x - 1)^2 = x^2 - 2x + 2.$$

Hence the parametric equations (7) describe the affine variety $\mathbf{V}(y - x^2 + 2x - 2)$.

In the above example, notice that the basic strategy was to eliminate the variable t so that we were left with an equation involving only x and y. This illustrates the role played by *elimination theory*, which will be studied in much greater detail in Chapter 3.

We will next discuss two examples of how geometry can be used to parametrize varieties. Let us start with the unit circle $x^2 + y^2 = 1$, which was parametrized in (4) via

$$x = \frac{1 - t^2}{1 + t^2},$$

$$y = \frac{2t}{1 + t^2}.$$

To see where this parametrization comes from, notice that each nonvertical line through $(-1, 0)$ will intersect the circle in a unique point (x, y):

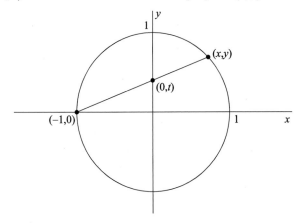

Each nonvertical line also meets the y-axis, and this is the point $(0, t)$ in the above picture.

This gives us a geometric parametrization of the circle: given t, draw the line connecting $(-1, 0)$ to $(0, t)$, and let (x, y) be the point where the line meets $x^2 + y^2 = 1$. Notice that the previous sentence really gives a parametrization: as t runs from $-\infty$ to ∞ on the vertical axis, the corresponding point (x, y) traverses all of the circle except for the point $(-1, 0)$.

It remains to find explicit formulas for x and y in terms of t. To do this, consider the *slope* of the line in the above picture. We can compute the slope in two ways, using either the points $(-1, 0)$ and $(0, t)$, or the points $(-1, 0)$ and (x, y). This gives us the equation

$$\frac{t - 0}{0 - (-1)} = \frac{y - 0}{x - (-1)},$$

which simplifies to become

$$t = \frac{y}{x + 1}.$$

Thus, $y = t(x + 1)$. If we substitute this into $x^2 + y^2 = 1$, we get

$$x^2 + t^2(x + 1)^2 = 1,$$

which gives the quadratic equation

(8) $(1 + t^2)x^2 + 2t^2 x + t^2 - 1 = 0.$

This equation gives the x-coordinates of where the line meets the circle, and it is quadratic since there are two points of intersection. One of the points is -1, so that $x+1$ is a factor of (8). It is now easy to find the other factor, and we can rewrite (8) as

$$(x + 1)((1 + t^2)x - (1 - t^2)) = 0.$$

Since the x-coordinate we want is given by the second factor, we obtain

$$x = \frac{1 - t^2}{1 + t^2}.$$

Furthermore, $y = t(x + 1)$ easily leads to

$$y = \frac{2t}{1 + t^2}$$

(you should check this), and we have now derived the parametrization given earlier. Note how the geometry tells us exactly what portion of the circle is covered.

For our second example, let us consider the twisted cubic $\mathbf{V}(y - x^2, z - x^3)$ from §2. This is a curve in 3-dimensional space, and by looking at the tangent lines to the curve, we will get an interesting surface. The idea is as follows. Given one point on the curve, we can draw the tangent line at that point:

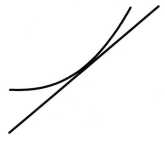

Now imagine taking the tangent lines for *all* points on the twisted cubic. This gives us the following surface:

This picture shows several of the tangent lines. The above surface is called the *tangent surface* of the twisted cubic.

To convert this geometric description into something more algebraic, notice that setting $x = t$ in $y - x^2 = z - x^3 = 0$ gives us a parametrization

$$x = t,$$
$$y = t^2,$$
$$z = t^3$$

of the twisted cubic. We will write this as $\mathbf{r}(t) = (t, t^2, t^3)$. Now fix a particular value of t, which gives us a point on the curve. From calculus, we know that the tangent vector to the curve at the point given by $\mathbf{r}(t)$ is $\mathbf{r}'(t) = (1, 2t, 3t^2)$. It follows that the tangent line is parametrized by

$$\mathbf{r}(t) + u\mathbf{r}'(t) = (t, t^2, t^3) + u(1, 2t, 3t^2) = (t + u, t^2 + 2tu, t^3 + 3t^2u),$$

where u is a parameter that moves along the tangent line. If we now allow t to vary, then we can parametrize the entire tangent surface by

$$x = t + u,$$
$$y = t^2 + 2tu,$$
$$z = t^3 + 3t^2u.$$

The parameters t and u have the following interpretations: t tells where we are on the curve, and u tells where we are on the tangent line. This parametrization was used to draw the picture of the tangent surface presented earlier.

A final question concerns the implicit representation of the tangent surface: how do we find its defining equation? This is a special case of the implicitization problem mentioned earlier and is equivalent to eliminating t and u from the above parametric equations. In Chapters 2 and 3, we will see that there is an algorithm for doing this, and, in particular, we will prove that the tangent surface to the twisted cubic is defined by the equation

$$x^3z - (3/4)x^2y^2 - (3/2)xyz + y^3 + (1/4)z^2 = 0.$$

We will end this section with an example from Computer Aided Geometric Design (CAGD). When creating complex shapes like automobile hoods or airplane wings, design engineers need curves and surfaces that are varied in shape, easy to describe, and quick to draw. Parametric equations involving polynomial and rational functions satisfy these requirements; there is a large body of literature on this topic.

For simplicity, let us suppose that a design engineer wants to describe a curve in the plane. Complicated curves are usually created by joining together simpler pieces, and for the pieces to join smoothly, the tangent directions must match up at the endpoints. Thus, for each piece, the designer needs to control the following geometric data:

- the starting and ending points of the curve;
- the tangent directions at the starting and ending points.

The *Bézier cubic*, introduced by Renault auto designer P. Bézier, is especially well suited for this purpose. A Bézier cubic is given parametrically by the equations

(9)
$$x = (1 - t)^3 x_0 + 3t(1 - t)^2 x_1 + 3t^2(1 - t)x_2 + t^3 x_3,$$
$$y = (1 - t)^3 y_0 + 3t(1 - t)^2 y_1 + 3t^2(1 - t)y_2 + t^3 y_3$$

for $0 \le t \le 1$, where $x_0, y_0, x_1, y_1, x_2, y_2, x_3, y_3$ are constants specified by the design engineer. Let us see how these constants correspond to the above geometric data.

If we evaluate the above formulas at $t = 0$ and $t = 1$, then we obtain

$$(x(0), y(0)) = (x_0, y_0),$$
$$(x(1), y(1)) = (x_3, y_3).$$

As t varies from 0 to 1, equations (9) describe a curve starting at (x_0, y_0) and ending at (x_3, y_3). This gives us half of the needed data. We will next use calculus to find the tangent directions when $t = 0$ and 1. We know that the tangent vector to (9) when $t = 0$ is $(x'(0), y'(0))$. To calculate $x'(0)$, we differentiate the first line of (9) to obtain

$$x' = -3(1 - t)^2 x_0 + 3((1 - t)^2 - 2t(1 - t))x_1 + 3(2t(1 - t) - t^2)x_2 + 3t^2 x_3.$$

Then substituting $t = 0$ yields

$$x'(0) = -3x_0 + 3x_1 = 3(x_1 - x_0),$$

and from here, it is straightforward to show that

(10)
$$(x'(0), y'(0)) = 3(x_1 - x_0, y_1 - y_0),$$
$$(x'(1), y'(1)) = 3(x_3 - x_2, y_3 - y_2).$$

Since $(x_1 - x_0, y_1 - y_0) = (x_1, y_1) - (x_0, y_0)$, it follows that $(x'(0), y'(0))$ is three times the vector from (x_0, y_0) to (x_1, y_1). Hence, by placing (x_1, y_1), the designer can control the tangent direction at the beginning of the curve. In a similar way, the placement of (x_2, y_2) controls the tangent direction at the end of the curve.

The points $(x_0, y_0), (x_1, y_1), (x_2, y_2)$ and (x_3, y_3) are called the *control points* of the Bézier cubic. They are usually labeled P_0, P_1, P_2, and P_3, and the convex quadrilateral they determine is called the *control polygon*. Here is a picture of a Bézier curve together with its control polygon:

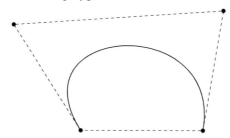

In the exercises, we will show that a Bézier cubic always lies inside its control polygon.

The data determining a Bézier cubic is thus easy to specify and has a strong geometric meaning. One issue not resolved so far is the *length* of the tangent vectors $(x'(0), y'(0))$ and $(x'(1), y'(1))$. According to (10), it is possible to change the points (x_1, y_1) and (x_2, y_2) without changing the tangent directions. For example, if we keep the same directions as in the previous picture, but lengthen the tangent vectors, then we get the following curve:

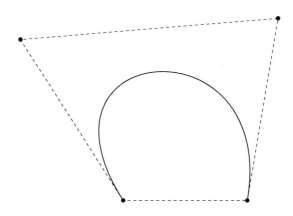

Thus, increasing the velocity at an endpoint makes the curve stay close to the tangent line for a longer distance. With practice and experience, a designer can become proficient in using Bézier cubics to create a wide variety of curves. It is interesting to note that the designer may never be aware of equations (9) that are used to describe the curve.

Besides CAGD, we should mention that Bézier cubics are also used in the page description language PostScript. The `curveto` command in PostScript has the coordinates of the control points as input and the Bézier cubic as output. This is how the above Bézier cubics were drawn—each curve was specified by a single `curveto` instruction in a PostScript file.

EXERCISES FOR §3

1. Parametrize all solutions of the linear equations

$$x + 2y - 2z + w = -1,$$
$$x + y + z - w = 2.$$

2. Use a trigonometric identity to show that

$$x = \cos(t),$$
$$y = \cos(2t)$$

parametrizes a portion of a parabola. Indicate exactly what portion of the parabola is covered.

3. Given $f \in k[x]$, find a parametrization of $\mathbf{V}(y - f(x))$.
4. Consider the parametric representation

$$x = \frac{t}{1+t},$$

$$y = 1 - \frac{1}{t^2}.$$

a. Find the equation of the affine variety determined by the above parametric equations.
b. Show that the above equations parametrize all points of the variety found in part (a) except for the point $(1, 1)$.

5. This problem will be concerned with the hyperbola $x^2 - y^2 = 1$.

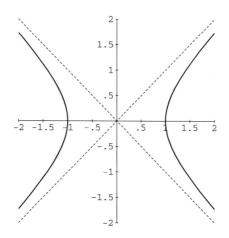

a. Just as trigonometric functions are used to parametrize the circle, hyperbolic functions are used to parametrize the hyperbola. Show that the point

$$x = \cosh(t),$$
$$y = \sinh(t)$$

always lies on $x^2 - y^2 = 1$. What portion of the hyperbola is covered?
b. Show that a straight line meets a hyperbola in 0, 1, or 2 points, and illustrate your answer with a picture. Hint: Consider the cases $x = a$ and $y = mx + b$ separately.
c. Adapt the argument given at the end of the section to derive a parametrization of the hyperbola. Hint: Consider nonvertical lines through the point $(-1, 0)$ on the hyperbola.
d. The parametrization you found in part (c) is undefined for two values of t. Explain how this relates to the asymptotes of the hyperbola.

6. The goal of this problem is to show that the sphere $x^2 + y^2 + z^2 = 1$ in 3-dimensional space can be parametrized by

$$x = \frac{2u}{u^2 + v^2 + 1},$$

$$y = \frac{2v}{u^2 + v^2 + 1},$$

$$z = \frac{u^2 + v^2 - 1}{u^2 + v^2 + 1}.$$

The idea is to adapt the argument given at the end of the section to 3-dimensional space.

a. Given a point $(u, v, 0)$ in the (x, y)-plane, draw the line from this point to the "north pole" $(0, 0, 1)$ of the sphere, and let (x, y, z) be the other point where the line meets the sphere. Draw a picture to illustrate this, and argue geometrically that mapping (u, v) to (x, y, z) gives a parametrization of the sphere minus the north pole.

b. Show that the line connecting $(0, 0, 1)$ to $(u, v, 0)$ is parametrized by $(tu, tv, 1 - t)$, where t is a parameter that moves along the line.

c. Substitute $x = tu, y = tv$ and $z = 1 - t$ into the equation for the sphere $x^2 + y^2 + z^2 = 1$. Use this to derive the formulas given at the beginning of the problem.

7. Adapt the argument of the previous exercise to parametrize the "sphere" $x_1^2 + \cdots + x_n^2 = 1$ in n-dimensional affine space. Hint: There will be $n - 1$ parameters.

8. Consider the curve defined by $y^2 = cx^2 - x^3$, where c is some constant. Here is a picture of the curve when $c > 0$:

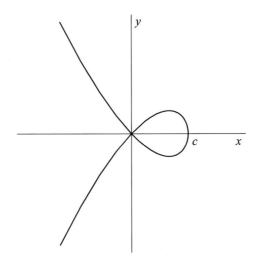

Our goal is to parametrize this curve.

a. Show that a line will meet this curve at either 0, 1, 2, or 3 points. Illustrate your answer with a picture. Hint: Let the equation of the line be either $x = a$ or $y = mx + b$.

b. Show that a nonvertical line through the origin meets the curve at exactly one other point when $m^2 \neq c$. Draw a picture to illustrate this, and see if you can come up with an intuitive explanation as to why this happens.

c. Now draw the vertical line $x = 1$. Given a point $(1, t)$ on this line, draw the line connecting $(1, t)$ to the origin. This will intersect the curve in a point (x, y). Draw a picture to illustrate this, and argue geometrically that this gives a parametrization of the entire curve.

d. Show that the geometric description from part (c) leads to the parametrization

$$x = c - t^2,$$
$$y = t(c - t^2).$$

9. The *strophoid* is a curve that was studied by various mathematicians, including Isaac Barrow (1630–1677), Jean Bernoulli (1667–1748), and Maria Agnesi (1718–1799). A trigonometric parametrization is given by

$$x = a\sin(t),$$
$$y = a\tan(t)(1 + \sin(t))$$

where a is a constant. If we let t vary in the range $-4.5 \le t \le 1.5$, we get the picture shown here.

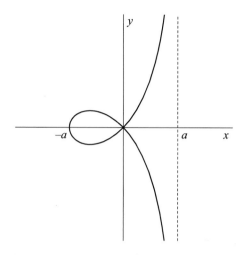

a. Find the equation in x and y that describes the strophoid. Hint: If you are sloppy, you will get the equation $(a^2 - x^2)y^2 = x^2(a + x)^2$. To see why this is not quite correct, see what happens when $x = -a$.

b. Find an algebraic parametrization of the strophoid.

10. Around 180 B.C.E., Diocles wrote the book *On Burning-Glasses*. One of the curves he considered was the *cissoid* and he used it to solve the problem of the duplication of the cube [see part (c) below]. The cissoid has the equation $y^2(a + x) = (a - x)^3$, where a is a constant. This gives the following curve in the plane:

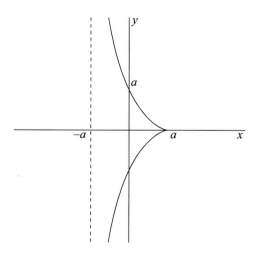

a. Find an algebraic parametrization of the cissoid.
b. Diocles described the cissoid using the following geometric construction. Given a circle of radius a (which we will take as centered at the origin), pick x between a and $-a$, and draw the line L connecting $(a, 0)$ to the point $P = (-x, \sqrt{a^2 - x^2})$ on the circle. This determines a point $Q = (x, y)$ on L:

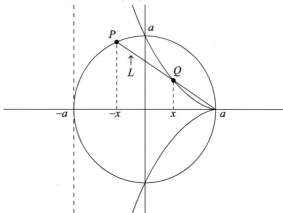

Prove that the cissoid is the locus of all such points Q.
c. The duplication of the cube is the classical Greek problem of trying to construct $\sqrt[3]{2}$ using ruler and compass. It is known that this is impossible given just a ruler and compass. Diocles showed that if in addition, you allow the use of the cissoid, then one can construct $\sqrt[3]{2}$. Here is how it works. Draw the line connecting $(-a, 0)$ to $(0, a/2)$. This line will meet the cissoid at a point (x, y). Then prove that

$$2 = \left(\frac{a - x}{y}\right)^3,$$

which shows how to construct $\sqrt[3]{2}$ using ruler, compass, and cissoid.

11. In this problem, we will derive the parametrization

$$x = t(u^2 - t^2),$$
$$y = u,$$
$$z = u^2 - t^2,$$

of the surface $x^2 - y^2 z^2 + z^3 = 0$ considered in the text.
a. Adapt the formulas in part (d) of Exercise 8 to show that the curve $x^2 = cz^2 - z^3$ is parametrized by

$$z = c - t^2,$$
$$x = t(c - t^2).$$

b. Now replace the c in part (a) by y^2, and explain how this leads to the above parametrization of $x^2 - y^2 z^2 + z^3 = 0$.
c. Explain why this parametrization covers the entire surface $\mathbf{V}(x^2 - y^2 z^2 + z^3)$. Hint: See part (c) of Exercise 8.

12. Consider the variety $V = \mathbf{V}(y - x^2, z - x^4) \subseteq \mathbb{R}^3$.
a. Draw a picture of V.

    b. Parametrize V in a way similar to what we did with the twisted cubic.

    c. Parametrize the tangent surface of V.

13. The general problem of finding the equation of a parametrized surface will be studied in Chapters 2 and 3. However, when the surface is a plane, methods from calculus or linear algebra can be used. For example, consider the plane in $\mathbb{R}^3$ parametrized by

$$x = 1 + u - v,$$
$$y = u + 2v,$$
$$z = -1 - u + v.$$

Find the equation of the plane determined this way. Hint: Let the equation of the plane be $ax + by + cz = d$. Then substitute in the above parametrization to obtain a system of equations for a, b, c, d. Another way to solve the problem would be to write the parametrization in vector form as $(1, 0, -1) + u(1, 1, -1) + v(-1, 2, 1)$. Then one can get a quick solution using the cross product.

14. This problem deals with convex sets and will be used in the next exercise to show that a Bézier cubic lies within its control polygon. A subset $C \subseteq \mathbb{R}^2$ is *convex* if for all $P, Q \in C$, the line segment joining P to Q also lies in C.

    a. If $P = \begin{pmatrix} x \\ y \end{pmatrix}$ and $Q = \begin{pmatrix} z \\ w \end{pmatrix}$ lie in a convex set C, then show that

$$t \begin{pmatrix} x \\ y \end{pmatrix} + (1 - t) \begin{pmatrix} z \\ w \end{pmatrix} \in C$$

    when $0 \le t \le 1$.

    b. If $P_i = \begin{pmatrix} x_i \\ y_i \end{pmatrix}$ lies in a convex set C for $1 \le i \le n$, then show that

$$\sum_{i=1}^{n} t_i \begin{pmatrix} x_i \\ y_i \end{pmatrix} \in C$$

    wherever $t_1, \ldots, t_n$ are nonnegative numbers such that $\sum_{i=1}^{n} t_i = 1$. Hint: Use induction on n.

15. Let a Bézier cubic be given by

$$x = (1 - t)^3 x_0 + 3t(1 - t)^2 x_1 + 3t^2(1 - t)x_2 + t^3 x_3,$$
$$y = (1 - t)^3 y_0 + 3t(1 - t)^2 y_1 + 3t^2(1 - t)y_2 + t^3 y_3.$$

    a. Show that the above equations can be written in vector form

$$\begin{pmatrix} x \\ y \end{pmatrix} = (1 - t)^3 \begin{pmatrix} x_0 \\ y_0 \end{pmatrix} + 3t(1 - t)^2 \begin{pmatrix} x_1 \\ y_1 \end{pmatrix} + 3t^2(1 - t) \begin{pmatrix} x_2 \\ y_2 \end{pmatrix} + t^3 \begin{pmatrix} x_3 \\ y_3 \end{pmatrix}.$$

    b. Use the previous exercise to show that a Bézier cubic always lies inside its control polygon. Hint: In the above equations, what is the sum of the coefficients?

16. One disadvantage of Bézier cubics is that curves like circles and hyperbolas cannot be described exactly by cubics. In this exercise, we will discuss a method similar to example (4) for parametrizing conic sections. Our treatment is based on BALL (1987) [see also GOLDMAN (2003), Section 5.7].

    A *conic section* is a curve in the plane defined by a second degree equation of the form $ax^2 + bxy + cy^2 + dx + ey + f = 0$. Conic sections include the familiar examples of circles, ellipses, parabolas, and hyperbolas. Now consider the curve parametrized by

$$x = \frac{(1-t)^2 x_1 + 2t(1-t)wx_2 + t^2 x_3}{(1-t)^2 + 2t(1-t)w + t^2},$$

$$y = \frac{(1-t)^2 y_1 + 2t(1-t)wy_2 + t^2 y_3}{(1-t)^2 + 2t(1-t)w + t^2}$$

for $0 \leq t \leq 1$. The constants $w, x_1, y_1, x_2, y_2, x_3, y_3$ are specified by the design engineer, and we will assume that $w \geq 0$. In Chapter 3, we will show that these equations parametrize a conic section. The goal of this exercise is to give a geometric interpretation for the quantities $w, x_1, y_1, x_2, y_2, x_3, y_3$.

a. Show that our assumption $w \geq 0$ implies that the denominator in the above formulas never vanishes.

b. Evaluate the above formulas at $t = 0$ and $t = 1$. This should tell you what x_1, y_1, x_3, y_3 mean.

c. Now compute $(x'(0), y'(0))$ and $(x'(1), y'(1))$. Use this to show that (x_2, y_2) is the intersection of the tangent lines at the start and end of the curve. Explain why $(x_1, y_1), (x_2, y_2),$ and (x_3, y_3) are called the *control points* of the curve.

d. Define the *control polygon* (it is actually a triangle in this case), and prove that the curve defined by the above equations always lies in its control polygon. Hint: Adapt the argument of the previous exercise. This gives the following picture:

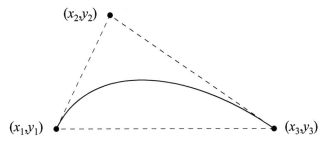

It remains to explain the constant w, which is called the *shape factor*. A hint should come from the answer to part (c), for note that w appears in the formulas for the tangent vectors when $t = 0$ and 1. So w somehow controls the "velocity," and a larger w should force the curve closer to (x_2, y_2). In the last two parts of the problem, we will determine exactly what w does.

e. Prove that

$$\begin{pmatrix} x(\frac{1}{2}) \\ y(\frac{1}{2}) \end{pmatrix} = \frac{1}{1+w} \left(\frac{1}{2} \begin{pmatrix} x_1 \\ y_1 \end{pmatrix} + \frac{1}{2} \begin{pmatrix} x_3 \\ y_3 \end{pmatrix} \right) + \frac{w}{1+w} \begin{pmatrix} x_2 \\ y_2 \end{pmatrix}.$$

Use this formula to show that $(x(\frac{1}{2}), y(\frac{1}{2}))$ lies on the line segment connecting (x_2, y_2) to the midpoint of the line between (x_1, y_1) and (x_3, y_3).

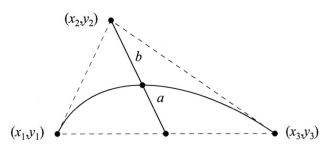

f. Notice that $(x(\frac{1}{2}), y(\frac{1}{2}))$ divides this line segment into two pieces, say of lengths a and b as indicated in the above picture. Then prove that

$$w = \frac{a}{b},$$

so that w tells us exactly where the curve crosses this line segment. Hint: Use the distance formula.

17. Use the formulas of the previous exercise to parametrize the arc of the circle $x^2 + y^2 = 1$ from $(1, 0)$ to $(0, 1)$. Hint: Use part (f) of Exercise 16 to show that $w = 1/\sqrt{2}$.

§4 Ideals

We next define the basic algebraic objects studied in this book.

Definition 1. A subset $I \subseteq k[x_1, \ldots, x_n]$ is an **ideal** if it satisfies:

(i) $0 \in I$.

(ii) If $f, g \in I$, then $f + g \in I$.

(iii) If $f \in I$ and $h \in k[x_1, \ldots, x_n]$, then $hf \in I$.

The goal of this section is to introduce the reader to some naturally occurring ideals and to see how ideals relate to affine varieties. The real importance of ideals is that they will give us a language for computing with affine varieties.

The first natural example of an ideal is the ideal generated by a finite number of polynomials.

Definition 2. Let $f_1, \ldots, f_s$ be polynomials in $k[x_1, \ldots, x_n]$. Then we set

$$\langle f_1, \ldots, f_s \rangle = \left\{ \sum_{i=1}^{s} h_i f_i \,\middle|\, h_1, \ldots, h_s \in k[x_1, \ldots, x_n] \right\}.$$

The crucial fact is that $\langle f_1, \ldots, f_s \rangle$ is an ideal.

Lemma 3. *If* $f_1, \ldots, f_s \in k[x_1, \ldots, x_n]$, *then* $\langle f_1, \ldots, f_s \rangle$ *is an ideal of* $k[x_1, \ldots, x_n]$. *We will call* $\langle f_1, \ldots, f_s \rangle$ *the* **ideal generated by** $f_1, \ldots, f_s$.

Proof. First, $0 \in \langle f_1, \ldots, f_s \rangle$ since $0 = \sum_{i=1}^{s} 0 \cdot f_i$. Next, suppose that $f = \sum_{i=1}^{s} p_i f_i$ and $g = \sum_{i=1}^{s} q_i f_i$, and let $h \in k[x_1, \ldots, x_n]$. Then the equations

$$f + g = \sum_{i=1}^{s} (p_i + q_i) f_i,$$

$$hf = \sum_{i=1}^{s} (hp_i) f_i$$

complete the proof that $\langle f_1, \ldots, f_s \rangle$ is an ideal. $\qquad\square$

The ideal $\langle f_1, \ldots, f_s \rangle$ has a nice interpretation in terms of polynomial equations. Given $f_1, \ldots, f_s \in k[x_1, \ldots, x_n]$, we get the system of equations

$$f_1 = 0,$$
$$\vdots$$
$$f_s = 0.$$

From these equations, one can derive others using algebra. For example, if we multiply the first equation by $h_1 \in k[x_1, \ldots, x_n]$, the second by $h_2 \in k[x_1, \ldots, x_n]$, etc., and then add the resulting equations, we obtain

$$h_1 f_1 + h_2 f_2 + \cdots + h_s f_s = 0,$$

which is a consequence of our original system. Notice that the left-hand side of this equation is exactly an element of the ideal $\langle f_1, \ldots, f_s \rangle$. Thus, we can think of $\langle f_1, \ldots, f_s \rangle$ as consisting of all "polynomial consequences" of the equations $f_1 = f_2 = \cdots = f_s = 0$.

To see what this means in practice, consider the example from §3 where we took

$$x = 1 + t,$$
$$y = 1 + t^2$$

and eliminated t to obtain

$$y = x^2 - 2x + 2$$

[see the discussion following equation (7) in §3]. Let us redo this example using the above ideas. We start by writing the equations as

(1)
$$x - 1 - t = 0,$$
$$y - 1 - t^2 = 0.$$

To cancel the terms involving t, we multiply the first equation by $x - 1 + t$ and the second by -1:

$$(x - 1)^2 - t^2 = 0,$$
$$-y + 1 + t^2 = 0,$$

and then add to obtain

$$(x - 1)^2 - y + 1 = x^2 - 2x + 2 - y = 0.$$

In terms of the ideal generated by equations (1), we can write this as

$$x^2 - 2x + 2 - y = (x - 1 + t)(x - 1 - t) + (-1)(y - 1 - t^2)$$
$$\in \langle x - 1 - t, y - 1 - t^2 \rangle.$$

Similarly, any other "polynomial consequence" of (1) leads to an element of this ideal.

We say that an ideal I is *finitely generated* if there exist $f_1, \ldots, f_s \in k[x_1, \ldots, x_n]$ such that $I = \langle f_1, \ldots, f_s \rangle$, and we say that $f_1, \ldots, f_s$, are a *basis* of I. In Chapter 2, we will prove the amazing fact that *every* ideal of $k[x_1, \ldots, x_n]$ is finitely generated (this is known as the Hilbert Basis Theorem). Note that a given ideal may have many different bases. In Chapter 2, we will show that one can choose an especially useful type of basis, called a Gröbner basis.

There is a nice analogy with linear algebra that can be made here. The definition of an ideal is similar to the definition of a subspace: both have to be closed under addition and multiplication, except that, for a subspace, we multiply by scalars, whereas for an ideal, we multiply by polynomials. Further, notice that the ideal generated by polynomials $f_1, \ldots, f_s$ is similar to the span of a finite number of vectors $v_1, \ldots, v_s$. In each case, one takes linear combinations, using field coefficients for the span and polynomial coefficients for the ideal. Relations with linear algebra are explored further in Exercise 6.

Another indication of the role played by ideals is the following proposition, which shows that a variety depends only on the *ideal* generated by its defining equations.

Proposition 4. *If $f_1, \ldots, f_s$ and $g_1, \ldots, g_t$ are bases of the same ideal in $k[x_1, \ldots, x_n]$, so that $\langle f_1, \ldots, f_s \rangle = \langle g_1, \ldots, g_t \rangle$, then we have $\mathbf{V}(f_1, \ldots, f_s) = \mathbf{V}(g_1, \ldots, g_t)$.*

Proof. The proof is very straightforward and is left as an exercise. $\qquad \square$

As an example, consider the variety $\mathbf{V}(2x^2 + 3y^2 - 11, x^2 - y^2 - 3)$. It is easy to show that $\langle 2x^2 + 3y^2 - 11, x^2 - y^2 - 3 \rangle = \langle x^2 - 4, y^2 - 1 \rangle$ (see Exercise 3), so that

$$\mathbf{V}(2x^2 + 3y^2 - 11, x^2 - y^2 - 3) = \mathbf{V}(x^2 - 4, y^2 - 1) = \{(\pm 2, \pm 1)\}$$

by the above proposition. Thus, by changing the basis of the ideal, we made it easier to determine the variety.

The ability to change the basis without affecting the variety is very important. Later in the book, this will lead to the observation that affine varieties are determined by *ideals*, not equations. (In fact, the correspondence between ideals and varieties is the main topic of Chapter 4.) From a more practical point of view, we will also see that Proposition 4, when combined with the Gröbner bases mentioned above, provides a powerful tool for understanding affine varieties.

We will next discuss how affine varieties give rise to an interesting class of ideals. Suppose we have an affine variety $V = \mathbf{V}(f_1, \ldots, f_s) \subseteq k^n$ defined by $f_1, \ldots, f_s \in k[x_1, \ldots, x_n]$. We know that $f_1, \ldots, f_s$ vanish on V, but are these the only ones? Are there other polynomials that vanish on V? For example, consider the twisted cubic studied in §2. This curve is defined by the vanishing of $y - x^2$ and $z - x^3$. From the parametrization (t, t^2, t^3) discussed in §3, we see that $z - xy$ and $y^2 - xz$ are two more polynomials that vanish on the twisted cubic. Are there other such polynomials? How do we find them all?

To study this question, we will consider the set of *all* polynomials that vanish on a given variety.

Definition 5. Let $V \subseteq k^n$ be an affine variety. Then we set

$$\mathbf{I}(V) = \{f \in k[x_1, \ldots, x_n] \mid f(a_1, \ldots, a_n) = 0 \text{ for all } (a_1, \ldots, a_n) \in V\}.$$

The crucial observation is that $\mathbf{I}(V)$ is an ideal.

Lemma 6. *If $V \subseteq k^n$ is an affine variety, then $\mathbf{I}(V) \subseteq k[x_1, \ldots, x_n]$ is an ideal. We will call $\mathbf{I}(V)$ the **ideal of** V.*

Proof. It is obvious that $0 \in \mathbf{I}(V)$ since the zero polynomial vanishes on all of k^n, and so, in particular, it vanishes on V. Next, suppose that $f, g \in \mathbf{I}(V)$ and $h \in k[x_1, \ldots, x_n]$.

Let $(a_1, \ldots, a_n)$ be an arbitrary point of V. Then

$$f(a_1, \ldots, a_n) + g(a_1, \ldots, a_n) = 0 + 0 = 0,$$
$$h(a_1, \ldots, a_n)f(a_1, \ldots, a_n) = h(a_1, \ldots, a_n) \cdot 0 = 0,$$

and it follows that $\mathbf{I}(V)$ is an ideal. $\square$

For an example of the ideal of a variety, consider the variety $\{(0,0)\}$ consisting of the origin in k^2. Then its ideal $\mathbf{I}(\{(0,0)\})$ consists of all polynomials that vanish at the origin, and we claim that

$$\mathbf{I}(\{(0,0)\}) = \langle x, y \rangle.$$

One direction of proof is trivial, for any polynomial of the form $A(x,y)x + B(x,y)y$ obviously vanishes at the origin. Going the other way, suppose that $f = \sum_{i,j} a_{ij}x^i y^j$ vanishes at the origin. Then $a_{00} = f(0,0) = 0$ and, consequently,

$$f = a_{00} + \sum_{i,j \neq 0,0} a_{ij}x^i y^j$$

$$= 0 + \left(\sum_{\substack{i,j \\ i>0}} a_{ij}x^{i-1}y^j\right)x + \left(\sum_{j>0} a_{0j}y^{j-1}\right)y \in \langle x, y \rangle.$$

Our claim is now proved.

For another example, consider the case when V is all of k^n. Then $\mathbf{I}(k^n)$ consists of polynomials that vanish everywhere, and, hence, by Proposition 5 of §1, we have

$$\mathbf{I}(k^n) = \{0\} \quad \text{when } k \text{ is infinite.}$$

(Here, "0" denotes the zero polynomial in $k[x_1, \ldots, x_n]$.) Note that Proposition 5 of §1 is equivalent to the above statement. In the exercises, we will discuss what happens when k is a finite field.

A more interesting example is given by the twisted cubic $V = \mathbf{V}(y - x^2, z - x^3)$ in $\mathbb{R}^3$. We claim that

$$\mathbf{I}(V) = \langle y - x^2, z - x^3 \rangle.$$

To prove this, we will first show that given a polynomial $f \in \mathbb{R}[x, y, z]$, we can write f in the form

$$(2) \qquad f = h_1(y - x^2) + h_2(z - x^3) + r,$$

where $h_1, h_2 \in \mathbb{R}[x, y, z]$ and r is a polynomial in the variable x alone. First, consider the case when f is a monomial $x^\alpha y^\beta z^\gamma$. Then the binomial theorem tells us that

$$\begin{aligned} x^\alpha y^\beta z^\gamma &= x^\alpha (x^2 + (y - x^2))^\beta (x^3 + (z - x^3))^\gamma \\ &= x^\alpha (x^{2\beta} + \text{terms involving } y - x^2)(x^{3\gamma} + \text{terms involving } z - x^3), \end{aligned}$$

and multiplying this out shows that

$$x^\alpha y^\beta z^\gamma = h_1(y - x^2) + h_2(z - x^3) + x^{\alpha + 2\beta + 3\gamma}$$

for some polynomials $h_1, h_2 \in \mathbb{R}[x, y, z]$. Thus, (2) is true in this case. Since an arbitrary $f \in \mathbb{R}[x, y, z]$ is an $\mathbb{R}$-linear combination of monomials, it follows that (2) holds in general.

We can now prove $\mathbf{I}(V) = \langle y - x^2, z - x^3 \rangle$. First, by the definition of the twisted cubic V, we have $y - x^2, z - x^3 \in \mathbf{I}(V)$, and since $\mathbf{I}(V)$ is an ideal, it follows that $h_1(y - x^2) + h_2(z - x^3) \in \mathbf{I}(V)$. This proves that $\langle y - x^2, z - x^3 \rangle \subseteq \mathbf{I}(V)$. To prove the opposite inclusion, let $f \in \mathbf{I}(V)$ and let

$$f = h_1(y - x^2) + h_2(z - x^3) + r$$

be the expression given by (2). To prove that r is zero, we will use the parametrization (t, t^2, t^3) of the twisted cubic. Since f vanishes on V, we obtain

$$0 = f(t, t^2, t^3) = 0 + 0 + r(t)$$

(recall that r is a polynomial in x alone). Since t can be any real number, $r \in \mathbb{R}[x]$ must be the zero polynomial by Proposition 5 of §1. But $r = 0$ shows that f has the desired form, and $\mathbf{I}(V) = \langle y - x^2, z - x^3 \rangle$ is proved.

What we did in (2) is reminiscent of the division of polynomials, except that we are dividing by two polynomials instead of one. In fact, (2) is a special case of the generalized division algorithm to be studied in Chapter 2.

A nice corollary of the above example is that given a polynomial $f \in \mathbb{R}[x, y, z]$, we have $f \in \langle y - x^2, z - x^3 \rangle$ if and only if $f(t, t^2, t^3)$ is identically zero. This gives us an algorithm for deciding whether a polynomial lies in the ideal. However, this method is dependent on the parametrization (t, t^2, t^3). Is there a way of deciding whether $f \in \langle y - x^2, z - x^3 \rangle$ without using the parametrization? In Chapter 2, we will answer this question positively using Gröbner bases and the generalized division algorithm.

The example of the twisted cubic is very suggestive. We started with the polynomials $y - x^2$ and $z - x^3$, used them to define an affine variety, took all functions vanishing on the variety, and got back the ideal generated by the two polynomials. It is natural to wonder if this happens in general. So take $f_1, \ldots, f_s \in k[x_1, \ldots, x_n]$. This gives us

$$\text{polynomials} \qquad \text{variety} \qquad \text{ideal}$$
$$f_1, \ldots, f_s \quad \longrightarrow \quad \mathbf{V}(f_1, \ldots, f_s) \quad \longrightarrow \quad \mathbf{I}(\mathbf{V}(f_1, \ldots, f_s)),$$

and the natural question to ask is whether $\mathbf{I}(\mathbf{V}(f_1, \ldots, f_s)) = \langle f_1, \ldots, f_s \rangle$? The answer, unfortunately, is not always yes. Here is the best answer we can give at this point.

Lemma 7. *Let* $f_1, \ldots, f_s \in k[x_1, \ldots, x_n]$. *Then* $\langle f_1, \ldots, f_s \rangle \subseteq \mathbf{I}(\mathbf{V}(f_1, \ldots, f_s))$, *although equality need not occur.*

Proof. Let $f \in \langle f_1, \ldots, f_s \rangle$, which means that $f = \sum_{i=1}^{s} h_i f_i$ for some polynomials $h_1, \ldots, h_s \in k[x_1, \ldots, x_n]$. Since $f_1, \ldots, f_s$ vanish on $\mathbf{V}(f_1, \ldots, f_s)$, so must $\sum_{i=1}^{s} h_i f_i$. Thus, f vanishes on $\mathbf{V}(f_1, \ldots, f_s)$, which proves $f \in \mathbf{I}(\mathbf{V}(f_1, \ldots, f_s))$.

For the second part of the lemma, we need an example where $\mathbf{I}(\mathbf{V}(f_1, \ldots, f_s))$ is strictly larger than $\langle f_1, \ldots, f_s \rangle$. We will show that the inclusion

$$\langle x^2, y^2 \rangle \subseteq \mathbf{I}(\mathbf{V}(x^2, y^2))$$

is not an equality. We first compute $\mathbf{I}(\mathbf{V}(x^2, y^2))$. The equations $x^2 = y^2 = 0$ imply that $\mathbf{V}(x^2, y^2) = \{(0,0)\}$. But an earlier example showed that the ideal of $\{(0,0)\}$ is $\langle x, y \rangle$, so that $\mathbf{I}(\mathbf{V}(x^2, y^2)) = \langle x, y \rangle$. To see that this is strictly larger than $\langle x^2, y^2 \rangle$, note that $x \notin \langle x^2, y^2 \rangle$ since for polynomials of the form $h_1(x, y)x^2 + h_2(x, y)y^2$, every monomial has total degree at least two. $\qquad \square$

For arbitrary fields, the relationship between $\langle f_1, \ldots, f_s \rangle$ and $\mathbf{I}(\mathbf{V}(f_1, \ldots, f_s))$ can be rather subtle (see the exercises for some examples). However, over an algebraically closed field like $\mathbb{C}$, there is a straightforward relation between these ideals. This will be explained when we prove the Nullstellensatz in Chapter 4.

Although for a general field, $\mathbf{I}(\mathbf{V}(f_1, \ldots, f_s))$ may not equal $\langle f_1, \ldots, f_s \rangle$, the ideal of a variety always contains enough information to determine the variety uniquely.

Proposition 8. *Let* V *and* W *be affine varieties in* k^n. *Then:*
(i) $V \subseteq W$ *if and only if* $\mathbf{I}(V) \supseteq \mathbf{I}(W)$.
(ii) $V = W$ *if and only if* $\mathbf{I}(V) = \mathbf{I}(W)$.

Proof. We leave it as an exercise to show that (ii) is an immediate consequence of (i). To prove (i), first suppose that $V \subseteq W$. Then any polynomial vanishing on W must vanish on V, which proves $\mathbf{I}(W) \subseteq \mathbf{I}(V)$. Next, assume that $\mathbf{I}(W) \subseteq \mathbf{I}(V)$. We know that W is the variety defined by some polynomials $g_1, \ldots, g_t \in k[x_1, \ldots, x_n]$. Then $g_1, \ldots, g_t \in \mathbf{I}(W) \subseteq \mathbf{I}(V)$, and hence the g_i's vanish on V. Since W consists of *all* common zeros of the g_i's, it follows that $V \subseteq W$. $\qquad \square$

There is a rich relationship between ideals and affine varieties; the material presented so far is just the tip of the iceberg. We will explore this relation further in Chapter 4. In particular, we will see that theorems proved about ideals have strong geometric implications. For now, let us list three questions we can pose concerning ideals in $k[x_1, \ldots, x_n]$:

- (Ideal Description) Can every ideal $I \subseteq k[x_1, \ldots, x_n]$ be written as $\langle f_1, \ldots, f_s \rangle$ for some $f_1, \ldots, f_s \in k[x_1, \ldots, x_n]$?
- (Ideal Membership) If $f_1, \ldots, f_s \in k[x_1, \ldots, x_n]$, is there an algorithm to decide whether a given $f \in k[x_1, \ldots, x_n]$ lies in $\langle f_1, \ldots, f_s \rangle$?
- (Nullstellensatz) Given $f_1, \ldots, f_s \in k[x_1, \ldots, x_n]$, what is the exact relation between $\langle f_1, \ldots, f_s \rangle$ and $\mathbf{I}(\mathbf{V}(f_1, \ldots, f_s))$?

In the chapters that follow, we will solve these problems completely (and we will explain where the name Nullstellensatz comes from), although we will need to be careful about which field we are working over.

EXERCISES FOR §4

1. Consider the equations

$$x^2 + y^2 - 1 = 0,$$
$$xy - 1 = 0$$

   which describe the intersection of a circle and a hyperbola.
   a. Use algebra to eliminate y from the above equations.
   b. Show how the polynomial found in part (a) lies in $\langle x^2 + y^2 - 1, xy - 1 \rangle$. Your answer should be similar to what we did in (1). Hint: Multiply the second equation by $xy + 1$.
2. Let $I \subseteq k[x_1, \ldots, x_n]$ be an ideal, and let $f_1, \ldots, f_s \in k[x_1, \ldots, x_n]$. Prove that the following statements are equivalent:
   (i) $f_1, \ldots, f_s \in I$.
   (ii) $\langle f_1, \ldots, f_s \rangle \subseteq I$.
   This fact is useful when you want to show that one ideal is contained in another.
3. Use the previous exercise to prove the following equalities of ideals in $\mathbb{Q}[x, y]$:
   a. $\langle x + y, x - y \rangle = \langle x, y \rangle$.
   b. $\langle x + xy, y + xy, x^2, y^2 \rangle = \langle x, y \rangle$.
   c. $\langle 2x^2 + 3y^2 - 11, x^2 - y^2 - 3 \rangle = \langle x^2 - 4, y^2 - 1 \rangle$.
   This illustrates that the same ideal can have many different bases and that different bases may have different numbers of elements.
4. Prove Proposition 4.
5. Show that $\mathbf{V}(x + xy, y + xy, x^2, y^2) = \mathbf{V}(x, y)$. Hint: See Exercise 3.
6. The word "basis" is used in various ways in mathematics. In this exercise, we will see that "a basis of an ideal," as defined in this section, is quite different from "a basis of a subspace," which is studied in linear algebra.
   a. First, consider the ideal $I = \langle x \rangle \subseteq k[x]$. As an ideal, I has a basis consisting of the one element x. But I can also be regarded as a subspace of $k[x]$, which is a vector space over k. Prove that any vector space basis of I over k is infinite. Hint: It suffices to find one basis that is infinite. Thus, allowing x to be multiplied by elements of $k[x]$ instead of just k is what enables $\langle x \rangle$ to have a finite basis.

b. In linear algebra, a basis must span and be linearly independent over k, whereas for an ideal, a basis is concerned only with spanning—there is no mention of any sort of independence. The reason is that once we allow polynomial coefficients, no independence is possible. To see this, consider the ideal $\langle x, y \rangle \subseteq k[x, y]$. Show that zero can be written as a linear combination of y and x with nonzero polynomial coefficients.

c. More generally, suppose that $f_1, \ldots, f_s$ is the basis of an ideal $I \subseteq k[x_1, \ldots, x_n]$. If $s \geq 2$ and $f_i \neq 0$ for all i, then show that for any i and j, zero can be written as a linear combination of f_i and f_j with nonzero polynomial coefficients.

d. A consequence of the lack of independence is that when we write an element $f \in \langle f_1, \ldots, f_s \rangle$ as $f = \sum_{i=1}^{s} h_i f_i$, the coefficients h_i are not unique. As an example, consider $f = x^2 + xy + y^2 \in \langle x, y \rangle$. Express f as a linear combination of x and y in two different ways. (Even though the h_i's are not unique, one can measure their lack of uniqueness. This leads to the interesting topic of syzygies.)

e. A basis $f_1, \ldots, f_s$ of an ideal I is said to be *minimal* if no proper subset of $f_1, \ldots, f_s$ is a basis of I. For example, x, x^2 is a basis of an ideal, but not a minimal basis since x generates the same ideal. Unfortunately, an ideal can have minimal bases consisting of different numbers of elements. To see this, show that x and $x + x^2, x^2$ are minimal bases of the same ideal of $k[x]$. Explain how this contrasts with the situation in linear algebra.

7. Show that $\mathbf{I}(\mathbf{V}(x^n, y^m)) = \langle x, y \rangle$ for any positive integers n and m.

8. The ideal $\mathbf{I}(V)$ of a variety has a special property not shared by all ideals. Specifically, we define an ideal I to be *radical* if whenever a power f^m of a polynomial f is in I, then f itself is in I. More succinctly, I is radical when $f \in I$ if and only if $f^m \in I$ for some positive integer m.

a. Prove that $\mathbf{I}(V)$ is always a radical ideal.

b. Prove that $\langle x^2, y^2 \rangle$ is not a radical ideal. This implies that $\langle x^2, y^2 \rangle \neq \mathbf{I}(V)$ for any variety $V \subseteq k^2$.

Radical ideals will play an important role in Chapter 4. In particular, the Nullstellensatz will imply that there is a one-to-one correspondence between varieties in $\mathbb{C}^n$ and radical ideals in $\mathbb{C}[x_1, \ldots, x_n]$.

9. Let $V = \mathbf{V}(y - x^2, z - x^3)$ be the twisted cubic. In the text, we showed that $\mathbf{I}(V) = \langle y - x^2, z - x^3 \rangle$.

a. Use the parametrization of the twisted cubic to show that $y^2 - xz \in \mathbf{I}(V)$.

b. Use the argument given in the text to express $y^2 - xz$ as a combination of $y - x^2$ and $z - x^3$.

10. Use the argument given in the discussion of the twisted cubic to show that $\mathbf{I}(\mathbf{V}(x - y)) = \langle x - y \rangle$. Your argument should be valid for any infinite field k.

11. Let $V \subseteq \mathbb{R}^3$ be the curve parametrized by (t, t^3, t^4).

a. Prove that V is an affine variety.

b. Adapt the method used in the case of the twisted cubic to determine $\mathbf{I}(V)$.

12. Let $V \subseteq \mathbb{R}^3$ be the curve parametrized by (t^2, t^3, t^4).

a. Prove that V is an affine variety.

b. Determine $\mathbf{I}(V)$.

This problem is quite a bit more challenging than the previous one—figuring out the proper analogue of equation (2) is not easy. Once we study the division algorithm in Chapter 2, this exercise will become much easier.

13. In Exercise 2 of §1, we showed that $x^2 y + y^2 x$ vanishes at all points of $\mathbb{F}_2^2$. More generally, let $I \subseteq \mathbb{F}_2[x, y]$ be the ideal of all polynomials that vanish at all points of $\mathbb{F}_2^2$. The goal of this exercise is to show that $I = \langle x^2 - x, y^2 - y \rangle$.

a. Show that $\langle x^2 - x, y^2 - y \rangle \subseteq I$.

b. Show that every $f \in \mathbb{F}_2[x, y]$ can be written as $f = A(x^2 - x) + B(y^2 - y) + axy + bx + cy + d$, where $A, B \in \mathbb{F}_2[x, y]$ and $a, b, c, d \in \mathbb{F}_2$. Hint: Write f in the form $\sum_i p_i(x)y^i$ and use the division algorithm (Proposition 2 of §5) to divide each p_i by $x^2 - x$. From this, you can write $f = A(x^2 - x) + q_1(y)x + q_2(y)$. Now divide q_1 and q_2 by $y^2 - y$. Again, this argument will become vastly simpler once we know the division algorithm from Chapter 2.

c. Show that $axy + bx + cy + d \in I$ if and only if $a = b = c = d = 0$.

d. Using parts (b) and (c), complete the proof that $I = \langle x^2 - x, y^2 - y \rangle$.

e. Express $x^2y + y^2x$ as a combination of $x^2 - x$ and $y^2 - y$. Hint: Remember that $2 = 1 + 1 = 0$ in $\mathbb{F}_2$.

14. This exercise is concerned with Proposition 8.
   a. Prove that part (ii) of the proposition follows from part (i).
   b. Prove the following corollary of the proposition: if V and W are affine varieties in k^n, then $V \subsetneq W$ if and only if $\mathbf{I}(V) \supsetneq \mathbf{I}(W)$.

15. In the text, we defined $\mathbf{I}(V)$ for a variety $V \subseteq k^n$. We can generalize this as follows: if $S \subseteq k^n$ is *any* subset, then we set

$$\mathbf{I}(S) = \{f \in k[x_1, \dots, x_n] \mid f(a_1, \dots, a_n) = 0 \text{ for all } (a_1, \dots, a_n) \in S\}.$$

   a. Prove that $\mathbf{I}(S)$ is an ideal.
   b. Let $X = \{(a, a) \in \mathbb{R}^2 \mid a \neq 1\}$. By Exercise 8 of §2, we know that X is not an affine variety. Determine $\mathbf{I}(X)$. Hint: What you proved in Exercise 8 of §2 will be useful. See also Exercise 10 of this section.
   c. Let $\mathbb{Z}^n$ be the points of $\mathbb{C}^n$ with integer coordinates. Determine $\mathbf{I}(\mathbb{Z}^n)$. Hint: See Exercise 6 of §1.

16. Here is more practice with ideals. Let I be an ideal in $k[x_1, \dots, x_n]$.
   a. Prove that $1 \in I$ if and only if $I = k[x_1, \dots, x_n]$.
   b. More generally, prove that I contains a nonzero constant if and only if $I = k[x_1, \dots, x_n]$.
   c. Suppose $f, g \in k[x_1, \dots, x_n]$ satisfy $f^2, g^2 \in I$. Prove that $(f + g)^3 \in I$. Hint: Expand $(f + g)^3$ using the Binomial Theorem.
   d. Now suppose $f, g \in k[x_1, \dots, x_n]$ satisfy $f^r, g^s \in I$. Prove that $(f + g)^{r+s-1} \in I$.

17. In the proof of Lemma 7, we showed that $x \notin \langle x^2, y^2 \rangle$ in $k[x, y]$.
   a. Prove that $xy \notin \langle x^2, y^2 \rangle$.
   b. Prove that $1, x, y, xy$ are the only monomials not contained in $\langle x^2, y^2 \rangle$.

18. In the text, we showed that $\mathbf{I}(\{(0, 0)\}) = \langle x, y \rangle$ in $k[x, y]$.
   a. Generalize this by proving that the origin $0 = (0, \dots, 0) \in k^n$ has the property that $\mathbf{I}(\{0\}) = \langle x_1, \dots, x_n \rangle$ in $k[x_1, \dots, x_n]$.
   b. What does part (a) say about polynomials in $k[x_1, \dots, x_n]$ with zero constant term?

19. One of the key ideas of this section is that a system of equations $f_1 = \cdots = f_s = 0$ gives the ideal $I = \langle f_1, \dots, f_s \rangle$ of polynomial consequences. Now suppose that the system has a consequence of the form $f = g$ and we take the mth power of each side to obtain $f^m = g^m$. In terms of the ideal I, this means that $f - g \in I$ should imply $f^m - g^m \in I$. Prove this by factoring $f^m - g^m$.

§5 Polynomials of One Variable

In this section, we will discuss polynomials of one variable and study the *division algorithm* from high school algebra. This simple algorithm has some surprisingly deep consequences—for example, we will use it to determine the structure of ideals of $k[x]$ and to explore the idea of a *greatest common divisor*. The theory developed will allow us to solve, in the special case of polynomials in $k[x]$, most of the problems raised in earlier sections. We will also begin to understand the important role played by algorithms.

By this point in their mathematics careers, most students have already seen a variety of algorithms, although the term "algorithm" may not have been used. Informally, an algorithm is a specific set of instructions for manipulating symbolic or numerical data. Examples are the differentiation formulas from calculus and the method of row reduction from linear algebra. An algorithm will have *inputs*, which are objects used by the algorithm, and *outputs*, which are the results of the algorithm. At each stage of execution, the algorithm must specify exactly what the next step will be.

When we are studying an algorithm, we will usually present it in "pseudocode," which will make the formal structure easier to understand. Pseudocode is similar to many common computer languages, and a brief discussion is given in Appendix B. Another reason for using pseudocode is that it indicates how the algorithm could be programmed on a computer. We should also mention that most of the algorithms in this book are implemented in Maple, *Mathematica*, and many other computer algebra systems. Appendix C has more details concerning these programs.

We begin by discussing the division algorithm for polynomials in $k[x]$. A crucial component of this algorithm is the notion of the "leading term" of a polynomial in one variable. The precise definition is as follows.

Definition 1. Given a nonzero polynomial $f \in k[x]$, let

$$f = c_0 x^m + c_1 x^{m-1} + \cdots + c_m,$$

where $c_i \in k$ and $c_0 \neq 0$ [thus, $m = \deg(f)$]. Then we say that $c_0 x^m$ is the **leading term** of f, written $\mathrm{LT}(f) = c_0 x^m$.

For example, if $f = 2x^3 - 4x + 3$, then $\mathrm{LT}(f) = 2x^3$. Notice also that if f and g are nonzero polynomials, then

(1) $\deg(f) \leq \deg(g) \iff \mathrm{LT}(f)$ divides $\mathrm{LT}(g)$.

We can now describe the division algorithm.

Proposition 2 (The Division Algorithm). *Let k be a field and let g be a nonzero polynomial in $k[x]$. Then every $f \in k[x]$ can be written as*

$$f = qg + r,$$

where $q, r \in k[x]$, and either $r = 0$ or $\deg(r) < \deg(g)$. Furthermore, q and r are unique, and there is an algorithm for finding q and r.

Proof. Here is the algorithm for finding q and r, presented in pseudocode:

> Input : g, f
> Output : q, r
>
> $q := 0; \ r := f$
> WHILE $r \neq 0$ AND LT(g) divides LT(r) DO
> $\qquad q := q + \text{LT}(r)/\text{LT}(g)$
> $\qquad r := r - (\text{LT}(r)/\text{LT}(g)) \, g$
> RETURN q, r

The WHILE...DO statement means doing the indented operations until the expression between the WHILE and DO becomes false. The statements $q := \ldots$ and $r := \ldots$ indicate that we are defining or redefining the values of q and r. Both q and r are *variables* in this algorithm—they change value at each step. We need to show that the algorithm terminates and that the final values of q and r have the required properties. (For a fuller discussion of pseudocode, see Appendix B.)

To see why this algorithm works, first note that $f = qg + r$ holds for the initial values of q and r, and that whenever we redefine q and r, the equality $f = qg + r$ remains true. This is because of the identity

$$f = qg + r = (q + \text{LT}(r)/\text{LT}(g)) \, g + (r - (\text{LT}(r)/\text{LT}(g)) \, g).$$

Next, note that the WHILE...DO statement terminates when "$r \neq 0$ and LT(g) divides LT(r)" is false, i.e., when either $r = 0$ or LT(g) does not divide LT(r). By (5), this last statement is equivalent to $\deg(r) < \deg(g)$. Thus, when the algorithm terminates, it produces q and r with the required properties.

We are not quite done; we still need to show that the algorithm terminates, i.e., that the expression between the WHILE and DO eventually becomes false (otherwise, we would be stuck in an infinite loop). The key observation is that $r - (\text{LT}(r)/\text{LT}(g)) \, g$ is either 0 or has smaller degree than r. To see why, suppose that

$$r = c_0 x^m + \cdots + c_m, \quad \text{LT}(r) = c_0 x^m,$$
$$g = d_0 x^\ell + \cdots + d_\ell, \quad \text{LT}(g) = d_0 x^\ell,$$

and suppose that $m \geq \ell$. Then

$$r - (\text{LT}(r)/\text{LT}(g)) \, g = (c_0 x^m + \cdots) - (c_0/d_0) x^{m-\ell} (d_0 x^\ell + \cdots),$$

and it follows that the degree of r must drop (or the whole expression may vanish). Since the degree is finite, it can drop at most finitely many times, which proves that the algorithm terminates.

To see how this algorithm corresponds to the process learned in high school, consider the following partially completed division:

$$
\begin{array}{r}
\frac{1}{2}x^2 \\[4pt]
2x+1\,\overline{\smash{\big)}\,x^3+2x^2+x+1} \\[2pt]
\underline{x^3+\tfrac{1}{2}x^2} \\[2pt]
\tfrac{3}{2}x^2+x+1
\end{array}
$$

Here, f and g are given by $f = x^3+2x^2+x+1$ and $g = 2x+1$, and more importantly, the current (but *not* final) values of q and r are $q = \frac{1}{2}x^2$ and $r = \frac{3}{2}x^2 + x + 1$. Now notice that the statements

$$q := q + \mathrm{LT}(r)/\mathrm{LT}(g),$$
$$r := r - \left(\mathrm{LT}(r)/\mathrm{LT}(g)\right)g$$

in the WHILE...DO loop correspond exactly to the next step in the above division.

The final step in proving the proposition is to show that q and r are unique. So suppose that $f = qg + r = q'g + r'$ where both r and r' have degree less than g (unless one or both are 0). If $r \neq r'$, then $\deg(r' - r) < \deg(g)$. On the other hand, since

(2) $$(q - q')g = r' - r,$$

we would have $q - q' \neq 0$, and consequently,

$$\deg(r' - r) = \deg((q - q')g) = \deg(q - q') + \deg(g) \geq \deg(g).$$

This contradiction forces $r = r'$, and then (2) shows that $q = q'$. This completes the proof of the proposition.   □

Most computer algebra systems implement the above algorithm [with some modifications—see VON ZUR GATHEN and GERHARD (2013)] for dividing polynomials.

A useful corollary of the division algorithm concerns the number of roots of a polynomial in one variable.

Corollary 3. *If k is a field and $f \in k[x]$ is a nonzero polynomial, then f has at most $\deg(f)$ roots in k.*

Proof. We will use induction on $m = \deg(f)$. When $m = 0$, f is a nonzero constant, and the corollary is obviously true. Now assume that the corollary holds for all polynomials of degree $m - 1$, and let f have degree m. If f has no roots in k, then we are done. So suppose a is a root in k. If we divide f by $x - a$, then Proposition 2 tells us that $f = q(x - a) + r$, where $r \in k$ since $x - a$ has degree one. To determine r, evaluate both sides at $x = a$, which gives $0 = f(a) = q(a)(a - a) + r = r$. It follows that $f = q(x - a)$. Note also that q has degree $m - 1$.

We claim that any root of f other than a is also a root of q. To see this, let $b \neq a$ be a root of f. Then $0 = f(b) = q(b)(b - a)$ implies that $q(b) = 0$ since k is a field. Since q has at most $m - 1$ roots by our inductive assumption, f has at most m roots in k. This completes the proof.   □

Corollary 3 was used to prove Proposition 5 in §1, which states that $\mathbf{I}(k^n) = \{0\}$ whenever k is infinite. This is an example of how a geometric fact can be the consequence of an algorithm.

We can also use Proposition 2 to determine the structure of all ideals of $k[x]$.

Corollary 4. *If k is a field, then every ideal of $k[x]$ can be written as $\langle f \rangle$ for some $f \in k[x]$. Furthermore, f is unique up to multiplication by a nonzero constant in k.*

Proof. Take an ideal $I \subseteq k[x]$. If $I = \{0\}$, then we are done since $I = \langle 0 \rangle$. Otherwise, let f be a nonzero polynomial of minimum degree contained in I. We claim that $\langle f \rangle = I$. The inclusion $\langle f \rangle \subseteq I$ is obvious since I is an ideal. Going the other way, take $g \in I$. By division algorithm (Proposition 2), we have $g = qf + r$, where either $r = 0$ or $\deg(r) < \deg(f)$. Since I is an ideal, $qf \in I$ and, thus, $r = g - qf \in I$. If r were not zero, then $\deg(r) < \deg(f)$, which would contradict our choice of f. Thus, $r = 0$, so that $g = qf \in \langle f \rangle$. This proves that $I = \langle f \rangle$.

To study uniqueness, suppose that $\langle f \rangle = \langle g \rangle$. Then $f \in \langle g \rangle$ implies that $f = hg$ for some polynomial h. Thus,

$$(3) \qquad\qquad \deg(f) = \deg(h) + \deg(g),$$

so that $\deg(f) \geq \deg(g)$. The same argument with f and g interchanged shows $\deg(f) \leq \deg(g)$, and it follows that $\deg(f) = \deg(g)$. Then (3) implies $\deg(h) = 0$, so that h is a nonzero constant. $\qquad\square$

In general, an ideal generated by one element is called a *principal ideal*. In view of Corollary 4, we say that $k[x]$ is a *principal ideal domain*, abbreviated PID.

The proof of Corollary 4 tells us that the generator of an ideal in $k[x]$ is the nonzero polynomial of minimum degree contained in the ideal. This description is not useful in practice, for it requires that we check the degrees of all polynomials (there are infinitely many) in the ideal. Is there a better way to find the generator? For example, how do we find a generator of the ideal

$$\langle x^4 - 1, x^6 - 1 \rangle \subseteq k[x]?$$

The tool needed to solve this problem is the greatest common divisor.

Definition 5. A **greatest common divisor** of polynomials $f, g \in k[x]$ is a polynomial h such that:

 (i) h divides f and g.
(ii) If p is another polynomial which divides f and g, then p divides h. When h has these properties, we write $h = \gcd(f, g)$.

Here are the main properties of gcd's.

Proposition 6. *Let $f, g \in k[x]$. Then:*

 (i) *$\gcd(f, g)$ exists and is unique up to multiplication by a nonzero constant in k.*
(ii) *$\gcd(f, g)$ is a generator of the ideal $\langle f, g \rangle$.*
(iii) *There is an algorithm for finding $\gcd(f, g)$.*

Proof. Consider the ideal $\langle f, g \rangle$. Since every ideal of $k[x]$ is principal (Corollary 4), there exists $h \in k[x]$ such that $\langle f, g \rangle = \langle h \rangle$. We claim that h is the gcd of f, g. To see this, first note that h divides f and g since $f, g \in \langle h \rangle$. Thus, the first part of Definition 5 is satisfied. Next, suppose that $p \in k[x]$ divides f and g. This means that $f = Cp$ and $g = Dp$ for some $C, D \in k[x]$. Since $h \in \langle f, g \rangle$, there are A, B such that $Af + Bg = h$. Substituting, we obtain

$$h = Af + Bg = ACp + BDp = (AC + BD)p,$$

which shows that p divides h. Thus, $h = \gcd(f, g)$.

This proves the existence of the gcd. To prove uniqueness, suppose that h' was another gcd of f and g. Then, by the second part of Definition 5, h and h' would each divide the other. This easily implies that h is a nonzero constant multiple of h'. Thus, part (i) of the corollary is proved, and part (ii) follows by the way we found h in the above paragraph.

The existence proof just given is not useful in practice. It depends on our ability to find a generator of $\langle f, g \rangle$. As we noted in the discussion following Corollary 4, this involves checking the degrees of infinitely many polynomials. Fortunately, there is a classic algorithm, known as the *Euclidean Algorithm*, which computes the gcd of two polynomials in $k[x]$. This is what part (iii) of the proposition is all about.

We will need the following notation. Let $f, g \in k[x]$, where $g \neq 0$, and write $f = qg + r$, where q and r are as in Proposition 2. Then we set $r = \mathrm{remainder}(f, g)$. We can now state the Euclidean Algorithm for finding $\gcd(f, g)$:

> Input : f, g
> Output : $h = \gcd(f, g)$
>
> $h := f$
> $s := g$
> WHILE $s \neq 0$ DO
> > $rem := \mathrm{remainder}(h, s)$
> > $h := s$
> > $s := rem$
> RETURN h

To see why this algorithm computes the gcd, write $f = qg + r$ as in Proposition 2. We claim that

$$(4) \qquad\qquad \gcd(f, g) = \gcd(f - qg, g) = \gcd(r, g).$$

To prove this, by part (ii) of the proposition, it suffices to show that the ideals $\langle f, g \rangle$ and $\langle f - qg, g \rangle$ are equal. We will leave this easy argument as an exercise.

We can write (4) in the form

$$\gcd(f, g) = \gcd(g, r).$$

Notice that $\deg(g) > \deg(r)$ or $r = 0$. If $r \neq 0$, we can make things yet smaller by repeating this process. Thus, we write $g = q'r + r'$ as in Proposition 2, and arguing as above, we obtain

$$\gcd(g, r) = \gcd(r, r'),$$

where $\deg(r) > \deg(r')$ or $r' = 0$. Continuing in this way, we get

(5) $$\gcd(f, g) = \gcd(g, r) = \gcd(r, r') = \gcd(r', r'') = \cdots,$$

where either the degrees drop

$$\deg(g) > \deg(r) > \deg(r') > \deg(r'') > \cdots,$$

or the process terminates when one of $r, r', r'', \ldots$ becomes 0.

We can now explain how the Euclidean Algorithm works. The algorithm has variables h and s, and we can see these variables in equation (5): the values of h are the first polynomial in each gcd, and the values of s are the second. You should check that in (5), going from one gcd to the next is exactly what is done in the WHILE...DO loop of the algorithm. Thus, at every stage of the algorithm, $\gcd(h, s) = \gcd(f, g)$.

The algorithm must terminate because the degree of s keeps dropping, so that at some stage, $s = 0$. When this happens, we have $\gcd(h, 0) = \gcd(f, g)$, and since $\langle h, 0 \rangle$ obviously equals $\langle h \rangle$, we have $\gcd(h, 0) = h$. Combining these last two equations, it follows that $h = \gcd(f, g)$ when $s = 0$. This proves that h is the gcd of f and g when the algorithm terminates, and the proof of Proposition 6 is now complete.   □

We should mention that there is also a version of the Euclidean Algorithm for finding the gcd of two integers. Most computer algebra systems have a command for finding the gcd of two polynomials (or integers) that uses a modified form of the Euclidean Algorithm [see VON ZUR GATHEN and GERHARD (2013) for more details].

For an example of how the Euclidean Algorithm works, let us compute the gcd of $x^4 - 1$ and $x^6 - 1$. First, we use the division algorithm:

$$x^4 - 1 = 0(x^6 - 1) + x^4 - 1,$$
$$x^6 - 1 = x^2(x^4 - 1) + x^2 - 1,$$
$$x^4 - 1 = (x^2 + 1)(x^2 - 1) + 0.$$

Then, by equation (5), we have

$$\gcd(x^4 - 1, x^6 - 1) = \gcd(x^6 - 1, x^4 - 1)$$
$$= \gcd(x^4 - 1, x^2 - 1) = \gcd(x^2 - 1, 0) = x^2 - 1.$$

Note that this gcd computation answers our earlier question of finding a generator for the ideal $\langle x^4 - 1, x^6 - 1 \rangle$. Namely, Proposition 6 and $\gcd(x^4 - 1, x^6 - 1) = x^2 - 1$ imply that

$$\langle x^4 - 1, x^6 - 1 \rangle = \langle x^2 - 1 \rangle.$$

At this point, it is natural to ask what happens for an ideal generated by three or more polynomials. How do we find a generator in this case? The idea is to extend the definition of gcd to more than two polynomials.

Definition 7. A **greatest common divisor** of polynomials $f_1, \dots, f_s \in k[x]$ is a polynomial h such that:

(i) h divides $f_1, \dots, f_s$.
(ii) If p is another polynomial which divides $f_1, \dots, f_s$, then p divides h.

When h has these properties, we write $h = \gcd(f_1, \dots, f_s)$.

Here are the main properties of these gcd's.

Proposition 8. *Let $f_1, \dots, f_s \in k[x]$, where $s \geq 2$. Then:*

(i) $\gcd(f_1, \dots, f_s)$ *exists and is unique up to multiplication by a nonzero constant in k.*
(ii) $\gcd(f_1, \dots, f_s)$ *is a generator of the ideal $\langle f_1, \dots, f_s \rangle$.*
(iii) *If $s \geq 3$, then $\gcd(f_1, \dots, f_s) = \gcd(f_1, \gcd(f_2, \dots, f_s))$.*
(iv) *There is an algorithm for finding $\gcd(f_1, \dots, f_s)$.*

Proof. The proofs of parts (i) and (ii) are similar to the proofs given in Proposition 6 and will be omitted. To prove part (iii), let $h = \gcd(f_2, \dots, f_s)$. We leave it as an exercise to show that

$$\langle f_1, h \rangle = \langle f_1, f_2, \dots, f_s \rangle.$$

By part (ii) of this proposition, we see that

$$\langle \gcd(f_1, h) \rangle = \langle \gcd(f_1, \dots, f_s) \rangle.$$

Then $\gcd(f_1, h) = \gcd(f_1, \dots, f_s)$ follows from the uniqueness part of Corollary 4, which proves what we want.

Finally, we need to show that there is an algorithm for finding $\gcd(f_1, \dots, f_s)$. The basic idea is to combine part (iii) with the Euclidean Algorithm. For example, suppose that we wanted to compute the gcd of four polynomials f_1, f_2, f_3, f_4. Using part (iii) of the proposition twice, we obtain

(6)
$$\gcd(f_1, f_2, f_3, f_4) = \gcd(f_1, \gcd(f_2, f_3, f_4))$$
$$= \gcd(f_1, \gcd(f_2, \gcd(f_3, f_4))).$$

Then if we use the Euclidean Algorithm three times [once for each gcd in the second line of (6)], we get the gcd of f_1, f_2, f_3, f_4. In the exercises, you will be asked to write pseudocode for an algorithm that implements this idea for an arbitrary number of polynomials. Proposition 8 is proved.   □

The gcd command in most computer algebra systems only handles two polynomials at a time. Thus, to work with more than two polynomials, you will need to use the method described in the proof of Proposition 8. For an example, consider the ideal

$$\langle x^3 - 3x + 2, x^4 - 1, x^6 - 1 \rangle \subseteq k[x].$$

We know that $\gcd(x^3 - 3x + 2, x^4 - 1, x^6 - 1)$ is a generator. Furthermore, you can check that

$$\gcd(x^3 - 3x + 2, x^4 - 1, x^6 - 1) = \gcd(x^3 - 3x + 2, \gcd(x^4 - 1, x^6 - 1))$$
$$= \gcd(x^3 - 3x + 2, x^2 - 1) = x - 1.$$

It follows that

$$\langle x^3 - 3x + 2, x^4 - 1, x^6 - 1 \rangle = \langle x - 1 \rangle.$$

More generally, given $f_1, \ldots, f_s \in k[x]$, it is clear that we now have an algorithm for finding a generator of $\langle f_1, \ldots, f_s \rangle$.

For another application of the algorithms developed here, consider the *ideal membership problem* from §4: Given $f_1, \ldots, f_s \in k[x]$, is there an algorithm for deciding whether a given polynomial $f \in k[x]$ lies in the ideal $\langle f_1, \ldots, f_s \rangle$? The answer is yes, and the algorithm is easy to describe. The first step is to use gcd's to find a generator h of $\langle f_1, \ldots, f_s \rangle$. Then, since $f \in \langle f_1, \ldots, f_s \rangle$ is equivalent to $f \in \langle h \rangle$, we need only use the division algorithm to write $f = qh + r$, where $\deg(r) < \deg(h)$. It follows that f is in the ideal if and only if $r = 0$. For example, suppose we wanted to know whether

$$x^3 + 4x^2 + 3x - 7 \in \langle x^3 - 3x + 2, x^4 - 1, x^6 - 1 \rangle.$$

We saw above that $x - 1$ is a generator of this ideal so that our question can be rephrased as the question whether

$$x^3 + 4x^2 + 3x - 7 \in \langle x - 1 \rangle.$$

Dividing, we find that

$$x^3 + 4x^2 + 3x - 7 = (x^2 + 5x + 8)(x - 1) + 1.$$

Hence $x^3 + 4x^2 + 3x - 7$ is *not* in the ideal $\langle x^3 - 3x + 2, x^4 - 1, x^6 - 1 \rangle$. In Chapter 2, we will solve the ideal membership problem for polynomials in $k[x_1, \ldots, x_n]$ using a similar strategy. We will first find a nice basis of the ideal (called a Gröbner basis) and then we will use a generalized division algorithm to determine whether or not a polynomial is in the ideal.

In the exercises, we will see that in the one-variable case, other problems posed in earlier sections can be solved algorithmically using the methods discussed here.

EXERCISES FOR §5

1. Over the complex numbers $\mathbb{C}$, Corollary 3 can be stated in a stronger form. Namely, prove that if $f \in \mathbb{C}[x]$ is a polynomial of degree $n > 0$, then f can be written in the form $f = c(x - a_1) \cdots (x - a_n)$, where $c, a_1, \ldots, a_n \in \mathbb{C}$ and $c \neq 0$. Hint: Use Theorem 7 of §1. Note that this result holds for *any* algebraically closed field.

2. Although Corollary 3 is simple to prove, it has some nice consequences. For example, consider the $n \times n$ Vandermonde determinant determined by $a_1, \ldots, a_n$ in a field k:

$$
\det
\begin{pmatrix}
1 & a_1 & a_1^2 & \cdots & a_1^{n-1} \\
1 & a_2 & a_2^2 & \cdots & a_2^{n-1} \\
\vdots & \vdots & \vdots & & \vdots \\
1 & a_n & a_n^2 & \cdots & a_n^{n-1}
\end{pmatrix}.
$$

Prove that this determinant is nonzero when the a_i's are distinct. Hint: If the determinant is zero, then the columns are linearly dependent. Show that the coefficients of the linear relation determine a polynomial of degree $\leq n - 1$ which has n roots. Then use Corollary 3.

3. The fact that every ideal of $k[x]$ is principal (generated by one element) is special to the case of polynomials in one variable. In this exercise we will see why. Namely, consider the ideal $I = \langle x, y \rangle \subseteq k[x, y]$. Prove that I is not a principal ideal. Hint: If $x = fg$, where $f, g \in k[x, y]$, then prove that f or g is a constant. It follows that the treatment of gcd's given in this section applies only to polynomials in one variable. One can compute gcd's for polynomials of ≥ 2 variables, but the theory involved is more complicated [see VON ZUR GATHEN and GERHARD (2013), Chapter 6].

4. If h is the gcd of $f, g \in k[x]$, then prove that there are $A, B \in k[x]$ such that $Af + Bg = h$.

5. If $f, g \in k[x]$, then prove that $\langle f - qg, g \rangle = \langle f, g \rangle$ for any q in $k[x]$. This will prove equation (4) in the text.

6. Given $f_1, \ldots, f_s \in k[x]$, let $h = \gcd(f_2, \ldots, f_s)$. Then use the equality $\langle h \rangle = \langle f_2, \ldots, f_s \rangle$ to show that $\langle f_1, h \rangle = \langle f_1, f_2, \ldots, f_s \rangle$. This equality is used in the proof of part (iii) of Proposition 8.

7. If you are allowed to compute the gcd of only two polynomials at a time (which is true for some computer algebra systems), give pseudocode for an algorithm that computes the gcd of polynomials $f_1, \ldots, f_s \in k[x]$, where $s > 2$. Prove that your algorithm works. Hint: See (6). This will complete the proof of part (iv) of Proposition 8.

8. Use a computer algebra system to compute the following gcd's:
   a. $\gcd(x^4 + x^2 + 1, x^4 - x^2 - 2x - 1, x^3 - 1)$.
   b. $\gcd(x^3 + 2x^2 - x - 2, x^3 - 2x^2 - x + 2, x^3 - x^2 - 4x + 4)$.

9. Use the method described in the text to decide whether $x^2 - 4$ is an element of the ideal $\langle x^3 + x^2 - 4x - 4, x^3 - x^2 - 4x + 4, x^3 - 2x^2 - x + 2 \rangle$.

10. Give pseudocode for an algorithm that has input $f, g \in k[x]$ and output $h, A, B \in k[x]$ where $h = \gcd(f, g)$ and $Af + Bg = h$. Hint: The idea is to add variables A, B, C, D to the algorithm so that $Af + Bg = h$ and $Cf + Dg = s$ remain true at every step of the algorithm. Note that the initial values of A, B, C, D are $1, 0, 0, 1$, respectively. You may find it useful to let quotient(f, g) denote the quotient of f on division by g, i.e., if the division algorithm yields $f = qg + r$, then $q = $ quotient(f, g).

11. In this exercise we will study the one-variable case of the *consistency problem* from §2. Given $f_1, \ldots, f_s \in k[x]$, this asks if there is an algorithm to decide whether $\mathbf{V}(f_1, \ldots, f_s)$ is nonempty. We will see that the answer is yes when $k = \mathbb{C}$.
    a. Let $f \in \mathbb{C}[x]$ be a nonzero polynomial. Then use Theorem 7 of §1 to show that $\mathbf{V}(f) = \emptyset$ if and only if f is constant.
    b. If $f_1, \ldots, f_s \in \mathbb{C}[x]$, prove $\mathbf{V}(f_1, \ldots, f_s) = \emptyset$ if and only if $\gcd(f_1, \ldots, f_s) = 1$.
    c. Describe (in words, not pseudocode) an algorithm for determining whether or not $\mathbf{V}(f_1, \ldots, f_s)$ is nonempty.
    When $k = \mathbb{R}$, the consistency problem is much more difficult. It requires giving an algorithm that tells whether a polynomial $f \in \mathbb{R}[x]$ has a real root.

12. This exercise will study the one-variable case of the *Nullstellensatz problem* from §4, which asks for the relation between $\mathbf{I}(\mathbf{V}(f_1, \ldots, f_s))$ and $\langle f_1, \ldots, f_s \rangle$ when $f_1, \ldots,$

$f_s \in \mathbb{C}[x]$. By using gcd's, we can reduce to the case of a single generator. So, in this problem, we will explicitly determine $\mathbf{I}(\mathbf{V}(f))$ when $f \in \mathbb{C}[x]$ is a nonconstant polynomial. Since we are working over the complex numbers, we know by Exercise 1 that f factors completely, i.e.,

$$f = c(x - a_1)^{r_1} \cdots (x - a_l)^{r_l},$$

where $a_1, \ldots, a_l \in \mathbb{C}$ are distinct and $c \in \mathbb{C} \setminus \{0\}$. Define the polynomial

$$f_{\mathrm{red}} = c(x - a_1) \cdots (x - a_l).$$

The polynomials f and f_{red} have the same roots, but their *multiplicities* may differ. In particular, all roots of f_{red} have multiplicity one. We call f_{red} the *reduced* or *square-free part* of f. The latter name recognizes that f_{red} is the square-free factor of f of largest degree.
  a. Show that $\mathbf{V}(f) = \{a_1, \ldots, a_l\}$.
  b. Show that $\mathbf{I}(\mathbf{V}(f)) = \langle f_{\mathrm{red}} \rangle$.
Whereas part (b) describes $\mathbf{I}(\mathbf{V}(f))$, the answer is not completely satisfactory because we need to factor f completely to find f_{red}. In Exercises 13, 14, and 15 we will show how to determine f_{red} without *any* factoring.
13. We will study the formal derivative of $f = c_0 x^n + c_1 x^{n-1} + \cdots + c_{n-1} x + c_n \in \mathbb{C}[x]$. The *formal derivative* is defined by the usual formulas from calculus:

$$f' = nc_0 x^{n-1} + (n-1)c_1 x^{n-2} + \cdots + c_{n-1} + 0.$$

Prove that the following rules of differentiation apply:

$$
\begin{aligned}
(af)' &= af' \quad \text{when } a \in \mathbb{C}, \\
(f + g)' &= f' + g', \\
(fg)' &= f'g + fg'.
\end{aligned}
$$

14. In this exercise we will use the differentiation properties of Exercise 13 to compute $\gcd(f, f')$ when $f \in \mathbb{C}[x]$.
  a. Suppose $f = (x - a)^r h$ in $\mathbb{C}[x]$, where $h(a) \neq 0$. Then prove that $f' = (x - a)^{r-1} h_1$, where $h_1 \in \mathbb{C}[x]$ does not vanish at a. Hint: Use the product rule.
  b. Let $f = c(x - a_1)^{r_1} \cdots (x - a_l)^{r_l}$ be the factorization of f, where $a_1, \ldots, a_l$ are distinct. Prove that f' is a product $f' = (x - a_1)^{r_1 - 1} \cdots (x - a_l)^{r_l - 1} H$, where $H \in \mathbb{C}[x]$ is a polynomial vanishing at none of $a_1, \ldots, a_l$.
  c. Prove that $\gcd(f, f') = (x - a_1)^{r_1 - 1} \cdots (x - a_l)^{r_l - 1}$.
15. Consider the square-free part f_{red} of a polynomial $f \in \mathbb{C}[x]$ defined in Exercise 12.
  a. Use Exercise 14 to prove that f_{red} is given by the formula

$$f_{\mathrm{red}} = \frac{f}{\gcd(f, f')}.$$

The virtue of this formula is that it allows us to find the square-free part without factoring f. This allows for much quicker computations.
  b. Use a computer algebra system to find the square-free part of the polynomial

$$x^{11} - x^{10} + 2x^8 - 4x^7 + 3x^5 - 3x^4 + x^3 + 3x^2 - x - 1.$$

16. Use Exercises 12 and 15 to describe (in words, not pseudocode) an algorithm whose input consists of polynomials $f_1, \ldots, f_s \in \mathbb{C}[x]$ and whose output consists of a basis of $\mathbf{I}(\mathbf{V}(f_1, \ldots, f_s))$. It is more difficult to construct such an algorithm when dealing with polynomials of more than one variable.
17. Find a basis for the ideal $\mathbf{I}(\mathbf{V}(x^5 - 2x^4 + 2x^2 - x, x^5 - x^4 - 2x^3 + 2x^2 + x - 1))$.

Chapter 2
Gröbner Bases

§1 Introduction

In Chapter 1, we have seen how the algebra of the polynomial rings $k[x_1, \ldots, x_n]$ and the geometry of affine algebraic varieties are linked. In this chapter, we will study the method of Gröbner bases, which will allow us to solve problems about polynomial ideals in an algorithmic or computational fashion. The method of Gröbner bases is also used in several powerful computer algebra systems to study specific polynomial ideals that arise in applications. In Chapter 1, we posed many problems concerning the algebra of polynomial ideals and the geometry of affine varieties. In this chapter and the next, we will focus on four of these problems.

Problems

a. The IDEAL DESCRIPTION PROBLEM: Does every ideal $I \subseteq k[x_1, \ldots, x_n]$ have a finite basis? In other words, can we write $I = \langle f_1, \ldots, f_s \rangle$ for $f_i \in k[x_1, \ldots, x_n]$?
b. The IDEAL MEMBERSHIP PROBLEM: Given $f \in k[x_1, \ldots, x_n]$ and an ideal $I = \langle f_1, \ldots, f_s \rangle$, determine if $f \in I$. Geometrically, this is closely related to the problem of determining whether $\mathbf{V}(f_1, \ldots, f_s)$ lies on the variety $\mathbf{V}(f)$.
c. The PROBLEM OF SOLVING POLYNOMIAL EQUATIONS: Find all common solutions in k^n of a system of polynomial equations

$$f_1(x_1, \ldots, x_n) = \cdots = f_s(x_1, \ldots, x_n) = 0.$$

This is the same as asking for the points in the affine variety $\mathbf{V}(f_1, \ldots, f_s)$.
d. The IMPLICITIZATION PROBLEM: Let $V \subseteq k^n$ be given parametrically as

$$x_1 = g_1(t_1, \ldots, t_m),$$
$$\vdots$$
$$x_n = g_n(t_1, \ldots, t_m).$$

© Springer International Publishing Switzerland 2015
D.A. Cox et al., *Ideals, Varieties, and Algorithms*, Undergraduate Texts
in Mathematics, DOI 10.1007/978-3-319-16721-3_2

If the g_i are polynomials (or rational functions) in the variables t_j, then V will be an affine variety or part of one. Find a system of polynomial equations (in the x_i) that defines the variety.

Some comments are in order. Problem (a) asks whether every polynomial ideal has a finite description via generators. Many of the ideals we have seen so far do have such descriptions—indeed, the way we have specified most of the ideals we have studied has been to give a finite generating set. However, there are other ways of constructing ideals that do not lead directly to this sort of description. The main example we have seen is the ideal of a variety, $\mathbf{I}(V)$. It will be useful to know that these ideals also have finite descriptions. On the other hand, in the exercises, we will see that if we allow *infinitely* many variables to appear in our polynomials, then the answer to Problem (a) is no.

Note that Problems (c) and (d) are, so to speak, inverse problems. In Problem (c), we ask for the set of solutions of a given system of polynomial equations. In Problem (d), on the other hand, we are given the solutions, and the problem is to find a system of equations with those solutions.

To begin our study of Gröbner bases, let us consider some special cases in which you have seen algorithmic techniques to solve the problems given above.

Example 1. When $n = 1$, we solved the ideal description problem in §5 of Chapter 1. Namely, given an ideal $I \subseteq k[x]$, we showed that $I = \langle g \rangle$ for some $g \in k[x]$ (see Corollary 4 of Chapter 1, §5). So ideals have an especially simple description in this case.

We also saw in §5 of Chapter 1 that the solution of the ideal membership problem follows easily from the division algorithm: given $f \in k[x]$, to check whether $f \in I = \langle g \rangle$, we divide g into f:

$$f = q \cdot g + r,$$

where $q, r \in k[x]$ and $r = 0$ or $\deg(r) < \deg(g)$. Then we proved that $f \in I$ if and only if $r = 0$. Thus, we have an algorithmic test for ideal membership in the case $n = 1$.

Example 2. Next, let n (the number of variables) be arbitrary, and consider the problem of solving a system of polynomial equations:

$$a_{11}x_1 + \cdots + a_{1n}x_n + b_1 = 0,$$

(1)
$$\vdots$$

$$a_{m1}x_1 + \cdots + a_{mn}x_n + b_m = 0,$$

where each polynomial is linear (total degree 1).

For example, consider the system

$$2x_1 + 3x_2 - x_3 = 0,$$
(2)
$$x_1 + x_2 - 1 = 0,$$
$$x_1 + x_3 - 3 = 0.$$

We row-reduce the matrix of the system to reduced row echelon form:

$$\begin{pmatrix} 1 & 0 & 1 & 3 \\ 0 & 1 & -1 & -2 \\ 0 & 0 & 0 & 0 \end{pmatrix}.$$

The form of this matrix shows that x_3 is a free variable, and setting $x_3 = t$ (any element of k), we have

$$x_1 = -t + 3,$$
$$x_2 = t - 2,$$
$$x_3 = t.$$

These are parametric equations for a line L in k^3. The original system of equations (2) presents L as an affine variety.

In the general case, one performs row operations on the matrix of (1)

$$\begin{pmatrix} a_{11} & \cdots & a_{1n} & -b_1 \\ \vdots & & \vdots & \vdots \\ a_{m1} & \cdots & a_{mn} & -b_m \end{pmatrix}$$

until it is in *reduced row echelon form* (where the first nonzero entry on each row is 1, and all other entries in the column containing a leading 1 are zero). Then we can find all solutions of the original system (1) by substituting values for the *free variables* in the reduced row echelon form system. In some examples there may be only one solution, or no solutions. This last case will occur, for instance, if the reduced row echelon matrix contains a row $(0\ldots0\,1)$, corresponding to the inconsistent equation $0 = 1$.

Example 3. Again, take n arbitrary, and consider the subset V of k^n parametrized by

$$x_1 = a_{11}t_1 + \cdots + a_{1m}t_m + b_1,$$
(3)
$$\vdots$$
$$x_n = a_{n1}t_1 + \cdots + a_{nm}t_m + b_n.$$

We see that V is an affine linear subspace of k^n since V is the image of the mapping $F : k^m \rightarrow k^n$ defined by

$$F(t_1,\ldots,t_m) = (a_{11}t_1 + \cdots + a_{1m}t_m + b_1,\ldots,a_{n1}t_1 + \cdots + a_{nm}t_m + b_n).$$

This is a linear mapping, followed by a translation. Let us consider the implicitization problem in this case. In other words, we seek a system of linear equations [as in (1)] whose solutions are the points of V.

For example, consider the affine linear subspace $V \subseteq k^4$ defined by

$$x_1 = t_1 + t_2 + 1,$$
$$x_2 = t_1 - t_2 + 3,$$
$$x_3 = 2t_1 - 2,$$
$$x_4 = t_1 + 2t_2 - 3.$$

We rewrite the equations by subtracting the x_i terms and constants from both sides and apply the row reduction algorithm to the corresponding matrix:

$$\begin{pmatrix} 1 & 1 & -1 & 0 & 0 & 0 & -1 \\ 1 & -1 & 0 & -1 & 0 & 0 & -3 \\ 2 & 0 & 0 & 0 & -1 & 0 & 2 \\ 1 & 2 & 0 & 0 & 0 & -1 & 3 \end{pmatrix}$$

(where the coefficients of the x_i have been placed after the coefficients of the t_j in each row). We obtain the reduced row echelon form:

$$\begin{pmatrix} 1 & 0 & 0 & 0 & -1/2 & 0 & 1 \\ 0 & 1 & 0 & 0 & 1/4 & -1/2 & 1 \\ 0 & 0 & 1 & 0 & -1/4 & -1/2 & 3 \\ 0 & 0 & 0 & 1 & -3/4 & 1/2 & 3 \end{pmatrix}.$$

Because the entries in the first two columns of rows 3 and 4 are zero, the last two rows of this matrix correspond to the following two equations with no t_j terms:

$$x_1 - (1/4)x_3 - (1/2)x_4 - 3 = 0,$$
$$x_2 - (3/4)x_3 + (1/2)x_4 - 3 = 0.$$

(Note that this system is also in echelon form.) These two equations define V in k^4.

The same method can be applied to find implicit equations for any affine linear subspace V given parametrically as in (3): one computes the reduced row echelon form of (3), and the rows involving only $x_1, \ldots, x_n$ give the equations for V. We thus have an algorithmic solution to the implicitization problem in this case.

Our goal in this chapter will be to develop extensions of the methods used in these examples to systems of polynomial equations of any degrees in any number of variables. What we will see is that a sort of "combination" of row-reduction and division of polynomials—the method of Gröbner bases mentioned at the outset—allows us to handle all these problems.

EXERCISES FOR §1

1. Determine whether the given polynomial is in the given ideal $I \subseteq \mathbb{R}[x]$ using the method of Example 1.
   a. $f(x) = x^2 - 3x + 2$, $I = \langle x - 2 \rangle$.
   b. $f(x) = x^5 - 4x + 1$, $I = \langle x^3 - x^2 + x \rangle$.
   c. $f(x) = x^2 - 4x + 4$, $I = \langle x^4 - 6x^2 + 12x - 8, 2x^3 - 10x^2 + 16x - 8 \rangle$.
   d. $f(x) = x^3 - 1$, $I = \langle x^9 - 1, x^5 + x^3 - x^2 - 1 \rangle$.

2. Find parametrizations of the affine varieties defined by the following sets of equations.
   a. In $\mathbb{R}^3$ or $\mathbb{C}^3$:

$$2x + 3y - z = 9,$$
$$x - y = 1,$$
$$3x + 7y - 2z = 17.$$

b. In $\mathbb{R}^4$ or $\mathbb{C}^4$:

$$x_1 + x_2 - x_3 - x_4 = 0,$$
$$x_1 - x_2 + x_3 = 0.$$

c. In $\mathbb{R}^3$ or $\mathbb{C}^3$:

$$y - x^3 = 0,$$
$$z - x^5 = 0.$$

3. Find implicit equations for the affine varieties parametrized as follows.
   a. In $\mathbb{R}^3$ or $\mathbb{C}^3$:

$$x_1 = t - 5,$$
$$x_2 = 2t + 1,$$
$$x_3 = -t + 6.$$

b. In $\mathbb{R}^4$ or $\mathbb{C}^4$:

$$x_1 = 2t - 5u,$$
$$x_2 = t + 2u,$$
$$x_3 = -t + u,$$
$$x_4 = t + 3u.$$

c. In $\mathbb{R}^3$ or $\mathbb{C}^3$:

$$x = t, \qquad y = t^4, \qquad z = t^7.$$

4. Let $x_1, x_2, x_3, \ldots$ be an infinite collection of independent variables indexed by the natural numbers. A *polynomial* with coefficients in a field k in the x_i is a finite linear combination of (finite) monomials $x_{i_1}^{e_1} \ldots x_{i_n}^{e_n}$. Let R denote the set of all polynomials in the x_i. Note that we can add and multiply elements of R in the usual way. Thus, R is the polynomial ring $k[x_1, x_2, \ldots]$ in infinitely many variables.
   a. Let $I = \langle x_1, x_2, x_3, \ldots \rangle$ be the set of polynomials of the form $x_{t_1} f_1 + \cdots + x_{t_m} f_m$, where $f_j \in R$. Show that I is an ideal in the ring R.
   b. Show, arguing by contradiction, that I has no finite generating set. Hint: It is not enough only to consider subsets of $\{x_i \mid i \geq 1\}$.
5. In this problem you will show that all polynomial parametric curves in k^2 are contained in affine algebraic varieties.
   a. Show that the number of distinct monomials $x^a y^b$ of total degree $\leq m$ in $k[x, y]$ is equal to $(m+1)(m+2)/2$. [Note: This is the binomial coefficient $\binom{m+2}{2}$.]
   b. Show that if $f(t)$ and $g(t)$ are polynomials of degree $\leq n$ in t, then for m large enough, the "monomials"

$$[f(t)]^a [g(t)]^b$$

with $a + b \leq m$ are linearly *dependent*.

c. Deduce from part (b) that if C is a curve in k^2 given parametrically by $x = f(t)$, $y = g(t)$ for $f(t), g(t) \in k[t]$, then C is contained in $\mathbf{V}(F)$ for some nonzero $F \in k[x, y]$.

d. Generalize parts (a), (b), and (c) to show that any polynomial parametric surface

$$x = f(t, u), \quad y = g(t, u), \quad z = h(t, u)$$

is contained in an algebraic surface $\mathbf{V}(F)$, where $F \in k[x, y, z]$ is nonzero.

§2 Orderings on the Monomials in $k[x_1, \ldots, x_n]$

If we examine the division algorithm in $k[x]$ and the row-reduction (Gaussian elimination) algorithm for systems of linear equations (or matrices) in detail, we see that a notion of *ordering of terms* in polynomials is a key ingredient of both (though this is not often stressed). For example, in dividing $f(x) = x^5 - 3x^2 + 1$ by $g(x) = x^2 - 4x + 7$ by the standard method, we would:

- Write the terms in the polynomials in decreasing order by degree in x.
- At the first step, the leading term (the term of highest degree) in f is $x^5 = x^3 \cdot x^2 = x^3 \cdot$ (leading term in g). Thus, we would subtract $x^3 \cdot g(x)$ from f to cancel the leading term, leaving $4x^4 - 7x^3 - 3x^2 + 1$.
- Then, we would repeat the same process on $f(x) - x^3 \cdot g(x)$, etc., until we obtain a polynomial of degree less than 2.

For the division algorithm on polynomials in one variable, we are dealing with the degree ordering on the one-variable monomials:

$$(1) \qquad\qquad \cdots > x^{m+1} > x^m > \cdots > x^2 > x > 1.$$

The success of the algorithm depends on working systematically with the leading terms in f and g, and not removing terms "at random" from f using arbitrary terms from g.

Similarly, in the row-reduction algorithm on matrices, in any given row, we systematically work with entries to the left first—leading entries are those nonzero entries farthest to the left on the row. On the level of linear equations, this is expressed by ordering the variables $x_1, \ldots, x_n$ as follows:

$$(2) \qquad\qquad x_1 > x_2 > \cdots > x_n.$$

We write the terms in our equations in decreasing order. Furthermore, in an echelon form system, the equations are listed with their leading terms in decreasing order. (In fact, the precise definition of an echelon form system could be given in terms of this ordering—see Exercise 8.)

From the above evidence, we might guess that a major component of any extension of division and row-reduction to arbitrary polynomials in several variables will be an ordering on the terms in polynomials in $k[x_1, \ldots, x_n]$. In this section, we will discuss the desirable properties such an ordering should have, and we will construct several different examples that satisfy our requirements. Each of these orderings will be useful in different contexts.

First, we note that we can reconstruct the monomial $x^\alpha = x_1^{\alpha_1} \cdots x_n^{\alpha_n}$ from the n-tuple of exponents $\alpha = (\alpha_1, \ldots, \alpha_n) \in \mathbb{Z}_{\geq 0}^n$. This observation establishes a one-to-one correspondence between the monomials in $k[x_1, \ldots, x_n]$ and $\mathbb{Z}_{\geq 0}^n$. Furthermore, any ordering $>$ we establish on the space $\mathbb{Z}_{\geq 0}^n$ will give us an ordering on monomials: if $\alpha > \beta$ according to this ordering, we will also say that $x^\alpha > x^\beta$.

There are many different ways to define orderings on $\mathbb{Z}_{\geq 0}^n$. For our purposes, most of these orderings will not be useful, however, since we will want our orderings to be compatible with the algebraic structure of polynomial rings.

To begin, since a polynomial is a sum of monomials, we would like to be able to arrange the terms in a polynomial unambiguously in descending (or ascending) order. To do this, we must be able to compare every pair of monomials to establish their proper relative positions. Thus, we will require that our orderings be *linear* or *total* orderings. This means that for every pair of monomials x^α and x^β, exactly one of the three statements

$$x^\alpha > x^\beta, \qquad x^\alpha = x^\beta, \qquad x^\beta > x^\alpha$$

should be true. A total order is also required to be *transitive*, so that $x^\alpha > x^\beta$ and $x^\beta > x^\gamma$ always imply $x^\alpha > x^\gamma$.

Next, we must take into account the effect of the sum and product operations on polynomials. When we add polynomials, after combining like terms, we may simply rearrange the terms present into the appropriate order, so sums present no difficulties. Products are more subtle, however. Since multiplication in a polynomial ring distributes over addition, it suffices to consider what happens when we multiply a monomial times a polynomial. If doing this changed the relative ordering of terms, significant problems could result in any process similar to the division algorithm in $k[x]$, in which we must identify the leading terms in polynomials. The reason is that the leading term in the product could be different from the product of the monomial and the leading term of the original polynomial.

Hence, we will require that all monomial orderings have the following additional property. If $x^\alpha > x^\beta$ and x^γ is any monomial, then we require that $x^\alpha x^\gamma > x^\beta x^\gamma$. In terms of the exponent vectors, this property means that if $\alpha > \beta$ in our ordering on $\mathbb{Z}_{\geq 0}^n$, then, for all $\gamma \in \mathbb{Z}_{\geq 0}^n$, $\alpha + \gamma > \beta + \gamma$.

With these considerations in mind, we make the following definition.

Definition 1. A **monomial ordering** $>$ on $k[x_1, \ldots, x_n]$ is a relation $>$ on $\mathbb{Z}_{\geq 0}^n$, or equivalently, a relation on the set of monomials x^α, $\alpha \in \mathbb{Z}_{\geq 0}^n$, satisfying:

(i) $>$ is a total (or linear) ordering on $\mathbb{Z}_{\geq 0}^n$.
(ii) If $\alpha > \beta$ and $\gamma \in \mathbb{Z}_{\geq 0}^n$, then $\alpha + \gamma > \beta + \gamma$.
(iii) $>$ is a well-ordering on $\mathbb{Z}_{\geq 0}^n$. This means that every nonempty subset of $\mathbb{Z}_{\geq 0}^n$ has a smallest element under $>$. In other words, if $A \subseteq \mathbb{Z}_{\geq 0}^n$ is nonempty, then there is $\alpha \in A$ such that $\beta > \alpha$ for every $\beta \neq \alpha$ in A.

Given a monomial ordering $>$, we say that $\alpha \geq \beta$ when either $\alpha > \beta$ or $\alpha = \beta$.

The following lemma will help us understand what the well-ordering condition of part (iii) of the definition means.

Lemma 2. *An order relation* $>$ *on* $\mathbb{Z}_{\geq 0}^n$ *is a well-ordering if and only if every strictly decreasing sequence in* $\mathbb{Z}_{\geq 0}^n$

$$\alpha(1) > \alpha(2) > \alpha(3) > \cdots$$

eventually terminates.

Proof. We will prove this in contrapositive form: $>$ is not a well-ordering if and only if there is an infinite strictly decreasing sequence in $\mathbb{Z}_{\geq 0}^n$.

If $>$ is not a well-ordering, then some nonempty subset $S \subseteq \mathbb{Z}_{\geq 0}^n$ has no least element. Now pick $\alpha(1) \in S$. Since $\alpha(1)$ is not the least element, we can find $\alpha(1) > \alpha(2)$ in S. Then $\alpha(2)$ is also not the least element, so that there is $\alpha(2) > \alpha(3)$ in S. Continuing this way, we get an infinite strictly decreasing sequence

$$\alpha(1) > \alpha(2) > \alpha(3) > \cdots.$$

Conversely, given such an infinite sequence, then $\{\alpha(1), \alpha(2), \alpha(3), \ldots\}$ is a non-empty subset of $\mathbb{Z}_{\geq 0}^n$ with no least element, and thus, $>$ is not a well-ordering. $\square$

The importance of this lemma will become evident in what follows. It will be used to show that various algorithms must terminate because some term strictly decreases (with respect to a fixed monomial order) at each step of the algorithm.

In §4, we will see that given parts (i) and (ii) in Definition 1, the well-ordering condition of part (iii) is equivalent to $\alpha \geq 0$ for all $\alpha \in \mathbb{Z}_{\geq 0}^n$.

For a simple example of a monomial order, note that the usual numerical order

$$\cdots > m + 1 > m > \cdots > 3 > 2 > 1 > 0$$

on the elements of $\mathbb{Z}_{\geq 0}$ satisfies the three conditions of Definition 1. Hence, the degree ordering (1) on the monomials in $k[x]$ is a monomial ordering, unique by Exercise 13.

Our first example of an ordering on n-tuples will be lexicographic order (or **lex** order, for short).

Definition 3 (Lexicographic Order). Let $\alpha = (\alpha_1, \ldots, \alpha_n)$ and $\beta = (\beta_1, \ldots, \beta_n)$ be in $\mathbb{Z}_{\geq 0}^n$. We say $\alpha >_{lex} \beta$ if the leftmost nonzero entry of the vector difference $\alpha - \beta \in \mathbb{Z}^n$ is positive. We will write $x^\alpha >_{lex} x^\beta$ if $\alpha >_{lex} \beta$.

Here are some examples:
a. $(1, 2, 0) >_{lex} (0, 3, 4)$ since $\alpha - \beta = (1, -1, -4)$.
b. $(3, 2, 4) >_{lex} (3, 2, 1)$ since $\alpha - \beta = (0, 0, 3)$.
c. The variables $x_1, \ldots, x_n$ are ordered in the usual way [see (2)] by the lex ordering:

$$(1, 0, \ldots, 0) >_{lex} (0, 1, 0, \ldots, 0) >_{lex} \cdots >_{lex} (0, \ldots, 0, 1).$$

so $x_1 >_{lex} x_2 >_{lex} \cdots >_{lex} x_n$.

In practice, when we work with polynomials in two or three variables, we will call the variables x, y, z rather than x_1, x_2, x_3. We will also assume that the alphabetical order $x > y > z$ on the variables is used to define the lexicographic ordering unless we explicitly say otherwise.

Lex order is analogous to the ordering of words used in dictionaries (hence the name). We can view the entries of an n-tuple $\alpha \in \mathbb{Z}_{\geq 0}^n$ as analogues of the letters in a word. The letters are ordered alphabetically:

$$a > b > \cdots > y > z.$$

Then, for instance,

$$\text{arrow} >_{lex} \text{arson}$$

since the third letter of "arson" comes after the third letter of "arrow" in alphabetical order, whereas the first two letters are the same in both. Since all elements $\alpha \in \mathbb{Z}_{\geq 0}^n$ have length n, this analogy only applies to words with a fixed number of letters.

For completeness, we must check that the lexicographic order satisfies the three conditions of Definition 1.

Proposition 4. *The lex ordering on $\mathbb{Z}_{\geq 0}^n$ is a monomial ordering.*

Proof. (i) That $>_{lex}$ is a total ordering follows directly from the definition and the fact that the usual numerical order on $\mathbb{Z}_{\geq 0}$ is a total ordering.

(ii) If $\alpha >_{lex} \beta$, then we have that the leftmost nonzero entry in $\alpha - \beta$, say $\alpha_i - \beta_i$, is positive. But $x^\alpha \cdot x^\gamma = x^{\alpha+\gamma}$ and $x^\beta \cdot x^\gamma = x^{\beta+\gamma}$. Then in $(\alpha + \gamma) - (\beta + \gamma) = \alpha - \beta$, the leftmost nonzero entry is again $\alpha_i - \beta_i > 0$.

(iii) Suppose that $>_{lex}$ were not a well-ordering. Then by Lemma 2, there would be an infinite strictly descending sequence

$$\alpha(1) >_{lex} \alpha(2) >_{lex} \alpha(3) >_{lex} \cdots$$

of elements of $\mathbb{Z}_{\geq 0}^n$. We will show that this leads to a contradiction.

Consider the first entries of the vectors $\alpha(i) \in \mathbb{Z}_{\geq 0}^n$. By the definition of the lex order, these first entries form a nonincreasing sequence of nonnegative integers. Since $\mathbb{Z}_{\geq 0}$ is well-ordered, the first entries of the $\alpha(i)$ must "stabilize" eventually. In other words, there exists an ℓ such that all the first entries of the $\alpha(i)$ with $i \geq \ell$ are equal.

Beginning at $\alpha(\ell)$, the second and subsequent entries come into play in determining the lex order. The second entries of $\alpha(\ell), \alpha(\ell+1), \ldots$ form a nonincreasing sequence. By the same reasoning as before, the second entries "stabilize" eventually as well. Continuing in the same way, we see that for some m, the $\alpha(m), \alpha(m+1), \ldots$ all are equal. This contradicts the fact that $\alpha(m) >_{lex} \alpha(m + 1)$. $\qquad\square$

It is important to realize that there are many lex orders, corresponding to how the variables are ordered. So far, we have used lex order with $x_1 > x_2 > \cdots > x_n$. But given *any* ordering of the variables $x_1, \ldots, x_n$, there is a corresponding lex order. For example, if the variables are x and y, then we get one lex order with $x > y$ and a second with $y > x$. In the general case of n variables, there are $n!$ lex orders. In what follows, the phrase "lex order" will refer to the one with $x_1 > \cdots > x_n$ unless otherwise stated.

In lex order, notice that a variable dominates *any* monomial involving only smaller variables, regardless of its total degree. Thus, for the lex order with $x > y > z$, we have $x >_{lex} y^5 z^3$. For some purposes, we may also want to take the total degrees of the monomials into account and order monomials of bigger degree first. One way to do this is the graded lexicographic order (or **grlex** order).

Definition 5 (Graded Lex Order). Let $\alpha, \beta \in \mathbb{Z}_{\geq 0}^n$. We say $\alpha >_{grlex} \beta$ if

$$|\alpha| = \sum_{i=1}^{n} \alpha_i > |\beta| = \sum_{i=1}^{n} \beta_i, \quad \text{or} \quad |\alpha| = |\beta| \text{ and } \alpha >_{lex} \beta.$$

We see that grlex orders by total degree first, then "break ties" using lex order. Here are some examples:

a. $(1,2,3) >_{grlex} (3,2,0)$ since $|(1,2,3)| = 6 > |(3,2,0)| = 5$.
b. $(1,2,4) >_{grlex} (1,1,5)$ since $|(1,2,4)| = |(1,1,5)|$ and $(1,2,4) >_{lex} (1,1,5)$.
c. The variables are ordered according to the lex order, i.e., $x_1 >_{grlex} \cdots >_{grlex} x_n$.

We will leave it as an exercise to show that the grlex ordering satisfies the three conditions of Definition 1. As in the case of lex order, there are $n!$ grlex orders on n variables, depending on how the variables are ordered.

Another (somewhat less intuitive) order on monomials is the graded reverse lexicographical order (or **grevlex** order). Even though this ordering "takes some getting used to," it has been shown that for some operations, the grevlex ordering is the most efficient for computations.

Definition 6 (Graded Reverse Lex Order). Let $\alpha, \beta \in \mathbb{Z}_{\geq 0}^n$. We say $\alpha >_{grevlex} \beta$ if

$$|\alpha| = \sum_{i=1}^{n} \alpha_i > |\beta| = \sum_{i=1}^{n} \beta_i, \text{ or } |\alpha| = |\beta| \text{ and the rightmost nonzero entry of } \alpha - \beta \in \mathbb{Z}^n \text{ is negative.}$$

Like grlex, grevlex orders by total degree, but it "breaks ties" in a different way. For example:

a. $(4,7,1) >_{grevlex} (4,2,3)$ since $|(4,7,1)| = 12 > |(4,2,3)| = 9$.
b. $(1,5,2) >_{grevlex} (4,1,3)$ since $|(1,5,2)| = |(4,1,3)|$ and $(1, 5, 2) - (4, 1, 3) = (-3, 4, -1)$.

You will show in the exercises that the grevlex ordering gives a monomial ordering. Note also that lex and grevlex give the same ordering on the variables. That is,

$$(1,0,\ldots,0) >_{grevlex} (0,1,\ldots,0) >_{grevlex} \cdots >_{grevlex} (0,\ldots,0,1)$$

or

$$x_1 >_{grevlex} x_2 >_{grevlex} \cdots >_{grevlex} x_n$$

Thus, grevlex is really different from the grlex order with the variables rearranged (as one might be tempted to believe from the name).

To explain the relation between grlex and grevlex, note that both use total degree in the same way. To break a tie, grlex uses lex order, so that it looks at the leftmost (or largest) variable and favors the *larger* power. In contrast, when grevlex finds the same total degree, it looks at the rightmost (or smallest) variable and favors the *smaller* power. In the exercises, you will check that this amounts to a "double-reversal" of lex order. For example,

$$x^5yz >_{grlex} x^4yz^2,$$

since both monomials have total degree 7 and $x^5yz >_{lex} x^4yz^2$. In this case, we also have

$$x^5yz >_{grevlex} x^4yz^2,$$

but for a different reason: x^5yz is larger because the smaller variable z appears to a smaller power.

As with lex and grlex, there are $n!$ grevlex orderings corresponding to how the n variables are ordered.

There are many other monomial orders besides the ones considered here. Some of these will be explored in the exercises for §4. Most computer algebra systems implement lex order, and most also allow other orders, such as grlex and grevlex. Once such an order is chosen, these systems allow the user to specify any of the $n!$ orderings of the variables. As we will see in §8 of this chapter and in later chapters, this facility becomes very useful when studying a variety of questions.

We will end this section with a discussion of how a monomial ordering can be applied to polynomials. If $f = \sum_\alpha a_\alpha x^\alpha$ is a nonzero polynomial in $k[x_1, \ldots, x_n]$ and we have selected a monomial ordering $>$, then we can order the monomials of f in an unambiguous way with respect to $>$. For example, consider the polynomial $f = 4xy^2z + 4z^2 - 5x^3 + 7x^2z^2 \in k[x, y, z]$. Then:

- With respect to lex order, we would reorder the terms of f in decreasing order as

$$f = -5x^3 + 7x^2z^2 + 4xy^2z + 4z^2.$$

- With respect to grlex order, we would have

$$f = 7x^2z^2 + 4xy^2z - 5x^3 + 4z^2.$$

- With respect to grevlex order, we would have

$$f = 4xy^2z + 7x^2z^2 - 5x^3 + 4z^2.$$

We will use the following terminology.

Definition 7. Let $f = \sum_\alpha a_\alpha x^\alpha$ be a nonzero polynomial in $k[x_1, \ldots, x_n]$ and let $>$ be a monomial order.

(i) The **multidegree** of f is

$$\mathrm{multideg}(f) = \max(\alpha \in \mathbb{Z}_{\geq 0}^n \mid a_\alpha \neq 0)$$

(the maximum is taken with respect to $>$).

(ii) The **leading coefficient** of f is

$$\mathrm{LC}(f) = a_{\mathrm{multideg}(f)} \in k.$$

(iii) The **leading monomial** of f is

$$\mathrm{LM}(f) = x^{\mathrm{multideg}(f)}$$

(with coefficient 1).

(iv) The **leading term** of f is

$$\mathrm{LT}(f) = \mathrm{LC}(f) \cdot \mathrm{LM}(f).$$

To illustrate, let $f = 4xy^2z + 4z^2 - 5x^3 + 7x^2z^2$ as before and let $>$ denote lex order. Then

$$\begin{aligned}
\mathrm{multideg}(f) &= (3,0,0), \\
\mathrm{LC}(f) &= -5, \\
\mathrm{LM}(f) &= x^3, \\
\mathrm{LT}(f) &= -5x^3.
\end{aligned}$$

In the exercises, you will show that the multidegree has the following useful properties.

Lemma 8. *Let $f, g \in k[x_1, \ldots, x_n]$ be nonzero polynomials. Then:*

(i) $\mathrm{multideg}(fg) = \mathrm{multideg}(f) + \mathrm{multideg}(g)$.

(ii) *If $f + g \neq 0$, then* $\mathrm{multideg}(f + g) \leq \max(\mathrm{multideg}(f), \mathrm{multideg}(g))$. *If, in addition,* $\mathrm{multideg}(f) \neq \mathrm{multideg}(g)$, *then equality occurs.*

Some books use different terminology. In EISENBUD (1999), the leading term $\mathrm{LT}(f)$ becomes the *initial term* $\mathrm{in}_>(f)$. A more substantial difference appears in BECKER and WEISPFENNING (1993), where the meanings of "monomial" and "term" are interchanged. For them, the leading term $\mathrm{LT}(f)$ is the *head monomial* $\mathrm{HM}(f)$, while the leading monomial $\mathrm{LM}(f)$ is the *head term* $\mathrm{HT}(f)$. Page 10 of KREUZER and ROBBIANO (2000) has a summary of the terminology used in different books. Our advice when reading other texts is to check the definitions carefully.

EXERCISES FOR §2

1. Rewrite each of the following polynomials, ordering the terms using the lex order, the grlex order, and the grevlex order, giving $\mathrm{LM}(f)$, $\mathrm{LT}(f)$, and $\mathrm{multideg}(f)$ in each case.

    a. $f(x, y, z) = 2x + 3y + z + x^2 - z^2 + x^3$.
    b. $f(x, y, z) = 2x^2 y^8 - 3x^5 yz^4 + xyz^3 - xy^4$.

2. Each of the following polynomials is written with its monomials ordered according to (exactly) one of lex, grlex, or grevlex order. Determine which monomial order was used in each case.
    a. $f(x, y, z) = 7x^2 y^4 z - 2xy^6 + x^2 y^2$.
    b. $f(x, y, z) = xy^3 z + xy^2 z^2 + x^2 z^3$.
    c. $f(x, y, z) = x^4 y^5 z + 2x^3 y^2 z - 4xy^2 z^4$.

3. Repeat Exercise 1 when the variables are ordered $z > y > x$.

4. Show that grlex is a monomial order according to Definition 1.

5. Show that grevlex is a monomial order according to Definition 1.

6. Another monomial order is the **inverse lexicographic** or **invlex** order defined by the following: for $\alpha, \beta \in \mathbb{Z}^n_{\geq 0}$, $\alpha >_{invlex} \beta$ if and only if the rightmost nonzero entry of $\alpha - \beta$ is positive. Show that invlex is equivalent to the lex order with the variables permuted in a certain way. (Which permutation?)

7. Let $>$ be any monomial order.
    a. Show that $\alpha \geq 0$ for all $\alpha \in \mathbb{Z}^n_{\geq 0}$. Hint: Proof by contradiction.
    b. Show that if x^α divides x^β, then $\alpha \leq \beta$. Is the converse true?
    c. Show that if $\alpha \in \mathbb{Z}^n_{\geq 0}$, then α is the smallest element of $\alpha + \mathbb{Z}^n_{\geq 0} = \{\alpha + \beta \mid \beta \in \mathbb{Z}^n_{\geq 0}\}$.

8. Write a precise definition of what it means for a system of linear equations to be in echelon form, using the ordering given in equation (2).

9. In this exercise, we will study grevlex in more detail. Let $>_{invlex}$, be the order given in Exercise 6, and define $>_{rinvlex}$ to be the reversal of this ordering, i.e., for $\alpha, \beta \in \mathbb{Z}^n_{\geq 0}$.

$$\alpha >_{rinvlex} \beta \iff \beta >_{invlex} \alpha.$$

Notice that rinvlex is a "double reversal" of lex, in the sense that we first reverse the order of the variables and then we reverse the ordering itself.
    a. Show that $\alpha >_{grevlex} \beta$ if and only if $|\alpha| > |\beta|$, or $|\alpha| = |\beta|$ and $\alpha >_{rinvlex} \beta$.
    b. Is rinvlex a monomial ordering according to Definition 1? If so, prove it; if not, say which properties fail.

10. In $\mathbb{Z}_{\geq 0}$ with the usual ordering, between any two integers, there are only a finite number of other integers. Is this necessarily true in $\mathbb{Z}^n_{\geq 0}$ for a monomial order? Is it true for the grlex order?

11. Let $>$ be a monomial order on $k[x_1, \ldots, x_n]$.
    a. Let $f \in k[x_1, \ldots, x_n]$ and let m be a monomial. Show that $\mathrm{LT}(m \cdot f) = m \cdot \mathrm{LT}(f)$.
    b. Let $f, g \in k[x_1, \ldots, x_n]$. Is $\mathrm{LT}(f \cdot g)$ necessarily the same as $\mathrm{LT}(f) \cdot \mathrm{LT}(g)$?
    c. If $f_i, g_i \in k[x_1, \ldots, x_n]$, $1 \leq i \leq s$, is $\mathrm{LM}(\sum_{i=1}^s f_i g_i)$ necessarily equal to $\mathrm{LM}(f_i) \cdot \mathrm{LM}(g_i)$ for some i?

12. Lemma 8 gives two properties of the multidegree.
    a. Prove Lemma 8. Hint: The arguments used in Exercise 11 may be relevant.
    b. Suppose that multideg$(f) = $ multideg(g) and $f + g \neq 0$. Give examples to show that multideg$(f + g)$ may or may not equal max(multideg(f), multideg(g)).

13. Prove that $1 < x < x^2 < x^3 < \cdots$ is the unique monomial order on $k[x]$.

§3 A Division Algorithm in $k[x_1, \ldots, x_n]$

In §1, we saw how the division algorithm could be used to solve the ideal membership problem for polynomials of one variable. To study this problem when there are more variables, we will formulate a division algorithm for polynomials in $k[x_1, \ldots, x_n]$ that extends the algorithm for $k[x]$. In the general case, the goal is to divide $f \in k[x_1, \ldots, x_n]$ by $f_1, \ldots, f_s \in k[x_1, \ldots, x_n]$. As we will see, this means expressing f in the form

$$f = q_1 f_1 + \cdots + q_s f_s + r,$$

where the "quotients" $q_1, \ldots, q_s$ and remainder r lie in $k[x_1, \ldots, x_n]$. Some care will be needed in deciding how to characterize the remainder. This is where we will use the monomial orderings introduced in §2. We will then see how the division algorithm applies to the ideal membership problem.

The basic idea of the algorithm is the same as in the one-variable case: we want to cancel the leading term of f (with respect to a fixed monomial order) by multiplying some f_i by an appropriate monomial and subtracting. Then this monomial becomes a term in the corresponding q_i. Rather than state the algorithm in general, let us first work through some examples to see what is involved.

Example 1. We will first divide $f = xy^2 + 1$ by $f_1 = xy + 1$ and $f_2 = y + 1$, using lex order with $x > y$. We want to employ the same scheme as for division of one-variable polynomials, the difference being that there are now several divisors and quotients. Listing the divisors f_1, f_2 and the quotients q_1, q_2 *vertically*, we have the following setup:

$$
\begin{array}{l}
q_1 : \\
q_2 : \\
\left.\begin{array}{r} xy + 1 \\ y + 1 \end{array}\right) \overline{xy^2 + 1}
\end{array}
$$

The leading terms $\mathrm{LT}(f_1) = xy$ and $\mathrm{LT}(f_2) = y$ both divide the leading term $\mathrm{LT}(f) = xy^2$. Since f_1 is listed first, we will use it. Thus, we divide xy into xy^2, leaving y, and then subtract $y \cdot f_1$ from f:

$$
\begin{array}{l}
q_1 : \quad y \\
q_2 : \\
\left.\begin{array}{r} xy + 1 \\ y + 1 \end{array}\right) \overline{\begin{array}{r} xy^2 + 1 \\ xy^2 + y \end{array}} \\
\hphantom{xxxxxxxxx} \overline{-y + 1}
\end{array}
$$

Now we repeat the same process on $-y + 1$. This time we must use f_2 since $\mathrm{LT}(f_1) = xy$ does not divide $\mathrm{LT}(-y + 1) = -y$. We obtain:

$$
\begin{array}{l}
q_1 : \quad y \\
q_2 : \quad -1 \\
\left.\begin{array}{r} xy + 1 \\ y + 1 \end{array}\right) \overline{\begin{array}{r} xy^2 + 1 \\ xy^2 + y \end{array}} \\
\hphantom{xxxxxxxxx} \overline{\begin{array}{r} -y + 1 \\ -y - 1 \end{array}} \\
\hphantom{xxxxxxxxxxxx} \overline{2}
\end{array}
$$

Since $\mathrm{LT}(f_1)$ and $\mathrm{LT}(f_2)$ do not divide 2, the remainder is $r = 2$ and we are done. Thus, we have written $f = xy^2 + 1$ in the form

$$xy^2 + 1 = y \cdot (xy + 1) + (-1) \cdot (y + 1) + 2.$$

Example 2. In this example, we will encounter an unexpected subtlety that can occur when dealing with polynomials of more than one variable. Let us divide $f = x^2y + xy^2 + y^2$ by $f_1 = xy - 1$ and $f_2 = y^2 - 1$. As in the previous example, we will use lex order with $x > y$. The first two steps of the algorithm go as usual, giving us the following partially completed division (remember that when both leading terms divide, we use f_1):

$$
\begin{array}{r}
q_1 : \quad x + y \\
q_2 : \\
\end{array}
$$

$$
\begin{array}{r}
xy - 1 \\
y^2 - 1
\end{array} \Bigg)
\begin{array}{l}
x^2y + xy^2 + y^2 \\
\underline{x^2y - x} \\
\quad xy^2 + x + y^2 \\
\quad \underline{xy^2 - y} \\
\qquad x + y^2 + y
\end{array}
$$

Note that neither $\mathrm{LT}(f_1) = xy$ nor $\mathrm{LT}(f_2) = y^2$ divides $\mathrm{LT}(x + y^2 + y) = x$. However, $x + y^2 + y$ is *not* the remainder since $\mathrm{LT}(f_2)$ divides y^2. Thus, if we move x to the remainder, we can continue dividing. (This is something that never happens in the one-variable case: once the leading term of the divisor no longer divides the leading term of what is at the bottom of the division, the algorithm terminates.)

To implement this idea, we create a remainder column r, to the right of the division, where we put the terms belonging to the remainder. Also, we call the polynomial at the bottom of division the *intermediate dividend.* Then we continue dividing until the intermediate dividend is zero. Here is the next step, where we move x to the remainder column (as indicated by the arrow):

$$
\begin{array}{r}
q_1 : \quad x + y \\
q_2 : \\
\end{array}
\qquad\qquad\qquad\qquad \underline{\quad r \quad}
$$

$$
\begin{array}{r}
xy - 1 \\
y^2 - 1
\end{array} \Bigg)
\begin{array}{l}
x^2y + xy^2 + y^2 \\
\underline{xy^2 - x} \\
\quad xy^2 + x + y^2 \\
\quad \underline{x^2y - y} \\
\qquad x + y^2 + y \\
\qquad \underline{\quad y^2 + y \quad} \longrightarrow \quad x
\end{array}
$$

Now we continue dividing. If we can divide by $\mathrm{LT}(f_1)$ or $\mathrm{LT}(f_2)$, we proceed as usual, and if neither divides, we move the leading term of the intermediate dividend to the remainder column.

Here is the rest of the division:

$$
\begin{array}{r}
q_1 : \quad x + y \\
q_2 : \quad 1
\end{array}
$$

$$
\begin{array}{rl}
\begin{matrix} xy - 1 \\ y^2 - 1 \end{matrix} \Big) & x^2 y + xy^2 + y^2 \\
& \underline{x^2 y - x} \\
& xy^2 + x + y^2 \\
& \underline{xy^2 - y} \\
& x + y^2 + y \\
& \underline{y^2 + y} \\
& \underline{y^2 - 1} \\
& y + 1 \\
\end{array}
$$

$$
\begin{array}{l}
y^2 + y \quad \longrightarrow \quad x \\
\\
1 \quad \longrightarrow \quad x + y \\
0 \quad \longrightarrow \quad x + y + 1
\end{array}
$$

Thus, the remainder is $x + y + 1$, and we obtain

$$(1) \qquad x^2 y + xy^2 + y^2 = (x + y) \cdot (xy - 1) + 1 \cdot (y^2 - 1) + x + y + 1.$$

Note that the remainder is a sum of monomials, none of which is divisible by the leading terms $\mathrm{LT}(f_1)$ or $\mathrm{LT}(f_2)$.

The above example is a fairly complete illustration of how the division algorithm works. It also shows us what property we want the remainder to have: none of its terms should be divisible by the leading terms of the polynomials by which we are dividing. We can now state the general form of the division algorithm.

Theorem 3 (Division Algorithm in $k[x_1, \ldots, x_n]$). *Let $>$ be a monomial order on $\mathbb{Z}_{\geq 0}^n$, and let $F = (f_1, \ldots, f_s)$ be an ordered s-tuple of polynomials in $k[x_1, \ldots, x_n]$. Then every $f \in k[x_1, \ldots, x_n]$ can be written as*

$$f = q_1 f_1 + \cdots + q_s f_s + r,$$

where $q_i, r \in k[x_1, \ldots, x_n]$, and either $r = 0$ or r is a linear combination, with coefficients in k, of monomials, none of which is divisible by any of $\mathrm{LT}(f_1), \ldots, \mathrm{LT}(f_s)$. We call r a **remainder** *of f on division by F. Furthermore, if $q_i f_i \neq 0$, then*

$$\mathrm{multideg}(f) \geq \mathrm{multideg}(q_i f_i).$$

Proof. We prove the existence of $q_1, \ldots, q_s$ and r by giving an algorithm for their construction and showing that it operates correctly on any given input. We recommend that the reader review the division algorithm in $k[x]$ given in Proposition 2 of Chapter 1, §5 before studying the following generalization:

Input : $f_1, \ldots, f_s, f$
Output : $q_1, \ldots, q_s, r$

$q_1 := 0; \ldots; q_s := 0; r := 0$
$p := f$
WHILE $p \neq 0$ DO
    $i := 1$
    divisionoccurred := false
    WHILE $i \leq s$ AND *divisionoccurred* = false DO
        IF LT(f_i) divides LT(p) THEN
            $q_i := q_i + \text{LT}(p)/\text{LT}(f_i)$
            $p := p - (\text{LT}(p)/\text{LT}(f_i))f_i$
            divisionoccurred := true
        ELSE
            $i := i + 1$
    IF *divisionoccurred* = false THEN
        $r := r + \text{LT}(p)$
        $p := p - \text{LT}(p)$
RETURN $q_1, \ldots, q_s, r$

We can relate this algorithm to the previous example by noting that the variable p represents the intermediate dividend at each stage, the variable r represents the column on the right-hand side, and the variables $q_1, \ldots, q_s$ are the quotients listed above the division. Finally, the boolean variable "divisionoccurred" tells us when some LT(f_i) divides the leading term of the intermediate dividend. You should check that each time we go through the main WHILE ... DO loop, precisely one of two things happens:

- (Division Step) If some LT(f_i) divides LT(p), then the algorithm proceeds as in the one-variable case.
- (Remainder Step) If no LT(f_i) divides LT(p), then the algorithm adds LT(p) to the remainder.

These steps correspond exactly to what we did in Example 2.

To prove that the algorithm works, we will first show that

(2) $$f = q_1 f_1 + \cdots + q_s f_s + p + r$$

holds at every stage. This is clearly true for the initial values of $q_1, \ldots, q_s, p$, and r. Now suppose that (2) holds at one step of the algorithm. If the next step is a Division Step, then some LT(f_i) divides LT(p), and the equality

$$q_i f_i + p = (q_i + \text{LT}(p)/\text{LT}(f_i))f_i + (p - (\text{LT}(p)/\text{LT}(f_i))f_i)$$

shows that $q_i f_i + p$ is unchanged. Since all other variables are unaffected, (2) remains true in this case. On the other hand, if the next step is a Remainder Step, then p and r will be changed, but the sum $p + r$ is unchanged since

$$p + r = (p - \mathrm{LT}(p)) + (r + \mathrm{LT}(p)).$$

As before, equality (2) is still preserved.

Next, notice that the algorithm comes to a halt when $p = 0$. In this situation, (2) becomes

$$f = q_1 f_1 + \cdots + q_s f_s + r.$$

Since terms are added to r only when they are divisible by none of the $\mathrm{LT}(f_i)$, it follows that $q_1, \ldots, q_s$ and r have the desired properties when the algorithm terminates.

Finally, we need to show that the algorithm does eventually terminate. The key observation is that each time we redefine the variable p, either its multidegree drops (relative to our term ordering) or it becomes 0. To see this, first suppose that during a Division Step, p is redefined to be

$$p' = p - \frac{\mathrm{LT}(p)}{\mathrm{LT}(f_i)} f_i.$$

By Lemma 8 of §2, we have

$$\mathrm{LT}\left(\frac{\mathrm{LT}(p)}{\mathrm{LT}(f_i)} f_i\right) = \frac{\mathrm{LT}(p)}{\mathrm{LT}(f_i)} \mathrm{LT}(f_i) = \mathrm{LT}(p),$$

so that p and $(\mathrm{LT}(p)/\mathrm{LT}(f_i)) f_i$ have the same leading term. Hence, their difference p' must have strictly smaller multidegree when $p' \neq 0$. Next, suppose that during a Remainder Step, p is redefined to be

$$p' = p - \mathrm{LT}(p).$$

Here, it is obvious that $\mathrm{multideg}(p') < \mathrm{multideg}(p)$ when $p' \neq 0$. Thus, in either case, the multidegree must decrease. If the algorithm never terminated, then we would get an infinite decreasing sequence of multidegrees. The well-ordering property of $>$, as stated in Lemma 2 of §2, shows that this cannot occur. Thus $p = 0$ must happen eventually, so that the algorithm terminates after finitely many steps.

It remains to study the relation between $\mathrm{multideg}(f)$ and $\mathrm{multideg}(q_i f_i)$. Every term in q_i is of the form $\mathrm{LT}(p)/\mathrm{LT}(f_i)$ for some value of the variable p. The algorithm starts with $p = f$, and we just finished proving that the multidegree of p decreases. This shows that $\mathrm{LT}(p) \leq \mathrm{LT}(f)$, and then it follows easily [using condition (ii) of the definition of a monomial order] that $\mathrm{multideg}(q_i f_i) \leq \mathrm{multideg}(f)$ when $q_i f_i \neq 0$ (see Exercise 4). This completes the proof of the theorem.   □

The algebra behind the division algorithm is very simple (there is nothing beyond high school algebra in what we did), which makes it surprising that this form of the algorithm was first isolated and exploited only within the past 50 years.

We will conclude this section by asking whether the division algorithm has the
same nice properties as the one-variable version. Unfortunately, the answer is not
pretty—the examples given below will show that the division algorithm is far from
perfect. In fact, the algorithm achieves its full potential only when coupled with the
Gröbner bases studied in §§5 and 6.

A first important property of the division algorithm in $k[x]$ is that the remainder is
uniquely determined. To see how this can fail when there is more than one variable,
consider the following example.

Example 4. Let us divide $f = x^2y + xy^2 + y^2$ by $f_1 = y^2 - 1$ and $f_2 = xy - 1$. We
will use lex order with $x > y$. This is the same as Example 2, except that we have
changed the order of the divisors. For practice, we suggest that the reader should do
the division. You should get the following answer:

$$
\begin{array}{l}
q_1 : \quad x + 1 \\
q_2 : \quad x \\
\end{array}
\qquad\qquad \overline{ r }
$$

$$
\begin{array}{r}
y^2 - 1 \\
xy - 1
\end{array}
\Big)
\begin{array}{l}
\overline{x^2y + xy^2 + y^2} \\
x^2y - x \\
\hline
\quad xy^2 + x + y^2 \\
\quad xy^2 - x \\
\hline
\qquad 2x + y^2 \\
\qquad\quad y^2 \qquad\qquad \longrightarrow \quad 2x \\
\qquad\quad y^2 - 1 \\
\hline
\qquad\qquad 1 \\
\hline
\qquad\qquad 0 \qquad\qquad \longrightarrow \quad 2x + 1
\end{array}
$$

This shows that

(3) $\qquad x^2y + xy^2 + y^2 = (x+1) \cdot (y^2 - 1) + x \cdot (xy - 1) + 2x + 1.$

If you compare this with equation (1), you will see that the remainder is different
from what we got in Example 2.

This shows that the remainder r is not uniquely characterized by the require-
ment that none of its terms be divisible by $\mathrm{LT}(f_1), \ldots, \mathrm{LT}(f_s)$. The situation is not
completely chaotic: if we follow the algorithm precisely as stated [most importantly,
testing $\mathrm{LT}(p)$ for divisibility by $\mathrm{LT}(f_1), \mathrm{LT}(f_2), \ldots$ in that order], then $q_1, \ldots, q_s$ and
r are uniquely determined. (See Exercise 11 for a more detailed discussion of how
to characterize the output of the algorithm.) However, Examples 2 and 4 show that
the *ordering* of the s-tuple of polynomials $(f_1, \ldots, f_s)$ definitely matters, both in the
number of steps the algorithm will take to complete the calculation and in the results. The q_i and r can change if we simply rearrange the f_i. (The q_i and r may also
change if we change the monomial ordering, but that is another story.)

One nice feature of the division algorithm in $k[x]$ is the way it solves the ideal
membership problem—recall Example 1 from §1. Do we get something similar

for several variables? One implication is an easy corollary of Theorem 3: if after division of f by $F = (f_1, \ldots, f_s)$ we obtain a remainder $r = 0$, then

$$f = q_1 f_1 + \cdots + q_s f_s,$$

so that $f \in \langle f_1, \ldots, f_s \rangle$. Thus $r = 0$ is a *sufficient* condition for ideal membership. However, as the following example shows, $r = 0$ is not a *necessary* condition for being in the ideal.

Example 5. Let $f_1 = xy - 1$, $f_2 = y^2 - 1 \in k[x, y]$ with the lex order. Dividing $f = xy^2 - x$ by $F = (f_1, f_2)$, the result is

$$xy^2 - x = y \cdot (xy - 1) + 0 \cdot (y^2 - 1) + (-x + y).$$

With $F = (f_2, f_1)$, however, we have

$$xy^2 - x = x \cdot (y^2 - 1) + 0 \cdot (xy - 1) + 0.$$

The second calculation shows that $f \in \langle f_1, f_2 \rangle$. Then the first calculation shows that even if $f \in \langle f_1, f_2 \rangle$, it is still possible to obtain a nonzero remainder on division by $F = (f_1, f_2)$.

Thus, we must conclude that the division algorithm given in Theorem 3 is an imperfect generalization of its one-variable counterpart. To remedy this situation, we turn to one of the lessons learned in Chapter 1. Namely, in dealing with a collection of polynomials $f_1, \ldots, f_s \in k[x_1, \ldots, x_n]$, it is frequently desirable to pass to the ideal I they generate. This allows the possibility of going from $f_1, \ldots, f_s$ to a different generating set for I. So we can still ask whether there might be a "good" generating set for I. For such a set, we would want the remainder r on division by the "good" generators to be uniquely determined and the condition $r = 0$ should be *equivalent* to membership in the ideal. In §6, we will see that Gröbner bases have exactly these "good" properties.

In the exercises, you will experiment with a computer algebra system to try to discover for yourself what properties a "good" generating set should have. We will give a precise definition of "good" in §5 of this chapter.

EXERCISES FOR §3

1. Compute the remainder on division of the given polynomial f by the ordered set F (by hand). Use the grlex order, then the lex order in each case.
   a. $f = x^7 y^2 + x^3 y^2 - y + 1$, $F = (xy^2 - x, x - y^3)$.
   b. Repeat part (a) with the order of the pair F reversed.
2. Compute the remainder on division:
   a. $f = xy^2 z^2 + xy - yz$, $F = (x - y^2, y - z^3, z^2 - 1)$.
   b. Repeat part (a) with the order of the set F permuted cyclically.
3. Using a computer algebra system, check your work from Exercises 1 and 2. (You may need to consult documentation to learn whether the system you are using has an explicit polynomial division command or you will need to perform the individual steps of the algorithm yourself.)

4. Let $f = q_1 f_1 + \cdots + q_s f_s + r$ be the output of the division algorithm.
   a. Complete the proof begun in the text that $\text{multideg}(f) \geq \text{multideg}(q_i f_i)$ provided that $q_i f_i \neq 0$.
   b. Prove that $\text{multideg}(f) \geq \text{multideg}(r)$ when $r \neq 0$.

The following problems investigate in greater detail the way the remainder computed by the division algorithm depends on the ordering and the form of the s-tuple of divisors $F = (f_1, \ldots, f_s)$. You may wish to use a computer algebra system to perform these calculations.

5. We will study the division of $f = x^3 - x^2 y - x^2 z + x$ by $f_1 = x^2 y - z$ and $f_2 = xy - 1$.
   a. Compute using grlex order:

$$r_1 = \text{remainder of } f \text{ on division by } (f_1, f_2).$$
$$r_2 = \text{remainder of } f \text{ on division by } (f_2, f_1).$$

   Your results should be *different*. Where in the division algorithm did the difference occur? (You may need to do a few steps by hand here.)
   b. Is $r = r_1 - r_2$ in the ideal $\langle f_1, f_2 \rangle$? If so, find an explicit expression $r = A f_1 + B f_2$. If not, say why not.
   c. Compute the remainder of r on division by (f_1, f_2). Why could you have predicted your answer before doing the division?
   d. Find another polynomial $g \in \langle f_1, f_2 \rangle$ such that the remainder on division of g by (f_1, f_2) is nonzero. Hint: $(xy + 1) \cdot f_2 = x^2 y^2 - 1$, whereas $y \cdot f_1 = x^2 y^2 - yz$.
   e. Does the division algorithm give us a solution for the ideal membership problem for the ideal $\langle f_1, f_2 \rangle$? Explain your answer.

6. Using the grlex order, find an element g of $\langle f_1, f_2 \rangle = \langle 2xy^2 - x, 3x^2 y - y - 1 \rangle \subseteq \mathbb{R}[x, y]$ whose remainder on division by (f_1, f_2) is nonzero. Hint: You can find such a g where the remainder is g itself.

7. Answer the question of Exercise 6 for $\langle f_1, f_2, f_3 \rangle = \langle x^4 y^2 - z, x^3 y^3 - 1, x^2 y^4 - 2z \rangle \subseteq \mathbb{R}[x, y, z]$. Find two different polynomials g (not constant multiples of each other).

8. Try to formulate a general pattern that fits the examples in Exercises 5(c)(d), 6, and 7. What condition on the leading term of the polynomial $g = A_1 f_1 + \cdots + A_s f_s$ would guarantee that there was a nonzero remainder on division by $(f_1, \ldots, f_s)$? What does your condition imply about the ideal membership problem?

9. The discussion around equation (2) of Chapter 1, §4 shows that every polynomial $f \in \mathbb{R}[x, y, z]$ can be written as

$$f = h_1(y - x^2) + h_2(z - x^3) + r,$$

   where r is a polynomial in x alone and $\mathbf{V}(y - x^2, z - x^3)$ is the twisted cubic curve in $\mathbb{R}^3$.
   a. Give a proof of this fact using the division algorithm. Hint: You need to specify carefully the monomial ordering to be used.
   b. Use the parametrization of the twisted cubic to show that $z^2 - x^4 y$ vanishes at every point of the twisted cubic.
   c. Find an explicit representation

$$z^2 - x^4 y = h_1(y - x^2) + h_2(z - x^3)$$

   using the division algorithm.

10. Let $V \subseteq \mathbb{R}^3$ be the curve parametrized by $(t, t^m, t^n), n, m \geq 2$.
   a. Show that V is an affine variety.
   b. Adapt the ideas in Exercise 9 to determine $\mathbf{I}(V)$.

11. In this exercise, we will characterize completely the expression

$$f = q_1 f_1 + \cdots + q_s f_s + r$$

that is produced by the division algorithm (among all the possible expressions for f of this form). Let $\mathrm{LM}(f_i) = x^{\alpha(i)}$ and define

$$\Delta_1 = \alpha(1) + \mathbb{Z}_{\geq 0}^n,$$
$$\Delta_2 = (\alpha(2) + \mathbb{Z}_{\geq 0}^n) \setminus \Delta_1,$$
$$\vdots$$
$$\Delta_s = (\alpha(s) + \mathbb{Z}_{\geq 0}^n) \setminus \left(\bigcup_{i=1}^{s-1} \Delta_i \right),$$
$$\overline{\Delta} = \mathbb{Z}_{\geq 0}^n \setminus \left(\bigcup_{i=1}^{s} \Delta_i \right).$$

(Note that $\mathbb{Z}_{\geq 0}^n$ is the disjoint union of the Δ_i and $\overline{\Delta}$.)

a. Show that $\beta \in \Delta_i$ if and only if $x^{\alpha(i)}$ divides x^β and no $x^{\alpha(j)}$ with $j < i$ divides x^β.
b. Show that $\gamma \in \overline{\Delta}$ if and only if no $x^{\alpha(i)}$ divides x^γ.
c. Show that in the expression $f = q_1 f_1 + \cdots + q_s f_s + r$ computed by the division algorithm, for every i, every monomial x^β in q_i satisfies $\beta + \alpha(i) \in \Delta_i$, and every monomial x^γ in r satisfies $\gamma \in \overline{\Delta}$.
d. Show that there is exactly one expression $f = q_1 f_1 + \cdots + q_s f_s + r$ satisfying the properties given in part (c).

12. Show that the operation of computing remainders on division by $F = (f_1, \ldots, f_s)$ is linear over k. That is, if the remainder on division of g_i by F is $r_i, i = 1, 2$, then, for any $c_1, c_2 \in k$, the remainder on division of $c_1 g_1 + c_2 g_2$ is $c_1 r_1 + c_2 r_2$. Hint: Use Exercise 11.

§4 Monomial Ideals and Dickson's Lemma

In this section, we will consider the ideal description problem of §1 for the special case of monomial ideals. This will require a careful study of the properties of these ideals. Our results will also have an unexpected application to monomial orderings.

To start, we define monomial ideals in $k[x_1, \ldots, x_n]$.

Definition 1. An ideal $I \subseteq k[x_1, \ldots, x_n]$ is a **monomial ideal** if there is a subset $A \subseteq \mathbb{Z}_{\geq 0}^n$ (possibly infinite) such that I consists of all polynomials which are finite sums of the form $\sum_{\alpha \in A} h_\alpha x^\alpha$, where $h_\alpha \in k[x_1, \ldots, x_n]$. In this case, we write $I = \langle x^\alpha \mid \alpha \in A \rangle$.

An example of a monomial ideal is given by $I = \langle x^4 y^2, x^3 y^4, x^2 y^5 \rangle \subseteq k[x, y]$. More interesting examples of monomial ideals will be given in §5.

We first need to characterize all monomials that lie in a given monomial ideal.

Lemma 2. Let $I = \langle x^\alpha \mid \alpha \in A \rangle$ be a monomial ideal. Then a monomial x^β lies in I if and only if x^β is divisible by x^α for some $\alpha \in A$.

Proof. If x^β is a multiple of x^α for some $\alpha \in A$, then $x^\beta \in I$ by the definition of ideal. Conversely, if $x^\beta \in I$, then $x^\beta = \sum_{i=1}^{s} h_i x^{\alpha(i)}$, where $h_i \in k[x_1, \ldots, x_n]$ and $\alpha(i) \in A$. If we expand each h_i as a sum of terms, we obtain

$$x^\beta = \sum_{i=1}^{s} h_i x^{\alpha(i)} = \sum_{i=1}^{s} \left(\sum_j c_{i,j} x^{\beta(i,j)} \right) x^{\alpha(i)} = \sum_{i,j} c_{i,j} x^{\beta(i,j)} x^{\alpha(i)}.$$

After collecting terms of the same multidegree, every term on the right side of the equation is divisible by some $x^{\alpha(i)}$. Hence, the left side x^β must have the same property. □

Note that x^β is divisible by x^α exactly when $x^\beta = x^\alpha \cdot x^\gamma$ for some $\gamma \in \mathbb{Z}_{\geq 0}^n$. This is equivalent to $\beta = \alpha + \gamma$. Thus, the set

$$\alpha + \mathbb{Z}_{\geq 0}^n = \{\alpha + \gamma \mid \gamma \in \mathbb{Z}_{\geq 0}^n\}$$

consists of the exponents of all monomials divisible by x^α. This observation and Lemma 2 allows us to draw pictures of the monomials in a given monomial ideal. For example, if $I = \langle x^4 y^2, x^3 y^4, x^2 y^5 \rangle$, then the exponents of the monomials in I form the set

$$((4,2) + \mathbb{Z}_{\geq 0}^2) \cup ((3,4) + \mathbb{Z}_{\geq 0}^2) \cup ((2,5) + \mathbb{Z}_{\geq 0}^2).$$

We can visualize this set as the union of the integer points in three translated copies of the first quadrant in the plane:

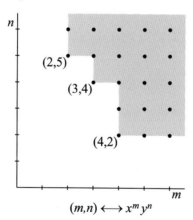

$$(m,n) \longleftrightarrow x^m y^n$$

Let us next show that whether a given polynomial f lies in a monomial ideal can be determined by looking at the monomials of f.

Lemma 3. *Let I be a monomial ideal, and let $f \in k[x_1, \ldots, x_n]$. Then the following are equivalent:*

(i) $f \in I$.
(ii) *Every term of f lies in I.*
(iii) *f is a k-linear combination of the monomials in I.*

Proof. The implications (iii) ⇒ (ii) ⇒ (i) and (ii) ⇒ (iii) are trivial. The proof of
(i) ⇒ (ii) is similar to what we did in Lemma 2 and is left as an exercise. □

An immediate consequence of part (iii) of the lemma is that a monomial ideal is
uniquely determined by its monomials. Hence, we have the following corollary.

Corollary 4. *Two monomial ideals are the same if and only if they contain the same
monomials.*

The main result of this section is that all monomial ideals of $k[x_1, \ldots, x_n]$ are
finitely generated.

Theorem 5 (Dickson's Lemma). *Let $I = \langle x^\alpha \mid \alpha \in A \rangle \subseteq k[x_1, \ldots, x_n]$ be a
monomial ideal. Then I can be written in the form $I = \langle x^{\alpha(1)}, \ldots, x^{\alpha(s)} \rangle$, where
$\alpha(1), \ldots, \alpha(s) \in A$. In particular, I has a finite basis.*

Proof. (By induction on n, the number of variables.) If $n = 1$, then I is generated by
the monomials x_1^α, where $\alpha \in A \subseteq \mathbb{Z}_{\geq 0}$. Let β be the smallest element of $A \subseteq \mathbb{Z}_{\geq 0}$.
Then $\beta \leq \alpha$ for all $\alpha \in A$, so that x_1^β divides all other generators x_1^α. From here,
$I = \langle x_1^\beta \rangle$ follows easily.

Now assume that $n > 1$ and that the theorem is true for $n - 1$. We will write the
variables as $x_1, \ldots, x_{n-1}, y$, so that monomials in $k[x_1, \ldots, x_{n-1}, y]$ can be written as
$x^\alpha y^m$, where $\alpha = (\alpha_1, \ldots, \alpha_{n-1}) \in \mathbb{Z}_{\geq 0}^{n-1}$ and $m \in \mathbb{Z}_{\geq 0}$.

Suppose that $I \subseteq k[x_1, \ldots, x_{n-1}, y]$ is a monomial ideal. To find generators for
I, let J be the ideal in $k[x_1, \ldots, x_{n-1}]$ generated by the monomials x^α for which
$x^\alpha y^m \in I$ for some $m \geq 0$. Since J is a monomial ideal in $k[x_1, \ldots, x_{n-1}]$,
our inductive hypothesis implies that finitely many of the x^α's generate J, say
$J = \langle x^{\alpha(1)}, \ldots, x^{\alpha(s)} \rangle$. The ideal J can be understood as the "projection" of I into
$k[x_1, \ldots, x_{n-1}]$.

For each i between 1 and s, the definition of J tells us that $x^{\alpha(i)} y^{m_i} \in I$ for some
$m_i \geq 0$. Let m be the largest of the m_i. Then, for each ℓ between 0 and $m - 1$,
consider the ideal $J_\ell \subseteq k[x_1, \ldots, x_{n-1}]$ generated by the monomials x^β such that
$x^\beta y^\ell \in I$. One can think of J_ℓ as the "slice" of I generated by monomials containing
y exactly to the ℓth power. Using our inductive hypothesis again, J_ℓ has a finite
generating set of monomials, say $J_\ell = \langle x^{\alpha_\ell(1)}, \ldots, x^{\alpha_\ell(s_\ell)} \rangle$.

We claim that I is generated by the monomials in the following list:

$$\text{from } J : x^{\alpha(1)} y^m, \ldots, x^{\alpha(s)} y^m,$$
$$\text{from } J_0 : x^{\alpha_0(1)}, \ldots, x^{\alpha_0(s_0)},$$
$$\text{from } J_1 : x^{\alpha_1(1)} y, \ldots, x^{\alpha_1(s_1)} y,$$
$$\vdots$$
$$\text{from } J_{m-1} : x^{\alpha_{m-1}(1)} y^{m-1}, \ldots, x^{\alpha_{m-1}(s_{m-1})} y^{m-1}.$$

First note that every monomial in I is divisible by one on the list. To see why, let
$x^\alpha y^p \in I$. If $p \geq m$, then $x^\alpha y^p$ is divisible by some $x^{\alpha(i)} y^m$ by the construction of J.
On the other hand, if $p \leq m - 1$, then $x^\alpha y^p$ is divisible by some $x^{\alpha_p(j)} y^p$ by the

construction of J_p. It follows from Lemma 2 that the above monomials generate an ideal having the same monomials as I. By Corollary 4, this forces the ideals to be the same, and our claim is proved.

To complete the proof, we need to show that the finite set of generators can be chosen from a given set of generators for the ideal. If we switch back to writing the variables as $x_1, \ldots, x_n$, then our monomial ideal is $I = \langle x^\alpha \mid \alpha \in A \rangle \subseteq k[x_1, \ldots, x_n]$. We need to show that I is generated by finitely many of the x^α's, where $\alpha \in A$. By the previous paragraph, we know that $I = \langle x^{\beta(1)}, \ldots, x^{\beta(s)} \rangle$ for some monomials $x^{\beta(i)}$ in I. Since $x^{\beta(i)} \in I = \langle x^\alpha : \alpha \in A \rangle$, Lemma 2 tells us that each $x^{\beta(i)}$ is divisible by $x^{\alpha(i)}$ for some $\alpha(i) \in A$. From here, it is easy to show that $I = \langle x^{\alpha(1)}, \ldots, x^{\alpha(s)} \rangle$ (see Exercise 6 for the details). This completes the proof. $\qquad\square$

To better understand how the proof of Theorem 5 works, let us apply it to the ideal $I = \langle x^4 y^2, x^3 y^4, x^2 y^5 \rangle$ discussed earlier in the section. From the picture of the exponents, you can see that the "projection" is $J = \langle x^2 \rangle \subseteq k[x]$. Since $x^2 y^5 \in I$, we have $m = 5$. Then we get the "slices" J_ℓ, $0 \le \ell \le 4 = m - 1$, generated by monomials containing y^ℓ:

$$J_0 = J_1 = \{0\},$$
$$J_2 = J_3 = \langle x^4 \rangle,$$
$$J_4 = \langle x^3 \rangle.$$

These "slices" are easy to see using the picture of the exponents. Then the proof of Theorem 5 gives $I = \langle x^2 y^5, x^4 y^2, x^4 y^3, x^3 y^4 \rangle$.

Theorem 5 solves the ideal description problem for monomial ideals, for it tells that such an ideal has a finite basis. This, in turn, allows us to solve the ideal membership problem for monomial ideals. Namely, if $I = \langle x^{\alpha(1)}, \ldots, x^{\alpha(s)} \rangle$, then one can easily show that a given polynomial f is in I if and only if the remainder of f on division by $x^{\alpha(1)}, \ldots, x^{\alpha(s)}$ is zero. See Exercise 8 for the details.

We can also use Dickson's Lemma to prove the following important fact about monomial orderings in $k[x_1, \ldots, x_n]$.

Corollary 6. *Let $>$ be a relation on $\mathbb{Z}_{\ge 0}^n$ satisfying:*

(i) $>$ *is a total ordering on $\mathbb{Z}_{\ge 0}^n$.*

(ii) *If $\alpha > \beta$ and $\gamma \in \mathbb{Z}_{\ge 0}^n$, then $\alpha + \gamma > \beta + \gamma$.*

Then $>$ is well-ordering if and only if $\alpha \ge 0$ for all $\alpha \in \mathbb{Z}_{\ge 0}^n$.

Proof. $\Rightarrow$: Assuming $>$ is a well-ordering, let α_0 be the smallest element of $\mathbb{Z}_{\ge 0}^n$. It suffices to show $\alpha_0 \ge 0$. This is easy: if $0 > \alpha_0$, then by hypothesis (ii), we can add α_0 to both sides to obtain $\alpha_0 > 2\alpha_0$, which is impossible since α_0 is the smallest element of $\mathbb{Z}_{\ge 0}^n$.

$\Leftarrow$: Assuming that $\alpha \ge 0$ for all $\alpha \in \mathbb{Z}_{\ge 0}^n$, let $A \subseteq \mathbb{Z}_{\ge 0}^n$ be nonempty. We need to show that A has a smallest element. Since $I = \langle x^\alpha \mid \alpha \in A \rangle$ is a monomial ideal, Dickson's Lemma gives us $\alpha(1), \ldots, \alpha(s) \in A$ so that $I = \langle x^{\alpha(1)}, \ldots, x^{\alpha(s)} \rangle$. Relabeling if necessary, we can assume that $\alpha(1) < \alpha(2) < \ldots < \alpha(s)$. We claim that $\alpha(1)$ is the smallest element of A. To prove this, take $\alpha \in A$. Then $x^\alpha \in I =$

$\langle x^{\alpha(1)}, \ldots, x^{\alpha(s)} \rangle$, so that by Lemma 2, x^{α} is divisible by some $x^{\alpha(i)}$. This tells us that $\alpha = \alpha(i) + \gamma$ for some $\gamma \in \mathbb{Z}_{\geq 0}^n$. Then $\gamma \geq 0$ and hypothesis (ii) imply that

$$\alpha = \alpha(i) + \gamma \geq \alpha(i) + 0 = \alpha(i) \geq \alpha(1).$$

Thus, $\alpha(1)$ is the least element of A.   □

As a result of this corollary, the definition of monomial ordering given in Definition 1 of §2 can be simplified. Conditions (i) and (ii) in the definition would be unchanged, but we could replace (iii) by the simpler condition that $\alpha \geq 0$ for all $\alpha \in \mathbb{Z}_{\geq 0}^n$. This makes it *much* easier to verify that a given ordering is actually a monomial ordering. See Exercises 9–11 for some examples.

Among all bases of a monomial ideal, there is one that is better than the others.

Proposition 7. *A monomial ideal $I \subseteq k[x_1, \ldots, x_n]$ has a basis $x^{\alpha(1)}, \ldots, x^{\alpha(s)}$ with the property that $x^{\alpha(i)}$ does not divide $x^{\alpha(j)}$ for $i \neq j$. Furthermore, this basis is unique and is called the* **minimal basis** *of I.*

Proof. By Theorem 5, I has a finite basis consisting of monomials. If one monomial in this basis divides another, then we can discard the other and still have a basis. Doing this repeatedly proves the existence of a minimal basis $x^{\alpha(1)}, \ldots, x^{\alpha(s)}$.

For uniqueness, assume that $x^{\beta(1)}, \ldots, x^{\beta(t)}$ is a second minimal basis of I. Then $x^{\alpha(1)} \in I$ and Lemma 2 imply that $x^{\beta(i)} \mid x^{\alpha(1)}$ for some i. Switching to the other basis, $x^{\beta(i)} \in I$ implies that $x^{\alpha(j)} \mid x^{\beta(i)}$ for some j. Thus $x^{\alpha(j)} \mid x^{\alpha(1)}$, which by minimality implies $j = 1$, and $x^{\alpha(1)} = x^{\beta(i)}$ follows easily. Continuing in this way, we see that the first basis is contained in the second. Then equality follows by interchanging the two bases.   □

EXERCISES FOR §4

1. Let $I \subseteq k[x_1, \ldots, x_n]$ be an ideal with the property that for every $f = \sum c_\alpha x^\alpha \in I$, every monomial x^α appearing in f is also in I. Show that I is a monomial ideal.
2. Complete the proof of Lemma 3 begun in the text.
3. Let $I = \langle x^6, x^2 y^3, xy^7 \rangle \subseteq k[x, y]$.
   a. In the (m, n)-plane, plot the set of exponent vectors (m, n) of monomials $x^m y^n$ appearing in elements of I.
   b. If we apply the division algorithm to an element $f \in k[x, y]$, using the generators of I as divisors, what terms can appear in the remainder?
4. Let $I \subseteq k[x, y]$ be the monomial ideal spanned over k by the monomials x^β corresponding to β in the shaded region shown at the top of the next page.
   a. Use the method given in the proof of Theorem 5 to find an ideal basis for I.
   b. Find a minimal basis for I in the sense of Proposition 7.
5. Suppose that $I = \langle x^\alpha \mid \alpha \in A \rangle$ is a monomial ideal, and let S be the set of all exponents that occur as monomials of I. For any monomial order $>$, prove that the smallest element of S with respect to $>$ must lie in A.

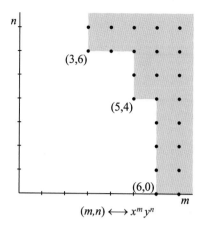

$(m,n) \longleftrightarrow x^m y^n$

6. Let $I = \langle x^\alpha \mid \alpha \in A \rangle$ be a monomial ideal, and assume that we have a finite basis $I = \langle x^{\beta(1)}, \ldots, x^{\beta(s)} \rangle$. In the proof of Dickson's Lemma, we observed that each $x^{\beta(i)}$ is divisible by $x^{\alpha(i)}$ for some $\alpha(i) \in A$. Prove that $I = \langle x^{\alpha(1)}, \ldots, x^{\alpha(s)} \rangle$.

7. Prove that Dickson's Lemma (Theorem 5) is equivalent to the following statement: given a nonempty subset $A \subseteq \mathbb{Z}_{\geq 0}^n$, there are finitely many elements $\alpha(1), \ldots, \alpha(s) \in A$ such that for every $\alpha \in A$, there exists some i and some $\gamma \in \mathbb{Z}_{\geq 0}^n$ such that $\alpha = \alpha(i) + \gamma$.

8. If $I = \langle x^{\alpha(1)}, \ldots, x^{\alpha(s)} \rangle$ is a monomial ideal, prove that a polynomial f is in I if and only if the remainder of f on division by $x^{\alpha(1)}, \ldots, x^{\alpha(s)}$ is zero. Hint: Use Lemmas 2 and 3.

9. Suppose we have the polynomial ring $k[x_1, \ldots, x_n, y_1, \ldots, y_m]$. Let us define a monomial order $>_{mixed}$ on this ring that mixes lex order for $x_1, \ldots x_n$, with grlex order for $y_1, \ldots, y_m$. If we write monomials in the $n + m$ variables as $x^\alpha y^\beta$, where $\alpha \in \mathbb{Z}_{\geq 0}^n$ and $\beta \in \mathbb{Z}_{\geq 0}^m$, then we define

$$x^\alpha y^\beta >_{mixed} x^\gamma y^\delta \iff x^\alpha >_{lex} x^\gamma \quad \text{or} \quad x^\alpha = x^\gamma \quad \text{and} \quad y^\beta >_{grlex} y^\delta.$$

Use Corollary 6 to prove that $>_{mixed}$ is a monomial order. This is an example of what is called a *product order*. It is clear that many other monomial orders can be created by this method.

10. In this exercise we will investigate a special case of a *weight order*. Let $\mathbf{u} = (u_1, \ldots, u_n)$ be a vector in $\mathbb{R}^n$ such that $u_1, \ldots, u_n$ are positive and linearly independent over $\mathbb{Q}$. We say that $\mathbf{u}$ is an *independent weight vector*. Then, for $\alpha, \beta \in \mathbb{Z}_{\geq 0}^n$, define

$$\alpha >_{\mathbf{u}} \beta \iff \mathbf{u} \cdot \alpha > \mathbf{u} \cdot \beta,$$

where the centered dot is the usual dot product of vectors. We call $>_{\mathbf{u}}$ the *weight order* determined by $\mathbf{u}$.

  a. Use Corollary 6 to prove that $>_{\mathbf{u}}$ is a monomial order. Hint: Where does your argument use the linear independence of $u_1, \ldots, u_n$?

  b. Show that $\mathbf{u} = (1, \sqrt{2})$ is an independent weight vector, so that $>_{\mathbf{u}}$ is a weight order on $\mathbb{Z}_{\geq 0}^2$.

  c. Show that $\mathbf{u} = (1, \sqrt{2}, \sqrt{3})$ is an independent weight vector, so that $>_{\mathbf{u}}$ is a weight order on $\mathbb{Z}_{\geq 0}^3$.

11. Another important weight order is constructed as follows. Let $\mathbf{u} = (u_1, \ldots, u_n)$ be in $\mathbb{Z}_{\geq 0}^n$, and fix a monomial order $>_\sigma$ (such as $>_{lex}$ or $>_{grevlex}$) on $\mathbb{Z}_{\geq 0}^n$. Then, for $\alpha, \beta \in \mathbb{Z}_{\geq 0}^n$, define $\alpha >_{\mathbf{u},\sigma} \beta$ if and only if

$$\mathbf{u}\cdot\alpha > \mathbf{u}\cdot\beta \quad \text{or} \quad \mathbf{u}\cdot\alpha = \mathbf{u}\cdot\beta \quad \text{and} \quad \alpha >_\sigma \beta.$$

We call $>_{\mathbf{u},\sigma}$ the *weight order* determined by $\mathbf{u}$ and $>_\sigma$.

a. Use Corollary 6 to prove that $>_{\mathbf{u},\sigma}$ is a monomial order.

b. Find $\mathbf{u} \in \mathbb{Z}^n_{\geq 0}$ so that the weight order $>_{\mathbf{u},lex}$ is the grlex order $>_{grlex}$.

c. In the definition of $>_{\mathbf{u},\sigma}$, the order $>_\sigma$ is used to break ties, and it turns out that ties will *always* occur when $n \geq 2$. More precisely, prove that given $\mathbf{u} \in \mathbb{Z}^n_{\geq 0}$, there are $\alpha \neq \beta$ in $\mathbb{Z}^n_{\geq 0}$ such that $\mathbf{u} \cdot \alpha = \mathbf{u} \cdot \beta$. Hint: Consider the linear equation $u_1 a_1 + \cdots + u_n a_n = 0$ over $\mathbb{Q}$. Show that there is a nonzero integer solution $(a_1, \dots, a_n)$, and then show that $(a_1, \dots, a_n) = \alpha - \beta$ for some $\alpha, \beta \in \mathbb{Z}^n_{\geq 0}$.

d. A useful example of a weight order is the *elimination order* introduced by BAYER and STILLMAN (1987b). Fix an integer $1 \leq l \leq n$ and let $\mathbf{u} = (1, \dots, 1, 0, \dots, 0)$, where there are l 1's and $n - l$ 0's. Then the l-th *elimination order* $>_l$ is the weight order $>_{\mathbf{u},grevlex}$. Prove that $>_l$ has the following property: if x^α is a monomial in which one of $x_1, \dots, x_l$ appears, then $x^\alpha >_l x^\beta$ for *any* monomial involving only $x_{l+1}, \dots, x_n$. Elimination orders play an important role in elimination theory, which we will study in the next chapter.

The weight orders described in Exercises 10 and 11 are only special cases of weight orders. In general, to determine a weight order, one starts with a vector $\mathbf{u}_1 \in \mathbb{R}^n$, whose entries may not be linearly independent over $\mathbb{Q}$. Then $\alpha > \beta$ if $\mathbf{u}_1 \cdot \alpha > \mathbf{u}_1 \cdot \beta$. But to break ties, one uses a second weight vector $\mathbf{u}_2 \in \mathbb{R}^n$. Thus, $\alpha > \beta$ also holds if $\mathbf{u}_1 \cdot \alpha = \mathbf{u}_1 \cdot \beta$ and $\mathbf{u}_2 \cdot \alpha > \mathbf{u}_2 \cdot \beta$. If there are still ties (when $\mathbf{u}_1 \cdot \alpha = \mathbf{u}_1 \cdot \beta$ and $\mathbf{u}_2 \cdot \alpha = \mathbf{u}_2 \cdot \beta$), then one uses a third weight vector $\mathbf{u}_3$, and so on. It can be proved that *every* monomial order on $\mathbb{Z}^n_{\geq 0}$ arises in this way. For a detailed treatment of weight orders and their relation to monomial orders, consult ROBBIANO (1986). See also Tutorial 10 of KREUZER and ROBBIANO (2000) or Section 1.2 of GREUEL and PFISTER (2008).

§5 The Hilbert Basis Theorem and Gröbner Bases

In this section, we will give a complete solution of the *ideal description problem* from §1. Our treatment will also lead to ideal bases with "good" properties relative to the division algorithm introduced in §3. The key idea we will use is that once we choose a monomial ordering, each nonzero $f \in k[x_1, \dots, x_n]$ has a unique leading term $\mathrm{LT}(f)$. Then, for any ideal I, we can define its *ideal of leading terms* as follows.

Definition 1. Let $I \subseteq k[x_1, \dots, x_n]$ be an ideal other than $\{0\}$, and fix a monomial ordering on $k[x_1, \dots, x_n]$. Then:

(i) We denote by $\mathrm{LT}(I)$ the set of leading terms of nonzero elements of I. Thus,

$$\mathrm{LT}(I) = \{cx^\alpha \mid \text{there exists } f \in I \setminus \{0\} \text{ with } \mathrm{LT}(f) = cx^\alpha\}.$$

(ii) We denote by $\langle \mathrm{LT}(I) \rangle$ the ideal generated by the elements of $\mathrm{LT}(I)$.

We have already seen that leading terms play an important role in the division algorithm. This brings up a subtle but important point concerning $\langle \mathrm{LT}(I) \rangle$. Namely, if we are given a finite generating set for I, say $I = \langle f_1, \dots, f_s \rangle$, then $\langle \mathrm{LT}(f_1), \dots, \mathrm{LT}(f_s) \rangle$ and $\langle \mathrm{LT}(I) \rangle$ may be *different* ideals. It is true that $\mathrm{LT}(f_i) \in$

$\mathrm{LT}(I) \subseteq \langle \mathrm{LT}(I) \rangle$ by definition, which implies $\langle \mathrm{LT}(f_1), \ldots, \mathrm{LT}(f_s) \rangle \subseteq \langle \mathrm{LT}(I) \rangle$. However, $\langle \mathrm{LT}(I) \rangle$ can be strictly larger. To see this, consider the following example.

Example 2. Let $I = \langle f_1, f_2 \rangle$, where $f_1 = x^3 - 2xy$ and $f_2 = x^2y - 2y^2 + x$, and use the grlex ordering on monomials in $k[x, y]$. Then

$$x \cdot (x^2y - 2y^2 + x) - y \cdot (x^3 - 2xy) = x^2,$$

so that $x^2 \in I$. Thus $x^2 = \mathrm{LT}(x^2) \in \langle \mathrm{LT}(I) \rangle$. However x^2 is not divisible by $\mathrm{LT}(f_1) = x^3$ or $\mathrm{LT}(f_2) = x^2y$, so that $x^2 \notin \langle \mathrm{LT}(f_1), \mathrm{LT}(f_2) \rangle$ by Lemma 2 of §4.

In the exercises for §3, you computed other examples of ideals $I = \langle f_1, \ldots, f_s \rangle$, where $\langle \mathrm{LT}(I) \rangle$ was strictly bigger than $\langle \mathrm{LT}(f_1), \ldots, \mathrm{LT}(f_s) \rangle$. The exercises at the end of the section will explore what this implies about the ideal membership problem.

We will now show that $\langle \mathrm{LT}(I) \rangle$ is a monomial ideal. This will allow us to apply the results of §4. In particular, it will follow that $\langle \mathrm{LT}(I) \rangle$ is generated by finitely many leading terms.

Proposition 3. *Let $I \subseteq k[x_1, \ldots, x_n]$ be an ideal different from $\{0\}$.*

(i) *$\langle \mathrm{LT}(I) \rangle$ is a monomial ideal.*
(ii) *There are $g_1, \ldots, g_t \in I$ such that $\langle \mathrm{LT}(I) \rangle = \langle \mathrm{LT}(g_1), \ldots, \mathrm{LT}(g_t) \rangle$.*

Proof. (i) The leading monomials $\mathrm{LM}(g)$ of elements $g \in I \setminus \{0\}$ generate the monomial ideal $\langle \mathrm{LM}(g) \mid g \in I \setminus \{0\} \rangle$. Since $\mathrm{LM}(g)$ and $\mathrm{LT}(g)$ differ by a nonzero constant, this ideal equals $\langle \mathrm{LT}(g) \mid g \in I \setminus \{0\} \rangle = \langle \mathrm{LT}(I) \rangle$ (see Exercise 4). Thus, $\langle \mathrm{LT}(I) \rangle$ is a monomial ideal.

(ii) Since $\langle \mathrm{LT}(I) \rangle$ is generated by the monomials $\mathrm{LM}(g)$ for $g \in I \setminus \{0\}$, Dickson's Lemma from §4 tells us that $\langle \mathrm{LT}(I) \rangle = \langle \mathrm{LM}(g_1), \ldots, \mathrm{LM}(g_t) \rangle$ for finitely many $g_1, \ldots, g_t \in I$. Since $\mathrm{LM}(g_i)$ differs from $\mathrm{LT}(g_i)$ by a nonzero constant, it follows that $\langle \mathrm{LT}(I) \rangle = \langle \mathrm{LT}(g_1), \ldots, \mathrm{LT}(g_t) \rangle$. This completes the proof. $\square$

We can now use Proposition 3 and the division algorithm to prove the existence of a finite generating set of *every* polynomial ideal, thus giving an affirmative answer to the ideal description problem from §1.

Theorem 4 (Hilbert Basis Theorem). *Every ideal $I \subseteq k[x_1, \ldots, x_n]$ has a finite generating set. In other words, $I = \langle g_1, \ldots, g_t \rangle$ for some $g_1, \ldots, g_t \in I$.*

Proof. If $I = \{0\}$, we take our generating set to be $\{0\}$, which is certainly finite. If I contains some nonzero polynomial, then a generating set $g_1, \ldots, g_t$ for I can be constructed as follows.

We first select one particular monomial order to use in the division algorithm and in computing leading terms. Then I has an ideal of leading terms $\langle \mathrm{LT}(I) \rangle$. By Proposition 3, there are $g_1, \ldots, g_t \in I$ such that $\langle \mathrm{LT}(I) \rangle = \langle \mathrm{LT}(g_1), \ldots, \mathrm{LT}(g_t) \rangle$. We claim that $I = \langle g_1, \ldots, g_t \rangle$.

It is clear that $\langle g_1, \ldots, g_t \rangle \subseteq I$ since each $g_i \in I$. Conversely, let $f \in I$ be any polynomial. If we apply the division algorithm from §3 to divide f by $(g_1, \ldots, g_t)$, then we get an expression of the form

$$f = q_1 g_1 + \cdots + q_t g_t + r$$

where no term of r is divisible by any of $\mathrm{LT}(g_1), \ldots, \mathrm{LT}(g_t)$. We claim that $r = 0$. To see this, note that

$$r = f - q_1 g_1 - \cdots - q_t g_t \in I.$$

If $r \neq 0$, then $\mathrm{LT}(r) \in \langle \mathrm{LT}(I) \rangle = \langle \mathrm{LT}(g_1), \ldots, \mathrm{LT}(g_t) \rangle$, and by Lemma 2 of §4, it follows that $\mathrm{LT}(r)$ must be divisible by some $\mathrm{LT}(g_i)$. This contradicts what it means to be a remainder, and, consequently, r must be zero. Thus,

$$f = q_1 g_1 + \cdots + q_t g_t + 0 \in \langle g_1, \ldots, g_t \rangle,$$

which shows that $I \subseteq \langle g_1, \ldots, g_t \rangle$. This completes the proof. □

Besides answering the ideal description question, the basis $\{g_1, \ldots, g_t\}$ used in the proof of Theorem 4 has the special property that $\langle \mathrm{LT}(I) \rangle = \langle \mathrm{LT}(g_1), \ldots, \mathrm{LT}(g_t) \rangle$. As we saw in Example 2, not all bases of an ideal behave this way. We will give these special bases the following name.

Definition 5. Fix a monomial order on the polynomial ring $k[x_1, \ldots, x_n]$. A finite subset $G = \{g_1, \ldots, g_t\}$ of an ideal $I \subseteq k[x_1, \ldots, x_n]$ different from $\{0\}$ is said to be a **Gröbner basis** (or **standard basis**) if

$$\langle \mathrm{LT}(g_1), \ldots, \mathrm{LT}(g_t) \rangle = \langle \mathrm{LT}(I) \rangle.$$

Using the convention that $\langle \emptyset \rangle = \{0\}$, we define the empty set $\emptyset$ to be the Gröbner basis of the zero ideal $\{0\}$.

Equivalently, but more informally, a set $\{g_1, \ldots, g_t\} \subseteq I$ is a Gröbner basis of I if and only if the leading term of any element of I is divisible by one of the $\mathrm{LT}(g_i)$ (this follows from Lemma 2 of §4—see Exercise 5). The proof of Theorem 4 also establishes the following result.

Corollary 6. *Fix a monomial order. Then every ideal $I \subseteq k[x_1, \ldots, x_n]$ has a Gröbner basis. Furthermore, any Gröbner basis for an ideal I is a basis of I.*

Proof. Given a nonzero ideal, the set $G = \{g_1, \ldots, g_t\}$ constructed in the proof of Theorem 4 is a Gröbner basis by definition. For the second claim, note that if $\langle \mathrm{LT}(I) \rangle = \langle \mathrm{LT}(g_1), \ldots, \mathrm{LT}(g_t) \rangle$, then the argument given in Theorem 4 shows that $I = \langle g_1, \ldots, g_t \rangle$, so that G is a basis for I. (A slightly different proof is given in Exercise 6.) □

In §6 we will study the properties of Gröbner bases in more detail, and, in particular, we will see how they give a solution of the ideal membership problem. Gröbner bases are the "good" generating sets we hoped for at the end of §3.

For some examples of Gröbner bases, first consider the ideal I from Example 2, which had the basis $\{f_1, f_2\} = \{x^3 - 2xy, x^2y - 2y^2 + x\}$. Then $\{f_1, f_2\}$ is *not* a Gröbner basis for I with respect to grlex order since we saw in Example 2 that

$x^2 \in \langle \mathrm{LT}(I) \rangle$, but $x^2 \notin \langle \mathrm{LT}(f_1), \mathrm{LT}(f_2) \rangle$. In §7 we will learn how to find a Gröbner basis of I.

Next, consider the ideal $J = \langle g_1, g_2 \rangle = \langle x+z, y-z \rangle$. We claim that g_1 and g_2 form a Gröbner basis using lex order in $\mathbb{R}[x, y, z]$. Thus, we must show that the leading term of every nonzero element of J lies in the ideal $\langle \mathrm{LT}(g_1), \mathrm{LT}(g_2) \rangle = \langle x, y \rangle$. By Lemma 2 of §4, this is equivalent to showing that the leading term of *any* nonzero element of J is divisible by either x or y.

To prove this, consider any $f = Ag_1 + Bg_2 \in J$. Suppose on the contrary that f is nonzero and $\mathrm{LT}(f)$ is divisible by neither x nor y. Then by the definition of lex order, f must be a polynomial in z alone. However, f vanishes on the linear subspace $L = \mathbf{V}(x+z, y-z) \subseteq \mathbb{R}^3$ since $f \in J$. It is easy to check that $(x, y, z) = (-t, t, t) \in L$ for any real number t. The only polynomial in z alone that vanishes at all of these points is the zero polynomial, which is a contradiction. It follows that $\langle g_1, g_2 \rangle$ is a Gröbner basis for J. In §6, we will learn a more systematic way to detect when a basis is a Gröbner basis.

Note, by the way, that the generators for the ideal J come from a row echelon matrix of coefficients:

$$\begin{pmatrix} 1 & 0 & 1 \\ 0 & 1 & -1 \end{pmatrix}.$$

This is no accident: for ideals generated by linear polynomials, a Gröbner basis for lex order is determined by the row echelon form of the matrix made from the coefficients of the generators (see Exercise 9).

Gröbner bases for ideals in polynomial rings were introduced by B. Buchberger in his PhD thesis BUCHBERGER (1965) and named by him in honor of W. Gröbner (1899–1980), Buchberger's thesis adviser. The closely related concept of "standard bases" for ideals in power series rings was discovered independently by H. Hironaka in HIRONAKA (1964). As we will see later in this chapter, Buchberger also developed the fundamental algorithms for working with Gröbner bases. Sometimes one sees the alternate spelling "Groebner bases," since this is how the command is spelled in some computer algebra systems.

We conclude this section with two applications of the Hilbert Basis Theorem. The first is an algebraic statement about the ideals in $k[x_1, \ldots, x_n]$. An **ascending chain** of ideals is a nested increasing sequence:

$$I_1 \subseteq I_2 \subseteq I_3 \subseteq \cdots.$$

For example, the sequence

$$(1) \qquad \langle x_1 \rangle \subseteq \langle x_1, x_2 \rangle \subseteq \cdots \subseteq \langle x_1, \ldots, x_n \rangle$$

forms a (finite) ascending chain of ideals. If we try to *extend* this chain by including an ideal with further generator(s), one of two alternatives will occur. Consider the ideal $\langle x_1, \ldots, x_n, f \rangle$ where $f \in k[x_1, \ldots, x_n]$. If $f \in \langle x_1, \ldots, x_n \rangle$, then we obtain $\langle x_1, \ldots, x_n \rangle$ again and nothing has changed. If, on the other hand, $f \notin \langle x_1, \ldots, x_n \rangle$, then we claim $\langle x_1, \ldots, x_n, f \rangle = k[x_1, \ldots, x_n]$. We leave the proof of this claim to the reader (Exercise 11 of this section). As a result, the ascending chain (1) can be continued in only two ways, either by repeating the last ideal *ad infinitum* or by appending $k[x_1, \ldots, x_n]$ and then repeating it *ad infinitum*. In either case, the ascending chain will have "stabilized" after a finite number of steps, in the sense that all the ideals after that point in the chain will be equal. Our next result shows that the same phenomenon occurs in *every* ascending chain of ideals in $k[x_1, \ldots, x_n]$.

Theorem 7 (The Ascending Chain Condition). *Let*

$$I_1 \subseteq I_2 \subseteq I_3 \subseteq \cdots$$

be an ascending chain of ideals in $k[x_1, \ldots, x_n]$. Then there exists an $N \geq 1$ such that

$$I_N = I_{N+1} = I_{N+2} = \cdots .$$

Proof. Given the ascending chain $I_1 \subseteq I_2 \subseteq I_3 \subseteq \cdots$, consider the set $I = \bigcup_{i=1}^{\infty} I_i$. We begin by showing that I is also an ideal in $k[x_1, \ldots, x_n]$. First, $0 \in I$ since $0 \in I_i$ for every i. Next, if $f, g \in I$, then, by definition, $f \in I_i$, and $g \in I_j$ for some i and j (possibly different). However, since the ideals I_i form an ascending chain, if we relabel so that $i \leq j$, then both f and g are in I_j. Since I_j is an ideal, the sum $f + g \in I_j$, hence, $\in I$. Similarly, if $f \in I$ and $r \in k[x_1, \ldots, x_n]$, then $f \in I_i$ for some i, and $r \cdot f \in I_i \subseteq I$. Hence, I is an ideal.

By the Hilbert Basis Theorem, the ideal I must have a finite generating set: $I = \langle f_1, \ldots, f_s \rangle$. But each of the generators is contained in some one of the I_j, say $f_i \in I_{j_i}$ for some $j_i, i = 1, \ldots, s$. We take N to be the maximum of the j_i. Then by the definition of an ascending chain $f_i \in I_N$ for all i. Hence we have

$$I = \langle f_1, \ldots, f_s \rangle \subseteq I_N \subseteq I_{N+1} \subseteq \cdots \subseteq I.$$

As a result the ascending chain stabilizes with I_N. All the subsequent ideals in the chain are equal. $\qquad \square$

The statement that every ascending chain of ideals in $k[x_1, \ldots, x_n]$ stabilizes is often called the **ascending chain condition**, or ACC for short. In Exercise 12 of this section, you will show that if we assume the ACC as hypothesis, then it follows that every ideal is finitely generated. Thus, the ACC is actually equivalent to the conclusion of the Hilbert Basis Theorem. We will use the ACC in a crucial way in §7, when we give Buchberger's algorithm for constructing Gröbner bases. We will also use the ACC in Chapter 4 to study the structure of affine varieties.

Our second consequence of the Hilbert Basis Theorem will be geometric. Up to this point, we have considered affine varieties as the sets of solutions of specific finite sets of polynomial equations:

$$\mathbf{V}(f_1, \ldots, f_s) = \{(a_1, \ldots, a_n) \in k^n \mid f_i(a_1, \ldots, a_n) = 0 \text{ for all } i\}.$$

The Hilbert Basis Theorem shows that, in fact, it also makes sense to speak of the affine variety defined by an *ideal* $I \subseteq k[x_1, \ldots, x_n]$.

Definition 8. Let $I \subseteq k[x_1, \ldots, x_n]$ be an ideal. We will denote by $\mathbf{V}(I)$ the set

$$\mathbf{V}(I) = \{(a_1, \ldots, a_n) \in k^n \mid f(a_1, \ldots, a_n) = 0 \text{ for all } f \in I\}.$$

Even though a nonzero ideal I always contains infinitely many different polynomials, the set $\mathbf{V}(I)$ can still be defined by a finite set of polynomial equations.

Proposition 9. $\mathbf{V}(I)$ *is an affine variety. In particular, if* $I = \langle f_1, \ldots, f_s \rangle$*, then* $\mathbf{V}(I) = \mathbf{V}(f_1, \ldots, f_s)$.

Proof. By the Hilbert Basis Theorem, $I = \langle f_1, \ldots, f_s \rangle$ for some finite generating set. We claim that $\mathbf{V}(I) = \mathbf{V}(f_1, \ldots, f_s)$. First, since the $f_i \in I$, if $f(a_1, \ldots, a_n) = 0$ for all $f \in I$, then $f_i(a_1, \ldots, a_n) = 0$, so $\mathbf{V}(I) \subseteq \mathbf{V}(f_1, \ldots, f_s)$. On the other hand, let $(a_1, \ldots, a_n) \in \mathbf{V}(f_1, \ldots, f_s)$ and let $f \in I$. Since $I = \langle f_1, \ldots, f_s \rangle$, we can write

$$f = \sum_{i=1}^{s} h_i f_i$$

for some $h_i \in k[x_1, \ldots, x_n]$. But then

$$f(a_1, \ldots, a_n) = \sum_{i=1}^{s} h_i(a_1, \ldots, a_n) f_i(a_1, \ldots, a_n)$$

$$= \sum_{i=1}^{s} h_i(a_1, \ldots, a_n) \cdot 0 = 0.$$

Thus, $\mathbf{V}(f_1, \ldots, f_s) \subseteq \mathbf{V}(I)$ and, hence, they are equal. $\square$

The most important consequence of this proposition is that *varieties are determined by ideals*. For example, in Chapter 1, we proved that $\mathbf{V}(f_1, \ldots, f_s) = \mathbf{V}(g_1, \ldots, g_t)$ whenever $\langle f_1, \ldots, f_s \rangle = \langle g_1, \ldots, g_t \rangle$ (see Proposition 4 of Chapter 1, §4). This proposition is an immediate corollary of Proposition 9. The relation between ideals and varieties will be explored in more detail in Chapter 4.

In the exercises, we will exploit Proposition 9 by showing that by using the right generating set for an ideal I, we can gain a better understanding of the variety $\mathbf{V}(I)$.

EXERCISES FOR §5

1. Let $I = \langle g_1, g_2, g_3 \rangle \subseteq \mathbb{R}[x, y, z]$, where $g_1 = xy^2 - xz + y$, $g_2 = xy - z^2$ and $g_3 = x - yz^4$. Using the lex order, give an example of $g \in I$ such that $\mathrm{LT}(g) \notin \langle \mathrm{LT}(g_1), \mathrm{LT}(g_2), \mathrm{LT}(g_3) \rangle$.
2. For the ideals and generators given in Exercises 5, 6, and 7 of §3, show that $\mathrm{LT}(I)$ is strictly bigger than $\langle \mathrm{LT}(f_1), \ldots, \mathrm{LT}(f_s) \rangle$. Hint: This should follow directly from what you did in those exercises.
3. To generalize the situation of Exercises 1 and 2, suppose that $I = \langle f_1, \ldots, f_s \rangle$ is an ideal such that $\langle \mathrm{LT}(f_1), \ldots, \mathrm{LT}(f_s) \rangle$ is strictly smaller than $\langle \mathrm{LT}(I) \rangle$.
   a. Prove that there is some $f \in I$ whose remainder on division by $f_1, \ldots, f_s$ is nonzero. Hint: First show that $\mathrm{LT}(f) \notin \langle \mathrm{LT}(f_1), \ldots, \mathrm{LT}(f_s) \rangle$ for some $f \in I$. Then use Lemma 2 of §4.

b. What does part (a) say about the ideal membership problem?

c. How does part (a) relate to the conjecture you were asked to make in Exercise 8 of §3?

4. If $I \subseteq k[x_1, \ldots, x_n]$ is an ideal, prove that $\langle \text{LT}(g) \mid g \in I \setminus \{0\} \rangle = \langle \text{LM}(g) \mid g \in I \setminus \{0\} \rangle$.

5. Let I be an ideal of $k[x_1, \ldots, x_n]$. Show that $G = \{g_1, \ldots, g_t\} \subseteq I$ is a Gröbner basis of I if and only if the leading term of any element of I is divisible by one of the $\text{LT}(g_i)$.

6. Corollary 6 asserts that a Gröbner basis is a basis, i.e., if $G = \{g_1, \ldots, g_t\} \subseteq I$ satisfies $\langle \text{LT}(I) \rangle = \langle \text{LT}(g_1), \ldots, \text{LT}(g_t) \rangle$, then $I = \langle g_1, \ldots, g_t \rangle$. We gave one proof of this in the proof of Theorem 4. Complete the following sketch to give a second proof. If $f \in I$, then divide f by $(g_1, \ldots, g_t)$. At each step of the division algorithm, the leading term of the polynomial under the division will be in $\langle \text{LT}(I) \rangle$ and, hence, will be divisible by one of the $\text{LT}(g_i)$. Hence, terms are never added to the remainder, so that $f = \sum_{i=1}^{t} a_i g_i$ when the algorithm terminates.

7. If we use grlex order with $x > y > z$, is $\{x^4y^2 - z^5, x^3y^3 - 1, x^2y^4 - 2z\}$ a Gröbner basis for the ideal generated by these polynomials? Why or why not?

8. Repeat Exercise 7 for $I = \langle x - z^2, y - z^3 \rangle$ using the lex order. Hint: The difficult part of this exercise is to determine exactly which polynomials are in $\langle \text{LT}(I) \rangle$.

9. Let $A = (a_{ij})$ be an $m \times n$ matrix with real entries in row echelon form and let $J \subseteq \mathbb{R}[x_1, \ldots, x_n]$ be an ideal generated by the linear polynomials $\sum_{j=1}^{n} a_{ij}x_j$ for $1 \leq i \leq m$. Show that the given generators form a Gröbner basis for J with respect to a suitable lexicographic order. Hint: Order the variables corresponding to the leading 1's before the other variables.

10. Let $I \subseteq k[x_1, \ldots, x_n]$ be a *principal ideal* (that is, I is generated by a single $f \in I$—see §5 of Chapter 1). Show that any finite subset of I containing a generator for I is a Gröbner basis for I.

11. Let $f \in k[x_1, \ldots, x_n]$. If $f \notin \langle x_1, \ldots, x_n \rangle$, then show $\langle x_1, \ldots, x_n, f \rangle = k[x_1, \ldots, x_n]$.

12. Show that if we take as hypothesis that every ascending chain of ideals in $k[x_1, \ldots, x_n]$ stabilizes, then the conclusion of the Hilbert Basis Theorem is a consequence. Hint: Argue by contradiction, assuming that some ideal $I \subseteq k[x_1, \ldots, x_n]$ has no finite generating set. The arguments you gave in Exercise 12 should not make any special use of properties of polynomials. Indeed, it is true that in any commutative ring R, the following two statements are equivalent:

  (i) Every ideal $I \subseteq R$ is finitely generated.

  (ii) Every ascending chain of ideals of R stabilizes.

13. Let

$$V_1 \supseteq V_2 \supseteq V_3 \supseteq \cdots$$

be a **descending chain** of affine varieties. Show that there is some $N \geq 1$ such that $V_N = V_{N+1} = V_{N+2} = \cdots$. Hint: Use the ACC and Exercise 14 of Chapter 1, §4.

14. Let $f_1, f_2, \ldots \in k[x_1, \ldots, x_n]$ be an infinite collection of polynomials. Prove that there is an integer N such that $f_i \in \langle f_1, \ldots, f_N \rangle$ for all $i \geq N + 1$. Hint: Use $f_1, f_2, \ldots$ to create an ascending chain of ideals.

15. Given polynomials $f_1, f_2, \ldots \in k[x_1, \ldots, x_n]$, let $\mathbf{V}(f_1, f_2, \ldots) \subseteq k^n$ be the solutions of the infinite system of equations $f_1 = f_2 = \cdots = 0$. Show that there is some N such that $\mathbf{V}(f_1, f_2, \ldots) = \mathbf{V}(f_1, \ldots, f_N)$.

16. In Chapter 1, §4, we defined the ideal $\mathbf{I}(V)$ of a variety $V \subseteq k^n$. In this section, we defined the variety of any ideal (see Definition 8). In particular, this means that $\mathbf{V}(\mathbf{I}(V))$ is a variety. Prove that $\mathbf{V}(\mathbf{I}(V)) = V$. Hint: See the proof of Lemma 7 of Chapter 1, §4.

17. Consider the variety $V = \mathbf{V}(x^2 - y, y + x^2 - 4) \subseteq \mathbb{C}^2$. Note that $V = \mathbf{V}(I)$, where $I = \langle x^2 - y, y + x^2 - 4 \rangle$.

  a. Prove that $I = \langle x^2 - y, x^2 - 2 \rangle$.

  b. Using the basis from part (a), prove that $\mathbf{V}(I) = \{(\pm\sqrt{2}, 2)\}$.

One reason why the second basis made V easier to understand was that $x^2 - 2$ could be *factored*. This implied that V "split" into two pieces. See Exercise 18 for a general statement.

18. When an ideal has a basis where some of the elements can be factored, we can use the factorization to help understand the variety.
   a. Show that if $g \in k[x_1, \dots, x_n]$ factors as $g = g_1 g_2$, then for any f, we have $\mathbf{V}(f, g) = \mathbf{V}(f, g_1) \cup \mathbf{V}(f, g_2)$.
   b. Show that in $\mathbb{R}^3$, $\mathbf{V}(y - x^2, xz - y^2) = \mathbf{V}(y - x^2, xz - x^4)$.
   c. Use part (a) to describe and/or sketch the variety from part (b).

§6 Properties of Gröbner Bases

As shown in §5, every nonzero ideal $I \subseteq k[x_1, \dots, x_n]$ has a Gröbner basis. In this section, we will study the properties of Gröbner bases and learn how to detect when a given basis is a Gröbner basis. We begin by showing that the undesirable behavior of the division algorithm in $k[x_1, \dots, x_n]$ noted in §3 does not occur when we divide by the elements of a Gröbner basis.

Let us first prove that the remainder is uniquely determined when we divide by a Gröbner basis.

Proposition 1. *Let $I \subseteq k[x_1, \dots, x_n]$ be an ideal and let $G = \{g_1, \dots, g_t\}$ be a Gröbner basis for I. Then given $f \in k[x_1, \dots, x_n]$, there is a unique $r \in k[x_1, \dots, x_n]$ with the following two properties:*

 (i) *No term of r is divisible by any of $\mathrm{LT}(g_1), \dots, \mathrm{LT}(g_t)$.*
 (ii) *There is $g \in I$ such that $f = g + r$.*

In particular, r is the remainder on division of f by G no matter how the elements of G are listed when using the division algorithm.

Proof. The division algorithm gives $f = q_1 g_1 + \cdots + q_t g_t + r$, where r satisfies (i). We can also satisfy (ii) by setting $g = q_1 g_1 + \cdots + q_t g_t \in I$. This proves the existence of r.

To prove uniqueness, suppose $f = g + r = g' + r'$ satisfy (i) and (ii). Then $r - r' = g' - g \in I$, so that if $r \neq r'$, then $\mathrm{LT}(r - r') \in \langle \mathrm{LT}(I) \rangle = \langle \mathrm{LT}(g_1), \dots, \mathrm{LT}(g_t) \rangle$. By Lemma 2 of §4, it follows that $\mathrm{LT}(r - r')$ is divisible by some $\mathrm{LT}(g_i)$. This is impossible since no term of r, r' is divisible by one of $\mathrm{LT}(g_1), \dots, \mathrm{LT}(g_t)$. Thus $r - r'$ must be zero, and uniqueness is proved.

The final part of the proposition follows from the uniqueness of r. □

The remainder r is sometimes called the *normal form* of f, and its uniqueness properties will be explored in Exercises 1 and 4. In fact, Gröbner bases can be characterized by the uniqueness of the remainder—see Theorem 5.35 of BECKER and WEISPFENNING (1993).

Although the remainder r is unique, even for a Gröbner basis, the "quotients" q_i produced by the division algorithm $f = q_1 g_1 + \cdots + q_t g_t + r$ can change if we list the generators in a different order. See Exercise 2 for an example.

As a corollary of Proposition 1, we get the following criterion for when a given polynomial lies in an ideal.

Corollary 2. *Let $G = \{g_1, \ldots, g_t\}$ be a Gröbner basis for an ideal $I \subseteq k[x_1, \ldots, x_n]$ and let $f \in k[x_1, \ldots, x_n]$. Then $f \in I$ if and only if the remainder on division of f by G is zero.*

Proof. If the remainder is zero, then we have already observed that $f \in I$. Conversely, given $f \in I$, then $f = f + 0$ satisfies the two conditions of Proposition 1. It follows that 0 is the remainder of f on division by G.   □

The property given in Corollary 2 is sometimes taken as the definition of a Gröbner basis, since one can show that it is true if and only if $\langle \mathrm{LT}(g_1), \ldots, \mathrm{LT}(g_t) \rangle = \langle \mathrm{LT}(I) \rangle$ (see Exercise 3). For this and similar conditions equivalent to being a Gröbner basis, see Proposition 5.38 of BECKER and WEISPFENNING (1993).

Using Corollary 2, we get an algorithm for solving the ideal membership problem from §1, *provided* that we know a Gröbner basis G for the ideal in question—we only need to compute a remainder with respect to G to determine whether $f \in I$. In §7, we will learn how to find Gröbner bases, and we will give a complete solution of the ideal membership problem in §8.

We will use the following notation for the remainder.

Definition 3. We will write $\overline{f}^F$ for the remainder on division of f by the ordered s-tuple $F = (f_1, \ldots, f_s)$. If F is a Gröbner basis for $\langle f_1, \ldots, f_s \rangle$, then we can regard F as a set (without any particular order) by Proposition 1.

For instance, with $F = (x^2 y - y^2, x^4 y^2 - y^2) \subseteq k[x, y]$, using the lex order, we have

$$\overline{x^5 y}^F = xy^3$$

since the division algorithm yields

$$x^5 y = (x^3 + xy)(x^2 y - y^2) + 0 \cdot (x^4 y^2 - y^2) + xy^3.$$

We will next discuss how to tell whether a given generating set of an ideal is a Gröbner basis. As we have indicated, the "obstruction" to $\{f_1, \ldots, f_s\}$ being a Gröbner basis is the possible occurrence of polynomial combinations of the f_i whose leading terms are not in the ideal generated by the $\mathrm{LT}(f_i)$. One way this can occur is if the leading terms in a suitable combination

$$ax^\alpha f_i - bx^\beta f_j$$

cancel, leaving only smaller terms. On the other hand, $ax^\alpha f_i - bx^\beta f_j \in I$, so its leading term is in $\langle \mathrm{LT}(I) \rangle$. You should check that this is what happened in Example 2 of §5. To study this cancellation phenomenon, we introduce the following special combinations.

Definition 4. Let $f, g \in k[x_1, \ldots, x_n]$ be nonzero polynomials.

(i) If multideg$(f) = \alpha$ and multideg$(g) = \beta$, then let $\gamma = (\gamma_1, \ldots, \gamma_n)$, where $\gamma_i = \max(\alpha_i, \beta_i)$ for each i. We call x^γ the **least common multiple** of $\mathrm{LM}(f)$ and $\mathrm{LM}(g)$, written $x^\gamma = \mathrm{lcm}(\mathrm{LM}(f), \mathrm{LM}(g))$.

(ii) The **S-polynomial** of f and g is the combination

$$S(f,g) = \frac{x^\gamma}{\mathrm{LT}(f)} \cdot f - \frac{x^\gamma}{\mathrm{LT}(g)} \cdot g.$$

(Note that we are inverting the leading coefficients here as well.)

For example, let $f = x^3y^2 - x^2y^3 + x$ and $g = 3x^4y + y^2$ in $\mathbb{R}[x,y]$ with the grlex order. Then $\gamma = (4,2)$ and

$$\begin{aligned} S(f,g) &= \frac{x^4y^2}{x^3y^2} \cdot f - \frac{x^4y^2}{3x^4y} \cdot g \\ &= x \cdot f - (1/3) \cdot y \cdot g \\ &= -x^3y^3 + x^2 - (1/3)y^3. \end{aligned}$$

An S-polynomial $S(f,g)$ is "designed" to produce cancellation of leading terms. See Exercise 7 for a precise description of the cancellation that occurs.

The following lemma shows that every cancellation of leading terms among polynomials of the same multidegree comes from the cancellation that occurs for S-polynomials.

Lemma 5. *Suppose we have a sum* $\sum_{i=1}^{s} p_i$, *where* $\mathrm{multideg}(p_i) = \delta \in \mathbb{Z}_{\geq 0}^{n}$ *for all i. If* $\mathrm{multideg}(\sum_{i=1}^{s} p_i) < \delta$, *then* $\sum_{i=1}^{s} p_i$ *is a linear combination, with coefficients in k, of the S-polynomials* $S(p_j, p_l)$ *for* $1 \leq j, l \leq s$. *Furthermore, each* $S(p_j, p_l)$ *has multidegree* $< \delta$.

Proof. Let $d_i = \mathrm{LC}(p_i)$, so that $d_i x^\delta$ is the leading term of p_i. Since the sum $\sum_{i=1}^{s} p_i$ has strictly smaller multidegree, it follows easily that $\sum_{i=1}^{s} d_i = 0$.

Next observe that since p_i and p_j have the same leading monomial, their S-polynomial reduces to

$$(1) \qquad\qquad S(p_i, p_j) = \frac{1}{d_i}p_i - \frac{1}{d_j}p_j.$$

It follows that

$$(2) \qquad \begin{aligned} \sum_{i=1}^{s-1} d_i S(p_i, p_s) &= d_1\left(\frac{1}{d_1}p_1 - \frac{1}{d_s}p_s\right) + d_2\left(\frac{1}{d_2}p_2 - \frac{1}{d_s}p_s\right) + \cdots \\ &= p_1 + p_2 + \cdots + p_{s-1} - \frac{1}{d_s}(d_1 + \cdots + d_{s-1})p_s. \end{aligned}$$

However, $\sum_{i=1}^{s} d_i = 0$ implies $d_1 + \cdots + d_{s-1} = -d_s$, so that (2) reduces to

$$\sum_{i=1}^{s-1} d_i S(p_i, p_s) = p_1 + \cdots + p_{s-1} + p_s.$$

Thus, $\sum_{i=1}^{s} p_i$ is a sum of S-polynomials of the desired form, and equation (1) makes it easy to see that $S(p_i, p_j)$ has multidegree $< \delta$. The lemma is proved. □

When $p_1, \ldots, p_s$ satisfy the hypothesis of Lemma 5, we get an equation of the form

$$\sum_{i=1}^{s} p_i = \sum_{j,l} c_{jl} S(p_j, p_l).$$

Let us consider where the cancellation occurs. In the sum on the left, every summand p_i has multidegree δ, so the cancellation occurs only after adding them up. However, in the sum on the right, each summand $c_{jl} S(p_j, p_l)$ has multidegree $< \delta$, so that the cancellation has already occurred. Intuitively, this means that all cancellation can be accounted for by S-polynomials.

Using S-polynomials and Lemma 5, we can now prove the following criterion of Buchberger for when a basis of an ideal is a Gröbner basis.

Theorem 6 (Buchberger's Criterion). *Let I be a polynomial ideal. Then a basis $G = \{g_1, \ldots, g_t\}$ of I is a Gröbner basis of I if and only if for all pairs $i \neq j$, the remainder on division of $S(g_i, g_j)$ by G (listed in some order) is zero.*

Proof. $\Rightarrow$: If G is a Gröbner basis, then since $S(g_i, g_j) \in I$, the remainder on division by G is zero by Corollary 2.

$\Leftarrow$: Let $f \in I$ be nonzero. We will show that $\mathrm{LT}(f) \in \langle \mathrm{LT}(g_1), \ldots, \mathrm{LT}(g_t) \rangle$ as follows. Write

$$f = \sum_{i=1}^{t} h_i g_i, \quad h_i \in k[x_1, \ldots, x_n].$$

From Lemma 8 of §2, it follows that

$$(3) \qquad \mathrm{multideg}(f) \leq \max(\mathrm{multideg}(h_i g_i) \mid h_i g_i \neq 0).$$

The strategy of the proof is to pick the most efficient representation of f, meaning that among all expressions $f = \sum_{i=1}^{t} h_i g_i$, we pick one for which

$$\delta = \max(\mathrm{multideg}(h_i g_i) \mid h_i g_i \neq 0)$$

is minimal. The minimal δ exists by the well-ordering property of our monomial ordering. By (3), it follows that $\mathrm{multideg}(f) \leq \delta$.

If equality occurs, then $\mathrm{multideg}(f) = \mathrm{multideg}(h_i g_i)$ for some i. This easily implies that $\mathrm{LT}(f)$ is divisible by $\mathrm{LT}(g_i)$. Then $\mathrm{LT}(f) \in \langle \mathrm{LT}(g_1), \ldots, \mathrm{LT}(g_t) \rangle$, which is what we want to prove.

It remains to consider the case when the minimal δ satisfies $\mathrm{multideg}(f) < \delta$. We will use $\overline{S(g_i, g_j)}^G = 0$ for $i \neq j$ to find a new expression for f that decreases δ. This will contradict the minimality of δ and complete the proof.

Given an expression $f = \sum_{i=1}^{t} h_i g_i$ with minimal δ, we begin by isolating the part of the sum where multidegree δ occurs:

$$
\text{(4)}\quad
\begin{aligned}
f &= \sum_{\text{multideg}(h_i g_i)=\delta} h_i g_i \;+\; \sum_{\text{multideg}(h_i g_i)<\delta} h_i g_i \\[2mm]
&= \sum_{\text{multideg}(h_i g_i)=\delta} \mathrm{LT}(h_i) g_i \;+\; \sum_{\text{multideg}(h_i g_i)=\delta} (h_i - \mathrm{LT}(h_i)) g_i \;+\; \sum_{\text{multideg}(h_i g_i)<\delta} h_i g_i.
\end{aligned}
$$

The monomials appearing in the second and third sums on the second line all have multidegree $< \delta$. Then multideg$(f) < \delta$ means that the first sum on the second line also has multidegree $< \delta$.

The key to decreasing δ is to rewrite the first sum in two stages: use Lemma 5 to rewrite the first sum in terms of S-polynomials, and then use $\overline{S(g_i, g_j)}^G = 0$ to rewrite the S-polynomials without cancellation.

To express the first sum on the second line of (4) using S-polynomials, note that

$$
\text{(5)}\qquad \sum_{\text{multideg}(h_i g_i)=\delta} \mathrm{LT}(h_i) g_i
$$

satisfies the hypothesis of Lemma 5 since each $p_i = \mathrm{LT}(h_i) g_i$ has multidegree δ and the sum has multidegree $< \delta$. Hence, the first sum is a linear combination with coefficients in k of the S-polynomials $S(p_i, p_j)$. In Exercise 8, you will verify that

$$
S(p_i, p_j) = x^{\delta - \gamma_{ij}} S(g_i, g_j),
$$

where $x^{\gamma_{ij}} = \mathrm{lcm}(\mathrm{LM}(g_i), \mathrm{LM}(g_j))$. It follows that the first sum (5) is a linear combination of $x^{\delta - \gamma_{ij}} S(g_i, g_j)$ for certain pairs (i, j).

Consider one of these S-polynomials $S(g_i, g_j)$. Since $\overline{S(g_i, g_j)}^G = 0$, the division algorithm (Theorem 3 of §3) gives an expression

$$
\text{(6)}\qquad S(g_i, g_j) = \sum_{l=1}^{t} A_l g_l,
$$

where $A_l \in k[x_1, \ldots, x_n]$ and

$$
\text{(7)}\qquad \mathrm{multideg}(A_l g_l) \le \mathrm{multideg}(S(g_i, g_j))
$$

when $A_l g_l \ne 0$. Now multiply each side of (6) by $x^{\delta - \gamma_{ij}}$ to obtain

$$
\text{(8)}\qquad x^{\delta - \gamma_{ij}} S(g_i, g_j) = \sum_{l=1}^{t} B_l g_l,
$$

where $B_l = x^{\delta - \gamma_{ij}} A_l$. Then (7) implies that when $B_l g_l \ne 0$, we have

$$
\text{(9)}\qquad \mathrm{multideg}(B_l g_l) \le \mathrm{multideg}(x^{\delta - \gamma_{ij}} S(g_i, g_j)) < \delta
$$

since $\mathrm{LT}(S(g_i, g_j)) < \mathrm{lcm}(\mathrm{LM}(g_i), \mathrm{LM}(g_j)) = x^{\gamma_{ij}}$ by Exercise 7.

It follows that the first sum (5) is a linear combination of certain $x^{\delta - \gamma_{ij}} S(g_i, g_j)$, each of which satisfies (8) and (9). Hence we can write the first sum as

$$(10) \qquad\qquad \sum_{\text{multideg}(h_i g_i) = \delta} \text{LT}(h_i) g_i = \sum_{l=1}^{t} \tilde{B}_l g_l$$

with the property that when $\tilde{B}_l g_l \neq 0$, we have

$$(11) \qquad\qquad \text{multideg}(\tilde{B}_l g_l) < \delta.$$

Substituting (10) into the second line of (4) gives an expression for f as a polynomial combination of the g_i's where *all* terms have multidegree $< \delta$. This contradicts the minimality of δ and completes the proof of the theorem. $\square$

The Buchberger criterion given in Theorem 6 is one of the key results about Gröbner bases. We have seen that Gröbner bases have many nice properties, but, so far, it has been difficult to determine if a basis of an ideal is a Gröbner basis (the examples we gave in §5 were rather trivial). Using the Buchberger criterion, also called the *S-pair criterion*, it is easy to show whether a given basis is a Gröbner basis. Furthermore, in §7, we will see that the S-pair criterion also leads naturally to an algorithm for computing Gröbner bases.

As an example of how to use Theorem 6, consider the ideal $I = \langle y - x^2, z - x^3 \rangle$ of the twisted cubic in $\mathbb{R}^3$. We claim that $G = \{y - x^2, z - x^3\}$ is a Gröbner basis for lex order with $y > z > x$. To prove this, consider the S-polynomial

$$S(y - x^2, z - x^3) = \frac{yz}{y}(y - x^2) - \frac{yz}{z}(z - x^3) = -zx^2 + yx^3.$$

Using the division algorithm, one finds that

$$-zx^2 + yx^3 = x^3 \cdot (y - x^2) + (-x^2) \cdot (z - x^3) + 0,$$

so that $\overline{S(y - x^2, z - x^3)}^G = 0$. Thus, by Theorem 6, G is a Gröbner basis for I. You can also check that G is *not* a Gröbner basis for lex order with $x > y > z$ (see Exercise 9).

EXERCISES FOR §6

1. Show that Proposition 1 can be strengthened slightly as follows. Fix a monomial ordering and let $I \subseteq k[x_1, \ldots, x_n]$ be an ideal. Suppose that $f \in k[x_1, \ldots, x_n]$.
   a. Show that f can be written in the form $f = g + r$, where $g \in I$ and no term of r is divisible by any element of $\text{LT}(I)$.
   b. Given two expressions $f = g + r = g' + r'$ as in part (a), prove that $r = r'$. Thus, r and g are uniquely determined.
   This result shows once a monomial order is fixed, we can define a unique "remainder of f on division by I." We will exploit this idea in Chapter 5.
2. In §5, we showed that $G = \{x + z, y - z\}$ is a Gröbner basis for lex order. Let us use this basis to study the uniqueness of the division algorithm.
   a. Divide xy by $x + z, y - z$.

b. Now interchange the two polynomials and divide xy by $y - z, x + z$.
You should get the same remainder (as predicted by Proposition 1), but the "quotients" should be different for the two divisions. This shows that the uniqueness of the remainder is the best one can hope for.

3. In Corollary 2, we showed that if $I = \langle g_1, \ldots, g_t \rangle$ and if $G = \{g_1, \ldots, g_t\}$ is a Gröbner basis for I, then $\overline{f}^G = 0$ for all $f \in I$. Prove the converse of this statement. Namely, show that if G is a basis for I with the property that $\overline{f}^G = 0$ for all $f \in I$, then G is a Gröbner basis for I.

4. Let G and G' be Gröbner bases for an ideal I with respect to the same monomial order in $k[x_1, \ldots, x_n]$. Show that $\overline{f}^G = \overline{f}^{G'}$ for all $f \in k[x_1, \ldots, x_n]$. Hence, the remainder on division by a Gröbner basis is even independent of which Gröbner basis we use, as long as we use one particular monomial order. Hint: See Exercise 1.

5. Compute $S(f, g)$ using the lex order.
   a. $f = 4x^2z - 7y^2$, $g = xyz^2 + 3xz^4$.
   b. $f = x^4y - z^2$, $g = 3xz^2 - y$.
   c. $f = x^7y^2z + 2ixyz$, $g = 2x^7y^2z + 4$.
   d. $f = xy + z^3$, $g = z^2 - 3^z$.

6. Does $S(f, g)$ depend on which monomial order is used? Illustrate your assertion with examples.

7. Prove that multideg$(S(f, g)) < \gamma$, where $x^\gamma = \text{lcm}(\text{LM}(f), \text{LM}(g))$. Explain why this inequality is a precise version of the claim that S-polynomials are designed to produce cancellation.

8. As in the proof of Theorem 6, suppose that $c_i x^{\alpha(i)} g_i$ and $c_j x^{\alpha(j)} g_j$ have multidegree δ. Prove that
$$S(x^{\alpha(i)} g_i, x^{\alpha(j)} g_j) = x^{\delta - \gamma_{ij}} S(g_i, g_j),$$
where $x^{\gamma_{ij}} = \text{lcm}(\text{LM}(g_i), \text{LM}(g_j))$.

9. Show that $\{y - x^2, z - x^3\}$ is not a Gröbner basis for lex order with $x > y > z$.

10. Using Theorem 6, determine whether the following sets G are Gröbner bases for the ideal they generate. You may want to use a computer algebra system to compute the S-polynomials and remainders.
    a. $G = \{x^2 - y, x^3 - z\}$ for grlex order.
    b. $G = \{x^2 - y, x^3 - z\}$ for invlex order (see Exercise 6 of §2).
    c. $G = \{xy^2 - xz + y, xy - z^2, x - yz^4\}$ for lex order.

11. Let $f, g \in k[x_1, \ldots, x_n]$ be polynomials such that $\text{LM}(f)$ and $\text{LM}(g)$ are *relatively prime* monomials and $\text{LC}(f) = \text{LC}(g) = 1$. Assume that f or g has at least two terms.
    a. Show that $S(f, g) = -(g - \text{LT}(g))f + (f - \text{LT}(f))g$.
    b. Deduce that $S(f, g) \neq 0$ and that the leading monomial of $S(f, g)$ is a multiple of either $\text{LM}(f)$ or $\text{LM}(g)$ in this case.

12. Let $f, g \in k[x_1, \ldots, x_n]$ be nonzero and x^α, x^β be monomials. Verify that
$$S(x^\alpha f, x^\beta g) = x^\gamma S(f, g)$$
where
$$x^\gamma = \frac{\text{lcm}(x^\alpha \text{LM}(f), x^\beta \text{LM}(g))}{\text{lcm}(\text{LM}(f), \text{LM}(g))}.$$
Be sure to prove that x^γ is a monomial. Also explain how this relates to Exercise 8.

13. Let $I \subseteq k[x_1, \ldots, x_n]$ be an ideal, and let G be a Gröbner basis of I.
    a. Show that $\overline{f}^G = \overline{g}^G$ if and only if $f - g \in I$. Hint: See Exercise 1.
    b. Use Exercise 1 to show that
$$\overline{f + g}^G = \overline{f}^G + \overline{g}^G.$$

c. Deduce that

$$\overline{fg}^{G} = \overline{f}^{G} \cdot \overline{g}^{G}.$$

We will return to an interesting consequence of these facts in Chapter 5.

§7 Buchberger's Algorithm

In Corollary 6 of §5, we saw that every ideal in $k[x_1, \ldots, x_n]$ has a Gröbner basis. Unfortunately, the proof given was nonconstructive in the sense that it did not tell us how to produce the Gröbner basis. So we now turn to the question: given an ideal $I \subseteq k[x_1, \ldots, x_n]$, how can we actually construct a Gröbner basis for I? To see the main ideas behind the method we will use, we return to the ideal of Example 2 from §5 and proceed as follows.

Example 1. Consider the ring $\mathbb{Q}[x, y]$ with grlex order, and let $I = \langle f_1, f_2 \rangle = \langle x^3 - 2xy, x^2y - 2y^2 + x \rangle$. Recall that $\{f_1, f_2\}$ is not a Gröbner basis for I since $\mathrm{LT}(S(f_1, f_2)) = -x^2 \notin \langle \mathrm{LT}(f_1), \mathrm{LT}(f_2) \rangle$.

To produce a Gröbner basis, one natural idea is to try first to extend the original generating set to a Gröbner basis by adding more polynomials in I. In one sense, this adds nothing new, and even introduces an element of redundancy. However, the extra information we get from a Gröbner basis more than makes up for this.

What new generators should we add? By what we have said about the S-polynomials in §6, the following should come as no surprise. We have $S(f_1, f_2) = -x^2 \in I$, and its remainder on division by $F = (f_1, f_2)$ is $-x^2$, which is nonzero. Hence, we should include that remainder in our generating set, as a new generator $f_3 = -x^2$. If we set $F = (f_1, f_2, f_3)$, we can use Theorem 6 of §6 to test if this new set is a Gröbner basis for I. We compute

$$S(f_1, f_2) = f_3, \text{ so}$$
$$\overline{S(f_1, f_2)}^{F} = 0,$$
$$S(f_1, f_3) = (x^3 - 2xy) - (-x)(-x^2) = -2xy, \text{ but}$$
$$\overline{S(f_1, f_3)}^{F} = -2xy \neq 0.$$

Thus, we must add $f_4 = -2xy$ to our generating set. If we let $F = (f_1, f_2, f_3, f_4)$, then by Exercise 12 we have

$$\overline{S(f_1, f_2)}^{F} = \overline{S(f_1, f_3)}^{F} = 0,$$
$$S(f_1, f_4) = y(x^3 - 2xy) - (-1/2)x^2(-2xy) = -2xy^2 = yf_4, \text{ so}$$
$$\overline{S(f_1, f_4)}^{F} = 0,$$
$$S(f_2, f_3) = (x^2y - 2y^2 + x) - (-y)(-x^2) = -2y^2 + x, \text{ but}$$
$$\overline{S(f_2, f_3)}^{F} = -2y^2 + x \neq 0.$$

Hence, we must also add $f_5 = -2y^2 + x$ to our generating set. Setting $F = (f_1, f_2, f_3, f_4, f_5)$, one can compute that

$$\overline{S(f_i, f_j)}^F = 0 \text{ for all } 1 \leq i < j \leq 5.$$

By Theorem 6 of §6, it follows that a grlex Gröbner basis for I is given by

$$\{f_1, f_2, f_3, f_4, f_5\} = \{x^3 - 2xy, x^2y - 2y^2 + x, -x^2, -2xy, -2y^2 + x\}.$$

The above example suggests that in general, one should try to extend a basis F to a Gröbner basis by successively adding nonzero remainders $\overline{S(f_i, f_j)}^F$ to F. This idea is a natural consequence of the S-pair criterion from §6 and leads to the following algorithm due to Buchberger for computing a Gröbner basis.

Theorem 2 (Buchberger's Algorithm). *Let $I = \langle f_1, \ldots, f_s \rangle \neq \{0\}$ be a polynomial ideal. Then a Gröbner basis for I can be constructed in a finite number of steps by the following algorithm:*

> Input : $F = (f_1, \ldots, f_s)$
> Output : a Gröbner basis $G = (g_1, \ldots, g_t)$ for I, with $F \subseteq G$
>
> $G := F$
> REPEAT
> $G' := G$
>   FOR each pair $\{p, q\}$, $p \neq q$ in G' DO
> $r := \overline{S(p, q)}^{G'}$
>   IF $r \neq 0$ THEN $G := G \cup \{r\}$
> UNTIL $G = G'$
> RETURN G

Proof. We begin with some frequently used notation. If $G = \{g_1, \ldots, g_t\}$, then $\langle G \rangle$ and $\langle \mathrm{LT}(G) \rangle$ will denote the following ideals:

$$\langle G \rangle = \langle g_1, \ldots, g_t \rangle,$$
$$\langle \mathrm{LT}(G) \rangle = \langle \mathrm{LT}(g_1), \ldots, \mathrm{LT}(g_t) \rangle.$$

Turning to the proof of the theorem, we first show that $G \subseteq I$ holds at every stage of the algorithm. This is true initially, and whenever we enlarge G, we do so by adding the remainder $r = \overline{S(p, q)}^{G'}$ for $p, q \in G' \subseteq G$. Thus, if $G \subseteq I$, then p, q and, hence, $S(p, q)$ are in I, and since we are dividing by $G' \subseteq I$, we get $G \cup \{r\} \subseteq I$. We also note that G contains the given basis F of I, so that G is actually a basis of I.

The algorithm terminates when $G = G'$, which means that $r = \overline{S(p, q)}^{G'} = 0$ for all $p, q \in G$. Hence G is a Gröbner basis of $\langle G \rangle = I$ by Theorem 6 of §6.

It remains to prove that the algorithm terminates. We need to consider what happens after each pass through the main loop. The set G consists of G' (the old G) together with the nonzero remainders of S-polynomials of elements of G'. Then

$$(1) \qquad\qquad \langle \mathrm{LT}(G') \rangle \subseteq \langle \mathrm{LT}(G) \rangle$$

since $G' \subseteq G$. Furthermore, if $G' \neq G$, we claim that $\langle \mathrm{LT}(G') \rangle$ is strictly smaller than $\langle \mathrm{LT}(G) \rangle$. To see this, suppose that a nonzero remainder r of an S-polynomial has been adjoined to G. Since r is a remainder on division by G', $\mathrm{LT}(r)$ is not divisible by the leading terms of elements of G', and thus $\mathrm{LT}(r) \notin \langle \mathrm{LT}(G') \rangle$ by Lemma 2 of §4. Yet $\mathrm{LT}(r) \in \langle \mathrm{LT}(G) \rangle$, which proves our claim.

By (1), the ideals $\langle \mathrm{LT}(G') \rangle$ from successive iterations of the loop form an ascending chain of ideals in $k[x_1, \ldots, x_n]$. Thus, the ACC (Theorem 7 of §5) implies that after a finite number of iterations the chain will stabilize, so that $\langle \mathrm{LT}(G') \rangle = \langle \mathrm{LT}(G) \rangle$ must happen eventually. By the previous paragraph, this implies that $G' = G$, so that the algorithm must terminate after a finite number of steps. $\qquad\square$

Taken together, the Buchberger criterion (Theorem 6 of §6) and the Buchberger algorithm (Theorem 2 above) provide an algorithmic basis for the theory of Gröbner bases. These contributions of Buchberger are central to the development of the subject. In §8, we will get our first hints of what can be done with these methods, and a large part of the rest of the book will be devoted to exploring their ramifications.

We should also point out the algorithm presented in Theorem 2 is only a rudimentary version of the Buchberger algorithm. It was chosen for what we hope will be its clarity for the reader, but it is not a very practical way to do the computation. Note (as a first improvement) that once a remainder $\overline{S(p,q)}^{G'} = 0$, that remainder will stay zero even if we adjoin further elements to the generating set G'. Thus, there is no reason to recompute those remainders on subsequent passes through the main loop. Indeed, if we add our new generators f_j one at a time, the only remainders that need to be checked are $\overline{S(f_i, f_j)}^{G'}$, where $i \leq j - 1$. It is a good exercise to revise the algorithm to take this observation into account. Other improvements of a deeper nature can also be made, but we will postpone considering them until §10.

Gröbner bases computed using the algorithm of Theorem 2 are often bigger than necessary. We can eliminate some unneeded generators by using the following fact.

Lemma 3. *Let G be a Gröbner basis of $I \subseteq k[x_1, \ldots, x_n]$. Let $p \in G$ be a polynomial such that $\mathrm{LT}(p) \in \langle \mathrm{LT}(G \setminus \{p\}) \rangle$. Then $G \setminus \{p\}$ is also a Gröbner basis for I.*

Proof. We know that $\langle \mathrm{LT}(G) \rangle = \langle \mathrm{LT}(I) \rangle$. If $\mathrm{LT}(p) \in \langle \mathrm{LT}(G \setminus \{p\}) \rangle$, then we have $\langle \mathrm{LT}(G \setminus \{p\}) \rangle = \langle \mathrm{LT}(G) \rangle$. By definition, it follows that $G \setminus \{p\}$ is also a Gröbner basis for I. $\qquad\square$

By adjusting constants to make all leading coefficients equal to 1 and removing any p with $\mathrm{LT}(p) \in \langle \mathrm{LT}(G \setminus \{p\}) \rangle$ from G, we arrive at what we will call a *minimal Gröbner basis*. We can construct a minimal Gröbner basis for a given nonzero ideal by applying the algorithm of Theorem 2 and then using Lemma 3 to eliminate any unneeded generators that might have been included.

To illustrate this procedure, we return to the ideal I studied in Example 1. Using grlex order, we found the Gröbner basis

$$f_1 = x^3 - 2xy,$$
$$f_2 = x^2y - 2y^2 + x,$$
$$f_3 = -x^2,$$
$$f_4 = -2xy,$$
$$f_5 = -2y^2 + x.$$

Since some of the leading coefficients are different from 1, the first step is to multiply the generators by suitable constants to make this true. Then note that $\text{LT}(f_1) = x^3 = -x \cdot \text{LT}(f_3)$. By Lemma 3, we can dispense with f_1 in the minimal Gröbner basis. Similarly, since $\text{LT}(f_2) = x^2y = -(1/2)x \cdot \text{LT}(f_4)$, we can also eliminate f_2. There are no further cases where the leading term of a generator divides the leading term of another generator. Hence,

$$\tilde{f}_3 = x^2, \qquad \tilde{f}_4 = xy, \qquad \tilde{f}_5 = y^2 - (1/2)x$$

is a minimal Gröbner basis for I.

When G is a minimal Gröbner basis, the leading terms $\text{LT}(p)$, $p \in G$, form the unique minimal basis of $\langle \text{LT}(I) \rangle$ by Proposition 7 of §4 (see Exercise 6). Unfortunately, the original ideal I may have many minimal Gröbner bases. For example, in the ideal I considered above, it is easy to check that

$$(2) \qquad \hat{f}_3 = x^2 + axy, \quad \tilde{f}_4 = xy, \quad \tilde{f}_5 = y^2 - (1/2)x$$

is also a minimal Gröbner basis, where $a \in \mathbb{Q}$ is any constant. Thus, we can produce infinitely many minimal Gröbner bases. Fortunately, we can single out one minimal basis that is better than the others. The definition is as follows.

Definition 4. A **reduced Gröbner basis** for a polynomial ideal I is a Gröbner basis G for I such that:

(i) $\text{LC}(p) = 1$ for all $p \in G$.
(ii) For all $p \in G$, no monomial of p lies in $\langle \text{LT}(G \setminus \{p\}) \rangle$.

Note that for the Gröbner bases given in (2), only the one with $a = 0$ is reduced. In general, reduced Gröbner bases have the following nice property.

Theorem 5. *Let $I \neq \{0\}$ be a polynomial ideal. Then, for a given monomial ordering, I has a reduced Gröbner basis, and the reduced Gröbner basis is unique.*

Proof. As noted above, all minimal Gröbner bases for I have the same leading terms. Now let G be a minimal Gröbner basis for I. We say that $g \in G$ is *fully reduced* for G provided that no monomial of g is in $\langle \text{LT}(G \setminus \{p\}) \rangle$. Observe that g is fully reduced for any other minimal Gröbner basis G' of I that contains g since G' and G have the same leading terms.

Next, given $g \in G$, let $g' = \overline{g}^{G \setminus \{g\}}$ and set $G' = (G \setminus \{g\}) \cup \{g'\}$. We claim that G' is a minimal Gröbner basis for I. To see this, first note that $\mathrm{LT}(g') = \mathrm{LT}(g)$, for when we divide g by $G \setminus \{g\}, \mathrm{LT}(g)$ goes to the remainder since it is not divisible by any element of $\mathrm{LT}(G \setminus \{g\})$. This shows that $\langle \mathrm{LT}(G') \rangle = \langle \mathrm{LT}(G) \rangle$. Since G' is clearly contained in I, we see that G' is a Gröbner basis, and minimality follows. Finally, note that g' is fully reduced for G' by construction.

Now, take the elements of G and apply the above process until they are all fully reduced. The Gröbner basis may change each time we do the process, but our earlier observation shows that once an element is fully reduced, it stays fully reduced since we never change the leading terms. Thus, we end up with a reduced Gröbner basis.

Finally, to prove uniqueness, suppose that G and $\widetilde{G}$ are reduced Gröbner bases for I. Then in particular, G and $\widetilde{G}$ are minimal Gröbner bases, and hence have the same leading terms, i.e., $\mathrm{LT}(G) = \mathrm{LT}(\widetilde{G})$. Thus, given $g \in G$, there is $\tilde{g} \in \widetilde{G}$ such that $\mathrm{LT}(g) = \mathrm{LT}(\tilde{g})$. If we can show that $g = \tilde{g}$, it will follow that $G = \widetilde{G}$, and uniqueness will be proved.

To show $g = \tilde{g}$, consider $g - \tilde{g}$. This is in I, and since G is a Gröbner basis, it follows that $\overline{g - \tilde{g}}^{G} = 0$. But we also know $\mathrm{LT}(g) = \mathrm{LT}(\tilde{g})$. Hence, these terms cancel in $g - \tilde{g}$, and the remaining terms are divisible by none of $\mathrm{LT}(G) = \mathrm{LT}(\widetilde{G})$ since G and $\widetilde{G}$ are reduced. This shows that $\overline{g - \tilde{g}}^{G} = g - \tilde{g}$, and then $g - \tilde{g} = 0$ follows. This completes the proof. $\qquad\square$

Many computer algebra systems implement a version of Buchberger's algorithm for computing Gröbner bases. These systems always compute a Gröbner basis whose elements are constant multiples of the elements in a reduced Gröbner basis. This means that they will give essentially the same answers for a given problem. Thus, answers can be easily checked from one system to the next.

Another consequence of the uniqueness in Proposition 5 is that we have an **ideal equality algorithm** for seeing when two sets of polynomials $\{f_1, \ldots, f_s\}$ and $\{g_1, \ldots, g_t\}$ generate the same ideal: simply fix a monomial order and compute a reduced Gröbner basis for $\langle f_1, \ldots, f_s \rangle$ and $\langle g_1, \ldots, g_t \rangle$. Then the ideals are equal if and only if the Gröbner bases are the same.

To conclude this section, we will indicate briefly some of the connections between Buchberger's algorithm and the row-reduction (Gaussian elimination) algorithm for systems of linear equations. The interesting fact here is that the row-reduction algorithm is essentially a special case of the general algorithm we have described. For concreteness, we will discuss the special case corresponding to the system of linear equations

$$
\begin{aligned}
3x - 6y - 2z \qquad\quad &= 0, \\
2x - 4y \qquad + 4w &= 0, \\
x - 2y - z - w &= 0.
\end{aligned}
$$

If we use row operations on the coefficient matrix to put it in row echelon form (which means that the leading 1's have been identified), then we get the matrix

$$(3) \qquad \begin{pmatrix} 1 & -2 & -1 & -1 \\ 0 & 0 & 1 & 3 \\ 0 & 0 & 0 & 0 \end{pmatrix}.$$

To get a *reduced* row echelon matrix, we need to make sure that each leading 1 is the only nonzero entry in its column. This leads to the matrix

$$(4) \qquad \begin{pmatrix} 1 & -2 & 0 & 2 \\ 0 & 0 & 1 & 3 \\ 0 & 0 & 0 & 0 \end{pmatrix}.$$

To translate these computations into algebra, let I be the ideal

$$I = \langle 3x - 6y - 2z, 2x - 4y + 4w, x - 2y - z - w \rangle \subseteq k[x, y, z, w]$$

corresponding to the original system of equations. We will use lex order with $x > y > z > w$. Then, in the exercises, you will verify that the linear forms determined by the row echelon matrix (3) give a minimal Gröbner basis

$$I = \langle x - 2y - z - w, z + 3w \rangle,$$

and you will also check that the reduced row echelon matrix (4) gives the reduced Gröbner basis

$$I = \langle x - 2y + 2w, z + 3w \rangle.$$

Recall from linear algebra that every matrix can be put in reduced row echelon form in a unique way. This can be viewed as a special case of the uniqueness of reduced Gröbner bases.

In the exercises, you will examine the relation between Buchberger's algorithm and the Euclidean Algorithm for finding the generator for the ideal $\langle f, g \rangle \subseteq k[x]$.

EXERCISES FOR §7

1. Check that $\overline{S(f_i, f_j)}^F = 0$ for all pairs $1 \le i < j \le 5$ in Example 1.
2. Use the algorithm given in Theorem 2 to find a Gröbner basis for each of the following ideals. You may wish to use a computer algebra system to compute the S-polynomials and remainders. Use the lex, then the grlex order in each case, and then compare your results.
   a. $I = \langle x^2 y - 1, xy^2 - x \rangle$.
   b. $I = \langle x^2 + y, x^4 + 2x^2 y + y^2 + 3 \rangle$. [What does your result indicate about the variety $\mathbf{V}(I)$?]
   c. $I = \langle x - z^4, y - z^5 \rangle$.
3. Find reduced Gröbner bases for the ideals in Exercise 2 with respect to lex and grlex.
4. Use the result of Exercise 7 of §4 to give an alternate proof that Buchberger's algorithm will always terminate after a finite number of steps.
5. Let G be a Gröbner basis of an ideal I with the property that $\mathrm{LC}(g) = 1$ for all $g \in G$. Prove that G is a minimal Gröbner basis if and only if no proper subset of G is a Gröbner basis of I.

6. The minimal basis of a monomial ideal was introduced in Proposition 7 of §4. Show that a Gröbner basis G of I is minimal if and only if $\mathrm{LC}(g) = 1$ for all $g \in G$ and $\mathrm{LT}(G)$ is the minimal basis of the monomial ideal $\langle \mathrm{LT}(I) \rangle$.

7. Fix a monomial order, and let G and $\widetilde{G}$ be minimal Gröbner bases for the ideal I.
   a. Prove that $\mathrm{LT}(G) = \mathrm{LT}(\widetilde{G})$.
   b. Conclude that G and $\widetilde{G}$ have the same number of elements.

8. Develop an algorithm that produces a reduced Gröbner basis (see Definition 4) for an ideal I, given as input an arbitrary Gröbner basis for I. Prove that your algorithm works.

9. Consider the ideal

$$I = \langle 3x - 6y - 2z, 2x - 4y + 4w, x - 2y - z - w \rangle \subseteq k[x, y, z, w]$$

mentioned in the text. We will use lex order with $x > y > z > w$.
   a. Show that the linear polynomials determined by the row echelon matrix (3) give a minimal Gröbner basis $I = \langle x - 2y - z - w, z + 3w \rangle$. Hint: Use Theorem 6 of §6.
   b. Show that the linear polynomials from the reduced row echelon matrix (4) give the reduced Gröbner basis $I = \langle x - 2y + 2w, z + 3w \rangle$.

10. Let $A = (a_{ij})$ be an $n \times m$ matrix with entries in k and let $f_i = a_{i1}x_1 + \cdots + a_{im}x_m$ be the linear polynomials in $k[x_1, \ldots, x_m]$ determined by the rows of A. Then we get the ideal $I = \langle f_1, \ldots, f_n \rangle$. We will use lex order with $x_1 > \cdots > x_m$. Now let $B = (b_{ij})$ be the reduced row echelon matrix determined by A and let $g_1, \ldots, g_t$ be the linear polynomials coming from the nonzero rows of B (so that $t \leq n$). We want to prove that $g_1, \ldots, g_t$ form the reduced Gröbner basis of I.
   a. Show that $I = \langle g_1, \ldots, g_t \rangle$. Hint: Show that the result of applying a row operation to A gives a matrix whose rows generate the same ideal.
   b. Use Theorem 6 of §6 to show that $g_1, \ldots, g_t$ form a Gröbner basis of I. Hint: If the leading 1 in the ith row of B is in the sth column, we can write $g_i = x_s + C$, where C is a linear polynomial involving none of the variables corresponding to leading 1's. If $g_j = x_t + D$ is written similarly, then you need to divide $S(g_i, g_j) = x_t C - x_s D$ by $g_1, \ldots, g_t$. Note that you will use only g_i and g_j in the division.
   c. Explain why $g_1, \ldots, g_t$ form the reduced Gröbner basis of I.

11. Show that the result of applying the Euclidean Algorithm in $k[x]$ to any pair of polynomials f, g is a reduced Gröbner basis for $\langle f, g \rangle$ (after dividing by a constant to make the leading coefficient equal to 1). Explain how the steps of the Euclidean Algorithm can be seen as special cases of the operations used in Buchberger's algorithm.

12. Fix $F = \{f_1, \ldots, f_s\}$ and let $r = \overline{f}^F$. Since dividing f by F gives r as remainder, adding r to the polynomials we divide by should reduce the remainder to zero. In other words, we should have $\overline{f}^{F \cup \{r\}} = 0$ when r comes last. Prove this as follows.
   a. When you divide f by $F \cup \{r\}$, consider the first place in the division algorithm where the intermediate dividend p is not divisible by any $\mathrm{LT}(f_i)$. Explain why $\mathrm{LT}(p) = \mathrm{LT}(r)$ and why the next intermediate dividend is $p - r$.
   b. From here on in the division algorithm, explain why the leading term of the intermediate dividend is always divisible by one of the $\mathrm{LT}(f_i)$. Hint: If this were false, consider the first time it fails. Remember that the terms of r are not divisible by any $\mathrm{LT}(f_i)$.
   c. Conclude that the remainder is zero, as desired.
   d. (For readers who did Exercise 11 of §3.) Give an alternate proof of $\overline{f}^{F \cup \{r\}} = 0$ using Exercise 11 of §3.

13. In the discussion following the proof of Theorem 2, we commented that if $\overline{S(f, g)}^{G'} = 0$, then remainder stays zero when we enlarge G'. More generally, if $\overline{f}^F = 0$ and F' is obtained from F by adding elements at the end, then $\overline{f}^{F'} = 0$. Prove this.

14. Suppose we have n points $V = \{(a_1, b_1), \ldots, (a_n, b_n)\} \subseteq k^2$ where $a_1, \ldots, a_n$ are distinct. This exercise will study the *Lagrange interpolation polynomial* defined by

$$h(x) = \sum_{i=1}^{n} b_i \prod_{j \neq i} \frac{x_j - a_j}{a_i - a_j} \in k[x].$$

We will also explain how $h(x)$ relates to the reduced Gröbner basis of $\mathbf{I}(V) \subseteq k[x, y]$.
a. Show that $h(a_i) = b_i$ for $i = 1, \ldots, n$ and explain why h has degree $\leq n - 1$.
b. Prove that $h(x)$ is the unique polynomial of degree $\leq n - 1$ satisfying $h(a_i) = b_i$ for $i = 1, \ldots, n$.
c. Prove that $\mathbf{I}(V) = \langle f(x), y - h(x) \rangle$, where $f(x) = \prod_{i=1}^{n}(x - a_i)$. Hint: Divide $g \in \mathbf{I}(V)$ by $f(x), y - h(x)$ using lex order with $y > x$.
d. Prove that $\{f(x), y - h(x)\}$ is the reduced Gröbner basis for $\mathbf{I}(V) \subseteq k[x, y]$ for lex order with $y > x$.

§8 First Applications of Gröbner Bases

In §1, we posed four problems concerning ideals and varieties. The first was the ideal description problem, which was solved by the Hilbert Basis Theorem in §5. Let us now consider the three remaining problems and see to what extent we can solve them using Gröbner bases.

The Ideal Membership Problem

If we combine Gröbner bases with the division algorithm, we get the following **ideal membership algorithm**: given an ideal $I = \langle f_1, \ldots, f_s \rangle$, we can decide whether a given polynomial f lies in I as follows. First, using a Gröbner basis algorithm (for instance, the one in Theorem 2 of §7), find a Gröbner basis $G = \{g_1, \ldots, g_t\}$ for I. Then Corollary 2 of §6 implies that

$$f \in I \text{ if and only if } \overline{f}^G = 0.$$

Example 1. Let $I = \langle f_1, f_2 \rangle = \langle xz - y^2, x^3 - z^2 \rangle \in \mathbb{C}[x, y, z]$, and use the grlex order. Let $f = -4x^2y^2z^2 + y^6 + 3z^5$. We want to know if $f \in I$.

The generating set given is not a Gröbner basis of I because $\mathrm{LT}(I)$ also contains polynomials such as $\mathrm{LT}(S(f_1, f_2)) = \mathrm{LT}(-x^2y^2 + z^3) = -x^2y^2$ that are not in the ideal $\langle \mathrm{LT}(f_1), \mathrm{LT}(f_2) \rangle = \langle xz, x^3 \rangle$. Hence, we begin by computing a Gröbner basis for I. Using a computer algebra system, we find a Gröbner basis

$$G = \{f_1, f_2, f_3, f_4, f_5\} = \{xz - y^2, x^3 - z^2, x^2y^2 - z^3, xy^4 - z^4, y^6 - z^5\}.$$

Note that this is a reduced Gröbner basis.

We may now test polynomials for membership in I. For example, dividing f above by G, we find

$$f = (-4xy^2z - 4y^4) \cdot f_1 + 0 \cdot f_2 + 0 \cdot f_3 + 0 \cdot f_4 + (-3) \cdot f_5 + 0.$$

Since the remainder is zero, we have $f \in I$.

For another example, consider $f = xy - 5z^2 + x$. Even without completely computing the remainder on division by G, we can see from the form of the elements in G that $f \notin I$. The reason is that $\mathrm{LT}(f) = xy$ is clearly not in the ideal $\langle \mathrm{LT}(G) \rangle = \langle xz, x^3, x^2y^2, xy^4, y^6 \rangle$. Hence, $\overline{f}^G \neq 0$, so that $f \notin I$.

This last observation illustrates the way the properties of an ideal are revealed by the form of the elements of a Gröbner basis.

The Problem of Solving Polynomial Equations

Next, we will investigate how the Gröbner basis technique can be applied to solve systems of polynomial equations in several variables. Let us begin by looking at some specific examples.

Example 2. Consider the equations

$$x^2 + y^2 + z^2 = 1,$$
(1)
$$x^2 + z^2 = y,$$
$$x = z$$

in $\mathbb{C}^3$. These equations determine $I = \langle x^2 + y^2 + z^2 - 1, x^2 + z^2 - y, x - z \rangle \subseteq \mathbb{C}[x, y, z]$, and we want to find all points in $\mathbf{V}(I)$. Proposition 9 of §5 implies that we can compute $\mathbf{V}(I)$ using *any* basis of I. So let us see what happens when we use a Gröbner basis.

Though we have no compelling reason as of yet to do so, we will compute a reduced Gröbner basis on I with respect to the lex order. The basis is

$$g_1 = x - z,$$
$$g_2 = y - 2z^2,$$
$$g_3 = z^4 + (1/2)z^2 - 1/4.$$

If we examine these polynomials closely, we find something remarkable. First, the polynomial g_3 depends on z alone. To find its roots, we solve for z^2 by the quadratic formula and take square roots. This gives four values of z:

$$z = \pm \frac{1}{2}\sqrt{\pm\sqrt{5} - 1}.$$

Next, when these values of z are substituted into the equations $g_2 = 0$ and $g_1 = 0$, those two equations can be solved uniquely for y and x, respectively. Thus, there are four solutions altogether of $g_1 = g_2 = g_3 = 0$, two real and two complex. Since $\mathbf{V}(I) = \mathbf{V}(g_1, g_2, g_3)$ by Proposition 9 of §5, we have found *all* solutions of the original equations (1).

Example 3. Next, we will consider the system of polynomial equations (2) from Chapter 1, §2, obtained by applying Lagrange multipliers to find the minimum and maximum values of $x^3 + 2xyz - z^2$ subject to the constraint $x^2 + y^2 + z^2 = 1$:

$$3x^2 + 2yz - 2x\lambda = 0,$$
$$2xz - 2y\lambda = 0,$$
$$2xy - 2z - 2z\lambda = 0,$$
$$x^2 + y^2 + z^2 - 1 = 0.$$

Again, we follow our general hunch and begin by computing a Gröbner basis for the ideal in $\mathbb{R}[x, y, z, \lambda]$ generated by the left-hand sides of the four equations, using the lex order with $\lambda > x > y > z$. We find a Gröbner basis:

(2)

$$\lambda - \frac{3}{2}x - \frac{3}{2}yz - \frac{167616}{3835}z^6 + \frac{36717}{590}z^4 - \frac{134419}{7670}z^2,$$
$$x^2 + y^2 + z^2 - 1,$$
$$xy - \frac{19584}{3835}z^5 + \frac{1999}{295}z^3 - \frac{6403}{3835}z,$$
$$xz + yz^2 - \frac{1152}{3835}z^5 - \frac{108}{295}z^3 + \frac{2556}{3835}z,$$
$$y^3 + yz^2 - y - \frac{9216}{3835}z^5 + \frac{906}{295}z^3 - \frac{2562}{3835}z,$$
$$y^2z - \frac{6912}{3835}z^5 + \frac{827}{295}z^3 - \frac{3839}{3835}z,$$
$$yz^3 - yz - \frac{576}{59}z^6 + \frac{1605}{118}z^4 - \frac{453}{118}z^2,$$
$$z^7 - \frac{1763}{1152}z^5 + \frac{655}{1152}z^3 - \frac{11}{288}z.$$

At first glance, this collection of polynomials looks horrendous. (The coefficients of the elements of Gröbner basis can be significantly messier than the coefficients of the original generating set.) However, on further observation, we see that once again the last polynomial depends only on the variable z. We have "eliminated" the other variables in the process of finding the Gröbner basis. (Miraculously) the equation obtained by setting this polynomial equal to zero has the roots

$$z = 0, \ \pm 1, \ \pm 2/3, \ \pm\sqrt{11}/8\sqrt{2}.$$

If we set z equal to each of these values in turn, the remaining equations can then be solved for y, x (and λ, though its values are essentially irrelevant for our purposes). We obtain the following solutions:

$$z = 0; \quad y = 0; \quad x = \pm 1,$$
$$z = 0; \quad y = \pm 1; \quad x = 0,$$
$$z = \pm 1; \quad y = 0; \quad x = 0,$$
$$z = 2/3; \quad y = 1/3; \quad x = -2/3,$$
$$z = -2/3; \quad y = -1/3; \quad x = -2/3,$$
$$z = \sqrt{11}/8\sqrt{2}; \quad y = -3\sqrt{11}/8\sqrt{2}; \quad x = -3/8,$$
$$z = -\sqrt{11}/8\sqrt{2}; \quad y = 3\sqrt{11}/8\sqrt{2}; \quad x = -3/8.$$

From here, it is easy to determine the minimum and maximum values.

Examples 2 and 3 indicate that finding a Gröbner basis for an ideal with respect to the lex order simplifies the form of the equations considerably. In particular, we seem to get equations where the variables are eliminated successively. Also, note that the *order* of elimination seems to correspond to the ordering of the variables. For instance, in Example 3, we had variables $\lambda > x > y > z$, and if you look back at the Gröbner basis (2), you will see that λ is eliminated first, x second, and so on.

A system of equations in this form is easy to solve, especially when the last equation contains only one variable. We can apply one-variable techniques to try and find its roots, then substitute back into the other equations in the system and solve for the other variables, using a procedure similar to the above examples. The reader should note the analogy between this procedure for solving polynomial systems and the method of "back-substitution" used to solve a linear system in triangular form.

We will study the process of elimination of variables from systems of polynomial equations intensively in Chapter 3. In particular, we will see why lex order gives a Gröbner basis that successively eliminates the variables.

The Implicitization Problem

Suppose that the parametric equations

(3)
$$x_1 = f_1(t_1, \ldots, t_m),$$
$$\vdots$$
$$x_n = f_n(t_1, \ldots, t_m),$$

define a subset of an algebraic variety V in k^n. For instance, this will always be the case if the f_i are rational functions in $t_1, \ldots, t_m$, as we will show in Chapter 3. How can we find polynomial equations in the x_i that define V? This problem can be solved using Gröbner bases, though a complete proof that this is the case will come only with the results of Chapter 3.

For simplicity, we will restrict our attention to the case where the f_i are actually *polynomials*. We begin with the affine variety in k^{m+n} defined by (3), namely

$$x_1 - f_1(t_1, \ldots, t_m) = 0,$$
$$\vdots$$
$$x_n - f_n(t_1, \ldots, t_m) = 0.$$

The basic idea is to eliminate the variables $t_1, \ldots, t_m$ from these equations. This should give us the equations for V.

Given what we saw in Examples 2 and 3, it makes sense to use a Gröbner basis to eliminate variables. We will take the lex order in $k[t_1, \ldots, t_m, x_1, \ldots, x_n]$ defined by the variable ordering

$$t_1 > \cdots > t_m > x_1 > \cdots > x_n.$$

Now suppose we have a Gröbner basis of the ideal $\tilde{I} = \langle x_1 - f_1, \ldots, x_n - f_n \rangle$. Since we are using lex order, we expect the Gröbner basis to have polynomials that eliminate variables, and $t_1, \ldots, t_m$ should be eliminated first since they are biggest in our monomial order. Thus, the Gröbner basis for $\tilde{I}$ should contain polynomials that only involve $x_1, \ldots, x_n$. These are our candidates for the equations of V.

The ideas just described will be explored in detail when we study elimination theory in Chapter 3. For now, we will content ourselves with some examples to see how this process works.

Example 4. Consider the parametric curve V:

$$x = t^4,$$
$$y = t^3,$$
$$z = t^2$$

in $\mathbb{C}^3$. We compute a Gröbner basis G of $I = \langle x - t^4, y - t^3, z - t^2 \rangle$ with respect to the lex order in $\mathbb{C}[t, x, y, z]$, and we find

$$G = \{t^2 - z, ty - z^2, tz - y, x - z^2, y^2 - z^3\}.$$

The last two polynomials depend only on x, y, z, so they define an affine variety of $\mathbb{C}^3$ containing our curve V. By the intuition on dimensions that we developed in Chapter 1, we would guess that two equations in $\mathbb{C}^3$ would define a curve (a 1-dimensional variety). The remaining question to answer is whether V is the entire intersection of the two surfaces

$$x - z^2 = 0, \quad y^2 - z^3 = 0.$$

Might there be other curves (or even surfaces) in the intersection? We will be able to show that the answer is no when we have established the general results in Chapter 3.

Example 5. Now consider the tangent surface of the twisted cubic in $\mathbb{R}^3$, which we studied in Chapter 1. This surface is parametrized by

$$x = t + u,$$
$$y = t^2 + 2tu,$$
$$z = t^3 + 3t^2u.$$

We compute a Gröbner basis G for this ideal relative to the lex order defined by $t > u > x > y > z$, and we find that G has 6 elements altogether. If you make the calculation, you will see that only one contains only x, y, z terms:

$$(4) \qquad x^3z - (3/4)x^2y^2 - (3/2)xyz + y^3 + (1/4)z^2 = 0.$$

The variety defined by this equation is a surface containing the tangent surface to the twisted cubic. However, it is possible that the surface given by (4) is strictly bigger than the tangent surface: there may be solutions of (4) that do not correspond to points on the tangent surface. We will return to this example in Chapter 3.

To summarize our findings in this section, we have seen that Gröbner bases and the division algorithm give a complete solution of the ideal membership problem. Furthermore, we have seen ways to produce solutions of systems of polynomial equations and to produce equations of parametrically given subsets of affine space. Our success in the examples given earlier depended on the fact that Gröbner bases, when computed using lex order, seem to eliminate variables in a very nice fashion. In Chapter 3, we will prove that this is always the case, and we will explore other aspects of what is called elimination theory.

EXERCISES FOR §8

In the following exercises, a computer algebra system should be used to perform the necessary calculations. (Most of the calculations would be very arduous if carried out by hand.)

1. Determine whether $f = xy^3 - z^2 + y^5 - z^3$ is in the ideal $I = \langle -x^3 + y, x^2y - z \rangle$.
2. Repeat Exercise 1 for $f = x^3z - 2y^2$ and $I = \langle xz - y, xy + 2z^2, y - z \rangle$.
3. By the method of Examples 2 and 3, find the points in $\mathbb{C}^3$ on the variety

$$\mathbf{V}(x^2 + y^2 + z^2 - 1, x^2 + y^2 + z^2 - 2x, 2x - 3y - z).$$

4. Repeat Exercise 3 for $\mathbf{V}(x^2y - z^3, 2xy - 4z - 1, z - y^2, x^3 - 4zy)$.
5. Recall from calculus that a *critical point* of a differentiable function $f(x, y)$ is a point where the partial derivatives $\frac{\partial f}{\partial x}$ and $\frac{\partial f}{\partial y}$ vanish simultaneously. When $f \in \mathbb{R}[x, y]$, it follows that the critical points can be found by applying our techniques to the system of polynomial equations

$$\frac{\partial f}{\partial x} = \frac{\partial f}{\partial y} = 0.$$

To see how this works, consider the function

$$f(x, y) = (x^2 + y^2 - 4)(x^2 + y^2 - 1) + (x - 3/2)^2 + (y - 3/2)^2.$$

   a. Find all critical points of $f(x, y)$.
   b. Classify your critical points as local maxima, local minima, or saddle points. Hint: Use the second derivative test.
6. Fill in the details of Example 5. In particular, compute the required Gröbner basis, and verify that this gives us (up to a constant multiple) the polynomial appearing on the left-hand side of equation (4).

7. Let the surface S in $\mathbb{R}^3$ be formed by taking the *union* of the straight lines joining pairs of points on the lines

$$\left\{\begin{array}{l} x = t \\ y = 0 \\ z = 1 \end{array}\right\}, \quad \left\{\begin{array}{l} x = 0 \\ y = 1 \\ z = t \end{array}\right\}$$

with the *same parameter value* (i.e., the same t). (This is a special example of a class of surfaces called *ruled surfaces*.)

a. Show that the surface S can be given parametrically as

$$x = ut,$$
$$y = 1 - u,$$
$$z = t + u(1 - t).$$

b. Using the method of Examples 4 and 5, find an (implicit) equation of a variety V containing the surface S.

c. Show $V = S$ (that is, show that every point of the variety V can be obtained by substituting some values for t, u in the equations of part (a). Hint: Try to "solve" the implicit equation of V for one variable as a function of the other two.

8. Some parametric curves and surfaces are algebraic varieties even when the given parametrizations involve transcendental functions such as sin and cos. In this problem, we will see that the parametric surface T,

$$x = (2 + \cos(t))\cos(u),$$
$$y = (2 + \cos(t))\sin(u),$$
$$z = \sin(t),$$

lies on an affine variety in $\mathbb{R}^3$.

a. Draw a picture of T. Hint: Use cylindrical coordinates.

b. Let $a = \cos(t), b = \sin(t), c = \cos(u), d = \sin(u)$, and rewrite the above equations as polynomial equations in a, b, c, d, x, y, z.

c. The pairs a, b and c, d in part (b) are not *independent* since there are additional polynomial identities

$$a^2 + b^2 - 1 = 0, \quad c^2 + d^2 - 1 = 0$$

stemming from the basic trigonometric identity. Form a system of five equations by adjoining the above equations to those from part (b) and compute a Gröbner basis for the corresponding ideal. Use the lex monomial ordering and the variable order

$$a > b > c > d > x > y > z.$$

There should be exactly one polynomial in your basis that depends only on x, y, z. This is the equation of a variety containing T.

9. Consider the parametric curve $K \subseteq \mathbb{R}^3$ given by

$$x = (2 + \cos(2s))\cos(3s),$$
$$y = (2 + \cos(2s))\sin(3s),$$
$$z = \sin(2s).$$

a. Express the equations of K as polynomial equations in $x, y, z, a = \cos(s), b = \sin(s)$. Hint: Trig identities.

b. By computing a Gröbner basis for the ideal generated by the equations from part (a) and $a^2 + b^2 - 1$ as in Exercise 8, show that K is (a subset of) an affine algebraic curve. Find implicit equations for a curve containing K.

c. Show that the equation of the surface from Exercise 8 is *contained* in the ideal generated by the equations from part (b). What does this result mean geometrically? (You
can actually reach the same conclusion by comparing the parametrizations of T and
K, without calculations.)

10. Use the method of Lagrange multipliers to find the point(s) on the surface defined by
$x^4 + y^2 + z^2 - 1 = 0$ that are closest to the point $(1, 1, 1)$ in $\mathbb{R}^3$. Hint: Proceed as in
Example 3. (You may need to "fall back" on a *numerical* method to solve the equations
you get.)

11. Suppose we have numbers a, b, c which satisfy the equations

$$a + b + c = 3,$$
$$a^2 + b^2 + c^2 = 5,$$
$$a^3 + b^3 + c^3 = 7.$$

a. Prove that $a^4 + b^4 + c^4 = 9$. Hint: Regard a, b, c as variables and show carefully that
$a^4 + b^4 + c^4 - 9 \in \langle a + b + c - 3, a^2 + b^2 + c^2 - 5, a^3 + b^3 + c^3 - 7 \rangle$.
b. Show that $a^5 + b^5 + c^5 \neq 11$.
c. What are $a^5 + b^5 + c^5$ and $a^6 + b^6 + c^6$? Hint: Compute remainders.

§9 Refinements of the Buchberger Criterion

The Buchberger criterion (Theorem 6 of §6) states that a basis $G = \{g_1, \ldots, g_t\}$ of
a polynomial ideal is a Gröbner basis provided that $\overline{S(g_i, g_j)}^G = 0$ for all $g_i, g_j \in G$.
In other words, if each of these S-polynomials has a representation

$$S(g_i, g_j) = \sum_{l=1}^{t} q_l g_l + 0$$

produced by the division algorithm, then G is a Gröbner basis of the ideal it generates. The goal of this section is to give two versions of the Buchberger criterion that
allow more flexibility in how the S-polynomials are represented.

Standard Representations

We first give a more general view of what it means to have zero remainder. The
definition is as follows.

Definition 1. Fix a monomial order and let $G = \{g_1, \ldots, g_t\} \subseteq k[x_1, \ldots, x_n]$. Given
$f \in k[x_1, \ldots, x_n]$, we say that f **reduces to zero modulo** G, written

$$f \to_G 0,$$

if f has a **standard representation**

$$f = A_1 g_1 + \cdots + A_t g_t, \quad A_i \in k[x_1, \ldots, x_n],$$

which means that whenever $A_i g_i \neq 0$, we have

$$\text{multideg}(f) \geq \text{multideg}(A_i g_i).$$

To understand the relation between Definition 1 and the division algorithm, we have the following lemma.

Lemma 2. *Let* $G = (g_1, \ldots, g_t)$ *be an ordered set of elements of* $k[x_1, \ldots, x_n]$ *and fix* $f \in k[x_1, \ldots, x_n]$. *Then* $\overline{f}^G = 0$ *implies* $f \to_G 0$, *though the converse is false in general.*

Proof. If $\overline{f}^G = 0$, then the division algorithm implies

$$f = q_1 g_1 + \cdots + q_t g_t + 0,$$

and by Theorem 3 of §3, whenever $q_i g_i \neq 0$, we have

$$\text{multideg}(f) \geq \text{multideg}(q_i g_i).$$

This shows that $f \to_G 0$. To see that the converse may fail, consider Example 5 from §3. If we divide $f = xy^2 - x$ by $G = (xy + 1, y^2 - 1)$, the division algorithm gives

$$xy^2 - x = y \cdot (xy + 1) + 0 \cdot (y^2 - 1) + (-x - y),$$

so that $\overline{f}^G = -x - y \neq 0$. Yet we can also write

$$xy^2 - x = 0 \cdot (xy + 1) + x \cdot (y^2 - 1),$$

and since

$$\text{multideg}(xy^2 - x) \geq \text{multideg}(x \cdot (y^2 - 1))$$

(in fact, they are equal), it follows that $f \to_G 0$. □

As an example of how Definition 1 can be used, let us state a more general version of the Gröbner basis criterion from §6.

Theorem 3. *A basis* $G = \{g_1, \ldots, g_t\}$ *for an ideal* I *is a Gröbner basis if and only if* $S(g_i, g_j) \to_G 0$ *for all* $i \neq j$.

Proof. If G is a Gröbner basis, then $S(g_i, g_j) \in I$ has zero remainder on division by G, hence $S(g_i, g_j) \to_G 0$ by Lemma 2. For the converse, Theorem 6 of §6 implies that G is a Gröbner basis when $\overline{S(g_i, g_j)}^G = 0$ for all $i \neq j$. But if you examine the proof, you will see that all we used was

$$S(g_i, g_j) = \sum_{l=1}^{t} A_l g_l,$$

where

$$\text{multideg}(A_l g_l) \leq \text{multideg}(S(g_i, g_j))$$

when $A_l g_l \neq 0$ (see (6) and (7) from §6). This is exactly what $S(g_i, g_j) \to_G 0$ means, and the theorem follows.   □

By Lemma 2, notice that Theorem 6 of §6 is a special case of Theorem 3. Using the notion of "standard representation" from Definition 1, Theorem 3 says that *a basis for an ideal I is a Gröbner basis if and only if all of its S-polynomials have standard representations.*

There are some situations where an S-polynomial is guaranteed to have a standard representation.

Proposition 4. *Given a finite set $G \subseteq k[x_1, \ldots, x_n]$, suppose that we have $f, g \in G$ such that the leading monomials of f and g are relatively prime. Then $S(f, g) \to_G 0$.*

Proof. For simplicity, we assume that f, g have been multiplied by appropriate constants to make $\text{LC}(f) = \text{LC}(g) = 1$. Write $f = \text{LM}(f) + p, g = \text{LM}(g) + q$. Since $\text{LM}(f)$ and $\text{LM}(g)$ are relatively prime, we know that $\text{lcm}(\text{LM}(f), \text{LM}(g)) = \text{LM}(f) \cdot \text{LM}(g)$. Hence, the S-polynomial $S(f, g)$ can be written

(1)
$$\begin{aligned} S(f, g) &= \text{LM}(g) \cdot f - \text{LM}(f) \cdot g \\ &= (g - q) \cdot f - (f - p) \cdot g \\ &= g \cdot f - q \cdot f - f \cdot g + p \cdot g \\ &= p \cdot g - q \cdot f. \end{aligned}$$

We claim that

(2) $$\text{multideg}(S(f, g)) = \max(\text{multideg}(p \cdot g), \text{multideg}(q \cdot f)).$$

Note that (1) and (2) imply $S(f, g) \to_G 0$ since $f, g \in G$. To prove (2), observe that in the last polynomial of (1), the leading monomials of $p \cdot g$ and $q \cdot f$ are distinct and, hence, cannot cancel. For if the leading monomials were the same, we would have

$$\text{LM}(p) \cdot \text{LM}(g) = \text{LM}(q) \cdot \text{LM}(f).$$

However this is impossible if $\text{LM}(f), \text{LM}(g)$ are relatively prime: from the last equation, $\text{LM}(g)$ would have to divide $\text{LM}(q)$, which is absurd since $\text{LM}(g) > \text{LM}(q)$. □

For an example of how this proposition works, let $G = (yz + y, x^3 + y, z^4)$ and use grlex order on $k[x, y, z]$. Since x^3 and z^4 are relatively prime, we have

$$S(x^3 + y, z^4) \to_G 0$$

by Proposition 4. However, using the division algorithm, it is easy to check that

$$S(x^3 + y, z^4) = yz^4 = (z^3 - z^2 + z - 1)(yz + y) + 0 \cdot (x^3 + y) + 0 \cdot z^4 + y.$$

so that

$$\overline{S(x^3 + y, z^4)}^G = y \neq 0.$$

This explains why we need Definition 1: Proposition 4 is false if we use the notion
of zero remainder coming from the division algorithm.

Another example of Proposition 4 is given by the ideal $I = \langle y - x^2, z - x^3 \rangle$. It
is easy to check that the given generators $f = y - x^2$ and $g = z - x^3$ do not form a
Gröbner basis for lex order with $x > y > z$. But if we switch to lex with $z > y > x$,
then the leading monomials are $\mathrm{LM}(f) = y$ and $\mathrm{LM}(g) = z$. Setting $G = \{f, g\}$,
Proposition 4 implies $S(f, g) \to_G 0$, so that G is a Gröbner basis of I by Theorem 3.
In §10, we see that Proposition 4 is part of a more efficient version of the Buchberger
algorithm.

LCM Representations

Our second version of the Buchberger criterion allows a yet more general way of
presenting S-polynomials. Recall from Exercise 7 of §6 that an S-polynomial $S(f, g)$
has leading term that is guaranteed to be strictly less than $\mathrm{lcm}(\mathrm{LM}(f), \mathrm{LM}(g))$.

Definition 5. Given nonzero polynomials $F = (f_1, \ldots, f_s)$, we say that

$$S(f_i, f_j) = \sum_{l=1}^{s} A_l f_l$$

is an **lcm representation** provided that

$$\mathrm{lcm}(\mathrm{LM}(f_i), \mathrm{LM}(f_j)) > \mathrm{LT}(A_l f_l) \text{ whenever } A_l f_l \neq 0.$$

To understand how lcm representations relate to standard representations, write
$S(f_i, f_j) = \sum_{l=1}^{s} A_l f_l$ and take l with $A_l f_l \neq 0$. Then consider the inequalities

(3) $\mathrm{lcm}(\mathrm{LM}(f_i), \mathrm{LM}(f_j)) > \mathrm{LT}(S(f_i, f_j)),$

(4) $\mathrm{lcm}(\mathrm{LM}(f_i), \mathrm{LM}(f_j)) > \mathrm{LT}(A_l f_l).$

Note that (3) is true by the definition of S-polynomial. In a standard representation,
we have (3) $\Rightarrow$ (4) since $\mathrm{LT}(S(f_i, f_j)) \geq \mathrm{LT}(A_l f_l)$. In an lcm representation, on the
other hand, we have (4), but we make no assumption about how $\mathrm{LT}(S(f_i, f_j))$ and
$\mathrm{LT}(A_l f_l)$ relate to each other.

The above discussion shows that every standard representation is also an lcm
representation. For an example of how the converse may fail, let $f_1 = xz + 1$, $f_2 = yz + 1$, $f_3 = xz + y - z + 1$. Using lex order with $x > y > z$, one can write

$$S(f_1, f_2) = (-1) \cdot f_1 + 0 \cdot f_2 + 1 \cdot f_3.$$

In Exercise 1, you will check that this is an lcm representation but not a standard
representation.

Here is a version of the Buchberger criterion that uses lcm representations.

Theorem 6. *A basis* $G = (g_1, \ldots, g_t)$ *for an ideal* I *is a Gröbner basis if and only if for every* $i \neq j$, *the S-polynomial* $S(g_i, g_j)$ *has an lcm representation.*

Proof. If G is a Gröbner basis, then every S-polynomial has a standard representation, hence an lcm representation. For the converse, we will look closely at the proof of Theorem 6 of §6, just as we did for Theorem 3.

We are assuming that $S(g_i, g_j)$ has an lcm representation

$$S(g_i, g_j) = \sum_{l=1}^{t} A_l g_l$$

with $x^{\gamma_{ij}} > \mathrm{LT}(A_l g_l)$ when $A_l g_l \neq 0$. Here, $x^{\gamma_{ij}} = \mathrm{lcm}(\mathrm{LM}(g_j), \mathrm{LM}(g_l))$. If we set $B_l = x^{\delta - \gamma_{ij}} A_l$, then

$$x^{\delta - \gamma_{ij}} S(g_i, g_j) = \sum_{l=1}^{t} B_l g_l,$$

where

$$\mathrm{multideg}(B_l g_l) = \mathrm{multideg}(x^{\delta - \gamma_{ij}}) + \mathrm{multideg}(A_l g_l) < (\delta - \gamma_{ij}) + \gamma_{ij} = \delta.$$

This gives the same inequality as (9) in the proof of Theorem 6 of §6. From here, the rest of the proof is identical to what we did in §6, and the theorem follows. □

We noted above any standard representation of $S(g_i, g_j)$ is an lcm representation. Thus Theorem 6 of §6 and Theorem 3 of this section follow from Theorem 6, since $S(g_i, g_j)$ has an lcm representation whenever it satisfies either $\overline{S(g_i, g_j)}^G = 0$ or $S(g_i, g_j) \to_G 0$. We will consider a further generalization of the Buchberger criterion in §10.

The ideas of this section are useful in elimination theory, which we will study in Chapter 3. Two of the central results are the *Extension Theorem* and the *Closure Theorem*. Standard representations appear in the proof of the Extension Theorem given in Chapter 3, §5, and lcm representations are used in the proof of the Closure Theorem given in Chapter 4, §7. We will also use Theorem 6 in the proof of the Nullstellensatz given in Chapter 4, §1.

EXERCISES FOR §9

1. Let $f_1 = xz + 1, f_2 = yz + 1$, and $f_3 = xz + y - z + 1$. For lex order with $x > y > z$, show that
$$S(f_1, f_2) = (-1) \cdot f_1 + 0 \cdot f_2 + 1 \cdot f_3.$$
Also show that this is an lcm representation but not a standard representation.
2. Consider the ideal $I = \langle x^2 + y + z - 1, x + y^2 + z - 1, x + y + z^2 - 1 \rangle \subseteq \mathbb{Q}[x, y, z]$.
   a. Show that the generators of I fail to be Gröbner basis for any lex order.
   b. Find a monomial order for which the leading terms of the generators are relatively prime.

   c. Explain why the generators automatically form a Gröbner basis for the monomial order you found in part (b).

3. The result of the previous exercise can be generalized as follows. Suppose that $I = \langle f_1, \ldots, f_s \rangle$ where $\mathrm{LM}(f_i)$ and $\mathrm{LM}(f_j)$ are relatively prime for all indices $i \neq j$. Prove that $\{f_1, \ldots, f_s\}$ is a Gröbner basis of I.

§10 Improvements on Buchberger's Algorithm

In designing useful mathematical software, attention must be paid not only to the *correctness* of the algorithms employed, but also to their *efficiency*. In this section, we will discuss two improvements on the basic Buchberger algorithm for computing Gröbner bases that can greatly speed up the calculations. Some version of these improvements has been built into most of the computer algebra systems that use Gröbner basis methods. The section will conclude with a brief discussion of the complexity of computing Gröbner bases. This is still an active area of research though, and there are as yet no definitive results in this direction.

The Buchberger algorithm presented in §7 computes remainders $\overline{S(f, g)}^G$ and adds them to G when they are nonzero. As you learned from doing examples by hand, these polynomial divisions are the most computationally intensive part of Buchberger's algorithm. Hence, one way to improve the efficiency of the algorithm would be to show that fewer S-polynomials $S(f, g)$ need to be considered. Any reduction of the number of divisions that need to be performed is all to the good.

Theorem 3 of §9 tells us that when checking for a Gröbner basis, we can replace $\overline{S(f, g)}^G = 0$ with $S(f, g) \to_G 0$. Thus, if we can predict in advance that certain S-polynomials are guaranteed to reduce to zero, then we can ignore them in the Buchberger algorithm.

We have already seen one example where reduction to zero is guaranteed, namely Proposition 4 of §9. This proposition is sufficiently important that we restate it here.

Proposition 1. *Given a finite set $G \subseteq k[x_1, \ldots, x_n]$, suppose that we have $f, g \in G$ such that*

$$\mathrm{lcm}(\mathrm{LM}(f), \mathrm{LM}(g)) = \mathrm{LM}(f) \cdot \mathrm{LM}(g).$$

This means that the leading monomials of f and g are relatively prime. Then $S(f, g) \to_G 0$.

Note that Proposition 1 gives a more efficient version of Theorem 3 of §9: to test for a Gröbner basis, we need only have $S(g_i, g_j) \to_G 0$ for those $i < j$ where $\mathrm{LM}(g_i)$ and $\mathrm{LM}(g_j)$ are not relatively prime. But before we apply this to improving Buchberger's algorithm, let us explore a second way to improve Theorem 3 of §9.

The basic idea is to better understand the role played by S-polynomials in the proof of Theorem 6 of §6. Since S-polynomials were constructed to cancel leading terms, this means we should study cancellation in greater generality. Hence, we will introduce the notion of a *syzygy* on the leading terms of $F = (f_1, \ldots, f_s)$. This word is used in astronomy to indicate an *alignment* of three planets or other heavenly bodies. The root is a Greek word meaning "yoke." In an astronomical syzygy, planets are "yoked together"; in a mathematical syzygy, it is polynomials that are "yoked."

Definition 2. Let $F = (f_1, \ldots, f_s)$. A **syzygy** on the leading terms $\text{LT}(f_1), \ldots, \text{LT}(f_s)$ of F is an s-tuple of polynomials $S = (h_1, \ldots, h_s) \in (k[x_1, \ldots, x_n])^s$ such that

$$\sum_{i=1}^{s} h_i \cdot \text{LT}(f_i) = 0.$$

We let $S(F)$ be the subset of $(k[x_1, \ldots, x_n])^s$ consisting of all syzygies on the leading terms of F.

For an example of a syzygy, consider $F = (x, x^2 + z, y + z)$. Then using the lex order, $S = (-x + y, 1, -x) \in (k[x, y, z])^3$ defines a syzygy in $S(F)$ since

$$(-x + y) \cdot \text{LT}(x) + 1 \cdot \text{LT}(x^2 + z) + (-x) \cdot \text{LT}(y + z) = 0.$$

Let $\mathbf{e}_i = (0, \ldots, 0, 1, 0, \ldots, 0) \in (k[x_1, \ldots, x_n])^s$, where the 1 is in the ith place. Then a syzygy $S \in S(F)$ can be written as $S = \sum_{i=1}^{s} h_i \mathbf{e}_i$. For an example of how to use this notation, consider the syzygies that come from S-polynomials. Namely, given a pair $\{f_i, f_j\} \subseteq F$ where $i < j$, let $x^\gamma = \text{lcm}(\text{LM}(f_i), \text{LM}(f_j))$. Then

(1) $$S_{ij} = \frac{x^\gamma}{\text{LT}(f_i)} \mathbf{e}_i - \frac{x^\gamma}{\text{LT}(f_j)} \mathbf{e}_j$$

gives a syzygy on the leading terms of F. In fact, the name S-polynomial is actually an abbreviation for "syzygy polynomial."

It is straightforward to check that the set of syzygies is closed under coordinate-wise sums, and under coordinate-wise multiplication by polynomials (see Exercise 1). An especially nice fact about $S(F)$ is that it has a finite *basis*—there is a finite collection of syzygies such that every other syzygy is a linear combination with polynomial coefficients of the basis syzygies.

However, before we can prove this, we need to learn a bit more about the structure of $S(F)$. We first define the notion of a *homogeneous* syzygy.

Definition 3. An element $S \in S(F)$ is **homogeneous of multidegree** α, where $\alpha \in \mathbb{Z}_{\geq 0}^n$, provided that

$$S = (c_1 x^{\alpha(1)}, \ldots, c_s x^{\alpha(s)}),$$

where $c_i \in k$ and $\alpha(i) + \text{multideg}(f_i) = \alpha$ whenever $c_i \neq 0$.

You should check that the syzygy S_{ij} given in (1) is homogeneous of multidegree γ (see Exercise 4). We can decompose syzygies into homogeneous ones as follows.

Lemma 4. *Every element of $S(F)$ can be written uniquely as a sum of homogeneous elements of $S(F)$.*

Proof. Let $S = (h_1, \ldots, h_s) \in S(F)$. Fix an exponent $\alpha \in \mathbb{Z}_{\geq 0}^n$, and let $h_{i\alpha}$ be the term of h_i (if any) such that $h_{i\alpha} f_i$ has multidegree α. Then we must have $\sum_{i=1}^{s} h_{i\alpha} \mathrm{LT}(f_i) = 0$ since the $h_{i\alpha} \mathrm{LT}(f_i)$ are the terms of multidegree α in the sum $\sum_{i=1}^{s} h_i \mathrm{LT}(f_i) = 0$. Then $S_\alpha = (h_{1\alpha}, \ldots, h_{s\alpha})$ is a homogeneous element of $S(F)$ of degree α and $S = \sum_\alpha S_\alpha$.

The proof of uniqueness will be left to the reader (see Exercise 5). □

We can now prove that the S_{ij}'s form a basis of all syzygies on the leading terms.

Proposition 5. *Given $F = (f_1, \ldots, f_s)$, every syzygy $S \in S(F)$ can be written as*

$$S = \sum_{i<j} u_{ij} S_{ij},$$

where $u_{ij} \in k[x_1, \ldots, x_n]$ and the syzygy S_{ij} is defined as in (1).

Proof. By Lemma 4, we can assume that S is homogeneous of multidegree α. Then S must have at least two nonzero components, say $c_i x^{\alpha(i)}$ and $c_j x^{\alpha(j)}$, where $i < j$. Then $\alpha(i) + \mathrm{multideg}(f_i) = \alpha(j) + \mathrm{multideg}(f_j) = \alpha$, which implies that $x^\gamma = \mathrm{lcm}(\mathrm{LM}(f_i), \mathrm{LM}(f_j))$ divides x^α. Since

$$S_{ij} = \frac{x^\gamma}{\mathrm{LT}(f_i)} \mathbf{e}_i - \frac{x^\gamma}{\mathrm{LT}(f_j)} \mathbf{e}_j,$$

an easy calculation shows that the ith component of

$$S - c_i \mathrm{LC}(f_i) x^{\alpha-\gamma} S_{ij}$$

must be zero, and the only other component affected is the jth. Hence we have produced a homogeneous syzygy with fewer nonzero components. Since a nonzero syzygy must have at least two nonzero components, continuing in this way will eventually enable us to write S as a combination of the S_{ij}'s, and we are done. □

This proposition explains our observation in §6 that S-polynomials account for all possible cancellation of leading terms.

We are now ready to state a more refined version of our algorithmic criterion for Gröbner bases.

Theorem 6. *A basis $G = (g_1, \ldots, g_t)$ for an ideal I is a Gröbner basis if and only if for every element $S = (H_1, \ldots, H_t)$ in a homogeneous basis for the syzygies $S(G)$, $S \cdot G = \sum_{i=1}^{t} H_i g_i$ can be written*

$$(2) \qquad\qquad S \cdot G = \sum_{i=1}^{t} A_i g_i,$$

where the multidegree α of S satisfies

$$(3) \qquad\qquad \alpha > \mathrm{multideg}(A_i g_i) \text{ whenever } A_i g_i \neq 0.$$

Proof. First assume that G is a Gröbner basis. Since S is a syzygy, it satisfies $\alpha > \text{multideg}(S \cdot G)$, and then any standard representation $S \cdot G = \sum_{i=1}^{t} A_i g_i$ has the desired property. For the converse, we will use the strategy (and notation) of the proof of Theorem 6 of §6. We start with $f = \sum_{i=1}^{t} h_i g_i$, where $\delta = \max(\text{multideg}(h_i g_i))$ is minimal among all ways of writing f in terms of G. As before, we need to show that $\text{multideg}(f) < \delta$ leads to a contradiction.

By (4) in §6, $\text{multideg}(f) < \delta$ implies that $\sum_{\text{multideg}(h_i g_i)=\delta} \text{LT}(h_i) g_i$ has strictly smaller multidegree. This therefore means that $\sum_{\text{multideg}(h_i g_i)=\delta} \text{LT}(h_i)\text{LT}(g_i) = 0$, so that

$$S = \sum_{\text{multideg}(h_i g_i)=\delta} \text{LT}(h_i)\mathbf{e}_i$$

is a syzygy in $S(G)$. Note also that S is homogeneous of multidegree δ. Our hypothesis then gives us a homogeneous basis $S_1, \ldots, S_m$ of $S(G)$ with the nice property that $S_j \cdot G$ satisfies (2) and (3) for all j. We can write S in the form

$$(4) \qquad\qquad S = u_1 S_1 + \cdots + u_m S_m.$$

By writing the u_j's as sums of terms and expanding, we see that (4) expresses S as a sum of homogeneous syzygies. Since S is homogeneous of multidegree δ, it follows from the uniqueness of Lemma 4 that we can discard all syzygies not of multidegree δ. Thus, in (4), we can assume that, for each j, either

$$u_j = 0, \text{ or } u_j S_j \text{ is homogeneous of multidegree } \delta.$$

Suppose that S_j has multidegree γ_j. If $u_j \neq 0$, then it follows that u_j can be written in the form $u_j = c_j x^{\delta - \gamma_j}$ for some $c_j \in k$. Thus, (4) can be written

$$S = \sum_j c_j x^{\delta - \gamma_j} S_j,$$

where the sum is over those j's with $u_j \neq 0$. If we take the dot product of each side with G, we obtain

$$(5) \qquad \sum_{\text{multideg}(h_i g_i)=\delta} \text{LT}(h_i) g_i = S \cdot G = \sum_j c_j x^{\delta - \gamma_j} S_j \cdot G.$$

Since S_j has multidegree γ_j, our hypothesis implies that $S_j \cdot G = \sum_{i=1}^{t} A_{ij} g_i$, where

$$\text{multideg}(A_{ij} g_i) < \gamma_j \text{ when } A_{ij} g_i \neq 0.$$

It follows that if we set $B_{ij} = x^{\delta - \gamma_j} A_{ij}$, then we have

$$x^{\delta - \gamma_j} S_j \cdot G = \sum_{i=1}^{t} B_{ij} g_i$$

where multideg$(B_{ij}g_i) < \delta$ when $B_{ij}g_i \neq 0$. Using this and (5), we can write the sum $\sum_{\text{multideg}(h_i g_i)=\delta} \text{LT}(h_i)g_i$ as

$$\sum_{\text{multideg}(h_i g_i)=\delta} \text{LT}(h_i)g_i = \sum_{l=1}^{t} \tilde{B}_l g_l,$$

where multideg$(\tilde{B}_l g_l) < \delta$ when $\tilde{B}_l g_l \neq 0$. This is exactly what we proved in (10) and (11) from §6. From here, the remainder of the proof is identical to what we did in §6. The theorem is proved. □

Note that Theorem 3 of §9 is a special case of this result. Namely, if we use the basis $\{S_{ij}\}$ for the syzygies $S(G)$, then the polynomials $S_{ij} \cdot G$ to be tested are precisely the S-polynomials $S(g_i, g_j)$.

A homogeneous syzygy S with the property that $S \cdot G \to_G 0$ is easily seen to satisfy (2) and (3) (Exercise 6). This gives the following corollary of Theorem 6.

Corollary 7. *A basis* $G = (g_1, \ldots, g_t)$ *for an ideal I is a Gröbner basis if and only if for every element* $S = (H_1, \ldots, H_t)$ *in a homogeneous basis for the syzygies $S(G)$,* $S \cdot G \to_G 0$.

To exploit the power of Theorem 6 and Corollary 7, we need to learn how to make small bases of $S(G)$. For an example of how a basis can be smaller than expected, consider $G = (x^2 y^2 + z, xy^2 - y, x^2 y + yz)$ and use lex order in $k[x, y, z]$. The basis formed by the three syzygies corresponding to the S-polynomials consists of

$$S_{12} = (1, -x, 0),$$
$$S_{13} = (1, 0, -y),$$
$$S_{23} = (0, x, -y).$$

However, we see that $S_{23} = S_{13} - S_{12}$. Thus S_{23} is *redundant* in the sense that it can be obtained from S_{12}, S_{13} by a linear combination. (Here, the coefficients are constants; in general, relations between syzygies may have polynomial coefficients.) It follows that $\{S_{12}, S_{13}\}$ is a smaller basis of $S(G)$.

We will show next that starting with the basis $\{S_{ij} \mid i < j\}$, there is a systematic way to predict when elements can be omitted.

Proposition 8. *Given* $G = (g_1, \ldots, g_t)$, *suppose that* $\mathcal{S} \subseteq \{S_{ij} \mid 1 \leq i < j \leq t\}$ *is a basis of $S(G)$. In addition, suppose we have distinct elements* $g_i, g_j, g_l \in G$ *such that*

$$\text{LT}(g_l) \text{ divides } \text{lcm}(\text{LT}(g_i), \text{LT}(g_j)).$$

If $S_{il}, S_{jl} \in \mathcal{S}$, *then* $\mathcal{S} \setminus \{S_{ij}\}$ *is also a basis of $S(G)$. (Note: If $i > j$, we set $S_{ij} = S_{ji}$.)*

Proof. For simplicity, assume that $i < j < l$. Set $x^{\gamma_{ij}} = \text{lcm}(\text{LM}(g_i), \text{LM}(g_j))$ and let $x^{\gamma_{il}}$ and $x^{\gamma_{jl}}$ be defined similarly. Then our hypothesis implies that $x^{\gamma_{il}}$ and $x^{\gamma_{jl}}$ both divide $x^{\gamma_{ij}}$. In Exercise 7, you will verify that

$$S_{ij} = \frac{x^{\gamma_{ij}}}{x^{\gamma_{il}}} S_{il} - \frac{x^{\gamma_{ij}}}{x^{\gamma_{jl}}} S_{jl},$$

and the proposition is proved. $\square$

To incorporate this proposition into an algorithm for creating Gröbner bases, we will use the ordered pairs (i,j) with $i < j$ to keep track of which syzygies we want. Since we sometimes will have an $i \neq j$ where we do not know which is larger, we will use the following notation: given $i \neq j$, define

$$[i,j] = \begin{cases} (i,j) & \text{if } i < j \\ (j,i) & \text{if } i > j. \end{cases}$$

We can now state an improved version of Buchberger's algorithm that takes into account the results proved so far.

Theorem 9. *Let $I = \langle f_1, \ldots, f_s \rangle$ be a polynomial ideal. Then a Gröbner basis of I can be constructed in a finite number of steps by the following algorithm:*

> Input : $F = (f_1, \ldots, f_s)$
> Output : a Gröbner basis G for $I = \langle f_1, \ldots, f_s \rangle$
>
> $B := \{(i,j) \mid 1 \leq i < j \leq s\}$
> $G := F$
> $t := s$
> WHILE $B \neq \emptyset$ DO
> > Select $(i,j) \in B$
> > IF $\text{lcm}(\text{LT}(f_i), \text{LT}(f_j)) \neq \text{LT}(f_i)\text{LT}(f_j)$ AND
> > > $\text{Criterion}(f_i, f_j, B) = \mathsf{false}$ THEN
> > > $r := \overline{S(f_i, f_j)}^G$
> > > IF $r \neq 0$ THEN
> > > > $t := t + 1;\ f_t := r$
> > > > $G := G \cup \{f_t\}$
> > > > $B := B \cup \{(i,t) \mid 1 \leq i \leq t - 1\}$
> > $B := B \setminus \{(i,j)\}$
> RETURN G

Here, $\text{Criterion}(f_i, f_j, B)$ is true provided that there is some $l \notin \{i,j\}$ for which the pairs $[i,l]$ and $[j,l]$ are **not** in B and $\text{LT}(f_l)$ divides $\text{lcm}(\text{LT}(f_i), \text{LT}(f_j))$. (Note that this criterion is based on Proposition 8.)

Proof. The basic idea of the algorithm is that B records the pairs (i,j) that remain to be considered. Furthermore, we only compute the remainder of those S-polynomials $S(g_i, g_j)$ for which neither Proposition 1 nor Proposition 8 applies.

To prove that the algorithm works, we first observe that at every stage of the algorithm, B has the property that if $1 \leq i < j \leq t$ and $(i,j) \notin B$, then

(6) $\qquad S(f_i, f_j) \to_G 0 \quad \text{or} \quad \text{Criterion}(f_i, f_j, B) \text{ holds.}$

Initially, this is true since B starts off as the set of all possible pairs. We must show that if (6) holds for some intermediate value of B, then it continues to hold when B changes, say to B'.

To prove this, assume that $(i,j) \notin B'$. If $(i,j) \in B$, then an examination of the algorithm shows that $B' = B \setminus \{(i,j)\}$. Now look at the step before we remove (i,j) from B. If $\text{lcm}(\text{LT}(f_i), \text{LT}(f_j)) = \text{LT}(f_i)\text{LT}(f_j)$, then $S(f_i, f_j) \to_G 0$ by Proposition 1, and (6) holds. Also if $\text{Criterion}(f_i, f_j, B)$ is true, then (6) clearly holds. Now suppose that both of these fail. In this case, the algorithm computes the remainder $r = \overline{S(f_i, f_j)}^G$. If $r = 0$, then $S(f_i, f_j) \to_G 0$ by Lemma 2, as desired. Finally, if $r \neq 0$, then we enlarge G to be $G' = G \cup \{r\}$, and we leave it as an exercise to show that $S(f_i, f_j) \to_{G'} 0$.

It remains to study the case when $(i,j) \notin B$. Here, (6) holds for B, and in Exercise 9, you will show that this implies that (6) also holds for B'.

Next, we need to show that G is a Gröbner basis when $B = \emptyset$. To prove this, let t be the length of G, and consider the set $\mathcal{I}$ consisting of all pairs (i,j) for $1 \leq i < j \leq t$ where $\text{Criterion}(f_i, f_j, B)$ was *false* when (i,j) was selected in the algorithm. We claim that $\mathcal{S} = \{S_{ij} \mid (i,j) \in \mathcal{I}\}$ is a basis of $S(G)$ with the property that $S_{ij} \cdot G = S(f_i, f_j) \to_G 0$ for all $S_{ij} \in \mathcal{S}$. This claim and Corollary 7 will prove that G is a Gröbner basis.

To prove our claim, note that $B = \emptyset$ implies that (6) holds for *all* pairs (i,j) for $1 \leq i < j \leq t$. It follows that $S(f_i, f_j) \to_G 0$ for all $(i,j) \in \mathcal{I}$. It remains to show that $\mathcal{S}$ is a basis of $S(G)$. To prove this, first notice that we can *order* the pairs (i,j) according to when they were removed from B in the algorithm (see Exercise 10 for the details of this ordering). Now go through the pairs in reverse order, starting with the last removed, and delete the pairs (i,j) for which $\text{Criterion}(f_i, f_j, B)$ was true at that point in the algorithm. After going through all pairs, those that remain are precisely the elements of $\mathcal{I}$. Let us show that at every stage of this process, the syzygies corresponding to the pairs (i,j) not yet deleted form a basis of $S(G)$. This is true initially because we started with all of the S_{ij}'s, which we know to be a basis. Further, if at some point we delete (i,j), then the definition of $\text{Criterion}(f_i, f_j, B)$ implies that there is some l where $\text{LT}(f_l)$ satisfies the lcm condition and $[i,l], [j,l] \notin B$. Thus, $[i,l]$ and $[j,l]$ were removed earlier from B, and hence S_{il} and S_{jl} are still in the set we are creating because we are going in reverse order. If follows from Proposition 8 that we still have a basis even after deleting S_{ij}.

Finally, we need to show that the algorithm terminates. As in the proof of the original algorithm (Theorem 2 of §7), G is always a basis of our ideal, and each time we enlarge G, the monomial ideal $\langle \text{LT}(G) \rangle$ gets strictly larger. By the ACC, it follows that at some point, G must stop growing, and thus, we eventually stop adding elements to B. Since every pass through the WHILE...DO loop removes an element of B, we must eventually get $B = \emptyset$, and the algorithm comes to an end. $\quad \square$

The algorithm given above is still not optimal, and several strategies have been found to improve its efficiency further. For example, in the division algorithm (Theorem 3 of §3), we allowed the divisors $f_1, \ldots, f_s$ to be listed in any order. In practice, some effort could be saved on average if we arranged the f_i so that their leading terms are listed in increasing order with respect to the chosen monomial ordering. Since the smaller $\mathrm{LT}(f_i)$ are more likely to be used during the division algorithm, listing them earlier means that fewer comparisons will be required. A second strategy concerns the step where we choose $(i,j) \in B$ in the algorithm of Theorem 9. BUCHBERGER (1985) suggests that there will often be some savings if we pick $(i,j) \in B$ such that $\mathrm{lcm}(\mathrm{LM}(f_i), \mathrm{LM}(f_j))$ is as *small* as possible. The corresponding S-polynomials will tend to yield any nonzero remainders (and new elements of the Gröbner basis) sooner in the process, so there will be more of a chance that subsequent remainders $\overline{S(f_i, f_j)}^G$ will be zero. This *normal selection strategy* is discussed in more detail in BECKER and WEISPFENNING (1993), BUCHBERGER (1985) and GEBAUER and MÖLLER (1988). Finally, there is the idea of *sugar*, which is a refinement of the normal selection strategy. Sugar and its variant *double sugar* can be found in GIOVINI, MORA, NIESI, ROBBIANO and TRAVERSO (1991).

In another direction, one can also modify the algorithm so that it will automatically produce a reduced Gröbner basis (as defined in §7). The basic idea is to systematically reduce G each time it is enlarged. Incorporating this idea also generally lessens the number of S-polynomials that must be divided in the course of the algorithm. For a further discussion of this idea, consult BUCHBERGER (1985).

We will discuss further ideas for computing Gröbner bases in Chapter 10.

Complexity Issues

We will end this section with a short discussion of the complexity of computing Gröbner bases. Even with the best currently known versions of the algorithm, it is still easy to generate examples of ideals for which the computation of a Gröbner basis takes a tremendously long time and/or consumes a huge amount of storage space. There are several reasons for this:

- The total degrees of intermediate polynomials that must be generated as the algorithm proceeds can be quite large.
- The coefficients of the elements of a Gröbner basis can be quite complicated rational numbers, even when the coefficients of the original ideal generators were small integers. See Example 3 of §8 or Exercise 13 of this section for some instances of this phenomenon.

For these reasons, a large amount of theoretical work has been done to try to establish uniform upper bounds on the degrees of the intermediate polynomials in Gröbner basis calculations when the degrees of the original generators are given. For some specific results in this area, see DUBÉ (1990) and MÖLLER and MORA (1984). The idea is to measure to what extent the Gröbner basis method will continue to be tractable as larger and larger problems are attacked.

The bounds on the degrees of the generators in a Gröbner basis are quite large, and it has been shown that large bounds are necessary. For instance, MAYR and MEYER (1982) give examples where the construction of a Gröbner basis for an ideal generated by polynomials of degree less than or equal to some d can involve polynomials of degree proportional to 2^{2^d}. As $d \to \infty$, 2^{2^d} grows *very* rapidly. Even when grevlex order is used (which often gives the smallest Gröbner bases—see below), the degrees can be quite large. For example, consider the polynomials

$$x^{n+1} - yz^{n-1}w, \quad xy^{n-1} - z^n, \quad x^nz - y^nw.$$

If we use grevlex order with $x > y > z > w$, then Mora [see LAZARD (1983)] showed that the reduced Gröbner basis contains the polynomial

$$z^{n^2+1} - y^{n^2}w.$$

The results led for a time to some pessimism concerning the ultimate practicality of the Gröbner basis method as a whole. Further work has shown, however, that for ideals in two and three variables, much more reasonable upper degree bounds are available [see, for example, LAZARD (1983) and WINKLER (1984)]. Furthermore, in any case the running time and storage space required by the algorithm seem to be much more manageable "on average" (and this tends to include most cases of geometric interest) than in the worst cases. There is also a growing realization that computing "algebraic" information (such as the primary decomposition of an ideal—see Chapter 4) should have greater complexity than computing "geometric" information (such as the dimension of a variety—see Chapter 9). A good reference for this is GIUSTI and HEINTZ (1993), and a discussion of a wide variety of complexity issues related to Gröbner bases can be found in BAYER and MUMFORD (1993). See also pages 616–619 of VON ZUR GATHEN and GERHARD (2013) for further discussion and references.

Finally, experimentation with changes of variables and varying the ordering of the variables often can reduce the difficulty of the computation drastically. BAYER and STILLMAN (1987a) have shown that in most cases, the grevlex order should produce a Gröbner basis with polynomials of the smallest total degree. In a different direction, it is tempting to consider changing the monomial ordering as the algorithm progresses in order to produce a Gröbner basis more efficiently. This idea was introduced in GRITZMANN and STURMFELS (1993) and has been taken up again in CABOARA and PERRY (2014).

EXERCISES FOR §10

1. Let $S = (c_1, \ldots, c_s)$ and $T = (d_1, \ldots, d_s) \in (k[x_1, \ldots, x_n])^s$ be syzygies on the leading terms of $F = (f_1, \ldots, f_s)$.
   a. Show that $S + T = (c_1 + d_1, \ldots, c_s + d_s)$ is also a syzygy.
   b. Show that if $g \in k[x_1, \ldots, x_n]$, then $g \cdot S = (gc_1, \ldots, gc_s)$ is also a syzygy.
2. Given any $G = (g_1, \ldots, g_s) \in (k[x_1, \ldots, x_n])^s$, we can define a syzygy on G to be an s-tuple $S = (h_1, \ldots, h_s) \in (k[x_1, \ldots, x_n])^s$ such that $\sum_i h_i g_i = 0$. [Note that the syzygies we studied in the text are syzygies on $\mathrm{LT}(G) = (\mathrm{LT}(g_1), \ldots, \mathrm{LT}(g_s))$.]

a. Show that if $G = (x^2 - y, xy - z, y^2 - xz)$, then $(z, -y, x)$ defines a syzygy on G.
b. Find another syzygy on G from part (a).
c. Show that if S, T are syzygies on G, and $g \in k[x_1, \ldots, x_n]$, then $S + T$ and gS are also syzygies on G.

3. Let M be an $m \times (m + 1)$ matrix of polynomials in $k[x_1, \ldots, x_n]$. Let I be the ideal generated by the determinants of all the $m \times m$ submatrices of M (such ideals are examples of *determinantal ideals*).
    a. Find a 2×3 matrix M such that the associated determinantal ideal of 2×2 submatrices is the ideal with generators G as in part (a) of Exercise 2.
    b. Explain the syzygy given in part (a) of Exercise 2 in terms of your matrix.
    c. Give a general way to produce syzygies on the generators of a determinantal ideal. Hint: Find ways to produce $(m + 1) \times (m + 1)$ matrices containing M, whose determinants are automatically zero.

4. Prove that the syzygy S_{ij} defined in (1) is homogeneous of multidegree γ.

5. Complete the proof of Lemma 4 by showing that the decomposition into homogeneous components is unique. Hint: First show that if $S = \sum_\alpha S'_\alpha$, where S'_α has multidegree α, then, for a fixed i, the ith components of the S'_α are either 0 or have multidegree equal to $\alpha - \text{multideg}(f_i)$ and, hence, give distinct terms as α varies.

6. Suppose that S is a homogeneous syzygy of multidegree α in $S(G)$.
    a. Prove that $S \cdot G$ has multidegree $< \alpha$.
    b. Use part (a) to show that Corollary 7 follows from Theorem 6.

7. Complete the proof of Proposition 8 by proving the formula expressing S_{ij} in terms of S_{il} and S_{jl}.

8. Let G be a finite subset of $k[x_1, \ldots, x_n]$ and let $f \in \langle G \rangle$. If $\overline{f}^G = r \neq 0$, then show that $f \to_{G'} 0$, where $G' = G \cup \{r\}$. This fact is used in the proof of Theorem 9.

9. In the proof of Theorem 9, we claimed that for every value of B, if $1 \leq i < j \leq t$ and $(i, j) \notin B$, then condition (6) was true. To prove this, we needed to show that if the claim held for B, then it held when B changed to some B'. The case when $(i, j) \notin B'$ but $(i, j) \in B$ was covered in the text. It remains to consider when $(i, j) \notin B' \cup B$. In this case, prove that (6) holds for B'. Hint: Note that (6) holds for B. There are two cases to consider, depending on whether B' is bigger or smaller than B. In the latter situation, $B' = B \setminus \{(l, m)\}$ for some $(l, m) \neq (i, j)$.

10. In this exercise, we will study the ordering on the set $\{(i, j) \mid 1 \leq i < j \leq t\}$ described in the proof of Theorem 9. Assume that $B = \emptyset$, and recall that t is the length of G when the algorithm stops.
    a. Show that any pair (i, j) with $1 \leq i < j \leq t$ was a member of B at some point during the algorithm.
    b. Use part (a) and $B = \emptyset$ to explain how we can order the set of *all* pairs according to when a pair was removed from B.

11. Consider $f_1 = x^3 - 2xy$ and $f_2 = x^2 y - 2y^2 + x$ and use grlex order on $k[x, y]$. These polynomials are taken from Example 1 of §7, where we followed Buchberger's algorithm to show how a Gröbner basis was produced. Redo this example using the algorithm of Theorem 9 and, in particular, keep track of how many times you have to use the division algorithm.

12. Consider the polynomials

$$x^{n+1} - yz^{n-1}w, \quad xy^{n-1} - z^n, \quad x^n z - y^n w,$$

and use grevlex order with $x > y > z > w$. Mora [see LAZARD (1983)] showed that the reduced Gröbner basis contains the polynomial

$$z^{n^2+1} - y^{n^2}w.$$

Prove that this is true when n is 3, 4, or 5. How big are the Gröbner bases?

13. In this exercise, we will look at some examples of how the term order can affect the length of a Gröbner basis computation and the complexity of the answer.

   a. Compute a Gröbner basis for $I = \langle x^5 + y^4 + z^3 - 1, x^3 + y^2 + z^2 - 1 \rangle$ using lex and grevlex orders with $x > y > z$. You will see that the Gröbner basis is much simpler when using grevlex.

   b. Compute a Gröbner basis for $I = \langle x^5 + y^4 + z^3 - 1, x^3 + y^3 + z^2 - 1 \rangle$ using lex and grevlex orders with $x > y > z$. This differs from the previous example by a single exponent, but the Gröbner basis for lex order is significantly nastier (one of its polynomials has 282 terms, total degree 25, and a largest coefficient of 170255391).

   c. Let $I = \langle x^4 - yz^2w, xy^2 - z^3, x^3z - y^3w \rangle$ be the ideal generated by the polynomials of Exercise 12 with $n = 3$. Using lex and grevlex orders with $x > y > z > w$, show that the resulting Gröbner bases are the same. So grevlex is not always better than lex, but in practice, it is usually a good idea to use grevlex whenever possible.

Chapter 3
Elimination Theory

This chapter will study systematic methods for eliminating variables from systems of polynomial equations. The basic strategy of elimination theory will be given in two main theorems: the Elimination Theorem and the Extension Theorem. We will prove these results using Gröbner bases and the classic theory of resultants. The geometric interpretation of elimination will also be explored when we discuss the Closure Theorem. Of the many applications of elimination theory, we will treat two in detail: the implicitization problem and the envelope of a family of curves.

§1 The Elimination and Extension Theorems

To get a sense of how elimination works, let us look at an example similar to those discussed at the end of Chapter 2. We will solve the system of equations

$$
\begin{aligned}
x^2 + y + z &= 1, \\
x + y^2 + z &= 1, \\
x + y + z^2 &= 1.
\end{aligned}
$$

(1)

If we let I be the ideal

(2) $\qquad I = \langle x^2 + y + z - 1, x + y^2 + z - 1, x + y + z^2 - 1 \rangle,$

then a Gröbner basis for I with respect to lex order is given by the four polynomials

$$
\begin{aligned}
g_1 &= x + y + z^2 - 1, \\
g_2 &= y^2 - y - z^2 + z, \\
g_3 &= 2yz^2 + z^4 - z^2, \\
g_4 &= z^6 - 4z^4 + 4z^3 - z^2.
\end{aligned}
$$

(3)

© Springer International Publishing Switzerland 2015
D.A. Cox et al., *Ideals, Varieties, and Algorithms*, Undergraduate Texts in Mathematics, DOI 10.1007/978-3-319-16721-3_3

It follows that equations (1) and (3) have the same solutions. However, since

$$g_4 = z^6 - 4z^4 + 4z^3 - z^2 = z^2(z-1)^2(z^2 + 2z - 1)$$

involves only z, we see that the possible z's are 0, 1, and $-1 \pm \sqrt{2}$. Substituting these values into $g_2 = y^2 - y - z^2 + z = 0$ and $g_3 = 2yz^2 + z^4 - z^2 = 0$, we can determine the possible y's, and then finally $g_1 = x + y + z^2 - 1 = 0$ gives the corresponding x's. In this way, one can check that equations (1) have exactly five solutions:

$$(1,\, 0,\, 0),\ (0,\, 1,\, 0),\ (0,\, 0,\, 1),$$
$$(-1 + \sqrt{2},\, -1 + \sqrt{2},\, -1 + \sqrt{2}),$$
$$(-1 - \sqrt{2},\, -1 - \sqrt{2},\, -1 - \sqrt{2}).$$

What enabled us to find these solutions? There were two things that made our success possible:

- (Elimination Step) We could find a consequence $g_4 = z^6 - 4z^4 + 4z^3 - z^2 = 0$ of our original equations which involved only z, i.e., we eliminated x and y from the system of equations.
- (Extension Step) Once we solved the simpler equation $g_4 = 0$ to determine the values of z, we could extend these solutions to solutions of the original equations.

The basic idea of *elimination theory* is that both the Elimination Step and the Extension Step can be done in great generality.

To see how the Elimination Step works, notice that our observation concerning g_4 can be written as

$$g_4 \in I \cap \mathbb{C}[z],$$

where I is the ideal given in equation (2). In fact, $I \cap \mathbb{C}[z]$ consists of *all* consequences of our equations which eliminate x and y. Generalizing this idea leads to the following definition.

Definition 1. Given $I = \langle f_1, \ldots, f_s \rangle \subseteq k[x_1, \ldots, x_n]$, the *l*-th **elimination ideal** I_l is the ideal of $k[x_{l+1}, \ldots, x_n]$ defined by

$$I_l = I \cap k[x_{l+1}, \ldots, x_n].$$

Thus, I_l consists of all consequences of $f_1 = \cdots = f_s = 0$ which eliminate the variables $x_1, \ldots, x_l$. In the exercises, you will verify that I_l is an ideal of $k[x_{l+1}, \ldots, x_n]$. Note that $I = I_0$ is the 0-th elimination ideal. Also observe that different orderings of the variables lead to different elimination ideals.

Using this language, we see that eliminating $x_1, \ldots, x_l$ means finding nonzero polynomials in the *l*-th elimination ideal I_l. Thus *a solution of the Elimination Step means giving a systematic procedure for finding elements of I_l*. With the proper term ordering, Gröbner bases allow us to do this instantly.

Theorem 2 (The Elimination Theorem). *Let $I \subseteq k[x_1, \ldots, x_n]$ be an ideal and let G be a Gröbner basis of I with respect to lex order where $x_1 > x_2 > \cdots > x_n$. Then, for every $0 \le l \le n$, the set*

$$G_l = G \cap k[x_{l+1}, \ldots, x_n]$$

is a Gröbner basis of the l-th elimination ideal I_l.

Proof. Fix l between 0 and n. Since $G_l \subseteq I_l$ by construction, it suffices to show that

$$\langle \mathrm{LT}(I_l) \rangle = \langle \mathrm{LT}(G_l) \rangle$$

by the definition of Gröbner basis. One inclusion is obvious, and to prove the other inclusion $\langle \mathrm{LT}(I_l) \rangle \subseteq \langle \mathrm{LT}(G_l) \rangle$, we need only to show that the leading term $\mathrm{LT}(f)$, for an arbitrary $f \in I_l$, is divisible by $\mathrm{LT}(g)$ for some $g \in G_l$.

To prove this, note that f also lies in I, which tells us that $\mathrm{LT}(f)$ is divisible by $\mathrm{LT}(g)$ for some $g \in G$ since G is a Gröbner basis of I. Since $f \in I_l$, this means that $\mathrm{LT}(g)$ involves only the variables $x_{l+1}, \ldots, x_n$. Now comes the crucial observation: since we are using lex order with $x_1 > \cdots > x_n$, any monomial involving $x_1, \ldots, x_l$ is greater than all monomials in $k[x_{l+1}, \ldots, x_n]$, so that $\mathrm{LT}(g) \in k[x_{l+1}, \ldots, x_n]$ implies $g \in k[x_{l+1}, \ldots, x_n]$. This shows that $g \in G_l$, and the theorem is proved. $\qquad\square$

For an example of how this theorem works, let us return to example (1) from the beginning of the section. Here, $I = \langle x^2 + y + z - 1, x + y^2 + z - 1, x + y + z^2 - 1 \rangle$, and a Gröbner basis with respect to lex order is given in (3). It follows from the Elimination Theorem that

$$I_1 = I \cap \mathbb{C}[y, z] = \langle y^2 - y - z^2 + z, 2yz^2 + z^4 - z^2, z^6 - 4z^4 + 4z^3 - z^2 \rangle,$$
$$I_2 = I \cap \mathbb{C}[z] = \langle z^6 - 4z^4 + 4z^3 - z^2 \rangle.$$

Thus, $g_4 = z^6 - 4z^4 + 4z^3 - z^2$ is not just some random way of eliminating x and y from our equations—it is the best possible way to do so since any other polynomial that eliminates x and y is a multiple of g_4.

The Elimination Theorem shows that a Gröbner basis for lex order eliminates not only the first variable, but also the first two variables, the first three variables, and so on. In some cases (such as the implicitization problem to be studied in §3), we only want to eliminate certain variables, and we do not care about the others. In such a situation, it is a bit of overkill to compute a Gröbner basis using lex order. This is especially true since lex order can lead to some very unpleasant Gröbner bases (see Exercise 13 of Chapter 2, §10 for an example). In the exercises, you will study versions of the Elimination Theorem that use more efficient monomial orderings than lex.

We next discuss the Extension Step. Suppose we have an ideal $I \subseteq k[x_1, \ldots, x_n]$. As in Chapter 2, we have the affine variety

$$\mathbf{V}(I) = \{(a_1, \ldots, a_n) \in k^n \mid f(a_1, \ldots, a_n) = 0 \text{ for all } f \in I\}.$$

To describe points of $\mathbf{V}(I)$, the basic idea is to build up solutions one coordinate at a time. Fix some l between 1 and n. This gives us the elimination ideal I_l, and we will call a solution $(a_{l+1}, \ldots, a_n) \in \mathbf{V}(I_l)$ a *partial solution* of the original system of equations. To extend $(a_{l+1}, \ldots, a_n)$ to a complete solution in $\mathbf{V}(I)$, we first need to add one more coordinate to the solution. This means finding a_l so that $(a_l, a_{l+1}, \ldots, a_n)$ lies in the variety $\mathbf{V}(I_{l-1})$ of the next elimination ideal. More concretely, suppose that $I_{l-1} = \langle g_1, \ldots, g_r \rangle$ in $k[x_l, x_{l+1}, \ldots, x_n]$. Then we want to find solutions $x_l = a_l$ of the equations

$$g_1(x_l, a_{l+1}, \ldots, a_n) = \cdots = g_r(x_l, a_{l+1}, \ldots, a_n) = 0.$$

Here we are dealing with polynomials of one variable x_l, and it follows that the possible a_l's are just the roots of the gcd of the above r polynomials.

The basic problem is that the above polynomials may not have a common root, i.e., there may be some partial solutions which do not extend to complete solutions. For a simple example, consider the equations

(4)
$$\begin{aligned} xy &= 1, \\ xz &= 1. \end{aligned}$$

Here, $I = \langle xy - 1, xz - 1 \rangle$, and an easy application of the Elimination Theorem shows that $y - z$ generates the first elimination ideal I_1. Thus, the partial solutions are given by (a, a), and these all extend to complete solutions $(1/a, a, a)$ *except* for the partial solution $(0, 0)$. To see what is going on geometrically, note that $y = z$ defines a plane in 3-dimensional space. Then the variety (4) is a hyperbola lying in this plane:

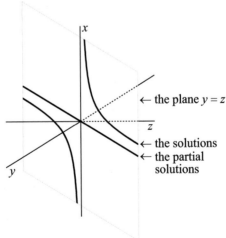

It is clear that the variety defined by (4) has no points lying over the partial solution $(0, 0)$. Pictures such as the one here will be studied in more detail in §2 when we study the geometric interpretation of eliminating variables. For now, our goal is to see if we can determine *in advance* which partial solutions extend to complete solutions.

Let us restrict our attention to the case where we eliminate just the first variable x_1. Thus, we want to know if a partial solution $(a_2, \ldots, a_n) \in \mathbf{V}(I_1)$ can be extended to a solution $(a_1, a_2, \ldots, a_n) \in \mathbf{V}(I)$. The following theorem tells us when this can be done.

Theorem 3 (The Extension Theorem). *Let* $I = \langle f_1, \ldots, f_s \rangle \subseteq \mathbb{C}[x_1, \ldots, x_n]$ *and let* I_1 *be the first elimination ideal of I. For each* $1 \leq i \leq s$, *write* f_i *in the form*

$$f_i = c_i(x_2, \ldots, x_n) \, x_1^{N_i} + \text{terms in which } x_1 \text{ has degree} < N_i,$$

where $N_i \geq 0$ *and* $c_i \in \mathbb{C}[x_2, \ldots, x_n]$ *is nonzero. Suppose that we have a partial solution* $(a_2, \ldots, a_n) \in \mathbf{V}(I_1)$. *If* $(a_2, \ldots, a_n) \notin \mathbf{V}(c_1, \ldots, c_s)$, *then there exists* $a_1 \in \mathbb{C}$ *such that* $(a_1, a_2, \ldots, a_n) \in \mathbf{V}(I)$.

We will give two proofs of this theorem, one using Gröbner bases in §5 and the other using resultants in §6. For the rest of the section, we will explain the Extension Theorem and discuss its consequences. A geometric interpretation will be given in §2.

A first observation is that the theorem is stated only for the field $k = \mathbb{C}$. To see why $\mathbb{C}$ is important, assume that $k = \mathbb{R}$ and consider the equations

$$
\begin{aligned}
x^2 &= y, \\
x^2 &= z.
\end{aligned}
\tag{5}
$$

Eliminating x gives $y = z$, so that we get the partial solutions (a, a) for all $a \in \mathbb{R}$. Since the leading coefficients of x in $x^2 - y$ and $x^2 - z$ never vanish, the Extension Theorem guarantees that (a, a) extends, *provided* we work over $\mathbb{C}$. Over $\mathbb{R}$, the situation is different. Here, $x^2 = a$ has no real solutions when a is negative, so that only those partial solutions with $a \geq 0$ extend to real solutions of (5). This shows that the Extension Theorem is false over $\mathbb{R}$.

Turning to the hypothesis $(a_2, \ldots, a_n) \notin \mathbf{V}(c_1, \ldots, c_s)$, note that the c_i's are the leading coefficients with respect to x_1 of the f_i's. Thus, $(a_2, \ldots, a_n) \notin \mathbf{V}(c_1, \ldots, c_s)$ says that the leading coefficients do not vanish simultaneously at the partial solution. To see why this condition is necessary, let us look at example (4). Here, the equations

$$
\begin{aligned}
xy &= 1, \\
xz &= 1
\end{aligned}
$$

have the partial solutions $(y, z) = (a, a)$. The only one that does not extend is $(0, 0)$, which is the partial solution where the leading coefficients y and z of x vanish. The Extension Theorem tells us that *the Extension Step can fail only when the leading coefficients vanish simultaneously.*

Finally, we should mention that the variety $\mathbf{V}(c_1, \ldots, c_s)$ where the leading coefficients vanish depends on the basis $\{f_1, \ldots, f_s\}$ of I: changing to a different basis may cause $\mathbf{V}(c_1, \ldots, c_s)$ to change. In Chapter 8, we will learn how to choose $(f_1, \ldots, f_s)$ so that $\mathbf{V}(c_1, \ldots, c_s)$ is as small as possible. We should also point out that if one works in projective space (to be defined in Chapter 8), then one can show that *all* partial solutions extend.

Although the Extension Theorem is stated only for the case of eliminating the first variable x_1, it can be used when eliminating any number of variables. For example, consider the equations

$$x^2 + y^2 + z^2 = 1,$$
(6)
$$xyz = 1.$$

A Gröbner basis for $I = \langle x^2 + y^2 + z^2 - 1, xyz - 1 \rangle$ with respect to lex order is

$$g_1 = y^4z^2 + y^2z^4 - y^2z^2 + 1,$$
$$g_2 = x + y^3z + yz^3 - yz.$$

By the Elimination Theorem, we obtain

$$I_1 = I \cap \mathbb{C}[y, z] = \langle g_1 \rangle,$$
$$I_2 = I \cap \mathbb{C}[z] = \{0\}.$$

Since $I_2 = \{0\}$, we have $\mathbf{V}(I_2) = \mathbb{C}$, and, thus, *every $c \in \mathbb{C}$ is a partial solution*. So we ask:

Which partial solutions $c \in \mathbb{C} = \mathbf{V}(I_2)$ extend to $(a, b, c) \in \mathbf{V}(I)$?

The idea is to extend c one coordinate at a time: first to (b, c), then to (a, b, c). To control which solutions extend, we will use the Extension Theorem at each step. The crucial observation is that I_2 is the first elimination ideal of I_1. This is easy to see here, and the general case is covered in the exercises. Thus, we will use the Extension Theorem once to go from $c \in \mathbf{V}(I_2)$ to $(b, c) \in \mathbf{V}(I_1)$, and a second time to go to $(a, b, c) \in \mathbf{V}(I)$. This will tell us exactly which c's extend.

To start, we apply the Extension Theorem to go from I_2 to $I_1 = \langle g_1 \rangle$. The coefficient of y^4 in g_1 is z^2, so that $c \in \mathbb{C} = \mathbf{V}(I_2)$ extends to (b, c) whenever $c \neq 0$. Note also that $g_1 = 0$ has *no* solution when $c = 0$. The next step is to go from I_1 to I; that is, to find a so that $(a, b, c) \in \mathbf{V}(I)$. If we substitute $(y, z) = (b, c)$ into (6), we get two equations in x, and it is not obvious that there is a common solution $x = a$. This is where the Extension Theorem shows its power. The leading coefficients of x in $x^2 + y^2 + z^2 - 1$ and $xyz - 1$ are 1 and yz, respectively. Since 1 never vanishes, the Extension Theorem *guarantees* that a always exists. We have thus proved that *all* partial solutions $c \neq 0$ extend to $\mathbf{V}(I)$.

The Extension Theorem is especially easy to use when one of the leading coefficients is constant. This case is sufficiently useful that we will state it as a separate corollary.

Corollary 4. *Let $I = \langle f_1, \ldots, f_s \rangle \subseteq \mathbb{C}[x_1, \ldots, x_n]$, and assume that for some i, f_i is of the form*

$$f_i = c_i x_1^{N_i} + \text{terms in which } x_1 \text{ has degree} < N_i,$$

where $c_i \in \mathbb{C}$ is nonzero and $N_i > 0$. If I_1 is the first elimination ideal of I and $(a_2, \ldots, a_n) \in \mathbf{V}(I_1)$, then there is $a_1 \in \mathbb{C}$ such that $(a_1, a_2, \ldots, a_n) \in \mathbf{V}(I)$.

Proof. This follows immediately from the Extension Theorem: since $c_i \neq 0$ in $\mathbb{C}$ implies $\mathbf{V}(c_1, \ldots, c_s) = \emptyset$, we have $(a_2, \ldots, a_n) \notin \mathbf{V}(c_1, \ldots, c_s)$ for all partial solutions $(a_2, \ldots, a_n)$. $\square$

We will end this section with an example of a system of equations that does not have nice solutions. Consider the equations

$$xy = 4,$$
$$y^2 = x^3 - 1.$$

Using lex order with $x > y$, the Gröbner basis is given by

$$g_1 = 16x - y^2 - y^4,$$
$$g_2 = y^5 + y^3 - 64,$$

but if we proceed as usual, we discover that $y^5 + y^3 - 64$ has *no* rational roots (in fact, it is *irreducible* over $\mathbb{Q}$, a concept we will discuss in Chapter 4, §2). One option is to compute the roots numerically. A variety of methods (such as the Newton-Raphson method) are available, and for $y^5 + y^3 - 64 = 0$, one obtains

$$y = 2.21363, \ -1.78719 \pm 1.3984i, \ \text{or} \ 0.680372 \pm 2.26969i.$$

These solutions can then be substituted into $g_1 = 16x - y^2 - y^4 = 0$ to determine the values of x. Thus, unlike the previous examples, we can only find numerical approximations to the solutions. See VON ZUR GATHEN and GERHARD (2013) for an introduction to finding the roots of a polynomial of one variable.

There are many interesting problems that arise when one tries to find numerical solutions of systems of polynomial equations. For further reading on this topic, we recommend LAZARD (1993) and MANOCHA (1994). The reader may also wish to consult COX, LITTLE and O'SHEA (2005) and DICKENSTEIN and EMIRIS (2005).

EXERCISES FOR §1

1. Let $I \subseteq k[x_1, \ldots, x_n]$ be an ideal.
   a. Prove that $I_l = I \cap k[x_{l+1}, \ldots, x_n]$ is an ideal of $k[x_{l+1}, \ldots, x_n]$.
   b. Prove that the ideal $I_{l+1} \subseteq k[x_{l+2}, \ldots, x_n]$ is the first elimination ideal of $I_l \subseteq k[x_{l+1}, \ldots, x_n]$. This observation allows us to use the Extension Theorem multiple times when eliminating more than one variable.
2. Consider the system of equations

$$x^2 + 2y^2 = 3,$$
$$x^2 + xy + y^2 = 3.$$

   a. If I is the ideal generated by these equations, find bases of $I \cap k[x]$ and $I \cap k[y]$.
   b. Find all solutions of the equations.
   c. Which of the solutions are *rational*, i.e., lie in $\mathbb{Q}^2$?
   d. What is the smallest field k containing $\mathbb{Q}$ such that all solutions lie in k^2?
3. Determine all solutions $(x, y) \in \mathbb{Q}^2$ of the system of equations

$$x^2 + 2y^2 = 2,$$
$$x^2 + xy + y^2 = 2.$$

   Also determine all solutions in $\mathbb{C}^2$.

4. Find bases for the elimination ideals I_1 and I_2 for the ideal I determined by the equations:

$$x^2 + y^2 + z^2 = 4,$$
$$x^2 + 2y^2 = 5,$$
$$xz = 1.$$

How many rational (i.e., in $\mathbb{Q}^3$) solutions are there?

5. In this exercise, we will prove a more general version of the Elimination Theorem. Fix an integer $1 \leq l \leq n$. We say that a monomial order $>$ on $k[x_1, \ldots, x_n]$ is of l-*elimination type* provided that any monomial involving one of $x_1, \ldots, x_l$ is greater than all monomials in $k[x_{l+1}, \ldots, x_n]$. Prove the following generalized Elimination Theorem. If I is an ideal in $k[x_1, \ldots, x_n]$ and G is a Gröbner basis of I with respect to a monomial order of l-elimination type, then $G \cap k[x_{l+1}, \ldots, x_n]$ is a Gröbner basis of the l-th elimination ideal $I_l = I \cap k[x_{l+1}, \ldots, x_n]$.

6. To exploit the generalized Elimination Theorem of Exercise 5, we need some interesting examples of monomial orders of l-elimination type. We will consider two such orders.

   a. Fix an integer $1 \leq l \leq n$, and define the order $>_l$ as follows: if $\alpha, \beta \in \mathbb{Z}^n_{\geq 0}$, then $\alpha >_l \beta$ if

   $$\alpha_1 + \cdots + \alpha_l > \beta_1 + \cdots + \beta_l, \text{ or } \alpha_1 + \cdots + \alpha_l = \beta_1 + \cdots + \beta_l \text{ and } \alpha >_{grevlex} \beta.$$

   This is the l-th *elimination order* of BAYER and STILLMAN (1987b). Prove that $>_l$ is a monomial order and is of l-elimination type. Hint: If you did Exercise 11 of Chapter 2, §4, then you have already done this problem.

   b. In Exercise 9 of Chapter 2, §4, we considered an example of a product order that mixed lex and grlex orders on different sets of variables. Explain how to create a product order that induces grevlex on both $k[x_1, \ldots, x_l]$ and $k[x_{l+1}, \ldots, x_n]$ and show that this order is of l-elimination type.

   c. If G is a Gröbner basis for $I \subseteq k[x_1, \ldots, x_n]$ for either of the monomial orders of parts (a) or (b), explain why $G \cap k[x_{l+1}, \ldots, x_n]$ is a Gröbner basis with respect to grevlex.

7. Consider the equations

$$t^2 + x^2 + y^2 + z^2 = 0,$$
$$t^2 + 2x^2 - xy - z^2 = 0,$$
$$t + y^3 - z^3 = 0.$$

We want to eliminate t. Let $I = \langle t^2 + x^2 + y^2 + z^2, t^2 + 2x^2 - xy - z^2, t + y^3 - z^3 \rangle$ be the corresponding ideal.

   a. Using lex order with $t > x > y > z$, compute a Gröbner basis for I, and then find a basis for $I \cap \mathbb{Q}[x, y, z]$. You should get four generators, one of which has total degree 12.

   b. Compute a grevlex Gröbner basis for $I \cap \mathbb{Q}[x, y, z]$. You will get a simpler set of two generators.

   c. Combine the answer to part (b) with the polynomial $t + y^3 - z^3$ and show that this gives a Gröbner basis for I with respect to the elimination order $>_1$ (this is $>_l$ with $l = 1$) of Exercise 6. Note that this Gröbner basis is much simpler than the one found in part (a). If you have access to a computer algebra system that knows elimination orders, then check your answer.

8. In equation (6), we showed that $z \neq 0$ could be specified arbitrarily. Hence, z can be regarded as a "parameter." To emphasize this point, show that there are formulas for x and y in terms of z. Hint: Use g_1 and the quadratic formula to get y in terms of z. Then use $xyz = 1$ to get x. The formulas you obtain give a "parametrization" of $\mathbf{V}(I)$ which is different from those studied in §3 of Chapter 1. Namely, in Chapter 1, we used parametrizations by

rational functions, whereas here, we have what is called a parametrization by *algebraic* functions. Note that x and y are *not* uniquely determined by z.

9. Consider the system of equations given by

$$x^5 + \frac{1}{x^5} = y,$$

$$x + \frac{1}{x} = z.$$

Let I be the ideal in $\mathbb{C}[x, y, z]$ determined by these equations.

a. Find a basis of $I_1 \subseteq \mathbb{C}[y, z]$ and show that $I_2 = \{0\}$.

b. Use the Extension Theorem to prove that each partial solution $c \in \mathbf{V}(I_2) = \mathbb{C}$ extends to a solution in $\mathbf{V}(I) \subseteq \mathbb{C}^3$.

c. Which partial solutions $(b, c) \in \mathbf{V}(I_1) \subseteq \mathbb{R}^2$ extend to solutions in $\mathbf{V}(I) \subseteq \mathbb{R}^3$? Explain why your answer does not contradict the Extension Theorem.

d. If we regard z as a "parameter" (see the previous problem), then solve for x and y as algebraic functions of z to obtain a "parametrization" of $\mathbf{V}(I)$.

§2 The Geometry of Elimination

In this section, we will give a geometric interpretation of the theorems from §1. The main idea is that elimination corresponds to projecting a variety onto a lower dimensional subspace. We will also discuss the Closure Theorem, which describes the relation between partial solutions and elimination ideals. For simplicity, we will work over the field $k = \mathbb{C}$.

Let us start by defining the projection of an affine variety. Suppose that we are given $V = \mathbf{V}(f_1, \ldots, f_s) \subseteq \mathbb{C}^n$. To eliminate the first l variables $x_1, \ldots, x_l$, we will consider the *projection map*

$$\pi_l : \mathbb{C}^n \longrightarrow \mathbb{C}^{n-l}$$

which sends $(a_1, \ldots, a_n)$ to $(a_{l+1}, \ldots, a_n)$. If we apply π_l to $V \subseteq \mathbb{C}^n$, then we get $\pi_l(V) \subseteq \mathbb{C}^{n-l}$. We can relate $\pi_l(V)$ to the l-th elimination ideal as follows.

Lemma 1. *With the above notation, let* $I_l = \langle f_1, \ldots, f_s \rangle \cap \mathbb{C}[x_{l+1}, \ldots, x_n]$ *be the l-th elimination ideal. Then, in $\mathbb{C}^{n-l}$, we have*

$$\pi_l(V) \subseteq \mathbf{V}(I_l).$$

Proof. Fix a polynomial $f \in I_l$. If $(a_1, \ldots, a_n) \in V$, then f vanishes at $(a_1, \ldots, a_n)$ since $f \in \langle f_1, \ldots, f_s \rangle$. But f involves only $x_{l+1}, \ldots, x_n$, so that we can write

$$f(a_{l+1}, \ldots, a_n) = f(\pi_l(a_1, \ldots, a_n)) = 0.$$

This shows that f vanishes at all points of $\pi_l(V)$.   □

As in §1, points of $\mathbf{V}(I_l)$ will be called *partial solutions*. Using the lemma, we can write $\pi_l(V)$ as follows:

$$\pi_l(V) = \{(a_{l+1}, \ldots, a_n) \in \mathbf{V}(I_l) \mid \text{there exist } a_1, \ldots, a_l \in \mathbb{C} \text{ such}$$
$$\text{that } (a_1, \ldots, a_l, a_{l+1}, \ldots, a_n) \in V\}.$$

Thus, $\pi_l(V)$ consists *exactly* of the partial solutions that extend to complete solutions. For an example, consider the variety V defined by equations (4) from §1:

(1)
$$xy = 1,$$
$$xz = 1.$$

Here, we have the following picture that simultaneously shows the solutions and the partial solutions:

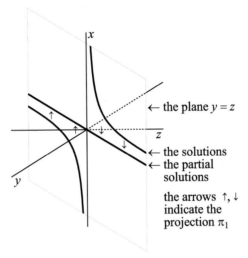

Note that $\mathbf{V}(I_1)$ is the line $y = z$ in the (y, z)-plane, and that

$$\pi_1(V) = \{(a, a) \in \mathbb{C}^2 \mid a \neq 0\}.$$

In particular, $\pi_1(V)$ is *not* an affine variety—it is missing the point $(0, 0)$.

The basic tool to understand the missing points is the Extension Theorem from §1. It only deals with π_1 (i.e., eliminating x_1), but gives us a good picture of what happens in this case. Stated geometrically, here is what the Extension Theorem says.

Theorem 2 (The Geometric Extension Theorem). *Given* $V = \mathbf{V}(f_1, \ldots, f_s) \subseteq \mathbb{C}^n$, *let* $c_i \in \mathbb{C}[x_2, \ldots, x_n]$ *be as in the Extension Theorem from §1. If* I_1 *is the first elimination ideal of* $\langle f_1, \ldots, f_s \rangle$, *then we have the equality in* $\mathbb{C}^{n-1}$

$$\mathbf{V}(I_1) = \pi_1(V) \cup (\mathbf{V}(c_1, \ldots, c_s) \cap \mathbf{V}(I_1)),$$

where $\pi_1 : \mathbb{C}^n \to \mathbb{C}^{n-1}$ *is projection onto the last* $n - 1$ *coordinates.*

Proof. The proof follows from Lemma 1 and the Extension Theorem. The details will be left as an exercise. □

This theorem tells us that $\pi_1(V)$ fills up the affine variety $\mathbf{V}(I_1)$, except possibly for a part that lies in $\mathbf{V}(c_1, \ldots, c_s)$. Unfortunately, it is not clear how big this part is, and sometimes $\mathbf{V}(c_1, \ldots, c_s)$ is unnaturally large. For example, one can show that the equations

$$
\begin{aligned}
(y-z)x^2 + xy &= 1, \\
(y-z)x^2 + xz &= 1
\end{aligned}
\tag{2}
$$

generate the same ideal as equations (1). Since $c_1 = c_2 = y - z$ generate the elimination ideal I_1, the Geometric Extension Theorem tells us *nothing* about the size of $\pi_1(V)$ in this case.

Nevertheless, we can still make the following strong statements about the relation between $\pi_l(V)$ and $\mathbf{V}(I_l)$.

Theorem 3 (The Closure Theorem). *Let* $V = \mathbf{V}(f_1, \ldots, f_s) \subseteq \mathbb{C}^n$ *and let* I_l *be the l-th elimination ideal of* $\langle f_1, \ldots, f_s \rangle$. *Then:*

(i) $\mathbf{V}(I_l)$ *is the smallest affine variety containing* $\pi_l(V) \subseteq \mathbb{C}^{n-l}$.

(ii) *When* $V \neq \emptyset$, *there is an affine variety* $W \subsetneq \mathbf{V}(I_l)$ *such that* $\mathbf{V}(I_l) \setminus W \subseteq \pi_l(V)$.

When we say "smallest variety" in part (i), we mean "smallest with respect to set-theoretic inclusion." Thus, $\mathbf{V}(I_l)$ being smallest means two things:

- $\pi_l(V) \subseteq \mathbf{V}(I_l)$
- If Z is *any other* affine variety in $\mathbb{C}^{n-l}$ containing $\pi_l(V)$, then $\mathbf{V}(I_l) \subseteq Z$.

In Chapter 4, we will express this by saying that $\mathbf{V}(I_l)$ is the *Zariski closure* of $\pi_l(V)$. This is where the theorem gets its name. Part (ii) of the theorem says that although $\pi_l(V)$ may not equal $\mathbf{V}(I_l)$, it fills up "most" of $\mathbf{V}(I_l)$ in the sense that what is missing lies in a strictly smaller affine variety.

We cannot yet prove the Closure Theorem, for it requires the Nullstellensatz and other tools from Chapter 4. The proof will be deferred until then. We will also say more about the variety $W \subsetneq \mathbf{V}(I_l)$ of part (ii) in Chapter 4.

The Closure Theorem gives us a partial description of $\pi_l(V)$ since it fills up $\mathbf{V}(I_l)$, except for some missing points that lie in a variety strictly smaller than $\mathbf{V}(I_l)$. Unfortunately, the missing points might not fill up all of the smaller variety. The precise structure of $\pi_l(V)$ can be described as follows: there are affine varieties $Z_i \subseteq W_i \subseteq \mathbb{C}^{n-l}$ for $1 \leq i \leq m$ such that

$$
\pi_l(V) = \bigcup_{i=1}^{m} (W_i \setminus Z_i).
$$

In general, a set of this form is called *constructible*. We will prove this in Chapter 4.

In §1, we saw that the nicest case of the Extension Theorem was when one of the leading coefficients c_i was a nonzero constant. Then the c_i's can *never* simultaneously vanish at a point $(a_2, \ldots, a_n)$, and, consequently, partial solutions *always* extend in this case. Thus, we have the following geometric version of Corollary 4 of §1.

Corollary 4. *Let* $V = \mathbf{V}(f_1,\ldots,f_s) \subseteq \mathbb{C}^n$, *and assume that for some* i, f_i *is of the form*

$$f_i = c_i x_1^{N_i} + \textit{terms in which } x_1 \textit{ has degree} < N_i,$$

where $c_i \in \mathbb{C}$ *is nonzero and* $N_i > 0$. *If* I_1 *is the first elimination ideal, then in* $\mathbb{C}^{n-1}$,

$$\pi_1(V) = \mathbf{V}(I_1),$$

where π_1 *is the projection on the last* $n-1$ *coordinates.*

A final point to make concerns fields. The Extension Theorem and the Closure Theorem (and their corollaries) are stated for the field of complex numbers $\mathbb{C}$. In §§5 and 6, we will see that the Extension Theorem actually holds for any algebraically closed field k, and in Chapter 4, we will show that the same is true for the Closure Theorem.

EXERCISES FOR §2

1. Prove the Geometric Extension Theorem (Theorem 2) using the Extension Theorem and Lemma 1.
2. In example (2), verify carefully that $\langle (y-z)x^2+xy-1, (y-z)x^2+xz-1 \rangle = \langle xy-1, xz-1 \rangle$. Also check that $y - z$ vanishes at all partial solutions in $\mathbf{V}(I_1)$.
3. In this problem, we will prove part (ii) of Theorem 3 in the special case when $I = \langle f_1, f_2, f_3 \rangle$, where

$$f_1 = yx^3 + x^2,$$
$$f_2 = y^3x^2 + y^2,$$
$$f_3 = yx^4 + x^2 + y^2.$$

   a. Find a Gröbner basis for I and show that $I_1 = \langle y^2 \rangle$.
   b. Let c_i be the coefficient of the highest power of x in f_i. Then explain why $W = \mathbf{V}(c_1, c_2, c_3) \cap \mathbf{V}(I_1)$ does not satisfy part (ii) of Theorem 3.
   c. Let $\tilde{I} = \langle f_1, f_2, f_3, c_1, c_2, c_3 \rangle$. Show that $\mathbf{V}(I) = \mathbf{V}(\tilde{I})$ and $\mathbf{V}(I_1) = \mathbf{V}(\tilde{I}_1)$.
   d. Let x^{N_i} be the highest power of x appearing in f_i and set $\tilde{f}_i = f_i - c_i x^{N_i}$. Show that $\tilde{I} = \langle \tilde{f}_1, \tilde{f}_2, \tilde{f}_3, c_1, c_2, c_3 \rangle$.
   e. Repeat part (b) for $\tilde{I}$ using the generators from part (d) to find $\widetilde{W} \subsetneq \mathbf{V}(I_1)$ that satisfies part (ii) of Theorem 3.
4. To see how the Closure Theorem can fail over $\mathbb{R}$, consider the ideal

$$I = \langle x^2 + y^2 + z^2 + 2, 3x^2 + 4y^2 + 4z^2 + 5 \rangle.$$

   Let $V = \mathbf{V}(I)$, and let π_1 be the projection taking (x, y, z) to (y, z).
   a. Working over $\mathbb{C}$, prove that $\mathbf{V}(I_1) = \pi_1(V)$.
   b. Working over $\mathbb{R}$, prove that $V = \emptyset$ and that $\mathbf{V}(I_1)$ is infinite. Thus, $\mathbf{V}(I_1)$ may be much larger than the smallest variety containing $\pi_1(V)$ when the field is not algebraically closed.
5. Suppose that $I \subseteq \mathbb{C}[x, y]$ is an ideal such that $I_1 \neq \{0\}$. Prove that $\mathbf{V}(I_1) = \pi_1(V)$, where $V = \mathbf{V}(I)$ and π_1 is the projection onto the y-axis. Hint: Use part (i) of the Closure Theorem. Also, the only varieties contained in $\mathbb{C}$ are either $\mathbb{C}$ or finite subsets of $\mathbb{C}$.

§3 Implicitization

In Chapter 1, we saw that a variety V can sometimes be described using parametric equations. The basic idea of the *implicitization problem* is to convert the parametrization into defining equations for V. The name "implicitization" comes from Chapter 1, where the equations defining V were called an "implicit representation" of V. However, some care is required in giving a precise formulation of implicitization. The problem is that the parametrization need not fill up all of the variety V—an example is given by equation (4) from Chapter 1, §3. So the implicitization problem really asks for the equations defining the *smallest* variety V containing the parametrization. In this section, we will use the elimination theory developed in §§1 and 2 to give a complete solution of the implicitization problem.

Furthermore, once the smallest variety V has been found, two other interesting questions arise. First, does the parametrization fill up all of V? Second, if there are missing points, how do we find them? As we will see, Gröbner bases and the Extension Theorem give us powerful tools for studying this situation.

To illustrate these issues in a specific case, let us look at the tangent surface to the twisted cubic in $\mathbb{R}^3$, first studied in Chapter 1, §3. Recall that this surface is given parametrically by

$$(1) \qquad \begin{aligned} x &= t + u, \\ y &= t^2 + 2tu, \\ z &= t^3 + 3t^2 u. \end{aligned}$$

In §8 of Chapter 2, we used these equations to show that the tangent surface lies on the variety V in $\mathbb{R}^3$ defined by

$$x^3 z - (3/4)x^2 y^2 - (3/2)xyz + y^3 + (1/4)z^2 = 0.$$

However, we do not know if V is the smallest variety containing the tangent surface and, thus, we cannot claim to have solved the implicitization problem. Furthermore, even if V is the smallest variety, we still do not know if the tangent surface fills it up completely. So there is a lot of work to do.

We begin our solution of the implicitization problem with the case of a *polynomial parametrization*, which is specified by the data

$$(2) \qquad \begin{aligned} x_1 &= f_1(t_1, \ldots, t_m), \\ &\ \ \vdots \\ x_n &= f_n(t_1, \ldots, t_m). \end{aligned}$$

Here, $f_1, \ldots, f_n$ are polynomials in $k[t_1, \ldots, t_m]$. We can think of this geometrically as the function

$$F : k^m \longrightarrow k^n$$

defined by

$$F(t_1, \ldots, t_m) = (f_1(t_1, \ldots, t_m), \ldots, f_n(t_1, \ldots, t_m)).$$

Then $F(k^m) \subseteq k^n$ is the subset of k^n parametrized by equations (2). Since $F(k^m)$ may not be an affine variety (examples will be given in the exercises), a solution of the implicitization problem means finding the smallest affine variety that contains $F(k^m)$.

We can relate implicitization to elimination as follows. Equations (2) define a variety $V = \mathbf{V}(x_1 - f_1, \ldots, x_n - f_n) \subseteq k^{m+n}$. Points of V can be written in the form

$$(t_1, \ldots, t_m, f_1(t_1, \ldots, t_m), \ldots, f_n(t_1, \ldots, t_m)),$$

which shows that V can be regarded as the *graph* of the function F. We also have two other functions

$$i : k^m \longrightarrow k^{m+n},$$
$$\pi_m : k^{m+n} \longrightarrow k^n$$

defined by

$$i(t_1, \ldots, t_m) = (t_1, \ldots, t_m, f_1(t_1, \ldots, t_m), \ldots, f_n(t_1, \ldots, t_m))$$

and

$$\pi_m(t_1, \ldots, t_m, x_1, \ldots, x_n) = (x_1, \ldots, x_n),$$

respectively. This gives us the following diagram of sets and maps:

(3)

Note that F is then the composition $F = \pi_m \circ i$. It is also straightforward to show that $i(k^m) = V$. Thus, we obtain

(4)
$$F(k^m) = \pi_m(i(k^m)) = \pi_m(V).$$

In more concrete terms, this says that the image of the parametrization is the projection of its graph. We can now use elimination theory to find the smallest variety containing $F(k^m)$.

Theorem 1 (Polynomial Implicitization). *If k is an infinite field, let $F : k^m \to k^n$ be the function determined by the polynomial parametrization (2). Let I be the ideal $I = \langle x_1 - f_1, \ldots, x_n - f_n \rangle \subseteq k[t_1, \ldots, t_m, x_1, \ldots, x_n]$ and let $I_m = I \cap k[x_1, \ldots, x_n]$ be the m-th elimination ideal. Then $\mathbf{V}(I_m)$ is the smallest variety in k^n containing $F(k^m)$.*

Proof. By equation (4) above and Lemma 1 of §2, $F(k^m) = \pi_m(V) \subseteq \mathbf{V}(I_m)$. Thus $\mathbf{V}(I_m)$ is a variety containing $F(k^m)$. To show it is the smallest, suppose that

$h \in k[x_1, \ldots, x_n]$ vanishes on $F(k^m)$. We show that $h \in I_m$ as follows. If we regard h as a polynomial in $k[t_1, \ldots, t_m, x_1, \ldots, x_n]$, then we can divide h by $x_1 - f_1, \ldots, x_n - f_n$ using lex order with $x_1 > \cdots > x_n > t_1 > \cdots > t_m$. This gives an equation

$$(5) \qquad h(x_1, \ldots, x_n) = q_1 \cdot (x_1 - f_1) + \cdots + q_n \cdot (x_n - f_n) + r(t_1, \ldots, t_m)$$

since $\mathrm{LT}(x_j - f_j) = x_j$. Given any $\mathbf{a} = (a_1, \ldots, a_m) \in k^m$, we substitute $t_i = a_i$ and $x_i = f_i(\mathbf{a})$ into the above equation to obtain

$$0 = h(f_1(\mathbf{a}), \ldots, f_n(\mathbf{a})) = 0 + \cdots + 0 + r(\mathbf{a}).$$

It follows that $r(\mathbf{a}) = 0$ for all $\mathbf{a} \in k^m$. Since k is infinite, Proposition 5 of Chapter 1, §1 implies that $r(t_1, \ldots, t_m)$ is the zero polynomial. Thus we obtain

$$h(x_1, \ldots, x_n) = q_1 \cdot (x_1 - f_1) + \cdots + q_n \cdot (x_n - f_n) \in I \cap k[x_1, \ldots, x_n] = I_m$$

since $I = \langle x_1 - f_1, \ldots, x_n - f_n \rangle$.

Now suppose that $Z = \mathbf{V}(h_1, \ldots, h_s) \subseteq k^n$ is variety of k^n such that $F(k^m) \subseteq Z$. Then each h_i vanishes on $F(k^m)$ and hence lies in I_m by the previous paragraph. Thus

$$\mathbf{V}(I_m) \subseteq \mathbf{V}(h_1, \ldots, h_s) = Z.$$

This proves that $\mathbf{V}(I_m)$ is the smallest variety of k^n containing $F(k^m)$. □

Theorem 1 gives the following **implicitization algorithm for polynomial parametrizations**: if we are given $x_i = f_i(t_1, \ldots, t_m)$ for polynomials $f_1, \ldots, f_n \in k[t_1, \ldots, t_m]$, consider the ideal $I = \langle x_1 - f_1, \ldots, x_n - f_n \rangle$ and compute a Gröbner basis with respect to a lexicographic ordering where every t_i is greater than every x_i. By the Elimination Theorem, the elements of the Gröbner basis not involving $t_1, \ldots, t_m$ form a basis of I_m, and by Theorem 1, they define the smallest variety in k^n containing the parametrization.

For an example of how this algorithm works, let us look at the tangent surface to the twisted cubic in $\mathbb{R}^3$, which is given by the polynomial parametrization (1). Thus, we need to consider the ideal

$$I = \langle x - t - u, y - t^2 - 2tu, z - t^3 - 3t^2u \rangle \subseteq \mathbb{R}[t, u, x, y, z].$$

Using lex order with $t > u > x > y > z$, a Gröbner basis for I is given by

$$g_1 = t + u - x,$$
$$g_2 = u^2 - x^2 + y,$$
$$g_3 = ux^2 - uy - x^3 + (3/2)xy - (1/2)z,$$
$$g_4 = uxy - uz - x^2y - xz + 2y^2,$$
$$g_5 = uxz - uy^2 + x^2z - (1/2)xy^2 - (1/2)yz,$$
$$g_6 = uy^3 - uz^2 - 2x^2yz + (1/2)xy^3 - xz^2 + (5/2)y^2z,$$
$$g_7 = x^3z - (3/4)x^2y^2 - (3/2)xyz + y^3 + (1/4)z^2.$$

The Elimination Theorem tells us that $I_2 = I \cap \mathbb{R}[x, y, z] = \langle g_7 \rangle$, and thus by Theorem 1, $\mathbf{V}(g_7)$ solves the implicitization problem for the tangent surface of the twisted cubic. The equation $g_7 = 0$ is exactly the one given at the start of this section, but now we know it defines the smallest variety in $\mathbb{R}^3$ containing the tangent surface.

But we still do not know if the tangent surface fills up *all* of $\mathbf{V}(g_7) \subseteq \mathbb{R}^3$. To answer this question, we must see whether all partial solutions $(x, y, z) \in \mathbf{V}(g_7) = \mathbf{V}(I_2)$ extend to $(t, u, x, y, z) \in \mathbf{V}(I)$. We will first work over $\mathbb{C}$ so that we can use the Extension Theorem. As usual, our strategy will be to add one coordinate at a time.

Let us start with $(x, y, z) \in \mathbf{V}(I_2) = \mathbf{V}(g_7)$. In §1, we observed that I_2 is the first elimination ideal of I_1. Further, the Elimination Theorem tells us that $I_1 = \langle g_2, \ldots, g_7 \rangle$. Then the Extension Theorem, in the form of Corollary 4 of §1, implies that (x, y, z) always extends to $(u, x, y, z) \in \mathbf{V}(I_1)$ since I_1 has a generator with a constant leading coefficient of u (we leave it to you to find which of $g_2, \ldots, g_7$ has this property). Going from $(u, x, y, z) \in \mathbf{V}(I_1)$ to $(t, u, x, y, z) \in \mathbf{V}(I)$ is just as easy: using Corollary 4 of §1 again, we can always extend since $g_1 = t + u - x$ has a constant leading coefficient of t. We have thus proved that the tangent surface to the twisted cubic equals $\mathbf{V}(g_7)$ in $\mathbb{C}^3$.

It remains to see what happens over $\mathbb{R}$. If we start with a real solution $(x, y, z) \in \mathbb{R}^3$ of $g_7 = 0$, we know that it extends to $(t, u, x, y, z) \in \mathbf{V}(I) \subseteq \mathbb{R}^5$. But are the parameters t and u real? This is not immediately obvious. However, if you look at the above Gröbner basis, you can check that t and u are real when $(x, y, z) \in \mathbb{R}^3$ (see Exercise 4). It follows that the tangent surface to the twisted cubic in $\mathbb{R}^3$ equals the variety defined by

$$x^3 z - (3/4)x^2 y^2 - (3/2)xyz + y^3 + (1/4)z^2 = 0.$$

In general, the question of whether a parametrization fills up all of its variety can be difficult to answer. Each case has to be analyzed separately. But as indicated by the example just completed, the combination of Gröbner bases and the Extension Theorem can shed considerable light on what is going on.

In our discussion of implicitization, we have thus far only considered polynomial parametrizations. The next step is to see what happens when we have a parametrization by rational functions. To illustrate the difficulties that can arise, consider the following rational parametrization:

$$(6) \qquad
\begin{aligned}
x &= \frac{u^2}{v}, \\[4pt]
y &= \frac{v^2}{u}, \\[4pt]
z &= u.
\end{aligned}$$

It is easy to check that the point (x, y, z) always lies on the surface $x^2 y = z^3$. Let us see what happens if we clear denominators in the above equations and apply the polynomial implicitization algorithm. We get the ideal

$$I = \langle vx - u^2, uy - v^2, z - u \rangle \subseteq k[u, v, x, y, z],$$

and we leave it as an exercise to show that $I_2 = I \cap k[x, y, z]$ is given by $I_2 = \langle z(x^2y - z^3) \rangle$. This implies that

$$\mathbf{V}(I_2) = \mathbf{V}(x^2y - z^3) \cup \mathbf{V}(z),$$

and, in particular, $\mathbf{V}(I_2)$ is *not* the smallest variety containing the parametrization. So the above ideal I is not what we want—simply "clearing denominators" is too naive. To find an ideal that works better, we will need to be more clever.

In the general situation of a rational parametrization, we have

(7)
$$x_1 = \frac{f_1(t_1, \ldots, t_m)}{g_1(t_1, \ldots, t_m)},$$

$$\vdots$$

$$x_n = \frac{f_n(t_1, \ldots, t_m)}{g_n(t_1, \ldots, t_m)},$$

where $f_1, g_1, \ldots, f_n, g_n$ are polynomials in $k[t_1, \ldots, t_m]$. The map F from k^m to k^n given by (7) may not be defined on all of k^m because of the denominators. But if we let $W = \mathbf{V}(g_1 g_2 \cdots g_n) \subseteq k^m$, then it is clear that

$$F(t_1, \ldots, t_m) = \left(\frac{f_1(t_1, \ldots, t_m)}{g_1(t_1, \ldots, t_m)}, \ldots, \frac{f_n(t_1, \ldots, t_m)}{g_n(t_1, \ldots, t_m)} \right)$$

defines a map

$$F : k^m \setminus W \longrightarrow k^n.$$

To solve the implicitization problem, we need to find the smallest variety of k^n containing $F(k^m \setminus W)$.

We can adapt diagram (3) to this case by writing

(8)

It is easy to check that $i(k^m \setminus W) \subseteq \mathbf{V}(I)$, where $I = \langle g_1 x_1 - f_1, \ldots, g_n x_n - f_n \rangle$ is the ideal obtained by "clearing denominators." The problem is that $\mathbf{V}(I)$ may *not* be the smallest variety containing $i(k^m \setminus W)$. In the exercises, you will see that (6) is such an example.

To avoid this difficulty, we will alter the ideal I slightly by using a new variable to control the denominators. Consider the polynomial ring $k[y, t_1, \ldots, t_m, x_1, \ldots, x_n]$, which gives us the affine space k^{1+m+n}. Let g be the product $g = g_1 \cdot g_2 \cdots g_n$, so that $W = \mathbf{V}(g)$. Then consider the ideal

$$J = \langle g_1 x_1 - f_1, \ldots, g_n x_n - f_n, 1 - gy \rangle \subseteq k[y, t_1, \ldots, t_m, x_1, \ldots, x_n].$$

Note that the equation $1 - gy = 0$ means that the denominators $g_1, \ldots, g_n$ never vanish on $\mathbf{V}(J)$. To adapt diagram (8) to this new situation, consider the maps

$$j : k^m \setminus W \longrightarrow k^{1+m+n},$$
$$\pi_{1+m} : k^{1+m+n} \longrightarrow k^n$$

defined by

$$j(t_1, \ldots, t_m) = \left(\frac{1}{g(t_1, \ldots, t_m)}, t_1, \ldots, t_m, \frac{f_1(t_1, \ldots, t_m)}{g_1(t_1, \ldots, t_m)}, \ldots, \frac{f_n(t_1, \ldots, t_m)}{g_n(t_1, \ldots, t_m)} \right)$$

and

$$\pi_{1+m}(y, t_1, \ldots, t_m, x_1, \ldots, x_n) = (x_1, \ldots, x_n),$$

respectively. Then we get the diagram

As before, we have $F = \pi_{1+m} \circ j$. The surprise is that $j(k^m \setminus W) = \mathbf{V}(J)$ in k^{1+m+n}. To see this, note that $j(k^m \setminus W) \subseteq \mathbf{V}(J)$ follows easily from the definitions of j and J. Going the other way, if $(y, t_1, \ldots, t_m, x_1, \ldots, x_n) \in \mathbf{V}(J)$, then $g(t_1, \ldots, t_m)y = 1$ implies that *none* of the g_i's vanish at $(t_1, \ldots, t_m)$ and, thus, $g_i(t_1, \ldots, t_m)x_i = f_i(t_1, \ldots, t_m)$ can be solved for $x_i = f_i(t_1, \ldots, t_m)/g_i(t_1, \ldots, t_m)$. Since $y = 1/g(t_1, \ldots, t_m)$, it follows that our point lies in $j(k^m \setminus W)$. This proves that $\mathbf{V}(J) \subseteq j(k^m \setminus W)$.

From $F = \pi_{1+m} \circ j$ and $j(k^m \setminus W) = \mathbf{V}(J)$, we obtain

$$(9) \qquad F(k^m \setminus W) = \pi_{1+m}(j(k^m \setminus W)) = \pi_{1+m}(\mathbf{V}(J)).$$

Thus, the image of the parametrization is the projection of the variety $\mathbf{V}(J)$. As with the polynomial case, we can now use elimination theory to solve the implicitization problem.

Theorem 2 (Rational Implicitization). *If k is an infinite field, let $F : k^m \setminus W \to k^n$ be the function determined by the rational parametrization (7). Let J be the ideal $J = \langle g_1 x_1 - f_1, \ldots, g_n x_n - f_n, 1 - gy \rangle \subseteq k[y, t_1, \ldots, t_m, x_1, \ldots, x_n]$, where $g = g_1 \cdot g_2 \cdots g_n$ and $W = \mathbf{V}(g)$. Also let $J_{1+m} = J \cap k[x_1, \ldots, x_n]$ be the $(1 + m)$-th elimination ideal. Then $\mathbf{V}(J_{1+m})$ is the smallest variety in k^n containing $F(k^m \setminus W)$.*

Proof. We will show that if $h \in k[x_1, \ldots, x_n]$ vanishes on $F(k^m \setminus W)$, then $h \in J_{1+m}$. In the proof of Theorem 1, we divided h by $x_i - f_i$ to obtain (5) . Here, we divide h by $g_i x_i - f_i$, except that we need to multiply h by a (possibly large) power of $g = g_1 \cdots g_n$ to make this work. In the exercises, you will show that (5) gets replaced with an equation

(10) $g^N h(x_1, \ldots, x_n) = q_1 \cdot (g_1 x_1 - f_1) + \cdots + q_n \cdot (g_n x_n - f_n) + r(t_1, \ldots, t_m)$

in $k[t_1, \ldots, t_m, x_1, \ldots, x_n]$, where N is a sufficiently large integer. Then, given any $\mathbf{a} = (a_1, \ldots, a_m) \in k^m \setminus W$, we have $g_i(\mathbf{a}) \neq 0$ for all i. Hence we can substitute $t_i = a_i$ and $x_i = f_i(\mathbf{a})/g_i(\mathbf{a})$ into (10) to obtain

$$0 = g(\mathbf{a})^N h(f_1(\mathbf{a})/g_1(\mathbf{a})), \ldots, f_n(\mathbf{a})/g_n(\mathbf{a})) = 0 + \cdots + 0 + r(\mathbf{a}).$$

Thus $r(\mathbf{a}) = 0$ for all $\mathbf{a} \in k^m \setminus W$. Since k is infinite, this implies that $r(t_1, \ldots, t_m)$ is the zero polynomial, as you will prove in the exercises. Hence

$$g^N h(x_1, \ldots, x_n) = q_1 \cdot (g_1 x_1 - f_1) + \cdots + q_n \cdot (g_n x_n - f_n),$$

so that $g^N h \in \langle g_1 x_1 - f_1, \ldots, g_n x_n - f_n \rangle \subseteq k[t_1, \ldots, t_m, x_1, \ldots, x_n]$. Combining this with the identity

$$h = g^N y^N h + h(1 - g^N y^N) = y^N (g^N h) + h(1 + gy + \cdots + g^{N-1} y^{N-1})(1 - gy),$$

we see that in the larger ring $k[y, t_1, \ldots, t_m, x_1, \ldots, x_n]$, we have

$$h(x_1, \ldots, x_n) \in \langle g_1 x_1 - f_1, \ldots, g_n x_n - f_n, 1 - gy \rangle \cap k[x_1, \ldots, x_n] = J_{1+m}$$

by the definition of J. From here, the proof is similar to what we did in Theorem 1. The exercises at the end of the section will take you through the details. $\square$

The interpretation of Theorem 2 is very nice: given the rational parametrization (7), consider the equations

$$g_1 x_1 = f_1,$$
$$\vdots$$
$$g_n x_n = f_n,$$
$$g_1 g_2 \cdots g_n y = 1.$$

These equations are obtained from (7) by "clearing denominators" and adding a final equation (in the new variable y) to prevent the denominators from vanishing. Then eliminating $y, t_1, \ldots, t_m$ gives us the equations we want.

Theorem 2 implies the following **implicitization algorithm for rational parametrizations**: if we have $x_i = f_i/g_i$ for polynomials $f_1, g_1, \ldots, f_n, g_n \in k[t_1, \ldots, t_m]$, consider the new variable y and set $J = \langle g_1 x_1 - f_1, \ldots, g_n x_n - f_n, 1 - gy \rangle$, where $g = g_1 \cdots g_n$. Compute a Gröbner basis with respect to a lexicographic ordering where y and every t_i are greater than every x_i. Then the elements of the Gröbner basis not involving $y, t_1, \ldots, t_m$ define the smallest variety in k^n containing the parametrization. This algorithm is due to KALKBRENER (1990).

Let us see how this algorithm works for example (6). Let w be the new variable, so that

$$J = \langle vx - u^2, uy - v^2, z - u, 1 - uvw \rangle \subseteq k[w, u, v, x, y, z].$$

One easily calculates that $J_3 = J \cap k[x, y, z] = \langle x^2 y - z^3 \rangle$, so that $\mathbf{V}(x^2 y - z^3)$ is the variety determined by the parametrization (6). In the exercises, you will study how much of $\mathbf{V}(x^2 y - z^3)$ is filled up by the parametrization.

We should also mention that in practice, resultants are often used to solve the implicitization problem. Implicitization for curves and surfaces is discussed in AN-DERSON, GOLDMAN and SEDERBERG (1984a) and (1984b). Another reference is CANNY and MANOCHA (1992), which shows how implicitization of parametric surfaces can be done using multipolynomial resultants. A more recent reference is DICKENSTEIN and EMIRIS (2005).

EXERCISES FOR §3

1. In diagram (3) in the text, prove carefully that $F = \pi_m \circ i$ and $i(k^m) = V$.
2. When $k = \mathbb{C}$, the conclusion of Theorem 1 can be strengthened. Namely, one can show that there is a variety $W \subsetneq \mathbf{V}(I_m)$ such that $\mathbf{V}(I_m) \setminus W \subseteq F(\mathbb{C}^m)$. Prove this using the Closure Theorem.
3. Give an example to show that Exercise 2 is false over $\mathbb{R}$. Hint: t^2 is always positive.
4. In the text, we proved that over $\mathbb{C}$, the tangent surface to the twisted cubic is defined by the equation

$$g_7 = x^3 z - (3/4)x^2 y^2 - (3/2)xyz + y^3 + (1/4)z^2 = 0.$$

We want to show that the same is true over $\mathbb{R}$. If (x, y, z) is a real solution of the above equation, then we proved (using the Extension Theorem) that there are $t, u \in \mathbb{C}$ such that

$$x = t + u,$$
$$y = t^2 + 2tu,$$
$$z = t^3 + 3t^2 u.$$

Use the Gröbner basis given in the text to show that t and u are real. This will prove that (x, y, z) is on the tangent surface in $\mathbb{R}^3$. Hint: First show that u is real.
5. In the parametrization of the tangent surface of the twisted cubic, show that the parameters t and u are uniquely determined by x, y, and z. Hint: The argument is similar to what you did in Exercise 4.
6. Let S be the parametric surface defined by

$$x = uv,$$
$$y = u^2,$$
$$z = v^2.$$

   a. Find the equation of the smallest variety V that contains S.
   b. Over $\mathbb{C}$, use the Extension Theorem to prove that $S = V$. Hint: The argument is similar to what we did for the tangent surface of the twisted cubic.
   c. Over $\mathbb{R}$, show that S only covers "half" of V. What parametrization would cover the other "half"?
7. Let S be the parametric surface

$$x = uv,$$
$$y = uv^2,$$
$$z = u^2.$$

    a. Find the equation of the smallest variety V that contains S.

    b. Over $\mathbb{C}$, show that V contains points which are not on S. Determine exactly which points of V are not on S. Hint: Use lexicographic order with $u > v > x > y > z$.

8. The *Enneper surface* is defined parametrically by

$$x = 3u + 3uv^2 - u^3,$$
$$y = 3v + 3u^2v - v^3,$$
$$z = 3u^2 - 3v^2.$$

    a. Find the equation of the smallest variety V containing the Enneper surface. It will be a very complicated equation!

    b. Over $\mathbb{C}$, use the Extension Theorem to prove that the above equations parametrize the entire surface V. Hint: There are a lot of polynomials in the Gröbner basis. Keep looking—you will find what you need.

9. The *Whitney umbrella surface* is given parametrically by

$$x = uv,$$
$$y = v,$$
$$z = u^2.$$

A picture of this surface is:

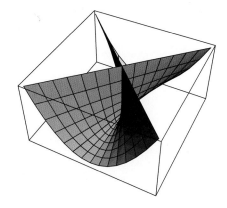

    a. Find the equation of the smallest variety containing the Whitney umbrella.

    b. Show that the parametrization fills up the variety over $\mathbb{C}$ but not over $\mathbb{R}$. Over $\mathbb{R}$, exactly what points are omitted?

    c. Show that the parameters u and v are not always uniquely determined by x, y, and z. Find the points where uniqueness fails and explain how your answer relates to the picture.

10. Consider the curve in $\mathbb{C}^n$ parametrized by $x_i = f_i(t)$, where $f_1, \dots, f_n$ are polynomials in $\mathbb{C}[t]$. This gives the ideal

$$I = \langle x_1 - f_1(t), \dots, x_n - f_n(t) \rangle \subseteq \mathbb{C}[t, x_1, \dots, x_n].$$

    a. Prove that the parametric equations fill up all of the variety $\mathbf{V}(I_1) \subseteq \mathbb{C}^n$.

    b. Show that the conclusion of part (a) may fail if we let $f_1 \dots, f_n$ be rational functions. Hint: See §3 of Chapter 1.

    c. Even if all of the f_i's are polynomials, show that the conclusion of part (a) may fail if we work over $\mathbb{R}$.

11. This problem is concerned with the proof of Theorem 2.
    a. Take $h \in k[x_1, \ldots, x_n]$ and let f_i, g_i be as in the theorem with $g = g_1 \cdots g_n$. Show that if N is sufficiently large, then there is $F \in k[t_1, \ldots, t_m, x_1, \ldots, x_n]$ such that $g^N h = F(t_1, \ldots, t_m, g_1 x_1, \ldots, g_n x_n)$.
    b. Divide F from part (a) by $x_1 - f_1, \ldots, x_n - f_n$. Then, in this division, replace x_i with $g_i x_i$ to obtain (10).
    c. Let k be an infinite field and let $f, g \in k[t_1, \ldots, t_m]$. Assume that $g \neq 0$ and that f vanishes on $k^m \setminus \mathbf{V}(g)$. Prove that f is the zero polynomial. Hint: Consider fg.
    d. Complete the proof of Theorem 2 using ideas from the proof of Theorem 1.

12. Consider the parametrization (6) given in the text. For simplicity, let $k = \mathbb{C}$. Also let $I = \langle vx - u^2, uy - v^2, z - u \rangle$ be the ideal obtained by "clearing denominators."
    a. Show that $I_2 = \langle z(x^2 y - z^3) \rangle$.
    b. Show that the smallest variety in $\mathbb{C}^5$ containing $i(\mathbb{C}^2 \setminus W)$ [see diagram (8)] is the variety $\mathbf{V}(vx - u^2, uy - v^2, z - u, x^2 y - z^3, vz - xy)$. Hint: Show that $i(\mathbb{C}^2 \setminus W) = \pi_1(\mathbf{V}(J))$, and then use the Closure Theorem.
    c. Show that $\{(0, 0, x, y, 0) \mid x, y \in \mathbb{C}\} \subseteq \mathbf{V}(I)$ and conclude that $\mathbf{V}(I)$ is *not* the smallest variety containing $i(\mathbb{C}^2 \setminus W)$.
    d. Determine exactly which portion of $x^2 y = z^3$ is parametrized by (6).

13. Given a rational parametrization as in (7), there is one case where the naive ideal $I = \langle g_1 x_1 - f_1, \ldots, g_n x_n - f_n \rangle$ obtained by "clearing denominators" gives the right answer. Suppose that $x_i = f_i(t)/g_i(t)$ where there is only one parameter t. We can assume that for each i, $f_i(t)$ and $g_i(t)$ are relatively prime in $k[t]$ (so in particular, they have no common roots). If $I \subseteq k[t, x_1, \ldots, x_n]$ is as above, then prove that $\mathbf{V}(I_1)$ is the smallest variety containing $F(k \setminus W)$, where as usual $g = g_1 \cdots g_n \in k[t]$ and $W = \mathbf{V}(g) \subseteq k$. Hint: In diagram (8), show that $i(k \setminus W) = \mathbf{V}(I)$, and adapt the proof of Theorem 2.

14. The *folium of Descartes* can be parametrized by

$$x = \frac{3t}{1 + t^3},$$

$$y = \frac{3t^2}{1 + t^3}.$$

    a. Find the equation of the folium. Hint: Use Exercise 13.
    b. Over $\mathbb{C}$ or $\mathbb{R}$, show that the above parametrization covers the entire curve.

15. In Exercise 16 to §3 of Chapter 1, we studied the parametric equations over $\mathbb{R}$

$$x = \frac{(1-t)^2 x_1 + 2t(1-t)w x_2 + t^2 x_3}{(1-t)^2 + 2t(1-t)w + t^2},$$

$$y = \frac{(1-t)^2 y_1 + 2t(1-t)w y_2 + t^2 y_3}{(1-t)^2 + 2t(1-t)w + t^2},$$

where $w, x_1, y_1, x_2, y_2, x_3, y_3$ are constants and $w > 0$. By eliminating t, show that these equations describe a portion of a conic section. Recall that a conic section is described by an equation of the form

$$ax^2 + bxy + cy^2 + dx + ey + f = 0.$$

Hint: In most computer algebra systems, the Gröbner basis command allows polynomials to have coefficients involving symbolic constants like $w, x_1, y_1, x_2, y_2, x_3, y_3$.

§4   Singular Points and Envelopes

In this section, we will discuss two topics from geometry:
- the *singular points* on a curve,
- the *envelope* of a family of curves.

Our goal is to show how geometry provides interesting equations that can be solved by the techniques studied in §§1 and 2.

   We will introduce some of the basic ideas concerning singular points and envelopes, but our treatment will be far from complete. One could write an entire book on these topics [see, for example, BRUCE and GIBLIN (1992)]. Also, our discussion of envelopes will not be fully rigorous. We will rely on some ideas from calculus to justify what is going on.

Singular Points

Suppose that we have a curve in the plane k^2 defined by $f(x, y) = 0$, where $f \in k[x, y]$. We expect that $\mathbf{V}(f)$ will have a well-defined tangent line at most points, although this may fail where the curve crosses itself or has a kink. Here are two examples:

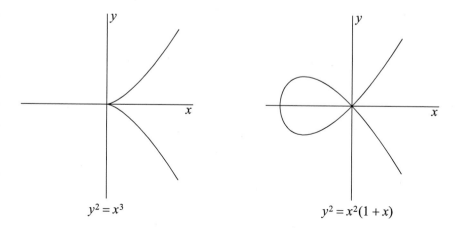

$$y^2 = x^3 \qquad\qquad\qquad y^2 = x^2(1 + x)$$

If we demand that a tangent line be unique and follow the curve on both sides of the point, then each of these curves has a point where there is no tangent. Intuitively, a *singular point* of $\mathbf{V}(f)$ is a point such as above where the tangent line fails to exist.

   To make this notion more precise, we first must give an algebraic definition of tangent line. We will use the following approach. Given a point $(a, b) \in \mathbf{V}(f)$, a line L through (a, b) is given parametrically by

(1)
$$\begin{aligned} x &= a + ct, \\ y &= b + dt. \end{aligned}$$

This line goes through (a, b) when $t = 0$. Notice also that $(c, d) \neq (0, 0)$ is a vector parallel to the line. Thus, by varying (c, d), we get all lines through (a, b). But how do we find the one that is *tangent* to $\mathbf{V}(f)$? Can we find it without using calculus?

Let us look at an example. Consider the line L

(2)
$$x = 1 + ct,$$
$$y = 1 + dt,$$

through the point $(1, 1)$ on the parabola $y = x^2$:

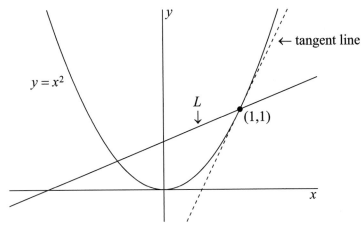

From calculus, we know that the tangent line has slope 2, which corresponds to the line with $d = 2c$ in the above parametrization. To find this line by algebraic means, we will study the *polynomial* that describes how the line meets the parabola. If we substitute (2) into the left-hand side of $y - x^2 = 0$, we get the polynomial

(3) $g(t) = 1 + dt - (1 + ct)^2 = -c^2 t^2 + (d - 2c)t = t(-c^2 t + d - 2c).$

The roots of g determine where the line intersects the parabola (be sure you understand this). If $d \neq 2c$, then g has two distinct roots when $c \neq 0$ and one root when $c = 0$. But if $d = 2c$, then g has a root of multiplicity 2. Thus, we can detect when the line (2) is tangent to the parabola by looking for a *multiple root*.

Based on this example, let us make the following definition.

Definition 1. Let m be a positive integer. Suppose that we have a point $(a, b) \in \mathbf{V}(f)$ and let L be the line through (a, b). Then L **meets** $\mathbf{V}(f)$ **with multiplicity** m at (a, b) if L can be parametrized as in (1) so that $t = 0$ is a root of multiplicity m of the polynomial $g(t) = f(a + ct, b + dt)$.

In this definition, note that $g(0) = f(a, b) = 0$, so that $t = 0$ is a root of g. Also, recall that $t = 0$ is a root of multiplicity m when $g = t^m h$, where $h(0) \neq 0$. One ambiguity with this definition is that a given line has many different parametrizations. So we need to check that the notion of multiplicity is independent of the parametrization. This will be covered in the exercises.

For an example of how this definition works, consider the line given by (2) above. It should be clear from (3) that the line meets the parabola $y = x^2$ with multiplicity 1 at $(1,1)$ when $d \neq 2c$ and with multiplicity 2 when $d = 2c$. Other examples will be given in the exercises.

We will use the notion of multiplicity to pick out the tangent line. To make this work, we will need the *gradient vector* of f, which is defined to be

$$\nabla f = \left(\frac{\partial f}{\partial x}, \frac{\partial f}{\partial y} \right).$$

We can now state our result.

Proposition 2. *Let $f \in k[x,y]$, and let $(a,b) \in \mathbf{V}(f)$.*

(i) *If $\nabla f(a,b) \neq (0,0)$, then there is a unique line through (a,b) which meets $\mathbf{V}(f)$ with multiplicity ≥ 2.*

(ii) *If $\nabla f(a,b) = (0,0)$, then every line through (a,b) meets $\mathbf{V}(f)$ with multiplicity ≥ 2.*

Proof. Let a line L through (a,b) be parametrized as in equation (1) and let $g(t) = f(a+ct, b+dt)$. Since $(a,b) \in \mathbf{V}(f)$, it follows that $t = 0$ is a root of g. The following observation will be proved in the exercises:

$$(4) \qquad t = 0 \text{ is a root of } g \text{ of multiplicity} \geq 2 \iff g'(0) = 0.$$

Using the chain rule, one sees that

$$g'(t) = \frac{\partial f}{\partial x}(a+ct, b+dt) \cdot c + \frac{\partial f}{\partial y}(a+ct, b+dt) \cdot d,$$

and thus

$$g'(0) = \frac{\partial f}{\partial x}(a,b) \cdot c + \frac{\partial f}{\partial y}(a,b) \cdot d.$$

If $\nabla f(a,b) = (0,0)$, then the above equation shows that $g'(0)$ always equals 0. By (4), it follows that L always meets $\mathbf{V}(f)$ at (a,b) with multiplicity ≥ 2. This proves the second part of the proposition. Turning to the first part, suppose that $\nabla f(a,b) \neq (0,0)$. We know that $g'(0) = 0$ if and only if

$$(5) \qquad \frac{\partial f}{\partial x}(a,b) \cdot c + \frac{\partial f}{\partial y}(a,b) \cdot d = 0.$$

This is a linear equation in the unknowns c and d. Since the coefficients $\frac{\partial}{\partial x}f(a,b)$ and $\frac{\partial}{\partial y}f(a,b)$ are not both zero, the solution space is 1-dimensional. Thus, there is $(c_0, d_0) \neq (0,0)$ such that (c,d) satisfies the above equation if and only if $(c,d) = \lambda(c_0, d_0)$ for some $\lambda \in k$. It follows that the (c,d)'s giving $g'(0) = 0$ all parametrize the same line L. This shows that there is a unique line which meets $\mathbf{V}(f)$ at (a,b) with multiplicity ≥ 2. Proposition 2 is proved. $\qquad\square$

Using Proposition 2, it is now obvious how to define the tangent line. From the second part of the proposition, it is also clear what a singular point should be.

Definition 3. Let $f \in k[x, y]$ and let $(a, b) \in \mathbf{V}(f)$.

(i) If $\nabla f(a, b) \neq (0, 0)$, then the **tangent line** of $\mathbf{V}(f)$ at (a, b) is the unique line through (a, b) which meets $\mathbf{V}(f)$ with multiplicity ≥ 2. We say that (a, b) is a **nonsingular point** of $\mathbf{V}(f)$.

(ii) If $\nabla f(a, b) = (0, 0)$, then we say that (a, b) is a **singular point** of $\mathbf{V}(f)$.

Over $\mathbb{R}$, the tangent line and the gradient have the following geometric interpretation. If the tangent to $\mathbf{V}(f)$ at (a, b) is parametrized by (1), then the vector (c, d) is parallel to the tangent line. But we also know from equation (5) that the dot product $\nabla f(a, b) \cdot (c, d)$ is zero, which means that $\nabla f(a, b)$ is perpendicular to (c, d). Thus, we have an algebraic proof of the theorem from calculus that *the gradient $\nabla f(a, b)$ is perpendicular to the tangent line of* $\mathbf{V}(f)$ *at* (a, b).

For any given curve $\mathbf{V}(f)$, we can compute the singular points as follows. The gradient ∇f is zero when $\frac{\partial}{\partial x} f$ and $\frac{\partial}{\partial y} f$ vanish simultaneously. Since we also have to be on $\mathbf{V}(f)$, we also need $f = 0$. It follows that the singular points of $\mathbf{V}(f)$ are determined by the equations

$$f = \frac{\partial f}{\partial x} = \frac{\partial f}{\partial y} = 0.$$

As an example, consider the curve $y^2 = x^2(1 + x)$ shown earlier. To find the singular points, we must solve

$$f = y^2 - x^2 - x^3 = 0,$$
$$\frac{\partial f}{\partial x} = -2x - 3x^2 = 0,$$
$$\frac{\partial f}{\partial y} = 2y = 0.$$

From these equations, it is easy to see that $(0, 0)$ is the only singular point of $\mathbf{V}(f)$. This agrees with the earlier picture.

Using the methods learned in §§1 and 2, we can tackle much more complicated problems. For example, later in this section we will determine the singular points of the curve defined by the sixth degree equation

$$0 = -1156 + 688x^2 - 191x^4 + 16x^6 + 544y + 30x^2y - 40x^4y$$
$$+ 225y^2 - 96x^2y^2 + 16x^4y^2 - 136y^3 - 32x^2y^3 + 16y^4.$$

The exercises will explore some other aspects of singular points. In Chapter 9, we will study singular and nonsingular points on an arbitrary affine variety.

Envelopes

In our discussion of envelopes, we will work over $\mathbb{R}$ to make the geometry easier to see. The best way to explain what we mean by envelope is to compute an example. Let $t \in \mathbb{R}$, and consider the circle in $\mathbb{R}^2$ defined by the equation

$$(6) \qquad (x - t)^2 + (y - t^2)^2 = 4.$$

Since (t, t^2) parametrizes a parabola, we can think of equation (6) as describing the *family* of circles of radius 2 in $\mathbb{R}^2$ whose centers lie on the parabola $y = x^2$. The picture is as follows:

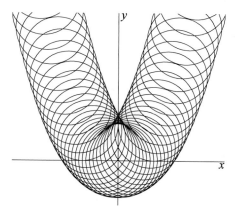

A Family of Circles in the Plane

Note that the "boundary" curve is *simultaneously tangent* to all the circles in the family. This is a special case of the *envelope* of a family of curves. The basic idea is that the envelope of a family of curves is a single curve that is tangent to all of the curves in the family. Our goal is to study envelopes and learn how to compute them. In particular, we want to find the equation of the envelope in the above example.

Before we can give a more careful definition of envelope, we must first understand the concept of a *family* of curves in $\mathbb{R}^2$.

Definition 4. Given a polynomial $F \in \mathbb{R}[x, y, t]$, fix a real number $t \in \mathbb{R}$. Then the variety in $\mathbb{R}^2$ defined by $F(x, y, t) = 0$ is denoted $\mathbf{V}(F_t)$, and the **family of curves** determined by F consists of the varieties $\mathbf{V}(F_t)$ as t varies over $\mathbb{R}$.

In this definition, we think of t as a *parameter* that tells us which curve in the family we are looking at. Strictly speaking, we should say "family of varieties" rather than "family of curves," but we will use the latter to emphasize the geometry of the situation.

For another example of a family and its envelope, consider the curves defined by

$$(7) \qquad F(x, y, t) = (x - t)^2 - y + t = 0.$$

Writing this as $y - t = (x - t)^2$, we see in the picture below that (7) describes the family $\mathbf{V}(F_t)$ of parabolas obtained by translating the standard parabola $y = x^2$ along the straight line $y = x$. In this case, the envelope is clearly the straight line that just touches each parabola in the family. This line has slope 1, and from here, it is easy to check that the envelope is given by $y = x - 1/4$ (the details are left as an exercise).

Not all envelopes are so easy to describe. The remarkable fact is that we can characterize the envelope in the following completely algebraic way.

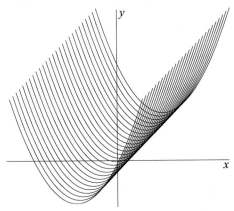

A Family of Parabolas in the Plane

Definition 5. Given a family $\mathbf{V}(F_t)$ of curves in $\mathbb{R}^2$, its **envelope** consists of all points $(x, y) \in \mathbb{R}^2$ with the property that

$$F(x, y, t) = 0,$$
$$\frac{\partial F}{\partial t}(x, y, t) = 0$$

for some $t \in \mathbb{R}$.

We need to explain how this definition corresponds to the intuitive idea of envelope. The argument given below is not rigorous, but it does explain where the condition on $\frac{\partial}{\partial t}F$ comes from. A complete treatment of envelopes requires a fair amount of theoretical machinery. We refer the reader to Chapter 5 of BRUCE and GIBLIN (1992) for more details.

Given a family $\mathbf{V}(F_t)$, we think of the envelope as a curve C with the property that at each point on the curve, C is tangent to one of the curves $\mathbf{V}(F_t)$ in the family. Suppose that C is parametrized by

$$x = f(t),$$
$$y = g(t).$$

We will assume that at time t, the point $(f(t), g(t))$ is on the curve $\mathbf{V}(F_t)$. This ensures that C meets all the members of the family. Algebraically, this means that

(8) $$F(f(t), g(t), t) = 0 \quad \text{for all } t \in \mathbb{R}.$$

But when is C *tangent* to $\mathbf{V}(F_t)$ at $(f(t), g(t))$? This is what is needed for C to be the envelope of the family. We know from calculus that the tangent vector to C is $(f'(t), g'(t))$. As for $\mathbf{V}(F_t)$, we have the gradient $\nabla F = \left(\frac{\partial}{\partial x} F, \frac{\partial}{\partial y} F \right)$, and from the first part of this section, we know that ∇F is perpendicular to the tangent line to $\mathbf{V}(F_t)$. Thus, for C to be tangent to $\mathbf{V}(F_t)$, the tangent $(f'(t), g'(t))$ must be perpendicular to the gradient ∇F. In terms of dot products, this means that $\nabla F \cdot (f'(t), g'(t)) = 0$ or, equivalently,

(9) $$\frac{\partial F}{\partial x}(f(t), g(t), t) \cdot f'(t) + \frac{\partial F}{\partial y}(f(t), g(t), t) \cdot g'(t) = 0.$$

We have thus shown that the envelope is determined by conditions (8) and (9). To relate this to Definition 5, differentiate (8) with respect to t. Using the chain rule, we get

$$\frac{\partial F}{\partial x}(f(t), g(t), t) \cdot f'(t) + \frac{\partial F}{\partial y}(f(t), g(t), t) \cdot g'(t) + \frac{\partial F}{\partial t}(f(t), g(t), t) = 0.$$

If we subtract equation (9) from this, we obtain

(10) $$\frac{\partial F}{\partial t}(f(t), g(t), t) = 0.$$

From (8) and (10), it follows that $(x, y) = (f(t), g(t))$ has exactly the property described in Definition 5.

As we will see later in the section, the above discussion of envelopes is rather naive. For us, the main consequence of Definition 5 is that the envelope of $\mathbf{V}(F_t)$ is determined by the equations

$$F(x, y, t) = 0,$$
$$\frac{\partial F}{\partial t}(x, y, t) = 0.$$

Note that x and y tell us where we are on the envelope and t tells us which curve in the family we are tangent to. Since these equations involve x, y, and t, we need to eliminate t to find the equation of the envelope. Thus, we can apply the theory from §§1 and 2 to study the envelope of a family of curves.

Let us see how this works in example (6). Here, $F = (x - t)^2 + (y - t^2)^2 - 4$, so that the envelope is described by the equations

(11) $$F = (x - t)^2 + (y - t^2)^2 - 4 = 0,$$
$$\frac{\partial F}{\partial t} = -2(x - t) - 4t(y - t^2) = 0.$$

Using lexicographic order with $t > x > y$, a Gröbner basis is given by the five polynomials

$$g_1 = -1156 + 688x^2 - 191x^4 + 16x^6 + 544y + 30x^2y - 40x^4y$$
$$+ 225y^2 - 96x^2y^2 + 16x^4y^2 - 136y^3 - 32x^2y^3 + 16y^4,$$

$$g_2 = (7327 - 1928y - 768y^2 - 896y^3 + 256y^4)t + 6929x - 2946x^3$$
$$+ 224x^5 + 2922xy - 1480x^3y + 128x^5y - 792xy^2 - 224x^3y^2$$
$$- 544xy^3 + 128x^3y^3 - 384xy^4,$$

$$g_3 = (431x - 12xy - 48xy^2 - 64xy^3)t + 952 - 159x^2 - 16x^4 + 320y$$
$$- 214x^2y + 32x^4y - 366y^2 - 32x^2y^2 - 80y^3 + 32x^2y^3 + 32y^4,$$

$$g_4 = (697 - 288x^2 + 108y - 336y^2 + 64y^3)t + 23x - 174x^3$$
$$+ 32x^5 + 322xy - 112x^3y + 32xy^2 + 32x^3y^2 - 96xy^3,$$

$$g_5 = 135t^2 + (26x + 40xy + 32xy^2)t - 128 + 111x^2$$
$$- 16x^4 + 64y + 8x^2y + 32y^2 - 16x^2y^2 - 16y^3.$$

We have written the Gröbner basis as polynomials in t with coefficients in $\mathbb{R}[x, y]$. The Elimination Theorem tells us that g_1 generates the first elimination ideal. Thus, the envelope lies on the curve $g_1 = 0$. Here is a picture of $\mathbf{V}(g_1)$ together with the parabola $y = x^2$:

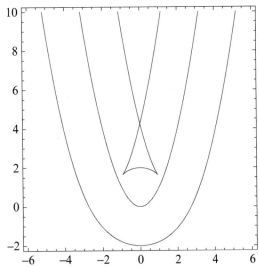

The surprise is the "triangular" portion of the graph that was somewhat unclear in the earlier picture of the family. By drawing some circles centered on the parabola, you can see how the triangle is still part of the envelope.

We have proved that the envelope lies on $\mathbf{V}(g_1)$, but the two may not be equal. In fact, there are two interesting questions to ask at this point:

- Is every point of $\mathbf{V}(g_1)$ on the envelope? This is the same as asking if every partial solution (x, y) of (11) extends to a complete solution (x, y, t).
- Given a point on the envelope, how many curves in the family are tangent to the envelope at the point? This asks how many t's are there for which (x, y) extends to (x, y, t).

Since the leading coefficient of t in g_5 is the constant 135, the Extension Theorem (in the form of Corollary 4 of §1) guarantees that *every* partial solution extends, provided we work over the complex numbers. Thus, t exists, but it might be complex. This illustrates the power and limitation of the Extension Theorem: it can guarantee that there is a solution, but it might lie in the wrong field.

In spite of this difficulty, the equation $g_5 = 0$ does have something useful to tell us: it is quadratic in t, so that a given (x, y) extends in at most *two* ways to a complete solution. Thus, *a point on the envelope of* (6) *is tangent to at most two circles in the family.* Can you see any points where there are two tangent circles?

To get more information on what is happening, let us look at the other polynomials in the Gröbner basis. Note that g_2, g_3, and g_4 involve t only to the first power. Thus, we can write them in the form

$$g_i = A_i(x, y)t + B_i(x, y), \quad i = 2, 3, 4.$$

If A_i does not vanish at (x, y) for one of $i = 2, 3, 4$, then we can solve $A_i t + B_i = 0$ to obtain

$$t = -\frac{B_i(x, y)}{A_i(x, y)}.$$

Thus, we see that t is real whenever x and y are. More importantly, this formula shows that t is uniquely determined when $A_i(x, y) \neq 0$. Thus, *a point on the envelope of* (6) *not in* $\mathbf{V}(A_2, A_3, A_4)$ *is tangent to exactly one circle in the family.*

It remains to understand where A_2, A_3, and A_4 vanish simultaneously. These polynomials might look complicated, but, using the techniques of §1, one can show that the real solutions of $A_2 = A_3 = A_4 = 0$ are

(12) $\qquad\qquad (x, y) = (0, 17/4)$ and $(\pm 0.936845, \ 1.63988)$.

Looking back at the picture of $\mathbf{V}(g_1)$, it appears that these are the singular points of $\mathbf{V}(g_1)$. Can you see the two circles tangent at these points? From the first part of this section, we know that the singular points of $\mathbf{V}(g_1)$ are determined by the equations $g_1 = \frac{\partial}{\partial x} g_1 = \frac{\partial}{\partial y} g_1 = 0$. Thus, to say that the singular points coincide with (12) means that

(13) $\qquad\qquad \mathbf{V}(A_2, A_3, A_4) = \mathbf{V}\left(g_1, \frac{\partial g_1}{\partial x}, \frac{\partial g_1}{\partial y}\right).$

To prove this, we will show that

(14)
$$g_1, \frac{\partial g_1}{\partial x}, \frac{\partial g_1}{\partial y} \in \langle A_2, A_3, A_4 \rangle,$$
$$A_2^2, A_3^2, A_4^2 \in \left\langle g_1, \frac{\partial g_1}{\partial x}, \frac{\partial g_1}{\partial y} \right\rangle.$$

The proof of (14) is a straightforward application of the ideal membership algorithm discussed in Chapter 2. For the first line, one computes a Gröbner basis of

$\langle A_2, A_3, A_4 \rangle$ and then applies the algorithm for the ideal membership problem to each of $g_1, \frac{\partial}{\partial x} g_1, \frac{\partial}{\partial y} g_1$ (see §8 of Chapter 2). The second line of (14) is treated similarly—the details will be left as an exercise.

Since (13) follows immediately from (14), we have proved that *a nonsingular point on* $\mathbf{V}(g_1)$ *is in the envelope of* (6) *and, at such a point, the envelope is tangent to exactly one circle in the family.* Also note that the singular points of $\mathbf{V}(g_1)$ are the most interesting points in the envelope, for they are the ones where there are two tangent circles. This last observation shows that singular points are not always bad—they can be a useful indication that something unusual is happening. An important part of algebraic geometry is devoted to the study of singular points.

In this example, equations (11) for the envelope were easy to write down. But to understand the equations, we had to use a Gröbner basis and the Elimination and Extension Theorems. Even though the Gröbner basis looked complicated, it told us exactly which points on the envelope were tangent to more than one circle. This illustrates nicely the power of the theory we have developed so far.

As we said earlier, our treatment of envelopes has been a bit naive. Evidence of this comes from the above example, which shows that the envelope can have singularities. How can the envelope be "tangent" to a curve in the family at a singular point? In the exercises, we will indicate another reason why our discussion has been too simple. We have also omitted the fascinating relation between the family of curves $\mathbf{V}(F_t) \subseteq \mathbb{R}^2$ and the surface $\mathbf{V}(F) \subseteq \mathbb{R}^3$ defined by $F(x, y, t) = 0$. We refer the reader to Chapter 5 of BRUCE and GIBLIN (1992) for a more complete treatment of these aspects of envelopes.

EXERCISES FOR §4

1. Let C be the curve in k^2 defined by $x^3 - xy + y^2 = 1$ and note that $(1,1) \in C$. Now consider the straight line parametrized by

$$x = 1 + ct,$$
$$y = 1 + dt.$$

Compute the multiplicity of this line when it meets C at $(1,1)$. What does this tell you about the tangent line? Hint: There will be two cases to consider.

2. In Definition 1, we need to show that the notion of multiplicity is independent of how the line is parametrized.
   a. Show that two parametrizations

$$x = a + ct, \quad x = a + c't,$$
$$y = b + dt, \quad y = b + d't,$$

   correspond to the same line if and only if there is a nonzero number $\lambda \in k$ such that $(c, d) = \lambda(c', d')$. Hint: In the parametrization $x = a + ct$, $y = b + dt$ of a line L, recall that L is parallel to the vector (c, d).
   b. Suppose that the two parametrizations of part (a) correspond to the same line L that meets $\mathbf{V}(f)$ at (a, b). Prove that the polynomials $g(t) = f(a + ct, b + dt)$ and $g'(t) = f(a + c't, b + d't)$ have the same multiplicity at $t = 0$. Hint: Use part (a) to relate g and g'. This will prove that the multiplicity of how L meets $\mathbf{V}(f)$ at (a, b) is well defined.

3. Consider the straight lines

$$x = t,$$
$$y = b + t.$$

These lines have slope 1 and y-intercept b. For which values of b is the line tangent to the circle $x^2 + y^2 = 2$? Draw a picture to illustrate your answer. Hint: Consider $g(t) = t^2 + (b+t)^2 - 2$. The roots of this quadratic determine the values of t where the line meets the circle.

4. If $(a, b) \in \mathbf{V}(f)$ and $\nabla f(a, b) \neq (0, 0)$, prove that the tangent line of $\mathbf{V}(f)$ at (a, b) is defined by the equation

$$\frac{\partial}{\partial x} f(a, b) \cdot (x - a) + \frac{\partial}{\partial y} f(a, b) \cdot (y - b) = 0.$$

5. Let $g \in k[t]$ be a polynomial such that $g(0) = 0$. Assume that $\mathbb{Q} \subset k$.
   a. Prove that $t = 0$ is a root of multiplicity ≥ 2 of g if and only if $g'(0) = 0$. Hint: Write $g(t) = th(t)$ and use the product rule.
   b. More generally, prove that $t = 0$ is a root of multiplicity $\geq m$ if and only if $g'(0) = g''(0) = \cdots = g^{(m-1)}(0) = 0$.
6. As in the Definition 1, let a line L be parametrized by (1), where $(a, b) \in \mathbf{V}(f)$. Also let $g(t) = f(a + ct, b + dt)$. Prove that L meets $\mathbf{V}(f)$ with multiplicity m if and only if $g'(0) = g''(0) = \cdots = g^{(m-1)}(0) = 0$ but $g^{(m)}(0) \neq 0$. Hint: Use the previous exercise.
7. In this exercise, we will study how a tangent line can meet a curve with multiplicity *greater* than 2. Let C be the curve defined by $y = f(x)$, where $f \in k[x]$. Thus, C is just the graph of f.
   a. Give an algebraic proof that the tangent line to C at $(a, f(a))$ is parametrized by

$$x = a + t,$$
$$y = f(a) + f'(a)t.$$

Hint: Consider $g(t) = f(a) + f'(a)t - f(a + t)$.
   b. Show that the tangent line at $(a, f(a))$ meets the curve with multiplicity ≥ 3 if and only if $f''(a) = 0$. Hint: Use the previous exercise.
   c. Show that the multiplicity is exactly 3 if and only if $f''(a) = 0$ but $f'''(a) \neq 0$.
   d. Over $\mathbb{R}$, a *point of inflection* is defined to be a point where $f''(x)$ changes sign. Prove that if the multiplicity is 3, then $(a, f(a))$ is a point of inflection.
8. In this problem, we will compute some singular points.
   a. Show that $(0, 0)$ is the *only* singular point of $y^2 = x^3$.
   b. In Exercise 8 of §3 of Chapter 1, we studied the curve $y^2 = cx^2 - x^3$, where c is some constant. Find all singular points of this curve and explain how your answer relates to the picture of the curve given in Chapter 1.
   c. Show that the circle $x^2 + y^2 = a^2$ in $\mathbb{R}^2$ has no singular points when $a > 0$.
9. One use of multiplicities is to show that one singularity is "worse" than another.
   a. For the curve $y^2 = x^3$, show that most lines through the origin meet the curve with multiplicity exactly 2.
   b. For $x^4 + 2xy^2 + y^3 = 0$, show that all lines through the origin meet the curve with multiplicity ≥ 3.

This suggests that the singularity at the origin is "worse" on the second curve. Using the ideas behind this exercise, one can define the notion of the *multiplicity* of a singular point.

10. We proved in the text that $(0, 0)$ is a singular point of the curve C defined by $y^2 = x^2(1 + x)$. But in the picture of C, it looks like there are two "tangent" lines through the origin. Can we use multiplicities to pick these out?

   a. Show that with two exceptions, all lines through the origin meet C with multiplicity 2. What are the lines that have multiplicity 3?

   b. Explain how your answer to part (a) relates to the picture of C in the text. Why should the "tangent" lines have higher multiplicity?

11. The four-leaved rose is defined in polar coordinates by the equation $r = \sin(2\theta)$:

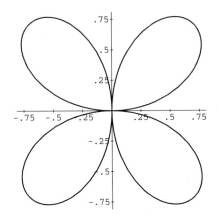

In Cartesian coordinates, this curve is defined by the equation $(x^2 + y^2)^3 = 4x^2y^2$.

   a. Show that most lines through the origin meet the rose with multiplicity 4 at the origin. Can you give a geometric explanation for this number?

   b. Find the lines through the origin that meet the rose with multiplicity > 4. Give a geometric explanation for the numbers you get.

12. Consider a surface $V(f) \subseteq k^3$ defined by $f \in k[x, y, z]$.

   a. Define what it means for $(a, b, c) \in V(f)$ to be a singular point.

   b. Determine all singular points of the sphere $x^2 + y^2 + z^2 = 1$. Does your answer make sense?

   c. Determine all singular points on the surface $V(x^2 - y^2z^2 + z^3)$. How does your answer relate to the picture of the surface drawn in §2 of Chapter 1?

13. Consider the family of curves given by $F = xy - t \in \mathbb{R}[x, y, t]$. Draw several of the curves $V(F_t)$. Be sure to include a picture of $V(F_0)$. What is the envelope of this family?

14. This problem will study the envelope of the family $F = (x - t)^2 - y + t$ considered in example (7).

   a. It is obvious that the envelope is a straight line of slope 1. Use elementary calculus to show that the line is $y = x - 1/4$.

   b. Use Definition 5 to compute the envelope.

   c. Find a parametrization of the envelope so that at time t, the point $(f(t), g(t))$ is on the parabola $V(F_t)$. Note that this is the kind of parametrization used in our discussion of Definition 5.

15. This problem is concerned with the envelope of example (6).

   a. Copy the picture in the text onto a sheet of paper and draw in the two tangent circles for each of the points in (12).

   b. For the point $(0, 4.25) = (0, 17/4)$, find the exact values of the t's that give the two tangent circles.

   c. Show that the exact values of the points (12) are given by

$$\left(0, \tfrac{17}{4}\right) \quad \text{and} \quad \left(\pm \sqrt{15 + 6\sqrt[3]{2} - 12\sqrt[3]{4}}, \tfrac{1}{4}(-1 + 6\sqrt[3]{2})\right).$$

Hint: Most computer algebra systems have commands to factor polynomials and solve cubic equations.

16. Consider the family determined by $F = (x - t)^2 + y^2 - (1/2)t^2$.
    a. Compute the envelope of this family.
    b. Draw a picture to illustrate your answer.

17. Consider the family of circles defined by $(x - t)^2 + (y - t^2)^2 = t^2$ in the plane $\mathbb{R}^2$.
    a. Compute the equation of the envelope of this family and show that the envelope is the union of two varieties.
    b. Use the Extension Theorem and a Gröbner basis to determine, for each point in the envelope, how many curves in the family are tangent to it. Draw a picture to illustrate your answer. Hint: You will use a different argument for each of the two curves making up the envelope.

18. Prove (14) using the hints given in the text. Also show that $A_2 \notin \langle g_1, \frac{\partial}{\partial x}g_1, \frac{\partial}{\partial y}g_1 \rangle$. This shows that the ideals $\langle g_1, \frac{\partial}{\partial x}g_1, \frac{\partial}{\partial y}g_1 \rangle$ and $\langle A_2, A_3, A_4 \rangle$ are not equal, even though they define the same variety.

19. In this exercise, we will show that our definition of envelope is too naive.
    a. Given a family of circles of radius 1 with centers lying on the x-axis, draw a picture to show that the envelope consists of the lines $y = \pm 1$.
    b. Use Definition 5 to compute the envelope of the family given by $F = (x-t)^2 + y^2 - 1$. Your answer should not be surprising.
    c. Use Definition 5 to find the envelope when the family is $F = (x - t^3)^2 + y^2 - 1$. Note that one of the curves in the family is part of the answer. This is because using t^3 allows the curves to "bunch up" near $t = 0$, which forces $\mathbf{V}(F_0)$ to appear in the envelope.

    In our intuitive discussion of envelope, recall that we assumed we could parametrize the envelope so that $(f(t), g(t))$ was in $\mathbf{V}(F_t)$ at time t. This presumes that the envelope is tangent to *different* curves in the family. Yet in the example given in part (c), part of the envelope lies in the *same* curve in the family. Thus, our treatment of envelope was too simple.

20. Suppose we have a family of curves in $\mathbb{R}^2$ determined by $F \in \mathbb{R}[x, y, t]$. Some of the curves $\mathbf{V}(F_t)$ may have singular points whereas others may not. Can we find the ones that have a singularity?
    a. By considering the equations $F = \frac{\partial}{\partial x}F = \frac{\partial}{\partial y}F = 0$ in $\mathbb{R}^3$ and using elimination theory, describe a procedure for determining those t's corresponding to curves in the family which have a singular point.
    b. Apply the method of part (a) to find the curves in the family of Exercise 13 that have singular points.

§5 Gröbner Bases and the Extension Theorem

The final task of Chapter 3 is to prove the Extension Theorem. We give two proofs, one using Gröbner bases in this section and a second using resultants in the next. The proofs are very different, which means that §§5 and 6 can be read independently of each other.

In the Extension Theorem, we have an ideal $I \subseteq k[x_1, \dots, x_n]$ with elimination ideal $I_1 = I \cap k[x_2, \dots, x_n]$. Given a partial solution $\mathbf{a} = (a_2, \dots, a_n) \in \mathbf{V}(I_1)$, the goal is to find $a_1 \in k$ such that $(a_1, \mathbf{a}) \in \mathbf{V}(I)$, i.e, to extend $\mathbf{a}$ to a solution of the original system.

Before beginning the proof, we introduce some notation. If $f \in k[x_1, \dots, x_n]$ is nonzero, we write f in the form

$$f = c_f(x_2, \ldots, x_n)x_1^N + \text{terms in which } x_1 \text{ has degree} < N,$$

where $N \geq 0$ and $c_f \in k[x_2, \ldots, x_n]$ is nonzero. We also define $\deg(f, x_1) = N$ and set $c_f = 0$ when $f = 0$.

Here are two properties of $\deg(f, x_1)$ and c_f that we will need. The proof will be left to the reader (Exercise 1).

Lemma 1. *Assume that* $f = \sum_{j=1}^{t} A_j g_j$ *is a standard representation for lex order with* $x_1 > \cdots > x_n$. *Then:*

(i) $\deg(f, x_1) \geq \deg(A_j g_j, x_1)$ *whenever* $A_j g_j \neq 0$.
(ii) $c_f = \sum_{\deg(A_j g_j, x_1) = N} c_{A_j} c_{g_j}$, *where* $N = \deg(f, x_1)$.

The first main result of this section tells us how lex Gröbner bases interact with nicely behaved partial solutions.

Theorem 2. *Let* $G = \{g_1, \ldots, g_t\}$ *be a Gröbner basis of* $I \subseteq k[x_1, \ldots, x_n]$ *for lex order with* $x_1 > \cdots > x_n$. *For each* $1 \leq j \leq t$, *let* $c_j = c_{g_j}$, *so that*

$$g_j = c_j(x_2, \ldots, x_n)x_1^{N_j} + \text{terms in which } x_1 \text{ has degree} < N_j,$$

where $N_j \geq 0$ *and* $c_j \in k[x_2, \ldots, x_n]$ *is nonzero. Assume* $\mathbf{a} = (a_2, \ldots, a_n) \in \mathbf{V}(I_1)$ *is a partial solution with the property that* $\mathbf{a} \notin \mathbf{V}(c_1, \ldots, c_t)$. *Then*

(1) $\{f(x_1, \mathbf{a}) \mid f \in I\} = \langle g_o(x_1, \mathbf{a}) \rangle \subseteq k[x_1],$

where $g_o \in G$ *satisfies* $c_o(\mathbf{a}) \neq 0$ *and* g_o *has minimal* x_1-*degree among all elements* $g_j \in G$ *with* $c_j(\mathbf{a}) \neq 0$. *Furthermore:*

(i) $\deg(g_o(x_1, \mathbf{a})) > 0$.
(ii) *If* $g_o(a_1, \mathbf{a}) = 0$ *for* $a_1 \in k$, *then* $(a_1, \mathbf{a}) \in \mathbf{V}(I)$.

In this theorem, we use the letter "o" for the index to indicate that g_o is *optimal* with respect to the partial solution $\mathbf{a} \in \mathbf{V}(I_1)$.

Proof. Choose an optimal $g_o \in G$. We first prove (i). If $\deg(g_o(x_1, \mathbf{a})) = 0$, then $\deg(g_o, x_1) = 0$ since $c_o(\mathbf{a}) \neq 0$. Thus $g_o \in I_1$ and $c_o = g_o$. But then $\mathbf{a} \in \mathbf{V}(I_1)$ implies $c_o(\mathbf{a}) = g_o(\mathbf{a}) = 0$, a contradiction. This proves that $\deg(g_o(x_1, \mathbf{a})) > 0$.

Next observe that (ii) is an immediate consequence of (1) since $g_o(a_1, \mathbf{a}) = 0$ and (1) imply that $f(a_1, \mathbf{a}) = 0$ for all $f \in I$, proving that $(a_1, \mathbf{a}) \in \mathbf{V}(I)$.

It remains to prove (1), which is the hardest part of the proof. We begin with the function

(2) $k[x_1, \ldots, x_n] \longrightarrow k[x_1]$

defined by evaluation at $\mathbf{a}$, i.e., $f(x_1, x_2, \ldots, x_n) \mapsto f(x_1, \mathbf{a})$. Since evaluation is compatible with addition and multiplication of polynomials, (2) is what we will call a *ring homomorphism* in Chapter 5, §2. In Exercise 2, you will show that the image of I under the ring homomorphism (2) is an ideal of $k[x_1]$. Furthermore, since I is generated by the $g_j \in G$, Exercise 3 implies that the image of I is generated by the $g_j(x_1, \mathbf{a})$. Hence (1) will follow once we show $g_j(x_1, \mathbf{a}) \in \langle g_0(x_1, \mathbf{a}) \rangle$ for all $g_j \in G$.

We will use a clever argument given in SCHAUENBURG (2007). The proof has two steps:

Step 1: Prove that $g_j(x_1, \mathbf{a}) = 0$ when $g_j \in G$ satisfies $\deg(g_j, x_1) < \deg(g_0, x_1)$.

Step 2: Prove that $g_j(x_1, \mathbf{a}) \in \langle g_0(x_1, \mathbf{a}) \rangle$ by induction on $\deg(g_j, x_1)$, with Step 1 as the base case.

For Step 1, we set $d_o = \deg(g_0, x_1)$. Our choice of g_0 implies that g_0 does not drop x_1-degree when evaluated at $\mathbf{a}$, but any $g_j \in G$ with $\deg(g_j, x_1) < d_o$ either drops x_1-degree or vanishes identically when evaluated at $\mathbf{a}$.

Suppose there are some $g_j \in G$ with $\deg(g_j, x_1) < d_o$ and $g_j(x_1, \mathbf{a}) \neq 0$. Among all such g_j's, let g_b be one that minimizes the degree drop when evaluated at $\mathbf{a}$. We use the letter "b" for the index to indicate that g_b is *bad* with respect to evaluation. Our goal is to show the existence of this bad g_b leads to a contradiction.

If we set $\delta = \deg(g_b, x_1) - \deg(g_b(x_1, \mathbf{a}))$, then g_b drops degree by δ when evaluated at $\mathbf{a}$, and any other $g_j \in G$ with $\deg(g_j, x_1) < d_o$ either vanishes identically or drops degree by at least δ when evaluated at $\mathbf{a}$.

Let $d_b = \deg(g_b, x_1)$, so that $\deg(g_b(x_1, \mathbf{a})) = d_b - \delta$. Then consider

$$S = c_o x_1^{d_o - d_b} g_b - c_b g_0 \in I$$
$$= c_o x_1^{d_o - d_b} (c_b x_1^{d_b} + \cdots) - c_b (c_o x_1^{d_o} + \cdots).$$

The second line makes it clear that $\deg(S, x_1) < d_o$. We will compute $\deg(S(x_1, \mathbf{a}))$ in two ways and show that this leads to a contradiction.

The first way to compute the degree is to evaluate directly at $\mathbf{a}$. This gives

$$S(x_1, \mathbf{a}) = c_o(\mathbf{a}) x_1^{d_o - d_b} g_b(x_1, \mathbf{a}) - c_b(\mathbf{a}) g_0(x_1, \mathbf{a}) = c_o(\mathbf{a}) x_1^{d_o - d_b} g_b(x_1, \mathbf{a})$$

since $c_b(\mathbf{a}) = 0$ (g_b drops degree when evaluated at $\mathbf{a}$). Then $c_o(\mathbf{a}) \neq 0$ (by the definition of g_0) implies that

$$(3) \qquad \deg(S(x_1, \mathbf{a})) = d_o - d_b + \deg(g_b(x_1, \mathbf{a})) = d_o - d_b + d_b - \delta = d_o - \delta.$$

The second way begins with a standard representation $S = \sum_{j=1}^{t} B_j g_j$. When we combine part (i) of Lemma 1 with $\deg(S, x_1) < d_o$, we get the inequality

$$(4) \qquad \deg(B_j, x_1) + \deg(g_j, x_1) = \deg(B_j g_j, x_1) \leq \deg(S, x_1) < d_o$$

when $B_j \neq 0$. Thus the g_j's that appear satisfy $\deg(g_j, x_1) < d_o$, so that either $g_j(x_1, \mathbf{a}) = 0$ or the x_1-degree of g_j drops by at least δ when we evaluate at $\mathbf{a}$. Hence

$$\deg(B_j(x_1, \mathbf{a})) + \deg(g_j(x_1, \mathbf{a})) \leq \deg(B_j, x_1) + \deg(g_j, x_1) - \delta < d_o - \delta,$$

where the final inequality uses (4). It follows that when we evaluate $S = \sum_{j=1}^{t} B_j g_j$ at $\mathbf{a}$, we obtain

$$\deg(S(x_1, \mathbf{a})) \leq \max(\deg(B_j(x_1, \mathbf{a})) + \deg(g_j(x_1, \mathbf{a}))) < d_o - \delta.$$

This contradicts (3), and Step 1 is proved.

For Step 2, we prove by induction on $\deg(g_j, x_1)$ that $g_j(x_1, \mathbf{a}) \in \langle g_0(x_1, \mathbf{a}) \rangle$ for all $g_j \in G$. The base case is when $\deg(g_j, x_1) < d_o$. Here, Step 1 implies that $g_j(x_1, \mathbf{a}) = 0$, which obviously lies in $\langle g_0(x_1, \mathbf{a}) \rangle$.

For the inductive step, fix $d \geq d_o$ and assume that the assertion is true for all $g_j \in G$ with $\deg(g_j, x_1) < d$. Then take $g_j \in G$ with $\deg(g_j, x_1) = d$ and consider the polynomial

$$S = c_0 g_j - c_j x_1^{d-d_o} g_0 \in I$$
$$= c_0(c_j x_1^d + \cdots) - c_j x_1^{d-d_o}(c_0 x_1^{d_o} + \cdots).$$

The second line makes it clear that $\deg(S, x_1) < d$.

Taking a standard representation $S = \sum_{\ell=1}^t B_\ell g_\ell$ and arguing as in (4) shows that $\deg(g_\ell, x_1) < d$ when $B_\ell \neq 0$. By our inductive hypothesis, we have $g_\ell(x_1, \mathbf{a}) \in \langle g_0(x_1, \mathbf{a}) \rangle$ when $B_\ell \neq 0$. Then

$$c_0 g_j = c_j x_1^{d-d_o} g_0 + S = c_j x_1^{d-d_o} g_0 + \sum_{\ell=1}^t B_\ell g_\ell$$

implies that

$$c_0(\mathbf{a}) g_j(x_1, \mathbf{a}) = c_j(\mathbf{a}) x_1^{d-d_o} g_0(x_1, \mathbf{a}) + \sum_{\ell=1}^t B_\ell(x_1, \mathbf{a}) g_\ell(x_1, \mathbf{a}) \in \langle g_0(x_1, \mathbf{a}) \rangle.$$

Since $c_0(\mathbf{a}) \neq 0$, we conclude that $g_j(x_1, \mathbf{a}) \in \langle g_0(x_1, \mathbf{a}) \rangle$. This proves the inductive step and completes the proof of Step 2. □

In concrete terms, equation (1) of Theorem 2 has the following interpretation. Suppose that $I = \langle f_1, \ldots, f_s \rangle$, so that $\mathbf{V}(I)$ is defined by $f_1 = \cdots = f_s = 0$. If $\mathbf{a} \in \mathbf{V}(I_1)$ is a partial solution satisfying the hypothesis of the theorem, then when we evaluate the system at $\mathbf{a}$, it reduces to the *single* equation $g_0(x_1, \mathbf{a}) = 0$, and we can find g_0 by computing a lex Gröbner basis of I.

For an example of how this works, consider $I = \langle x^2 y + xz + 1, xy - xz^2 + z - 1 \rangle$ in $\mathbb{C}[x, y, z]$. A Gröbner basis G of I for lex order with $x > y > z$ consists of the four polynomials

(5)
$$g_1 = x^2 z^2 + x + 1,$$
$$g_2 = xy - xz^2 + z - 1 = (y - z^2)x + z - 1,$$
$$g_3 = xz^3 - xz^2 - y + z^2 + z - 1 = (z^3 - z^2)x - y + z^2 + z - 1,$$
$$g_4 = y^2 - 2yz^2 - yz + y + 2z^4 - z^3.$$

Thus $I_1 = \langle g_4 \rangle = \langle y^2 - 2yz^2 - yz + y + 2z^4 - z^3 \rangle \subseteq \mathbb{C}[y, z]$. In the notation of Theorem 2,

$$c_1 = z^2, \ c_2 = y - z^2, \ c_3 = z^2(z - 1), \ c_4 = g_4.$$

Since

$$\mathbf{V}(c_1, c_2, c_3, c_4) = \mathbf{V}(z^2, y - z^2, z^2(z-1), y^2 - 2yz^2 - yz + y + 2z^4 - z^3) = \{(0,0)\},$$

we see that if $(b, c) \in \mathbf{V}(I_1) \subseteq \mathbb{C}^2$ is a partial solution different from $(0,0)$, then

$$\{f(x, b, c) \mid f \in I\} = \langle g_0(x, b, c) \rangle \subseteq \mathbb{C}[x],$$

where $g_0 \in G$ is the polynomial from Theorem 2.

Let us compute g_0 for $(1, 1) \in \mathbf{V}(I_1)$. Here, $c_1(1, 1) \neq 0$, while c_2, c_3, c_4 all vanish at $(1, 1)$. Hence $g_0 = g_1$. Note that

$$g_2(x, 1, 1) = g_3(x, 1, 1) = g_4(x, 1, 1) = 0,$$

as predicted by Step 1 of the proof of Theorem 2 since $\deg(g_i, x) < 2 = \deg(g_1, x)$ for $i = 2, 3, 4$. Since $g_0 = g_1 = x^2 z^2 + x + 1$, we obtain

$$\{f(x, 1, 1) \mid f \in I\} = \langle g_1(x, 1, 1) \rangle = \langle x^2 + x + 1 \rangle.$$

For the partial solution $(0, \frac{1}{2}) \in \mathbf{V}(I_1)$, the situation is a bit different. Here, (5) gives

$$g_1(x, 0, \tfrac{1}{2}) = \tfrac{1}{4}x^2 + x + 1,$$
$$g_2(x, 0, \tfrac{1}{2}) = -\tfrac{1}{4}x - \tfrac{1}{2},$$
$$g_3(x, 0, \tfrac{1}{2}) = -\tfrac{1}{8}x - \tfrac{1}{4},$$
$$g_4(x, 0, \tfrac{1}{2}) = 0.$$

The third polynomial is a constant multiple of the second, so that g_0 can be taken to be g_2 or g_3. Then part (ii) of Theorem 2 tells us that when trying to extend $(0, \frac{1}{2}) \in \mathbf{V}(I_1)$ to a solution of the original system, *there is only one equation to solve*, namely $g_0(x, 0, \frac{1}{2}) = 0$. In particular, we can ignore the quadratic polynomial $g_1(x, 0, \frac{1}{2})$ since it is a multiple of $g_0(x, 0, \frac{1}{2})$ by equation (1) of Theorem 2.

We are now ready to give our first proof of the Extension Theorem. In the original statement of the theorem, we worked over the field $\mathbb{C}$. Here we prove a more general version that holds for any algebraically closed field.

Theorem 3 (The Extension Theorem). *Let $I = \langle f_1, \ldots, f_s \rangle \subseteq k[x_1, \ldots, x_n]$ and let I_1 be the first elimination ideal of I. For each $1 \leq i \leq s$, write f_i in the form*

$$f_i = c_i(x_2, \ldots, x_n) x_1^{N_i} + \text{terms in which } x_1 \text{ has degree} < N_i,$$

where $N_i \geq 0$ and $c_i \in k[x_2, \ldots, x_n]$ is nonzero. Suppose that we have a partial solution $(a_2, \ldots, a_n) \in \mathbf{V}(I_1)$. If $(a_2, \ldots, a_n) \notin \mathbf{V}(c_1, \ldots, c_s)$ and k is algebraically closed, then there exists $a_1 \in k$ such that $(a_1, \ldots, a_n) \in \mathbf{V}(I)$.

Proof. Let $G = \{g_1, \ldots, g_t\}$ be a lex Gröbner basis of I with $x_1 > \cdots > x_n$ and set $\mathbf{a} = (a_2, \ldots, a_n)$. We first show that there is $g_j \in G$ such that $c_{g_j}(\mathbf{a}) \neq 0$. Here, we are using the notation introduced in the discussion leading up to Lemma 1.

Our hypothesis implies that $c_i(\mathbf{a}) \neq 0$ for some i. Take a standard representation $f_i = \sum_{j=1}^{t} A_j g_j$ of f_i. Since $c_i = c_{f_i}$ and $N_i = \deg(f_i, x_1)$, part (ii) of Lemma 1 yields

$$c_i = \sum_{\deg(A_j g_j, x_1) = N_i} c_{A_j} c_{g_j}.$$

Then $c_i(\mathbf{a}) \neq 0$ implies that $c_{g_j}(\mathbf{a}) \neq 0$ for at least one g_j appearing in the sum.

Hence we can apply Theorem 2. This gives $g_0 \in G$ with $\deg(g_0(x_1, \mathbf{a})) > 0$ by part (i) of the theorem. Then there is $a_1 \in k$ with $g_0(a_1, \mathbf{a}) = 0$ since k is algebraically closed. By part (ii) of Theorem 2, $(a_1, \mathbf{a}) \in \mathbf{V}(I)$, and the Extension Theorem is proved.   □

EXERCISES FOR §5

1. As in Lemma 1, let $f = \sum_{j=1}^{t} A_j g_j$ be a standard representation and set $N = \deg(f, x_1)$.
    a. Prove that $N \geq \deg(A_j g_j, x_1)$ when $A_j g_j \neq 0$. Hint: Recall that $\text{multideg}(f) \geq \text{multideg}(A_j g_j)$ when $A_j g_j \neq 0$. Then explain why the first entry in $\text{multideg}(f)$ is $\deg(f, x_1)$.
    b. Prove that

    $$c_f = \sum_{\deg(A_j g_j, x_1) = N} c_{A_j} c_{g_j}.$$

    Hint: Use part (a) and compare the coefficients of x_1^N in $f = \sum_{j=1}^{t} A_j g_j$.

2. Suppose that k is a field and $\varphi : k[x_1, \ldots, x_n] \to k[x_1]$ is a ring homomorphism that is the identity on k and maps x_1 to x_1. Given an ideal $I \subseteq k[x_1, \ldots, x_n]$, prove that $\varphi(I) \subseteq k[x_1]$ is an ideal. (In the proof of Theorem 2, we use this result when φ is the map that evaluates x_i at a_i for $2 \leq i \leq n$.)

3. In the proof of Theorem 2, show that (1) follows from the assertion that $g_j(x_1, \mathbf{a}) \in \langle g_0(x_1, \mathbf{a}) \rangle$ for all $g_j \in G$.

4. This exercise will explore the example $I = \langle x^2 y + xz + 1, xy - xz^2 + z - 1 \rangle$ discussed in the text.
    a. Show that the partial solution $(b, c) = (0, 0)$ does not extend to a solution $(a, 0, 0) \in \mathbf{V}(I)$.
    b. In the text, we showed that $g_0 = g_1$ for the partial solution $(1, 1)$. Show that $g_0 = g_3$ works for all partial solutions different from $(1, 1)$ and $(0, 0)$.

5. Evaluation at $\mathbf{a}$ is sometimes called *specialization*. Given $I \subseteq k[x_1, \ldots, x_n]$ with lex Gröbner basis $G = \{g_1, \ldots, g_t\}$, we get the specialized basis $\{g_1(x_1, \mathbf{a}), \ldots, g_t(x_1, \mathbf{a})\}$. Discarding the polynomials the specialize to zero, we get $G' = \{g_j(x_1, \mathbf{a}) \mid g_j(x_1, \mathbf{a}) \neq 0\}$.
    a. Show that G' is a basis of the ideal $\{f(x_1, \mathbf{a}) \mid f \in I\} \subseteq k[x_1]$.
    b. If in addition $\mathbf{a} \in \mathbf{V}(I_1)$ is a partial solution satisfying the hypothesis of Theorem 2, prove that G' is a Gröbner basis of $\{f(x_1, \mathbf{a}) \mid f \in I\}$.
    The result of part (b) is an example of a *specialization theorem* for Gröbner bases. We will study the specialization of Gröbner bases in more detail in Chapter 6.

6. Show that Theorem 2 remains true if we replace lex order for $x_1 > \cdots > x_n$ with any monomial order for which x_1 is greater than all monomials in $x_2, \ldots, x_n$. This is an order of 1-elimination type in the terminology of Exercise 5 of §1. Hint: You will need to show that Lemma 1 of this section holds for such monomial orders.

7. Use the strategy explained in the discussion following Theorem 2 to find all solutions of the system of equations given in Example 3 of Chapter 2, §8.

§6 Resultants and the Extension Theorem

So far, we have presented elimination theory from the point of view of Gröbner bases. Here, we will prove the Extension Theorem using a classical approach to elimination theory based on the theory of resultants.

We introduce the concept of resultant by asking when two polynomials in $k[x]$ have a common factor. This might seem far removed from elimination, but we will see the connection by the end of the section, where we use resultants to construct elements of elimination ideals and give a second proof of the Extension Theorem. Resultants will reappear in §7 of Chapter 8 when we prove Bezout's Theorem.

This section can be read independently of §5.

Resultants

We begin with the question of whether two polynomials $f, g \in k[x]$ have a *common factor*, which means a polynomial $h \in k[x]$ of degree > 0 that divides f and g. One approach would be to compute the gcd of f and g using the Euclidean Algorithm from Chapter 1. A drawback is that the Euclidean Algorithm requires divisions in the field k. As we will see later, this is something we want to avoid when doing elimination. Is there a way of determining whether a common factor exists without doing any divisions in k? Here is a first answer.

Lemma 1. *Let $f, g \in k[x]$ be polynomials of degrees $l > 0$ and $m > 0$, respectively. Then f and g have a common factor in $k[x]$ if and only if there are polynomials $A, B \in k[x]$ such that:*

(i) *A and B are not both zero.*
(ii) *A has degree at most $m - 1$ and B has degree at most $l - 1$.*
(iii) *$Af + Bg = 0$.*

Proof. First, assume that f and g have a common factor $h \in k[x]$. Then $f = hf_1$ and $g = hg_1$, where $f_1, g_1 \in k[x]$. Note that f_1 has degree at most $l - 1$, and similarly $\deg(g_1) \le m - 1$. Then

$$g_1 \cdot f + (-f_1) \cdot g = g_1 \cdot hf_1 - f_1 \cdot hg_1 = 0.$$

and, thus, $A = g_1$ and $B = -f_1$ have the required properties.

Conversely, suppose that A and B have the above three properties. By (i), we may assume $B \ne 0$. If f and g have no common factor, then their gcd is 1, so we can find polynomials $\tilde{A}, \tilde{B} \in k[x]$ such that $\tilde{A}f + \tilde{B}g = 1$ (see Proposition 6 of Chapter 1, §5). Now multiply by B and use $Bg = -Af$:

$$B = B(\tilde{A}f + \tilde{B}g) = \tilde{A}Bf + \tilde{B}Bg = \tilde{A}Bf + \tilde{B}(-Af) = (\tilde{A}B - \tilde{B}A)f.$$

Since B is nonzero, this equation shows that B has degree at least l, contradicting (ii). Hence, A and B must have a common factor of positive degree. $\qquad\square$

The answer given by Lemma 1 may not seem very satisfactory, for we still need to decide whether the required A and B exist. Remarkably, we can use *linear algebra* to answer this last question. The idea is to turn $Af + Bg = 0$ into a system of linear equations. We begin by writing

$$A = u_0 x^{m-1} + \cdots + u_{m-1},$$
$$B = v_0 x^{l-1} + \cdots + v_{l-1},$$

where for now we will regard the $l + m$ coefficients $u_0, \ldots, u_{m-1}, v_0, \ldots, v_{l-1}$ as unknowns. Our goal is to find $u_i, v_i \in k$, not all zero, so that the equation

(1) $Af + Bg = 0$

holds. Note that this will automatically give us A and B as required in Lemma 1.

To get a system of linear equations, let us also write out f and g:

$$f = c_0 x^l + \cdots + c_l, \quad c_0 \neq 0,$$
$$g = d_0 x^m + \cdots + d_m, \quad d_0 \neq 0,$$

where $c_i, d_i \in k$. If we substitute the formulas for f, g, A, and B into equation (1) and compare the coefficients of powers of x, then we get the following system of linear equations with unknowns u_i, v_i and coefficients c_i, d_i in k:

(2)
$$
\begin{array}{llll}
c_0 u_0 & + \quad d_0 v_0 & = 0 & \text{coefficient of } x^{l+m-1} \\
c_1 u_0 + c_0 u_1 & + \quad d_1 v_0 + d_0 v_1 & = 0 & \text{coefficient of } x^{l+m-2} \\
\quad \ddots & \qquad \ddots & \ \vdots & \qquad \vdots \\
c_l u_{m-1} & + \qquad d_m v_{l-1} & = 0 & \text{coefficient of } x^0.
\end{array}
$$

Since there are $l + m$ linear equations and $l + m$ unknowns, we know from linear algebra that there is a nonzero solution if and only if the coefficient matrix has zero determinant. This leads to the following definition.

Definition 2. Given nonzero polynomials $f, g \in k[x]$ of degree l, m, respectively, write them as

$$f = c_0 x^l + \cdots + c_l, \quad c_0 \neq 0,$$
$$g = d_0 x^m + \cdots + d_m, \quad d_0 \neq 0.$$

If $l, m > 0$, then the **Sylvester matrix** of f and g with respect to x, denoted $\mathrm{Syl}(f, g, x)$, is the coefficient matrix of the system of equations given in (2). Thus, $\mathrm{Syl}(f, g, x)$ is the following $(l + m) \times (l + m)$ matrix:

$$\text{Syl}(f,g,x) = \begin{pmatrix} c_0 & & & & d_0 & & & \\ c_1 & c_0 & & & d_1 & d_0 & & \\ c_2 & c_1 & \ddots & & d_2 & d_1 & \ddots & \\ \vdots & & \ddots & c_0 & \vdots & & \ddots & d_0 \\ & \vdots & & c_1 & & \vdots & & d_1 \\ c_l & & & & d_m & & & \vdots \\ & c_l & & \vdots & & d_m & & \vdots \\ & & \ddots & & & & \ddots & \\ & & & c_l & & & & d_m \end{pmatrix},$$

$$\underbrace{\hspace{3cm}}_{m \text{ columns}} \quad \underbrace{\hspace{3cm}}_{l \text{ columns}}$$

where the empty spaces are filled by zeros. The **resultant** of f and g with respect to x, denoted $\text{Res}(f,g,x)$, is the determinant of the Sylvester matrix. Thus,

$$\text{Res}(f,g,x) = \det(\text{Syl}(f,g,x)).$$

Finally, when f,g do not both have positive degree, we define:

(3)
$$\begin{aligned} \text{Res}(c_0,g,x) &= c_0^m, \text{ when } c_0 \in k \setminus \{0\},\ m > 0, \\ \text{Res}(f,d_0,x) &= d_0^l, \text{ when } d_0 \in k \setminus \{0\},\ l > 0, \\ \text{Res}(c_0,d_0,x) &= 1, \text{ when } c_0, d_0 \in k \setminus \{0\}. \end{aligned}$$

To understand the formula for $\text{Res}(c_0,g,x)$, notice that $\text{Syl}(c_0,g,x)$ reduces to an $m \times m$ diagonal matrix with c_0's on the diagonal, and similarly for $\text{Res}(f,d_0,x)$.

From this definition, we get the following properties of the resultant. A polynomial is called an *integer polynomial* provided that all of its coefficients are integers.

Proposition 3. *Given nonzero* $f, g \in k[x]$, *the resultant* $\text{Res}(f,g,x) \in k$ *is an integer polynomial in the coefficients of* f *and* g. *Furthermore,* f *and* g *have a common factor in* $k[x]$ *if and only if* $\text{Res}(f,g,x) = 0$.

Proof. The determinant of an $s \times s$ matrix $A = (a_{ij})_{1 \le i,j \le s}$ is given by the formula

$$\det(A) = \sum_{\substack{\sigma \text{ a permutation} \\ \text{of } \{1,\dots,s\}}} \text{sgn}(\sigma)\, a_{1\sigma(1)} \cdot a_{2\sigma(2)} \cdots a_{s\sigma(s)},$$

where $\text{sgn}(\sigma)$ is $+1$ if σ interchanges an even number of pairs of elements of $\{1,\dots,s\}$ and -1 if σ interchanges an odd number of pairs (see Appendix A, §4 for more details). This shows that the determinant is an integer polynomial (in fact, the coefficients are ± 1) in its entries, and the first statement of the proposition then follows immediately from the definition of resultant when f, g have positive degree. The formulas (3) take care of the remaining cases.

The second statement is easy to prove when f, g have positive degree: the resultant is zero $\Leftrightarrow$ the coefficient matrix of equations (2) has zero determinant $\Leftrightarrow$ equations (2) have a nonzero solution. We observed earlier that this is equivalent to the existence of A and B as in Lemma 1, and then Lemma 1 completes the proof in this case.

When f or g is a nonzero constant, $\text{Res}(f, g, x) \neq 0$ by (3), and f and g cannot have a common factor since by definition common factors have positive degree. $\square$

As an example, let us see if $f = 2x^2 + 3x + 1$ and $g = 7x^2 + x + 3$ have a common factor in $\mathbb{Q}[x]$. One computes that

$$\text{Res}(f, g, x) = \det \begin{pmatrix} 2 & 0 & 7 & 0 \\ 3 & 2 & 1 & 7 \\ 1 & 3 & 3 & 1 \\ 0 & 1 & 0 & 3 \end{pmatrix} = 153 \neq 0,$$

so that there is no common factor.

Here is an important consequence of Proposition 3.

Corollary 4. *If $f, g \in \mathbb{C}[x]$ are nonzero, then $\text{Res}(f, g, x) = 0$ if and only if f and g have a common root in $\mathbb{C}$. Furthermore, we can replace $\mathbb{C}$ with any algebraically closed field k.*

Proof. Since $\mathbb{C}$ is algebraically closed, f and g have a common factor in $\mathbb{C}[x]$ of positive degree if and only if they have a common root in $\mathbb{C}$. Then we are done by Proposition 3. This argument clearly works over any algebraically closed field. $\square$

One disadvantage to using resultants is that large determinants are hard to compute. In the exercises, we will explain an alternate method for computing resultants that is similar to the Euclidean Algorithm. Most computer algebra systems have a resultant command that implements this algorithm.

To link resultants to elimination, let us compute the resultant of the polynomials $f = xy - 1$ and $g = x^2 + y^2 - 4$. Regarding f and g as polynomials in x whose coefficients are polynomials in y, we get

$$\text{Res}(f, g, x) = \det \begin{pmatrix} y & 0 & 1 \\ -1 & y & 0 \\ 0 & -1 & y^2 - 4 \end{pmatrix} = y^4 - 4y^2 + 1.$$

This eliminates x, but how does this relate to elimination that we did in §§1 and 2? In particular, is $\text{Res}(f, g, x) = y^4 - 4y^2 + 1$ in the first elimination ideal $\langle f, g \rangle \cap k[y]$? The full answer will come later in the section and will use the following result.

Proposition 5. *Given nonzero $f, g \in k[x]$, there are polynomials $A, B \in k[x]$ such that*

$$Af + Bg = \text{Res}(f, g, x).$$

Furthermore, if at least one of f, g has positive degree, then the coefficients of A and B are integer polynomials in the coefficients of f and g.

Proof. The definition of resultant was based on the equation $Af + Bg = 0$. In this proof, we will apply the same methods to the equation

$$(4) \qquad\qquad\qquad \tilde{A}f + \tilde{B}g = 1.$$

The reason for using $\tilde{A}, \tilde{B}$ rather than A, B will soon become apparent.

The proposition is trivially true if $\mathrm{Res}(f, g, x) = 0$ (simply choose $A = B = 0$). If $f = c_0 \in k$ and $m = \deg(g) > 0$, then by (3), we have

$$\mathrm{Res}(f, g, x) = c_0^m = c_0^{m-1} \cdot f + 0 \cdot g,$$

and the case $l = \deg(f) > 0$, $g = d_0 \in k$ is handled similarly.

Hence we may assume that f, g have positive degree and satisfy $\mathrm{Res}(f, g, x) \neq 0$. Now let

$$
\begin{aligned}
f &= c_0 x^l + \cdots + c_l, \quad c_0 \neq 0, \\
g &= d_0 x^m + \cdots + d_m, \quad d_0 \neq 0, \\
\tilde{A} &= u_0 x^{m-1} + \cdots + u_{m-1}, \\
\tilde{B} &= v_0 x^{l-1} + \cdots + v_{l-1},
\end{aligned}
$$

where $u_0, \ldots, u_{m-1}, v_0, \ldots, v_{l-1}$ are unknowns in k. Equation (4) holds if and only if substituting these formulas into (4) gives an equality of polynomials. Comparing coefficients of powers of x, we conclude that (4) is equivalent to the following system of linear equations with unknowns u_i, v_i and coefficients c_i, d_i in k:

$$(5)
\begin{array}{llll}
c_0 u_0 & + \quad d_0 v_0 & = 0 & \text{coefficient of } x^{l+m-1} \\
c_1 u_0 + c_0 u_1 & + \quad d_1 v_0 + d_0 v_1 & = 0 & \text{coefficient of } x^{l+m-2} \\
\quad \ddots & \quad\quad \ddots & \ \vdots & \qquad\qquad \vdots \\
c_l u_{m-1} & + \quad d_m v_{l-1} = 1 & & \text{coefficient of } x^0.
\end{array}
$$

These equations are the same as (2) except for the 1 on the right-hand side of the last equation. Thus, the coefficient matrix is the Sylvester matrix of f and g, and then $\mathrm{Res}(f, g, x) \neq 0$ guarantees that (5) has a unique solution in k.

In this situation, we use *Cramer's Rule* to give a *formula* for the unique solution. Cramer's Rule states that the ith unknown is a ratio of two determinants, where the denominator is the determinant of the coefficient matrix and the numerator is the determinant of the matrix where the ith column of the coefficient matrix has been replaced by the right-hand side of the equation. For a more precise statement of Cramer's rule, the reader should consult Appendix A, §4. In our case, Cramer's rule gives formulas for the u_i's and v_i's. For example, the first unknown u_0 is given by the formula

$$u_0 = \frac{1}{\text{Res}(f,g,x)} \det \begin{pmatrix} 0 & & & & d_0 & & \\ 0 & c_0 & & & \vdots & \ddots & \\ \vdots & \vdots & \ddots & & \vdots & & d_0 \\ 0 & c_l & & c_0 & d_m & & \vdots \\ \vdots & & \ddots & \vdots & & \ddots & \vdots \\ 1 & & & c_l & & & d_m \end{pmatrix}.$$

Since a determinant is an integer polynomial in its entries, it follows that

$$u_0 = \frac{\text{an integer polynomial in } c_i, d_i}{\text{Res}(f,g,x)}.$$

All of the u_i's and v_i's can be written this way. Since $\tilde{A} = u_0 x^{m-1} + \cdots + u_{m-1}$, we can pull out the common denominator $\text{Res}(f,g,x)$ and write $\tilde{A}$ in the form

$$\tilde{A} = \frac{1}{\text{Res}(f,g,x)} A,$$

where $A \in k[x]$ and the coefficients of A are integer polynomials in c_i, d_i. Similarly, we can write

$$\tilde{B} = \frac{1}{\text{Res}(f,g,x)} B,$$

where $B \in k[x]$ has the same properties as A. Since $\tilde{A}$ and $\tilde{B}$ satisfy $\tilde{A}f + \tilde{B}g = 1$, we can multiply through by $\text{Res}(f,g,x)$ to obtain

$$Af + Bg = \text{Res}(f,g,x).$$

Since A and B have the required kind of coefficients, the proposition is proved. $\square$

Most courses in linear algebra place little emphasis on Cramer's rule, mainly because Gaussian elimination is much more efficient than Cramer's rule from a computational point of view. But for theoretical uses, where one needs to worry about the *form* of the solution, Cramer's rule is very important.

We can now explain the relation between the resultant and the gcd. Given $f, g \in k[x]$, $\text{Res}(f,g,x) \neq 0$ tells us that f and g have no common factor, and hence their gcd is 1. Then Proposition 6 of Chapter 1, §5 says that there are $\tilde{A}$ and $\tilde{B}$ such that $\tilde{A}f + \tilde{B}g = 1$. As the above formulas for $\tilde{A}$ and $\tilde{B}$ make clear, the coefficients of $\tilde{A}$ and $\tilde{B}$ have a denominator given by the resultant. Then clearing these denominators leads to $Af + Bg = \text{Res}(f,g,x)$.

To see this more explicitly, let us return to the case of $f = xy - 1$ and $g = x^2 + y^2 - 4$. If we regard these as polynomials in x, then we computed that $\text{Res}(f,g,x) = y^4 - 4y^2 + 1 \neq 0$. Thus, their gcd is 1, and we leave it as an exercise to check that

$$-\left(\frac{y}{y^4 - 4y^2 + 1}x + \frac{1}{y^4 - 4y^2 + 1}\right)f + \frac{y^2}{y^4 - 4y^2 + 1}g = 1.$$

This equation takes place in $k(y)[x]$, i.e., the coefficients are *rational functions* in y, because the gcd theory from Chapter 1, §5 requires *field* coefficients. If we want to work in $k[x, y]$, we must clear denominators, which leads to the equation

$$(6) \qquad -(yx + 1)f + y^2 g = y^4 - 4y^2 + 1.$$

This, of course, is just a special case of Proposition 5. Hence, we can regard the resultant as a "denominator-free" version of the gcd.

The Extension Theorem

Our final task is to use resultants to give a second proof of the Extension Theorem. We begin with nonzero polynomials $f, g \in k[x_1, \ldots, x_n]$ of degree l, m in x_1, written

$$(7) \qquad \begin{aligned} f &= c_0 x_1^l + \cdots + c_l, \quad c_0 \neq 0, \\ g &= d_0 x_1^m + \cdots + d_m, \quad d_0 \neq 0, \end{aligned}$$

where now $c_i, d_i \in k[x_2, \ldots, x_n]$. We set $x_1\text{-deg}(f) = l$ and $x_1\text{-deg}(g) = m$. By Proposition 3,

$$\mathrm{Res}(f, g, x_1) = \det(\mathrm{Syl}(f, g, x_1))$$

is a polynomial in $k[x_2, \ldots, x_n]$ since $c_i, d_i \in k[x_2, \ldots, x_n]$. If at least one of f, g has positive x_1-degree, then Proposition 5 implies that

$$\mathrm{Res}(f, g, x_1) = Af + Bg$$

with $A, B \in k[x_1, \ldots, x_n]$. It follows that

$$(8) \qquad h = \mathrm{Res}(f, g, x_1) \in \langle f, g \rangle \cap k[x_2, \ldots, x_n].$$

Hence $\mathrm{Res}(f, g, x_1)$ lies in the first elimination ideal of $\langle f, g \rangle$. In particular, this answers the question posed before Proposition 5.

We will prove the Extension Theorem by studying the interaction between resultants and partial solutions. Substituting $\mathbf{a} = (a_2, \ldots, a_n) \in k^{n-1}$ into $h = \mathrm{Res}(f, g, x_1)$ gives a *specialization* of the resultant. However, this need not equal the resultant of the specialized polynomials $f(x_1, \mathbf{a})$ and $g(x_1, \mathbf{a})$. See Exercises 12 and 13 for some examples. Fortunately, there is one situation where the relation between the two resultants is easy to state.

Proposition 6. *Let nonzero $f, g \in k[x_1, \ldots, x_n]$ have x_1-degree l, m, respectively, and assume that $\mathbf{a} = (a_2, \ldots, a_n) \in k^{n-1}$ satisfies the following:*
(i) $f(x_1, \mathbf{a}) \in k[x_1]$ *has degree l.*
(ii) $g(x_1, \mathbf{a}) \in k[x_1]$ *is nonzero of degree $p \leq m$.*
If c_0 is as in (7), then the polynomial $h = \mathrm{Res}(f, g, x_1) \in k[x_2, \ldots, x_n]$ satisfies

$$h(\mathbf{a}) = c_0(\mathbf{a})^{m-p} \, \mathrm{Res}(f(x_1, \mathbf{a}), g(x_1, \mathbf{a}), x_1).$$

Proof. We first consider the case when $l, m > 0$. If we substitute $\mathbf{a} = (a_2, \ldots, a_n)$ for $x_2, \ldots, x_n$ in the determinantal formula for $h = \mathrm{Res}(f, g, x_1)$, we obtain

$$
(9) \qquad h(\mathbf{a}) = \det
\begin{pmatrix}
c_0(\mathbf{a}) & & & & d_0(\mathbf{a}) & & \\
\vdots & \ddots & & & \vdots & \ddots & \\
\vdots & & c_0(\mathbf{a}) & & \vdots & & d_0(\mathbf{a}) \\
c_l(\mathbf{a}) & & \vdots & & d_m(\mathbf{a}) & & \vdots \\
& \ddots & \vdots & & & \ddots & \vdots \\
& & c_l(\mathbf{a}) & & & & d_m(\mathbf{a})
\end{pmatrix}.
$$

$$
\underbrace{}_{m \text{ columns}} \quad \underbrace{}_{l \text{ columns}}
$$

If $g(x_1, \mathbf{a})$ has degree $p = m$, then our assumptions imply that

$$
\begin{aligned}
f(x_1, \mathbf{a}) &= c_0(\mathbf{a})x_1^l + \cdots + c_l(\mathbf{a}), & c_0(\mathbf{a}) \neq 0, \\
g(x_1, \mathbf{a}) &= d_0(\mathbf{a})x_1^m + \cdots + d_m(\mathbf{a}), & d_0(\mathbf{a}) \neq 0.
\end{aligned}
$$

Hence the above determinant is the resultant of $f(x_1, \mathbf{a})$ and $g(x_1, \mathbf{a})$, so that

$$
h(\mathbf{a}) = \mathrm{Res}(f(x_1, \mathbf{a}), g(x_1, \mathbf{a}), x_1).
$$

This proves the proposition when $p = m$. When $p < m$, the determinant (9) is no longer the resultant of $f(x_1, \mathbf{a})$ and $g(x_1, \mathbf{a})$ (it has the wrong size). Here, we get the desired resultant by repeatedly expanding by minors along the first row. We leave the details to the reader (see Exercise 14).

Next suppose that $l = 0$ and $m > 0$. Then $f(x_1, \mathbf{a}) = c_0(\mathbf{a}) \neq 0$, so by (3),

$$
\mathrm{Res}(f(x_1, \mathbf{a}), g(x_1, \mathbf{a}), x_1) =
\begin{cases}
c_0(\mathbf{a})^p & p > 0 \\
1 & p = 0.
\end{cases}
$$

Since the determinant (9) reduces to $h(\mathbf{a}) = c_0(\mathbf{a})^m$ when $l = 0$, the desired equation holds in this case. The proof when $l > 0$ and $m = 0$ is similar. Finally, when $l = m = 0$, our hypotheses imply that $c_0(\mathbf{a}), d_0(\mathbf{a}) \neq 0$, and then our desired equation reduces to $1 = 1$ by (3). $\qquad\square$

Over the complex numbers (or, more generally, over any algebraically closed field), we have the following corollary of Proposition 6.

Corollary 7. *Let nonzero $f, g \in \mathbb{C}[x_1, \ldots, x_n]$ have x_1-degree l, m, respectively, and assume that $\mathbf{a} = (a_2, \ldots, a_n) \in \mathbb{C}^{n-1}$ satisfies the following:*

(i) *$f(x_1, \mathbf{a}) \in k[x_1]$ has degree l or $g(x_1, \mathbf{a}) \in k[x_1]$ has degree m.*
(ii) *$\mathrm{Res}(f, g, x_1)(\mathbf{a}) = 0$.*

Then there exists $a_1 \in \mathbb{C}$ such that $f(a_1, \mathbf{a}) = g(a_1, \mathbf{a}) = 0$. Furthermore, we can replace $\mathbb{C}$ with any algebraically closed field k.

Proof. We give the proof in the case of an algebraically closed field k. Switching f and g if necessary, we may assume that $f(x_1, \mathbf{a})$ has degree l, so that $c_0(\mathbf{a}) \neq 0$. If $g(x_1, \mathbf{a}) \neq 0$, then Proposition 6 implies that

$$0 = \text{Res}(f, g, x_1)(\mathbf{a}) = c_0(\mathbf{a})^{m-p} \text{Res}(f(x_1, \mathbf{a}), g(x_1, \mathbf{a}), x_1),$$

where p is the degree of $g(x_1, \mathbf{a})$. Thus $\text{Res}(f(x_1, \mathbf{a}), g(x_1, \mathbf{a}), x_1) = 0$. Since $f(x_1, \mathbf{a}), g(x_1, \mathbf{a})$ are nonzero, the desired $a_1 \in k$ exists by Corollary 4.

It remains to study what happens when $g(x_1, \mathbf{a}) = 0$. If $l > 0$, then $f(x_1, \mathbf{a})$ has positive degree, so any root $a_1 \in k$ of $f(x_1, \mathbf{a})$ will work. On the other hand, if $l = 0$, then (3) implies that $\text{Res}(f, g, x_1) = c_0^m$, which does not vanish at $\mathbf{a}$ since $c_0(\mathbf{a}) \neq 0$. Hence this case cannot occur. $\qquad\square$

We can now prove the Extension Theorem. In §1, we stated the result over $\mathbb{C}$. Here is a more general version that holds over any algebraically closed field k.

Theorem 8 (The Extension Theorem). *Let $I = \langle f_1, \ldots, f_s \rangle \subseteq k[x_1, \ldots, x_n]$ and let I_1 be the first elimination ideal of I. For each $1 \leq i \leq s$, write f_i in the form*

$$f_i = c_i(x_2, \ldots, x_n) x_1^{N_i} + \text{ terms in which } x_1 \text{ has degree} < N_i,$$

where $N_i \geq 0$ and $c_i \in k[x_2, \ldots, x_n]$ is nonzero. Suppose that we have a partial solution $(a_2, \ldots, a_n) \in \mathbf{V}(I_1)$. If $(a_2, \ldots, a_n) \notin \mathbf{V}(c_1, \ldots, c_s)$ and k is algebraically closed, then there exists $a_1 \in k$ such that $(a_1, a_2, \ldots, a_n) \in \mathbf{V}(I)$.

Proof. The first part of the argument repeats material from the proof of Theorem 2 of §5. We reproduce it verbatim for the convenience of readers who skipped §5.

Set $\mathbf{a} = (a_2, \ldots, a_n)$ and consider the function

(10) $$k[x_1, \ldots, x_n] \longrightarrow k[x_1]$$

defined by evaluation at $\mathbf{a}$, i.e., $f(x_1, x_2, \ldots, x_n) \mapsto f(x_1, \mathbf{a})$. Since evaluation is compatible with addition and multiplication of polynomials, (10) is what we will call a *ring homomorphism* in Chapter 5, §2. In Exercise 15, you will show that the image of I under the ring homomorphism (10) is an ideal of $k[x_1]$. By Corollary 4 of Chapter 1, §5, it follows that the image of I is generated by a single polynomial $u(x_1) \in k[x_1]$. Thus

$$\{f(x_1, \mathbf{a}) \mid f \in I\} = \langle u(x_1) \rangle.$$

In particular, there is $f^* \in I$ such that $f^*(x_1, \mathbf{a}) = u_1(x_1)$. Then we can rewrite the above equation as

(11) $$\{f(x_1, \mathbf{a}) \mid f \in I\} = \langle f^*(x_1, \mathbf{a}) \rangle.$$

Since $\mathbf{a} \notin \mathbf{V}(c_1, \ldots, c_s)$, we have $c_i(\mathbf{a}) \neq 0$ for some i, and then $f_i(x_1, \mathbf{a})$ has positive x_1-degree since $\mathbf{a} \in \mathbf{V}(I_1)$. By (11), it follows that $f^*(x_1, \mathbf{a})$ is nonzero. We now apply our results to the polynomials $f_i, f^* \in I$. Since $f_i(x_1, \mathbf{a})$ has positive x_1-degree, (8) implies that

$$h = \mathrm{Res}(f_i, f^*, x_1) \in \langle f_i, f_* \rangle \cap k[x_2, \ldots, x_n] \subseteq I \cap k[x_2, \ldots, x_n] = I_1.$$

It follows that $h(\mathbf{a}) = 0$ since $\mathbf{a} \in \mathbf{V}(I_1)$. Since k is algebraically closed and f_i does not drop degree in x_1 when evaluated at $\mathbf{a}$, Corollary 7 gives us $a_1 \in k$ with $f_i(a_1, \mathbf{a}) = f^*(a_1, \mathbf{a}) = 0$. By (11), we see that $f(a_1, \mathbf{a}) = 0$ for all $f \in I$, so that $(a_1, \mathbf{a}) = (a_1, a_2, \ldots, a_n) \in \mathbf{V}(I)$. This completes the proof of the Extension Theorem.   □

   In addition to the resultant of two polynomials discussed here, the resultant of three or more polynomials can be defined. Readers interested in *multipolynomial resultants* should consult MACAULAY (1902) or VAN DER WAERDEN (1931). A modern introduction to this theory can be found in COX, LITTLE and O'SHEA (2005). A more treatment of resultants is presented in JOUANOLOU (1991), and a vast generalization of the concept of resultant is discussed in GELFAND, KAPRANOV and ZELEVINSKY (1994).

EXERCISES FOR §6

1. Compute the resultant of $x^5 - 3x^4 - 2x^3 + 3x^2 + 7x + 6$ and $x^4 + x^2 + 1$. Do these polynomials have a common factor in $\mathbb{Q}[x]$? Explain your reasoning.
2. If $f, g \in \mathbb{Z}[x]$, explain why $\mathrm{Res}(f, g, x) \in \mathbb{Z}$.
3. Assume that f has degree l and g has degree m. Here are some properties of the resultant. We assume that at least one of l, m is positive.
   a. Prove that the resultant has the symmetry property

   $$\mathrm{Res}(f, g, x) = (-1)^{lm} \mathrm{Res}(g, f, x).$$

   Be sure to take the cases $l = 0$ or $m = 0$ into account. Hint: A determinant changes sign if you switch two columns.
   b. If $\lambda, \mu \in k$ are nonzero, then prove that

   $$\mathrm{Res}(\lambda f, \mu g, x) = \lambda^m \mu^l \, \mathrm{Res}(f, g, x).$$

   Again, be sure to consider the cases when $l = 0$ or $m = 0$.
   c. When $l = m = 0$, show that part (a) is still true but part (b) can fail.
4. In §3, we mentioned that resultants are sometimes used to solve implicitization problems. For a simple example of how this works, consider the curve parametrized by

   $$u = \frac{t^2}{1 + t^2}, \quad v = \frac{t^3}{1 + t^2}.$$

   To get an implicit equation, form the equations

   $$u(1 + t^2) - t^2 = 0, \quad v(1 + t^2) - t^3 = 0$$

   and use an appropriate resultant to eliminate t. Then compare your result to the answer obtained by the methods of §3. (Note that Exercise 13 of §3 is relevant.)
5. Consider the polynomials $f = 2x^2 + 3x + 1$ and $g = 7x^2 + x + 3$.
   a. Use the Euclidean Algorithm (by hand, not computer) to find the gcd of these polynomials.
   b. Find polynomials $\tilde{A}, \tilde{B} \in \mathbb{Q}[x]$ such that $\tilde{A}f + \tilde{B}g = 1$. Hint: Use the calculations you made in part (a).

c. In the equation you found in part (b), clear the denominators. How does this answer relate to the resultant?

6. Let $f = xy - 1$ and $g = x^2 + y^2 - 4$. We will regard f and g as polynomials in x with coefficients in $k(y)$.

   a. With f and g as above, set up the system of equations (5) that describes $\tilde{A}f + \tilde{B}g = 1$. Hint: $\tilde{A}$ is linear and $\tilde{B}$ is constant. Thus, you should have three equations in three unknowns.

   b. Use Cramer's rule to solve the system of equations obtained in part (a). Hint: The denominator is the resultant.

   c. What equation do you get when you clear denominators in part (b)? Hint: See equation (6) in the text.

7. The last part of Proposition 5 requires that at least one of $f, g \in k[x]$ have positive degree in x. Working in $\mathbb{Q}[x]$, explain why $\text{Res}(2, 2, x)$ shows that this requirement is necessary.

8. The discussion of resultants in the text assumes that the polynomials f, g are nonzero. For some purposes, however, it is useful to let $f = 0$ or $g = 0$. We define

$$\text{Res}(0, g, x) = \text{Res}(f, 0, x) = 0$$

for any $f, g \in k[x]$. This definition will play an important role in Exercises 10 and 11. On the other hand, it is not compatible with some of the results proved in the text. Explain why $\text{Res}(1, 0, x) = 0$ implies that the "nonzero" hypothesis is essential in Corollary 4.

9. Let $f = c_0 x^l + \cdots + c_l$ and $g = d_0 x^m + \cdots + d_m$ be nonzero polynomials in $k[x]$, and assume that $c_0, d_0 \neq 0$ and $l \geq m$.

   a. Let $\tilde{f} = f - (c_0/d_0)x^{l-m}g$, so that $\deg(\tilde{f}) \leq l - 1$. If $\deg(\tilde{f}) = l - 1$, then prove that

$$\text{Res}(f, g, x) = (-1)^m d_0 \text{Res}(\tilde{f}, g, x).$$

   Hint: Use column operations on the Sylvester matrix. You will subtract c_0/d_0 times the first m columns in the g part from the columns in the f part. Then expand by minors along the first row. [See Section 11.4 of Dummit and Foote (2004) for a description of expansion by minors, also called cofactor expansion.]

   b. Let $\tilde{f}$ be as in part (a), but this time we allow the possibility that the degree of $\tilde{f}$ could be strictly smaller than $l - 1$. If $\tilde{f} \neq 0$, prove that

$$\text{Res}(f, g, x) = (-1)^{m(l-\deg(\tilde{f}))} d_0^{l-\deg(\tilde{f})} \text{Res}(\tilde{f}, g, x).$$

   Hint: The exponent $l - \deg(\tilde{f})$ tells you how many times to expand by minors.

   c. Now use the division algorithm to write $f = qg + r$ in $k[x]$, where $\deg(r) < \deg(g)$ or $r = 0$. In the former case, use part (b) to prove that

$$\text{Res}(f, g, x) = (-1)^{m(l-\deg(r))} d_0^{l-\deg(r)} \text{Res}(r, g, x).$$

   d. If the remainder r vanishes, Exercise 8 implies that $\text{Res}(r, g, x) = \text{Res}(0, g, x) = 0$. Explain why the formula of part (c) correctly computes $\text{Res}(f, g, x)$ in this case. Hint: Proposition 3.

10. In this exercise and the next, we will modify the Euclidean Algorithm to give an algorithm for computing resultants. The basic idea is the following: to find the gcd of f and g, we used the division algorithm to write $f = qg + r, g = q'r + r'$, etc. In equation (5) of Chapter 1, §5, the equalities

$$\gcd(f, g) = \gcd(g, r) = \gcd(r, r') = \cdots$$

enabled us to compute the gcd since the degrees were decreasing. Use Exercises 3 and 9 to prove the following "resultant" version of the first two equalities above:

$$\mathrm{Res}(f,g,x) = (-1)^{\deg(g)(\deg(f)-\deg(r))} d_0^{\deg(f)-\deg(r)} \mathrm{Res}(r,g,x)$$

$$= (-1)^{\deg(f)\deg(g)} d_0^{\deg(f)-\deg(r)} \mathrm{Res}(g,r,x)$$

$$= (-1)^{\deg(f)\deg(g)+\deg(r)(\deg(g)-\deg(r'))} d_0^{\deg(f)-\deg(r)} d_0'^{\deg(g)-\deg(r')} \mathrm{Res}(r',r,x)$$

$$= (-1)^{\deg(f)\deg(g)+\deg(g)\deg(r)} d_0^{\deg(f)-\deg(r)} d_0'^{\deg(g)-\deg(r')} \mathrm{Res}(r,r',x)$$

where d_0 (resp. d_0') is the leading coefficient of g (resp. r). Continuing in this way, we can reduce to the case where the second polynomial is constant, and then we can use (3) to compute the resultant.

11. To turn the previous exercises into an algorithm, we will use pseudocode and two functions: let $r = \mathrm{remainder}(f,g)$ be the remainder on division of f by g and let $\mathrm{LC}(f)$ be the leading coefficient of f. We can now state the algorithm for finding $\mathrm{Res}(f,g,x)$:

> Input : $f,\ g \in k[x] \setminus \{0\}$
> Output : $res = \mathrm{Res}(f,g,x)$
>
> $h := f$
> $s := g$
> $res := 1$
> WHILE $\deg(s) > 0$ DO
> $r := \mathrm{remainder}(h,s)$
> $res := (-1)^{\deg(h)\deg(s)} \mathrm{LC}(s)^{\deg(h)-\deg(r)} res$
> $h := s$
> $s := r$
> IF $h = 0$ or $s = 0$ THEN $res := 0$ ELSE
> IF $\deg(h) > 0$ THEN $res := s^{\deg(h)} res$
> RETURN res

Prove that this algorithm computes the resultant of f and g. Hint: Use (3) and Exercises 3, 8, 9, and 10, and follow the proof of Proposition 6 of Chapter 1, §5.

12. In the discussion leading up to Proposition 6, we claimed that the specialization of a resultant need not be the resultant of the specialized polynomials. Let us work out some examples.

     a. Let $f = x^2 y + 3x - 1$ and $g = 6x^2 + y^2 - 4$. Compute $h = \mathrm{Res}(f,g,x)$ and show that $h(0) = -180$. But if we set $y = 0$ in f and g, we get the polynomials $3x - 1$ and $6x^2 - 4$. Check that $\mathrm{Res}(3x-1, 6x^2-4, x) = -30$. Thus, $h(0)$ is *not* a resultant—it is off by a factor of 6. Note why equality fails: $h(0)$ is a 4×4 determinant, whereas $\mathrm{Res}(3x-1, 6x^2-4, x)$ is a 3×3 determinant.

     b. Now let $f = x^2 y + 3xy - 1$ and $g = 6x^2 + y^2 - 4$. Compute $h = \mathrm{Res}(f,g,x)$ and verify that $h(0) = 36$. Setting $y = 0$ in f and g gives polynomials -1 and $6x^2 - 4$. Use (3) to show that the resultant of these polynomials is 1. Thus, $h(0)$ is off by a factor of 36.

When the degree of f drops by 1 [in part (a)], we get an extra factor of 6, and when it drops by 2 [in part (b)], we get an extra factor of $36 = 6^2$. And the leading coefficient of x in g is 6. In Exercise 14, we will see that this is no accident.

13. Let $f = x^2 y + x - 1$ and $g = x^2 y + x + y^2 - 4$. If $h = \mathrm{Res}(f,g,x) \in \mathbb{C}[y]$, show that $h(0) = 0$. But if we substitute $y = 0$ into f and g, we get $x - 1$ and $x - 4$. Show that these polynomials have a nonzero resultant. Thus, $h(0)$ is *not* a resultant.

14. In this problem you will complete the proof of Proposition 6 by determining what happens to a resultant when specializing causes the degree of one of the polynomials to

drop. Let $f, g \in k[x_1, \ldots, x_n]$ and set $h = \mathrm{Res}(f, g, x_1)$. If $\mathbf{a} = (a_2, \ldots, a_n) \in k^{n-1}$, let $f(x_1, \mathbf{a})$ be the polynomial in $k[x_1]$ obtained by substituting in $\mathbf{a}$. As in (7), let $c_0, d_0 \in k[x_2, \ldots, x_n]$ be the leading coefficients of x_1 in f, g, respectively. We will assume that $c_0(\mathbf{a}) \neq 0$ and $d_0(\mathbf{a}) = 0$, and our goal is to see how $h(\mathbf{a})$ relates to $\mathrm{Res}(f(x_1, \mathbf{a}), g(x_1, \mathbf{a}), x_1)$.

a. First suppose that the degree of g drops by exactly 1, which means that $d_1(\mathbf{a}) \neq 0$. In this case, prove that

$$h(\mathbf{a}) = c_0(\mathbf{a}) \cdot \mathrm{Res}(f(x_1, \mathbf{a}), g(x_1, \mathbf{a}), x_1).$$

Hint: $h(\mathbf{a})$ is given by the following determinant:

$$\underbrace{}_{m \text{ columns}} \underbrace{}_{l \text{ columns}}$$

The determinant is the wrong size to be the resultant of $f(x_1, \mathbf{a})$ and $g(x_1, \mathbf{a})$. If you expand by minors along the first row [see Section 11.4 of DUMMIT and FOOTE (2004)], the desired result will follow.

b. Now let us do the general case. Suppose that the degree of $g(x_1, \mathbf{a})$ is $m - p$, where $p \geq 1$. Then prove that

$$h(\mathbf{a}) = c_0(\mathbf{a})^p \cdot \mathrm{Res}(f(x_1, \mathbf{a}), g(x_1, \mathbf{a}), x_1).$$

Hint: Expand by minors p times. Note how this formula explains Exercise 12.

15. Suppose that k is a field and $\varphi : k[x_1, \ldots, x_n] \rightarrow k[x_1]$ is a ring homomorphism that is the identity on k and maps x_1 to x_1. Given an ideal $I \subseteq k[x_1, \ldots, x_n]$, prove that $\varphi(I) \subseteq k[x_1]$ is an ideal. (In the proof of Theorem 8, we use this result when φ is the map that evaluates x_i at a_i for $2 \leq i \leq n$.)

16. If $f = c_0 x^l + \cdots + c_l \in k[x]$, where $c_0 \neq 0$ and $l > 0$, then the *discriminant* of f is

$$\mathrm{disc}(f) = \frac{(-1)^{l(l-1)/2}}{c_0} \mathrm{Res}(f, f', x),$$

where f' is the formal derivative of f from Exercise 13 of Chapter 1, §5. When $k = \mathbb{C}$, prove that f has a multiple root if and only if $\mathrm{disc}(f) = 0$. Hint: See Exercise 14 of Chapter 1, §5.

17. Use the previous exercise to determine whether or not $6x^4 - 23x^3 + 32x^2 - 19x + 4$ has a multiple root in $\mathbb{C}$. What is the multiple root?

18. Compute the discriminant of the quadratic polynomial $f = ax^2 + bx + c$. Explain how your answer relates to the quadratic formula, and, without using Exercise 16, prove that f has a multiple root if and only if its discriminant vanishes.

19. Suppose that $f, g \in \mathbb{C}[x]$ are polynomials of positive degree. The goal of this problem is to construct a polynomial whose roots are all sums of a root of f plus a root of g.

   a. Show that a complex number $\gamma \in \mathbb{C}$ can be written $\gamma = \alpha + \beta$, where $f(\alpha) = g(\beta) = 0$, if and only if the equations $f(x) = g(y - x) = 0$ have a solution with $y = \gamma$.

   b. Using Proposition 6, show that γ is a root of $\mathrm{Res}(f(x), g(y - x), x)$ if and only if $\gamma = \alpha + \beta$, where $f(\alpha) = g(\beta) = 0$.

   c. Construct a polynomial with coefficients in $\mathbb{Q}$ which has $\sqrt{2} + \sqrt{3}$ as a root. Hint: What are f and g in this case?

   d. Modify your construction to create a polynomial whose roots are all differences of a root of f minus a root of g.

20. Suppose that $f, g \in \mathbb{C}[x]$ are polynomials of positive degree. If all of the roots of f are nonzero, adapt the argument of Exercise 19 to construct a polynomial whose roots are all products of a root of f times a root of g.

21. Suppose that $I = \langle f, g \rangle \subseteq \mathbb{C}[x, y]$ and assume that $\mathrm{Res}(f, g, x) \neq 0$. Prove that $\mathbf{V}(I_1) = \pi_1(V)$, where $V = \mathbf{V}(I)$ and π_1 is projection onto the y-axis. Hint: See Exercise 5 of §2. This exercise is due to GALLET, RAHKOOY and ZAFEIRAKOPOULOS (2013).

Chapter 4
The Algebra–Geometry Dictionary

In this chapter, we will explore the correspondence between ideals and varieties. In §§1 and 2, we will prove the Nullstellensatz, a celebrated theorem which identifies exactly which ideals correspond to varieties. This will allow us to construct a "dictionary" between geometry and algebra, whereby any statement about varieties can be translated into a statement about ideals (and conversely). We will pursue this theme in §§3 and 4, where we will define a number of natural algebraic operations on ideals and study their geometric analogues. In keeping with the computational emphasis of the book, we will develop algorithms to carry out the algebraic operations. In §§5 and 6, we will study the more important algebraic and geometric concepts arising out of the Hilbert Basis Theorem: notably the possibility of decomposing a variety into a union of simpler varieties and the corresponding algebraic notion of writing an ideal as an intersection of simpler ideals. In §7, we will prove the Closure Theorem from Chapter 3 using the tools developed in this chapter.

§1 Hilbert's Nullstellensatz

In Chapter 1, we saw that a variety $V \subseteq k^n$ can be studied by passing to the ideal

$$\mathbf{I}(V) = \{f \in k[x_1, \ldots, x_n] \mid f(a) = 0 \text{ for all } a \in V\}$$

of all polynomials vanishing on V. Hence, we have a map

$$
\begin{array}{ccc}
\text{affine varieties} & & \text{ideals} \\
V & \longrightarrow & \mathbf{I}(V).
\end{array}
$$

Conversely, given an ideal $I \subseteq k[x_1, \ldots, x_n]$, we can define the set

$$\mathbf{V}(I) = \{a \in k^n \mid f(a) = 0 \text{ for all } f \in I\}.$$

© Springer International Publishing Switzerland 2015
D.A. Cox et al., *Ideals, Varieties, and Algorithms*, Undergraduate Texts in Mathematics, DOI 10.1007/978-3-319-16721-3_4

The Hilbert Basis Theorem assures us that $\mathbf{V}(I)$ is actually an affine variety, for it tells us that there exists a finite set of polynomials $f_1, \ldots, f_s \in I$ such that $I = \langle f_1, \ldots, f_s \rangle$, and we proved in Proposition 9 of Chapter 2, §5 that $\mathbf{V}(I)$ is the set of common roots of these polynomials. Thus, we have a map

$$
\begin{array}{ccc}
\text{ideals} & & \text{affine varieties} \\
I & \longrightarrow & \mathbf{V}(I).
\end{array}
$$

These two maps give us a correspondence between ideals and varieties. In this chapter, we will explore the nature of this correspondence.

The first thing to note is that this correspondence (more precisely, the map $\mathbf{V}$) is not one-to-one: different ideals can give the same variety. For example, $\langle x \rangle$ and $\langle x^2 \rangle$ are different ideals in $k[x]$ which have the same variety $\mathbf{V}(x) = \mathbf{V}(x^2) = \{0\}$. More serious problems can arise if the field k is not algebraically closed. For example, consider the three polynomials 1, $1 + x^2$, and $1 + x^2 + x^4$ in $\mathbb{R}[x]$. These generate different ideals

$$
I_1 = \langle 1 \rangle = \mathbb{R}[x], \quad I_2 = \langle 1 + x^2 \rangle, \quad I_3 = \langle 1 + x^2 + x^4 \rangle,
$$

but each polynomial has no real roots, so that the corresponding varieties are all empty:

$$
\mathbf{V}(I_1) = \mathbf{V}(I_2) = \mathbf{V}(I_3) = \emptyset.
$$

Examples of polynomials in two variables without real roots include $1 + x^2 + y^2$ and $1 + x^2 + y^4$. These give different ideals in $\mathbb{R}[x, y]$ which correspond to the empty variety.

Does this problem of having different ideals represent the empty variety go away if the field k is algebraically closed? It does in the one-variable case when the ring is $k[x]$. To see this, recall from §5 of Chapter 1 that any ideal I in $k[x]$ can be generated by a single polynomial because $k[x]$ is a principal ideal domain. So we can write $I = \langle f \rangle$ for some polynomial $f \in k[x]$. Then $\mathbf{V}(I)$ is the set of roots of f; i.e., the set of $a \in k$ such that $f(a) = 0$. But since k is algebraically closed, every nonconstant polynomial in $k[x]$ has a root. Hence, the only way that we could have $\mathbf{V}(I) = \emptyset$ would be to have f be a nonzero constant. In this case, $1/f \in k$. Thus, $1 = (1/f) \cdot f \in I$, which means that $g = g \cdot 1 \in I$ for all $g \in k[x]$. This shows that $I = k[x]$ is the only ideal of $k[x]$ that represents the empty variety when k is algebraically closed.

A wonderful thing now happens: the same property holds when there is more than one variable. In any polynomial ring, algebraic closure is enough to guarantee that the only ideal which represents the empty variety is the entire polynomial ring itself. This is the *Weak Nullstellensatz*, which is the basis of (and is equivalent to) one of the most celebrated mathematical results of the late nineteenth century, Hilbert's Nullstellensatz. Such is its impact that, even today, one customarily uses the original German name *Nullstellensatz*: a word formed, in typical German fashion, from three simpler words: Null (= Zero), Stellen (= Places), Satz (= Theorem).

Theorem 1 (The Weak Nullstellensatz). *Let k be an algebraically closed field and let $I \subseteq k[x_1, \ldots, x_n]$ be an ideal satisfying $\mathbf{V}(I) = \emptyset$. Then $I = k[x_1, \ldots, x_n]$.*

Proof. Our proof is inspired by GLEBSKY (2012). We will prove the theorem in contrapositive form:

$$I \subsetneq k[x_1, \ldots, x_n] \implies \mathbf{V}(I) \neq \emptyset.$$

We will make frequent use of the standard equivalence $I = k[x_1, \ldots, x_n] \Leftrightarrow 1 \in I$. This is part (a) of Exercise 16 from Chapter 1, §4.

Given $a \in k$ and $f \in k[x_1, \ldots, x_n]$, let $\bar{f} = f(x_1, \ldots, x_{n-1}, a) \in k[x_1, \ldots, x_{n-1}]$. Similar to Exercise 2 of Chapter 3, §5 and Exercise 15 of Chapter 3, §6, the set

$$I_{x_n=a} = \{\bar{f} \mid f \in I\}$$

is an ideal of $k[x_1, \ldots, x_{n-1}]$. The key step in the proof is the following claim.

Claim. If k is algebraically closed and $I \subsetneq k[x_1, \ldots, x_n]$ is a proper ideal, then there is $a \in k$ such that $I_{x_n=a} \subsetneq k[x_1, \ldots, x_{n-1}]$.

Once we prove the claim, an easy induction gives elements $a_1, \ldots, a_n \in k$ such that $I_{x_n=a_n,\ldots,x_1=a_1} \subsetneq k$. But the only ideals of k are $\{0\}$ and k (Exercise 3), so that $I_{x_n=a_n,\ldots,x_1=a_1} = \{0\}$. This implies $(a_1, \ldots, a_n) \in \mathbf{V}(I)$. We conclude that $\mathbf{V}(I) \neq \emptyset$, and the theorem will follow.

To prove the claim, there are two cases, depending on the size of $I \cap k[x_n]$.

Case 1. $I \cap k[x_n] \neq \{0\}$. Let $f \in I \cap k[x_n]$ be nonzero, and note that f is nonconstant, since otherwise $1 \in I \cap k[x_n] \subseteq I$, contradicting $I \neq k[x_1, \ldots, x_n]$.

Since k is algebraically closed, $f = c \prod_{i=1}^{r} (x_n - b_i)^{m_i}$ where $c, b_1, \ldots, b_r \in k$ and $c \neq 0$. Suppose that $I_{x_n=b_i} = k[x_1, \ldots, x_{n-1}]$ for all i. Then for all i there is $B_i \in I$ with $B_i(x_1, \ldots, x_{n-1}, b_i) = 1$. This implies that

$$1 = B_i(x_1, \ldots, x_{n-1}, b_i) = B_i(x_1, \ldots, x_{n-1}, x_n - (x_n - b_i)) = B_i + A_i(x_n - b_i)$$

for some $A_i \in k[x_1, \ldots, x_n]$. Since this holds for $i = 1, \ldots, r$, we obtain

$$1 = \prod_{i=1}^{r}(A_i(x_n - b_i) + B_i)^{m_i} = A\prod_{i=1}^{r}(x_n - b_i)^{m_i} + B,$$

where $A = \prod_{i=1}^{r} A_i^{m_i}$ and $B \in I$. This and $\prod_{i=1}^{r}(x_n - b_i)^{m_i} = c^{-1}f \in I$ imply that $1 \in I$, which contradicts $I \neq k[x_1, \ldots, x_n]$. Thus $I_{x_n=b_i} \neq k[x_1, \ldots, x_{n-1}]$ for some i. This b_i is the desired a.

Case 2. $I \cap k[x_n] = \{0\}$. Let $\{g_1, \ldots, g_t\}$ be a Gröbner basis of I for lex order with $x_1 > \cdots > x_n$ and write

(1) $$g_i = c_i(x_n)x^{\alpha_i} + \text{terms} < x^{\alpha_i},$$

where $c_i(x_n) \in k[x_n]$ is nonzero and x^{α_i} is a monomial in $x_1, \ldots, x_{n-1}$.

Now pick $a \in k$ such that $c_i(a) \neq 0$ for all i. This is possible since algebraically closed fields are infinite by Exercise 4. It is easy to see that the polynomials

$$\bar{g}_i = g_i(x_1, \ldots, x_{n-1}, a)$$

form a basis of $I_{x_n=a}$ (Exercise 5). Substituting $x_n = a$ into equation (1), one easily sees that $\mathrm{LT}(\bar{g}_i) = c_i(a)x^{\alpha_i}$ since $c_i(a) \neq 0$. Also note that $x^{\alpha_i} \neq 1$, since otherwise $g_i = c_i \in I \cap k[x_n] = \{0\}$, yet $c_i \neq 0$. This shows that $\mathrm{LT}(\bar{g}_i)$ is nonconstant for all i.

We claim that the $\bar{g}_i$ form a Gröbner basis of $I_{x_n=a}$. Assuming the claim, it follows that $1 \notin I_{x_n=a}$ since no $\mathrm{LT}(\bar{g}_i)$ can divide 1. Thus $I_{x_n=a} \neq k[x_1, \ldots, x_{n-1}]$, which is what we want to show.

To prove the claim, take $g_i, g_j \in G$ and consider the polynomial

$$S = c_j(x_n) \frac{x^\gamma}{x^{\alpha_i}} g_i - c_i(x_n) \frac{x^\gamma}{x^{\alpha_j}} g_j,$$

where $x^\gamma = \mathrm{lcm}(x^{\alpha_i}, x^{\alpha_j})$. By construction, $x^\gamma > \mathrm{LT}(S)$ (be sure you understand this). Since $S \in I$, it has a standard representation $S = \sum_{l=1}^{t} A_l g_l$. Then evaluating at $x_n = a$ gives

$$c_j(a) \frac{x^\gamma}{x^{\alpha_i}} \bar{g}_i - c_i(a) \frac{x^\gamma}{x^{\alpha_j}} \bar{g}_j = \bar{S} = \sum_{l=1}^{t} \bar{A}_l \bar{g}_l.$$

Since $\mathrm{LT}(\bar{g}_i) = c_i(a)x^{\alpha_i}$, we see that $\bar{S}$ is the S-polynomial $S(\bar{g}_i, \bar{g}_j)$ up to the nonzero constant $c_i(a)c_j(a)$. Then

$$x^\gamma > \mathrm{LT}(S) \geq \mathrm{LT}(A_l g_l), \quad A_l g_l \neq 0$$

implies that

$$x^\gamma > \mathrm{LT}(\bar{A}_l \bar{g}_l), \quad \bar{A}_l \bar{g}_l \neq 0$$

(Exercise 6). Since $x^\gamma = \mathrm{lcm}(\mathrm{LM}(\bar{g}_i), \mathrm{LM}(\bar{g}_j))$, it follows that $S(\bar{g}_i, \bar{g}_j)$ has an lcm representation for all i, j and hence is a Gröbner basis by Theorem 6 of Chapter 2, §9. This proves the claim and completes the proof of the theorem. □

In the special case when $k = \mathbb{C}$, the Weak Nullstellensatz may be thought of as the "Fundamental Theorem of Algebra for multivariable polynomials"—every system of polynomials that generates an ideal strictly smaller than $\mathbb{C}[x_1, \ldots, x_n]$ has a common zero in $\mathbb{C}^n$.

The Weak Nullstellensatz also allows us to solve the *consistency problem* from §2 of Chapter 1. Recall that this problem asks whether a system

$$f_1 = 0,$$
$$f_2 = 0,$$
$$\vdots$$
$$f_s = 0$$

of polynomial equations has a common solution in $\mathbb{C}^n$. The polynomials fail to have a common solution if and only if $\mathbf{V}(f_1, \ldots, f_s) = \emptyset$. By the Weak Nullstellensatz, the latter holds if and only if $1 \in \langle f_1, \ldots, f_s \rangle$. Thus, to solve the consistency problem, we need to be able to determine whether 1 belongs to an ideal. This is made easy by the observation that for any monomial ordering, $\{1\}$ *is the only reduced Gröbner basis of the ideal* $\langle 1 \rangle = k[x_1, \ldots, x_n]$.

To see this, let $\{g_1, \ldots, g_t\}$ be a Gröbner basis of $I = \langle 1 \rangle$. Thus, $1 \in \langle \mathrm{LT}(I) \rangle = \langle \mathrm{LT}(g_1), \ldots, \mathrm{LT}(g_t) \rangle$, and then Lemma 2 of Chapter 2, §4 implies that 1 is divisible by some $\mathrm{LT}(g_i)$, say $\mathrm{LT}(g_1)$. This forces $\mathrm{LT}(g_1)$ to be constant. Then every other $\mathrm{LT}(g_i)$ is a multiple of that constant, so that $g_2, \ldots, g_t$ can be removed from the Gröbner basis by Lemma 3 of Chapter 2, §7. Finally, since $\mathrm{LT}(g_1)$ is constant, g_1 itself is constant since every nonconstant monomial is > 1 (see Corollary 6 of Chapter 2, §4). We can multiply by an appropriate constant to make $g_1 = 1$. Our reduced Gröbner basis is thus $\{1\}$.

To summarize, we have the following **consistency algorithm**: if we have polynomials $f_1, \ldots, f_s \in \mathbb{C}[x_1, \ldots, x_n]$, we compute a reduced Gröbner basis of the ideal they generate with respect to any ordering. If this basis is $\{1\}$, the polynomials have no common zero in $\mathbb{C}^n$; if the basis is not $\{1\}$, they must have a common zero. Note that this algorithm works over any algebraically closed field.

If we are working over a field k which is not algebraically closed, then the consistency algorithm still works in one direction: if $\{1\}$ is a reduced Gröbner basis of $\langle f_1, \ldots, f_s \rangle$, then the equations $f_1 = \cdots = f_s = 0$ have no common solution. The converse is not true, as shown by the examples preceding the statement of the Weak Nullstellensatz.

Inspired by the Weak Nullstellensatz, one might hope that the correspondence between ideals and varieties is one-to-one provided only that one restricts to algebraically closed fields. Unfortunately, our earlier example $\mathbf{V}(x) = \mathbf{V}(x^2) = \{0\}$ works over *any* field. Similarly, the ideals $\langle x^2, y \rangle$ and $\langle x, y \rangle$ (and, for that matter, $\langle x^n, y^m \rangle$ where n and m are integers greater than one) are different but define the same variety: namely, the single point $\{(0,0)\} \subseteq k^2$. These examples illustrate a basic reason why different ideals can define the same variety (equivalently, that the map $\mathbf{V}$ can fail to be one-to-one): namely, a power of a polynomial vanishes on the same set as the original polynomial. The Hilbert Nullstellensatz states that over an algebraically closed field, this is the *only* reason that different ideals can give the same variety: if a polynomial f vanishes at all points of some variety $\mathbf{V}(I)$, then some power of f must belong to I.

Theorem 2 (Hilbert's Nullstellensatz). *Let k be an algebraically closed field. If $f, f_1, \ldots, f_s \in k[x_1, \ldots, x_n]$, then $f \in \mathbf{I}(\mathbf{V}(f_1, \ldots, f_s))$ if and only if*

$$f^m \in \langle f_1, \ldots, f_s \rangle$$

for some integer $m \geq 1$.

Proof. Given a nonzero polynomial f which vanishes at every common zero of the polynomials $f_1, \ldots, f_s$, we must show that there exists an integer $m \geq 1$ and polynomials $A_1, \ldots, A_s$ such that

$$f^m = \sum_{i=1}^{s} A_i f_i.$$

The most direct proof is based on an ingenious trick. Consider the ideal

$$\tilde{I} = \langle f_1, \ldots, f_s, 1 - yf \rangle \subseteq k[x_1, \ldots, x_n, y],$$

where $f, f_1, \ldots, f_s$ are as above. We claim that

$$\mathbf{V}(\tilde{I}) = \emptyset.$$

To see this, let $(a_1, \ldots, a_n, a_{n+1}) \in k^{n+1}$. Either

- $(a_1, \ldots, a_n)$ is a common zero of $f_1, \ldots, f_s$, or
- $(a_1, \ldots, a_n)$ is not a common zero of $f_1, \ldots, f_s$.

In the first case $f(a_1, \ldots, a_n) = 0$ since f vanishes at any common zero of $f_1, \ldots, f_s$. Thus, the polynomial $1 - yf$ takes the value $1 - a_{n+1}f(a_1, \ldots, a_n) = 1 \neq 0$ at the point $(a_1, \ldots, a_n, a_{n+1})$. In particular, $(a_1, \ldots, a_n, a_{n+1}) \notin \mathbf{V}(\tilde{I})$. In the second case, for some $i, 1 \leq i \leq s$, we must have $f_i(a_1, \ldots, a_n) = 0$. Thinking of f_i as a function of $n + 1$ variables which does not depend on the last variable, we have $f_i(a_1, \ldots, a_n, a_{n+1}) \neq 0$. In particular, we again conclude that $(a_1, \ldots, a_n, a_{n+1}) \notin \mathbf{V}(\tilde{I})$. Since $(a_1, \ldots, a_n, a_{n+1}) \in k^{n+1}$ was arbitrary, we obtain $\mathbf{V}(\tilde{I}) = \emptyset$, as claimed.

Now apply the Weak Nullstellensatz to conclude that $1 \in \tilde{I}$. Hence

$$(2) \qquad 1 = \sum_{i=1}^{s} p_i(x_1, \ldots, x_n, y) f_i + q(x_1, \ldots, x_n, y)(1 - yf)$$

for some polynomials $p_i, q \in k[x_1, \ldots, x_n, y]$. Now set $y = 1/f(x_1, \ldots, x_n)$. Then relation (2) above implies that

$$(3) \qquad 1 = \sum_{i=1}^{s} p_i(x_1, \ldots, x_n, 1/f) f_i.$$

Multiply both sides of this equation by a power f^m, where m is chosen sufficiently large to clear all the denominators. This yields

$$(4) \qquad f^m = \sum_{i=1}^{s} A_i f_i,$$

for some polynomials $A_i \in k[x_1, \ldots, x_n]$, which is what we had to show. $\square$

EXERCISES FOR §1

1. Recall that $V(y - x^2, z - x^3)$ is the twisted cubic in $\mathbb{R}^3$.
   a. Show that $V((y - x^2)^2 + (z - x^3)^2)$ is also the twisted cubic.
   b. Show that any variety $V(I) \subseteq \mathbb{R}^n$, $I \subseteq \mathbb{R}[x_1, \ldots, x_n]$, can be defined by a single equation (and hence by a principal ideal).
2. Let $J = \langle x^2 + y^2 - 1, y - 1 \rangle$. Find $f \in I(V(J))$ such that $f \notin J$.
3. Prove that $\{0\}$ and k are the only ideals of a field k.
4. Prove that an algebraically closed field k must be infinite. Hint: Given n elements $a_1, \ldots, a_n$ of a field k, can you write down a nonconstant polynomial $f \in k[x]$ with the property that $f(a_i) = 1$ for all i?
5. In the proof of Theorem 1, prove that $I_{x_n=a} = \langle \bar{g}_1, \ldots, \bar{g}_t \rangle$.
6. In the proof of Theorem 1, let x^δ be a monomial in $x_1, \ldots, x_{n-1}$ satisfying $x^\delta > \mathrm{LT}(f)$ for some $f \in k[x_1, \ldots, x_n]$. Prove that $x^\delta > \mathrm{LT}(\bar{f})$, where $\bar{f} = f(x_1, \ldots, x_{n-1}, a)$.
7. In deducing Hilbert's Nullstellensatz from the Weak Nullstellensatz, we made the substitution $y = 1/f(x_1, \ldots, x_n)$ to deduce relations (3) and (4) from (2). Justify this rigorously. Hint: In what set is $1/f$ contained?
8. The purpose of this exercise is to show that if k is any field that is not algebraically closed, then *any* variety $V \subseteq k^n$ can be defined by a *single* equation.
   a. If $g = a_0 x^n + a_1 x^{n-1} + \cdots + a_{n-1} x + a_n$ is a polynomial of degree n in x, define the *homogenization* g^h of g with respect to some variable y to be the polynomial $g^h = a_0 x^n + a_1 x^{n-1} y + \cdots + a_{n-1} x y^{n-1} + a_n y^n$. Show that g has a root in k if and only if there is $(a, b) \in k^2$ such that $(a, b) \neq (0, 0)$ and $g^h(a, b) = 0$. Hint: Show that $g^h(a, b) = b^n g^h(a/b, 1)$ when $b \neq 0$.
   b. If k is not algebraically closed, show that there exists $f \in k[x, y]$ such that the variety defined by $f = 0$ consists of just the origin $(0, 0) \in k^2$. Hint: Choose a polynomial in $k[x]$ with no root in k and consider its homogenization.
   c. If k is not algebraically closed, show that for each integer $l > 0$ there exists $f \in k[x_1, \ldots, x_l]$ such that the only solution of $f = 0$ is the origin $(0, \ldots, 0) \in k^l$. Hint: Use induction on l and part (b) above.
   d. If $W = V(g_1, \ldots, g_s)$ is any variety in k^n, where k is not algebraically closed, then show that W can be defined by a single equation. Hint: Consider the polynomial $f(g_1, \ldots, g_s)$ where f is as in part (c).
9. Let k be an arbitrary field and let S be the subset of all polynomials in $k[x_1, \ldots, x_n]$ that have no zeros in k^n. If I is any ideal in $k[x_1, \ldots, x_n]$ such that $I \cap S = \emptyset$, show that $V(I) \neq \emptyset$. Hint: When k is not algebraically closed, use the previous exercise.
10. In Exercise 1, we encountered two ideals in $\mathbb{R}[x, y]$ that give the same nonempty variety. Show that one of these ideals is contained in the other. Can you find two ideals in $\mathbb{R}[x, y]$, neither contained in the other, which give the same nonempty variety? Can you do the same for $\mathbb{R}[x]$?

§2 Radical Ideals and the Ideal–Variety Correspondence

To further explore the relation between ideals and varieties, it is natural to recast Hilbert's Nullstellensatz in terms of ideals. Can we characterize the kinds of ideals that appear as the ideal of a variety? In other words, can we identify those ideals that consist of *all* polynomials which vanish on some variety V? The key observation is contained in the following simple lemma.

Lemma 1. *Let V be a variety. If $f^m \in \mathbf{I}(V)$, then $f \in \mathbf{I}(V)$.*

Proof. Let $a \in V$. If $f^m \in \mathbf{I}(V)$, then $(f(a))^m = 0$. But this can happen only if $f(a) = 0$. Since $a \in V$ was arbitrary, we must have $f \in \mathbf{I}(V)$. $\square$

Thus, an ideal consisting of *all* polynomials which vanish on a variety V has the property that if some power of a polynomial belongs to the ideal, then the polynomial itself must belong to the ideal. This leads to the following definition.

Definition 2. An ideal I is **radical** if $f^m \in I$ for some integer $m \geq 1$ implies that $f \in I$.

Rephrasing Lemma 1 in terms of radical ideals gives the following statement.

Corollary 3. $\mathbf{I}(V)$ *is a radical ideal.*

On the other hand, Hilbert's Nullstellensatz tells us that the only way that an arbitrary ideal I can fail to be the ideal of all polynomials vanishing on $\mathbf{V}(I)$ is for I to contain powers f^m of polynomials f which are not in I—in other words, for I to fail to be a radical ideal. This suggests that there is a one-to-one correspondence between affine varieties and radical ideals. To clarify this and get a sharp statement, it is useful to introduce the operation of taking the radical of an ideal.

Definition 4. Let $I \subseteq k[x_1, \ldots, x_n]$ be an ideal. The **radical** of I, denoted $\sqrt{I}$, is the set

$$\{f \mid f^m \in I \text{ for some integer } m \geq 1\}.$$

Note that we always have $I \subseteq \sqrt{I}$ since $f \in I$ implies $f^1 \in I$ and, hence, $f \in \sqrt{I}$ by definition. It is an easy exercise to show that an ideal I is radical if and only if $I = \sqrt{I}$. A somewhat more surprising fact is that the radical of an ideal is always an ideal. To see what is at stake here, consider, for example, the ideal $J = \langle x^2, y^3 \rangle \subseteq k[x, y]$. Although neither x nor y belongs to J, it is clear that $x \in \sqrt{J}$ and $y \in \sqrt{J}$. Note that $(x \cdot y)^2 = x^2 y^2 \in J$ since $x^2 \in J$; thus, $x \cdot y \in \sqrt{J}$. It is less obvious that $x + y \in \sqrt{J}$. To see this, observe that

$$(x + y)^4 = x^4 + 4x^3 y + 6x^2 y^2 + 4xy^3 + y^4 \in J$$

because $x^4, 4x^3 y, 6x^2 y^2 \in J$ (they are all multiples of x^2) and $4xy^3, y^4 \in J$ (because they are multiples of y^3). Thus, $x + y \in \sqrt{J}$. By way of contrast, neither xy nor $x + y$ belong to J.

Lemma 5. *If I is an ideal in $k[x_1, \ldots, x_n]$, then $\sqrt{I}$ is an ideal in $k[x_1, \ldots, x_n]$ containing I. Furthermore, $\sqrt{I}$ is a radical ideal.*

Proof. We have already shown that $I \subseteq \sqrt{I}$. To show $\sqrt{I}$ is an ideal, suppose $f, g \in \sqrt{I}$. Then there are positive integers m and l such that $f^m, g^l \in I$. In the binomial expansion of $(f + g)^{m+l-1}$ every term has a factor $f^i g^j$ with $i + j = m + l - 1$. Since either $i \geq m$ or $j \geq l$, either f^i or g^j is in I, whence $f^i g^j \in I$ and every term in the binomial expansion is in I. Hence, $(f + g)^{m+l-1} \in I$ and, therefore, $f + g \in \sqrt{I}$. Finally, suppose $f \in \sqrt{I}$ and $h \in k[x_1, \ldots, x_n]$. Then $f^m \in I$ for some integer $m \geq 1$. Since I is an ideal, we have $(h \cdot f)^m = h^m f^m \in I$. Hence, $hf \in \sqrt{I}$. This shows that $\sqrt{I}$ is an ideal. In Exercise 4, you will show that $\sqrt{I}$ is a radical ideal. $\square$

We are now ready to state the ideal-theoretic form of the Nullstellensatz.

Theorem 6 (The Strong Nullstellensatz). *Let k be an algebraically closed field. If I is an ideal in $k[x_1, \ldots, x_n]$, then*

$$\mathbf{I}(\mathbf{V}(I)) = \sqrt{I}.$$

Proof. We certainly have $\sqrt{I} \subseteq \mathbf{I}(\mathbf{V}(I))$ because $f \in \sqrt{I}$ implies that $f^m \in I$ for some m. Hence, f^m vanishes on $\mathbf{V}(I)$, which implies that f vanishes on $\mathbf{V}(I)$. Thus, $f \in \mathbf{I}(\mathbf{V}(I))$.

Conversely, take $f \in \mathbf{I}(\mathbf{V}(I))$. Then, by definition, f vanishes on $\mathbf{V}(I)$. By Hilbert's Nullstellensatz, there exists an integer $m \geq 1$ such that $f^m \in I$. But this means that $f \in \sqrt{I}$. Since f was arbitrary, $\mathbf{I}(\mathbf{V}(I)) \subseteq \sqrt{I}$, and we are done. $\square$

It has become a custom, to which we shall adhere, to refer to Theorem 6 as *the* Nullstellensatz with no further qualification. The most important consequence of the Nullstellensatz is that it allows us to set up a "dictionary" between geometry and algebra. The basis of the dictionary is contained in the following theorem.

Theorem 7 (The Ideal–Variety Correspondence). *Let k be an arbitrary field.*

  (i) *The maps*

$$\text{affine varieties} \xrightarrow{\ \mathbf{I}\ } \text{ideals}$$

  and

$$\text{ideals} \xrightarrow{\ \mathbf{V}\ } \text{affine varieties}$$

  are inclusion-reversing, i.e., if $I_1 \subseteq I_2$ are ideals, then $\mathbf{V}(I_1) \supseteq \mathbf{V}(I_2)$ and, similarly, if $V_1 \subseteq V_2$ are varieties, then $\mathbf{I}(V_1) \supseteq \mathbf{I}(V_2)$.
  (ii) *For any variety V,*

$$\mathbf{V}(\mathbf{I}(V)) = V,$$

  so that $\mathbf{I}$ is always one-to-one. On the other hand, any ideal I satisfies

$$\mathbf{V}(\sqrt{I}) = \mathbf{V}(I).$$

 (iii) *If k is algebraically closed, and if we restrict to radical ideals, then the maps*

$$\text{affine varieties} \xrightarrow{\ \mathbf{I}\ } \text{radical ideals}$$

  and

$$\text{radical ideals} \xrightarrow{\ \mathbf{V}\ } \text{affine varieties}$$

  are inclusion-reversing bijections which are inverses of each other.

Proof. (i) The proof will be covered in the exercises.

(ii) Let $V = \mathbf{V}(f_1, \ldots, f_s)$ be an affine variety in k^n. Since every $f \in \mathbf{I}(V)$ vanishes on V, the inclusion $V \subseteq \mathbf{V}(\mathbf{I}(V))$ follows directly from the definition of $\mathbf{V}$. Going the other way, note that $f_1, \ldots, f_s \in \mathbf{I}(V)$ by the definition of $\mathbf{I}$, and, thus, $\langle f_1, \ldots, f_s \rangle \subseteq \mathbf{I}(V)$. Since $\mathbf{V}$ is inclusion-reversing, it follows that $\mathbf{V}(\mathbf{I}(V)) \subseteq \mathbf{V}(\langle f_1, \ldots, f_s \rangle) = V$. This proves that $\mathbf{V}(\mathbf{I}(V)) = V$, and, consequently, $\mathbf{I}$ is one-to-one since it has a left inverse. The final assertion of part (ii) is left as an exercise.

(iii) Since $\mathbf{I}(V)$ is radical by Corollary 3, we can think of $\mathbf{I}$ as a function which takes varieties to radical ideals. Furthermore, we already know $\mathbf{V}(\mathbf{I}(V)) = V$ for any variety V. It remains to prove $\mathbf{I}(\mathbf{V}(I)) = I$ whenever I is a radical ideal. This is easy: the Nullstellensatz tells us $\mathbf{I}(\mathbf{V}(I)) = \sqrt{I}$, and I being radical implies $\sqrt{I} = I$ (see Exercise 4). This gives the desired equality. Hence, $\mathbf{V}$ and $\mathbf{I}$ are inverses of each other and, thus, define bijections between the set of radical ideals and affine varieties. The theorem is proved. □

As a consequence of this theorem, any question about varieties can be rephrased as an algebraic question about radical ideals (and conversely), provided that we are working over an algebraically closed field. This ability to pass between algebra and geometry will give us considerable power.

In view of the Nullstellensatz and the importance it assigns to radical ideals, it is natural to ask whether one can compute generators for the radical from generators of the original ideal. In fact, there are three questions to ask concerning an ideal $I = \langle f_1, \ldots, f_s \rangle$:

- (Radical Generators) Is there an algorithm which produces a set $\{g_1, \ldots, g_m\}$ of polynomials such that $\sqrt{I} = \langle g_1, \ldots, g_m \rangle$?
- (Radical Ideal) Is there an algorithm which will determine whether I is radical?
- (Radical Membership) Given $f \in k[x_1, \ldots, x_n]$, is there an algorithm which will determine whether $f \in \sqrt{I}$?

The existence of these algorithms follows from the work of HERMANN (1926) [see also MINES, RICHMAN, and RUITENBERG (1988) and SEIDENBERG (1974, 1984) for more modern expositions]. More practical algorithms for finding radicals follow from the work of GIANNI, TRAGER and ZACHARIAS (1988), KRICK and LOGAR (1991), and EISENBUD, HUNEKE and VASCONCELOS (1992). These algorithms have been implemented in CoCoA, Singular, and Macaulay2, among others. See, for example, Section 4.5 of GREUEL and PFISTER (2008).

For now, we will settle for solving the more modest *radical membership problem*. To test whether $f \in \sqrt{I}$, we could use the ideal membership algorithm to check whether $f^m \in I$ for all integers $m > 0$. This is not satisfactory because we might have to go to very large powers of m, and it will never tell us if $f \notin \sqrt{I}$ (at least, not until we work out *a priori* bounds on m). Fortunately, we can adapt the proof of Hilbert's Nullstellensatz to give an algorithm for determining whether $f \in \sqrt{\langle f_1, \ldots, f_s \rangle}$.

Proposition 8 (Radical Membership). *Let k be an arbitrary field and let $I = \langle f_1, \ldots, f_s \rangle \subseteq k[x_1, \ldots, x_n]$ be an ideal. Then $f \in \sqrt{I}$ if and only if the constant* *polynomial* 1 *belongs to the ideal* $\tilde{I} = \langle f_1,\ldots,f_s, 1 - yf \rangle \subseteq k[x_1,\ldots,x_n, y]$, *in which case* $\tilde{I} = k[x_1,\ldots,x_n, y]$.

Proof. From equations (2), (3), and (4) in the proof of Hilbert's Nullstellensatz in §1, we see that $1 \in \tilde{I}$ implies $f^m \in I$ for some m, which, in turn, implies $f \in \sqrt{I}$. Going the other way, suppose that $f \in \sqrt{I}$. Then $f^m \in I \subseteq \tilde{I}$ for some m. But we also have $1 - yf \in \tilde{I}$, and, consequently,

$$1 = y^m f^m + (1 - y^m f^m) = y^m \cdot f^m + (1 - yf) \cdot (1 + yf + \cdots + y^{m-1} f^{m-1}) \in \tilde{I},$$

as desired. □

Proposition 8, together with our earlier remarks on determining whether 1 belongs to an ideal (see the discussion of the consistency problem in §1), immediately leads to the following **radical membership algorithm**: to determine if $f \in \sqrt{\langle f_1,\ldots,f_s \rangle} \subseteq k[x_1,\ldots,x_n]$, we compute a reduced Gröbner basis of the ideal $\langle f_1,\ldots,f_s, 1 - yf \rangle \subseteq k[x_1,\ldots,x_n, y]$ with respect to some ordering. If the result is $\{1\}$, then $f \in \sqrt{I}$. Otherwise, $f \notin \sqrt{I}$.

As an example, consider the ideal $I = \langle xy^2 + 2y^2, x^4 - 2x^2 + 1 \rangle$ in $k[x, y]$. Let us test if $f = y - x^2 + 1$ lies in $\sqrt{I}$. Using lex order on $k[x, y, z]$, one checks that the ideal

$$\tilde{I} = \langle xy^2 + 2y^2, x^4 - 2x^2 + 1, 1 - z(y - x^2 + 1) \rangle \subseteq k[x, y, z]$$

has reduced Gröbner basis $\{1\}$. It follows that $y - x^2 + 1 \in \sqrt{I}$ by Proposition 8. Using the division algorithm, we can check what power of $y - x^2 + 1$ lies in I:

$$\overline{y - x^2 + 1}^G = y - x^2 + 1,$$

$$\overline{(y - x^2 + 1)^2}^G = -2x^2y + 2y,$$

$$\overline{(y - x^2 + 1)^3}^G = 0,$$

where $G = \{x^4 - 2x^2 + 1, y^2\}$ is a Gröbner basis of I with respect to lex order and $\overline{p}^G$ is the remainder of p on division by G. As a consequence, we see that $(y - x^2 + 1)^3 \in I$, but no lower power of $y - x^2 + 1$ is in I (in particular, $y - x^2 + 1 \notin I$).

We can also see what is happening in this example geometrically. As a set, $\mathbf{V}(I) = \{(\pm 1, 0)\}$, but (speaking somewhat imprecisely) every polynomial in I vanishes to order at least 2 at each of the two points in $\mathbf{V}(I)$. This is visible from the form of the generators of I if we factor them:

$$xy^2 + 2y^2 = y^2(x + 2) \quad \text{and} \quad x^4 - 2x^2 + 1 = (x^2 - 1)^2.$$

Even though $f = y - x^2 + 1$ also vanishes at $(\pm 1, 0)$, f only vanishes to order 1 there. We must take a higher power of f to obtain an element of I.

We will end this section with a discussion of the one case where we can compute the radical of an ideal, which is when we are dealing with a principal ideal $I = \langle f \rangle$. A nonconstant polynomial f is said to be *irreducible* if it has the property that whenever $f = g \cdot h$ for some polynomials g and h, then either g or h is a constant. As noted in §2 of Appendix A, any nonconstant polynomial f can always be written as a product of irreducible polynomials. By collecting the irreducible polynomials which differ by constant multiples of one another, we can write f in the form

$$f = cf_1^{a_1} \cdots f_r^{a_r}, \quad c \in k,$$

where the f_i's, $1 \le i \le r$, are *distinct* irreducible polynomials, meaning that f_i and f_j are not constant multiples of one another whenever $i \ne j$. Moreover, this expression for f is *unique* up to reordering the f_i's and up to multiplying the f_i's by constant multiples. (This *unique factorization* is Theorem 2 from Appendix A, §2.)

If we have f expressed as a product of irreducible polynomials, then it is easy to write down the radical of the principal ideal generated by f.

Proposition 9. *Let $f \in k[x_1, \ldots, x_n]$ and $I = \langle f \rangle$ be the principal ideal generated by f. If $f = cf_1^{a_1} \cdots f_r^{a_r}$ is the factorization of f into a product of distinct irreducible polynomials, then*

$$\sqrt{I} = \sqrt{\langle f \rangle} = \langle f_1 f_2 \cdots f_r \rangle.$$

Proof. We first show that $f_1 f_2 \cdots f_r$ belongs to $\sqrt{I}$. Let N be an integer strictly greater than the maximum of $a_1, \ldots, a_r$. Then

$$(f_1 f_2 \cdots f_r)^N = f_1^{N-a_1} f_2^{N-a_2} \cdots f_r^{N-a_r} f$$

is a polynomial multiple of f. This shows that $(f_1 f_2 \cdots f_r)^N \in I$, which implies that $f_1 f_2 \cdots f_r \in \sqrt{I}$. Thus $\langle f_1 f_2 \cdots f_r \rangle \subseteq \sqrt{I}$.

Conversely, suppose that $g \in \sqrt{I}$. Then there exists a positive integer M such that $g^M \in I = \langle f \rangle$, so that g^M is a multiple of f and hence a multiple of each irreducible factor f_i of f. Thus, f_i is an irreducible factor of g^M. However, the unique factorization of g^M into distinct irreducible polynomials is the Mth power of the factorization of g. It follows that each f_i is an irreducible factor of g. This implies that g is a polynomial multiple of $f_1 f_2 \cdots f_r$ and, therefore, g is contained in the ideal $\langle f_1 f_2 \cdots f_r \rangle$. The proposition is proved. ☐

In view of Proposition 9, we make the following definition:

Definition 10. If $f \in k[x_1, \ldots, x_n]$ is a polynomial, we define the **reduction** of f, denoted f_{red}, to be the polynomial such that $\langle f_{\text{red}} \rangle = \sqrt{\langle f \rangle}$. A polynomial is said to be **reduced** (or **square-free**) if $f = f_{\text{red}}$.

Thus, f_{red} is the polynomial f with repeated factors "stripped away." So, for example, if $f = (x + y^2)^3 (x - y)$, then $f_{\text{red}} = (x + y^2)(x - y)$. Note that f_{red} is only unique up to a constant factor in k.

The usefulness of Proposition 9 is mitigated by the requirement that f be factored into irreducible factors. We might ask if there is an algorithm to compute f_{red} from f without factoring f first. It turns out that such an algorithm exists.

To state the algorithm, we will need the notion of a greatest common divisor of two polynomials.

Definition 11. Let $f, g \in k[x_1, \ldots, x_n]$. Then $h \in k[x_1, \ldots, x_n]$ is called a **greatest common divisor** of f and g, and denoted $h = \gcd(f, g)$, if

(i) h divides f and g.

(ii) If p is any polynomial that divides both f and g, then p divides h.

It is easy to show that $\gcd(f, g)$ exists and is unique up to multiplication by a nonzero constant in k (see Exercise 9). Unfortunately, the one-variable algorithm for finding the gcd (i.e., the Euclidean Algorithm) does not work in the case of several variables. To see this, consider the polynomials xy and xz in $k[x, y, z]$. Clearly, $\gcd(xy, xz) = x$. However, no matter what term ordering we use, dividing xy by xz gives 0 plus remainder xy and dividing xz by xy gives 0 plus remainder xz. As a result, neither polynomial "reduces" with respect to the other and there is no next step to which to apply the analogue of the Euclidean Algorithm.

Nevertheless, there is an algorithm for calculating the gcd of two polynomials in several variables. We defer a discussion of it until the next section after we have studied intersections of ideals. For the purposes of our discussion here, let us assume that we have such an algorithm. We also remark that given polynomials $f_1, \ldots, f_s \in k[x_1, \ldots, x_n]$, one can define $\gcd(f_1, f_2, \ldots, f_s)$ exactly as in the one-variable case. There is also an algorithm for computing $\gcd(f_1, f_2, \ldots, f_s)$.

Using this notion of gcd, we can now give a formula for computing the radical of a principal ideal.

Proposition 12. *Suppose that k is a field containing the rational numbers $\mathbb{Q}$ and let $I = \langle f \rangle$ be a principal ideal in $k[x_1, \ldots, x_n]$. Then $\sqrt{I} = \langle f_{\mathrm{red}} \rangle$, where*

$$f_{\mathrm{red}} = \frac{f}{\gcd\left(f, \frac{\partial f}{\partial x_1}, \frac{\partial f}{\partial x_2}, \ldots, \frac{\partial f}{\partial x_n}\right)}.$$

Proof. Writing f as in Proposition 9, we know that $\sqrt{I} = \langle f_1 f_2 \cdots f_r \rangle$. Thus, it suffices to show that

(1)
$$\gcd\left(f, \frac{\partial f}{\partial x_1}, \ldots, \frac{\partial f}{\partial x_n}\right) = f_1^{a_1-1} f_2^{a_2-1} \cdots f_r^{a_r-1}.$$

We first use the product rule to note that

$$\frac{\partial f}{\partial x_j} = f_1^{a_1-1} f_2^{a_2-1} \cdots f_r^{a_r-1}\left(a_1 \frac{\partial f_1}{\partial x_j} f_2 \cdots f_r + \cdots + a_r f_1 \cdots f_{r-1} \frac{\partial f_r}{\partial x_j}\right).$$

This proves that $f_1^{a_1-1} f_2^{a_2-1} \cdots f_r^{a_r-1}$ divides the gcd. It remains to show that for each i, there is some $\frac{\partial f}{\partial x_j}$ which is not divisible by $f_i^{a_i}$.

Write $f = f_i^{a_i} h_i$, where h_i is not divisible by f_i. Since f_i is nonconstant, some variable x_j must appear in f_i. The product rule gives us

$$\frac{\partial f}{\partial x_j} = f_i^{a_i-1}\left(a_1 \frac{\partial f_i}{\partial x_j} h_i + f_i \frac{\partial h_i}{\partial x_j}\right).$$

If this expression is divisible by $f_i^{a_i}$, then $\frac{\partial f_i}{\partial x_j} h_i$ must be divisible by f_i. Since f_i is irreducible and does not divide h_i, this forces f_i to divide $\frac{\partial f_i}{\partial x_j}$. In Exercise 13, you will show that $\frac{\partial f_i}{\partial x_j}$ is nonzero since $\mathbb{Q} \subseteq k$ and x_j appears in f_i. As $\frac{\partial f_i}{\partial x_j}$ also has smaller total degree than f_i, it follows that f_i cannot divide $\frac{\partial f_i}{\partial x_j}$. Consequently, $\frac{\partial f}{\partial x_j}$ is not divisible by $f_i^{a_i}$, which proves (1), and the proposition follows.   □

It is worth remarking that for fields which do not contain $\mathbb{Q}$, the above formula for f_{red} may fail (see Exercise 13).

EXERCISES FOR §2

1. Given a field k (not necessarily algebraically closed), show that $\sqrt{\langle x^2, y^2 \rangle} = \langle x, y \rangle$ and, more generally, show that $\sqrt{\langle x^n, y^m \rangle} = \langle x, y \rangle$ for any positive integers n and m.

2. Let f and g be distinct nonconstant polynomials in $k[x, y]$ and let $I = \langle f^2, g^3 \rangle$. Is it necessarily true that $\sqrt{I} = \langle f, g \rangle$? Explain.

3. Show that $\langle x^2 + 1 \rangle \subseteq \mathbb{R}[x]$ is a radical ideal, but that $\mathbf{V}(x^2 + 1)$ is the empty variety.

4. Let I be an ideal in $k[x_1, \ldots, x_n]$, where k is an arbitrary field.

   a. Show that $\sqrt{I}$ is a radical ideal.
   b. Show that I is radical if and only if $I = \sqrt{I}$.
   c. Show that $\sqrt{\sqrt{I}} = \sqrt{I}$.

5. Prove that $\mathbf{I}$ and $\mathbf{V}$ are inclusion-reversing and that $\mathbf{V}(\sqrt{I}) = \mathbf{V}(I)$ for any ideal I.

6. Let I be an ideal in $k[x_1, \ldots, x_n]$.

   a. In the special case when $\sqrt{I} = \langle f_1, f_2 \rangle$, with $f_i^{m_i} \in I$, prove that $f^{m_1 + m_2 - 1} \in I$ for all $f \in \sqrt{I}$.
   b. Now prove that for any I, there exists a single integer m such that $f^m \in I$ for all $f \in \sqrt{I}$. Hint: Write $\sqrt{I} = \langle f_1, \ldots, f_s \rangle$.

7. Determine whether the following polynomials lie in the following radicals. If the answer is yes, what is the smallest power of the polynomial that lies in the ideal?

   a. Is $x + y \in \sqrt{\langle x^3, y^3, xy(x + y) \rangle}$?
   b. Is $x^2 + 3xz \in \sqrt{\langle x + z, x^2 y, x - z^2 \rangle}$?

8. Let $f_1 = y^2 + 2xy - 1$ and $f_2 = x^2 + 1$. Prove that $\langle f_1, f_2 \rangle$ is not a radical ideal. Hint: What is $f_1 + f_2$?

9. Given $f, g \in k[x_1, \ldots, x_n]$, use unique factorization to prove that $\gcd(f, g)$ exists. Also prove that $\gcd(f, g)$ is unique up to multiplication by a nonzero constant of k.

10. Prove the following ideal-theoretic characterization of $\gcd(f, g)$: given polynomials f, g, h in $k[x_1, \ldots, x_n]$, then $h = \gcd(f, g)$ if and only if h is a generator of the smallest principal ideal containing $\langle f, g \rangle$ (i.e., if $\langle h \rangle \subseteq J$, whenever J is a principal ideal such that $J \supseteq \langle f, g \rangle$).

11. Find a basis for the ideal

$$\sqrt{\langle x^5 - 2x^4 + 2x^2 - x, x^5 - x^4 - 2x^3 + 2x^2 + x - 1 \rangle}.$$

Compare with Exercise 17 of Chapter 1, §5.

12. Let $f = x^5 + 3x^4 y + 3x^3 y^2 - 2x^4 y^2 + x^2 y^3 - 6x^3 y^3 - 6x^2 y^4 + x^3 y^4 - 2xy^5 + 3x^2 y^5 + 3xy^6 + y^7 \in \mathbb{Q}[x, y]$. Compute $\sqrt{\langle f \rangle}$.

13. A field k has *characteristic zero* if it contains the rational numbers $\mathbb{Q}$; otherwise, k has *positive characteristic*.

    a. Let k be the field $\mathbb{F}_2$ from Exercise 1 of Chapter 1, §1. If $f = x_1^2 + \cdots + x_n^2 \in \mathbb{F}_2[x_1, \ldots, x_n]$, then show that $\frac{\partial f}{\partial x_i} = 0$ for all i. Conclude that the formula given in Proposition 12 may fail when the field is $\mathbb{F}_2$.

b. Let k be a field of characteristic zero and let $f \in k[x_1, \ldots, x_n]$ be nonconstant. If the variable x_j appears in f, then prove that $\frac{\partial f}{\partial x_j} \neq 0$. Also explain why $\frac{\partial f}{\partial x_j}$ has smaller total degree than f.

14. Let $J = \langle xy, (x-y)x \rangle$. Describe $\mathbf{V}(J)$ and show that $\sqrt{J} = \langle x \rangle$.

15. Prove that $I = \langle xy, xz, yz \rangle$ is a radical ideal. Hint: If you divide $f \in k[x, y, z]$ by xy, xz, yz, what does the remainder look like? What does f^m look like?

16. Let $I \subseteq k[x_1, \ldots, x_n]$ be an ideal. Assume that I has a Gröbner basis $G = \{g_1, \ldots, g_t\}$ such that for all i, $\mathrm{LT}(g_i)$ is square-free in the sense of Definition 10.
    a. If $f \in \sqrt{I}$, prove that $\mathrm{LT}(f)$ is divisible by $\mathrm{LT}(g_i)$ for some i. Hint: $f^m \in I$.
    b. Prove that I is radical. Hint: Use part (a) to show that G is a Gröbner basis of $\sqrt{I}$.

17. This exercise continues the line of thought begun in Exercise 16.
    a. Prove that a monomial ideal in $k[x_1, \ldots, x_n]$ is radical if and only if its minimal generators are square-free.
    b. Given an ideal $I \subseteq k[x_1, \ldots, x_n]$, prove that if $\langle \mathrm{LT}(I) \rangle$ is radical, then I is radical.
    c. Give an example to show that the converse of part (b) can fail.

§3 Sums, Products, and Intersections of Ideals

Ideals are algebraic objects and, as a result, there are natural algebraic operations we can define on them. In this section, we consider three such operations: sum, intersection, and product. These are binary operations: to each pair of ideals, they associate a new ideal. We shall be particularly interested in two general questions which arise in connection with each of these operations. The first asks how, given generators of a pair of ideals, one can compute generators of the new ideals which result on applying these operations. The second asks for the geometric significance of these algebraic operations. Thus, the first question fits the general computational theme of this book; the second, the general thrust of this chapter. We consider each of the operations in turn.

Sums of Ideals

Definition 1. If I and J are ideals of the ring $k[x_1, \ldots, x_n]$, then the **sum** of I and J, denoted $I + J$, is the set

$$I + J = \{f + g \mid f \in I \text{ and } g \in J\}.$$

Proposition 2. If I and J are ideals in $k[x_1, \ldots, x_n]$, then $I + J$ is also an ideal in $k[x_1, \ldots, x_n]$. In fact, $I + J$ is the smallest ideal containing I and J. Furthermore, if $I = \langle f_1, \ldots, f_r \rangle$ and $J = \langle g_1, \ldots, g_s \rangle$, then $I + J = \langle f_1, \ldots, f_r, g_1, \ldots, g_s \rangle$.

Proof. Note first that $0 = 0 + 0 \in I + J$. Suppose $h_1, h_2 \in I + J$. By the definition of $I + J$, there exist $f_1, f_2 \in I$ and $g_1, g_2 \in J$ such that $h_1 = f_1 + g_1, h_2 = f_2 + g_2$. Then, after rearranging terms slightly, $h_1 + h_2 = (f_1 + f_2) + (g_1 + g_2)$. But $f_1 + f_2 \in I$ because I is an ideal and, similarly, $g_1 + g_2 \in J$, whence $h_1 + h_2 \in I + J$. To check closure under multiplication, let $h \in I + J$ and $p \in k[x_1, \ldots, x_n]$ be any polynomial. Then, as above, there exist $f \in I$ and $g \in J$ such that $h = f + g$. But then $p \cdot h = p \cdot (f + g) = p \cdot f + p \cdot g$. Now $p \cdot f \in I$ and $p \cdot g \in J$ because I and J are ideals. Consequently, $p \cdot h \in I + J$. This shows that $I + J$ is an ideal.

If H is an ideal which contains I and J, then H must contain all elements $f \in I$ and $g \in J$. Since H is an ideal, H must contain all $f + g$, where $f \in I, g \in J$. In particular, $H \supseteq I + J$. Therefore, every ideal containing I and J contains $I + J$ and, thus, $I + J$ must be the smallest such ideal.

Finally, if $I = \langle f_1, \ldots, f_r \rangle$ and $J = \langle g_1, \ldots, g_s \rangle$, then $\langle f_1, \ldots, f_r, g_1, \ldots, g_s \rangle$ is an ideal containing I and J, so that $I + J \subseteq \langle f_1, \ldots, f_r, g_1, \ldots, g_s \rangle$. The reverse inclusion is obvious, so that $I + J = \langle f_1, \ldots, f_r, g_1, \ldots, g_s \rangle$. □

The following corollary is an immediate consequence of Proposition 2.

Corollary 3. *If $f_1, \ldots, f_r \in k[x_1, \ldots, x_n]$, then*

$$\langle f_1, \ldots, f_r \rangle = \langle f_1 \rangle + \cdots + \langle f_r \rangle.$$

To see what happens geometrically, let $I = \langle x^2 + y \rangle$ and $J = \langle z \rangle$ be ideals in $\mathbb{R}[x, y, z]$. We have sketched $\mathbf{V}(I)$ and $\mathbf{V}(J)$ on the next page. Then $I + J = \langle x^2 + y, z \rangle$ contains both $x^2 + y$ and z. Thus, the variety $\mathbf{V}(I + J)$ must consist of those points where both $x^2 + y$ and z vanish, i.e., it must be the intersection of $\mathbf{V}(I)$ and $\mathbf{V}(J)$.

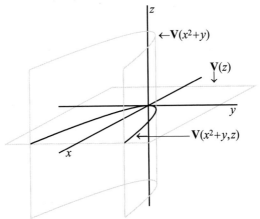

The same line of reasoning generalizes to show that addition of ideals corresponds geometrically to taking intersections of varieties.

Theorem 4. *If I and J are ideals in $k[x_1, \ldots, x_n]$, then $\mathbf{V}(I + J) = \mathbf{V}(I) \cap \mathbf{V}(J)$.*

Proof. If $a \in \mathbf{V}(I + J)$, then $a \in \mathbf{V}(I)$ because $I \subseteq I + J$; similarly, $a \in \mathbf{V}(J)$. Thus, $a \in \mathbf{V}(I) \cap \mathbf{V}(J)$ and we conclude that $\mathbf{V}(I + J) \subseteq \mathbf{V}(I) \cap \mathbf{V}(J)$.

To get the opposite inclusion, suppose $a \in \mathbf{V}(I) \cap \mathbf{V}(J)$. Let h be any polynomial in $I + J$. Then there exist $f \in I$ and $g \in J$ such that $h = f + g$. We have $f(a) = 0$ because $a \in \mathbf{V}(I)$ and $g(a) = 0$ because $a \in \mathbf{V}(J)$. Thus, $h(a) = f(a) + g(a) = 0 + 0 = 0$. Since h was arbitrary, we conclude that $a \in \mathbf{V}(I + J)$. Hence, $\mathbf{V}(I + J) \supseteq \mathbf{V}(I) \cap \mathbf{V}(J)$. □

An analogue of Theorem 4 stated in terms of generators was given in Lemma 2 of Chapter 1, §2.

Products of Ideals

In Lemma 2 of Chapter 1, §2, we encountered the fact that an ideal generated by the products of the generators of two other ideals corresponds to the union of varieties:

$$\mathbf{V}(f_1,\ldots,f_r) \cup \mathbf{V}(g_1,\ldots,g_s) = \mathbf{V}(f_i g_j, 1 \le i \le r, 1 \le j \le s).$$

Thus, for example, the variety $\mathbf{V}(xz, yz)$ corresponding to an ideal generated by the product of the generators of the ideals, $\langle x, y\rangle$ and $\langle z\rangle$ in $k[x,y,z]$ is the union of $\mathbf{V}(x,y)$ (the z-axis) and $\mathbf{V}(z)$ [the (x,y)-plane]. This suggests the following definition.

Definition 5. If I and J are two ideals in $k[x_1,\ldots,x_n]$, then their **product**, denoted $I \cdot J$, is defined to be the ideal generated by all polynomials $f \cdot g$ where $f \in I$ and $g \in J$.

Thus, the product $I \cdot J$ of I and J is the set

$$I \cdot J = \{f_1 g_1 + \cdots + f_r g_r \mid f_1,\ldots,f_r \in I,\ g_1,\ldots,g_r \in J,\ r \text{ a positive integer}\}.$$

To see that this is an ideal, note that $0 = 0 \cdot 0 \in I \cdot J$. Moreover, it is clear that $h_1, h_2 \in I \cdot J$ implies that $h_1 + h_2 \in I \cdot J$. Finally, if $h = f_1 g_1 + \cdots + f_r g_r \in I \cdot J$ and p is any polynomial, then

$$ph = (pf_1)g_1 + \cdots + (pf_r)g_r \in I \cdot J$$

since $pf_i \in I$ for all i, $1 \le i \le r$. Note that the set of products would not be an ideal because it would not be closed under addition. The following easy proposition shows that computing a set of generators for $I \cdot J$ given sets of generators for I and J is completely straightforward.

Proposition 6. Let $I = \langle f_1,\ldots,f_r\rangle$ and $J = \langle g_1,\ldots,g_s\rangle$. Then $I \cdot J$ is generated by the set of all products of generators of I and J:

$$I \cdot J = \langle f_i g_j \mid 1 \le i \le r, 1 \le j \le s\rangle.$$

Proof. It is clear that the ideal generated by products $f_i g_j$ of the generators is contained in $I \cdot J$. To establish the opposite inclusion, note that any polynomial in $I \cdot J$ is a sum of polynomials of the form fg with $f \in I$ and $g \in J$. But we can write f and g in terms of the generators $f_1,\ldots,f_r$ and $g_1,\ldots,g_s$, respectively, as

$$f = a_1 f_1 + \cdots + a_r f_r, \quad g = b_1 g_1 + \cdots + b_s g_s$$

for appropriate polynomials $a_1,\ldots,a_r,b_1,\ldots,b_s$. Thus, fg, and consequently any sum of polynomials of this form, can be written as a sum $\sum_{ij} c_{ij} f_i g_j$, where $c_{ij} \in k[x_1,\ldots,x_n]$. □

The following proposition guarantees that the product of ideals does indeed correspond geometrically to the operation of taking the union of varieties.

Theorem 7. *If I and J are ideals in* $k[x_1, \ldots, x_n]$, *then* $\mathbf{V}(I \cdot J) = \mathbf{V}(I) \cup \mathbf{V}(J)$.

Proof. Let $a \in \mathbf{V}(I \cdot J)$. Then $g(a)h(a) = 0$ for all $g \in I$ and all $h \in J$. If $g(a) = 0$ for all $g \in I$, then $a \in \mathbf{V}(I)$. If $g(a) \neq 0$ for some $g \in I$, then we must have $h(a) = 0$ for all $h \in J$. In either event, $a \in \mathbf{V}(I) \cup \mathbf{V}(J)$.

Conversely, suppose $a \in \mathbf{V}(I) \cup \mathbf{V}(J)$. Either $g(a) = 0$ for all $g \in I$ or $h(a) = 0$ for all $h \in J$. Thus, $g(a)h(a) = 0$ for all $g \in I$ and $h \in J$. Thus, $f(a) = 0$ for all $f \in I \cdot J$ and, hence, $a \in \mathbf{V}(I \cdot J)$.   □

In what follows, we will often write the product of ideals as IJ rather than $I \cdot J$.

Intersections of Ideals

The operation of forming the intersection of two ideals is, in some ways, even more primitive than the operations of addition and multiplication.

Definition 8. The **intersection** $I \cap J$ of two ideals I and J in $k[x_1, \ldots, x_n]$ is the set of polynomials which belong to both I and J.

As in the case of sums, the set of ideals is closed under intersections.

Proposition 9. *If I and J are ideals in* $k[x_1, \ldots, x_n]$, *then* $I \cap J$ *is also an ideal.*

Proof. Note that $0 \in I \cap J$ since $0 \in I$ and $0 \in J$. If $f, g \in I \cap J$, then $f + g \in I$ because $f, g \in I$. Similarly, $f + g \in J$ and, hence, $f + g \in I \cap J$. Finally, to check closure under multiplication, let $f \in I \cap J$ and h be any polynomial in $k[x_1, \ldots, x_n]$. Since $f \in I$ and I is an ideal, we have $h \cdot f \in I$. Similarly, $h \cdot f \in J$ and, hence, $h \cdot f \in I \cap J$.   □

Note that we always have $IJ \subseteq I \cap J$ since elements of IJ are sums of polynomials of the form fg with $f \in I$ and $g \in J$. But the latter belongs to both I (since $f \in I$) and J (since $g \in J$). However, IJ can be strictly contained in $I \cap J$. For example, if $I = J = \langle x, y \rangle$, then $IJ = \langle x^2, xy, y^2 \rangle$ is strictly contained in $I \cap J = I = \langle x, y \rangle$ ($x \in I \cap J$, but $x \notin IJ$).

Given two ideals and a set of generators for each, we would like to be able to compute a set of generators for the intersection. This is much more delicate than the analogous problems for sums and products of ideals, which were entirely straightforward. To see what is involved, suppose I is the ideal in $\mathbb{Q}[x, y]$ generated by the polynomial $f = (x + y)^4(x^2 + y)^2(x - 5y)$ and let J be the ideal generated by the polynomial $g = (x + y)(x^2 + y)^3(x + 3y)$. We leave it as an (easy) exercise to check that

$$I \cap J = \langle (x + y)^4(x^2 + y)^3(x - 5y)(x + 3y) \rangle.$$

This computation is easy precisely because we were given factorizations of f and g into irreducible polynomials. In general, such factorizations may not be available. So any algorithm which allows one to compute intersections will have to be powerful enough to circumvent this difficulty.

Nevertheless, there is a nice trick that reduces the computation of intersections to computing the intersection of an ideal with a subring (i.e., eliminating variables), a problem which we have already solved. To state the theorem, we need a little notation: if I is an ideal in $k[x_1, \ldots, x_n]$ and $f(t) \in k[t]$ a polynomial in the single variable t, then $f(t)I$ denotes the ideal in $k[x_1, \ldots, x_n, t]$ generated by the set of polynomials $\{f(t) \cdot h \mid h \in I\}$. This is a little different from our usual notion of product in that the ideal I and the ideal generated by $f(t)$ in $k[t]$ lie in different rings: in fact, the ideal $I \subseteq k[x_1, \ldots, x_n]$ is *not* an ideal in $k[x_1, \ldots, x_n, t]$ because it is not closed under multiplication by t. When we want to stress that a polynomial $h \in k[x_1, \ldots, x_n]$ involves only the variables $x_1, \ldots, x_n$, we write $h = h(x)$. Along the same lines, if we are considering a polynomial g in $k[x_1, \ldots, x_n, t]$ and we want to emphasize that it can involve the variables $x_1, \ldots, x_n$ as well as t, we will write $g = g(x, t)$. In terms of this notation, $f(t)I = \langle f(t)h(x) \mid h(x) \in I \rangle$. So, for example, if $f(t) = t^2 - t$ and $I = \langle x, y \rangle$, then the ideal $f(t)I$ in $k[x, y, t]$ contains $(t^2 - t)x$ and $(t^2 - t)y$. In fact, it is not difficult to see that $f(t)I$ is generated as an ideal by $(t^2 - t)x$ and $(t^2 - t)y$. This is a special case of the following assertion.

Lemma 10.

(i) *If I is generated as an ideal in $k[x_1, \ldots, x_n]$ by $p_1(x), \ldots, p_r(x)$, then $f(t)I$ is generated as an ideal in $k[x_1, \ldots, x_n, t]$ by $f(t) \cdot p_1(x), \ldots, f(t) \cdot p_r(x)$.*
(ii) *If $g(x, t) \in f(t)I$ and a is any element of the field k, then $g(x, a) \in I$.*

Proof. To prove the first assertion, note that any polynomial $g(x, t) \in f(t)I$ can be expressed as a sum of terms of the form $h(x, t) \cdot f(t) \cdot p(x)$ for $h \in k[x_1, \ldots, x_n, t]$ and $p \in I$. But because I is generated by $p_1, \ldots, p_r$ the polynomial $p(x)$ can be expressed as a sum of terms of the form $q_i(x)p_i(x)$, $1 \le i \le r$. In other words,

$$p(x) = \sum_{i=1}^{r} q_i(x)p_i(x).$$

Hence,

$$h(x, t) \cdot f(t) \cdot p(x) = \sum_{i=1}^{r} h(x, t)q_i(x)f(t)p_i(x).$$

Now, for each i, $1 \le i \le r$, $h(x, t) \cdot q_i(x) \in k[x_1, \ldots, x_n, t]$. Thus, $h(x, t) \cdot f(t) \cdot p(x)$ belongs to the ideal in $k[x_1, \ldots, x_n, t]$ generated by $f(t) \cdot p_1(x), \ldots, f(t) \cdot p_r(x)$. Since $g(x, t)$ is a sum of such terms,

$$g(x, t) \in \langle f(t) \cdot p_1(x), \ldots, f(t) \cdot p_r(x) \rangle,$$

which establishes (i). The second assertion follows immediately upon substituting $a \in k$ for t. $\qquad\square$

Theorem 11. *Let I, J be ideals in $k[x_1, \ldots, x_n]$. Then*

$$I \cap J = (tI + (1 - t)J) \cap k[x_1, \ldots, x_n].$$

Proof. Note that $tI + (1 - t)J$ is an ideal in $k[x_1, \ldots, x_n, t]$. To establish the desired equality, we use the usual strategy of proving containment in both directions.

Suppose $f \in I \cap J$. Since $f \in I$, we have $t \cdot f \in tI$. Similarly, $f \in J$ implies $(1 - t) \cdot f \in (1 - t)J$. Thus, $f = t \cdot f + (1 - t) \cdot f \in tI + (1 - t)J$. Since $I, J \subseteq k[x_1, \ldots, x_n]$, we have $f \in (tI + (1 - t)J) \cap k[x_1, \ldots, x_n]$. This shows that $I \cap J \subseteq (tI + (1 - t)J) \cap k[x_1, \ldots, x_n]$.

To establish the opposite containment, take $f \in (tI + (1 - t)J) \cap k[x_1, \ldots, x_n]$. Then $f(x) = g(x, t) + h(x, t)$, where $g(x, t) \in tI$ and $h(x, t) \in (1 - t)J$. First set $t = 0$. Since every element of tI is a multiple of t, we have $g(x, 0) = 0$. Thus, $f(x) = h(x, 0)$ and hence, $f(x) \in J$ by Lemma 10. On the other hand, set $t = 1$ in the relation $f(x) = g(x, t) + h(x, t)$. Since every element of $(1 - t)J$ is a multiple of $1 - t$, we have $h(x, 1) = 0$. Thus, $f(x) = g(x, 1)$ and, hence, $f(x) \in I$ by Lemma 10. Since f belongs to both I and J, we have $f \in I \cap J$. Thus, $I \cap J \supseteq (tI + (1 - t)J) \cap k[x_1, \ldots, x_n]$ and this completes the proof. $\square$

The above result and the Elimination Theorem (Theorem 2 of Chapter 3, §1) lead to the following **algorithm for computing intersections of ideals**: if $I = \langle f_1, \ldots, f_r \rangle$ and $J = \langle g_1, \ldots, g_s \rangle$ are ideals in $k[x_1, \ldots, x_n]$, we consider the ideal

$$\langle tf_1, \ldots, tf_r, (1 - t)g_1, \ldots, (1 - t)g_s \rangle \subseteq k[x_1, \ldots, x_n, t]$$

and compute a Gröbner basis with respect to lex order in which t is greater than the x_i. The elements of this basis which do not contain the variable t will form a basis (in fact, a Gröbner basis) of $I \cap J$. For more efficient calculations, one could also use one of the orders described in Exercises 5 and 6 of Chapter 3, §1. An algorithm for intersecting three or more ideals is described in Proposition 6.19 of BECKER and WEISPFENNING (1993).

As a simple example of the above procedure, suppose we want to compute the intersection of the ideals $I = \langle x^2 y \rangle$ and $J = \langle xy^2 \rangle$ in $\mathbb{Q}[x, y]$. We consider the ideal

$$tI + (1 - t)J = \langle tx^2 y, (1 - t)xy^2 \rangle = \langle tx^2 y, txy^2 - xy^2 \rangle$$

in $\mathbb{Q}[t, x, y]$. Computing the S-polynomial of the generators, we obtain $tx^2 y^2 - (tx^2 y^2 - x^2 y^2) = x^2 y^2$. It is easily checked that $\{tx^2 y, txy^2 - xy^2, x^2 y^2\}$ is a Gröbner basis of $tI + (1 - t)J$ with respect to lex order with $t > x > y$. By the Elimination Theorem, $\{x^2 y^2\}$ is a (Gröbner) basis of $(tI + (1 - t)J) \cap \mathbb{Q}[x, y]$. Thus,

$$I \cap J = \langle x^2 y^2 \rangle.$$

As another example, we invite the reader to apply the algorithm for computing intersections of ideals to give an alternate proof that the intersection $I \cap J$ of the ideals

$$I = \langle (x + y)^4 (x^2 + y)^2 (x - 5y) \rangle \quad \text{and} \quad J = \langle (x + y)(x^2 + y)^3 (x + 3y) \rangle$$

in $\mathbb{Q}[x, y]$ is

$$I \cap J = \langle (x + y)^4 (x^2 + y)^3 (x - 5y)(x + 3y) \rangle.$$

These examples above are rather simple in that our algorithm applies to ideals which are not necessarily principal, whereas the examples given here involve intersections of principal ideals. We shall see a somewhat more complicated example in the exercises.

We can generalize both of the examples above by introducing the following definition.

Definition 12. A polynomial $h \in k[x_1, \ldots, x_n]$ is called a **least common multiple** of $f, g \in k[x_1, \ldots, x_n]$ and denoted $h = \text{lcm}(f, g)$ if

(i) f divides h and g divides h.
(ii) If f and g both divide a polynomial p, then h divides p.

For example,

$$\text{lcm}(x^2 y, xy^2) = x^2 y^2$$

and

$$\text{lcm}((x + y)^4 (x^2 + y)^2 (x - 5y), (x + y)(x^2 + y)^3 (x + 3y))$$
$$= (x + y)^4 (x^2 + y)^3 (x - 5y)(x + 3y).$$

More generally, suppose $f, g \in k[x_1, \ldots, x_n]$ and let $f = c f_1^{a_1} \ldots f_r^{a_r}$ and $g = c' g_1^{b_1} \ldots g_s^{b_s}$ be their factorizations into distinct irreducible polynomials. It may happen that some of the irreducible factors of f are constant multiples of those of g. In this case, let us suppose that we have rearranged the order of the irreducible polynomials in the expressions for f and g so that for some l, $1 \le l \le \min(r, s)$, f_i is a constant (nonzero) multiple of g_i for $1 \le i \le l$ and for all $i, j > l, f_i$ is not a constant multiple of g_j. Then it follows from unique factorization that

(1) $$\text{lcm}(f, g) = f_1^{\max(a_1, b_1)} \cdots f_l^{\max(a_l, b_l)} \cdot g_{l+1}^{b_{l+1}} \cdots g_s^{b_s} \cdot f_{l+1}^{a_{l+1}} \cdots f_r^{a_r}.$$

[In the case that f and g share no common factors, we have $\text{lcm}(f, g) = f \cdot g$.] This, in turn, implies the following result.

Proposition 13.

(i) *The intersection $I \cap J$ of two principal ideals $I, J \subseteq k[x_1, \ldots, x_n]$ is a principal ideal.*
(ii) *If $I = \langle f \rangle$, $J = \langle g \rangle$ and $I \cap J = \langle h \rangle$, then*

$$h = \text{lcm}(f, g).$$

Proof. The proof will be left as an exercise. $\qquad\square$

This result, together with our algorithm for computing the intersection of two ideals immediately gives an **algorithm for computing the least common multiple** of two polynomials: to compute the least common multiple of two polynomials

$f, g \in k[x_1, \ldots, x_n]$, we compute the intersection $\langle f \rangle \cap \langle g \rangle$ using our algorithm for computing the intersection of ideals. Proposition 13 assures us that this intersection is a principal ideal (in the exercises, we ask you to prove that the intersection of principal ideals is principal) and that any generator of it is a least common multiple of f and g.

This algorithm for computing least common multiples allows us to clear up a point which we left unfinished in §2: namely, the computation of the greatest common divisor of two polynomials f and g. The crucial observation is the following.

Proposition 14. *Let $f, g \in k[x_1, \ldots, x_n]$. Then*

$$\text{lcm}(f, g) \cdot \gcd(f, g) = fg.$$

Proof. This follows by expressing f and g as products of distinct irreducibles and then using the remarks preceding Proposition 13, especially equation (1). You will provide the details in Exercise 5.   □

It follows immediately from Proposition 14 that

$$(2) \qquad\qquad\qquad \gcd(f, g) = \frac{f \cdot g}{\text{lcm}(f, g)}.$$

This gives an **algorithm for computing the greatest common divisor** of two polynomials f and g. Namely, we compute $\text{lcm}(f, g)$ using our algorithm for the least common multiple and divide it into the product of f and g using the division algorithm.

We should point out that the gcd algorithm just described is rather cumbersome. In practice, more efficient algorithms are used [see DAVENPORT, SIRET and TOURNIER (1993)].

Having dealt with the computation of intersections, we now ask what operation on varieties corresponds to the operation of intersection on ideals. The following result answers this question.

Theorem 15. *If I and J are ideals in $k[x_1, \ldots, x_n]$, then $\mathbf{V}(I \cap J) = \mathbf{V}(I) \cup \mathbf{V}(J)$.*

Proof. Let $a \in \mathbf{V}(I) \cup \mathbf{V}(J)$. Then $a \in \mathbf{V}(I)$ or $a \in \mathbf{V}(J)$. This means that either $f(a) = 0$ for all $f \in I$ or $f(a) = 0$ for all $f \in J$. Thus, certainly, $f(a) = 0$ for all $f \in I \cap J$. Hence, $a \in \mathbf{V}(I \cap J)$. Hence, $\mathbf{V}(I) \cup \mathbf{V}(J) \subseteq \mathbf{V}(I \cap J)$.

On the other hand, note that since $IJ \subseteq I \cap J$, we have $\mathbf{V}(I \cap J) \subseteq \mathbf{V}(IJ)$. But $\mathbf{V}(IJ) = \mathbf{V}(I) \cup \mathbf{V}(J)$ by Theorem 7, and we immediately obtain the reverse inequality.   □

Thus, the intersection of two ideals corresponds to the same variety as the product. In view of this and the fact that the intersection is much more difficult to compute than the product, one might legitimately question the wisdom of bothering with the intersection at all. The reason is that intersection behaves much better with respect to the operation of taking radicals: the product of radical ideals need not be a radical ideal (consider IJ where $I = J$), but the intersection of radical ideals is always a radical ideal. The latter fact is a consequence of the next proposition.

Proposition 16. *If I, J are any ideals, then*

$$\sqrt{I \cap J} = \sqrt{I} \cap \sqrt{J}.$$

Proof. If $f \in \sqrt{I \cap J}$, then $f^m \in I \cap J$ for some integer $m > 0$. Since $f^m \in I$, we have $f \in \sqrt{I}$. Similarly, $f \in \sqrt{J}$. Thus, $\sqrt{I \cap J} \subseteq \sqrt{I} \cap \sqrt{J}$.

For the reverse inclusion, take $f \in \sqrt{I} \cap \sqrt{J}$. Then, there exist integers $m, p > 0$ such that $f^m \in I$ and $f^p \in J$. Thus $f^{m+p} = f^m f^p \in I \cap J$, so $f \in \sqrt{I \cap J}$. □

EXERCISES FOR §3

1. Show that in $\mathbb{Q}[x, y]$, we have

$$\langle (x+y)^4 (x^2+y)^2 (x-5y) \rangle \cap \langle (x+y)(x^2+y)^3 (x+3y) \rangle = \langle (x+y)^4 (x^2+y)^3 (x-5y)(x+3y) \rangle.$$

2. Prove formula (1) for the least common multiple of two polynomials f and g.

3. Prove assertion (i) of Proposition 13. In other words, show that the intersection of two principal ideals is principal.

4. Prove assertion (ii) of Proposition 13. In other words, show that the least common multiple of two polynomials f and g in $k[x_1, \ldots, x_n]$ is the generator of the ideal $\langle f \rangle \cap \langle g \rangle$.

5. Prove Proposition 14. In other words, show that the least common multiple of two polynomials times the greatest common divisor of the same two polynomials is the product of the polynomials. Hint: Use the remarks following the statement of Proposition 14.

6. Let $I_1, \ldots, I_r$ and J be ideals in $k[x_1, \ldots, x_n]$. Show the following:
   a. $(I_1 + I_2)J = I_1 J + I_2 J$.
   b. $(I_1 \cdots I_r)^m = I_1^m \cdots I_r^m$.

7. Let I and J be ideals in $k[x_1, \ldots, x_n]$, where k is an arbitrary field. Prove the following:
   a. If $I^\ell \subseteq J$ for some integer $\ell > 0$, then $\sqrt{I} \subseteq \sqrt{J}$.
   b. $\sqrt{I + J} = \sqrt{\sqrt{I} + \sqrt{J}}$.

8. Let
$$f = x^4 + x^3 y + x^3 z^2 - x^2 y^2 + x^2 yz^2 - xy^3 - xy^2 z^2 - y^3 z^2$$
and
$$g = x^4 + 2x^3 z^2 - x^2 y^2 + x^2 z^4 - 2xy^2 z^2 - y^2 z^4.$$
   a. Use a computer algebra program to compute generators for $\langle f \rangle \cap \langle g \rangle$ and $\sqrt{\langle f \rangle \langle g \rangle}$.
   b. Use a computer algebra program to compute $\gcd(f, g)$.
   c. Let $p = x^2 + xy + xz + yz$ and $q = x^2 - xy - xz + yz$. Use a computer algebra program to calculate $\langle f, g \rangle \cap \langle p, q \rangle$.

9. For an arbitrary field, show that $\sqrt{IJ} = \sqrt{I \cap J}$. Give an example to show that the product of radical ideals need not be radical. Also give an example to show that $\sqrt{IJ}$ can differ from $\sqrt{I}\sqrt{J}$.

10. If I is an ideal in $k[x_1, \ldots, x_n]$ and $\langle f(t) \rangle$ is an ideal in $k[t]$, show that the ideal $f(t)I$ defined in the text is the product of the ideal $\tilde{I}$ generated by all elements of I in $k[x_1, \ldots, x_n, t]$ and the ideal $\langle f(t) \rangle$ generated by $f(t)$ in $k[x_1, \ldots, x_n, t]$.

11. Two ideals I and J of $k[x_1, \ldots, x_n]$ are said to be *comaximal* if and only if $I + J = k[x_1, \ldots, x_n]$.
   a. Show that if $k = \mathbb{C}$, then I and J are comaximal if and only if $\mathbf{V}(I) \cap \mathbf{V}(J) = \emptyset$. Give an example to show that this is false in general.
   b. Show that if I and J are comaximal, then $IJ = I \cap J$.

c. Is the converse to part (b) true? That is, if $IJ = I \cap J$, does it necessarily follow that I and J are comaximal? Proof or counterexample?

d. If I and J are comaximal, show that I and J^2 are comaximal. In fact, show that I^r and J^s are comaximal for all positive integers r and s.

e. Let $I_1, \ldots, I_r$ be ideals in $k[x_1, \ldots, x_n]$ and suppose that I_i and $J_i = \bigcap_{j \neq i} I_j$ are comaximal for all i. Show that

$$I_1^m \cap \cdots \cap I_r^m = (I_1 \cdots I_r)^m = (I_1 \cap \cdots \cap I_r)^m$$

for all positive integers m.

12. Let I, J be ideals in $k[x_1, \ldots, x_n]$ and suppose that $I \subseteq \sqrt{J}$. Show that $I^m \subseteq J$ for some integer $m > 0$. Hint: You will need to use the Hilbert Basis Theorem.

13. Let A be an $m \times n$ constant matrix and suppose that $x = Ay$, where we are thinking of $x \in k^m$ and $y \in k^n$ as column vectors of variables. Define a map

$$\alpha_A : k[x_1, \ldots, x_m] \longrightarrow k[y_1, \ldots, y_n]$$

by sending $f \in k[x_1, \ldots, x_m]$ to $\alpha_A(f) \in k[y_1, \ldots, y_n]$, where $\alpha_A(f)$ is the polynomial defined by $\alpha_A(f)(y) = f(Ay)$.

a. Show that α_A is k-linear, i.e., show that $\alpha_A(rf + sg) = r\alpha_A(f) + s\alpha_A(g)$ for all $r, s \in k$ and all $f, g \in k[x_1, \ldots, x_n]$.

b. Show that $\alpha_A(f \cdot g) = \alpha_A(f) \cdot \alpha_A(g)$ for all $f, g \in k[x_1, \ldots, x_n]$. (As we will see in Definition 8 of Chapter 5, §2, a map between rings which preserves addition and multiplication and also preserves the multiplicative identity is called a *ring homomorphism*. Since it is clear that $\alpha_A(1) = 1$, this shows that α_A is a ring homomorphism.)

c. Show that the set $\{f \in k[x_1, \ldots, x_m] \mid \alpha_A(f) = 0\}$ is an ideal in $k[x_1, \ldots, x_m]$. [This set is called the *kernel* of α_A and denoted $\ker(\alpha_A)$.]

d. If I is an ideal in $k[x_1, \ldots, x_n]$, show that the set $\alpha_A(I) = \{\alpha_A(f) \mid f \in I\}$ need not be an ideal in $k[y_1, \ldots, y_n]$. [We will often write $\langle \alpha_A(I) \rangle$ to denote the ideal in $k[y_1, \ldots, y_n]$ generated by the elements of $\alpha_A(I)$—it is called the *extension* of I to $k[y_1, \ldots, y_n]$.]

e. If I' is an ideal in $k[y_1, \ldots, y_n]$, set $\alpha_A^{-1}(I') = \{f \in k[x_1, \ldots, x_m] \mid \alpha_A(f) \in I'\}$. Show that $\alpha_A^{-1}(I')$ is an ideal in $k[x_1, \ldots, x_m]$ (often called the *contraction* of I').

14. Let A and α_A be as above and let $K = \ker(\alpha_A)$. Let I and J be ideals in $k[x_1, \ldots, x_m]$. Show that:

a. $I \subseteq J$ implies $\langle \alpha_A(I) \rangle \subseteq \langle \alpha_A(J) \rangle$.

b. $\langle \alpha_A(I + J) \rangle = \langle \alpha_A(I) \rangle + \langle \alpha_A(J) \rangle$.

c. $\langle \alpha_A(IJ) \rangle = \langle \alpha_A(I) \rangle \langle \alpha_A(J) \rangle$.

d. $\langle \alpha_A(I \cap J) \rangle \subseteq \langle \alpha_A(I) \rangle \cap \langle \alpha_A(J) \rangle$, with equality if $I \supseteq K$ or $J \supseteq K$ and α_A is onto.

e. $\langle \alpha_A(\sqrt{I}) \rangle \subseteq \sqrt{\langle \alpha_A(I) \rangle}$ with equality if $I \supseteq K$ and α_A is onto.

15. Let A, α_A, and $K = \ker(\alpha_A)$ be as above. Let I' and J' be ideals in $k[y_1, \ldots, y_n]$. Show that:

a. $I' \subseteq J'$ implies $\alpha_A^{-1}(I') \subseteq \alpha_A^{-1}(J')$.

b. $\alpha_A^{-1}(I' + J') \supseteq \alpha_A^{-1}(I') + \alpha_A^{-1}(J')$, with equality if α_A is onto.

c. $\alpha_A^{-1}(I'J') \supseteq (\alpha_A^{-1}(I'))(\alpha_A^{-1}(J'))$, with equality if α_A is onto and the right-hand side contains K.

d. $\alpha_A^{-1}(I' \cap J') = \alpha_A^{-1}(I') \cap \alpha_A^{-1}(J')$.

e. $\alpha_A^{-1}(\sqrt{I'}) = \sqrt{\alpha_A^{-1}(I')}$.

§4 Zariski Closures, Ideal Quotients, and Saturations

We have already encountered a number of examples of sets which are not varieties. Such sets arose very naturally in Chapter 3, where we saw that the projection of a variety need not be a variety, and in the exercises in Chapter 1, where we saw that the (set-theoretic) difference of varieties can fail to be a variety.

Whether or not a set $S \subseteq k^n$ is an affine variety, the set

$$\mathbf{I}(S) = \{f \in k[x_1, \ldots, x_n] \mid f(a) = 0 \text{ for all } a \in S\}$$

is an ideal in $k[x_1, \ldots, x_n]$ (check this!). In fact, it is radical. By the ideal–variety correspondence, $\mathbf{V}(\mathbf{I}(S))$ is a variety. The following proposition states that this variety is the smallest variety that contains the set S.

Proposition 1. *If $S \subseteq k^n$, the affine variety $\mathbf{V}(\mathbf{I}(S))$ is the smallest variety that contains S [in the sense that if $W \subseteq k^n$ is any affine variety containing S, then $\mathbf{V}(\mathbf{I}(S)) \subseteq W$].*

Proof. If $W \supseteq S$, then $\mathbf{I}(W) \subseteq \mathbf{I}(S)$ because $\mathbf{I}$ is inclusion-reversing. But then $\mathbf{V}(\mathbf{I}(W)) \supseteq \mathbf{V}(\mathbf{I}(S))$ because $\mathbf{V}$ also reverses inclusions. Since W is an affine variety, $\mathbf{V}(\mathbf{I}(W)) = W$ by Theorem 7 from §2, and the result follows. $\qquad\square$

This proposition leads to the following definition.

Definition 2. The **Zariski closure** of a subset of affine space is the smallest affine algebraic variety containing the set. If $S \subseteq k^n$, the Zariski closure of S is denoted $\overline{S}$ and is equal to $\mathbf{V}(\mathbf{I}(S))$.

We note the following properties of Zariski closure.

Lemma 3. *Let S and T be subsets of k^n. Then:*
  (i) $\mathbf{I}(\overline{S}) = \mathbf{I}(S)$.
  (ii) *If $S \subseteq T$, then $\overline{S} \subseteq \overline{T}$.*
  (iii) $\overline{S \cup T} = \overline{S} \cup \overline{T}$.

Proof. For (i), the inclusion $\mathbf{I}(\overline{S}) \subseteq \mathbf{I}(S)$ follows from $S \subseteq \overline{S}$. Going the other way, $f \in \mathbf{I}(S)$ implies $S \subseteq \mathbf{V}(f)$. Then $S \subseteq \overline{S} \subseteq \mathbf{V}(f)$ by Definition 2, so that $f \in \mathbf{I}(\overline{S})$.

The proofs of (ii) and (iii) will be covered in the exercises. $\qquad\square$

A natural example of Zariski closure is given by elimination ideals. We can now prove the first assertion of the Closure Theorem (Theorem 3 of Chapter 3, §2).

Theorem 4 (The Closure Theorem, first part). *Assume k is algebraically closed. Let $V = \mathbf{V}(f_1, \ldots, f_s) \subseteq k^n$, and let $\pi_l : k^n \to k^{n-l}$ be projection onto the last $n - l$ coordinates. If I_l is the l-th elimination ideal $I_l = \langle f_1, \ldots, f_s \rangle \cap k[x_{l+1}, \ldots, x_n]$, then $\mathbf{V}(I_l)$ is the Zariski closure of $\pi_l(V)$.*

Proof. In view of Proposition 1, we must show that $\mathbf{V}(I_l) = \mathbf{V}(\mathbf{I}(\pi_l(V)))$. By Lemma 1 of Chapter 3, §2, we have $\pi_l(V) \subseteq \mathbf{V}(I_l)$. Since $\mathbf{V}(\mathbf{I}(\pi_l(V)))$ is the smallest variety containing $\pi_l(V)$, it follows immediately that $\mathbf{V}(\mathbf{I}(\pi_l(V))) \subseteq \mathbf{V}(I_l)$.

To get the opposite inclusion, suppose $f \in \mathbf{I}(\pi_l(V))$, i.e., $f(a_{l+1}, \ldots, a_n) = 0$ for all $(a_{l+1}, \ldots, a_n) \in \pi_l(V)$. Then, considered as an element of $k[x_1, x_2, \ldots, x_n]$, we certainly have $f(a_1, a_2, \ldots, a_n) = 0$ for all $(a_1, \ldots, a_n) \in V$. By Hilbert's Nullstellensatz, $f^N \in \langle f_1, \ldots, f_s \rangle$ for some integer N. Since f does not depend on $x_1, \ldots, x_l$, neither does f^N, and we have $f^N \in \langle f_1, \ldots, f_s \rangle \cap k[x_{l+1}, \ldots, x_n] = I_l$. Thus, $f \in \sqrt{I_l}$, which implies $\mathbf{I}(\pi_l(V)) \subseteq \sqrt{I_l}$. It follows that $\mathbf{V}(I_l) = \mathbf{V}(\sqrt{I_l}) \subseteq \mathbf{V}(\mathbf{I}(\pi_l(V)))$, and the theorem is proved. $\qquad\square$

The conclusion of Theorem 4 can be stated as $\mathbf{V}(I_l) = \overline{\pi_l(V)}$. In general, if V is a variety, then we say that a subset $S \subseteq V$ is *Zariski dense in V* if $V = \overline{S}$, i.e., V is the Zariski closure of S. is Thus Theorem 4 tells us that $\pi_l(V)$ is Zariski dense in $\mathbf{V}(I_l)$ when the field is algebraically closed.

One context in which we encountered sets that were not varieties was in taking the difference of varieties. For example, let $V = \mathbf{V}(I)$ where $I \subseteq k[x, y, z]$ is the ideal $\langle xz, yz \rangle$ and $W = \mathbf{V}(J)$ where $J = \langle z \rangle$. Then we have already seen that V is the union of the (x, y)-plane and the z-axis. Since W is the (x, y)-plane, $V \setminus W$ is the z-axis with the origin removed [because the origin also belongs to the (x, y)-plane]. We have seen in Chapter 1 that this is not a variety. The z-axis [i.e., $\mathbf{V}(x, y)$] is the Zariski closure of $V \setminus W$.

We could ask if there is a general way to compute the ideal corresponding to the Zariski closure $\overline{V \setminus W}$ of the difference of two varieties V and W. The answer is affirmative, but it involves two new algebraic constructions on ideals called *ideal quotients* and *saturations*.

We begin with the first construction.

Definition 5. If I, J are ideals in $k[x_1, \ldots, x_n]$, then $I : J$ is the set

$$\{f \in k[x_1, \ldots, x_n] \mid fg \in I \text{ for all } g \in J\}$$

and is called the **ideal quotient** (or **colon ideal**) of I by J.

So, for example, in $k[x, y, z]$ we have

$$
\begin{aligned}
\langle xz, yz \rangle : \langle z \rangle &= \{f \in k[x, y, z] \mid f \cdot z \in \langle xz, yz \rangle\} \\
&= \{f \in k[x, y, z] \mid f \cdot z = Axz + Byz\} \\
&= \{f \in k[x, y, z] \mid f = Ax + By\} \\
&= \langle x, y \rangle.
\end{aligned}
$$

Proposition 6. *If I, J are ideals in $k[x_1, \ldots, x_n]$, then the ideal quotient $I : J$ is an ideal in $k[x_1, \ldots, x_n]$ and $I : J$ contains I.*

Proof. To show $I : J$ contains I, note that because I is an ideal, if $f \in I$, then $fg \in I$ for all $g \in k[x_1, \ldots, x_n]$ and, hence, certainly $fg \in I$ for all $g \in J$. To show that $I : J$

is an ideal, first note that $0 \in I : J$ because $0 \in I$. Let $f_1, f_2 \in I : J$. Then $f_1 g$ and $f_2 g$ are in I for all $g \in J$. Since I is an ideal $(f_1 + f_2)g = f_1 g + f_2 g \in I$ for all $g \in J$. Thus, $f_1 + f_2 \in I : J$. To check closure under multiplication is equally straightforward: if $f \in I : J$ and $h \in k[x_1, \dots, x_n]$, then $fg \in I$ and, since I is an ideal, $hfg \in I$ for all $g \in J$, which means that $hf \in I : J$. □

The algebraic properties of ideal quotients and methods for computing them will be discussed later in the section. For now, we want to explore the relation between ideal quotients and the Zariski closure of a difference of varieties.

Proposition 7.

(i) *If I and J are ideals in $k[x_1, \dots, x_n]$, then*

$$\mathbf{V}(I) = \mathbf{V}(I + J) \cup \mathbf{V}(I : J).$$

(ii) *If V and W are varieties k^n, then*

$$V = (V \cap W) \cup \overline{(V \setminus W)}.$$

(iii) *In the situation of* (i), *we have*

$$\overline{\mathbf{V}(I) \setminus \mathbf{V}(J)} \subseteq \mathbf{V}(I : J).$$

Proof. We begin with (ii). Since V contains $V \setminus W$ and V is a variety, the smallest variety containing $V \setminus W$ must be contained in V. Hence, $\overline{V \setminus W} \subseteq V$. Since $V \cap W \subseteq V$, we have $(V \cap W) \cup \overline{(V \setminus W)} \subseteq V$.

To get the reverse containment, note that $V = (V \cap W) \cup (V \setminus W)$. Since $V \setminus W \subseteq \overline{V \setminus W}$, the desired inclusion $V \subseteq (V \cap W) \cup \overline{V \setminus W}$ follows immediately.

For (iii), we first claim that $I : J \subseteq \mathbf{I}(\mathbf{V}(I) \setminus \mathbf{V}(J))$. For suppose that $f \in I : J$ and $a \in \mathbf{V}(I) \setminus \mathbf{V}(J)$. Then $fg \in I$ for all $g \in J$. Since $a \in \mathbf{V}(I)$, we have $f(a)g(a) = 0$ for all $g \in J$. Since $a \notin \mathbf{V}(J)$, there is some $g \in J$ such that $g(a) \neq 0$. Hence, $f(a) = 0$ for all $a \in \mathbf{V}(I) \setminus \mathbf{V}(J)$. Thus, $f \in \mathbf{I}(\mathbf{V}(I) \setminus \mathbf{V}(J))$, which proves the claim. Since $\mathbf{V}$ reverses inclusions, we have $\mathbf{V}(I : J) \supseteq \mathbf{V}(\mathbf{I}(\mathbf{V}(I) \setminus \mathbf{V}(J))) = \overline{\mathbf{V}(I) \setminus \mathbf{V}(J)}$.

Finally, for (i), note that $\mathbf{V}(I + J) = \mathbf{V}(I) \cap \mathbf{V}(J)$ by Theorem 4 of §3. Then applying (ii) with $V = \mathbf{V}(I)$ and $W = \mathbf{V}(J)$ gives

$$\mathbf{V}(I) = \mathbf{V}(I + J) \cup \overline{\mathbf{V}(I) \setminus \mathbf{V}(J)} \subseteq \mathbf{V}(I + J) \cup \mathbf{V}(I : J),$$

where the inclusion follows from (iii). But $I \subseteq I + J$ and $I \subseteq I : J$ imply that

$$\mathbf{V}(I + J) \subseteq \mathbf{V}(I) \quad \text{and} \quad \mathbf{V}(I : J) \subseteq \mathbf{V}(I).$$

These inclusions give $\mathbf{V}(I + J) \cup \mathbf{V}(I : J) \subseteq \mathbf{V}(I)$, and then we are done. □

In Proposition 7, note that $\mathbf{V}(I + J)$ from part (i) matches up with $V \cap W$ in part (ii) since $\mathbf{V}(I + J) = \mathbf{V}(I) \cap \mathbf{V}(J)$. So it is natural to ask if $\mathbf{V}(I : J)$ in part (i) matches up with $\overline{V \setminus W}$ in part (ii). This is equivalent to asking if the inclusion $\overline{\mathbf{V}(I) \setminus \mathbf{V}(J)} \subseteq \mathbf{V}(I : J)$ in part (iii) is an equality.

Unfortunately, this can fail, even when the field is algebraically closed. To see what can go wrong, let $I = \langle x^2(y-1) \rangle$ and $J = \langle x \rangle$ in the polynomial ring $\mathbb{C}[x, y]$. Then one easily checks that

$$\mathbf{V}(I) = \mathbf{V}(x) \cup \mathbf{V}(y-1) = \mathbf{V}(J) \cup \mathbf{V}(y-1) \subseteq \mathbb{C}^2,$$

which is the union of the y-axis and the line $y = 1$. It follows without difficulty that $\overline{\mathbf{V}(I) \setminus \mathbf{V}(J)} = \mathbf{V}(y-1)$. However, the ideal quotient is

$$
\begin{aligned}
I : J = \langle x^2(y-1) \rangle : \langle x \rangle &= \{ f \in \mathbb{C}[x, y] \mid f \cdot x = Ax^2(y-1) \} \\
&= \{ f \in \mathbb{C}[x, y] \mid f = Ax(y-1) \} = \langle x(y-1) \rangle.
\end{aligned}
$$

Then $\mathbf{V}(I : J) = \mathbf{V}(x(y-1)) = \mathbf{V}(x) \cup \mathbf{V}(y-1)$, which is strictly bigger than $\overline{\mathbf{V}(I) \setminus \mathbf{V}(J)} = \mathbf{V}(y-1)$. In other words, the inclusion in part (iii) of Proposition 7 can be strict, even over an algebraically closed field.

However, if we replace J with J^2, then a computation similar to the above gives $I : J^2 = \langle y-1 \rangle$, so that $\mathbf{V}(I : J^2) = \overline{\mathbf{V}(I) \setminus \mathbf{V}(J)}$. In general, higher powers may be required, which leads to our second algebraic construction on ideals.

Definition 8. If I, J are ideals in $k[x_1, \ldots, x_n]$, then $I : J^\infty$ is the set

$$\{ f \in k[x_1, \ldots, x_n] \mid \text{for all } g \in J, \text{ there is } N \geq 0 \text{ such that } fg^N \in I \}$$

and is called the **saturation** of I with respect to J.

Proposition 9. *If I, J are ideals in $k[x_1, \ldots, x_n]$, then the saturation $I : J^\infty$ is an ideal in $k[x_1, \ldots, x_n]$. Furthermore:*

(i) $I \subseteq I : J \subseteq I : J^\infty$.
(ii) $I : J^\infty = I : J^N$ *for all sufficiently large N.*
(iii) $\sqrt{I : J^\infty} = \sqrt{I : J}$.

Proof. First observe that $J_1 \subseteq J_2$ implies $I : J_2 \subseteq I : J_1$. Since $J^{N+1} \subseteq J^N$ for all N, we obtain the ascending chain of ideals

$$(1) \qquad\qquad I \subseteq I : J \subseteq I : J^2 \subseteq I : J^3 \subseteq \cdots .$$

By the ACC, there is N such that $I : J^N = I : J^{N+1} = \cdots$. We claim that $I : J^\infty = I : J^N$. One inclusion is easy, for if $f \in I : J^N$ and $g \in J$, then $g^N \in J^N$. Hence, $fg^N \in I$, proving that $f \in I : J^\infty$. For the other inclusion, take $f \in I : J^\infty$ and let $J = \langle g_1, \ldots, g_s \rangle$. By Definition 8, f times a power of g_i lies in I. If M is the largest such power, then $fg_i^M \in I$ for $i = 1, \ldots, s$. In the exercises, you will show that

$$J^{sM} \subseteq \langle g_1^M, \ldots, g_s^M \rangle.$$

This implies $fJ^{sM} \subseteq I$, so that $f \in I : J^{sM}$. Then $f \in I : J^N$ since (1) stabilizes at N.

Part (ii) follows from the claim just proved, and $I : J^\infty = I : J^N$ implies that $I : J^\infty$ is an ideal by Proposition 6. Note also that part (i) follows from (1) and part (ii) .

For part (iii), we first show $\sqrt{I:J^\infty} \subseteq \sqrt{I}:J$. This is easy, for $f \in \sqrt{I:J^\infty}$ implies $f^m \in I:J^\infty$ for some m. Given $g \in J$, it follows that $f^m g^N \in I$ for some N. Then $(fg)^M \in I$ for $M = \max(m, N)$, so that $fg \in \sqrt{I}$. Since this holds for all $g \in J$, we conclude that $f \in \sqrt{I}:J$.

For the opposite inclusion, take $f \in \sqrt{I}:J$ and write $J = \langle g_1, \ldots, g_s \rangle$. Then $fg_i \in \sqrt{I}$, so we can find M with $f^M g_i^M \in I$ for all i. The argument from (ii) implies $f^M J^{sM} \subseteq I$, so

$$f^M \in I:J^{sM} \subseteq I:J^\infty.$$

It follows that $f \in \sqrt{I:J^\infty}$, and the proof is complete. □

Later in the section we will discuss further algebraic properties of saturations and how to compute them. For now, we focus on their relation to geometry.

Theorem 10. *Let I and J be ideals in $k[x_1, \ldots, x_n]$. Then:*

(i) $\mathbf{V}(I) = \mathbf{V}(I+J) \cup \mathbf{V}(I:J^\infty)$.
(ii) $\overline{\mathbf{V}(I) \setminus \mathbf{V}(J)} \subseteq \mathbf{V}(I:J^\infty)$.
(iii) *If k is algebraically closed, then* $\mathbf{V}(I:J^\infty) = \overline{\mathbf{V}(I) \setminus \mathbf{V}(J)}$.

Proof. In the exercises, you will show that (i) and (ii) follow by easy modifications of the proofs of parts (i) and (iii) of Proposition 7.

For (iii), suppose that k is algebraically closed. We first show that

$$(2) \qquad\qquad \mathbf{I}(\mathbf{V}(I) \setminus \mathbf{V}(J)) \subseteq \sqrt{I}:J.$$

Let $f \in \mathbf{I}(\mathbf{V}(I) \setminus \mathbf{V}(J))$. If $g \in J$, then fg vanishes on $\mathbf{V}(I)$ because f vanishes on $\mathbf{V}(I) \setminus \mathbf{V}(J)$ and g on $\mathbf{V}(J)$. Thus, $fg \in \mathbf{I}(\mathbf{V}(I))$, so $fg \in \sqrt{I}$ by the Nullstellensatz. Since this holds for all $g \in J$, we have $f \in \sqrt{I}:J$, as claimed.

Since $\mathbf{V}$ is inclusion-reversing, (2) implies

$$\mathbf{V}(\sqrt{I}:J) \subseteq \mathbf{V}(\mathbf{I}(\mathbf{V}(I) \setminus \mathbf{V}(J))) = \overline{\mathbf{V}(I) \setminus \mathbf{V}(J)}.$$

However, we also have

$$\mathbf{V}(I:J^\infty) = \mathbf{V}(\sqrt{I:J^\infty}) = \mathbf{V}(\sqrt{I}:J),$$

where the second equality follows from part (iii) of Proposition 9. Combining the last two displays, we obtain

$$\mathbf{V}(I:J^\infty) \subseteq \overline{\mathbf{V}(I) \setminus \mathbf{V}(J)}.$$

Then (iii) follows immediately from this inclusion and (ii). □

When k is algebraically closed, Theorem 10 and Theorem 4 of §3 imply that the decomposition

$$\mathbf{V}(I) = \mathbf{V}(I+J) \cup \mathbf{V}(I:J^\infty)$$

is *precisely* the decomposition

$$\mathbf{V}(I) = (\mathbf{V}(I) \cap \mathbf{V}(J)) \cup (\overline{\mathbf{V}(I) \setminus \mathbf{V}(J)})$$

from part (ii) of Proposition 7. This shows that the saturation $I : J^\infty$ is the ideal-theoretic analog of the Zariski closure $\overline{\mathbf{V}(I) \setminus \mathbf{V}(J)}$.

In some situations, saturations can be replaced with ideal quotients. For example, the proof of Theorem 10 yields the following corollary when the ideal I is radical.

Corollary 11. *Let I and J be ideals in $k[x_1, \ldots, x_n]$. If k is algebraically closed and I is radical, then*

$$\mathbf{V}(I : J) = \overline{\mathbf{V}(I) \setminus \mathbf{V}(J)}.$$

You will prove this in the exercises. Another nice fact (also covered in the exercises) is that if k is arbitrary and V and W are varieties in k^n, then

$$\mathbf{I}(V) : \mathbf{I}(W) = \mathbf{I}(V \setminus W).$$

The following proposition takes care of some simple properties of ideal quotients and saturations.

Proposition 12. *Let I and J be ideals in $k[x_1, \ldots, x_n]$. Then:*

(i) $I : k[x_1, \ldots, x_n] = I : k[x_1, \ldots, x_n]^\infty = I$.
(ii) $J \subseteq I$ *if and only if* $I : J = k[x_1, \ldots, x_n]$.
(iii) $J \subseteq \sqrt{I}$ *if and only if* $I : J^\infty = k[x_1, \ldots, x_n]$.

Proof. The proof is left as an exercise.   □

When the field is algebraically closed, the reader is urged to translate parts (i) and (iii) of the proposition into terms of varieties (upon which they become clear).

The following proposition will help us compute ideal quotients and saturations.

Proposition 13. *Let I and $J_1, \ldots, J_r$ be ideals in $k[x_1, \ldots, x_n]$. Then:*

(3)
$$I : \left(\sum_{i=1}^{r} J_i \right) = \bigcap_{i=1}^{r} (I : J_i),$$

(4)
$$I : \left(\sum_{i=1}^{r} J_i \right)^\infty = \bigcap_{i=1}^{r} (I : J_i^\infty).$$

Proof. We again leave the (straightforward) proofs to the reader.   □

If f is a polynomial and I an ideal, we will often write $I : f$ instead of $I : \langle f \rangle$, and similarly $I : f^\infty$ instead of $I : \langle f \rangle^\infty$. Note that (3) and (4) imply that

(5) $$I : \langle f_1, f_2, \ldots, f_r \rangle = \bigcap_{i=1}^{r} (I : f_i) \quad \text{and} \quad I : \langle f_1, f_2, \ldots, f_r \rangle^\infty = \bigcap_{i=1}^{r} (I : f_i^\infty).$$

We now turn to the question of how to compute generators of the ideal quotient $I : J$ and saturation $I : J^\infty$, given generators of I and J. Inspired by (5), we begin with the case when J is generated by a single polynomial.

Theorem 14. *Let I be an ideal and g an element of* $k[x_1, \ldots, x_n]$. *Then:*

(i) *If* $\{h_1, \ldots, h_p\}$ *is a basis of the ideal* $I \cap \langle g \rangle$, *then* $\{h_1/g, \ldots, h_p/g\}$ *is a basis of* $I : g$.

(ii) *If* $\{f_1, \ldots, f_s\}$ *is a basis of I and* $\tilde{I} = \langle f_1, \ldots, f_s, 1 - yg \rangle \subseteq k[x_1, \ldots, x_n, y]$, *where* y *is a new variable, then*

$$I : g^\infty = \tilde{I} \cap k[x_1, \ldots, x_n].$$

Furthermore, if G is a lex Gröbner basis of $\tilde{I}$ *for* $y > x_1 > \cdots > x_n$, *then* $G \cap k[x_1, \ldots, x_n]$ *is a basis of* $I : g^\infty$.

Proof. For (i), observe that if $h \in \langle g \rangle$, then $h = bg$ for some polynomial $b \in k[x_1, \ldots, x_n]$. Thus, if $f \in \langle h_1/g, \ldots, h_p/g \rangle$, then

$$hf = bgf \in \langle h_1, \ldots, h_p \rangle = I \cap \langle g \rangle \subseteq I.$$

Thus, $f \in I : g$. Conversely, suppose $f \in I : g$. Then $fg \in I$. Since $fg \in \langle g \rangle$, we have $fg \in I \cap \langle g \rangle$. If $I \cap \langle g \rangle = \langle h_1, \ldots, h_p \rangle$, this means $fg = \sum r_i h_i$ for some polynomials r_i. Since each $h_i \in \langle g \rangle$, each h_i/g is a polynomial, and we conclude that $f = \sum r_i(h_i/g)$, whence $f \in \langle h_1/g, \ldots, h_p/g \rangle$.

The first assertion of (ii) is left as an exercise. Then the Elimination Theorem from Chapter 3, §1 implies that $G \cap k[x_1, \ldots, x_n]$ is a Gröbner basis of $I : g^\infty$. □

This theorem, together with our procedure for computing intersections of ideals and equation (5), immediately leads to an **algorithm for computing a basis of an ideal quotient**: given $I = \langle f_1, \ldots, f_r \rangle$ and $J = \langle g_1, \ldots, g_s \rangle$, to compute a basis of $I : J$, we first compute a basis for $I : g_i$ for each i. In view of Theorem 14, this means computing a basis $\{h_1, \ldots, h_p\}$ of $\langle f_1, \ldots, f_r \rangle \cap \langle g_i \rangle$. Recall that we do this via the algorithm for computing intersections of ideals from §3. Using the division algorithm, we divide each of basis element h_j by g_i to get a basis for $I : g_i$ by part (i) of Theorem 14. Finally, we compute a basis for $I : J$ by applying the intersection algorithm $s - 1$ times, computing first a basis for $I : \langle g_1, g_2 \rangle = (I : g_1) \cap (I : g_2)$, then a basis for $I : \langle g_1, g_2, g_3 \rangle = (I : \langle g_1, g_2 \rangle) \cap (I : g_3)$, and so on.

Similarly, we have an **algorithm for computing a basis of a saturation**: given $I = \langle f_1, \ldots, f_r \rangle$ and $J = \langle g_1, \ldots, g_s \rangle$, to compute a basis of $I : J^\infty$, we first compute a basis for $I : g_i^\infty$ for each i using the method described in part (ii) of Theorem 14. Then by (5), we need to intersect the ideals $I : g_i^\infty$, which we do as above by applying the intersection algorithm $s - 1$ times.

EXERCISES FOR §4

1. Find the Zariski closure of the following sets:
   a. The projection of the hyperbola $\mathbf{V}(xy - 1)$ in $\mathbb{R}^2$ onto the x-axis.
   b. The boundary of the first quadrant in $\mathbb{R}^2$.
   c. The set $\{(x, y) \in \mathbb{R}^2 \mid x^2 + y^2 \le 4\}$.
2. Complete the proof of Lemma 3. Hint: For part (iii), use Lemma 2 from Chapter 1, §2.
3. Let $f = (x + y)^2(x - y)(x + z^2)$ and $g = (x + z^2)^3(x - y)(z + y)$. Compute generators for $\langle f \rangle : \langle g \rangle$.

4. Let I and J be ideals in $k[x_1, \ldots, x_n]$. Show that if I is radical, then $I : J$ is radical and $I : J = I : \sqrt{J} = I : J^\infty$.

5. As in the proof of Proposition 9, assume $J = \langle g_1, \ldots, g_s \rangle$. Prove that $J^{sM} \subseteq \langle g_1^M, \ldots, g_s^M \rangle$. Hint: See the proof of Lemma 5 of §2.

6. Prove parts (i) and (ii) of Theorem 10. Hint: Adapt the proofs of parts (i) and (iii) of Proposition 7.

7. Prove Corollary 11. Hint: Combine Theorem 10 and the Exercise 4. Another approach would be look closely at the proof of Theorem 10 when I is radical.

8. Let $V, W \subseteq k^n$ be varieties. Prove that $\mathbf{I}(V) : \mathbf{I}(W) = \mathbf{I}(V \setminus W)$.

9. Prove Proposition 12 and find geometric interpretations of parts (i) and (iii)

10. Prove Proposition 13 and find a geometric interpretation of (4).

11. Prove $I : g^\infty = \tilde{I} \cap k[x_1, \ldots, x_n]$ from part (ii) of Theorem 14. Hint: See the proof of Proposition 8 of §2.

12. Show that Proposition 8 of §2 is a corollary of Proposition 12 and Theorem 14.

13. An example mentioned in the text used $I = \langle x^2(y-1) \rangle$ and $J = \langle x \rangle$. Compute $I : J^\infty$ and explain how your answer relates to the discussion in the text.

14. Let $I, J \subseteq k[x_1, \ldots, x_n]$ be ideals. Prove that $I : J^\infty = I : J^N$ if and only if $I : J^N = I : J^{N+1}$. Then use this to describe an algorithm for computing the saturation $I : J^\infty$ based on the algorithm for computing ideal quotients.

15. Show that N can be arbitrarily large in $I : J^\infty = I : J^N$. Hint: Look at $I = \langle x^N(y-1) \rangle$.

16. Let $I, J, K \subseteq k[x_1, \ldots, x_n]$ be ideals. Prove the following:
    a. $IJ \subseteq K$ if and only if $I \subseteq K : J$.
    b. $(I : J) : K = I : JK$.

17. Given ideals $I_1, \ldots, I_r, J \subseteq k[x_1, \ldots, x_n]$, prove that $\left(\bigcap_{i=1}^r I_i \right) : J = \bigcap_{i=1}^r (I_i : J)$. Then prove a similar result for saturations and give a geometric interpretation.

18. Let A be an $m \times n$ constant matrix and suppose that $x = Ay$. where we are thinking of $x \in k^m$ and $y \in k^n$ as column vectors of variables. As in Exercise 13 of §3, define a map

$$\alpha_A : k[x_1, \ldots, x_m] \longrightarrow k[y_1, \ldots, y_n]$$

by sending $f \in k[x_1, \ldots, x_m]$ to $\alpha_A(f) \in k[y_1, \ldots, y_n]$, where $\alpha_A(f)$ is the polynomial defined by $\alpha_A(f)(y) = f(Ay)$.
    a. Show that $\alpha_A(I : J) \subseteq \alpha_A(I) : \alpha_A(J)$ with equality if $I \supseteq \ker(\alpha_A)$ and α_A is onto.
    b. Show that $\alpha_A^{-1}(I' : J') = \alpha_A^{-1}(I') : \alpha_A^{-1}(J')$ when α_A is onto.

§5   Irreducible Varieties and Prime Ideals

We have already seen that the union of two varieties is a variety. For example, in Chapter 1 and in the last section, we considered $\mathbf{V}(xz, yz)$, which is the union of a line and a plane. Intuitively, it is natural to think of the line and the plane as "more fundamental" than $\mathbf{V}(xz, yz)$. Intuition also tells us that a line or a plane are "irreducible" or "indecomposable" in some sense: they do not obviously seem to be a union of finitely many simpler varieties. We formalize this notion as follows.

Definition 1. An affine variety $V \subseteq k^n$ is **irreducible** if whenever V is written in the form $V = V_1 \cup V_2$, where V_1 and V_2 are affine varieties, then either $V_1 = V$ or $V_2 = V$.

Thus, $\mathbf{V}(xz, yz)$ is not an irreducible variety. On the other hand, it is not completely clear when a variety is irreducible. If this definition is to correspond to our geometric intuition, it is clear that a point, a line, and a plane ought to be irreducible. For that matter, the twisted cubic $\mathbf{V}(y - x^2, z - x^3)$ in $\mathbb{R}^3$ appears to be irreducible. But how do we prove this? The key is to capture this notion algebraically: if we can characterize ideals which correspond to irreducible varieties, then perhaps we stand a chance of establishing whether a variety is irreducible.

The following notion turns out to be the right one.

Definition 2. An ideal $I \subseteq k[x_1, \ldots, x_n]$ is **prime** if whenever $f, g \in k[x_1, \ldots, x_n]$ and $fg \in I$, then either $f \in I$ or $g \in I$.

If we have set things up right, an irreducible variety will correspond to a prime ideal and conversely. The following theorem assures us that this is indeed the case.

Proposition 3. *Let $V \subseteq k^n$ be an affine variety. Then V is irreducible if and only if $\mathbf{I}(V)$ is a prime ideal.*

Proof. First, assume that V is irreducible and let $fg \in \mathbf{I}(V)$. Set $V_1 = V \cap \mathbf{V}(f)$ and $V_2 = V \cap \mathbf{V}(g)$; these are affine varieties because an intersection of affine varieties is a variety. Then $fg \in \mathbf{I}(V)$ easily implies that $V = V_1 \cup V_2$. Since V is irreducible, we have either $V = V_1$ or $V = V_2$. Say the former holds, so that $V = V_1 = V \cap \mathbf{V}(f)$. This implies that f vanishes on V, so that $f \in \mathbf{I}(V)$. Thus, $\mathbf{I}(V)$ is prime.

Next, assume that $\mathbf{I}(V)$ is prime and let $V = V_1 \cup V_2$. Suppose that $V \neq V_1$. We claim that $\mathbf{I}(V) = \mathbf{I}(V_2)$. To prove this, note that $\mathbf{I}(V) \subseteq \mathbf{I}(V_2)$ since $V_2 \subseteq V$. For the opposite inclusion, first note that $\mathbf{I}(V) \subsetneq \mathbf{I}(V_1)$ since $V_1 \subsetneq V$. Thus, we can pick $f \in \mathbf{I}(V_1) \setminus \mathbf{I}(V)$. Now take any $g \in \mathbf{I}(V_2)$. Since $V = V_1 \cup V_2$, it follows that fg vanishes on V, and, hence, $fg \in \mathbf{I}(V)$. But $\mathbf{I}(V)$ is prime, so that f or g lies in $\mathbf{I}(V)$. We know that $f \notin \mathbf{I}(V)$ and, thus, $g \in \mathbf{I}(V)$. This proves $\mathbf{I}(V) = \mathbf{I}(V_2)$, whence $V = V_2$ because $\mathbf{I}$ is one-to-one. Thus, V is an irreducible variety. □

It is an easy exercise to show that every prime ideal is radical. Then, using the ideal-variety correspondence between radical ideals and varieties, we get the following corollary of Proposition 3.

Corollary 4. *When k is algebraically closed, the functions $\mathbf{I}$ and $\mathbf{V}$ induce a one-to-one correspondence between irreducible varieties in k^n and prime ideals in $k[x_1, \ldots, x_n]$.*

As an example of how to use Proposition 3, let us prove that the ideal $\mathbf{I}(V)$ of the twisted cubic is prime. Suppose that $fg \in \mathbf{I}(V)$. Since the curve is parametrized by (t, t^2, t^3), it follows that, for all t,

$$f(t, t^2, t^3)g(t, t^2, t^3) = 0.$$

This implies that $f(t, t^2, t^3)$ or $g(t, t^2, t^3)$ must be the zero polynomial, so that f or g vanishes on V. Hence, f or g lies in $\mathbf{I}(V)$, proving that $\mathbf{I}(V)$ is a prime ideal.

By the proposition, the twisted cubic is an irreducible variety in $\mathbb{R}^3$. One proves that a straight line is irreducible in the same way: first parametrize it, then apply the above argument.

In fact, the above argument holds much more generally.

Proposition 5. *If k is an infinite field and $V \subseteq k^n$ is a variety defined parametrically*

$$x_1 = f_1(t_1, \ldots, t_m),$$
$$\vdots$$
$$x_n = f_n(t_1, \ldots, t_m),$$

where $f_1, \ldots, f_n$ are polynomials in $k[t_1, \ldots, t_m]$, then V is irreducible.

Proof. As in §3 of Chapter 3, we let $F : k^m \to k^n$ be defined by

$$F(t_1, \ldots, t_m) = (f_1(t_1, \ldots, t_m), \ldots, f_n(t_1, \ldots, t_m)).$$

Saying that V is defined parametrically by the above equations means that V is the Zariski closure of $F(k^m)$. In particular, $\mathbf{I}(V) = \mathbf{I}(F(k^m))$.

For any polynomial $g \in k[x_1, \ldots, x_n]$, the function $g \circ F$ is a polynomial in $k[t_1, \ldots, t_m]$. In fact, $g \circ F$ is the polynomial obtained by "plugging the polynomials $f_1, \ldots, f_n$ into g":

$$g \circ F = g(f_1(t_1, \ldots, t_m), \ldots, f_n(t_1, \ldots, t_m)).$$

Because k is infinite, $\mathbf{I}(V) = \mathbf{I}(F(k^m))$ is the set of polynomials in $k[x_1, \ldots, x_n]$ whose composition with F is the zero polynomial in $k[t_1, \ldots, t_m]$:

$$\mathbf{I}(V) = \{g \in k[x_1, \ldots, x_n] \mid g \circ F = 0\}.$$

Now suppose that $gh \in \mathbf{I}(V)$. Then $(gh) \circ F = (g \circ F)(h \circ F) = 0$. (Make sure you understand this.) But, if the product of two polynomials in $k[t_1, \ldots, t_m]$ is the zero polynomial, one of them must be the zero polynomial. Hence, either $g \circ F = 0$ or $h \circ F = 0$. This means that either $g \in \mathbf{I}(V)$ or $h \in \mathbf{I}(V)$. This shows that $\mathbf{I}(V)$ is a prime ideal and, therefore, that V is irreducible. $\qquad\qquad\square$

With a little care, the above argument extends still further to show that any variety defined by a *rational* parametrization is irreducible.

Proposition 6. *If k is an infinite field and V is a variety defined by the rational parametrization*

$$x_1 = \frac{f_1(t_1, \ldots, t_m)}{g_1(t_1, \ldots, t_m)},$$
$$\vdots$$
$$x_n = \frac{f_n(t_1, \ldots, t_m)}{g_n(t_1, \ldots, t_m)},$$

where $f_1, \ldots, f_n, g_1, \ldots, g_n \in k[t_1, \ldots, t_m]$, then V is irreducible.

Proof. Set $W = \mathbf{V}(g_1 g_2 \cdots g_n)$ and let $F : k^m \setminus W \to k^n$ defined by

$$F(t_1, \ldots, t_m) = \left(\frac{f_1(t_1, \ldots, t_m)}{g_1(t_1, \ldots, t_m)}, \ldots, \frac{f_n(t_n, \ldots, t_m)}{g_n(t_1, \ldots, t_m)} \right).$$

Then V is the Zariski closure of $F(k^m \setminus W)$, which implies that $\mathbf{I}(V)$ is the set of $h \in k[x_1, \ldots, x_n]$ such that the function $h \circ F$ is zero for all $(t_1, \ldots, t_m) \in k^m \setminus W$. The difficulty is that $h \circ F$ need not be a polynomial, and we, thus, cannot directly apply the argument in the latter part of the proof of Proposition 5.

We can get around this difficulty as follows. Let $h \in k[x_1, \ldots, x_n]$. Since

$$g_1(t_1, \ldots, t_m) g_2(t_1, \ldots, t_m) \cdots g_n(t_1, \ldots, t_m) \neq 0$$

for any $(t_1, \ldots, t_m) \in k^m \setminus W$, the function $(g_1 g_2 \cdots g_n)^N (h \circ F)$ is equal to zero at precisely those values of $(t_1, \ldots, t_m) \in k^m \setminus W$ for which $h \circ F$ is equal to zero. Moreover, if we let N be the total degree of $h \in k[x_1, \ldots, x_n]$, then we leave it as an exercise to show that $(g_1 g_2 \cdots g_n)^N (h \circ F)$ is a polynomial in $k[t_1, \ldots, t_m]$. We deduce that $h \in \mathbf{I}(V)$ if and only if $(g_1 g_2 \cdots g_n)^N (h \circ F)$ is zero for all $t \in k^m \setminus W$. But, by Exercise 11 of Chapter 3, §3, this happens if and only if $(g_1 g_2 \cdots g_n)^N (h \circ F)$ is the zero polynomial in $k[t_1, \ldots, t_m]$. Thus, we have shown that

$$h \in \mathbf{I}(V) \quad \text{if and only if} \quad (g_1 g_2 \cdots g_n)^N (h \circ F) = 0 \in k[t_1, \ldots, t_m].$$

Now, we continue our proof that $\mathbf{I}(V)$ is prime. Suppose $p, q \in k[x_1, \ldots, x_n]$ satisfy $p \cdot q \in \mathbf{I}(V)$. If the total degrees of p and q are M and N, respectively, then the total degree of $p \cdot q$ is $M + N$. Thus, $(g_1 g_2 \cdots g_n)^{M+N} (p \circ F) \cdot (q \circ F) = 0$. But the former is a product of the polynomials $(g_1 g_2 \cdots g_n)^M (p \circ F)$ and $(g_1 g_2 \cdots g_n)^N (q \circ F)$ in $k[t_1, \ldots, t_m]$. Hence one of them must be the zero polynomial. In particular, either $p \in \mathbf{I}(V)$ or $q \in \mathbf{I}(V)$. This shows that $\mathbf{I}(V)$ is a prime ideal and, therefore, that V is an irreducible variety. $\square$

The simplest variety in k^n given by a parametrization consists of a single point, $\{(a_1, \ldots, a_n)\}$. In the notation of Proposition 5, it is given by the parametrization in which each f_i is the constant polynomial $f_i(t_1, \ldots, t_m) = a_i$, $1 \leq i \leq n$. It is clearly irreducible and it is easy to check that $\mathbf{I}(\{(a_1, \ldots, a_n)\}) = \langle x_1 - a_1, \ldots, x_n - a_n \rangle$ (see Exercise 7), which implies that the latter is prime. The ideal $\langle x_1 - a_1, \ldots, x_n - a_n \rangle$ has another distinctive property: it is maximal in the sense that the only ideal which strictly contains it is the whole ring $k[x_1, \ldots, x_n]$. Such ideals are important enough to merit special attention.

Definition 7. An ideal $I \subseteq k[x_1, \ldots, x_n]$ is said to be **maximal** if $I \neq k[x_1, \ldots, x_n]$ and any ideal J containing I is such that either $J = I$ or $J = k[x_1, \ldots, x_n]$.

In order to streamline statements, we make the following definition.

Definition 8. An ideal $I \subseteq k[x_1, \ldots, x_n]$ is said to be **proper** if I is not equal to $k[x_1, \ldots, x_n]$.

Thus, an ideal is maximal if it is proper and no other proper ideal strictly contains it. We now show that any ideal of the form $\langle x_1 - a_1, \ldots, x_n - a_n \rangle$ is maximal.

Proposition 9. *If k is any field, an ideal $I \subseteq k[x_1, \ldots, x_n]$ of the form*

$$I = \langle x_1 - a_1, \ldots, x_n - a_n \rangle,$$

where $a_1, \ldots, a_n \in k$, is maximal.

Proof. Suppose that J is some ideal strictly containing I. Then there must exist $f \in J$ such that $f \notin I$. We can use the division algorithm to write f as $A_1(x_1 - a_1) + \cdots + A_n(x_n - a_n) + b$ for some $b \in k$. Since $A_1(x_1 - a_1) + \cdots + A_n(x_n - a_n) \in I$ and $f \notin I$, we must have $b \neq 0$. However, since $f \in J$ and since $A_1(x_1 - a_1) + \cdots + A_n(x_n - a_n) \in I \subseteq J$, we also have

$$b = f - (A_1(x_1 - a_1) + \cdots + A_n(x_n - a_n)) \in J.$$

Since b is nonzero, $1 = 1/b \cdot b \in J$, so $J = k[x_1, \ldots, x_n]$. $\qquad\square$

Since

$$\mathbf{V}(x_1 - a_1, \ldots, x_n - a_n) = \{(a_1, \ldots, a_n)\},$$

every point $(a_1, \ldots, a_n) \in k^n$ corresponds to a maximal ideal of $k[x_1, \ldots, x_n]$, namely $\langle x_1 - a_1, \ldots, x_n - a_n \rangle$. The converse does not hold if k is not algebraically closed. In the exercises, we ask you to show that $\langle x^2 + 1 \rangle$ is maximal in $\mathbb{R}[x]$. The latter does not correspond to a point of $\mathbb{R}$. The following result, however, holds in any polynomial ring.

Proposition 10. *If k is any field, a maximal ideal in $k[x_1, \ldots, x_n]$ is prime.*

Proof. Suppose that I is a proper ideal which is not prime and let $fg \in I$, where $f \notin I$ and $g \notin I$. Consider the ideal $\langle f \rangle + I$. This ideal strictly contains I because $f \notin I$. Moreover, if we were to have $\langle f \rangle + I = k[x_1, \ldots, x_n]$, then $1 = cf + h$ for some polynomial c and some $h \in I$. Multiplying through by g would give $g = cfg + hg \in I$ which would contradict our choice of g. Thus, $I + \langle f \rangle$ is a proper ideal containing I, so that I is not maximal. $\qquad\square$

Note that Propositions 9 and 10 together imply that $\langle x_1 - a_1, \ldots, x_n - a_n \rangle$ is prime in $k[x_1, \ldots, x_n]$ even if k is not infinite. Over an algebraically closed field, it turns out that every maximal ideal corresponds to some point of k^n.

Theorem 11. *If k is an algebraically closed field, then every maximal ideal of $k[x_1, \ldots, x_n]$ is of the form $\langle x_1 - a_1, \ldots, x_n - a_n \rangle$ for some $a_1, \ldots, a_n \in k$.*

Proof. Let $I \subseteq k[x_1, \ldots, x_n]$ be maximal. Since $I \neq k[x_1, \ldots, x_n]$, we have $\mathbf{V}(I) \neq \emptyset$ by the Weak Nullstellensatz (Theorem 1 of §1). Hence, there is some point $(a_1, \ldots, a_n) \in \mathbf{V}(I)$. This means that every $f \in I$ vanishes at $(a_1, \ldots, a_n)$, so that $f \in \mathbf{I}(\{(a_1, \ldots, a_n)\})$. Thus, we can write

$$I \subseteq \mathbf{I}(\{(a_1, \ldots, a_n)\}).$$

We have already observed that $\mathbf{I}(\{(a_1, \ldots, a_n)\}) = \langle x_1 - a_1, \ldots, x_n - a_n \rangle$ (see Exercise 7), and, thus, the above inclusion becomes

$$I \subseteq \langle x_1 - a_1, \ldots, x_n - a_n \rangle \subsetneqq k[x_1, \ldots, x_n].$$

Since I is maximal, it follows that $I = \langle x_1 - a_1, \ldots, x_n - a_n \rangle$. □

Note the proof of Theorem 11 uses the Weak Nullstellensatz. It is not difficult to see that it is, in fact, equivalent to the Weak Nullstellensatz.

We have the following easy corollary of Theorem 11.

Corollary 12. *If k is an algebraically closed field, then there is a one-to-one correspondence between points of k^n and maximal ideals of $k[x_1, \ldots, x_n]$.*

Thus, we have extended our algebra–geometry dictionary. Over an algebraically closed field, every nonempty irreducible variety corresponds to a proper prime ideal, and conversely. Every point corresponds to a maximal ideal, and conversely.

We can use Zariski closure to characterize when a variety is irreducible.

Proposition 13. *A variety V is irreducible if and only if for every variety $W \subsetneqq V$, the difference $V \setminus W$ is Zariski dense in V.*

Proof. First assume that V is irreducible and take $W \subsetneqq V$. Then Proposition 7 of §4 gives the decomposition $V = W \cup \overline{V \setminus W}$. Since V is irreducible and $V \neq W$, this forces $V = \overline{V \setminus W}$.

For the converse, suppose that $V = V_1 \cup V_2$. If $V_1 \subsetneqq V$, then $\overline{V \setminus V_1} = V$. But $V \setminus V_1 \subseteq V_2$, so that $\overline{V \setminus V_1} \subseteq V_2$. This implies $V \subseteq V_2$, and $V = V_2$ follows. □

Let us make a final comment about terminology. Some references, such as HARTSHORNE (1977), use the term "variety" for what we call an irreducible variety and say "algebraic set" instead of variety. When reading other books on algebraic geometry, be sure to check the definitions!

EXERCISES FOR §5

1. If $h \in k[x_1, \ldots, x_n]$ has total degree N and if F is as in Proposition 6, show that $(g_1 g_2 \ldots g_n)^N (h \circ F)$ is a polynomial in $k[t_1, \ldots, t_m]$.
2. Show that a prime ideal is radical.
3. Show that an ideal I is prime if and only if for any ideals J and K such that $JK \subseteq I$, either $J \subseteq I$ or $K \subseteq I$.
4. Let $I_1, \ldots, I_n$ be ideals and P a prime ideal containing $\bigcap_{i=1}^{n} I_i$. Then prove that $P \supseteq I_i$ for some i. Further, if $P = \bigcap_{i=1}^{n} I_i$, show that $P = I_i$ for some i.
5. Express $f = x^2 z - 6y^4 + 2xy^3 z$ in the form $f = f_1(x, y, z)(x + 3) + f_2(x, y, z)(y - 1) + f_3(x, y, z)(z - 2)$ for some $f_1, f_2, f_3 \in k[x, y, z]$.

6. Let k be an infinite field.
   a. Show that any straight line in k^n is irreducible.
   b. Prove that any linear subspace of k^n is irreducible. Hint: Parametrize and use Proposition 5.
7. Show that
$$\mathbf{I}(\{(a_1, \ldots, a_n)\}) = \langle x_1 - a_1, \ldots, x_n - a_n \rangle.$$
8. Show the following:
   a. $\langle x^2 + 1 \rangle$ is maximal in $\mathbb{R}[x]$.
   b. If $I \subseteq \mathbb{R}[x_1, \ldots, x_n]$ is maximal, show that $\mathbf{V}(I)$ is either empty or a point in $\mathbb{R}^n$. Hint: Examine the proof of Theorem 11.
   c. Give an example of a maximal ideal I in $\mathbb{R}[x_1, \ldots, x_n]$ for which $\mathbf{V}(I) = \emptyset$. Hint: Consider the ideal $\langle x_1^2 + 1, x_2, \ldots, x_n \rangle$.
9. Suppose that k is a field which is not algebraically closed.
   a. Show that if $I \subseteq k[x_1, \ldots, x_n]$ is maximal, then $\mathbf{V}(I)$ is either empty or a point in k^n. Hint: Examine the proof of Theorem 11.
   b. Show that there exists a maximal ideal I in $k[x_1, \ldots, x_n]$ for which $\mathbf{V}(I) = \emptyset$. Hint: See the previous exercise.
   c. Conclude that if k is not algebraically closed, there is always a maximal ideal of $k[x_1, \ldots, x_n]$ which is not of the form $\langle x_1 - a_1, \ldots, x_n - a_n \rangle$.
10. Prove that Theorem 11 implies the Weak Nullstellensatz.
11. If $f \in \mathbb{C}[x_1, \ldots, x_n]$ is irreducible, then $\mathbf{V}(f)$ is irreducible. Hint: Show that $\langle f \rangle$ is prime.
12. Prove that if I is any proper ideal in $\mathbb{C}[x_1, \ldots, x_n]$, then $\sqrt{I}$ is the intersection of all maximal ideals containing I. Hint: Use Theorem 11.
13. Let $f_1, \ldots, f_n \in k[x_1]$ be polynomials of one variable and consider the ideal
$$I = \langle f_1(x_1), x_2 - f_2(x_1), \ldots, x_n - f_n(x_1) \rangle \subseteq k[x_1, \ldots, x_n].$$

We also assume that $\deg(f_1) = m > 0$.
   a. Show that every $f \in k[x_1, \ldots, x_n]$ can be written uniquely as $f = q + r$ where $q \in I$ and $r \in k[x_1]$ with either $r = 0$ or $\deg(r) < m$. Hint: Use lex order with x_1 last.
   b. Let $f \in k[x_1]$. Use part (a) to show that $f \in I$ if and only if f is divisible by f_1 in $k[x_1]$.
   c. Prove that I is prime if and only if $f_1 \in k[x_1]$ is irreducible.
   d. Prove that I is radical if and only if $f_1 \in k[x_1]$ is square-free.
   e. Prove that $\sqrt{I} = \langle (f_1)_{\text{red}} \rangle + I$, where $(f_1)_{\text{red}}$ is defined in §2.

§6   Decomposition of a Variety into Irreducibles

In the last section, we saw that irreducible varieties arise naturally in many contexts. It is natural to ask whether an arbitrary variety can be built up out of irreducibles. In this section, we explore this and related questions.

We begin by translating the *ascending chain condition* (ACC) for ideals (see §5 of Chapter 2) into the language of varieties.

Proposition 1 (The Descending Chain Condition). *Any descending chain of varieties*
$$V_1 \supseteq V_2 \supseteq V_3 \supseteq \cdots$$
in k^n must stabilize, meaning that there exists a positive integer N such that $V_N = V_{N+1} = \cdots$.

Proof. Passing to the corresponding ideals gives an ascending chain of ideals

$$\mathbf{I}(V_1) \subseteq \mathbf{I}(V_2) \subseteq \mathbf{I}(V_3) \subseteq \cdots .$$

By the ascending chain condition for ideals (see Theorem 7 of Chapter 2, §5), there exists N such that $\mathbf{I}(V_N) = \mathbf{I}(V_{N+1}) = \cdots$. Since $\mathbf{V}(\mathbf{I}(V)) = V$ for any variety V, we have $V_N = V_{N+1} = \cdots$. $\qquad\square$

We can use Proposition 1 to prove the following basic result about the structure of affine varieties.

Theorem 2. *Let $V \subseteq k^n$ be an affine variety. Then V can be written as a finite union*

$$V = V_1 \cup \cdots \cup V_m,$$

where each V_i is an irreducible variety.

Proof. Assume that V is an affine variety which cannot be written as a finite union of irreducibles. Then V is not irreducible, so that $V = V_1 \cup V_1'$, where $V \neq V_1$ and $V \neq V_1'$. Further, one of V_1 and V_1' must not be a finite union of irreducibles, for otherwise V would be of the same form. Say V_1 is not a finite union of irreducibles. Repeating the argument just given, we can write $V_1 = V_2 \cup V_2'$, where $V_1 \neq V_2, V_1 \neq V_2'$, and V_2 is not a finite union of irreducibles. Continuing in this way gives us an infinite sequence of affine varieties

$$V \supseteq V_1 \supseteq V_2 \supseteq \cdots$$

with

$$V \neq V_1 \neq V_2 \neq \cdots .$$

This contradicts Proposition 1. $\qquad\square$

As a simple example of Theorem 2, consider the variety $\mathbf{V}(xz, yz)$ which is a union of a line (the z-axis) and a plane [the (x, y)-plane], both of which are irreducible by Exercise 6 of §5. For a more complicated example of the decomposition of a variety into irreducibles, consider the variety

$$V = \mathbf{V}(xz - y^2, x^3 - yz).$$

A sketch of this variety appears at the top of the next page. The picture suggests that this variety is not irreducible. It appears to be a union of two curves. Indeed, since both $xz - y^2$ and $x^3 - yz$ vanish on the z-axis, it is clear that the z-axis $\mathbf{V}(x, y)$ is contained in V. What about the other curve $V \setminus \mathbf{V}(x, y)$?

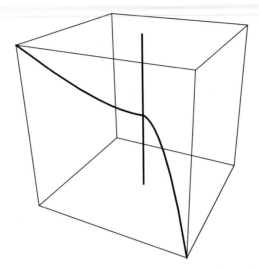

By Corollary 11 of §4, this suggests looking at the ideal quotient

$$\langle xz - y^2, x^3 - yz \rangle : \langle x, y \rangle.$$

(At the end of the section we will see that $\langle xz - y^2, x^3 - yz \rangle$ is a radical ideal.) We can compute this quotient using our algorithm for computing ideal quotients (make sure you review this algorithm). By equation (5) of §4, the above is equal to

$$(I : x) \cap (I : y),$$

where $I = \langle xz - y^2, x^3 - yz \rangle$. To compute $I : x$, we first compute $I \cap \langle x \rangle$ using our algorithm for computing intersections of ideals. Using lex order with $z > y > x$, we obtain

$$I \cap \langle x \rangle = \langle x^2 z - xy^2, x^4 - xyz, x^3 y - xz^2, x^5 - xy^3 \rangle.$$

We can omit $x^5 - xy^3$ since it is a combination of the first and second elements in the basis. Hence

$$
I : x = \left\langle \frac{x^2 z - xy^2}{x}, \frac{x^4 - xyz}{x}, \frac{x^3 y - xz^2}{x} \right\rangle
$$

(1)

$$
= \langle xz - y^2, x^3 - yz, x^2 y - z^2 \rangle
$$

$$
= I + \langle x^2 y - z^2 \rangle.
$$

Similarly, to compute $I : \langle y \rangle$, we compute

$$I \cap \langle y \rangle = \langle xyz - y^3, x^3 y - y^2 z, x^2 y^2 - yz^2 \rangle,$$

which gives

$$I : y = \left\langle \frac{xyz - y^3}{y}, \frac{x^3y - y^2z}{y}, \frac{x^2y^2 - yz^2}{y} \right\rangle$$
$$= \langle xz - y^2, x^3 - yz, x^2y - z^2 \rangle$$
$$= I + \langle x^2y - z^2 \rangle$$
$$= I : x.$$

(Do the computations using a computer algebra system.) Since $I : x = I : y$, we have

$$I : \langle x, y \rangle = \langle xz - y^2, x^3 - yz, x^2y - z^2 \rangle.$$

The variety $W = \mathbf{V}(xz - y^2, x^3 - yz, x^2y - z^2)$ turns out to be an irreducible curve. To see this, note that it can be parametrized as (t^3, t^4, t^5) [it is clear that $(t^3, t^4, t^5) \in W$ for any t—we leave it as an exercise to show every point of W is of this form], so that W is irreducible by Proposition 5 of the last section. It then follows easily that

$$V = \mathbf{V}(I) = \mathbf{V}(x, y) \cup \mathbf{V}(I : \langle x, y \rangle) = \mathbf{V}(x, y) \cup W$$

(see Proposition 7 of §4), which gives decomposition of V into irreducibles.

Both in the above example and the case of $\mathbf{V}(xz, yz)$, it appears that the decomposition of a variety into irreducible pieces is unique. It is natural to ask whether this is true in general. It is clear that, to avoid trivialities, we must rule out decompositions in which the same irreducible piece appears more than once, or in which one irreducible piece contains another. This is the aim of the following definition.

Definition 3. Let $V \subseteq k^n$ be an affine variety. A decomposition

$$V = V_1 \cup \cdots \cup V_m,$$

where each V_i is an irreducible variety, is called a **minimal decomposition** (or, sometimes, an **irredundant union**) if $V_i \not\subseteq V_j$ for $i \neq j$. Also, we call the V_i the **irreducible components** of V.

With this definition, we can now prove the following uniqueness result.

Theorem 4. *Let $V \subseteq k^n$ be an affine variety. Then V has a minimal decomposition*

$$V = V_1 \cup \cdots \cup V_m$$

(so each V_i is an irreducible variety and $V_i \not\subseteq V_j$ for $i \neq j$). Furthermore, this minimal decomposition is unique up to the order in which $V_1, \ldots, V_m$ are written.

Proof. By Theorem 2, V can be written in the form $V = V_1 \cup \cdots \cup V_m$, where each V_i is irreducible. Further, if a V_i lies in some V_j for $i \neq j$, we can drop V_i, and V will be the union of the remaining V_j's for $j \neq i$. Repeating this process leads to a minimal decomposition of V.

To show uniqueness, suppose that $V = V_1' \cup \cdots \cup V_l'$ is another minimal decomposition of V. Then, for each V_i in the first decomposition, we have

$$V_i = V_i \cap V = V_i \cap (V_1' \cup \cdots \cup V_l') = (V_i \cap V_1') \cup \cdots \cup (V_i \cap V_l').$$

Since V_i is irreducible, it follows that $V_i = V_i \cap V_j'$ for some j, i.e., $V_i \subseteq V_j'$. Applying the same argument to V_j' (using the V_i's to decompose V) shows that $V_j' \subseteq V_k$ for some k, and, thus,

$$V_i \subseteq V_j' \subseteq V_k.$$

By minimality, $i = k$, and it follows that $V_i = V_j'$. Hence, every V_i appears in $V = V_1' \cup \cdots \cup V_l'$, which implies $m \leq l$. A similar argument proves $l \leq m$, and $m = l$ follows. Thus, the V_i''s are just a permutation of the V_i's, and uniqueness is proved.   □

The uniqueness part of Theorem 4 guarantees that the irreducible components of V are well-defined. We remark that the uniqueness is false if one does not insist that the union be finite. (A plane P is the union of all the points on it. It is also the union of some line in P and all the points not on the line—there are infinitely many lines in P with which one could start.) This should alert the reader to the fact that although the proof of Theorem 4 is easy, it is far from vacuous: one makes subtle use of finiteness (which follows, in turn, from the Hilbert Basis Theorem).

Here is a result that relates irreducible components to Zariski closure.

Proposition 5. *Let V, W be varieties with $W \subsetneq V$. Then $V \setminus W$ is Zariski dense in V if and only if W contains no irreducible component of V.*

Proof. Suppose that $V = V_1 \cup \cdots \cup V_m$ as in Theorem 4 and that $V_i \nsubseteq W$ for all i. This implies $V_i \cap W \subsetneq V_i$, and since V_i is irreducible, we deduce $\overline{V_i \setminus (V_i \cap W)} = V_i$ by Proposition 13 of §5. Then

$$\overline{V \setminus W} = \overline{(V_1 \cup \cdots \cup V_m) \setminus W} = \overline{(V_1 \setminus (V_1 \cap W)) \cup \cdots \cup (V_m \setminus (V_m \cap W))}$$
$$= \overline{V_1 \setminus (V_1 \cap W)} \cup \cdots \cup \overline{V_m \setminus (V_m \cap W)}$$
$$= V_1 \cup \cdots \cup V_m = V,$$

where the second line uses Lemma 3 of §4. The other direction of the proof will be covered in the exercises.   □

Theorems 2 and 4 can also be expressed purely algebraically using the one-to-one correspondence between radical ideals and varieties.

Theorem 6. *If k is algebraically closed, then every radical ideal in $k[x_1, \ldots, x_n]$ can be written uniquely as a finite intersection of prime ideals $P_1 \cap \cdots \cap P_r$, where $P_i \nsubseteq P_j$ for $i \neq j$. (As in the case of varieties, we often call such a presentation of a radical ideal a **minimal decomposition** or an **irredundant intersection**).*

Proof. Theorem 6 follows immediately from Theorems 2 and 4 and the ideal–variety correspondence.   □

We can also use ideal quotients from §4 to describe the prime ideals that appear in the minimal representation of a radical ideal.

Theorem 7. *If k is algebraically closed and I is a proper radical ideal such that*

$$I = \bigcap_{i=1}^{r} P_i$$

is its minimal decomposition as an intersection of prime ideals, then the P_i's are precisely the proper prime ideals that occur in the set $\{I:f \mid f \in k[x_1, \ldots, x_n]\}$.

Proof. First, note that since I is proper, each P_i is also a proper ideal (this follows from minimality).

For any $f \in k[x_1, \ldots, x_n]$, we have

$$I:f = \left(\bigcap_{i=1}^{r} P_i\right):f = \bigcap_{i=1}^{r}(P_i:f)$$

by Exercise 17 of §4. Note also that for any prime ideal P, either $f \in P$, in which case $P:f = \langle 1 \rangle$, or $f \notin P$, in which case $P:f = P$ (see Exercise 3).

Now suppose that $I:f$ is a proper prime ideal. By Exercise 4 of §5, the above formula for $I:f$ implies that $I:f = P_i:f$ for some i. Since $P_i:f = P_i$ or $\langle 1 \rangle$ by the above observation, it follows that $I:f = P_i$.

To see that every P_i can arise in this way, fix i and pick $f \in \left(\bigcap_{j \neq i} P_j\right) \setminus P_i$; such an f exists because $\bigcap_{i=1}^{r} P_i$ is minimal. Then $P_i:f = P_i$ and $P_j:f = \langle 1 \rangle$ for $j \neq i$. If we combine this with the above formula for $I:f$, then it follows that $I:f = P_i$. □

We should mention that Theorems 6 and 7 hold for any field k, although the proofs in the general case are different (see Corollary 10 of §8).

For an example of these theorems, consider the ideal $I = \langle xz - y^2, x^3 - yz \rangle$. Recall that the variety $V = \mathbf{V}(I)$ was discussed earlier in this section. For the time being, let us assume that I is radical (eventually we will see that this is true). Can we write I as an intersection of prime ideals?

We start with the geometric decomposition

$$V = \mathbf{V}(x, y) \cup W$$

proved earlier, where $W = \mathbf{V}(xz - y^2, x^3 - yz, x^2y - z^2)$. This suggests that

$$I = \langle x, y \rangle \cap \langle xz - y^2, x^3 - yz, x^2y - z^2 \rangle,$$

which is straightforward to prove by the techniques we have learned so far (see Exercise 4). Also, from equation (1), we know that $I:x = \langle xz - y^2, x^3 - yz, x^2y - z^2 \rangle$. Thus,

$$I = \langle x, y \rangle \cap (I:x).$$

To represent $\langle x, y \rangle$ as an ideal quotient of I, let us think geometrically. The idea is to remove W from V. Of the three equations defining W, the first two give V. So it makes sense to use the third one, $x^2y - z^2$, and one can check that $I : (x^2y - z^2) = \langle x, y \rangle$ (see Exercise 4). Thus,

$$(2) \qquad\qquad I = (I : (x^2y - z^2)) \cap (I : x).$$

It remains to show that $I : (x^2y - z^2)$ and $I : x$ are prime ideals. The first is easy since $I : (x^2y - z^2) = \langle x, y \rangle$ is obviously prime. As for the second, we have already seen that $W = \mathbf{V}(xz - y^2, x^3 - yz, x^2y - z^2)$ is irreducible and, in the exercises, you will show that $\mathbf{I}(W) = \langle xz - y^2, x^3 - yz, x^2y - z^2 \rangle = I : x$. It follows from Proposition 3 of §5 that $I : x$ is a prime ideal. This completes the proof that (2) is the minimal representation of I as an intersection of prime ideals. Finally, since I is an intersection of prime ideals, we see that I is a radical ideal (see Exercise 1).

The arguments used in this example are special to the case $I = \langle xz - y^2, x^3 - yz \rangle$. It would be nice to have more general methods that could be applied to any ideal. Theorems 2, 4, 6, and 7 tell us that certain decompositions exist, but the proofs give no indication of how to find them. The problem is that the proofs rely on the Hilbert Basis Theorem, which is intrinsically nonconstructive. Based on what we have seen in §§5 and 6, the following questions arise naturally:

- (Primality) Is there an algorithm for deciding if a given ideal is prime?
- (Irreducibility) Is there an algorithm for deciding if a given affine variety is irreducible?
- (Decomposition) Is there an algorithm for finding the minimal decomposition of a given variety or radical ideal?

The answer to all three questions is *yes*, and descriptions of the algorithms can be found in the works of HERMANN (1926), MINES, RICHMAN, and RUITEN-BERG (1988), and SEIDENBERG (1974, 1984). As in §2, the algorithms in these articles are very inefficient. However, the work of GIANNI, TRAGER and ZACHARIAS (1988) and EISENBUD, HUNEKE and VASCONCELOS (1992) has led to more efficient algorithms. See also Chapter 8 of BECKER and WEISPFENNING (1993) and §4.4 of ADAMS and LOUSTAUNAU (1994).

EXERCISES FOR §6

1. Show that the intersection of any collection of prime ideals is radical.
2. Show that an irredundant intersection of at least two prime ideals is never prime.
3. If $P \subseteq k[x_1, \ldots, x_n]$ is a prime ideal, then prove that $P : f = P$ iff $f \notin P$ and $P : f = \langle 1 \rangle$ if $f \in P$.
4. Let $I = \langle xz - y^2, x^3 - yz \rangle$.
   a. Show that $I : (x^2y - z^2) = \langle x, y \rangle$.
   b. Show that $I : (x^2y - z^2)$ is prime.
   c. Show that $I = \langle x, y \rangle \cap \langle xz - y^2, x^3 - yz, z^2 - x^2y \rangle$.
5. Let $J = \langle xz - y^2, x^3 - yz, z^2 - x^2y \rangle \subseteq k[x, y, z]$, where k is infinite.
   a. Show that every point of $W = \mathbf{V}(J)$ is of the form (t^3, t^4, t^5) for some $t \in k$.

b. Show that $J = \mathbf{I}(W)$. Hint: Compute a Gröbner basis for J using lex order with $z > y > x$ and show that every $f \in k[x, y, z]$ can be written in the form

$$f = g + a + bz + xA(x) + yB(x) + y^2 C(x),$$

where $g \in J, a, b \in k$ and $A, B, C \in k[x]$.
6. Complete the proof of Proposition 5. Hint: $V_i \subseteq W$ implies $V \setminus W \subseteq V \setminus V_i$.
7. Translate Theorem 7 and its proof into geometry.
8. Let $I = \langle xz - y^2, z^3 - x^5 \rangle \subseteq \mathbb{Q}[x, y, z]$.
    a. Express $\mathbf{V}(I)$ as a finite union of irreducible varieties. Hint: The parametrizations (t^3, t^4, t^5) and $(t^3, -t^4, t^5)$ will be useful.
    b. Express I as an intersection of prime ideals which are ideal quotients of I and conclude that I is radical.
9. Let V, W be varieties in k^n with $V \subseteq W$. Show that each irreducible component of V is contained in some irreducible component of W.
10. Let $f \in \mathbb{C}[x_1, \ldots, x_n]$ and let $f = f_1^{a_1} f_2^{a_2} \cdots f_r^{a_r}$ be the decomposition of f into irreducible factors. Show that $\mathbf{V}(f) = \mathbf{V}(f_1) \cup \cdots \cup \mathbf{V}(f_r)$ is the decomposition of $\mathbf{V}(f)$ into irreducible components and $\mathbf{I}(\mathbf{V}(f)) = \langle f_1 f_2 \cdots f_r \rangle$. Hint: See Exercise 11 of §5.

§7 Proof of the Closure Theorem

This section will complete the proof of the Closure Theorem from Chapter 3, §2. We will use many of the tools introduced in this chapter, including the Nullstellensatz, Zariski closures, saturations, and irreducible components.

We begin by recalling the basic situation. Let k be an algebraically closed field, and let $\pi_l : k^n \to k^{n-l}$ is projection onto the last $n - l$ components. If $V = \mathbf{V}(I)$ is an affine variety in k^n, then we get the l-th elimination ideal $I_l = I \cap k[x_{l+1}, \ldots, x_n]$. The first part of the Closure Theorem, which asserts that $\mathbf{V}(I_l)$ is the Zariski closure of $\pi_l(V)$ in k^{n-l}, was proved earlier in Theorem 4 of §4.

The remaining part of the Closure Theorem tells us that $\pi_l(V)$ fills up "most" of $\mathbf{V}(I_l)$. Here is the precise statement.

Theorem 1 (The Closure Theorem, second part). *Let k be algebraically closed, and let $V = \mathbf{V}(I) \subseteq k^n$. Then there is an affine variety $W \subseteq \mathbf{V}(I_l)$ such that*

$$\mathbf{V}(I_l) \setminus W \subseteq \pi_l(V) \text{ and } \overline{\mathbf{V}(I_l) \setminus W} = \mathbf{V}(I_l).$$

This is slightly different from the Closure Theorem stated in §2 of Chapter 3. There, we assumed $V \neq \emptyset$ and asserted that $\mathbf{V}(I_l) \setminus W \subseteq \pi_l(V)$ for some $W \subsetneq \mathbf{V}(I_l)$. In Exercise 1 you will prove that Theorem 1 implies the version in Chapter 3.

The proof of Theorem 1 will use the following notation. Rename $x_{l+1}, \ldots, x_n$ as $y_{l+1}, \ldots, y_n$ and write $k[x_1, \ldots, x_l, y_{l+1}, \ldots, y_n]$ as $k[\mathbf{x}, \mathbf{y}]$ for $\mathbf{x} = (x_1, \ldots, x_l)$ and $\mathbf{y} = (y_{l+1}, \ldots, y_n)$. Also fix a monomial order $>$ on $k[\mathbf{x}, \mathbf{y}]$ with the property that $\mathbf{x}^\alpha > \mathbf{x}^\beta$ implies $\mathbf{x}^\alpha > \mathbf{x}^\beta \mathbf{y}^\gamma$ for all γ. The product order described in Exercise 9 of Chapter 2, §4 is an example of such a monomial order. Another example is given by lex order with $x_1 > \cdots > x_l > y_{l+1} > \cdots > y_n$.

An important tool in proving Theorem 1 is the following result.

Theorem 2. *Fix a field k. Let $I \subseteq k[\mathbf{x}, \mathbf{y}]$ be an ideal and let $G = \{g_1, \ldots, g_t\}$ be a Gröbner basis for I with respect to a monomial order as above. For $1 \leq i \leq t$ with $g_i \notin k[\mathbf{y}]$, write g_i in the form*

(1) $g_i = c_i(\mathbf{y}) \mathbf{x}^{\alpha_i} + \text{terms} < \mathbf{x}^{\alpha_i}.$

Finally, assume that $\mathbf{b} = (a_{l+1}, \ldots, a_n) \in \mathbf{V}(I_l) \subseteq k^{n-l}$ is a partial solution such that $c_i(\mathbf{b}) \neq 0$ for all $g_i \notin k[\mathbf{y}]$. Then:

(i) *The set*

$$\overline{G} = \{g_i(\mathbf{x}, \mathbf{b}) \mid g_i \notin k[\mathbf{y}]\} \subseteq k[\mathbf{x}]$$

   is a Gröbner basis of the ideal $\{f(\mathbf{x}, \mathbf{b}) \mid f \in I\}$.

(ii) *If k is algebraically closed, then there exists $\mathbf{a} = (a_1, \ldots, a_l) \in k^l$ such that $(\mathbf{a}, \mathbf{b}) \in V = \mathbf{V}(I)$.*

Proof. Given $f \in k[\mathbf{x}, \mathbf{y}]$, we set

$$\bar{f} = f(\mathbf{x}, \mathbf{b}) \in k[\mathbf{x}].$$

In this notation, $\overline{G} = \{\bar{g}_i \mid g_i \notin k[\mathbf{y}]\}$. Also observe that $\bar{g}_i = 0$ when $g_i \in k[\mathbf{y}]$ since $\mathbf{b} \in \mathbf{V}(I_l)$. If we set $\bar{I} = \{\bar{f} \mid f \in I\}$, then it is an easy exercise to show that

$$\bar{I} = \langle \overline{G} \rangle \subseteq k[\mathbf{x}].$$

In particular, $\bar{I}$ is an ideal of $k[\mathbf{x}]$.

To prove that $\overline{G}$ is a Gröbner basis of $\bar{I}$, take $g_i, g_j \in G \setminus k[\mathbf{y}]$ and consider the polynomial

$$S = c_j(\mathbf{y}) \frac{\mathbf{x}^\gamma}{\mathbf{x}^{\alpha_i}} g_i - c_i(\mathbf{y}) \frac{\mathbf{x}^\gamma}{\mathbf{x}^{\alpha_j}} g_j,$$

where $\mathbf{x}^\gamma = \text{lcm}(\mathbf{x}^{\alpha_i}, \mathbf{x}^{\alpha_j})$. Our chosen monomial order has the property that $\text{LT}(g_i) = \text{LT}(c_i(\mathbf{y})) \mathbf{x}^{\alpha_i}$, and it follows easily that $\mathbf{x}^\gamma > \text{LT}(S)$. Since $S \in I$, it has a standard representation $S = \sum_{k=1}^t A_k g_k$. Then evaluating at $\mathbf{b}$ gives

$$c_j(\mathbf{b}) \frac{\mathbf{x}^\gamma}{\mathbf{x}^{\alpha_i}} \bar{g}_i - c_i(\mathbf{b}) \frac{\mathbf{x}^\gamma}{\mathbf{x}^{\alpha_j}} \bar{g}_j = \overline{S} = \sum_{\bar{g}_k \in \overline{G}} \overline{A}_k \bar{g}_k$$

since $\bar{g}_i = 0$ for $g_i \in k[\mathbf{y}]$.

Then $c_i(\mathbf{b}), c_j(\mathbf{b}) \neq 0$ imply that $\overline{S}$ is the S-polynomial $S(\bar{g}_i, \bar{g}_j)$ up to the nonzero constant $c_i(\mathbf{b})c_j(\mathbf{b})$. Since

$$\mathbf{x}^\gamma > \text{LT}(S) \geq \text{LT}(A_k g_k), \quad A_k g_k \neq 0,$$

it follows that

$$\mathbf{x}^\gamma > \text{LT}(\overline{A}_k \bar{g}_k), \quad \overline{A}_k \bar{g}_k \neq 0,$$

by Exercise 3 of Chapter 2, §9. Hence $S(\bar{g}_i, \bar{g}_j)$ has an lcm representation as defined in Chapter 2, §9. By Theorem 6 of that section, we conclude that $\overline{G}$ is a Gröbner basis of $\bar{I}$, as claimed.

For part (ii), note that by construction, every element of $\overline{G}$ has positive total degree in the $\mathbf{x}$ variables, so that $\bar{g}_i$ is nonconstant for every i. It follows that $1 \notin \bar{I}$ since $\overline{G}$ is a Gröbner basis of $\bar{I}$. Hence $\bar{I} \subsetneq k[\mathbf{x}]$, so that by the Nullstellensatz, there exists $\mathbf{a} \in k^l$ such that $\bar{g}_i(\mathbf{a}) = 0$ for all $\bar{g}_i \in \overline{G}$, i.e., $g_i(\mathbf{a}, \mathbf{b}) = 0$ for all $g_i \in G \setminus k[\mathbf{y}]$. Since $\bar{g}_i = 0$ when $g_i \in k[\mathbf{y}]$, it follows that $g_i(\mathbf{a}, \mathbf{b}) = 0$ for all $g_i \in G$. Hence $(\mathbf{a}, \mathbf{b}) \in V = \mathbf{V}(I)$. $\square$

Part (ii) of Theorem 2 is related to the Extension Theorem from Chapter 3. Compared to the Extension theorem, part (ii) is simultaneously stronger (the Extension Theorem assumes $l = 1$, i.e., just one variable is eliminated) and weaker [part (ii) requires the nonvanishing of *all* relevant leading coefficients, while the Extension Theorem requires just one].

For our purposes, Theorem 2 has the following important corollary.

Corollary 3. *With the same notation as Theorem 2, we have*

$$\mathbf{V}(I_l) \setminus \mathbf{V}\big(\textstyle\prod_{g_i \in G \setminus k[\mathbf{y}]} c_i\big) \subseteq \pi_l(V).$$

Proof. Take $\mathbf{b} \in \mathbf{V}(I_l) \setminus \mathbf{V}\big(\prod_{g_i \in G \setminus k[\mathbf{y}]} c_i\big)$. Then $\mathbf{b} \in \mathbf{V}(I_l)$ and $c_i(\mathbf{b}) \neq 0$ for all $g_i \in G \setminus k[\mathbf{y}]$. By Theorem 2, there is $\mathbf{a} \in k^l$ such that $(\mathbf{a}, \mathbf{b}) \in V = \mathbf{V}(I)$. In other words, $\mathbf{b} \in \pi_l(V)$, and the corollary follows. $\square$

Since $A \setminus B = A \setminus (A \cap B)$, Corollary 3 implies that the intersection

$$W = \mathbf{V}(I_l) \cap \mathbf{V}\big(\textstyle\prod_{g_i \in G \setminus k[\mathbf{y}]} c_i\big) \subseteq \mathbf{V}(I_l)$$

has the property that $\mathbf{V}(I_l) \setminus W \subseteq \pi_l(V)$. If $\mathbf{V}(I_l) \setminus W$ is also Zariski dense in $\mathbf{V}(I_l)$, then $W \subseteq \mathbf{V}(I_l)$ satisfies the conclusion of the Closure Theorem.

Hence, to complete the proof of the Closure Theorem, we need to explore what happens when the difference $\mathbf{V}(I_l) \setminus \mathbf{V}\big(\prod_{g_i \in G \setminus k[\mathbf{y}]} c_i\big)$ is not Zariski dense in $\mathbf{V}(I_l)$. The following proposition shows that in this case, the original variety $V = \mathbf{V}(I)$ decomposes into varieties coming from strictly bigger ideals.

Proposition 4. *Assume that k is algebraically closed and the Gröbner basis G is reduced. If $\mathbf{V}(I_l) \setminus \mathbf{V}\big(\prod_{g_i \in G \setminus k[\mathbf{y}]} c_i\big)$ is not Zariski dense in $\mathbf{V}(I_l)$, then there is some $g_i \in G \setminus k[\mathbf{y}]$ whose c_i has the following two properties:*
(i) $V = \mathbf{V}(I + \langle c_i \rangle) \cup \mathbf{V}(I : c_i^\infty)$.
(ii) $I \subsetneq I + \langle c_i \rangle$ *and* $I \subsetneq I : c_i^\infty$.

Proof. For (i), we have $V = \mathbf{V}(I) = \mathbf{V}(I + \langle c_i \rangle) \cup \mathbf{V}(I : c_i^\infty)$ by Theorem 10 of §4.

For (ii), we first show that $I \subsetneq I + \langle c_i \rangle$ for all $g_i \in G \setminus k[\mathbf{y}]$. To see why, suppose that $c_i \in I$ for some i. Since G is a Gröbner basis of I, $\mathrm{LT}(c_i)$ is divisible by some $\mathrm{LT}(g_j)$, and then $g_j \in k[\mathbf{y}]$ since the monomial order eliminates the $\mathbf{x}$ variables.

Hence $g_j \neq g_i$. But then (1) implies that $\mathrm{LT}(g_j)$ divides $\mathrm{LT}(g_i) = \mathrm{LT}(c_i)\mathbf{x}^{\alpha_i}$, which contradicts our assumption that G is reduced. Hence $c_i \notin I$, and $I \subsetneq I + \langle c_i \rangle$ follows.

Now suppose that $I = I : c_i^\infty$ for all i with $g_i \in G \setminus k[\mathbf{y}]$. In Exercise 4, you will show that this implies $I_l : c_i^\infty = I_l$ for all i. Hence

$$\mathbf{V}(I_l) = \mathbf{V}(I_l : c_i^\infty) = \overline{\mathbf{V}(I_l) \setminus \mathbf{V}(c_i)} = \overline{\mathbf{V}(I_l) \setminus (\mathbf{V}(I_l) \cap \mathbf{V}(c_i))},$$

where the second equality uses Theorem 10 of §4. If follows that $\mathbf{V}(I_l) \cap \mathbf{V}(c_i)$ contains no irreducible component of $\mathbf{V}(I_l)$ by Proposition 5 of §6. Since this holds for all i, the finite union

$$\bigcup_{g_i \in G \setminus k[\mathbf{y}]} \mathbf{V}(I_l) \cap \mathbf{V}(c_i) = \mathbf{V}(I_l) \cap \bigcup_{g_i \in G \setminus k[\mathbf{y}]} \mathbf{V}(c_i) = \mathbf{V}(I_l) \cap \mathbf{V}\left(\textstyle\prod_{g_i \in G \setminus k[\mathbf{y}]} c_i\right)$$

also contains no irreducible component of $\mathbf{V}(I_l)$ (see Exercise 5). By the same proposition from §6, we conclude that the difference

$$\mathbf{V}(I_l) \setminus \left(\mathbf{V}(I_l) \cap \mathbf{V}\left(\textstyle\prod_{g_i \in G \setminus k[\mathbf{y}]} c_i\right)\right) = \mathbf{V}(I_l) \setminus \mathbf{V}\left(\textstyle\prod_{g_i \in G \setminus k[\mathbf{y}]} c_i\right)$$

is Zariski dense in $\mathbf{V}(I_l)$. This contradiction shows that $I \subsetneq I : c_i^\infty$ for some i and completes the proof of the proposition. $\square$

In the situation of Proposition 4, we have a decomposition of V into two pieces. The next step is to show that if we can find a W that works for each piece, then we can find a W what works for V. Here is the precise result.

Proposition 5. *Let k be algebraically closed. Suppose that a variety $V = \mathbf{V}(I)$ can be written $V = \mathbf{V}(I^{(1)}) \cup \mathbf{V}(I^{(2)})$ and that we have varieties*

$$W_1 \subseteq \mathbf{V}(I_l^{(1)}) \quad and \quad W_2 \subseteq \mathbf{V}(I_l^{(2)})$$

such that $\overline{\mathbf{V}(I_l^{(i)}) \setminus W_i} = \mathbf{V}(I_l^{(i)})$ *and* $\mathbf{V}(I_l^{(i)}) \setminus W_i \subseteq \pi_l(\mathbf{V}(I^{(i)}))$ *for* $i = 1, 2$. *Then* $W = W_1 \cup W_2$ *is a variety contained in V that satisfies*

$$\overline{\mathbf{V}(I_l) \setminus W} = \mathbf{V}(I_l) \quad and \quad \mathbf{V}(I_l) \setminus W \subseteq \pi_l(V).$$

Proof. For simplicity, set $V_i = \mathbf{V}(I^{(i)})$, so that $V = V_1 \cup V_2$. The first part of the Closure Theorem proved in §4 implies that $\mathbf{V}(I_l) = \overline{\pi_l(V)}$ and $\mathbf{V}(I_l^{(i)}) = \overline{\pi_l(V_i)}$. Hence

$$\mathbf{V}(I_l) = \overline{\pi_l(V)} = \overline{\pi_l(V_1 \cup V_2)} = \overline{\pi_l(V_1) \cup \pi_l(V_2)} = \overline{\pi_l(V_1)} \cup \overline{\pi_l(V_2)}$$
$$= \mathbf{V}(I_l^{(1)}) \cup \mathbf{V}(I_l^{(2)}),$$

where the last equality of the first line uses Lemma 3 of §4.

Now let $W_i \subseteq \mathbf{V}(I_l^{(i)})$ be as in the statement of the proposition. By Proposition 5 of §6, we know that W_i contains no irreducible component of $\mathbf{V}(I_l^{(i)})$. As you will prove in Exercise 5, this implies that the union $W = W_1 \cup W_2$ contains no irreducible component of $\mathbf{V}(I_l) = \mathbf{V}(I_l^{(1)}) \cup \mathbf{V}(I_l^{(2)})$. Using Proposition 5 of §6 again, we deduce that $\mathbf{V}(I_l) \setminus W$ is Zariski dense in $\mathbf{V}(I_l)$. Since we also have

$$\mathbf{V}(I_l) \setminus W = \left(\mathbf{V}(I_l^{(1)}) \cup \mathbf{V}(I_l^{(2)})\right) \setminus (W_1 \cup W_2) \subseteq \left(\mathbf{V}(I_l^{(1)}) \setminus W_1\right) \cup \left(\mathbf{V}(I_l^{(2)}) \setminus W_2\right)$$
$$\subseteq \pi_l(V_1) \cup \pi_l(V_2) = \pi_l(V),$$

the proof of the proposition is complete. $\square$

The final ingredient we need for the proof of the Closure Theorem is the following maximum principle for ideals.

Proposition 6 (Maximum Principle for Ideals). *Given a nonempty collection of ideals $\{I_\alpha\}_{\alpha \in \mathcal{A}}$ in a polynomial ring $k[x_1, \ldots, x_n]$, there exists $\alpha_0 \in \mathcal{A}$ such that for all $\beta \in \mathcal{A}$, we have*

$$I_{\alpha_0} \subseteq I_\beta \implies I_{\alpha_0} = I_\beta.$$

In other words, I_{α_0} is maximal with respect to inclusion among the I_α for $\alpha \in \mathcal{A}$.

Proof. This is an easy consequence of the ascending chain condition (Theorem 7 of Chapter 2, §5). The proof will be left as an exercise. $\square$

We are now ready to prove the second part of the Closure Theorem.

Proof of Theorem 1. Suppose the theorem fails for some ideal $I \subseteq k[x_1, \ldots, x_n]$, i.e., there is no affine variety $W \subsetneq \mathbf{V}(I)$ such that

$$\mathbf{V}(I_l) \setminus W \subseteq \pi_l(\mathbf{V}(I)) \text{ and } \overline{\mathbf{V}(I_l) \setminus W} = \mathbf{V}(I_l).$$

Our goal is to derive a contradiction.

Among all ideals for which the theorem fails, the maximum principle of Proposition 6 guarantees that there is a maximal such ideal, i.e., there is an ideal I such that the theorem fails for I but holds for every strictly larger ideal $I \subsetneq J$.

Let us apply our results to I. By Corollary 3, we know that

$$\mathbf{V}(I_l) \setminus \mathbf{V}\left(\prod_{g_i \in G \setminus k[\mathbf{y}]} c_i\right) \subseteq \pi_l(V).$$

Since the theorem fails for I, $\mathbf{V}(I) \setminus \mathbf{V}\left(\prod_{g_i \in G \setminus k[\mathbf{y}]} c_i\right)$ cannot be Zariski dense in $\mathbf{V}(I_l)$. Therefore, by Proposition 4, there is some i such that

$$I \subsetneq I^{(1)} = I + \langle c_i \rangle, \quad I \subsetneq I^{(2)} = I : c_i^\infty$$

and

$$\mathbf{V}(I) = \mathbf{V}(I^{(1)}) \cup \mathbf{V}(I^{(2)}).$$

Our choice of I guarantees that the theorem holds for the strictly larger ideals $I^{(1)}$ and $I^{(2)}$. The resulting affine varieties $W_i \subseteq \mathbf{V}(I_l^{(i)})$, $i = 1, 2$, satisfy the hypothesis of Proposition 5, and then the proposition implies that $W = W_1 \cup W_2 \subseteq \mathbf{V}(I)$ satisfies the theorem for I. This contradicts our choice of I, and we are done. $\square$

The proof of the Closure Theorem just given is nonconstructive. Fortunately, in practice it is straightforward to find $W \subseteq \mathbf{V}(I_l)$ with the required properties. We will give two examples and then describe a general procedure.

The first example is very simple. Consider the ideal

$$I = \langle yx^2 + yx + 1 \rangle \subseteq \mathbb{C}[x, y].$$

We use lex order with $x > y$, and $I_1 = \{0\}$ since $g_1 = yx^2 + yx + 1$ is a Gröbner basis for I. In the notation of Theorem 2, we have $c_1 = y$, and then Corollary 3 implies that

$$\mathbf{V}(I_1) \setminus \mathbf{V}(c_1) = \mathbb{C} \setminus \mathbf{V}(y) = \mathbb{C} \setminus \{0\} \subseteq \pi_1(\mathbf{V}(I)) = \mathbb{C}.$$

Hence, we can take $W = \{0\}$ in Theorem 1 since $\mathbb{C} \setminus \{0\}$ is Zariski dense in $\mathbb{C}$.

The second example, taken from SCHAUENBURG (2007), uses the ideal

$$I = \langle xz + y - 1, w + y + z - 2, z^2 \rangle \subseteq \mathbb{C}[w, x, y, z].$$

It is straightforward to check that $V = \mathbf{V}(I)$ is the line $V = \mathbf{V}(w - 1, y - 1, z) \subseteq \mathbb{C}^4$, which projects to the point $\pi_2(V) = \mathbf{V}(y - 1, z) \subseteq \mathbb{C}^2$ when we eliminate w and x. Thus, $W = \emptyset$ satisfies Theorem 1 in this case.

Here is a systematic way to discover that $W = \emptyset$. A lex Gröbner basis of I for $w > x > y > z$ consists of

$$g_1 = w + y + z - 2, \ g_2 = xz + y - 1, \ g_3 = y^2 - 2y + 1, \ g_4 = yz - z, \ g_5 = z^2.$$

Eliminating w and x gives $I_2 = \langle g_3, g_4, g_5 \rangle$, and one sees easily that

$$\mathbf{V}(I_2) = \mathbf{V}(y - 1, z).$$

Since $g_1 = 1 \cdot w + y + z - 2$ and $g_2 = z \cdot x + y - 1$, we have $c_1 = 1$ and $c_2 = z$. If we set

$$J = \langle c_1 c_2 \rangle = \langle z \rangle,$$

then Corollary 3 implies that $\mathbf{V}(I_2) \setminus \mathbf{V}(J) \subseteq \pi_2(V)$. However, $\mathbf{V}(I_2) \setminus \mathbf{V}(J) = \emptyset$, so the difference is not Zariski dense in $\mathbf{V}(I_2)$.

In this situation, we use the decomposition of $\mathbf{V}(I)$ guaranteed to exist by Proposition 4. Note that $I = I : c_1^\infty$ since $c_1 = 1$. Hence we use $c_2 = z$ in the proposition. This gives the two ideals

$$I^{(1)} = I + \langle c_2 \rangle = \langle xz + y - 1, w + y + z - 2, z^2, z \rangle = \langle w - 1, y - 1, z \rangle,$$
$$I^{(2)} = I : c_2^\infty = I : z^\infty = \langle 1 \rangle \text{ since } z^2 \in I.$$

Now we start over with $I^{(1)}$ and $I^{(2)}$.

For $I^{(1)}$, observe that $\{w-1, y-1, z\}$ is a Gröbner basis of $I^{(1)}$, and only $g_1^{(1)} = w-1 \notin \mathbb{C}[y, z]$. The coefficient of w is $c_1^{(1)} = 1$, and then Corollary 3 applied to $I^{(1)}$ gives

$$\mathbf{V}(I_2^{(1)}) \setminus \mathbf{V}(1) \subseteq \pi_2(\mathbf{V}(I^{(1)})).$$

Since $\mathbf{V}(1) = \emptyset$, we can pick $W_1 = \emptyset$ for $I^{(1)}$ in Theorem 1.

Applying the same systematic process to $I^{(2)} = \langle 1 \rangle$, we see that there are *no* $g_i \notin \mathbb{C}[y, z]$. Thus Corollary 3 involves the product over the empty set. By convention (see Exercise 7) the empty product is 1. Then Corollary 3 tells us that we can pick $W_2 = \emptyset$ for $I^{(2)}$ in Theorem 1. By Proposition 5, it follows that Theorem 1 holds for the ideal I with

$$W = W_1 \cup W_2 = \emptyset \cup \emptyset = \emptyset.$$

To do this in general, we use the following recursive algorithm to produce the desired subset W:

Input : an ideal $I \subseteq k[\mathbf{x}, \mathbf{y}]$ with variety $V = \mathbf{V}(I)$
Output : $\mathrm{FindW}(I) = W \subseteq \mathbf{V}(I_l)$ with $\mathbf{V}(I_l) \setminus W \subseteq \pi_l(V)$, $\overline{\mathbf{V}(I_l) \setminus W} = \mathbf{V}(I_l)$

$G :=$ reduced Gröbner basis for I for a monomial order as in Theorem 2
$c_i :=$ coefficient in $g_i = c_i(\mathbf{y})\mathbf{x}^{\alpha_i} +$ terms $< \mathbf{x}^{\alpha_i}$ when $g_i \in G \setminus k[\mathbf{y}]$
$I_l := I \cap k[\mathbf{y}] = \langle G \cap k[\mathbf{y}] \rangle$
$J := \langle \prod_{g_i \in G \setminus k[\mathbf{y}]} c_i \rangle$
IF $\overline{\mathbf{V}(I_l) \setminus \mathbf{V}(J)} = \mathbf{V}(I_l)$ THEN
$\qquad \mathrm{FindW}(I) := \mathbf{V}(I_l) \cap \mathbf{V}(J)$
ELSE
$\qquad$ Select $g_i \in G \setminus k[\mathbf{y}]$ with $I \subsetneq I : c_i^{\infty}$
$\qquad \mathrm{FindW}(I) := \mathrm{FindW}(I + \langle c_i \rangle) \cup \mathrm{FindW}(I : c_i^{\infty})$
RETURN $\mathrm{FindW}(I)$

The function FindW takes the input ideal I and computes the ideals I_l and $J = \langle \prod_{g_i \in G \setminus k[\mathbf{y}]} c_i \rangle$. The IF statement asks whether $\mathbf{V}(I_l) \setminus \mathbf{V}(J)$ is Zariski dense in $\mathbf{V}(I_l)$. If the answer is yes, then $\mathbf{V}(I_l) \cap \mathbf{V}(J)$ has the desired property by Corollary 3, which is why $\mathrm{FindW}(I) = \mathbf{V}(I_l) \cap \mathbf{V}(J)$ in this case. In the exercises, you will describe an algorithm for determining whether $\overline{\mathbf{V}(I_l) \setminus \mathbf{V}(J)} = \mathbf{V}(I_l)$.

When $\mathbf{V}(I_l) \setminus \mathbf{V}(J)$ fails to be Zariski dense in $\mathbf{V}(I_l)$, Proposition 4 guarantees that we can find c_i such that the ideals

$$I^{(1)} = I + \langle c_i \rangle \quad \text{and} \quad I^{(2)} = I : c_i^{\infty}$$

are strictly larger than I and satisfy $V = \mathbf{V}(I) = \mathbf{V}(I^{(1)}) \cup \mathbf{V}(I^{(2)})$. Then, as in the second example above, we repeat the process on the two new ideals, which means computing $\mathrm{FindW}(I^{(1)})$ and $\mathrm{FindW}(I^{(2)})$. By Proposition 5, the union of these varieties works for I, which explains the last line of FindW.

We say that FindW is *recursive* since it calls itself. We leave it as an exercise to show that the maximum principle from Proposition 6 implies that FindW always terminates in finitely many steps. When it does, correctness follows from the above discussion.

We end this section by using the Closure Theorem to give a precise description of the projection $\pi_l(V) \subseteq k^{n-l}$ of an affine variety $V \subseteq k^n$.

Theorem 7. *Let k be algebraically closed and let $V \subseteq k^n$ be an affine variety. Then there are affine varieties $Z_i \subseteq W_i \subseteq k^{n-l}$ for $1 \le i \le p$ such that*

$$\pi_l(V) = \bigcup_{i=1}^{p}(W_i \setminus Z_i).$$

Proof. We assume $V \ne \emptyset$. First let $W_1 = \mathbf{V}(I_l)$. By the Closure Theorem, there is a variety $Z_1 \subsetneq W_1$ such that $W_1 \setminus Z_1 \subseteq \pi_l(V)$. Then, back in k^n, consider the set

$$V_1 = V \cap \{(a_1, \ldots, a_n) \in k^n \mid (a_{l+1}, \ldots, a_n) \in Z_1\}.$$

One easily checks that V_1 is an affine variety (see Exercise 10) and furthermore, $V_1 \subsetneq V$ since otherwise we would have $\pi_l(V) \subseteq Z_1$, which would imply $W_1 \subseteq Z_1$ by Zariski closure. Moreover, you will check in Exercise 10 that

(2) $\pi_l(V) = (W_1 \setminus Z_1) \cup \pi_l(V_1)$.

If $V_1 = \emptyset$, then we are done. If V_1 is nonempty, let W_2 be the Zariski closure of $\pi_l(V_1)$. Applying the Closure Theorem to V_1, we get $Z_2 \subsetneq W_2$ with $W_2 \backslash Z_2 \subset \pi_l(V_1)$. Then, repeating the above construction, we get the variety

$$V_2 = V_1 \cap \{(a_1, \ldots, a_n) \in k^n \mid (a_{l+1}, \ldots, a_n) \in Z_2\}$$

such that $V_2 \subsetneq V_1$ and

$$\pi_l(V) = (W_1 \setminus Z_1) \cup (W_2 \setminus Z_2) \cup \pi_l(V_2).$$

If $V_2 = \emptyset$, we are done, and if not, we repeat this process again to obtain W_3, Z_3 and $V_3 \subsetneq V_2$. Continuing in this way, we must eventually have $V_N = \emptyset$ for some N, since otherwise we would get an infinite descending chain of varieties

$$V \supsetneq V_1 \supsetneq V_2 \supsetneq \cdots,$$

which would contradict Proposition 1 of §6. Once we have $V_N = \emptyset$, the desired formula for $\pi_l(V)$ follows easily.   ☐

In general, a set of the form described in Theorem 7 is called *constructible*.

As a simple example of Theorem 7, consider $I = \langle xy + z - 1, y^2 z^2 \rangle \subseteq \mathbb{C}[x, y, z]$ and set $V = \mathbf{V}(I) \subseteq \mathbb{C}^3$. We leave it as an exercise to show that

$$\mathbf{V}(I_1) = \mathbf{V}(z) \cup \mathbf{V}(y, z - 1) = \mathbf{V}(z) \cup \{(0, 1)\}$$

and that $W = \mathbf{V}(y,z) = \{(0,0)\}$ satisfies $\mathbf{V}(I_1) \setminus W \subseteq \pi_1(V)$. However, we also have $\pi_1(V) \subseteq \mathbf{V}(I_1)$, and since $xy + z - 1 \in I$, no point of V has vanishing y and z coordinates. It follows that $\pi_1(V) \subseteq \mathbf{V}(I_1) \setminus \{(0,0)\}$. Hence

$$\pi_1(V) = \mathbf{V}(I_1) \setminus \{(0,0)\} = (\mathbf{V}(z) \setminus \{(0,0)\}) \cup \{(0,1)\}.$$

This gives an explicit representation of $\pi_1(V)$ as a constructible set. You will work out another example of Theorem 7 in the exercises. More substantial examples can be found in SCHAUENBURG (2007), which also describes an algorithm for writing $\pi_l(V)$ as a constructible set. Another approach is described in ULLRICH (2006).

EXERCISES FOR §7

1. Prove that Theorem 3 of Chapter 3, §2 follows from Theorem 1 of this section. Hint: Show that the W from Theorem 1 satisfies $W \subsetneq \mathbf{V}(I_l)$ when $V \neq \emptyset$.
2. In the notation of Theorem 2, prove that $\bar{I} = \langle \overline{G} \rangle$ for $\bar{I} = \{\bar{f} \mid f \in I\}$.
3. Given sets A and B, prove that $A \setminus B = A \setminus (A \cap B)$.
4. In the proof of Proposition 4, prove that $I = I : c_i^\infty$ implies that $I_l = I_l : c_i^\infty$.
5. This exercise will explore some properties of irreducible components needed in the proofs of Propositions 4 and 5.
   a. Let $W_1, \ldots, W_r$ be affine varieties contained in a variety V and assume that for each $1 \leq i \leq r$, no irreducible component of V is contained in W_i. Prove that the same is true for $\bigcup_{i=1}^r W_i$. (This fact is used in the proof of Proposition 4.)
   b. Let $W_i \subseteq V_i$ be affine varieties for $i = 1,2$ such that W_i contains no irreducible component of V_i. Prove that $W = W_1 \cup W_2$ contains no irreducible component of $V = V_1 \cup V_2$. (This fact is used in the proof of Proposition 5.)
6. Prove Proposition 6. Hint: Assume that the proposition is false for some nonempty collection of ideals $\{I_\alpha\}_{\alpha \in \mathcal{A}}$ and show that this leads to a contradiction of the ascending chain condition.
7. In this exercise we will see why it is reasonable to make the convention that the empty product is 1. Let R be a commutative ring with 1 and let $\mathcal{A}$ be a finite set such that for every $\alpha \in \mathcal{A}$, we have $r_\alpha \in R$. Then we get the product

$$\prod_{\alpha \in \mathcal{A}} r_\alpha.$$

Although $\mathcal{A}$ is unordered, the product is well-defined since R is commutative.
   a. Assume $\mathcal{B}$ is finite and disjoint from $\mathcal{A}$ such that for every $\beta \in \mathcal{B}$, we have $r_\beta \in R$. Prove that

$$\prod_{\gamma \in \mathcal{A} \cup \mathcal{B}} r_\gamma = \left(\prod_{\alpha \in \mathcal{A}} r_\alpha \right) \left(\prod_{\beta \in \mathcal{B}} r_\beta \right).$$

   b. It is likely that the proof you gave in part (a) assumed that $\mathcal{A}$ and $\mathcal{B}$ are nonempty. Explain why $\prod_{\alpha \in \emptyset} r_\alpha = 1$ makes the above formula work in all cases.
   c. In a similar way, define $\sum_{\alpha \in \mathcal{A}} r_\alpha$ and explain why $\sum_{\alpha \in \emptyset} r_\alpha = 0$ is needed to make the analog of part (a) true for sums.
8. The goal of this exercise is to describe an algorithm for deciding whether $\overline{\mathbf{V}(I) \setminus \mathbf{V}(g)} = \mathbf{V}(I)$ when the field k is algebraically closed.
   a. Prove that $\overline{\mathbf{V}(I) \setminus \mathbf{V}(g)} = \mathbf{V}(I)$ is equivalent to $I : g^\infty \subseteq \sqrt{I}$. Hint: Use the Nullstellensatz and Theorem 10 of §4. Also remember that $I \subseteq I : g^\infty$.
   b. Use Theorem 14 of §4 and the Radical Membership Algorithm from §2 to describe an algorithm for deciding whether $I : g^\infty \subseteq \sqrt{I}$.

9. Give a proof of the termination of FindW that uses the maximum principle stated in Proposition 6. Hint: Consider the set of all ideals in $k[\mathbf{x}, \mathbf{y}]$ for which FindW does not terminate.

10. This exercise is concerned with the proof of Theorem 7.
    a. Verify that $V_1 = V \cap \{(a_1, \ldots, a_n) \in k^n \mid (a_{l+1}, \ldots, a_n) \in Z_1\}$ is an affine variety.
    b. Verify that $\pi_l(V) = (W_1 \setminus Z_1) \cup \pi_l(V_1)$.

11. As in the text, let $V = \mathbf{V}(I)$ for $I = \langle xy + z - 1, y^2 z^2 \rangle \subseteq \mathbb{C}[x, y, z]$. Show that

$$\mathbf{V}(I_1) = \mathbf{V}(z) \cup \mathbf{V}(y, z - 1) = \mathbf{V}(z) \cup \{(0, 1)\}$$

and that $W = \mathbf{V}(y, z) = \{(0, 0)\}$ satisfies $\mathbf{V}(I_1) \setminus W \subseteq \pi_1(V)$.

12. Let $V = \mathbf{V}(y - xz) \subseteq \mathbb{C}^3$. Theorem 7 tells us that $\pi_1(V) \subseteq \mathbb{C}^2$ is a constructible set. Find an explicit decomposition of $\pi_1(V)$ of the form given by Theorem 7. Hint: Your answer will involve W_1, Z_1 and W_2.

13. Proposition 6 is the maximum principle for ideals. The geometric analog is the *minimum principle* for varieties, which states that among any nonempty collection of varieties in k^n, there is a variety in the collection which is minimal with respect to inclusion. More precisely, this means that if we are given varieties V_α, $\alpha \in \mathcal{A}$, where $\mathcal{A}$ is a nonempty set, then there is some $\alpha_0 \in \mathcal{A}$ with the property that for all $\beta \in \mathcal{A}$, $V_\beta \subseteq V_{\alpha_0}$ implies $V_\beta = V_{\alpha_0}$. Prove the minimum principle. Hint: Use Proposition 1 of §6.

14. Apply the minimum principle of Exercise 13 to give a different proof of Theorem 7. Hint: Consider the collection of all varieties $V \subseteq k^n$ for which $\pi_l(V)$ is not constructible. By the minimum principle, there is a variety V such that $\pi_l(V)$ is not constructible but $\pi_l(W)$ is constructible for every variety $W \subsetneq V$. Show how the proof of Theorem 7 up to (2) can be used to obtain a contradiction and thereby prove the theorem.

§8 Primary Decomposition of Ideals

In view of the decomposition theorem proved in §6 for radical ideals, it is natural to ask whether an arbitrary ideal I (not necessarily radical) can be represented as an intersection of simpler ideals. In this section, we will prove the Lasker-Noether decomposition theorem, which describes the structure of I in detail.

There is no hope of writing an arbitrary ideal I as an intersection of prime ideals (since an intersection of prime ideals is always radical). The next thing that suggests itself is to write I as an intersection of powers of prime ideals. This does not quite work either: consider the ideal $I = \langle x, y^2 \rangle$ in $\mathbb{C}[x, y]$. Any prime ideal containing I must contain x and y and, hence, must equal $\langle x, y \rangle$ (since $\langle x, y \rangle$ is maximal). Thus, if I were to be an intersection of powers of prime ideals, it would have to be a power of $\langle x, y \rangle$ (see Exercise 1 for the details).

The concept we need is a bit more subtle.

Definition 1. An ideal I in $k[x_1, \ldots, x_n]$ is **primary** if $fg \in I$ implies either $f \in I$ or some power $g^m \in I$ for some $m > 0$.

It is easy to see that prime ideals are primary. Also, you can check that the ideal $I = \langle x, y^2 \rangle$ discussed above is primary (see Exercise 1).

Lemma 2. *If I is a primary ideal, then $\sqrt{I}$ is prime and is the smallest prime ideal containing I.*

Proof. See Exercise 2. $\qquad\square$

In view of this lemma, we make the following definition.

Definition 3. If I is primary and $\sqrt{I} = P$, then we say that I is P-**primary**.

We can now prove that every ideal is an intersection of primary ideals.

Theorem 4. *Every ideal* $I \subseteq k[x_1, \dots, x_n]$ *can be written as a finite intersection of primary ideals.*

Proof. We first define an ideal I to be *irreducible* if $I = I_1 \cap I_2$ implies that $I = I_1$ or $I = I_2$. We claim that every ideal is an intersection of finitely many irreducible ideals. The argument is an "ideal" version of the proof of Theorem 2 from §6. One uses the ACC rather than the DCC—we leave the details as an exercise.

Next we claim that an irreducible ideal is primary. Note that this will prove the theorem. To see why the claim is true, suppose that I is irreducible and that $fg \in I$ with $f \notin I$. We need to prove that some power of g lies in I. Consider the saturation $I : g^\infty$. By Proposition 9 of §4, we know that $I : g^\infty = I : g^N$ once N is sufficiently large. We will leave it as an exercise to show that $(I + \langle g^N \rangle) \cap (I + \langle f \rangle) = I$. Since I is irreducible, it follows that $I = I + \langle g^N \rangle$ or $I = I + \langle f \rangle$. The latter cannot occur since $f \notin I$, so that $I = I + \langle g^N \rangle$. This proves that $g^N \in I$. $\qquad\square$

As in the case of varieties, we can define what it means for a decomposition to be minimal.

Definition 5. A **primary decomposition** of an ideal I is an expression of I as an intersection of primary ideals: $I = \bigcap_{i=1}^r Q_i$. It is called **minimal** or **irredundant** if the $\sqrt{Q_i}$ are all distinct and $Q_i \not\supseteq \bigcap_{j \neq i} Q_j$.

To prove the existence of a minimal decomposition, we will need the following lemma, the proof of which we leave as an exercise.

Lemma 6. *If I, J are primary and $\sqrt{I} = \sqrt{J}$, then $I \cap J$ is primary.*

We can now prove the first part of the Lasker-Noether decomposition theorem.

Theorem 7 (Lasker-Noether). *Every ideal* $I \subseteq k[x_1, \dots, x_n]$ *has a minimal primary decomposition.*

Proof. By Theorem 4, we know that there is a primary decomposition $I = \bigcap_{i=1}^r Q_i$. Suppose that Q_i and Q_j have the same radical for some $i \neq j$. Then, by Lemma 6, $Q = Q_i \cap Q_j$ is primary, so that in the decomposition of I, we can replace Q_i and Q_j by the single ideal Q. Continuing in this way, eventually all of the Q_i's will have distinct radicals.

Next, suppose that some Q_i contains $\bigcap_{j \neq i} Q_j$. Then we can omit Q_i, and I will be the intersection of the remaining Q_j's for $j \neq i$. Continuing in this way, we can reduce to the case where $Q_i \not\supseteq \bigcap_{j \neq i} Q_j$ for all i. $\qquad\square$

Unlike the case of varieties (or radical ideals), a minimal primary decomposition need not be unique. In the exercises, you will verify that the ideal $\langle x^2, xy \rangle \subseteq k[x, y]$ has the two distinct minimal decompositions

$$\langle x^2, xy \rangle = \langle x \rangle \cap \langle x^2, xy, y^2 \rangle = \langle x \rangle \cap \langle x^2, y \rangle.$$

Although $\langle x^2, xy, y^2 \rangle$ and $\langle x^2, y \rangle$ are distinct, note that they have the same radical. To prove that this happens in general, we will use ideal quotients from §4. We start by computing some ideal quotients of a primary ideal.

Lemma 8. *If I is primary with* $\sqrt{I} = P$ *and* $f \in k[x_1, \dots, x_n]$, *then:*

(i) *If* $f \in I$, *then* $I : f = \langle 1 \rangle$.
(ii) *If* $f \notin I$, *then* $I : f$ *is P-primary.*
(iii) *If* $f \notin P$, *then* $I : f = I$.

Proof. See Exercise 7.   □

The second part of the Lasker-Noether theorem tells us that the *radicals* of the ideals in a minimal decomposition are uniquely determined.

Theorem 9 (Lasker-Noether). *Let* $I = \bigcap_{i=1}^{r} Q_i$ *be a minimal primary decomposition of a proper ideal* $I \subseteq k[x_1, \dots, x_n]$ *and let* $P_i = \sqrt{Q_i}$. *Then the* P_i *are precisely the proper prime ideals occurring in the set* $\{ \sqrt{I : f} \mid f \in k[x_1, \dots, x_n] \}$.

Remark. In particular, the P_i are independent of the primary decomposition of I. We say that the P_i *belong* to I.

Proof. The proof is very similar to the proof of Theorem 7 from §6. The details are covered in Exercises 8–10.   □

In §6, we proved a decomposition theorem for radical ideals over an algebraically closed field. Using the Lasker–Noether theorems, we can now show that these results hold over an arbitrary field k.

Corollary 10. *Let* $I = \bigcap_{i=1}^{r} Q_i$ *be a minimal primary decomposition of a proper radical ideal* $I \subseteq k[x_1, \dots, x_n]$. *Then the* Q_i *are prime and are precisely the proper prime ideals occurring in the set* $\{ I : f \mid f \in k[x_1, \dots, x_n] \}$.

Proof. See Exercise 12.   □

The two Lasker–Noether theorems do not tell the full story of a minimal primary decomposition $I = \bigcap_{i=1}^{r} Q_i$. For example, if P_i is minimal in the sense that no P_j is strictly contained in P_i, then one can show that Q_i is uniquely determined. Thus there is a uniqueness theorem for *some* of the Q_i's [see Chapter 4 of ATIYAH and MACDONALD (1969) for the details]. We should also mention that the conclusion of Theorem 9 can be strengthened: one can show that the P_i's are precisely the proper prime ideals in the set $\{ I : f \mid f \in k[x_1, \dots, x_n] \}$ [see Chapter 7 of ATIYAH and MACDONALD (1969)].

Finally, it is natural to ask if a primary decomposition can be done constructively. More precisely, given $I = \langle f_1, \ldots, f_s \rangle$, we can ask the following:

- (Primary Decomposition) Is there an algorithm for finding bases for the primary ideals Q_i in a minimal primary decomposition of I?
- (Associated Primes) Can we find bases for the associated primes $P_i = \sqrt{Q_i}$?

If you look in the references given at the end of §6, you will see that the answer to these questions is *yes*. Primary decomposition has been implemented in CoCoA, Macaulay2, Singular, and Maple.

EXERCISES FOR §9

1. Consider the ideal $I = \langle x, y^2 \rangle \subseteq \mathbb{C}[x, y]$.
   a. Prove that $\langle x, y \rangle^2 \subsetneq I \subsetneq \langle x, y \rangle$, and conclude that I is not a prime power.
   b. Prove that I is primary.
2. Prove Lemma 2.
3. This exercise is concerned with the proof of Theorem 4. Let $I \subseteq k[x_1, \ldots, x_n]$ be an ideal.
   a. Using the hints given in the text, prove that I is a finite intersection of irreducible ideals.
   b. Suppose that $fg \in I$ and $I : g^\infty = I : g^N$. Then prove that $(I + \langle g^N \rangle) \cap (I + \langle f \rangle) = I$. Hint: Elements of $(I + \langle g^N \rangle) \cap (I + \langle f \rangle)$ can be written as $a + bg^N = c + df$, where $a, c \in I$ and $b, d \in k[x_1, \ldots, x_n]$. Now multiply through by g and use $I : g^N = I : g^{N+1}$.
4. In the proof of Theorem 4, we showed that every irreducible ideal is primary. Surprisingly, the converse is false. Let I be the ideal $\langle x^2, xy, y^2 \rangle \subseteq k[x, y]$.
   a. Show that I is primary.
   b. Show that $I = \langle x^2, y \rangle \cap \langle x, y^2 \rangle$ and conclude that I is not irreducible.
5. Prove Lemma 6. Hint: Proposition 16 from §3 will be useful.
6. Let I be the ideal $\langle x^2, xy \rangle \subseteq \mathbb{Q}[x, y]$.
   a. Prove that
   $$I = \langle x \rangle \cap \langle x^2, xy, y^2 \rangle = \langle x \rangle \cap \langle x^2, y \rangle$$
   are two distinct minimal primary decompositions of I.
   b. Prove that for any $a \in \mathbb{Q}$,
   $$I = \langle x \rangle \cap \langle x^2, y - ax \rangle$$
   is a minimal primary decomposition of I. Thus I has infinitely many distinct minimal primary decompositions.
7. Prove Lemma 8.
8. Prove that an ideal is proper if and only if its radical is.
9. Use Exercise 8 to show that the primes belonging to a proper ideal are also proper.
10. Prove Theorem 9. Hint: Adapt the proof of Theorem 7 from §6. The extra ingredient is that you will need to take radicals. Proposition 16 from §3 will be useful. You will also need to use Exercise 9 and Lemma 8.
11. Let $P_1, \ldots, P_r$ be the prime ideals belonging to I.
    a. Prove that $\sqrt{I} = \bigcap_{i=1}^r P_i$. Hint: Use Proposition 16 from §3.
    b. Show that $\sqrt{I} = \bigcap_{i=1}^r P_i$ need not be a minimal decomposition of $\sqrt{I}$. Hint: Exercise 4.
12. Prove Corollary 10. Hint: Use Proposition 9 of §4 to show that $I : f$ is radical.

§9   Summary

The table on the next page summarizes the results of this chapter. In the table, it is supposed that all ideals are radical and that the field is algebraically closed.

ALGEBRA		GEOMETRY
radical ideals		varieties
I	$\longrightarrow$	$\mathbf{V}(I)$
$\mathbf{I}(V)$	$\longleftarrow$	V
addition of ideals		intersection of varieties
$I + J$	$\longrightarrow$	$\mathbf{V}(I) \cap \mathbf{V}(J)$
$\sqrt{\mathbf{I}(V) + \mathbf{I}(W)}$	$\longleftarrow$	$V \cap W$
product of ideals		union of varieties
IJ	$\longrightarrow$	$\mathbf{V}(I) \cup \mathbf{V}(J)$
$\sqrt{\mathbf{I}(V)\mathbf{I}(W)}$	$\longleftarrow$	$V \cup W$
intersection of ideals		union of varieties
$I \cap J$	$\longrightarrow$	$\mathbf{V}(I) \cup \mathbf{V}(J)$
$\mathbf{I}(V) \cap \mathbf{I}(W)$	$\longleftarrow$	$V \cup W$
ideal quotients		difference of varieties
$I : J$	$\longrightarrow$	$\overline{\mathbf{V}(I) \setminus \mathbf{V}(J)}$
$\mathbf{I}(V) : \mathbf{I}(W)$	$\longleftarrow$	$\overline{V \setminus W}$
elimination of variables		projection of varieties
$I \cap k[x_{l+1}, \ldots, x_n]$	$\longleftrightarrow$	$\overline{\pi_l(\mathbf{V}(I))}$
prime ideal	$\longleftrightarrow$	irreducible variety
minimal decomposition		minimal decomposition
$I = P_1 \cap \cdots \cap P_m$	$\longrightarrow$	$\mathbf{V}(I) = \mathbf{V}(P_1) \cup \cdots \cup \mathbf{V}(P_m)$
$\mathbf{I}(V) = \mathbf{I}(V_1) \cap \cdots \cap \mathbf{I}(V_m)$	$\longleftarrow$	$V = V_1 \cup \cdots \cup V_m$
maximal ideal	$\longleftrightarrow$	point of affine space
ascending chain condition	$\longleftrightarrow$	descending chain condition

Chapter 5
Polynomial and Rational Functions on a Variety

One of the unifying themes of modern mathematics is that in order to understand any class of mathematical objects, one should also study *mappings* between those objects, and especially the mappings which preserve some property of interest. For instance, in linear algebra after studying vector spaces, you also studied the properties of *linear mappings* between vector spaces (mappings that preserve the vector space operations of sum and scalar product).

This chapter will consider mappings between varieties, and the results of our investigation will form another chapter of the "algebra–geometry dictionary" that we started in Chapter 4. The algebraic properties of polynomial and rational functions on a variety yield many insights into the geometric properties of the variety itself. This chapter will also serve as an introduction to (and motivation for) the idea of a quotient ring. The chapter will end with a discussion of *Noether normalization*.

§1 Polynomial Mappings

We will begin our study of functions between varieties by reconsidering two examples that we have encountered previously. First, recall the tangent surface of the twisted cubic curve in $\mathbb{R}^3$. As in equation (1) of Chapter 3, §3, we describe this surface parametrically:

$$
\begin{aligned}
x &= t + u, \\
y &= t^2 + 2tu, \\
z &= t^3 + 3t^2u.
\end{aligned}
\tag{1}
$$

In functional language, giving the parametric representation (1) is equivalent to defining a mapping

$$\phi : \mathbb{R}^2 \longrightarrow \mathbb{R}^3$$

© Springer International Publishing Switzerland 2015
D.A. Cox et al., *Ideals, Varieties, and Algorithms*, Undergraduate Texts in Mathematics, DOI 10.1007/978-3-319-16721-3_5

by

(2) $\phi(t, u) = (t + u, t^2 + 2tu, t^3 + 3t^2 u).$

The domain of ϕ is an affine variety $V = \mathbb{R}^2$ and the image of ϕ is the tangent surface S.

We saw in §3 of Chapter 3 that S is the same as the affine variety

$$W = \mathbf{V}(x^3 z - (3/4)x^2 y^2 - (3/2)xyz + y^3 + (1/4)z^2).$$

Hence, our parametrization gives what we might call a *polynomial mapping* between V and W. (The adjective "polynomial" refers to the fact that the components of ϕ are polynomials in t and u.)

Second, in the discussion of the geometry of elimination of variables from systems of equations in §2 of Chapter 3, we considered the projection mappings

$$\pi_l : \mathbb{C}^n \longrightarrow \mathbb{C}^{n-l}$$

defined by

$$\pi_l(a_1, \ldots, a_n) = (a_{l+1}, \ldots, a_n).$$

If we have a variety $V = \mathbf{V}(I) \subseteq \mathbb{C}^n$, then we can also restrict π_l to V and, as we know, $\pi_l(V)$ will be contained in the affine variety $W = \mathbf{V}(I_l)$, where $I_l = I \cap \mathbb{C}[x_{l+1}, \ldots, x_n]$, the l-th elimination ideal of I. Hence, we can consider $\pi_l : V \to W$ as a mapping of varieties. Here too, by the definition of π_l we see that the components of π_l are polynomials in the coordinates in the domain.

Definition 1. Let $V \subseteq k^m$, $W \subseteq k^n$ be varieties. A function $\phi : V \to W$ is said to be a **polynomial mapping** (or **regular mapping**) if there exist polynomials $f_1, \ldots, f_n \in k[x_1, \ldots, x_m]$ such that

$$\phi(a_1, \ldots, a_m) = (f_1(a_1, \ldots, a_m), \ldots, f_n(a_1, \ldots, a_m))$$

for all $(a_1, \ldots, a_m) \in V$. We say that the n-tuple of polynomials

$$(f_1, \ldots, f_n) \in (k[x_1, \ldots, x_m])^n$$

represents ϕ. The f_i are the **components** of this representation.

To say that ϕ is a polynomial mapping from $V \subseteq k^m$ to $W \subseteq k^n$ represented by $(f_1, \ldots, f_n)$ means that $(f_1(a_1, \ldots, a_m), \ldots, f_n(a_1, \ldots, a_m))$ must satisfy the defining equations of W for all points $(a_1, \ldots, a_m) \in V$. For example, consider $V = \mathbf{V}(y - x^2, z - x^3) \subseteq k^3$ (the twisted cubic) and $W = \mathbf{V}(y^3 - z^2) \subseteq k^2$. Then the projection $\pi_1 : k^3 \to k^2$ represented by (y, z) gives a polynomial mapping $\pi_1 : V \to W$. This is true because every point in $\pi_1(V) = \{(a^2, a^3) \mid a \in k\}$ satisfies the defining equation of W.

Of particular interest is the case $W = k$, where ϕ simply becomes a scalar polynomial function defined on the variety V. One reason to consider polynomial functions from V to k is that a general polynomial mapping $\phi : V \to k^n$ is constructed by using any n polynomial functions $\phi : V \to k$ as the components. Hence, if we understand functions $\phi : V \to k$, we understand how to construct all mappings $\phi : V \to k^n$ as well.

To begin our study of polynomial functions, note that, for $V \subseteq k^m$, Definition 1 says that a mapping $\phi : V \to k$ is a polynomial function if *there exists* a polynomial $f \in k[x_1, \ldots, x_m]$ representing ϕ. In fact, we usually specify a polynomial function by giving an explicit polynomial representative. Thus, finding a representative is not actually the key issue. What we will see next, however, is that the cases where a representative is uniquely determined are very rare. For example, consider the variety $V = \mathbf{V}(y - x^2) \subseteq \mathbb{R}^2$. The polynomial $f = x^3 + y^3$ represents a polynomial function from V to $\mathbb{R}$. However, $g = x^3 + y^3 + (y - x^2)$, $h = x^3 + y^3 + (x^4 y - x^6)$, and $F = x^3 + y^3 + A(x, y)(y - x^2)$ for any $A(x, y)$ define *the same polynomial function* on V. Indeed, since $\mathbf{I}(V)$ is the set of polynomials which are zero at every point of V, adding any element of $\mathbf{I}(V)$ to f does not change the values of the polynomial at the points of V. The general pattern is the same.

Proposition 2. *Let $V \subseteq k^m$ be an affine variety. Then:*

(i) *f and $g \in k[x_1, \ldots, x_m]$ represent the same polynomial function on V if and only if $f - g \in \mathbf{I}(V)$.*

(ii) *$(f_1, \ldots, f_n)$ and $(g_1, \ldots, g_n)$ represent the same polynomial mapping from V to k^n if and only if $f_i - g_i \in \mathbf{I}(V)$ for each i, $1 \le i \le n$.*

Proof. (i) If $f - g = h \in \mathbf{I}(V)$, then for any point $p = (a_1, \ldots, a_m) \in V$, $f(p) - g(p) = h(p) = 0$. Hence, f and g represent the same function on V. Conversely, if f and g represent the same function, then, at every $p \in V$, $f(p) - g(p) = 0$. Thus, $f - g \in \mathbf{I}(V)$ by definition. Part (ii) follows directly from (i). $\quad\square$

It follows that the correspondence between polynomials in $k[x_1, \ldots, x_m]$ and polynomial functions $V \to k$ is one-to-one only in the case where $\mathbf{I}(V) = \{0\}$. In Exercise 7, you will show that $\mathbf{I}(V) = \{0\}$ if and only if k is infinite and $V = k^m$.

There are two ways of dealing with this potential ambiguity in describing polynomial functions on a variety:

- In rough terms, we can "lump together" *all* the polynomials $f \in k[x_1, \ldots, x_m]$ that represent the same function on V and think of that collection as a "new object" in its own right. We can then take the collection of polynomials as our description of the function on V.
- Alternatively, we can systematically look for the simplest possible individual polynomial that represents each function on V and work with those "standard representative" polynomials exclusively.

Each of these approaches has its own advantages, and we will consider both of them in detail in later sections of this chapter. We will conclude this section by looking at two further examples to show the kinds of properties of varieties that can be revealed by considering polynomial functions.

Definition 3. We denote by $k[V]$ the collection of polynomial functions $\phi : V \to k$.

Since k is a field, we can define a sum and a product function for any pair of functions $\phi, \psi : V \to k$ by adding and multiplying images. For each $p \in V$,

$$(\phi + \psi)(p) = \phi(p) + \psi(p),$$
$$(\phi \cdot \psi)(p) = \phi(p) \cdot \psi(p).$$

Furthermore, if we pick specific representatives $f, g \in k[x_1, \ldots, x_m]$ for ϕ, ψ, respectively, then by definition the polynomial sum $f + g$ represents $\phi + \psi$ and the polynomial product $f \cdot g$ represents $\phi \cdot \psi$. It follows that $\phi + \psi$ and $\phi \cdot \psi$ are polynomial functions on V.

Thus, we see that $k[V]$ has sum and product operations constructed using the sum and product operations in $k[x_1, \ldots, x_m]$. All of the usual properties of sums and products of polynomials also hold for functions in $k[V]$. Thus, $k[V]$ is another example of a *commutative ring*. (See Appendix A for the precise definition.) We will return to this point in §2.

Now we are ready to start exploring what $k[V]$ can tell us about the geometric properties of a variety V. First, recall from §5 of Chapter 4 that a variety $V \subseteq k^m$ is said to be *reducible* if it can be written as the union of two nonempty proper subvarieties: $V = V_1 \cup V_2$, where $V_1 \neq V$ and $V_2 \neq V$. For example, the variety $V = \mathbf{V}(x^3 + xy^2 - xz, yx^2 + y^3 - yz)$ in k^3 is reducible since, from the factorizations of the defining equations, we can decompose V as $V = \mathbf{V}(x^2 + y^2 - z) \cup \mathbf{V}(x, y)$. We would like to demonstrate that geometric properties such as reducibility can be "read off" from a sufficiently good algebraic description of $k[V]$. To see this, let

(3) $f = x^2 + y^2 - z, \ g = 2x^2 - 3y^4z \in k[x, y, z]$

and let ϕ, ψ be the corresponding elements of $k[V]$.

Note that neither ϕ nor ψ is identically zero on V. For example, at $(0, 0, 5) \in V$, $\phi(0, 0, 5) = f(0, 0, 5) = -5 \neq 0$. Similarly, at $(1, 1, 2) \in V$, $\psi(1, 1, 2) = g(1, 1, 2) = -4 \neq 0$. However, the product function $\phi \cdot \psi$ is zero at every point of V. The reason is that

$$\begin{aligned} f \cdot g &= (x^2 + y^2 - z)(2x^2 - 3y^4z) \\ &= 2x(x^3 + xy^2 - xz) - 3y^3z(x^2y + y^3 - yz) \\ &\in \langle x^3 + xy^2 - xz, x^2y + y^3 - yz \rangle. \end{aligned}$$

Hence $f \cdot g \in \mathbf{I}(V)$, so the corresponding polynomial function $\phi \cdot \psi$ on V is identically zero.

The product of two nonzero elements of a field or of two nonzero polynomials in $k[x_1, \ldots, x_n]$ is never zero. In general, a commutative ring R is said to be an *integral domain* if whenever $a \cdot b = 0$ in R, either $a = 0$ or $b = 0$. Hence, for the variety V in the above example, we see that $k[V]$ is not an integral domain. Furthermore, the existence of $\phi \neq 0$ and $\psi \neq 0$ in $k[V]$ such that $\phi \cdot \psi = 0$ is a direct consequence of the reducibility of V, since f in (3) is zero on $V_1 = \mathbf{V}(x^2 + y^2 - z)$ but not on $V_2 = \mathbf{V}(x, y)$, and similarly g is zero on V_2 but not on V_1. This is why $f \cdot g = 0$ at every point of $V = V_1 \cup V_2$. Hence, we see a relation between the geometric properties of V and the algebraic properties of $k[V]$.

The general case of this relation can be stated as follows.

Proposition 4. *Let $V \subseteq k^n$ be an affine variety. The following are equivalent:*

(i) *V is irreducible.*

(ii) $\mathbf{I}(V)$ *is a prime ideal.*

(iii) $k[V]$ *is an integral domain.*

Proof. (i) $\Leftrightarrow$ (ii) is Proposition 3 of Chapter 4, §5.

To show (iii) $\Rightarrow$ (i), suppose that $k[V]$ is an integral domain but that V is reducible. By Definition 1 of Chapter 4, §5, this means that we can write $V = V_1 \cup V_2$, where V_1 and V_2 are proper, nonempty subvarieties of V. Let $f_1 \in k[x_1, \dots, x_n]$ be a polynomial that vanishes on V_1 but not identically on V_2 and, similarly, let f_2 be identically zero on V_2, but not on V_1. (Such polynomials must exist since V_1 and V_2 are varieties and neither is contained in the other.) Hence, neither f_1 nor f_2 represents the zero function in $k[V]$. However, the product $f_1 \cdot f_2$ vanishes at all points of $V_1 \cup V_2 = V$. Hence, the product function is zero in $k[V]$. This is a contradiction to our hypothesis that $k[V]$ was an integral domain. Hence, V is irreducible.

Finally, for (i) $\Rightarrow$ (iii), suppose that $k[V]$ is not an integral domain. Then there must be polynomials $f, g \in k[x_1, \dots, x_n]$ such that neither f nor g vanishes identically on V but their product does. In Exercise 9, you will check that we get a decomposition of V as a union of subvarieties:

$$V = (V \cap \mathbf{V}(f)) \cup (V \cap \mathbf{V}(g)).$$

You will also show in Exercise 9 that, under these hypotheses, neither $V \cap \mathbf{V}(f)$ nor $V \cap \mathbf{V}(g)$ is all of V. This contradicts our assumption that V is irreducible. $\quad\square$

Next we will consider another example of the kind of information about varieties revealed by polynomial mappings. The variety $V \subseteq \mathbb{C}^3$ defined by

(4)
$$x^2 + 2xz + 2y^2 + 3y = 0,$$
$$xy + 2x + z = 0,$$
$$xz + y^2 + 2y = 0$$

is the intersection of three quadric surfaces.

To study V, we compute a Gröbner basis for the ideal generated by the polynomials in (4) using lex order with $y > z > x$. The result is

(5)
$$g_1 = y - x^2,$$
$$g_2 = z + x^3 + 2x.$$

Geometrically, by the results of Chapter 3, §2, we know that the projection of V on the x-axis is onto since the two polynomials in (5) have constant leading coefficients. Furthermore, for each value of x in $\mathbb{C}$, there are unique y, z satisfying equations (4).

We can rephrase this observation using the maps

$$\pi : V \longrightarrow \mathbb{C}, \quad (x, y, z) \longmapsto x,$$
$$\phi : \mathbb{C} \longrightarrow V, \quad x \longmapsto (x, x^2, -x^3 - 2x).$$

Note that (5) guarantees that ϕ takes values in V. Both ϕ and π are visibly poly-nomial mappings. We claim that these maps establish a one-to-one correspondence between the points of the variety V and the points of the variety $\mathbb{C}$.

Our claim will follow if we can show that π and ϕ are inverses of each other. To verify this last claim, we first check that $\pi \circ \phi = \mathrm{id}_{\mathbb{C}}$. This is actually quite clear since

$$(\pi \circ \phi)(x) = \pi(x, x^2, -x^3 - 2x) = x.$$

On the other hand, if $(x, y, z) \in V$, then

$$(\phi \circ \pi)(x, y, z) = (x, x^2, -x^3 - 2x).$$

By (5), we have $y - x^2, z + x^3 + 2x \in \mathbf{I}(V)$ and it follows that $\phi \circ \pi$ defines the *same* mapping on V as $\mathrm{id}_V(x, y, z) = (x, y, z)$.

The conclusion we draw from this example is that $V \subseteq \mathbb{C}^3$ and $\mathbb{C}$ are "isomor-phic" varieties in the sense that there is a one-to-one, onto, polynomial mapping from V to $\mathbb{C}$, with a polynomial inverse. Even though our two varieties are defined by different equations and are subsets of different ambient spaces, they are "the same" in a certain sense. In addition, the Gröbner basis calculation leading to equa-tion (5) shows that $\mathbb{C}[V] = \mathbb{C}[x]$, in the sense that every $\psi \in \mathbb{C}[V]$ can be (uniquely) expressed by substituting for y and z from (5) to yield a polynomial in x alone. Of course, if we use x as the coordinate on $W = \mathbb{C}$, then $\mathbb{C}[W] = \mathbb{C}[x]$ as well, and we obtain the *same* collection of functions on our two isomorphic varieties.

Thus, the collection of polynomial functions on an affine variety can detect ge-ometric properties such as reducibility or irreducibility. In addition, knowing the structure of $k[V]$ can also furnish information leading toward the beginnings of a *classification* of varieties, a topic we have not broached before. We will return to these questions later in the chapter, once we have developed several different tools to analyze the algebraic properties of $k[V]$.

EXERCISES FOR §1

1. Let V be the twisted cubic in $\mathbb{R}^3$ and let $W = \mathbf{V}(v - u - u^2)$ in $\mathbb{R}^2$. Show that $\phi(x, y, z) = (xy, z + x^2 y^2)$ defines a polynomial mapping from V to W. Hint: The easiest way is to use a parametrization of V.
2. Let $V = \mathbf{V}(y - x)$ in $\mathbb{R}^2$ and let $\phi : \mathbb{R}^2 \to \mathbb{R}^3$ be the polynomial mapping represented by $\phi(x, y) = (x^2 - y, y^2, x - 3y^2)$. The image of V under ϕ is a variety in $\mathbb{R}^3$. Find a system of equations defining the image of ϕ.
3. Given a polynomial function $\phi : V \to k$, we define a *level set* or *fiber* of ϕ to be

$$\phi^{-1}(c) = \{(a_1, \ldots, a_m) \in V \mid \phi(a_1, \ldots, a_m) = c\},$$

where $c \in k$ is fixed. In this exercise, we will investigate how level sets can be used to analyze and reconstruct a variety. We will assume that $k = \mathbb{R}$, and we will work with the surface

$$V = \mathbf{V}(x^2 - y^2 z^2 + z^3) \subseteq \mathbb{R}^3.$$

   a. Let ϕ be the polynomial function represented by $f(x, y, z) = z$. The image of ϕ is all of $\mathbb{R}$ in this case. For each $c \in \mathbb{R}$, explain why the level set $\phi^{-1}(c)$ is the affine variety defined by the equations:

$$x^2 - y^2 z^2 + z^3 = 0,$$
$$z - c = 0.$$

b. Eliminate z between these equations to find the equation of the intersection of V with the plane $z = c$. Explain why your equation defines a hyperbola in the plane $z = c$ if $c \neq 0$, and the y-axis if $c = 0$. (Refer to the sketch of V in §3 of Chapter 1, and see if you can visualize the way these hyperbolas lie on V.)

c. Let $\pi : V \to \mathbb{R}$ be the polynomial mapping $\pi(x, y, z) = x$. Describe the level sets $\pi^{-1}(c)$ in V geometrically for $c = -1, 0, 1$.

d. Do the same for the level sets of $\sigma : V \to \mathbb{R}$ given by $\sigma(x, y, z) = y$.

e. Construct a one-to-one polynomial mapping $\psi : \mathbb{R} \to V$ and identify the image as a subvariety of V. Hint: The y-axis.

4. Let $V = \mathbf{V}(z^2 - (x^2 + y^2 - 1)(4 - x^2 - y^2))$ in $\mathbb{R}^3$ and let $\pi : V \to \mathbb{R}^2$ be the vertical projection $\pi(x, y, z) = (x, y)$.

a. What is the maximum number of points in $\pi^{-1}(a, b)$ for $(a, b) \in \mathbb{R}^2$?

b. For which subsets $R \subseteq \mathbb{R}^2$ does $(a, b) \in R$ imply $\pi^{-1}(a, b)$ consists of two points, one point, no points?

c. Using part (b) describe and/or sketch V.

5. Show that $\phi_1(x, y, z) = (2x^2 + y^2, z^2 - y^3 + 3xz)$ and $\phi_2(x, y, z) = (2y + xz, 3y^2)$ represent the same polynomial mapping from the twisted cubic in $\mathbb{R}^3$ to $\mathbb{R}^2$.

6. Consider the mapping $\phi : \mathbb{R}^2 \to \mathbb{R}^5$ defined by $\phi(u, v) = (u, v, u^2, uv, v^2)$.

a. The image of ϕ is a variety S known as an affine Veronese surface. Find implicit equations for S.

b. Show that the projection $\pi : S \to \mathbb{R}^2$ defined by $\pi(x_1, x_2, x_3, x_4, x_5) = (x_1, x_2)$ is the inverse mapping of $\phi : \mathbb{R}^2 \to S$. What does this imply about S and $\mathbb{R}^2$?

7. This problem characterizes the varieties for which $\mathbf{I}(V) = \{0\}$.

a. Show that if k is an infinite field and $V \subseteq k^n$ is a variety, then $\mathbf{I}(V) = \{0\}$ if and only if $V = k^n$.

b. On the other hand, show that if k is finite, then $\mathbf{I}(V)$ is never equal to $\{0\}$. Hint: See Exercise 4 of Chapter 1, §1.

8. Let $V = \mathbf{V}(xy, xz) \subseteq \mathbb{R}^3$.

a. Show that neither of the polynomial functions $f = y^2 + z^3, g = x^2 - x$ is identically zero on V, but that their product is identically zero on V.

b. Find $V_1 = V \cap \mathbf{V}(f)$ and $V_2 = V \cap \mathbf{V}(g)$ and show that $V = V_1 \cup V_2$.

9. Let V be an irreducible variety and let ϕ, ψ be functions in $k[V]$ represented by polynomials f, g, respectively. Assume that $\phi \cdot \psi = 0$ in $k[V]$ and that neither ϕ nor ψ is the zero function on V.

a. Show that $V = (V \cap \mathbf{V}(f)) \cup (V \cap \mathbf{V}(g))$.

b. Show that neither $V \cap \mathbf{V}(f)$ nor $V \cap \mathbf{V}(g)$ is all of V and deduce a contradiction.

10. In this problem, we will see that there are no nonconstant polynomial mappings from $V = \mathbb{R}$ to $W = \mathbf{V}(y^2 - x^3 + x) \subseteq \mathbb{R}^2$. Thus, these varieties are not isomorphic (i.e., they are not "the same" in the sense introduced in this section).

a. Suppose $\phi : \mathbb{R} \to W$ is a polynomial mapping represented by $\phi(t) = (a(t), b(t))$ where $a(t), b(t) \in \mathbb{R}[t]$. Explain why it must be true that $b(t)^2 = a(t)(a(t)^2 - 1)$.

b. Explain why the two factors on the right of the equation in part (a) must be relatively prime in $\mathbb{R}[t]$.

c. Using the unique factorizations of a and b into products of powers of irreducible polynomials, show that $b^2 = ac^2$ for some polynomial $c \in \mathbb{R}[t]$ relatively prime to a.

d. From part (c) it follows that $c^2 = a^2 - 1$. Deduce from this equation that c, a, and, hence, b must be constant polynomials. Hint: $a^2 - c^2 = 1$.

§2 Quotients of Polynomial Rings

The construction of $k[V]$ given in §1 is a special case of what is called the quotient of $k[x_1, \ldots, x_n]$ modulo an ideal I. From the word *quotient*, you might guess that the issue is to define a division operation, but this is *not* the case. Instead, forming the quotient will indicate the sort of "lumping together" of polynomials that we mentioned in §1 when describing the elements $\phi \in k[V]$. The quotient construction is a fundamental tool in commutative algebra and algebraic geometry, so if you pursue these subjects further, the acquaintance you make with quotient rings here will be valuable.

To begin, we introduce some new terminology.

Definition 1. Let $I \subseteq k[x_1, \ldots, x_n]$ be an ideal, and let $f, g \in k[x_1, \ldots, x_n]$. We say f and g are **congruent modulo** I, written

$$f \equiv g \mod I,$$

if $f - g \in I$.

For instance, if $I = \langle x^2 - y^2, x + y^3 + 1 \rangle \subseteq k[x, y]$, then $f = x^4 - y^4 + x$ and $g = x + x^5 + x^4 y^3 + x^4$ are congruent modulo I since

$$f - g = x^4 - y^4 - x^5 - x^4 y^3 - x^4$$
$$= (x^2 + y^2)(x^2 - y^2) - (x^4)(x + y^3 + 1) \in I.$$

The most important property of the congruence relation is given by the following proposition.

Proposition 2. *Let $I \subseteq k[x_1, \ldots, x_n]$ be an ideal. Then congruence modulo I is an equivalence relation on $k[x_1, \ldots, x_n]$.*

Proof. Congruence modulo I is reflexive since $f - f = 0 \in I$ for every $f \in k[x_1, \ldots, x_n]$. To prove symmetry, suppose that $f \equiv g \mod I$. Then $f - g \in I$, which implies that $g - f = (-1)(f - g) \in I$ as well. Hence, $g \equiv f \mod I$ also. Finally, we need to consider transitivity. If $f \equiv g \mod I$ and $g \equiv h \mod I$, then $f - g, g - h \in I$. Since I is closed under addition, we have $f - h = f - g + g - h \in I$ as well. Hence, $f \equiv h \mod I$. ☐

An equivalence relation on a set S partitions S into a collection of disjoint subsets called equivalence classes. For any $f \in k[x_1, \ldots, x_n]$, the class of f is the set

$$[f] = \{g \in k[x_1, \ldots, x_n] \mid g \equiv f \mod I\}.$$

The definition of congruence modulo I and Proposition 2 makes sense for *every* ideal $I \subseteq k[x_1, \ldots, x_n]$. In the special case that $I = \mathbf{I}(V)$ is the ideal of the variety V, then by Proposition 2 of §1, it follows that $f \equiv g \mod \mathbf{I}(V)$ if and only if f and g define the same function on V. In other words, the "lumping together" of polynomials that define the same function on a variety V is accomplished by passing to the equivalence classes for the congruence relation modulo $\mathbf{I}(V)$. More formally, we have the following proposition.

Proposition 3. *The distinct polynomial functions $\phi : V \to k$ are in one-to-one correspondence with the equivalence classes of polynomials under congruence modulo $\mathbf{I}(V)$.*

Proof. This is a corollary of Proposition 2 of §1 and the (easy) proof is left to the reader as an exercise. □

We are now ready to introduce the quotients mentioned in the title of this section.

Definition 4. The **quotient** of $k[x_1, \ldots, x_n]$ modulo I, written $k[x_1, \ldots, x_n]/I$, is the set of equivalence classes for congruence modulo I:

$$k[x_1, \ldots, x_n]/I = \{[f] \mid f \in k[x_1, \ldots, x_n]\}.$$

For instance, take $k = \mathbb{R}$, $n = 1$, and $I = \langle x^2 - 2 \rangle$. We may ask whether there is some way to describe all the equivalence classes for congruence modulo I. By the division algorithm, every $f \in \mathbb{R}[x]$ can be written as $f = q \cdot (x^2 - 2) + r$, where $r = ax + b$ for some $a, b \in \mathbb{R}$. By the definition, $f \equiv r \bmod I$ since $f - r = q \cdot (x^2 - 2) \in I$. Thus, every element of $\mathbb{R}[x]$ belongs to one of the equivalence classes $[ax + b]$, and $\mathbb{R}[x]/I = \{[ax + b] \mid a, b \in \mathbb{R}\}$. In §3, we will extend the idea used in this example to a method for dealing with $k[x_1, \ldots, x_n]/I$ in general.

Because $k[x_1, \ldots, x_n]$ is a ring, given any two classes $[f], [g] \in k[x_1, \ldots, x_n]/I$, we can attempt to define sum and product operations *on classes* by using the corresponding operations on elements of $k[x_1, \ldots, x_n]$. That is, we can try to define

$$
\begin{aligned}
(1) \qquad & [f] + [g] = [f + g] \quad \text{(sum in } k[x_1, \ldots, x_n]\text{)}, \\
& [f] \cdot [g] = [f \cdot g] \quad \text{(product in } k[x_1, \ldots, x_n]\text{)}.
\end{aligned}
$$

We must check, however, that these formulas actually make sense. We need to show that if we choose different $f' \in [f]$ and $g' \in [g]$, then the class $[f' + g']$ is the same as the class $[f + g]$. Similarly, we need to check that $[f' \cdot g'] = [f \cdot g]$.

Proposition 5. *The operations defined in equations* (1) *yield the same classes in $k[x_1, \ldots, x_n]/I$ on the right-hand sides no matter which $f' \in [f]$ and $g' \in [g]$ we use. (We say that the operations on classes given in* (1) *are* **well-defined** *on classes.)*

Proof. If $f' \in [f]$ and $g' \in [g]$, then $f' = f + A$ and $g' = g + B$, where $A, B \in I$. Hence,

$$f' + g' = (f + A) + (g + B) = (f + g) + (A + B).$$

Since we also have $A + B \in I$ (I is an ideal), it follows that $f' + g' \equiv f + g \bmod I$, so $[f' + g'] = [f + g]$. Similarly,

$$f' \cdot g' = (f + A) \cdot (g + B) = fg + Ag + fB + AB.$$

Since $A, B \in I$, we have $Ag + fB + AB \in I$. Thus, $f' \cdot g' \equiv f \cdot g \bmod I$, which implies that $[f' \cdot g'] = [f \cdot g]$. □

To illustrate this result, consider the sum and product operations in $\mathbb{R}[x]/\langle x^2 - 2\rangle$. As we saw earlier, the classes $[ax + b], a, b \in \mathbb{R}$ form a complete list of the elements of $\mathbb{R}[x]/\langle x^2 - 2\rangle$. The sum operation is defined by $[ax + b] + [cx + d] = [(a + c)x + (b + d)]$. Note that this amounts to the usual vector sum on ordered pairs of real numbers. The product operation is also easily understood. We have

$$[ax + b] \cdot [cx + d] = [acx^2 + (ad + bc)x + bd]$$
$$= [(ad + bc)x + (bd + 2ac)],$$

as we can see by dividing the quadratic polynomial in the first line by $x^2 - 2$ and using the remainder as our representative of the class of the product.

Once we know that the operations in (1) are well-defined, it follows immediately that all of the axioms for a commutative ring are satisfied in $k[x_1, \ldots, x_n]/I$. This is so because the sum and product in $k[x_1, \ldots, x_n]/I$ are defined in terms of the corresponding operations in $k[x_1, \ldots, x_n]$, where we know that the axioms do hold. For example, to check that sums are associative in $k[x_1, \ldots, x_n]/I$, we argue as follows: if $[f], [g], [h] \in k[x_1, \ldots, x_n]/I$, then

$$([f] + [g]) + [h] = [f + g] + [h]$$
$$= [(f + g) + h] \quad \text{[by (1)]}$$
$$= [f + (g + h)] \quad \text{(by associativity in } k[x_1, \ldots, x_n])$$
$$= [f] + [g + h]$$
$$= [f] + ([g] + [h]).$$

Similarly, commutativity of addition, associativity, and commutativity of multiplication, and the distributive law all follow because polynomials satisfy these properties. The additive identity is $[0] \in k[x_1, \ldots, x_n]/I$, and the multiplicative identity is $[1] \in k[x_1, \ldots, x_n]/I$. To summarize, we have sketched the proof of the following theorem.

Theorem 6. *Let I be an ideal in $k[x_1, \ldots, x_n]$. The quotient $k[x_1, \ldots, x_n]/I$ is a commutative ring under the sum and product operations given in (1).*

Next, given a variety V, let us relate the quotient ring $k[x_1, \ldots, x_n]/\mathbf{I}(V)$ to the ring $k[V]$ of polynomial functions on V. It turns out that these two rings are "the same" in the following sense.

Theorem 7. *The one-to-one correspondence between the elements of $k[V]$ and the elements of $k[x_1, \ldots, x_n]/\mathbf{I}(V)$ given in Proposition 3 preserves sums and products.*

Proof. Let $\Phi : k[x_1, \ldots, x_n]/\mathbf{I}(V) \to k[V]$ be the mapping defined by $\Phi([f]) = \phi$, where ϕ is the polynomial function represented by f. Since every element of $k[V]$ is represented by some polynomial, we see that Φ is onto. To see that Φ is also one-to-one, suppose that $\Phi([f]) = \Phi([g])$. Then by Proposition 3, $f \equiv g \bmod \mathbf{I}(V)$. Hence, $[f] = [g]$ in $k[x_1, \ldots, x_n]/\mathbf{I}(V)$.

To study sums and products, let $[f], [g] \in k[x_1, \ldots, x_n]/\mathbf{I}(V)$. Then $\Phi([f]+[g]) = \Phi([f+g])$ by the definition of sum in the quotient ring. If f represents the polynomial function ϕ and g represents ψ, then $f + g$ represents $\phi + \psi$. Hence,

$$\Phi([f + g]) = \phi + \psi = \Phi([f]) + \Phi([g]).$$

Thus, Φ preserves sums. Similarly,

$$\Phi([f] \cdot [g]) = \Phi([f \cdot g]) = \phi \cdot \psi = \Phi([f]) \cdot \Phi([g]).$$

Thus, Φ preserves products as well.

The inverse correspondence $\Psi = \Phi^{-1}$ also preserves sums and products by a similar argument, and the theorem is proved. □

The result of Theorem 7 illustrates a basic notion from abstract algebra. The following definition tells us what it means for two rings to be essentially the same.

Definition 8. Let R, S be commutative rings.

(i) A mapping $\Phi : R \to S$ is said to be a **ring isomorphism** if:
  a. Φ preserves sums: $\Phi(r + r') = \Phi(r) + \Phi(r')$ for all $r, r' \in R$.
  b. Φ preserves products: $\Phi(r \cdot r') = \Phi(r) \cdot \Phi(r')$ for all $r, r' \in R$.
  c. Φ is one-to-one and onto.
(ii) Two rings R, S are **isomorphic** if there exists an isomorphism $\Phi : R \to S$. We write $R \cong S$ to denote that R is isomorphic to S.
(iii) A mapping $\Phi : R \to S$ is a **ring homomorphism** if Φ satisfies properties (a) and (b) of (i), but not necessarily property (c), and if, in addition, Φ maps the multiplicative identity $1 \in R$ to $1 \in S$.

In general, a "homomorphism" is a mapping that preserves algebraic structure. A ring homomorphism $\Phi : R \to S$ is a mapping that preserves the addition and multiplication operations in the ring R.

From Theorem 7, we get a ring isomorphism $k[V] \cong k[x_1, \ldots, x_n]/\mathbf{I}(V)$. A natural question to ask is what happens if we replace $\mathbf{I}(V)$ by some other ideal I which defines V. [From Chapter 4, we know that there are *lots* of ideals I such that $V = \mathbf{V}(I)$.] Could it be true that *all* the quotient rings $k[x_1, \ldots, x_n]/I$ are isomorphic to $k[V]$? The following example shows that the answer to this question is *no*. Let $V = \{(0,0)\}$. We saw in Chapter 1, §4 that $\mathbf{I}(V) = \mathbf{I}(\{(0,0)\}) = \langle x, y \rangle$. By Theorem 7, we know that $k[x, y]/\mathbf{I}(V) \cong k[V]$.

Our first claim is that the quotient ring $k[x, y]/\mathbf{I}(V)$ is isomorphic to the field k. The easiest way to see this is to note that a polynomial function on the one-point set $\{(0,0)\}$ can be represented by a constant since the function will have only one function value. Alternatively, we can derive the same fact algebraically by constructing a mapping

$$\Phi : k[x,y]/\mathbf{I}(V) \longrightarrow k$$

by setting $\Phi([f]) = f(0,0)$ (the constant term of the polynomial). We will leave it as an exercise to show that Φ is a well-defined ring isomorphism.

Now, let $I = \langle x^3+y^2, 3y^4\rangle \subseteq k[x,y]$. It is easy to check that $\mathbf{V}(I) = \{(0,0)\} = V$. We ask whether $k[x,y]/I$ is also isomorphic to k. A moment's thought shows that this is not so. For instance, consider the class $[y] \in k[x,y]/I$. Note that $y \notin I$, a fact which can be checked by finding a Gröbner basis for I (use any monomial order) and computing a remainder. In the ring $k[x,y]/I$, this shows that $[y] \neq [0]$. But we also have $[y]^4 = [y^4] = [0]$ since $y^4 \in I$. Thus, there is an element of $k[x,y]/I$ which is not zero itself, but whose fourth power is zero. In a field, this is impossible. We conclude that $k[x,y]/I$ is not a field. But this says that $k[x,y]/\mathbf{I}(V)$ and $k[x,y]/I$ cannot be isomorphic rings since one is a field and the other is not. (See Exercise 8.)

In a commutative ring R, an element $a \in R$ such that $a^n = 0$ for some $n \geq 1$ is said to be *nilpotent*. The example just given is actually quite representative of the kind of difference that can appear when we compare $k[x_1,\ldots,x_n]/\mathbf{I}(V)$ with $k[x_1,\ldots,x_n]/I$ for another ideal I with $\mathbf{V}(I) = V$. If I is not a radical ideal, there will be elements $f \in \sqrt{I}$ which are not in I itself. Thus, in $k[x_1,\ldots,x_n]/I$, we will have $[f] \neq [0]$, whereas $[f]^n = [0]$ for the $n > 1$ such that $f^n \in I$. The ring $k[x_1,\ldots,x_n]/I$ *will* have nonzero nilpotent elements, whereas $k[x_1,\ldots,x_n]/\mathbf{I}(V)$ never does: $\mathbf{I}(V)$ is always a radical ideal, so $[f]^n = 0$ if and only if $[f] = 0$.

Since a quotient $k[x_1,\ldots,x_n]/I$ is a commutative ring in its own right, we can study other facets of its ring structure as well, and, in particular, we can consider ideals in $k[x_1,\ldots,x_n]/I$. The definition is the same as the definition of ideals in $k[x_1,\ldots,x_n]$.

Definition 9. A subset I of a commutative ring R is said to be an **ideal** in R if it satisfies

(i) $0 \in I$ (where 0 is the zero element of R).
(ii) If $a,b \in I$, then $a + b \in I$.
(iii) If $a \in I$ and $r \in R$, then $r \cdot a \in I$.

There is a close relation between ideals in the quotient $k[x_1,\ldots,x_n]/I$ and ideals in $k[x_1,\ldots,x_n]$.

Proposition 10. *Let I be an ideal in $k[x_1,\ldots,x_n]$. The ideals in the quotient ring $k[x_1,\ldots,x_n]/I$ are in one-to-one correspondence with the ideals of $k[x_1,\ldots,x_n]$ containing I (i.e., the ideals J satisfying $I \subseteq J \subseteq k[x_1,\ldots,x_n]$).*

Proof. First, we give a way to produce an ideal in $k[x_1,\ldots,x_n]/I$ corresponding to each J containing I in $k[x_1,\ldots,x_n]$. Given an ideal J in $k[x_1,\ldots,x_n]$ containing I, let J/I denote the set $\{[j] \in k[x_1,\ldots,x_n]/I \mid j \in J\}$. We claim that J/I is an ideal in $k[x_1,\ldots,x_n]/I$. To prove this, first note that $[0] \in J/I$ since $0 \in J$. Next, let $[j],[k] \in J/I$. Then $[j] + [k] = [j+k]$ by the definition of the sum in $k[x_1,\ldots,x_n]/I$. Since $j,k \in J$, we have $j+k \in J$ as well. Hence, $[j] + [k] \in J/I$. Finally, if $[j] \in J/I$ and $[r] \in k[x_1,\ldots,x_n]/I$, then $[r] \cdot [j] = [r \cdot j]$ by the definition of the product in

$k[x_1,\ldots,x_n]/I$. But $r \cdot j \in J$ since J is an ideal in $k[x_1,\ldots,x_n]$. Hence, $[r] \cdot [j] \in J/I$. As a result, J/I is an ideal in $k[x_1,\ldots,x_n]/I$.

If $\tilde{J} \subseteq k[x_1,\ldots,x_n]/I$ is an ideal, we next show how to produce an ideal $J \subseteq k[x_1,\ldots,x_n]$ which contains I. Let $J = \{j \in k[x_1,\ldots,x_n] \mid [j] \in \tilde{J}\}$. Then we have $I \subseteq J$ since $[i] = [0] \in \tilde{J}$ for any $i \in I$. It remains to show that J is an ideal of $k[x_1,\ldots,x_n]$. First note that $0 \in I \subseteq J$. Furthermore, if $j,k \in J$, then $[j],[k] \in \tilde{J}$ implies that $[j] + [k] = [j+k] \in \tilde{J}$. It follows that $j + k \in J$. Finally, if $j \in J$ and $r \in k[x_1,\ldots,x_n]$, then $[j] \in \tilde{J}$, so $[r][j] = [rj] \in \tilde{J}$. But this says $rj \in J$, and, hence, J is an ideal in $k[x_1,\ldots,x_n]$.

We have thus shown that there are correspondences between the two collections of ideals:

$$\{J \mid I \subseteq J \subseteq k[x_1,\ldots,x_n]\} \qquad \{\tilde{J} \subseteq k[x_1,\ldots,x_n]/I\}$$

(2)
$$J \longrightarrow J/I = \{[j] \mid j \in J\}$$
$$J = \{j \mid [j] \in \tilde{J}\} \longleftarrow \tilde{J}.$$

We leave it as an exercise to prove that each of these arrows is the inverse of the other. This gives the desired one-to-one correspondence. □

For example, consider the ideal $I = \langle x^2 - 4x + 3 \rangle$ in $R = \mathbb{R}[x]$. We know from Chapter 1 that R is a principal ideal domain, i.e., every ideal in R is generated by a single polynomial. The ideals containing I are precisely the ideals generated by polynomials that *divide* $x^2 - 4x + 3 = (x-1)(x-3)$. Hence, the quotient ring R/I has exactly four ideals in this case:

ideals in R/I	ideals in R containing I
$\{[0]\}$	I
$\langle [x-1] \rangle$	$\langle x-1 \rangle$
$\langle [x-3] \rangle$	$\langle x-3 \rangle$
R/I	R

As in another example earlier in this section, we can compute in R/I by computing remainders with respect to $x^2 - 4x + 3$.

As a corollary of Proposition 10, we deduce the following result about ideals in quotient rings, parallel to the Hilbert Basis Theorem from Chapter 2.

Corollary 11. *Every ideal in the quotient ring $k[x_1,\ldots,x_n]/I$ is finitely generated.*

Proof. Let $\tilde{J}$ be any ideal in $k[x_1,\ldots,x_n]/I$. By Proposition 10, $\tilde{J} = \{[j] \mid j \in J\}$ for an ideal J in $k[x_1,\ldots,x_n]$ containing I. Then the Hilbert Basis Theorem implies that $J = \langle f_1,\ldots,f_s \rangle$ for some $f_i \in k[x_1,\ldots,x_n]$. But then for any $j \in J$, we have $j = h_1 f_1 + \cdots + h_s f_s$ for some $h_i \in k[x_1,\ldots,x_n]$. Hence,

$$[j] = [h_1 f_1 + \cdots + h_s f_s]$$
$$= [h_1][f_1] + \cdots + [h_s][f_s].$$

As a result, the classes $[f_1],\ldots,[f_s]$ generate $\tilde{J}$ in $k[x_1,\ldots,x_n]/I$. □

In the next section, we will discuss a more constructive method to study the quotient rings $k[x_1, \ldots, x_n]/I$ and their algebraic properties.

EXERCISES FOR §2

1. Let $I = \langle f_1, \ldots, f_s \rangle \subseteq k[x_1, \ldots, x_n]$. Describe an algorithm for determining whether $f \equiv g \bmod I$ using techniques from Chapter 2.
2. Prove Proposition 3.
3. Prove Theorem 6. This means showing that the other axioms for a commutative ring are satisfied by $k[x_1, \ldots, x_n]/I$.
4. In this problem, we will give an algebraic construction of a field containing $\mathbb{Q}$ in which 2 has a square root. Note that the field of real numbers is one such field. However, our construction will not make use of the *limit* process necessary, for example, to make sense of an infinite decimal expansion such as the usual expansion $\sqrt{2} = 1.414\ldots$. Instead, we will work with the polynomial $x^2 - 2$.
   a. Show that every $f \in \mathbb{Q}[x]$ is congruent modulo the ideal $I = \langle x^2 - 2 \rangle \subseteq \mathbb{Q}[x]$ to a unique polynomial of the form $ax + b$, where $a, b \in \mathbb{Q}$.
   b. Show that the class of x in $\mathbb{Q}[x]/I$ is a square root of 2 in the sense that $[x]^2 = [2]$ in $\mathbb{Q}[x]/I$.
   c. Show that $F = \mathbb{Q}[x]/I$ is a field. Hint: Using Theorem 6, the only thing left to prove is that every nonzero element of F has a multiplicative inverse in F.
   d. Find a subfield of F isomorphic to $\mathbb{Q}$.
5. In this problem, we will consider the addition and multiplication operations in the quotient ring $\mathbb{R}[x]/\langle x^2 + 1 \rangle$.
   a. Show that every $f \in \mathbb{R}[x]$ is congruent modulo $I = \langle x^2 + 1 \rangle$ to a unique polynomial of the form $ax + b$, where $a, b \in \mathbb{R}$.
   b. Construct formulas for the addition and multiplication rules in $\mathbb{R}[x]/\langle x^2 + 1 \rangle$ using these polynomials as the standard representatives for classes.
   c. Do we know another way to describe the ring $\mathbb{R}[x]/\langle x^2 + 1 \rangle$ (that is, another well-known ring isomorphic to this one)? Hint: What is $[x]^2$?
6. Show that $\mathbb{R}[x]/\langle x^2 - 4x + 3 \rangle$ is not an integral domain.
7. One can define a quotient ring R/I for any ideal I in a commutative ring R. The general construction is the same as what we did for $k[x_1, \ldots, x_n]/I$. Here is a simple example.
   a. Let $I = \langle p \rangle$ in $R = \mathbb{Z}$, where p is a prime number. Show that the relation of congruence modulo p, defined by

   $$m \equiv n \bmod p \iff p \text{ divides } m - n$$

   is an equivalence relation on $\mathbb{Z}$, and list the different equivalence classes. We will denote the set of equivalence classes by $\mathbb{Z}/\langle p \rangle$.
   b. Construct sum and product operations in $\mathbb{Z}/\langle p \rangle$ by the analogue of equation (1) and then prove that they are well-defined by adapting the proof of Proposition 5.
   c. Explain why $\mathbb{Z}/\langle p \rangle$ is a commutative ring under the operations you defined in part (b).
   d. Show that the finite field $\mathbb{F}_p$ introduced in Chapter 1 is isomorphic as a ring to $\mathbb{Z}/\langle p \rangle$.
8. In this problem, we study how ring homomorphisms interact with multiplicative inverses.
   a. Show that every ring isomorphism $\Phi : R \to S$ takes the multiplicative identity in R to the multiplicative identity in S, i.e., $\Phi(1) = 1$.
   b. Show that if $r \in R$ has a multiplicative inverse, then for any ring homomorphism $\Phi : R \to S$, $\Phi(r^{-1})$ is a multiplicative inverse for $\Phi(r)$ in the ring S.
   c. Show that if R and S are isomorphic as rings and R is a field, then S is also a field.
9. Prove that the map $f \mapsto f(0,0)$ induces a ring isomorphism $k[x, y]/\langle x, y \rangle \cong k$. Hint: An efficient proof can be given using Exercise 16.

10. This problem illustrates one important use of nilpotent elements in rings. Let $R = k[x, t]$ and let $I = \langle t^2 \rangle \subseteq R$.
    a. Show that $[t]$ is a nilpotent element in R/I and find the smallest power of $[t]$ which is equal to zero.
    b. Show that every class in R/I has a unique representative of the form $a + b\varepsilon$, where $a, b \in k[x]$ and ε is shorthand for $[t]$.
    c. Given $a + b\varepsilon \in R/I$ and $f(x) \in k[x]$, we can define an element $f(a + b\varepsilon) \in R/I$ by substituting $x = a + b\varepsilon$ into $f(x)$. For instance, with $a + b\varepsilon = 2 + \varepsilon$ and $f(x) = x^2$, we obtain $(2 + \varepsilon)^2 = 4 + 4\varepsilon + \varepsilon^2 = 4 + 4\varepsilon$. Show that

$$f(a + b\varepsilon) = f(a) + f'(a) \cdot b\varepsilon,$$

    where f' is the formal derivative of f defined in Exercise 13 of Chapter 1, §5.
    d. Suppose $\varepsilon = [t] \in R/\langle t^3 \rangle$ and $\frac{1}{2} \in k$. Find an analogous formula for $f(a + b\varepsilon)$.
11. Let R be a commutative ring. Show that the set of nilpotent elements of R forms an ideal in R. Hint: To show that the sum of two nilpotent elements is also nilpotent, use the binomial expansion of $(a + b)^m$ for a suitable exponent m.
12. This exercise will show that the two mappings given in (2) are inverses of each other.
    a. If $I \subseteq J$ is an ideal of $k[x_1, \ldots, x_n]$, show that $J = \{f \in k[x_1, \ldots, x_n] \mid [f] \in J/I\}$, where $J/I = \{[j] \mid j \in J\}$. Explain how your proof uses the assumption $I \subseteq J$.
    b. If $\tilde{J}$ is an ideal of $k[x_1, \ldots, x_n]/I$, show that $\tilde{J} = \{[f] \in k[x_1, \ldots, x_n]/I \mid f \in J\}$, where $J = \{j \mid [j] \in \tilde{J}\}$.
13. Let R and S be commutative rings and let $\Phi : R \to S$ be a ring homomorphism.
    a. If $J \subseteq S$ is an ideal, show that $\Phi^{-1}(J)$ is an ideal in R.
    b. If Φ is an isomorphism of rings, show that there is a one-to-one, inclusion-preserving correspondence between the ideals of R and the ideals of S.
14. This problem studies the ideals in some quotient rings.
    a. Let $I = \langle x^3 - x \rangle \subseteq R = \mathbb{R}[x]$. Determine the ideals in the quotient ring R/I using Proposition 10. Draw a diagram indicating the inclusions among these ideals.
    b. How does your answer change if $I = \langle x^3 + x \rangle$?
15. This problem considers some special quotient rings of $\mathbb{R}[x, y]$.
    a. Let $I = \langle x^2, y^2 \rangle \subseteq \mathbb{R}[x, y]$. Describe the ideals in $\mathbb{R}[x, y]/I$. Hint: Use Proposition 10.
    b. Is $\mathbb{R}[x, y]/\langle x^3, y \rangle$ isomorphic to $\mathbb{R}[x, y]/\langle x^2, y^2 \rangle$?
16. Let $\Phi : k[x_1, \ldots, x_n] \to S$ be a ring homomorphism. The set $\{r \in k[x_1, \ldots, x_n] \mid \Phi(r) = 0 \in S\}$ is called the *kernel* of Φ, written $\ker(\Phi)$.
    a. Show that $\ker(\Phi)$ is an ideal in $k[x_1, \ldots, x_n]$.
    b. Show that the mapping ν from $k[x_1, \ldots, x_n]/\ker(\Phi)$ to S defined by $\nu([r]) = \Phi(r)$ is well-defined in the sense that $\nu([r]) = \nu([r'])$ whenever $r \equiv r' \bmod \ker(\Phi)$.
    c. Show that ν is a ring homomorphism.
    d. (The Isomorphism Theorem) Assume that Φ is onto. Show that ν is a one-to-one and onto ring homomorphism. As a result, we have $S \cong k[x_1, \ldots, x_n]/\ker(\Phi)$ when $\Phi : k[x_1, \ldots, x_n] \to S$ is onto.
17. Use Exercise 16 to give a more concise proof of Theorem 7. Consider the mapping $\Phi : k[x_1, \ldots, x_n] \to k[V]$ that takes a polynomial to the element of $k[V]$ that it represents. Hint: What is the kernel of Φ?
18. Prove that $k[x_1, \ldots, x_n]/I$ has no nonzero nilpotent elements if and only if I is radical.
19. An $m \times n$ constant matrix A gives a map $\alpha_A : k[x_1, \ldots, x_m] \to k[y_1, \ldots, y_n]$ defined by $\alpha_A(f)(y) = f(Ay)$ for $f \in k[x_1, \ldots, x_m]$. Exercise 13 of Chapter 4, §3 shows that α_A is a ring homomorphism. If $m = n$ and A is invertible, prove that α_A is a ring isomorphism.

§3 Algorithmic Computations in $k[x_1, \ldots, x_n]/I$

In this section, we will use the division algorithm to produce simple representatives of equivalence classes for congruence modulo I, where $I \subseteq k[x_1, \ldots, x_n]$ is an ideal. These representatives will enable us to develop an explicit method for computing the sum and product operations in a quotient ring $k[x_1, \ldots, x_n]/I$. As an added dividend, we will derive an easily checked criterion to determine when a system of polynomial equations over $\mathbb{C}$ has only finitely many solutions. We will also describe a strategy for finding the solutions of such a system.

The basic idea that we will use is a direct consequence of the fact that the remainder on division of a polynomial f by a Gröbner basis G for an ideal I is uniquely determined by the polynomial f. (This was Proposition 1 of Chapter 2, §6.) Furthermore, we have the following basic observations reinterpreting the result of the division and the form of the remainder.

Proposition 1. *Fix a monomial ordering on $k[x_1, \ldots, x_n]$ and let $I \subseteq k[x_1, \ldots, x_n]$ be an ideal. As in Chapter 2, §5, $\langle \mathrm{LT}(I) \rangle$ will denote the ideal generated by the leading terms of elements of I.*

(i) *Every $f \in k[x_1, \ldots, x_n]$ is congruent modulo I to a unique polynomial r which is a k-linear combination of the monomials in the complement of $\langle \mathrm{LT}(I) \rangle$.*

(ii) *The elements of $\{x^\alpha \mid x^\alpha \notin \langle \mathrm{LT}(I) \rangle\}$ are "linearly independent modulo I," i.e., if we have*

$$\sum_\alpha c_\alpha x^\alpha \equiv 0 \bmod I,$$

where the x^α are all in the complement of $\langle \mathrm{LT}(I) \rangle$, then $c_\alpha = 0$ for all α.

Proof. (i) Let G be a Gröbner basis for I and let $f \in k[x_1, \ldots, x_n]$. By the division algorithm, the remainder $r = \overline{f}^G$ satisfies $f = q + r$, where $q \in I$. Hence, $f - r = q \in I$, so $f \equiv r \bmod I$. The division algorithm also tells us that r is a k-linear combination of the monomials $x^\alpha \notin \langle \mathrm{LT}(I) \rangle$. The uniqueness of r follows from Proposition 1 of Chapter 2, §6.

(ii) The argument to establish this part of the proposition is essentially the same as the proof of the uniqueness of the remainder in Proposition 1 of Chapter 2, §6. We leave it to the reader to carry out the details. $\square$

Historically, this was the first application of Gröbner bases. Buchberger's thesis BUCHBERGER (1965) concerned the question of finding "standard sets of representatives" for the classes in quotient rings. We also note that if $I = \mathbf{I}(V)$ for a variety V, Proposition 1 gives standard representatives for the polynomial functions $\phi \in k[V]$.

Example 2. Let $I = \langle xy^3 - x^2, x^3 y^2 - y \rangle$ in $\mathbb{R}[x, y]$ and use graded lex order. We find that

$$G = \{x^3 y^2 - y, x^4 - y^2, xy^3 - x^2, y^4 - xy\}$$

is a Gröbner basis for I. Hence, $\langle \mathrm{LT}(I) \rangle = \langle x^3 y^2, x^4, xy^3, y^4 \rangle$. As in Chapter 2, §4, we can draw a diagram in $\mathbb{Z}_{\geq 0}^2$ to represent the exponent vectors of the monomials in $\langle \mathrm{LT}(I) \rangle$ and its complement as follows. The vectors

$$\alpha(1) = (3, 2),$$
$$\alpha(2) = (4, 0),$$
$$\alpha(3) = (1, 3),$$
$$\alpha(4) = (0, 4)$$

are the exponent vectors of the generators of $\langle \mathrm{LT}(I) \rangle$. Thus, the elements of

$$((3,2) + \mathbb{Z}_{\geq 0}^2) \cup ((4,0) + \mathbb{Z}_{\geq 0}^2) \cup ((1,3) + \mathbb{Z}_{\geq 0}^2) \cup ((0,4) + \mathbb{Z}_{\geq 0}^2)$$

are the exponent vectors of monomials in $\langle \mathrm{LT}(I) \rangle$. As a result, we can represent the monomials in $\langle \mathrm{LT}(I) \rangle$ by the integer points in the shaded region in $\mathbb{Z}_{\geq 0}^2$ given below:

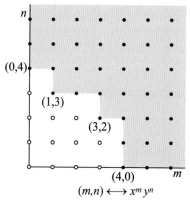

$$(m,n) \longleftrightarrow x^m y^n$$

Given any $f \in \mathbb{R}[x, y]$, Proposition 1 implies that the remainder $\overline{f}^G$ will be a $\mathbb{R}$-linear combination of the 12 monomials $1, x, x^2, x^3, y, xy, x^2y, x^3y, y^2, xy^2, x^2y^2, y^3$ *not* contained in the shaded region. Note that in this case the remainders all belong to a finite-dimensional vector subspace of $\mathbb{R}[x, y]$.

We may also ask what happens if we use a different monomial order in $\mathbb{R}[x, y]$ with the same ideal. If we use lex order instead of grlex, with the variables ordered $y > x$, we find that a Gröbner basis in this case is

$$G = \{y - x^7, x^{12} - x^2\}.$$

Hence, for this monomial order, $\langle \mathrm{LT}(I) \rangle = \langle y, x^{12} \rangle$, and $\langle \mathrm{LT}(I) \rangle$ contains all the monomials with exponent vectors in the shaded region shown below:

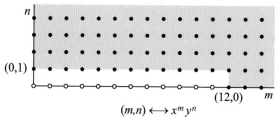

$$(m,n) \longleftrightarrow x^m y^n$$

Thus, for every $f \in \mathbb{R}[x, y]$, we see that $\overline{f}^G \in \mathrm{Span}(1, x, x^2, \ldots, x^{11})$.

Note that $\langle \mathrm{LT}(I) \rangle$ and the remainders can be completely different depending on which monomial order we use. In both cases, however, the possible remainders form the elements of a 12-dimensional vector space. The fact that the dimension is the same in both cases is no accident, as we will soon see. No matter what monomial order we use, for a given ideal I, we will always find the same *number* of monomials in the complement of $\langle \mathrm{LT}(I) \rangle$ (in the case that this number is finite).

Example 3. For the ideal considered in Example 2, there were only finitely many monomials in the complement of $\langle \mathrm{LT}(I) \rangle$. This is actually a very special situation. For instance, consider $I = \langle x - z^2, y - z^3 \rangle \subseteq k[x, y, z]$. Using lex order, the given generators for I already form a Gröbner basis, so that $\langle \mathrm{LT}(I) \rangle = \langle x, y \rangle$. The set of possible remainders modulo I is thus the set of all k-linear combinations of the powers of z. In this case, we recognize I as the ideal of a twisted cubic curve in k^3. As a result of Proposition 1, we see that every polynomial function on the twisted cubic can be uniquely represented by a polynomial in $k[z]$. Hence, the space of possible remainders is not finite-dimensional and $\mathbf{V}(I)$ is a curve. What can you say about $\mathbf{V}(I)$ for the ideal in Example 2?

In any case, we can use Proposition 1 in the following way to describe a portion of the algebraic structure of the quotient ring $k[x_1, \ldots, x_n]/I$.

Proposition 4. *Let $I \subseteq k[x_1, \ldots, x_n]$ be an ideal. Then $k[x_1, \ldots, x_n]/I$ is isomorphic as a k-vector space to $S = \mathrm{Span}(x^\alpha \mid x^\alpha \notin \langle \mathrm{LT}(I) \rangle)$.*

Proof. By Proposition 1, the mapping $\Phi : k[x_1, \ldots, x_n]/I \to S$ defined by $\Phi([f]) = \overline{f}^G$ defines a one-to-one correspondence between the classes in $k[x_1, \ldots, x_n]/I$ and the elements of S. Hence, it remains to check that Φ preserves the vector space operations. Consider the sum operation in $k[x_1, \ldots, x_n]/I$ introduced in §2. If $[f], [g]$ are elements of $k[x_1, \ldots, x_n]/I$, then using Proposition 1, we can "standardize" our polynomial representatives by computing remainders with respect to a Gröbner basis G for I. By Exercise 13 of Chapter 2, §6, we have $\overline{f + g}^G = \overline{f}^G + \overline{g}^G$, so that if

$$\overline{f}^G = \sum_\alpha c_\alpha x^\alpha \quad \text{and} \quad \overline{g}^G = \sum_\alpha d_\alpha x^\alpha,$$

where the sum is over those α with $x^\alpha \notin \langle \mathrm{LT}(I) \rangle$, then

$$(1) \qquad\qquad \overline{f + g}^G = \sum_\alpha (c_\alpha + d_\alpha) x^\alpha.$$

It follows that with the standard representatives, the sum operation in $k[x_1, \ldots, x_n]/I$ is the same as the vector sum in the k-vector space $S = \mathrm{Span}(x^\alpha \mid x^\alpha \notin \langle \mathrm{LT}(I) \rangle)$. Further, if $c \in k$, we leave it as an exercise to prove that $\overline{c \cdot f}^G = c \cdot \overline{f}^G$ (this is an easy consequence of the uniqueness part of Proposition 1). It follows that

$$\overline{c \cdot f}^G = \sum_\alpha c c_\alpha x^\alpha,$$

which shows that multiplication by c in $k[x_1, \ldots, x_n]/I$ is the same as scalar multiplication in S. Thus Φ is linear and hence is a vector space isomorphism.   □

The product operation in $k[x_1, \ldots, x_n]/I$ is slightly less straightforward. The reason for this is clear, however, if we consider an example. Let I be the ideal

$$I = \langle y + x^2 - 1, xy - 2y^2 + 2y \rangle \subseteq \mathbb{R}[x, y].$$

If we compute a Gröbner basis for I using lex order with $x > y$, then we get

(2) $$G = \{x^2 + y - 1, xy - 2y^2 + 2y, y^3 - (7/4)y^2 + (3/4)y\}.$$

Thus, $\langle \mathrm{LT}(I) \rangle = \langle x^2, xy, y^3 \rangle$, and $\{1, x, y, y^2\}$ forms a basis for the vector space of remainders modulo I. Consider the classes of $f = 3y^2 + x$ and $g = x - y$ in $\mathbb{R}[x, y]/I$. The product of $[f]$ and $[g]$ is represented by $f \cdot g = 3xy^2 + x^2 - 3y^3 - xy$. However, this polynomial cannot be the standard representative of the product function because it contains monomials that are in $\langle \mathrm{LT}(I) \rangle$. Hence, we should divide by G, and the remainder $\overline{f \cdot g}^G$ will be the standard representative of the product. We have

$$\overline{3xy^2 + x^2 - 3y^3 - xy}^G = (-11/4)y^2 - (5/4)y + 1,$$

which is in $\mathrm{Span}(1, x, y, y^2)$ as we expect.

The above discussion gives a completely algorithmic way to handle computations in $k[x_1, \ldots, x_n]/I$. To summarize, we have proved the following result.

Proposition 5. *Let I be an ideal in $k[x_1, \ldots, x_n]$ and let G be a Gröbner basis of I with respect to any monomial order. For each $[f] \in k[x_1, \ldots, x_n]/I$, we get the standard representative $\overline{f} = \overline{f}^G$ in $S = \mathrm{Span}(x^\alpha \mid x^\alpha \notin \langle \mathrm{LT}(I) \rangle)$. Then:*
(i) *$[f] + [g]$ is represented by $\overline{f} + \overline{g}$.*
(ii) *$[f] \cdot [g]$ is represented by $\overline{\overline{f} \cdot \overline{g}} \in S$.*

We will conclude this section by using the ideas we have developed to give an algorithmic criterion to determine when a variety in $\mathbb{C}^n$ contains only a finite number of points or, equivalently, to determine when a system of polynomial equations has only a finite number of solutions in $\mathbb{C}^n$. Here is a general version of the result.

Theorem 6 (Finiteness Theorem). *Let $I \subseteq k[x_1, \ldots, x_n]$ be an ideal and fix a monomial ordering on $k[x_1, \ldots, x_n]$. Consider the following five statements:*
(i) *For each i, $1 \leq i \leq n$, there is some $m_i \geq 0$ such that $x_i^{m_i} \in \langle \mathrm{LT}(I) \rangle$.*
(ii) *Let G be a Gröbner basis for I. Then for each i, $1 \leq i \leq n$, there is some $m_i \geq 0$ such that $x_i^{m_i} = \mathrm{LM}(g)$ for some $g \in G$.*
(iii) *The set $\{x^\alpha \mid x^\alpha \notin \langle \mathrm{LT}(I) \rangle\}$ is finite.*
(iv) *The k-vector space $k[x_1, \ldots, x_n]/I$ is finite-dimensional.*
(v) *$\mathbf{V}(I) \subseteq k^n$ is a finite set.*

Then (i)–(iv) are equivalent and they all imply (v). Furthermore, if k is algebraically closed, then (i)–(v) are all equivalent.

Proof. We prove that (i)–(iv) are equivalent by showing (i) $\Leftrightarrow$ (ii), (i) $\Leftrightarrow$ (iii), and (iii) $\Leftrightarrow$ (iv).

(i) $\Leftrightarrow$ (ii) Assume $x_i^{m_i} \in \langle \mathrm{LT}(I) \rangle$. Since G is a Gröbner basis of I, $\langle \mathrm{LT}(I) \rangle = \langle \mathrm{LT}(g) \mid g \in G \rangle$. By Lemma 2 of Chapter 2, §4, there is some $g \in G$, such that $\mathrm{LT}(g)$ divides $x_i^{m_i}$. But this implies that $\mathrm{LM}(g)$ is a power of x_i, as claimed. The opposite implication follows directly from the definition of $\langle \mathrm{LT}(I) \rangle$ since $G \subseteq I$.

(i) $\Leftrightarrow$ (iii) If some power $x_i^{m_i} \in \langle \mathrm{LT}(I) \rangle$ for each i, then the monomials $x_1^{\alpha_1} \cdots x_n^{\alpha_n}$ for which some $\alpha_i \geq m_i$ are all in $\langle \mathrm{LT}(I) \rangle$. The monomials in the complement of $\langle \mathrm{LT}(I) \rangle$ must have $0 \leq \alpha_i \leq m_i - 1$ for each i. As a result, the number of monomials in the complement of $\langle \mathrm{LT}(I) \rangle$ is at most $m_1 \cdot m_2 \cdots m_n$. For the opposite implication, assume that the complement consists of $N < \infty$ monomials. Then, for each i, at least one of the $N + 1$ monomials $1, x_i, x_i^2, \ldots, x_i^N$ must lie in $\langle \mathrm{LT}(I) \rangle$.

(iii) $\Leftrightarrow$ (iv) follows from Proposition 4.

To complete the proof, we will show (iv) $\Rightarrow$ (v) and, assuming k is algebraically closed, (v) $\Rightarrow$ (i).

(iv) $\Rightarrow$ (v) To show that $V = \mathbf{V}(I)$ is finite, it suffices to show that for each i there can be only finitely many distinct i-th coordinates for the points of V. Fix i and consider the classes $[x_i^j]$ in $k[x_1, \ldots, x_n]/I$, where $j = 0, 1, 2, \ldots$. Since $k[x_1, \ldots, x_n]/I$ is finite-dimensional, the $[x_i^j]$ must be linearly dependent in $k[x_1, \ldots, x_n]/I$. Thus, there exist constants c_j (not all zero) and some m such that

$$\sum_{j=0}^{m} c_j [x_i^j] = \left[\sum_{j=0}^{m} c_j x_i^j \right] = [0].$$

However, this implies that $\sum_{j=0}^{m} c_j x_i^j \in I$. Since a nonzero polynomial can have only finitely many roots in k, this shows that the points of V have only finitely many different i-th coordinates.

(v) $\Rightarrow$ (i) Assume k is algebraically closed and $V = \mathbf{V}(I)$ is finite. If $V = \emptyset$, then $1 \in I$ by the Weak Nullstellensatz. In this case, we can take $m_i = 0$ for all i. If V is nonempty, then for a fixed i, let $a_1, \ldots, a_m \in k$ be the distinct i-th coordinates of points in V. Form the one-variable polynomial

$$f(x_i) = \prod_{j=1}^{m} (x_i - a_j).$$

By construction, f vanishes at every point in V, so $f \in \mathbf{I}(V)$. By the Nullstellensatz, there is some $N \geq 1$ such that $f^N \in I$. But this says that the leading monomial of f^N is in $\langle \mathrm{LT}(I) \rangle$. Examining our expression for f, we see that $x_i^{mN} \in \langle \mathrm{LT}(I) \rangle$. $\qquad\square$

Over an algebraically closed field k, Theorem 6 shows how we can characterize "zero-dimensional" varieties (varieties containing only finitely many points) using the properties of $k[x_1, \ldots, x_n]/I$. In Chapter 9, we will take up the question of assigning a *dimension* to a general variety, and some of the ideas introduced in Theorem 6 will be useful. The same holds for Example 3.

A judicious choice of monomial ordering can sometimes lead to a very easy determination that a variety is finite. For example, consider the ideal

$$I = \langle x^5 + y^3 + z^2 - 1, x^2 + y^3 + z - 1, x^4 + y^5 + z^6 - 1 \rangle.$$

Using grlex, we see that $x^5, y^3, z^6 \in \langle \mathrm{LT}(I) \rangle$ since those are the leading monomials of the three generators. By (i) $\Rightarrow$ (v) of Theorem 6, we know that $\mathbf{V}(I)$ is finite (even without computing a Gröbner basis). If we actually wanted to determine which points were in $\mathbf{V}(I)$, we would need to do elimination, for instance, by computing a lexicographic Gröbner basis. We will say more about finding solutions below.

We also have a quantitative estimate for the number of solutions when the criteria of Theorem 6 are satisfied.

Proposition 7. *Let $I \subseteq k[x_1, \ldots, x_n]$ be an ideal such that for each i, some power $x_i^{m_i} \in \langle \mathrm{LT}(I) \rangle$, and set $V = \mathbf{V}(I)$. Then:*

(i) *The number of points of V is at most $\dim k[x_1, \ldots, x_n]/I$ (where "dim" means dimension as a vector space over k).*

(ii) *The number of points of V is at most $m_1 \cdot m_2 \cdots m_n$.*

(iii) *If I is radical and k is algebraically closed, then equality holds in part (i), i.e., the number of points in V is exactly $\dim k[x_1, \ldots, x_n]/I$.*

Proof. We first show that given distinct points $p_1, \ldots, p_m \in k^n$, there is a polynomial $f_1 \in k[x_1, \ldots, x_n]$ with $f_1(p_1) = 1$ and $f_1(p_2) = \cdots = f_1(p_m) = 0$. To prove this, note that if $a \neq b \in k^n$, then they must differ at some coordinate, say the j-th, and it follows that $g = (x_j - b_j)/(a_j - b_j)$ satisfies $g(a) = 1, g(b) = 0$. If we apply this observation to each pair $p_1 \neq p_i, i \geq 2$, we get polynomials g_i such that $g_i(p_1) = 1$ and $g_i(p_i) = 0$ for $i \geq 2$. Then $f_1 = g_2 \cdot g_3 \cdots g_m$ has the desired property. In this argument, there is nothing special about p_1. If we apply the same argument with p_1 replaced by each of $p_1, \ldots, p_m$ in turn, we get polynomials $f_1, \ldots, f_m$ such that $f_i(p_i) = 1$ and $f_i(p_j) = 0$ for $i \neq j$.

Now we can prove the proposition. By Theorem 6, we know that V is finite. Write $V = \{p_1, \ldots, p_m\}$, where the p_i are distinct.

(i) Let $f_1, \ldots, f_m$ be as above. If we can prove that $[f_1], \ldots, [f_m] \in k[x_1, \ldots, x_n]/I$ are linearly independent, then

$$(3) \qquad\qquad m \leq \dim k[x_1, \ldots, x_n]/I$$

will follow, proving (i).

To prove linear independence, suppose that $\sum_{i=1}^{m} a_i[f_i] = [0]$ in $k[x_1, \ldots, x_n]/I$, where $a_i \in k$. Back in $k[x_1, \ldots, x_n]$, this means that $g = \sum_{i=1}^{m} a_i f_i \in I$, so that g vanishes at all points of $V = \{p_1, \ldots, p_m\}$. Then, for $1 \leq j \leq m$, we have

$$0 = g(p_j) = \sum_{i=1}^{m} a_i f_i(p_j) = 0 + a_j f_j(p_j) = a_j,$$

and linear independence follows.

(ii) Proposition 4 implies that dim $k[x_1, \ldots, x_n]/I$ is the number of monomials in the set $\{x^\alpha \mid x^\alpha \notin \langle \mathrm{LT}(I) \rangle\}$. The proof of (i) $\Leftrightarrow$ (iii) from Theorem 6 shows that this number is at most $m_1 \cdots m_n$. Then the desired bound follows from (i).

(iii) Now assume that k is algebraically closed and I is a radical ideal. To prove that equality holds in (3), it suffices to show that $[f_1], \ldots, [f_m]$ form a basis of $k[x_1, \ldots, x_n]/I$. Since we just proved linear independence, we only need to show that they span. Thus, let $[g] \in k[x_1, \ldots, x_n]/I$ be arbitrary, and set $a_i = g(p_i)$. Then consider $h = g - \sum_{i=1}^m a_i f_i$. One easily computes $h(p_j) = 0$ for all j, so that $h \in \mathbf{I}(V)$. By the Nullstellensatz, $\mathbf{I}(V) = \mathbf{I}(\mathbf{V}(I)) = \sqrt{I}$ since k is algebraically closed, and since I is radical, we conclude that $h \in I$. Thus $[h] = [0]$ in $k[x_1, \ldots, x_n]/I$, which implies $[g] = \sum_{i=1}^m a_i[f_i]$. The proposition is now proved. $\square$

For an example of Proposition 7, consider the ideal $I = \langle xy^3 - x^2, x^3 y^2 - y \rangle$ in $\mathbb{R}[x, y]$ from Example 2 at the beginning of this section. Using grlex, we found $x^4, y^4 \in \langle \mathrm{LT}(I) \rangle$, so that $\mathbf{V}(I) \subseteq \mathbb{R}^2$ has at most $4 \cdot 4 = 16$ points by part (ii) of Proposition 7. Yet Example 2 also shows that $\mathbb{R}[x, y]/I$ has dimension 12 over $\mathbb{R}$. Thus part (i) of the proposition gives the better bound of 12. Note that the bound of 12 also holds if we work over $\mathbb{C}$.

In Example 2, we discovered that switching to lex order with $y > x$ gives the Gröbner basis $G = \{y - x^7, x^{12} - x^2\}$ for I. Then part (ii) of Proposition 7 gives the bound of $1 \cdot 12 = 12$, which agrees with the bound given by part (i). By solving the equations from G, we see that $\mathbf{V}(I)$ actually contains only 11 distinct points in this case:

$$(4) \qquad \mathbf{V}(I) = \{(0,0)\} \cup \{(\zeta, \zeta^7) \mid \zeta^{10} = 1\}.$$

(Recall that there are 10 distinct 10-th roots of unity in $\mathbb{C}$.)

For any ideal I, we have $\mathbf{V}(I) = \mathbf{V}(\sqrt{I})$. Thus, when $\mathbf{V}(I)$ is finite, Proposition 7 shows how to find the exact number of solutions over an algebraically closed field, provided we know $\sqrt{I}$. Computing radicals was discussed in §2 of Chapter 4, though when I satisfies the conditions of Theorem 6, $\sqrt{I}$ is simple to compute. See Proposition (2.7) of COX, LITTLE and O'SHEA (2005) for the details.

Thus far in this section, we have studied the *finiteness of solutions* in Theorem 6 and the *number of solutions* in Proposition 7. We now combine this with earlier results to describe a method for *finding the solutions*. For simplicity, we work over the complex numbers $\mathbb{C}$. Consider a system of equations

$$(5) \qquad f_1 = \cdots = f_s = 0$$

whose ideal $I \subseteq \mathbb{C}[x_1, \ldots, x_n]$ satisfies the conditions of Theorem 6. There are no solutions when $I = \mathbb{C}[x_1, \ldots, x_n]$, so we will assume that I is a proper ideal.

Let G be a reduced Gröbner basis of I for lex order with $x_1 > \cdots > x_n$. By assumption, for each i, there is $g \in G$ such that $\mathrm{LT}(g) = x_i^{m_i}$. Note that $g \in \mathbb{C}[x_i, \ldots, x_n]$ since we are using lex order. By Theorem 6, it follows that (5) has only finitely many solutions. We will find the solutions by working one variable at a time, using the "back-substitution" method discussed earlier in the book.

We start with $G \cap \mathbb{C}[x_n]$, which is nonempty by the previous paragraph. Since G is reduced, this intersection consists of a single polynomial, say $G \cap \mathbb{C}[x_n] = \{g\}$. The first step in solving (5) is to find the solutions of $g(x_n) = 0$, i.e., the roots of g. These exist since $\mathbb{C}$ is algebraically closed. This gives partial solutions of (5).

Now we work backwards. Suppose that $\mathbf{a} = (a_{i+1}, \ldots, a_n)$ is a partial solution of (5). This means that $\mathbf{a} \in \mathbf{V}(I_i)$, $I_i = I \cap \mathbb{C}[x_{i+1}, \ldots, x_n]$. The goal to find all $a_i \in \mathbb{C}$ such that $(a_i, \mathbf{a}) = (a_i, a_{i+1}, \ldots, a_n)$ is a partial solution in $\mathbf{V}(I_{i-1})$ for $I_{i-1} = I \cap \mathbb{C}[x_i, \ldots, x_n]$. By the Elimination Theorem, $G_{i-1} = G \cap \mathbb{C}[x_i, \ldots, x_n]$ is a basis of I_{i-1}. Since G_{i-1} contains a polynomial with $x_i^{m_i}$ as leading term, the Extension Theorem (specifically, Corollary 4 of Chapter 3, §1) implies that the partial solution $\mathbf{a}$ extends to $(a_i, \mathbf{a}) = (a_i, a_{i+1}, \ldots, a_n) \in \mathbf{V}(I_{i-1})$.

We can find the possible a_i's as follows. Let $g_1, \ldots, g_\ell$ be the polynomials G_{i-1} in which x_i actually appears. When we evaluate G_{i-1} at $\mathbf{a}$, any polynomial lying in $\mathbb{C}[x_{i+1}, \ldots, x_n]$ will vanish identically, leaving us with the equations

(6) $$g_1(x_i, \mathbf{a}) = \cdots = g_\ell(x_i, \mathbf{a}) = 0.$$

The desired a_i's are the solutions of this system of equations.

Now comes a key observation: there is only *one* polynomial in (6) to worry about. To find this polynomial, write each g_j in the form

$$g_j = c_j(x_{i+1}, \ldots, x_n)x_i^{N_j} + \text{terms in which } x_i \text{ has degree} < N_j,$$

where $N_j \geq 0$ and $c_j \in k[x_{i+1}, \ldots, x_n]$ is nonzero. Among all g_j for which $c_j(\mathbf{a}) \neq 0$, choose g_* with minimal degree in x_i. Then Theorem 2 of Chapter 3, §5 implies that the system (6) is equivalent to the *single* equation

$$g_*(x_i, \mathbf{a}) = 0.$$

Thus the roots of $g_*(x_i, \mathbf{a})$ give all possible ways of extending $\mathbf{a}$ to a partial solution in $\mathbf{V}(I_{i-1})$. Continuing in this way, we get all solutions of the original system (5).

It is very satisfying to see how our theory gives a complete answer to solving equations in the zero-dimensional case. In practice, things are more complicated because of complexity issues related to computing Gröbner bases and numerical issues related to finding roots of univariate polynomials.

EXERCISES FOR §3

1. Complete the proof of part (ii) of Proposition 1.
2. In Proposition 5, we stated a method for computing $[f] \cdot [g]$ in $k[x_1, \ldots, x_n]/I$. Could we simply compute $\overline{f \cdot g}^G$ rather than first computing the remainders of f and g separately?
3. Let $I = \langle x^4 y - z^6, x^2 - y^3 z, x^3 z^2 - y^3 \rangle$ in $k[x, y, z]$.
   a. Using lex order, find a Gröbner basis G of I and a collection of monomials that spans the space of remainders modulo G.
   b. Repeat part (a) for grlex order. How do your sets of monomials compare?
4. Use the division algorithm and the uniqueness part of Proposition 1 to prove that $\overline{c \cdot f}^G = c \cdot \overline{f}^G$ whenever $f \in k[x_1, \ldots, x_n]$ and $c \in k$.

5. Let $I = \langle y + x^2 - 1, xy - 2y^2 + 2y \rangle \subseteq \mathbb{R}[x, y]$. This is the ideal used in the example following Proposition 4.

   a. Construct a vector space isomorphism $\mathbb{R}[x, y]/I \cong \mathbb{R}^4$.

   b. Using the lexicographic Gröbner basis given in (2), compute a "multiplication table" for the elements $\{[1], [x], [y], [y^2]\}$ in $\mathbb{R}[x, y]/I$. Hint: Express each product as a linear combination of these four classes.

   c. Is $\mathbb{R}[x, y]/I$ a field? Why or why not?

   d. Compute $\mathbf{V}(I)$. Hint: It has four points.

   e. The lex Gröbner basis computed in (2) shows that $m_1 = 2$ and $m_2 = 3$ are the smallest powers of x and y, respectively, contained in $\langle \mathrm{LT}(I) \rangle$. What bound does part (ii) of Proposition 7 give for the number of points in V? Does part (i) of the proposition give a better bound?

6. Let $V = \mathbf{V}(x_3 - x_1^2, x_4 - x_1 x_2, x_2 x_4 - x_1 x_5, x_4^2 - x_3 x_5) \subseteq \mathbb{C}^5$.

   a. Using any convenient monomial order, determine a collection of monomials spanning the space of remainders modulo a Gröbner basis for the ideal generated by the defining equations of V.

   b. For which i is there some $m_i \geq 0$ such that $x_i^{m_i} \in \langle \mathrm{LT}(I) \rangle$?

   c. Is V a finite set? Why or why not?

7. Let I be any ideal in $k[x_1, \ldots, x_n]$.

   a. Suppose that the set $\{x^\alpha \mid x^\alpha \notin \langle \mathrm{LT}(I) \rangle\}$ is finite with d elements for some choice of monomial order. Show that the dimension of $k[x_1, \ldots, x_n]/I$ as a k-vector space is equal to d.

   b. Deduce from part (a) that the number of monomials in the complement of $\langle \mathrm{LT}(I) \rangle$ is independent of the choice of the monomial order, when that number is finite.

8. Suppose that $I \subseteq \mathbb{C}[x_1, \ldots, x_n]$ is a radical ideal with a Gröbner basis $\{g_1, \ldots, g_n\}$ such that $\mathrm{LT}(g_i) = x_i^{m_i}$ for each i. Prove that $\mathbf{V}(I)$ contains *exactly* $m_1 \cdot m_2 \cdots m_n$ points.

9. Most computer algebra systems contain routines for simplifying radical expressions. For example, instead of writing

$$r = \frac{1}{x + \sqrt{2} + \sqrt{3}},$$

most systems would allow you to rationalize the denominator and rewrite r as a quotient of polynomials in x, where $\sqrt{2}$ and $\sqrt{3}$ appear in the coefficients only in the numerator. The idea behind one method used here is as follows.

   a. Explain why r can be seen as a rational function in x, whose coefficients are elements of the quotient ring $R = \mathbb{Q}[y_1, y_2]/\langle y_1^2 - 2, y_2^2 - 3 \rangle$. Hint: See Exercise 4 from §2 of this chapter.

   b. Compute a Gröbner basis G for $I = \langle y_1^2 - 2, y_2^2 - 3 \rangle$ and construct a multiplication table for the classes of the monomials spanning the possible remainders modulo G (which should be $\{[1], [y_1], [y_2], [y_1 y_2]\}$).

   c. Now, to rationalize the denominator of r, we can try to solve the following equation

$$(x[1] + [y_1] + [y_2]) \cdot (a_0[1] + a_1[y_1] + a_2[y_2] + a_3[y_1 y_2]) = [1],$$

where a_0, a_1, a_2, a_3 are rational functions of x with rational number coefficients. Multiply out the above equation using your table from part (b), match coefficients, and solve the resulting linear equations for a_0, a_1, a_2, a_3. Then

$$a_0[1] + a_1[y_1] + a_2[y_2] + a_3[y_1 y_2]$$

gives the rationalized expression for r.

10. In this problem, we will establish a fact about the number of monomials of total degree less than or equal to d in $k[x_1, \ldots, x_n]$ and relate this to the intuitive notion of the dimension of the variety $V = k^n$.

    a. Explain why every monomial in $k[x_1, \ldots, x_n]$ is in the complement of $\langle \mathrm{LT}(\, \mathbf{I}(V))\rangle$ for $V = k^n$.

    b. Show that for all $d, n \geq 0$, the number of distinct monomials of total degree less than or equal to d in $k[x_1, \ldots, x_n]$ is the binomial coefficient $\binom{n+d}{n}$. (This generalizes part (a) of Exercise 5 in Chapter 2, §1.)

    c. When n is fixed, explain why this number of monomials grows like d^n as $d \to \infty$. Note that the *exponent* n is the same as the intuitive dimension of the variety $V = k^n$, for which $k[V] = k[x_1, \ldots, x_n]$.

11. In this problem, we will compare what happens with the monomials not in $\langle \mathrm{LT}(I)\rangle$ in two examples where $\mathbf{V}(I)$ is not finite, and one where $\mathbf{V}(I)$ is finite.

    a. Consider the variety $\mathbf{V}(I) \subseteq \mathbb{C}^3$, where $I = \langle x^2 + y, x - y^2 + z^2, xy - z\rangle$. Compute a Gröbner basis for I using lex order, and, for $1 \leq d \leq 10$, tabulate the number of monomials of total degree $\leq d$ that are not in $\langle \mathrm{LT}(I)\rangle$. Note that by Theorem 6, $\mathbf{V}(I)$ is a finite subset of $\mathbb{C}^3$. Hint: It may be helpful to try to visualize or sketch a 3-dimensional analogue of the diagrams in Example 2 for this ideal.

    b. Repeat the calculations of part (a) for $J = \langle x^2 + y, x - y^2 + z^2\rangle$. Here, $\mathbf{V}(J)$ is not finite. How does the behavior of the number of monomials of total degree $\leq d$ in the complement of $\langle \mathrm{LT}(J)\rangle$ (as a function of d) differ from the behavior in part (a)?

    c. Let $H(d)$ be the number of monomials of total degree $\leq d$ in the complement of $\langle \mathrm{LT}(J)\rangle$. Can you guess a power ℓ such that $H(d)$ will grow roughly like d^ℓ as d grows?

    d. Now repeat parts (b) and (c) for the ideal $K = \langle x^2 + y\rangle$.

    e. Using the intuitive notion of the dimension of a variety that we developed in Chapter 1, can you see a pattern here? We will return to these questions in Chapter 9.

12. Let k be any field, and suppose $I \subseteq k[x_1, \ldots, x_n]$ has the property that $k[x_1, \ldots, x_n]/I$ is a finite-dimensional vector space over k.

    a. Prove that $\dim k[x_1, \ldots, x_n]/\sqrt{I} \leq \dim k[x_1, \ldots, x_n]/I$. Hint: Show that $I \subseteq \sqrt{I}$ induces a map of quotient rings $k[x_1, \ldots, x_n]/I \to k[x_1, \ldots, x_n]/\sqrt{I}$ which is onto.

    b. Show that the number of points in $\mathbf{V}(I)$ is at most $\dim k[x_1, \ldots, x_n]/\sqrt{I}$.

    c. Give an example to show that equality need not hold in part (b) when k is not algebraically closed.

    d. Assume that k is algebraically closed. Strengthen part (iii) of Proposition 7 by showing that I is radical if and only if $\dim k[x_1, \ldots, x_n]/I$ equals the number of points in $\mathbf{V}(I)$. Hint: Use part (a).

13. A polynomial in $k[x_1, \ldots, x_n]$ of the form $x^\alpha - x^\beta$ is called a *binomial*, and an ideal generated by binomials is a *binomial ideal*. Let $I \subseteq \mathbb{C}[x_1, \ldots, x_n]$ be a binomial ideal.

    a. Prove that I has a reduced lex Gröbner basis consisting of binomials. Hint: What is the S-polynomial of two binomials?

    b. Assume that $\mathbf{V}(I) \subseteq \mathbb{C}^n$ is finite. Prove that every coordinate of a solution is either zero or a root of unity. Hint: Follow the solution procedure described at the end of the section. Note that a root of a root of unity is again a root of unity.

    c. Explain how part (b) relates to equation (4) in the text.

§4 The Coordinate Ring of an Affine Variety

In this section, we will apply the algebraic tools developed in §§2 and 3 to study the ring $k[V]$ of polynomial functions on an affine variety $V \subseteq k^n$. Using the isomorphism $k[V] \cong k[x_1, \ldots, x_n]/\mathbf{I}(V)$ from §2, we will frequently identify $k[V]$ with the quotient ring $k[x_1, \ldots, x_n]/\mathbf{I}(V)$. Thus, given a polynomial $f \in k[x_1, \ldots, x_n]$, we let $[f]$ denote the polynomial function in $k[V]$ represented by f.

In particular, each variable x_i gives a polynomial function $[x_i] : V \to k$ whose value at a point $p \in V$ is the i-th coordinate of p. We call $[x_i] \in k[V]$ the i-th *coordinate function* on V. Then the isomorphism $k[V] \cong k[x_1, \ldots, x_n]/\mathbf{I}(V)$ shows that the coordinate functions generate $k[V]$ in the sense that any polynomial function on V is a k-linear combination of products of the $[x_i]$. This explains the following terminology.

Definition 1. The **coordinate ring** of an affine variety $V \subseteq k^n$ is the ring $k[V]$.

Many results from previous sections of this chapter can be rephrased in terms of the coordinate ring. For example:

- Proposition 4 from §1: A variety is irreducible if and only if its coordinate ring is an integral domain.
- Theorem 6 from §3: Over an algebraically closed field k, a variety is finite if and only if its coordinate ring is finite-dimensional as a k-vector space.

In the "algebra–geometry" dictionary of Chapter 4, we related varieties in k^n to ideals in $k[x_1, \ldots, x_n]$. One theme of Chapter 5 is that this dictionary still works if we replace k^n and $k[x_1, \ldots, x_n]$ by a general variety V and its coordinate ring $k[V]$. For this purpose, we introduce the following definitions.

Definition 2. Let $V \subseteq k^n$ be an affine variety.
(i) For any ideal $J = \langle \phi_1, \ldots, \phi_s \rangle \subseteq k[V]$, we define
$$\mathbf{V}_V(J) = \{(a_1, \ldots, a_n) \in V \mid \phi(a_1, \ldots, a_n) = 0 \text{ for all } \phi \in J\}.$$

We call $\mathbf{V}_V(J)$ a **subvariety** of V.
(ii) For each subset $W \subseteq V$, we define
$$\mathbf{I}_V(W) = \{\phi \in k[V] \mid \phi(a_1, \ldots, a_n) = 0 \text{ for all } (a_1, \ldots, a_n) \in W\}.$$

For instance, let $V = \mathbf{V}(z - x^2 - y^2) \subseteq \mathbb{R}^3$. If we take $J = \langle [x] \rangle \subseteq \mathbb{R}[V]$, then
$$W = \mathbf{V}_V(J) = \langle (0, a, a^2) \mid a \in \mathbb{R} \rangle \subseteq V$$

is a subvariety of V. Note that this is the same as $\mathbf{V}(z - x^2 - y^2, x)$ in $\mathbb{R}^3$. Similarly, if we let $W = \{(1, 1, 2)\} \subseteq V$, then we leave it as an exercise to show that
$$\mathbf{I}_V(W) = \langle [x - 1], [y - 1] \rangle.$$

Given a fixed affine variety V, we can use $\mathbf{I}_V$ and $\mathbf{V}_V$ to relate subvarieties of V to ideals in $k[V]$. The first result we get is the following.

Proposition 3. *Let $V \subseteq k^n$ be an affine variety.*
(i) *For each ideal $J \subseteq k[V]$, $W = \mathbf{V}_V(J)$ is an affine variety in k^n contained in V.*
(ii) *For each subset $W \subseteq V$, $\mathbf{I}_V(W)$ is an ideal of $k[V]$.*
(iii) *If $J \subseteq k[V]$ is an ideal, then $J \subseteq \sqrt{J} \subseteq \mathbf{I}_V(\mathbf{V}_V(J))$.*
(iv) *If $W \subseteq V$ is a subvariety, then $W = \mathbf{V}_V(\mathbf{I}_V(W))$.*

Proof. To prove (i), we will use the one-to-one correspondence of Proposition 10 of §2 between the ideals of $k[V]$ and the ideals in $k[x_1, \ldots, x_n]$ containing $\mathbf{I}(V)$. Let $\tilde{J} = \{f \in k[x_1, \ldots, x_n] \mid [f] \in J\} \subseteq k[x_1, \ldots, x_n]$ be the ideal corresponding to $J \subseteq k[V]$. Then $\mathbf{V}(\tilde{J}) \subseteq V$, since $\mathbf{I}(V) \subseteq \tilde{J}$. But we also have $\mathbf{V}(\tilde{J}) = \mathbf{V}_V(J)$ by definition since the elements of $\tilde{J}$ represent the functions in J on V. Thus, W (considered as a subset of k^n) is an affine variety in its own right.

The proofs of (ii), (iii), and (iv) are similar to arguments given in earlier chapters and the details are left as an exercise. Note that the definition of the radical of an ideal is the same in $k[V]$ as it is in $k[x_1, \ldots, x_n]$. □

We can also show that the radical ideals in $k[V]$ correspond to the radical ideals in $k[x_1, \ldots, x_n]$ containing $\mathbf{I}(V)$.

Proposition 4. *An ideal $J \subseteq k[V]$ is radical if and only if the corresponding ideal $\tilde{J} = \{f \in k[x_1, \ldots, x_n] \mid [f] \in J\} \subseteq k[x_1, \ldots, x_n]$ is radical.*

Proof. Assume J is radical, and let $f \in k[x_1, \ldots, x_n]$ satisfy $f^m \in \tilde{J}$ for some $m \geq 1$. Then $[f^m] = [f]^m \in J$. Since J is a radical ideal, this implies that $[f] \in J$. Hence, $f \in \tilde{J}$, so $\tilde{J}$ is also a radical ideal. Conversely, if $\tilde{J}$ is radical and $[f]^m \in J$, then $[f^m] \in J$, so $f^m \in \tilde{J}$. Since $\tilde{J}$ is radical, this shows that $f \in \tilde{J}$. Hence, $[f] \in J$ and J is radical. □

Rather than discuss the complete "ideal–variety" correspondence (as we did in Chapter 4), we will confine ourselves to the following result which highlights some of the important properties of the correspondence.

Theorem 5. *Let k be an algebraically closed field and let $V \subseteq k^n$ be an affine variety.*

(i) (**The Nullstellensatz in $k[V]$**) *If J is any ideal in $k[V]$, then*

$$\mathbf{I}_V(\mathbf{V}_V(J)) = \sqrt{J} = \{[f] \in k[V] \mid [f]^m \in J\}.$$

(ii) *The correspondences*

$$\left\{ \begin{array}{c} \text{affine subvarieties} \\ W \subseteq V \end{array} \right\} \underset{\mathbf{V}_V}{\overset{\mathbf{I}_V}{\rightleftarrows}} \left\{ \begin{array}{c} \text{radical ideals} \\ J \subseteq k[V] \end{array} \right\}$$

are inclusion-reversing bijections and are inverses of each other.

(iii) *Under the correspondence given in (ii), points of V correspond to maximal ideals of $k[V]$.*

Proof. (i) Let J be an ideal of $k[V]$. By the correspondence of Proposition 10 of §2, J corresponds to the ideal $\tilde{J} \subseteq k[x_1, \ldots, x_n]$ as in the proof of Proposition 4, where $\mathbf{V}(\tilde{J}) = \mathbf{V}_V(J)$. As a result, if $[f] \in \mathbf{I}_V(\mathbf{V}_V(J))$, then $f \in \mathbf{I}(\mathbf{V}(\tilde{J}))$. By the Nullstellensatz in k^n, $\mathbf{I}(\mathbf{V}(\tilde{J})) = \sqrt{\tilde{J}}$, so $f^m \in \tilde{J}$ for some $m \geq 1$. But then, $[f^m] = [f]^m \in J$, so $[f] \in \sqrt{J}$ in $k[V]$. We have shown that $\mathbf{I}_V(\mathbf{V}_V(J)) \subseteq \sqrt{J}$. Since the opposite inclusion holds for any ideal, our Nullstellensatz in $k[V]$ is proved.

(ii) follows from (i) as in Chapter 4.

(iii) is proved in the same way as Theorem 11 of Chapter 4, §5.   □

When k is algebraically closed, the Weak Nullstellensatz also holds in $k[V]$. You will prove this in Exercise 16.

Next, we return to the general topic of a *classification* of varieties that we posed in §1. What should it mean for two affine varieties to be "isomorphic"? One reasonable answer is given in the following definition.

Definition 6. Let $V \subseteq k^m$ and $W \subseteq k^n$ be affine varieties. We say that V and W are **isomorphic** if there exist polynomial mappings $\alpha : V \to W$ and $\beta : W \to V$ such that $\alpha \circ \beta = \mathrm{id}_W$ and $\beta \circ \alpha = \mathrm{id}_V$. (For any variety V, we write id_V for the identity mapping from V to itself. This is always a polynomial mapping.)

Intuitively, varieties that are isomorphic should share properties such as irreducibility, dimension, etc. In addition, subvarieties of V should correspond to subvarieties of W, and so forth. For instance, saying that a variety $W \subseteq k^n$ is isomorphic to $V = k^m$ implies that there is a one-to-one and onto polynomial mapping $\alpha : k^m \to W$ with a polynomial inverse. Thus, we have a polynomial *parametrization* of W with especially nice properties! Here is an example, inspired by a technique used in geometric modeling, which illustrates the usefulness of this idea.

Example 7. Let us consider the two surfaces

$$Q_1 = \mathbf{V}(x^2 - xy - y^2 + z^2) = \mathbf{V}(f_1),$$
$$Q_2 = \mathbf{V}(x^2 - y^2 + z^2 - z) = \mathbf{V}(f_2)$$

in $\mathbb{R}^3$. (These might be boundary surfaces of a solid region in a shape we were designing, for example.) To study the *intersection curve* $C = \mathbf{V}(f_1, f_2)$ of the two surfaces, we could proceed as follows. Neither Q_1 nor Q_2 is an especially simple surface, so the intersection curve is fairly difficult to visualize directly. However, as usual, we are *not limited* to using the particular equations f_1, f_2 to define the curve! It is easy to check that $C = \mathbf{V}(f_1, f_1 + cf_2)$, where $c \in \mathbb{R}$ is any nonzero real number. Hence, the surfaces $F_c = \mathbf{V}(f_1 + cf_2)$ also contain C. These surfaces, together with Q_2, are often called the elements of the *pencil* of surfaces determined by Q_1 and Q_2. (A pencil of varieties is a one-parameter family of varieties, parametrized by the points of k. In the above case, the parameter is $c \in \mathbb{R}$.)

If we can find a value of c making the surface F_c particularly simple, then understanding the curve C will be correspondingly easier. Here, if we take $c = -1$, then F_{-1} is defined by

$$0 = f_1 - f_2$$
$$= z - xy.$$

The surface $Q = F_{-1} = \mathbf{V}(z - xy)$ is much easier to understand because it is *isomorphic as a variety* to $\mathbb{R}^2$ [as is the graph of any polynomial function $f(x, y)$]. To see this, note that we have polynomial mappings:

$$\alpha : \mathbb{R}^2 \longrightarrow Q,$$
$$(x, y) \longmapsto (x, y, xy),$$
$$\pi : Q \longrightarrow \mathbb{R}^2,$$
$$(x, y, z) \longmapsto (x, y),$$

which satisfy $\alpha \circ \pi = \mathrm{id}_Q$ and $\pi \circ \alpha = \mathrm{id}_{\mathbb{R}^2}$.

Hence, curves on Q can be reduced to plane curves in the following way. To study C, we can project to the curve $\pi(C) \subset \mathbb{R}^2$, and we obtain the equation

$$x^2 y^2 + x^2 - xy - y^2 = 0$$

for $\pi(C)$ by substituting $z = xy$ in either f_1 or f_2. Note that π and α restrict to give isomorphisms between C and $\pi(C)$, so we have not really lost anything by projecting in this case.

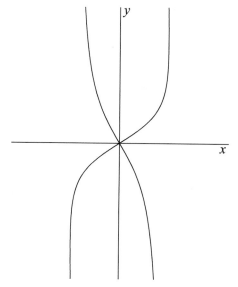

In particular, each point (a, b) on $\pi(C)$ corresponds to exactly one point (a, b, ab) on C. In the exercises, you will show that $\pi(C)$ can also be parametrized as

(1)
$$x = \frac{-t^2 + t + 1}{t^2 + 1},$$
$$y = \frac{-t^2 + t + 1}{t(t + 2)}.$$

From this we can also obtain a parametrization of C via the mapping α.

Given the above example, it is natural to ask how we can tell whether two varieties are isomorphic. One way is to consider the relation between their coordinate rings

$$k[V] \cong k[x_1,\ldots,x_m]/\mathbf{I}(V) \quad \text{and} \quad k[W] \cong k[y_1,\ldots,y_n]/\mathbf{I}(W).$$

The fundamental observation is that if we have a polynomial mapping $\alpha : V \to W$, then every polynomial function $\phi : W \to k$ in $k[W]$ gives us another polynomial function $\phi \circ \alpha : V \to k$ in $k[V]$. This will give us a map from $k[W]$ to $k[V]$ with the following properties.

Proposition 8. *Let V and W be varieties (possibly in different affine spaces).*

(i) *Let $\alpha : V \to W$ be a polynomial mapping. Then for every polynomial function $\phi : W \to k$, the composition $\phi \circ \alpha : V \to k$ is also a polynomial function. Furthermore, the map $\alpha^* : k[W] \to k[V]$ defined by $\alpha^*(\phi) = \phi \circ \alpha$ is a ring homomorphism which is the identity on the constant functions $k \subseteq k[W]$. (Note that α^* "goes in the opposite direction" from α since α^* maps functions on W to functions on V. For this reason we call α^* the **pullback mapping** on functions.)*
(ii) *Conversely, let $\Phi : k[W] \to k[V]$ be a ring homomorphism which is the identity on constants. Then there is a unique polynomial mapping $\alpha : V \to W$ such that $\Phi = \alpha^*$.*

Proof. (i) Suppose that $V \subseteq k^m$ has coordinates $x_1,\ldots,x_m$ and $W \subseteq k^n$ has coordinates $y_1,\ldots,y_n$. Then $\phi : W \to k$ can be represented by a polynomial $f(y_1,\ldots,y_n)$, and $\alpha : V \to W$ can be represented by an n-tuple of polynomials:

$$\alpha(x_1,\ldots,x_m) = (h_1(x_1,\ldots,x_m),\ldots,h_n(x_1,\ldots,x_m)).$$

We compute $\phi \circ \alpha$ by substituting $\alpha(x_1,\ldots,x_m)$ into ϕ. Thus,

$$(\phi \circ \alpha)(x_1,\ldots,x_m) = f(h_1(x_1,\ldots,x_m),\ldots,h_n(x_1,\ldots,x_m)),$$

which is a polynomial in $x_1,\ldots,x_m$. Hence, $\phi \circ \alpha$ is a polynomial function on V.

It follows that we can define $\alpha^* : k[W] \to k[V]$ by the formula $\alpha^*(\phi) = \phi \circ \alpha$. To show that α^* is a ring homomorphism, let ψ be another element of $k[W]$, represented by a polynomial $g(y_1,\ldots,y_n)$. Then

$$\begin{aligned}(\alpha^*(\phi+\psi))(x_1,\ldots,x_m) = {}& f(h_1(x_1,\ldots,x_m),\ldots,h_n(x_1,\ldots,x_m)) + \\ & g(h_1(x_1,\ldots,x_m),\ldots,h_n(x_1,\ldots,x_m)) \\ = {}& \alpha^*(\phi)(x_1,\ldots,x_m) + \alpha^*(\psi)(x_1,\ldots,x_m).\end{aligned}$$

Hence, $\alpha^*(\phi+\psi) = \alpha^*(\phi) + \alpha^*(\psi)$, and $\alpha^*(\phi \cdot \psi) = \alpha^*(\phi) \cdot \alpha^*(\psi)$ is proved similarly. Thus, α^* is a ring homomorphism.

Finally, consider $[a] \in k[W]$ for some $a \in k$. Then $[a]$ is a constant function on W with value a, and it follows that $\alpha^*([a]) = [a] \circ \alpha$ is constant on V, again with value a. Thus, $\alpha^*([a]) = [a]$, so that α^* is the identity on constants.

(ii) Now let $\Phi : k[W] \to k[V]$ be a ring homomorphism which is the identity on the constants. We need to show that Φ comes from some polynomial mapping $\alpha : V \to W$. Since $W \subseteq k^n$ has coordinates $y_1,\ldots,y_n$, we get coordinate functions $[y_i] \in k[W]$. Then $\Phi([y_i]) \in k[V]$, and since $V \subseteq k^m$ has coordinates $x_1,\ldots,x_m$, we can write $\Phi([y_i]) = [h_i(x_1, \ldots, x_m)] \in k[V]$ for some polynomial $h_i \in k[x_1, \ldots, x_m]$. Then consider the polynomial mapping

$$\alpha = (h_1(x_1, \ldots, x_m), \ldots, h_n(x_1, \ldots, x_m)).$$

We need to show that α maps V to W and that $\Phi = \alpha^*$.

Given any polynomial $F \in k[y_1, \ldots, y_n]$, we first claim that

(2) $$[F \circ \alpha] = \Phi([F])$$

in $k[V]$. To prove this, note that

$$[F \circ \alpha] = [F(h_1, \ldots, h_n)] = F([h_1], \ldots, [h_n]) = F(\Phi([y_1]), \ldots, \Phi([y_n])),$$

where the second equality follows from the definition of sum and product in $k[V]$, and the third follows from $[h_i] = \Phi([y_i])$. But $[F] = [F(y_1, \ldots, y_n)]$ is a k-linear combination of products of the $[y_i]$, so that

$$F(\Phi([y_1]), \ldots, \Phi([y_n])) = \Phi([F(y_1, \ldots, y_n)]) = \Phi([F])$$

since Φ is a ring homomorphism which is the identity on k (see Exercise 10). Equation (2) follows immediately.

We can now prove that α maps V to W. Given a point $(c_1, \ldots, c_m) \in V$, we must show that $\alpha(c_1, \ldots, c_m) \in W$. If $F \in \mathbf{I}(W)$, then $[F] = 0$ in $k[W]$, and since Φ is a ring homomorphism, we have $\Phi([F]) = 0$ in $k[V]$. By (2), this implies that $[F \circ \alpha]$ is the zero function on V. In particular,

$$[F \circ \alpha](c_1, \ldots, c_m) = F(\alpha(c_1, \ldots, c_m)) = 0.$$

Since F was an arbitrary element of $\mathbf{I}(W)$, this shows $\alpha(c_1, \ldots, c_m) \in W$, as desired.

Once we know α maps V to W, equation (2) implies that $[F] \circ \alpha = \Phi([F])$ for any $[F] \in k[W]$. Since $\alpha^*([F]) = [F] \circ \alpha$, this proves $\Phi = \alpha^*$. It remains to show that α is uniquely determined. So suppose we have $\beta : V \to W$ such that $\Phi = \beta^*$. If β is represented by

$$\beta(x_1, \ldots, x_m) = (\tilde{h}_1(x_1, \ldots, x_m), \ldots, \tilde{h}_n(x_1, \ldots, x_m)),$$

then note that $\beta^*([y_i]) = [y_i] \circ \beta = [\tilde{h}_i(x_1, \ldots, x_m)]$. A similar computation gives $\alpha^*([y_i]) = [h_i(x_1, \ldots, x_m)]$, and since $\alpha^* = \Phi = \beta^*$, we have $[h_i] = [\tilde{h}_i]$ for all i. Then h_i and $\tilde{h}_i$ give the same polynomial function on V, and, hence, $\alpha = (h_1, \ldots, h_n)$ and $\beta = (\tilde{h}_1, \ldots, \tilde{h}_n)$ define the same mapping on V. This shows $\alpha = \beta$, and uniqueness is proved.   □

Now suppose that $\alpha : V \to W$ and $\beta : W \to V$ are inverse polynomial mappings. Then $\alpha \circ \beta = \mathrm{id}_W$, where $\mathrm{id}_W : W \to W$ is the identity map. By general properties of functions, this implies $(\alpha \circ \beta)^*(\phi) = \mathrm{id}_W^*(\phi) = \phi \circ \mathrm{id}_W = \phi$ for all $\phi \in k[W]$. However, we also have

$$(\alpha \circ \beta)^*(\phi) = \phi \circ (\alpha \circ \beta) = (\phi \circ \alpha) \circ \beta$$

(3)

$$= \alpha^*(\phi) \circ \beta = \beta^*(\alpha^*(\phi)) = (\beta^* \circ \alpha^*)(\phi).$$

Hence, $(\alpha \circ \beta)^* = \beta^* \circ \alpha^* = \mathrm{id}_{k[W]}$ as a mapping from $k[W]$ to itself. Similarly, one can show that $(\beta \circ \alpha)^* = \alpha^* \circ \beta^* = \mathrm{id}_{k[V]}$. This proves the first half of the following theorem.

Theorem 9. *Two affine varieties $V \subseteq k^m$ and $W \subseteq k^n$ are isomorphic if and only if there is an isomorphism $k[V] \cong k[W]$ of coordinate rings which is the identity on constant functions.*

Proof. The above discussion shows that if V and W are isomorphic varieties, then $k[V] \cong k[W]$ as rings. Proposition 8 shows that the isomorphism is the identity on constants.

For the converse, we must show that if we have a ring isomorphism $\Phi : k[W] \to k[V]$ which is the identity on k, then Φ and Φ^{-1} "come from" inverse polynomial mappings between V and W. By part (ii) of Proposition 8, we know that $\Phi = \alpha^*$ for some $\alpha : V \to W$ and $\Phi^{-1} = \beta^*$ for $\beta : W \to V$. We need to show that α and β are inverse mappings. First consider the composite map $\alpha \circ \beta : W \to W$. This is clearly a polynomial map, and, using the argument from (3), we see that for any $\phi \in k[W]$,

(4) $$(\alpha \circ \beta)^*(\phi) = \beta^*(\alpha^*(\phi)) = \Phi^{-1}(\Phi(\phi)) = \phi.$$

Since the identity map $\mathrm{id}_W : W \to W$ is a polynomial map on W, and we saw above that $\mathrm{id}_W^*(\phi) = \phi$ for all $\phi \in k[W]$, from (4), we conclude that $(\alpha \circ \beta)^* = \mathrm{id}_W^*$, and then $\alpha \circ \beta = \mathrm{id}_W$ follows from the uniqueness statement of part (ii) of Proposition 8. In a similar way, one proves that $\beta \circ \alpha = \mathrm{id}_V$, and hence α and β are inverse mappings. This completes the proof of the theorem. $\square$

We conclude with several examples to illustrate isomorphisms of varieties and the corresponding isomorphisms of their coordinate rings.

Let A be an invertible $n \times n$ matrix with entries in k and consider the linear mapping $L_A : k^n \to k^n$ defined by $L_A(x) = Ax$, where Ax is matrix multiplication. This is easily seen to be an isomorphism of varieties by considering $L_{A^{-1}}$. (An isomorphism of a variety with itself is often called an *automorphism* of the variety.) It follows from Theorem 9 that $L_A^* : k[x_1, \ldots, x_n] \to k[x_1, \ldots, x_n]$ is a ring isomorphism. In Exercise 9, you will show that if V is any subvariety of k^n, then $L_A(V)$ is a subvariety of k^n isomorphic to V since L_A restricts to give an isomorphism of V onto $L_A(V)$. For example, the curve we studied in the final example of §1 of this chapter was obtained from the "standard" twisted cubic curve in $\mathbb{C}^3$ by an invertible linear mapping. Refer to equation (5) of §1 and see if you can identify the mapping L_A that was used.

Next, let $f(x, y) \in k[x, y]$ and consider the *graph* of the polynomial function on k^2 given by f [that is, the variety $V = \mathbf{V}(z - f(x, y)) \subseteq k^3$]. Generalizing what we said concerning the variety $\mathbf{V}(z - xy)$ in analyzing the curve given in Example 7, it will always be the case that a graph V is isomorphic as a variety to k^2. The reason is that the projection on the (x, y)-plane $\pi : V \to k^2$, and the parametrization of the graph given by $\alpha : k^2 \to V$, $\alpha(x,y) = (x,y,f(x,y))$ are inverse mappings. The isomorphism of coordinate rings corresponding to α just consists of substituting $z = f(x,y)$ into every polynomial function $F(x,y,z)$ on V.

Finally, consider the curve $V = \mathbf{V}(y^5 - x^2)$ in $\mathbb{R}^2$.

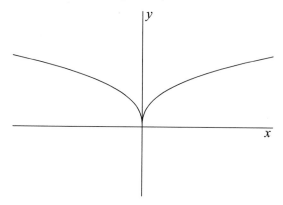

We claim that V is *not* isomorphic to $\mathbb{R}$ as a variety, even though there is a one-to-one polynomial mapping from V to $\mathbb{R}$ given by projecting V onto the x-axis. The reason lies in the coordinate ring of V, $\mathbb{R}[V] = \mathbb{R}[x,y]/\langle y^5 - x^2 \rangle$. If there were an isomorphism $\alpha : \mathbb{R} \to V$, then the "pullback" $\alpha^* : \mathbb{R}[V] \to \mathbb{R}[u]$ would be a ring isomorphism given by

$$\alpha^*([x]) = g(u),$$
$$\alpha^*([y]) = h(u),$$

where $g(u), h(u) \in \mathbb{R}[u]$ are polynomials. Since $y^5 - x^2$ represents the zero function on V, we must have $\alpha^*([y^5 - x^2]) = (h(u))^5 - (g(u))^2 = 0$ in $\mathbb{R}[u]$.

We may assume that $g(0) = h(0) = 0$ since the parametrization α can be "arranged" so that $\alpha(0) = (0,0) \in V$. But then let us examine the possible polynomial solutions

$$g(u) = c_1 u + c_2 u^2 + \cdots, \quad h(u) = d_1 u + d_2 u^2 + \cdots$$

of the equation $(g(u))^2 = (h(u))^5$. Since $(h(u))^5$ contains no power of u lower than u^5, the same must be true of $(g(u))^2$. However,

$$(g(u))^2 = c_1^2 u^2 + 2c_1 c_2 u^3 + (c_2^2 + 2c_1 c_3)u^4 + 2(c_1 c_4 + c_2 c_3)u^5 + \cdots .$$

The coefficient of u^2 must be zero, which implies $c_1 = 0$. The coefficient of u^4 must also be zero, which implies $c_2 = 0$ as well. Since $c_1, c_2 = 0$, the smallest power of u that can appear in $(g(u))^2$ is u^6, which implies that $d_1 = 0$ also.

It follows that u cannot be in the image of α^* since the image of α^* consists of polynomials in $g(u)$ and $h(u)$. This is a contradiction since α^* was supposed to be a ring isomorphism *onto* $\mathbb{R}[u]$. Thus, our two varieties are not isomorphic. In the exercises, you will derive more information about $\mathbb{R}[V]$ by the method of §3 to yield another proof that $\mathbb{R}[V]$ is not isomorphic to a polynomial ring in one variable.

EXERCISES FOR §4

1. Let C be the twisted cubic curve in k^3.
   a. Show that C is a subvariety of the surface $S = \mathbf{V}(xz - y^2)$.
   b. Find an ideal $J \subseteq k[S]$ such that $C = \mathbf{V}_S(J)$.

2. Let $V \subseteq \mathbb{C}^n$ be a nonempty affine variety.
   a. Let $\phi \in \mathbb{C}[V]$. Show that $\mathbf{V}_V(\phi) = \emptyset$ if and only if ϕ is invertible in $\mathbb{C}[V]$ (which means that there is some $\psi \in \mathbb{C}[V]$ such that $\phi\psi = [1]$ in $\mathbb{C}[V]$).
   b. Is the statement of part (a) true if we replace $\mathbb{C}$ by $\mathbb{R}$? If so, prove it; if not, give a counterexample.

3. Prove parts (ii), (iii), and (iv) of Proposition 3.

4. Let $V = \mathbf{V}(y - x^n, z - x^m)$, where m, n are any integers ≥ 1. Show that V is isomorphic as a variety to k by constructing explicit inverse polynomial mappings $\alpha : k \to V$ and $\beta : V \to k$.

5. Show that any surface in k^3 with a defining equation of the form $x - f(y, z) = 0$ or $y - g(x, z) = 0$ is isomorphic as a variety to k^2.

6. Let V be a variety in k^n defined by a single equation of the form $x_n - f(x_1, \ldots, x_{n-1}) = 0$. Show that V is isomorphic as a variety to k^{n-1}.

7. In this exercise, we will derive the parametrization (1) for the projected curve $\pi(C)$ from Example 7.
   a. Show that every hyperbola in $\mathbb{R}^2$ whose asymptotes are horizontal and vertical and which passes through the points $(0, 0)$ and $(1, 1)$ is defined by an equation of the form

   $$xy + tx - (t + 1)y = 0$$

   for some $t \in \mathbb{R}$.
   b. Using a computer algebra system, compute a Gröbner basis for the ideal generated by the equation of $\pi(C)$, and the above equation of the hyperbola. Use lex order with the variables ordered $x > y > t$.
   c. The Gröbner basis will contain one polynomial depending on y, t only. By collecting powers of y and factoring, show that this polynomial has $y = 0$ as a double root, $y = 1$ as a single root, and one root which depends on t, namely $y = \frac{-t^2 + t + 1}{t(t + 2)}$.
   d. Now consider the other elements of the basis and show that for the "movable" root from part (c) there is a unique corresponding x value given by the first equation in (1).
   The method sketched in Exercise 7 probably seems exceedingly *ad hoc*, but it is an example of a general pattern that can be developed with some more machinery concerning algebraic curves. Using the complex projective plane to be introduced in Chapter 8, it can be shown that $\pi(C)$ is contained in a projective algebraic curve with *three* singular points similar to the one at $(0, 0)$ in the sketch. Using the family of conics passing through all three singular points and any one additional point, we can give a rational parametrization for *any* irreducible quartic curve with three singular points as in this example. However, one can show that nonsingular quartic curves have no such parametrizations.

8. Let $Q_1 = \mathbf{V}(x^2 + y^2 + z^2 - 1)$, and $Q_2 = \mathbf{V}((x - 1/2)^2 - 3y^2 - 2z^2)$ in $\mathbb{R}^3$.
   a. Using the idea of Example 7 and Exercise 5, find a surface in the pencil defined by Q_1 and Q_2 that is isomorphic as a variety to $\mathbb{R}^2$.
   b. Describe and/or sketch the intersection curve $Q_1 \cap Q_2$.

9. Let $\alpha : V \to W$ and $\beta : W \to V$ be inverse polynomial mappings between two isomorphic varieties V and W. Let $U = \mathbf{V}_V(I)$ for some ideal $I \subseteq k[V]$. Show that $\alpha(U)$ is a subvariety of W and explain how to find an ideal $J \subseteq k[W]$ such that $\alpha(U) = \mathbf{V}_W(J)$.

10. Let $\Phi : k[V] \to k[W]$ be a ring homomorphism of coordinate rings which is the identity on constants. Suppose that $V \subseteq k^m$ with coordinates $x_1, \ldots, x_m$. If $F \in k[x_1, \ldots, x_m]$, then prove that $\Phi([F]) = F(\Phi([x_1]), \ldots, \Phi([x_m]))$. Hint: Express $[F]$ as a k-linear combination of products of the $[x_i]$.

11. Recall the example following Definition 2 where $V = \mathbf{V}(z - x^2 - y^2) \subseteq \mathbb{R}^3$.
    a. Show that the subvariety $W = \{(1, 1, 2)\} \subseteq V$ is equal to $\mathbf{V}_V([x-1], [y-1])$. Explain why this implies that $\langle [x-1], [y-1] \rangle \subseteq \mathbf{I}_V(W)$.
    b. Prove that $\langle [x-1], [y-1] \rangle = \mathbf{I}_V(W)$. Hint: Show that V is isomorphic to $\mathbb{R}^2$ and use Exercise 9.

12. Let $V = \mathbf{V}(y^2 - 3x^2 z + 2) \subseteq \mathbb{R}^3$ and let L_A be the linear mapping on $\mathbb{R}^3$ defined by the matrix

$$A = \begin{pmatrix} 2 & 0 & 1 \\ 1 & 1 & 0 \\ 0 & 1 & 1 \end{pmatrix}.$$

    a. Verify that L_A is an isomorphism from $\mathbb{R}^3$ to $\mathbb{R}^3$.
    b. Find the equation of the image of V under L_A.

13. In this exercise, we will rotate the twisted cubic in $\mathbb{R}^3$.
    a. Find the matrix A of the linear mapping on $\mathbb{R}^3$ that rotates every point through an angle of $\pi/6$ counterclockwise about the z-axis.
    b. What are the equations of the image of the standard twisted cubic curve under the linear mapping defined by the rotation matrix A?

14. This exercise will outline another proof that $V = \mathbf{V}(y^5 - x^2) \subseteq \mathbb{R}^2$ is not isomorphic to $\mathbb{R}$ as a variety. This proof will use the algebraic structure of $\mathbb{R}[V]$. We will show that there is no ring isomorphism from $\mathbb{R}[V]$ to $\mathbb{R}[t]$. (Note that $\mathbb{R}[t]$ is the coordinate ring of $\mathbb{R}$.)
    a. Using the techniques of §3, explain how each element of $\mathbb{R}[V]$ can be uniquely represented by a polynomial of the form $a(y) + b(y)x$, where $a, b \in \mathbb{R}[y]$.
    b. Express the product $(a + bx)(a' + b'x)$ in $\mathbb{R}[V]$ in the form given in part (a).
    c. Aiming for a contradiction, suppose that there were some ring isomorphism $\Phi : \mathbb{R}[t] \to \mathbb{R}[V]$. Since Φ is assumed to be onto, $x = \Phi(f(t))$ and $y = \Phi(g(t))$ for some polynomials f, g. Using the unique factorizations of f, g and the product formula from part (b), deduce a contradiction.

15. Let $V \subseteq \mathbb{R}^3$ be the tangent surface of the twisted cubic curve.
    a. Show that the usual parametrization of V sets up a one-to-one correspondence between the points of V and the points of $\mathbb{R}^2$. Hint: Recall the discussion of V in Chapter 3, §3. In light of part (a), it is natural to ask whether V is *isomorphic* to $\mathbb{R}^2$. We will show that the answer to this question is *no*.
    b. Show that V is singular at each point on the twisted cubic curve by using the method of Exercise 12 of Chapter 3, §4. (The tangent surface has what is called a "cuspidal edge" along this curve.)
    c. Show that if $\alpha : \mathbb{R}^2 \to V$ is *any* polynomial parametrization of V, and $\alpha(a, b)$ is contained in the twisted cubic itself, then the derivative matrix of α must have rank strictly less than 2 at (a, b) (in other words, the columns of the derivative matrix must be linearly dependent there). (Note: α need not be the standard parametrization, although the statement will be true also for that parametrization.)
    d. Now suppose that the polynomial parametrization α has a polynomial inverse mapping $\beta : V \to \mathbb{R}^2$. Using the chain rule from multivariable calculus, show that part (c) gives a contradiction if we consider (a, b) such that $\alpha(a, b)$ is on the twisted cubic.

16. Let k be algebraically closed. Prove the Weak Nullstellensatz for $k[V]$, which asserts that for any ideal $J \subseteq k[V]$, $\mathbf{V}_V(J) = \emptyset$ if and only if $J = k[V]$. Also explain how this relates to Exercise 2 when $J = \langle \phi \rangle$.

17. Here is some practice with isomorphisms.
    a. Let $f : R \to S$ be a ring isomorphism. Prove that R is an integral domain if and only if S is an integral domain.
    b. Let $\phi : V \to W$ be an isomorphism of affine varieties. Prove that V is irreducible if and only if W is irreducible. Hint: Combine part (a) with Theorem 9 of this section and Proposition 4 of §1.

18. Let A be an invertible $n \times n$ matrix with entries in k and consider the map $L_A : k^n \to k^n$ from the discussion following Theorem 9.

  a. Prove that L_A^* is the ring homomorphism denoted α_A in Exercise 13 of Chapter 4, §3.

  b. Explain how the discussion in the text gives a proof of Exercise 19 of §2.

§5 Rational Functions on a Variety

The ring of integers can be embedded in many fields. The *smallest* of these is the field of rational numbers $\mathbb{Q}$ because $\mathbb{Q}$ is formed by constructing fractions $\frac{m}{n}$, where $m, n \in \mathbb{Z}$ and $n \neq 0$. Nothing more than integers was used. Similarly, the polynomial ring $k[x_1, \ldots, x_n]$ is included as a subring in the field of *rational functions*

$$k(x_1, \ldots, x_n) = \left\{ \frac{f(x_1, \ldots, x_n)}{g(x_1, \ldots, x_n)} \;\middle|\; f, g \in k[x_1, \ldots, x_n], \; g \neq 0 \right\}.$$

Generalizing these examples, if R is any integral domain, then we can form what is called the *field of fractions*, or *quotient field* of R, denoted $FF(R)$. The elements of $FF(R)$ are thought of as "fractions" r/s, where $r, s \in R$ and $s \neq 0$. Two of these fractions r/s and r'/s' represent the same element in the field of fractions if $rs' = r's$. We add and multiply elements of $FF(R)$ as we do rational numbers or rational functions:

$$r/s + t/u = (ru + ts)/su \quad \text{and} \quad r/s \cdot t/u = rt/su.$$

The assumption that R is an integral domain ensures that the denominators of the sum and product will be nonzero. You will check in Exercise 1 that these operations are well-defined and that $FF(R)$ satisfies all the axioms of a field. Furthermore, $FF(R)$ contains the subset $\{r/1 \mid r \in R\}$, which is a subring isomorphic to R itself. Hence, the terminology "field of fractions, or quotient field of R" is fully justified.

Now if $V \subseteq k^n$ is an *irreducible* variety, then we have seen in §1 that the coordinate ring $k[V]$ is an integral domain. The field of fractions $FF(k[V])$ is given the following name.

Definition 1. Let V be an irreducible affine variety in k^n. We call $FF(k[V])$ the **function field** (or **field of rational functions**) of V, and we denote this field by $k(V)$.

Note the consistency of our notation. We use $k[x_1, \ldots, x_n]$ for a polynomial ring and $k[V]$ for the coordinate ring of V. Similarly, we use $k(x_1, \ldots, x_n)$ for a rational function field and $k(V)$ for the function field of V.

We can write the function field $k(V)$ of $V \subseteq k^n$ explicitly as

$$k(V) = \{\phi/\psi \mid \phi, \psi \in k[V], \; \psi \neq 0\}$$
$$= \{[f]/[g] \mid f, g \in k[x_1, \ldots, x_n], \; g \notin \mathbf{I}(V)\}.$$

As with any rational function, we must be careful to avoid zeros of the denominator if we want a well-defined function value in k. Thus, an element $\phi/\psi \in k(V)$ defines a function only on the complement of $\mathbf{V}_V(\psi)$.

The most basic example of the function field of a variety is given by $V = k^n$. In this case, we have $k[V] = k[x_1, \ldots, x_n]$ and, hence,

$$k(V) = k(x_1, \ldots, x_n).$$

We next consider some more complicated examples.

Example 2. In §4, we showed that the curve

$$V = \mathbf{V}(y^5 - x^2) \subseteq \mathbb{R}^2$$

is not isomorphic to $\mathbb{R}$ because the coordinate rings of V and $\mathbb{R}$ are not isomorphic. Let us see what we can say about the function field of V. To begin, note that by the method of §3, we can represent the elements of $\mathbb{R}[V]$ by remainders modulo $G = \{y^5 - x^2\}$, which is a Gröbner basis for $\mathbf{I}(V)$ with respect to lex order with $x > y$ in $\mathbb{R}[x, y]$. Then $\mathbb{R}[V] = \{a(y) + xb(y) \mid a, b \in \mathbb{R}[y]\}$ as a real vector space, and multiplication is defined by

$$(1) \qquad (a + xb) \cdot (c + xd) = (ac + y^5 \cdot bd) + x(ad + bc).$$

In Exercise 2, you will show that V is irreducible, so that $\mathbb{R}[V]$ is an integral domain.

Now, using this description of $\mathbb{R}[V]$, we can also describe the function field $\mathbb{R}(V)$ as follows. If $c + xd \neq 0$ in $\mathbb{R}[V]$, then in the function field we can write

$$\begin{aligned}
\frac{a + xb}{c + xd} &= \frac{a + xb}{c + xd} \cdot \frac{c - xd}{c - xd} \\
&= \frac{(ac - y^5bd) + x(bc - ad)}{c^2 - y^5d^2} \\
&= \frac{ac - y^5bd}{c^2 - y^5d^2} + x\frac{bc - ad}{c^2 - y^5d^2}.
\end{aligned}$$

This is an element of $\mathbb{R}(y) + x\mathbb{R}(y)$. Conversely, it is clear that every element of $\mathbb{R}(y) + x\mathbb{R}(y)$ defines an element of $\mathbb{R}(V)$. Hence, the field $\mathbb{R}(V)$ can be identified with the set of functions $\mathbb{R}(y) + x\mathbb{R}(y)$, where the addition and multiplication operations are defined as before in $\mathbb{R}[V]$, only using rational functions of y rather than polynomials.

Now consider the mappings:

$$\begin{aligned}
\alpha : V &\longrightarrow \mathbb{R}, & (x, y) &\longmapsto x/y^2, \\
\beta : \mathbb{R} &\longrightarrow V, & u &\longmapsto (u^5, u^2).
\end{aligned}$$

Note that α is defined except at $(0, 0) \in V$, whereas β is a polynomial parametrization of V. As in §4, we can use α and β to define mappings "going in the opposite direction" on functions. However, since α itself is defined as a rational function, we will not stay within $\mathbb{R}[V]$ if we compose α with a function in $\mathbb{R}[u]$. Hence, we will consider the maps

$$\alpha^* : \mathbb{R}(u) \longrightarrow \mathbb{R}(V), \quad f(u) \longmapsto f(x/y^2),$$
$$\beta^* : \mathbb{R}(V) \longrightarrow \mathbb{R}(u), \quad a(y) + xb(y) \longmapsto a(u^2) + u^5 b(u^2).$$

We claim that α^* and β^* are inverse ring isomorphisms. That α^* and β^* preserve sums and products follows by the argument given in the proof of Proposition 8 from §4. To check that α^* and β^* are inverses, first we have that for any $f(u) \in \mathbb{R}(u)$, $\alpha^*(f) = f(x/y^2)$. Hence, $\beta^*(\alpha^*(f)) = f(u^5/(u^2)^2) = f(u)$. Therefore, $\beta^* \circ \alpha^*$ is the identity on $\mathbb{R}(u)$. Similarly, if $a(y) + xb(y) \in \mathbb{R}(V)$, then $\beta^*(a + xb) = a(u^2) + u^5 b(u^2)$, so

$$\alpha^*(\beta^*(a + xb)) = a((x/y^2)^2) + (x/y^2)^5 b((x/y^2)^2)$$
$$= a(x^2/y^4) + (x^5/y^{10})b(x^2/y^4).$$

However, in $\mathbb{R}(V), x^2 = y^5$, so $x^2/y^4 = y$, and $x^5/y^{10} = xy^{10}/y^{10} = x$. Hence, $\alpha^* \circ \beta^*$ is the identity on $\mathbb{R}(V)$. Thus, α^*, β^* define ring isomorphisms between the *function fields* $\mathbb{R}(V)$ and $\mathbb{R}(u)$.

Example 2 shows that it is possible for two varieties to have the same (i.e., isomorphic) function fields, even when they are not isomorphic. It also gave us an example of a rational mapping between two varieties. Before we give a precise definition of a rational mapping, let us look at another example.

Example 3. Let $Q = \mathbf{V}(x^2 + y^2 - z^2 - 1)$, a hyperboloid of one sheet in $\mathbb{R}^3$, and let $W = \mathbf{V}(x + 1)$, the plane $x = -1$. Let $p = (1, 0, 0) \in Q$. For any $q \in Q \setminus \{p\}$, we construct the line L_q joining p and q, and we define a mapping ϕ to W by setting

$$\phi(q) = L_q \cap W$$

if the line intersects W. (If the line does not intersect W, then $\phi(q)$ is undefined.) We can find an algebraic formula for ϕ as follows. If $q = (x_0, y_0, z_0) \in Q$, then L_q is given in parametric form by

$$x = 1 + t(x_0 - 1),$$
(2) $$y = ty_0,$$
$$z = tz_0.$$

At $\phi(q) = L_q \cap W$, we must have $1 + t(x_0 - 1) = -1$, so $t = \frac{-2}{x_0 - 1}$. From (2), it follows that

(3) $$\phi(q) = \left(-1, \frac{-2y_0}{x_0 - 1}, \frac{-2z_0}{x_0 - 1}\right).$$

This shows that ϕ is defined on all of Q except for the points on the two lines

$$\mathbf{V}_Q(x - 1) = Q \cap \mathbf{V}(x - 1) = \{(1, t, t) \mid t \in \mathbb{R}\} \cup \{(1, t, -t) \mid t \in \mathbb{R}\}.$$

We will call $\phi : Q \setminus \mathbf{V}_Q(x-1) \to W$ a *rational mapping* on Q since the components of ϕ are rational functions. [We can think of them as elements of $\mathbb{R}(Q)$ if we like.]

Going in the other direction, if $(-1, a, b) \in W$, then the line L through $p = (1, 0, 0)$ and $(-1, a, b)$ can be parametrized by

$$x = 1 - 2t,$$
$$y = ta,$$
$$z = tb,$$

Computing the intersections with Q, we find

$$L \cap Q = \left\{ (1, 0, 0), \left(\frac{a^2 - b^2 - 4}{a^2 - b^2 + 4}, \frac{4a}{a^2 - b^2 + 4}, \frac{4b}{a^2 - b^2 + 4} \right) \right\}.$$

Thus, if we let H denote the hyperbola $\mathbf{V}_W(a^2 - b^2 + 4)$, then we can define a second rational mapping

$$\psi : W \setminus H \longrightarrow Q$$

by

(4) $$\psi(-1, a, b) = \left(\frac{a^2 - b^2 - 4}{a^2 - b^2 + 4}, \frac{4a}{a^2 - b^2 + 4}, \frac{4b}{a^2 - b^2 + 4} \right).$$

From the geometric descriptions of ϕ and ψ, $\phi \circ \psi$ is the identity mapping on the subset $W \setminus H \subseteq W$. Similarly, we see that $\psi \circ \phi$ is the identity on $Q \setminus \mathbf{V}_Q(x-1)$. Also, using the formulas from equations (3) and (4), it can be checked that $\phi^* \circ \psi^*$ and $\psi^* \circ \phi^*$ are the identity mappings on the function fields. (We should mention that as in the second example, Q and W are *not* isomorphic varieties. However, this is not an easy fact to prove given what we know.)

We now introduce some general terminology that was implicit in the above examples.

Definition 4. Let $V \subseteq k^m$ and $W \subseteq k^n$ be irreducible affine varieties. A **rational mapping** from V to W is a function ϕ represented by

(5) $$\phi(x_1, \ldots, x_m) = \left(\frac{f_1(x_1, \ldots, x_m)}{g_1(x_1, \ldots, x_m)}, \ldots, \frac{f_n(x_1, \ldots, x_m)}{g_n(x_1, \ldots, x_m)} \right),$$

where $f_i/g_i \in k(x_1, \ldots, x_m)$ satisfy:
 (i) ϕ is defined at some point of V.
 (ii) For every $(a_1, \ldots, a_m) \in V$ where ϕ is defined, $\phi(a_1, \ldots, a_m) \in W$.

Note that a rational mapping ϕ from V to W may fail to be a *function* from V to W in the usual sense because, as we have seen in the examples, ϕ may not be defined everywhere on V. For this reason, many authors use a special notation to indicate a rational mapping:

$$\phi : V \dashrightarrow W.$$

We will follow this convention as well. By condition (i), the set of points of V when the rational mapping ϕ in (5) is defined includes $V \setminus \mathbf{V}_V(g_1 \cdots g_n) = V \setminus (\mathbf{V}_V(g_1) \cup \cdots \cup \mathbf{V}_V(g_n))$, where $\mathbf{V}_V(g_1 \cdots g_n)$ is a proper subvariety of V.

Because rational mappings are not defined everywhere on their domains, we must exercise some care in studying them. In particular, we will need the following precise definition of when two rational mappings are to be considered equal.

Definition 5. Let $\phi, \psi : V \dashrightarrow W$ be rational mappings represented by

$$\phi = \left(\frac{f_1}{g_1}, \ldots, \frac{f_n}{g_n}\right) \quad \text{and} \quad \psi = \left(\frac{f_1'}{g_1'}, \ldots, \frac{f_n'}{g_n'}\right).$$

Then we say that $\phi = \psi$ if for each i, $1 \le i \le n$,

$$f_i g_i' - f_i' g_i \in \mathbf{I}(V).$$

We have the following geometric criterion for the equality of rational mappings.

Proposition 6. *Two rational mappings $\phi, \psi : V \dashrightarrow W$ are equal if and only if there is a proper subvariety $V' \subseteq V$ such that ϕ and ψ are defined on $V \setminus V'$ and $\phi(p) = \psi(p)$ for all $p \in V \setminus V'$.*

Proof. We will assume that $\phi = (f_1/g_1, \ldots, f_n/g_n)$ and $\psi = (f_1'/g_1', \ldots, f_n'/g_n')$. First, suppose that ϕ and ψ are equal as in Definition 5 and let $V_1 = \mathbf{V}_V(g_1 \cdots g_n)$ and $V_2 = \mathbf{V}_V(g_1' \cdots g_n')$. By hypothesis, V_1 and V_2 are proper subvarieties of V, and since V is irreducible, it follows that $V' = V_1 \cup V_2$ is also a proper subvariety of V. Then ϕ and ψ are defined on $V \setminus V'$, and since $f_i g_i' - f_i' g_i \in \mathbf{I}(V)$, it follows that f_i/g_i and f_i'/g_i' give the same function on $V \setminus V'$. Hence, the same is true for ϕ and ψ.

Conversely, suppose that ϕ and ψ are defined and equal (as functions) on $V \setminus V'$. This implies that for each i, we have $f_i/g_i = f_i'/g_i'$ on $V \setminus V'$. Then $f_i g_i' - f_i' g_i$ vanishes on $V \setminus V'$, which shows that $V = \mathbf{V}(f_i g_i' - f_i' g_i) \cup V'$. Since V is irreducible and V' is a proper subvariety, this forces $V = \mathbf{V}(f_i g_i' - f_i' g_i)$. Thus, $f_i g_i' - f_i' g_i \in \mathbf{I}(V)$, as desired. $\qquad\square$

As an example, recall from Example 3 that we had rational maps $\phi : Q \dashrightarrow W$ and $\psi : W \dashrightarrow Q$ such that $\phi \circ \psi$ was the identity on $W \setminus H \subseteq W$. By Proposition 6, this proves that $\phi \circ \psi$ equals the identity map id_W in the sense of Definition 5.

We also need to be careful in dealing with the composition of rational mappings.

Definition 7. Given $\phi : V \dashrightarrow W$ and $\psi : W \dashrightarrow Z$, we say that $\psi \circ \phi$ is **defined** if there is a point $p \in V$ such that ϕ is defined at p and ψ is defined at $\phi(p)$.

When a composition $\psi \circ \phi$ is defined, it gives us a rational mapping as follows.

Proposition 8. *Let $\phi : V \dashrightarrow W$ and $\psi : W \dashrightarrow Z$ be rational mappings such that $\psi \circ \phi$ is defined. Then there is a proper subvariety $V' \subsetneq V$ such that:*
(i) ϕ is defined on $V \setminus V'$ and ψ is defined on $\phi(V \setminus V')$.
(ii) $\psi \circ \phi : V \dashrightarrow Z$ is a rational mapping defined on $V \setminus V'$.

Proof. Suppose that ϕ and ψ are represented by

$$\phi(x_1,\ldots,x_m) = \left(\frac{f_1(x_1,\ldots,x_m)}{g_1(x_1,\ldots,x_m)},\ldots,\frac{f_n(x_1,\ldots,x_m)}{g_n(x_1,\ldots,x_m)}\right).$$

$$\psi(y_1,\ldots,y_n) = \left(\frac{f_1'(y_1,\ldots,y_n)}{g_1'(y_1,\ldots,y_n)},\ldots,\frac{f_l'(y_1,\ldots,y_n)}{g_l'(y_1,\ldots,y_n)}\right).$$

Then the j-th component of $\psi \circ \phi$ is

$$\frac{f_j'(f_1/g_1,\ldots,f_n/g_n)}{g_j'(f_1/g_1,\ldots,f_n/g_n)},$$

which is clearly a rational function in $x_1,\ldots,x_m$. To get a quotient of polynomials, we can write this as

$$\frac{P_j}{Q_j} = \frac{(g_1\cdots g_n)^M f_j'(f_1/g_1\cdots,f_n/g_n)}{(g_1\cdots g_n)^M g_j'(f_1/g_1,\cdots,f_n/g_n)},$$

when M is sufficiently large.

Now set

$$V' = \mathbf{V}_V([Q_1\cdots Q_l g_1\cdots g_n]) \subseteq V.$$

It should be clear that ϕ is defined on $V \setminus V'$ and ψ is defined on $\phi(V \setminus V')$. It remains to show that $V' \neq V$. But by assumption, there is $p \in V$ such that $\phi(p)$ and $\psi(\phi(p))$ are defined. This means that $g_i(p) \neq 0$ for $1 \leq i \leq n$ and

$$g_j'(f_1(p)/g_1(p),\ldots,f_n(p)/g_n(p)) \neq 0$$

for $1 \leq j \leq l$. It follows that $Q_j(p) \neq 0$ and consequently, $p \in V \setminus V'$. □

In the exercises, you will work out an example to show how $\psi \circ \phi$ can fail to be defined. Basically, this happens when the domain of definition of ψ lies outside the image of ϕ.

Examples 2 and 3 illustrate the following alternative to the notion of isomorphism of varieties.

Definition 9. (i) Two irreducible varieties $V \subseteq k^m$ and $W \subseteq k^n$ are **birationally equivalent** if there exist rational mappings $\phi : V \dashrightarrow W$ and $\psi : W \dashrightarrow V$ such that $\phi \circ \psi$ is defined (as in Definition 7) and equal to the identity map id_W (as in Definition 5), and similarly for $\psi \circ \phi$.

(ii) A **rational variety** is a variety that is birationally equivalent to k^n for some n.

Just as isomorphism of varieties can be detected from the coordinate rings, birational equivalence can be detected from the function fields.

Theorem 10. *Two irreducible varieties V and W are birationally equivalent if and only if there is an isomorphism of function fields $k(V) \cong k(W)$ which is the identity on k. (By definition, two fields are isomorphic if they are isomorphic as commutative rings.)*

Proof. The proof is similar to what we did in Theorem 9 of §4. Suppose first that V and W are birationally equivalent via $\phi : V \dashrightarrow W$ and $\psi : W \dashrightarrow V$. We will define a pullback mapping $\phi^* : k(W) \to k(V)$ by the rule $\phi^*(f) = f \circ \phi$ and show that ϕ^* is an isomorphism. Unlike the polynomial case, it is not obvious that $\phi^*(f) = f \circ \phi$ exists for all $f \in k(W)$—we need to prove that $f \circ \phi$ is defined at some point of V.

We first show that our assumption $\phi \circ \psi = \mathrm{id}_W$ implies the existence of a proper subvariety $W' \subseteq W$ such that

(6)
$$\psi \text{ is defined on } W \setminus W',$$
$$\phi \text{ is defined on } \psi(W \setminus W'),$$
$$\phi \circ \psi \text{ is the identity function on } W \setminus W'.$$

To prove this, we first use Proposition 8 to find a proper subvariety $W_1 \subseteq W$ such that ψ is defined on $W \setminus W_1$ and ϕ is defined on $\psi(W \setminus W_1)$. Also, from Proposition 6, we get a proper subvariety $W_2 \subseteq W$ such that $\phi \circ \psi$ is the identity function on $W \setminus W_2$. Since W is irreducible, $W' = W_1 \cup W_2$ is a proper subvariety, and it follows easily that (6) holds for this choice of W'.

Given $f \in k(W)$, we can now prove that $f \circ \phi$ is defined. If f is defined on $W \setminus W'' \subseteq W$, then we can pick $q \in W \setminus (W' \cup W'')$ since W is irreducible. From (6), we get $p = \psi(q) \in V$ such that $\phi(p)$ is defined, and since $\phi(p) = q \notin W''$, we also know that f is defined at $\phi(p)$, i.e., $f \circ \phi$ is defined at p. By Definition 4, $\phi^*(f) = f \circ \phi$ exists as an element of $k(V)$.

This proves that we have a map $\phi^* : k(W) \to k(V)$, and ϕ^* is a ring homomorphism by the proof of Proposition 8 from §4. Similarly, we get a ring homomorphism $\psi^* : k(V) \to k(W)$. To show that these maps are inverses of each other, let us look at

$$(\psi^* \circ \phi^*)(f) = f \circ \phi \circ \psi$$

for $f \in k(W)$. Using the above notation, we see that $f \circ \phi \circ \psi$ equals f as a function on $W \setminus (W' \cup W'')$, so that $f \circ \phi \circ \psi = f$ in $k(W)$ by Proposition 6. This shows that $\psi^* \circ \phi^*$ is the identity on $k(W)$, and a similar argument shows that $\phi^* \circ \psi^* = \mathrm{id}_{k(V)}$. Thus, $\phi^* : k(W) \to k(V)$ is an isomorphism of fields. We leave it to the reader to show that ϕ^* is the identity on the constant functions $k \subseteq k(W)$.

The proof of the converse implication is left as an exercise for the reader. Once again the idea is basically the same as in the proof of Theorem 9 of §4. $\square$

In the exercises, you will prove that two irreducible varieties are birationally equivalent if there are "big" subsets (complements of proper subvarieties) that can be put in one-to-one correspondence by rational mappings. For example, the curve $V = \mathbf{V}(y^5 - x^2)$ from Example 2 is birationally equivalent to $W = \mathbb{R}$. You should check that $V \setminus \{(0,0)\}$ and $W \setminus \{0\}$ are in a one-to-one correspondence via the rational mappings f and g from equation (1). The birational equivalence between the hyperboloid and the plane in Example 3 works similarly. This example also shows that outside of the "big" subsets, birationally equivalent varieties may be quite different (you will check this in Exercise 14).

As we see from these examples, birational equivalence of irreducible varieties is a *weaker* equivalence relation than isomorphism. By this we mean that the set of varieties birationally equivalent to a given variety will contain many different non-isomorphic varieties. Nevertheless, in the history of algebraic geometry, the classification of varieties up to birational equivalence has received more attention than classification up to isomorphism, perhaps because constructing rational functions on a variety is easier than constructing polynomial functions. There are reasonably complete classifications of irreducible varieties of dimensions 1 and 2 up to birational equivalence, and, recently, significant progress has been made in dimension ≥ 3 with the so-called *minimal model program*. The birational classification of varieties remains an area of active research in algebraic geometry.

EXERCISES FOR §5

1. Let R be an integral domain, and let $FF(R)$ be the field of fractions of R as described in the text.
   a. Show that addition is well-defined in $FF(R)$. This means that if $r/s = r'/s'$ and $t/u = t'/u'$, then you must show that $(ru + ts)/su = (r'u' + t's')/s'u'$. Hint: Remember what it means for two elements of $FF(R)$ to be equal.
   b. Show that multiplication is well-defined in $FF(R)$.
   c. Show that the field axioms are satisfied for $FF(R)$.

2. As in Example 2, let $V = \mathbf{V}(y^5 - x^2) \subseteq \mathbb{R}^2$.
   a. Show that $y^5 - x^2$ is irreducible in $\mathbb{R}[x, y]$ and prove that $\mathbf{I}(V) = \langle y^5 - x^2 \rangle$.
   b. Conclude that $\mathbb{R}[V]$ is an integral domain.

3. Show that the singular cubic curve $\mathbf{V}(y^2 - x^3)$ is a rational variety (birationally equivalent to k) by adapting what we did in Example 2.

4. Consider the singular cubic curve $V_c = \mathbf{V}(y^2 - cx^2 + x^3)$ studied in Exercise 8 of Chapter 1, §3. Using the parametrization given there, prove that V_c is a rational variety and find subvarieties $V_c' \subseteq V_c$ and $W \subseteq \mathbb{R}$ such that your rational mappings define a one-to-one correspondence between $V_c \setminus V_c'$ and $\mathbb{R} \setminus W$. Hint: Recall that t in the parametrization of V_c is the slope of a line passing through $(0, 0)$.

5. Verify that the curve $\pi(C)$ from Exercise 7 of §4 is a rational variety. Hint: To define a rational inverse of the parametrization we derived in that exercise, you need to solve for t as a function of x and y on the curve. The equation of the hyperbola may be useful.

6. In Example 3, verify directly that (3) and (4) define inverse rational mappings from the hyperboloid of the one sheet to the plane.

7. Let $S = \mathbf{V}(x^2 + y^2 + z^2 - 1)$ in $\mathbb{R}^3$ and let $W = \mathbf{V}(z)$ be the (x, y)-plane. In this exercise, we will show that S and W are birationally equivalent varieties, via an explicit mapping called *stereographic projection*. See also Exercise 6 of Chapter 1, §3.
   a. Derive parametric equations as in (2) for the line L_q in $\mathbb{R}^3$ passing through the north pole $(0, 0, 1)$ of S and a general point $q = (x_0, y_0, z_0) \neq (0, 0, 1)$ in S.
   b. Using the line from part (a) show that $\phi(q) = L_q \cap W$ defines a rational mapping $\phi : S \dashrightarrow \mathbb{R}^2$. This mapping is the stereographic projection mentioned above.
   c. Show that the rational parametrization of S given in Exercise 6 of Chapter 1, §3 is the inverse mapping of ϕ.
   d. Deduce that S and W are birationally equivalent varieties and find subvarieties $S' \subseteq S$ and $W' \subseteq W$ such that ϕ and ψ put $S \setminus S'$ and $W \setminus W'$ into one-to-one correspondence.

8. In Exercise 10 of §1, you showed that there were no nonconstant polynomial mappings from $\mathbb{R}$ to $V = \mathbf{V}(y^2 - x^3 + x)$. In this problem, you will show that there are no nonconstant *rational* mappings either, so V is not birationally equivalent to $\mathbb{R}$. In the process, we

will need to consider polynomials with complex coefficients, so the proof will actually show that $\mathbf{V}(y^2 - x^3 + x) \subseteq \mathbb{C}^2$ is not birationally equivalent to $\mathbb{C}$ either. The proof will be by contradiction.

a. Start by assuming that $\alpha : \mathbb{R} \dashrightarrow V$ is a nonconstant rational mapping defined by $\alpha(t) = (a(t)/b(t), c(t)/d(t))$ with a and b relatively prime, c and d relatively prime, and b, d monic. By substituting into the equation of V, show that $b^3 = d^2$ and $c^2 = a^3 - ab^2$.

b. Deduce that $a, b, a+b$, and $a-b$ are all squares of polynomials in $\mathbb{C}[t]$. In other words, show that $a = A^2, b = B^2, a + b = C^2$ and $a - b = D^2$ for some $A, B, C, D \in \mathbb{C}[t]$.

c. Show that the polynomials $A, B \in \mathbb{C}[t]$ from part (b) are nonconstant and relatively prime and that $A^4 - B^4$ is the square of a polynomial in $\mathbb{C}[t]$.

d. The key step of the proof is to show that such polynomials cannot exist using *infinite descent*. Suppose that $A, B \in \mathbb{C}[t]$ satisfy the conclusions of part (c). Prove that there are polynomials $A_1, B_1, C_1 \in \mathbb{C}[t]$ such that

$$A - B = A_1^2$$
$$A + B = B_1^2$$
$$A^2 + B^2 = C_1^2.$$

e. Prove that the polynomials A_1, B_1 from part (d) are relatively prime and nonconstant and that their degrees satisfy

$$\max(\deg(A_1), \deg(B_1)) \leq \tfrac{1}{2} \max(\deg(A), \deg(B)).$$

Also show that $A_1^4 - (\sqrt{i}B_1)^4 = A_1^4 + B_1^4$ is the square of a polynomial in $\mathbb{C}[t]$. Conclude that $A_1, \sqrt{i}B_1$ satisfy the conclusions of part (c).

f. Conclude that if such a pair A, B exists, then one can repeat parts (d) and (e) infinitely many times with decreasing degrees at each step (this is the "infinite descent"). Explain why this is impossible and conclude that our original polynomials a, b, c, d must be constant.

9. Let V be an irreducible variety and let $f \in k(V)$. If we write $f = \phi/\psi$, where $\phi, \psi \in k[V]$, then we know that f is defined on $V \setminus \mathbf{V}_V(\psi)$. What is interesting is that f might make sense on a larger set. In this exercise, we will work out how this can happen on the variety $V = \mathbf{V}(xz - yw) \subseteq \mathbb{C}^4$.

a. Prove that $xz - yw \in \mathbb{C}[x, y, z, w]$ is irreducible. Hint: Look at the total degrees of its factors.

b. Use unique factorization in $\mathbb{C}[x, y, z, w]$ to prove that $\langle xz - yw \rangle$ is a prime ideal.

c. Conclude that V is irreducible and that $\mathbf{I}(V) = \langle xz - yw \rangle$.

d. Let $f = [x]/[y] \in \mathbb{C}(V)$ so that f is defined on $V \setminus \mathbf{V}_V([y])$. Show that $\mathbf{V}_V([y])$ is the union of planes $\{(0, 0, z, w) \mid z, w \in \mathbb{C}\} \cup \{(x, 0, 0, w) \mid x, w \in \mathbb{C}\}$.

e. Show that $f = [w]/[z]$ and conclude that f is defined everywhere outside of the plane $\{(x, 0, 0, w) \mid x, w \in \mathbb{C}\}$.

Note that what made this possible was that we had two fundamentally different ways of representing the rational function f. This is part of why rational functions are subtle to deal with.

10. Consider the rational mappings $\phi : \mathbb{R} \dashrightarrow \mathbb{R}^3$ and $\psi : \mathbb{R}^3 \dashrightarrow \mathbb{R}$ defined by

$$\phi(t) = (t, 1/t, t^2) \quad \text{and} \quad \psi(x, y, z) = \frac{x + yz}{x - yz}.$$

Show that $\psi \circ \phi$ is not defined.

11. Complete the proof of Theorem 10 by showing that if V and W are irreducible varieties and $k(V) \cong k(W)$ is an isomorphism of their function fields which is the identity on constants, then there are inverse rational mappings $\phi : V \dashrightarrow W$ and $\psi : W \dashrightarrow V$. Hint: Follow the proof of Theorem 9 from §4.

12. Suppose that $\phi : V \dashrightarrow W$ is a rational mapping defined on $V \setminus V'$. If $W' \subseteq W$ is a subvariety, then prove that

$$V'' = V' \cup \{p \in V \setminus V' \mid \phi(p) \in W'\}$$

is a subvariety of V. Hint: Find equations for V'' by substituting the rational functions representing ϕ into the equations for W' and setting the numerators of the resulting functions equal to zero.

13. Suppose that V and W are birationally equivalent varieties via $\phi : V \dashrightarrow W$ and $\psi : W \dashrightarrow V$. As mentioned in the text after the proof of Theorem 10, this means that V and W have "big" subsets that are the same. More precisely, there are proper subvarieties $V_1 \subseteq V$ and $W_1 \subseteq W$ such that ϕ and ψ induce inverse bijections between subsets $V \setminus V_1$ and $W \setminus W_1$. Note that Exercises 4 and 7 involved special cases of this result.

   a. Let $V' \subseteq V$ be the subvariety that satisfies the properties given in (6) for $\phi \circ \psi$. Similarly, we get $W' \subseteq W$ that satisfies the analogous properties for $\psi \circ \phi$. Let

$$\mathcal{V} = \{p \in V \setminus V' \mid \phi(p) \in W \setminus W'\},$$
$$\mathcal{W} = \{q \in W \setminus W' \mid \psi(q) \in V \setminus V'\}.$$

   Show that we have bijections $\phi : \mathcal{V} \to \mathcal{W}$ and $\psi : \mathcal{W} \to \mathcal{V}$ which are inverses of each other.

   b. Use Exercise 12 to prove that $\mathcal{V} = V \setminus V_1$ and $\mathcal{W} = W \setminus W_1$ for proper subvarieties V_1 and W_1.

   Parts (a) and (b) give the desired one-to-one correspondence between "big" subsets of V and W.

14. In Example 3, we had rational mappings $\phi : Q \dashrightarrow W$ and $\psi : W \dashrightarrow Q$.

   a. Show that ϕ and ψ induce inverse bijections $\phi : Q \setminus \mathbf{V}_Q(x-1) \to W \setminus H$ and $\psi : W \setminus H \to Q \setminus \mathbf{V}_Q(x-1)$, where $H = \mathbf{V}_W(a^2 - b^2 + 4)$.

   b. Show that H and $\mathbf{V}_Q(x-1)$ are very different varieties that are neither isomorphic nor birationally equivalent.

§6 Relative Finiteness and Noether Normalization

A major theme of this book is the relation between geometric objects, specifically affine varieties, and algebraic objects, notably ideals in polynomial rings. In this chapter, we learned about other algebraic objects, namely the coordinate ring $k[V]$ of a variety V and the quotient ring $k[x_1, \ldots, x_n]/I$ of an ideal $I \subseteq k[x_1, \ldots, x_n]$. But the rings $k[V]$ and $k[x_1, \ldots, x_n]/I$ have an additional structure that deserves explicit recognition.

Definition 1. A k-algebra is a ring which contains the field k as a subring. Also:

(i) A k-algebra is **finitely generated** if it contains finitely many elements such that every element can be expressed as a polynomial (with coefficients in k) in these finitely many elements.

(ii) A **homomorphism** of k-algebras is a ring homomorphism which is the identity
on elements of k.

   In addition to being a ring, a k-algebra is a vector space over k (usually infinite
dimensional) where addition is defined by the addition in the ring, and scalar multiplication is just multiplication by elements of the subring k. Examples of k-algebras
include the coordinate ring of a nonempty variety or the quotient ring of a proper
polynomial ideal. We need to assume $V \neq \emptyset$ in order to guarantee that $k \subseteq k[V]$,
and similarly $I \subsetneq k[x_1, \dots, x_n]$ ensures that $k[x_1, \dots, x_n]/I$ contains a copy of k.
   The k-algebra $k[x_1, \dots, x_n]/I$ is finitely generated, and every finitely generated
k-algebra is isomorphic (as a k-algebra) to the quotient of a polynomial ring by
an ideal (see Exercise 1). We can also characterize which k-algebras correspond
to coordinate rings of varieties. A k-algebra is said to be *reduced* if it contains no
nilpotent elements (i.e., no nonzero element r such that $r^m = 0$ for some integer
$m > 0$). The coordinate ring of a variety is reduced, because the ideal of a variety
is radical (see Exercise 18 of §2). In the exercises, you will prove that when k is
algebraically closed, every reduced, finitely generated k-algebra is isomorphic to
the coordinate ring of a variety.
   One advantage of k-algebras over ideals is that two varieties are isomorphic if
and only if their coordinate rings are isomorphic as k-algebras (Theorem 9 of §4).
When we study affine varieties up to isomorphism, it no longer makes sense to talk
about the ideal of a variety, since isomorphic varieties may live in affine spaces
of different dimensions and hence have ideals in polynomial rings with different
numbers of variables. Yet the coordinate rings of these varieties are essentially the
same, since they are isomorphic as k-algebras.
   In this section, we will explore two related aspects of the structure of finitely
generated k-algebras.

Relative Finiteness

Given an ideal $I \subseteq k[x_1, \dots, x_n]$, the Finiteness Theorem (Theorem 6 of §3) characterizes when the k-algebra $k[x_1, \dots, x_n]/I$ has finite dimension as a vector space
over k. To generalize this result, suppose that we allow the ideal to depend on a set
of parameters $y_1, \dots, y_m$. Thus, we consider an ideal

$$I \subseteq k[x_1, \dots, x_n, y_1, \dots, y_m],$$

and we assume that $I \cap k[y_1, \dots, y_m] = \{0\}$ since we do not the ideal to impose
any restrictions on the parameters. One consequence of this assumption is that the
natural map

$$k[y_1, \dots, y_m] \longrightarrow k[x_1, \dots, x_n, y_1, \dots, y_m]/I$$

is easily seen to be one-to-one. This allows us to regard $k[y_1, \dots, y_m]$ as a subring of
the quotient ring $k[x_1, \dots, x_n, y_1, \dots, y_m]/I$.

To adapt the Finiteness Theorem to this situation, we need to know what it means for $k[x_1, \ldots, x_n, y_1, \ldots, y_m]/I$ to be "finite" over the subring $k[y_1, \ldots, y_m]$. We have the following general notion of finiteness for a subring of a commutative ring.

Definition 2. Given a commutative ring S and a subring $R \subseteq S$, we say that S is **finite over R** if there are finitely many elements $s_1, \ldots, s_\ell \in S$ such that every $s \in S$ can be written in the form
$$s = a_1 s_1 + \cdots + a_\ell s_\ell, \quad a_1, \ldots, a_\ell \in R.$$

For example, consider $k[x, y]/\langle x^2 - y^2 \rangle$. It is easy to see that $k[y] \cap \langle x^2 - y^2 \rangle = \{0\}$, so that we can consider $k[y]$ as a subring of $k[x, y]/\langle x^2 - y^2 \rangle$. Moreover, $k[x, y]/\langle x^2 - y^2 \rangle$ is finite over $k[y]$ because every element can be expressed as a $k[y]$-linear combination of the images of 1 and x in $k[x, y]/\langle x^2 - y^2 \rangle$.

On the other hand, if we consider the quotient ring $k[x, y]/\langle xy \rangle$, we again have $k[y] \cap \langle xy \rangle = \{0\}$, so that we can again consider $k[y]$ as a subring of $k[x, y]/\langle xy \rangle$. However, $k[x, y]/\langle xy \rangle$ is *not* finite over $k[y]$ because none of the images of $1, x, x^2, x^3, \ldots$ can be expressed as a $k[y]$-linear combination of the others. Similarly, in the exercises we ask you to show that $k[x, y]/\langle xy - 1 \rangle$ is not finite over the subring $k[y]$.

Here is an important consequence of finiteness.

Proposition 3. *Assume that S is finite over R. Then every $s \in S$ satisfies an equation of the form*

(1) $$s^\ell + a_1 s^{\ell-1} + \cdots + a_\ell = 0, \quad a_1, \ldots, a_\ell \in R.$$

Proof. By assumption, there exist $s_1, \ldots s_\ell \in S$ such that every $s \in S$ can be written in the form
$$s = a_1 s_1 + \cdots + a_\ell s_\ell, \quad a_i \in R.$$

Now fix an element $s \in S$. Thinking of the set $s_1, \ldots, s_\ell$ as akin to a basis, we can define an $\ell \times \ell$ matrix $A = (a_{ij})$ with entries in the ring R that represents multiplication by s on $s_1, \ldots, s_\ell$:
$$s \cdot s_i = a_{i1} s_1 + \cdots + a_{i\ell} s_\ell, \quad 1 \leq i \leq \ell, \, a_{ij} \in R.$$

If we write $\mathbf{v}$ for the transpose $(s_1, \ldots, s_\ell)^t$, then A has the property that $A\mathbf{v} = s\mathbf{v}$.

The characteristic polynomial of A is $\det(A - xI_\ell)$, where I_ℓ is the $\ell \times \ell$ identity matrix. The coefficient of x is $(-1)^\ell$, which allows us to write
$$\det(A - xI_\ell) = (-1)^\ell(x^\ell + a_1 x^{\ell-1} + \cdots + a_\ell).$$

Since A has entries in R and the determinant of a matrix is a polynomial in its entries (see §4 of Appendix A), it follows that $a_i \in R$ for all i. Hence the proposition will follow once we prove that $\det(A - sI_\ell) = 0$. If we were doing linear algebra over a field, then this would follow immediately since $A\mathbf{v} = s\mathbf{v}$ would imply that s was an eigenvalue of A.

However, standard linear algebra does not apply here since we are working with rings $R \subseteq S$. Fortunately, the argument is not difficult. Let $B = A - sI_\ell$ and let C be the transpose of the matrix of cofactors of B as defined in Appendix A, §4. The formula given there shows that C has entries in R and satisfies $CB = \det(B)I_\ell$.

The equation $A\mathbf{v} = s\mathbf{v}$ implies that $B\mathbf{v} = (A - sI_\ell)\mathbf{v} = \mathbf{0}$, where $\mathbf{0}$ is the $\ell \times 1$ column vector with all entries 0. Then

$$\det(B)\mathbf{v} = (\det(B)I_\ell)\mathbf{v} = (CB)\mathbf{v} = C(B\mathbf{v}) = C\mathbf{0} = \mathbf{0}.$$

Thus $\det(B)s_i = 0$ for all $1 \leq i \leq \ell$. Every element of S, in particular $1 \in S$, is a linear combination of the s_i with coefficients in R. Thus $1 = b_1 s_1 + \cdots + b_\ell s_\ell$, $b_i \in R$. Multiplying this equation by $\det(B)$, we obtain $\det(B) = \det(B) \cdot 1 = 0$, and the proposition follows.   □

The proof above is often called the "determinant trick." It is a good example of an elementary proof that is not obvious.

In general, an element $s \in S$ is *integral* over a subring R if s satisfies an equation of the form (1). Thus Proposition 3 can be restated as saying that if S is finite over R, then every element of S is integral over R.

We can now state the relative version of the Finiteness Theorem from §3.

Theorem 4 (Relative Finiteness Theorem). *Let $I \subseteq k[x_1, \ldots, x_n, y_1, \ldots, y_m]$ be an ideal such that $I \cap k[y_1, \ldots, y_m] = \{0\}$ and fix an ordering of n-elimination type (as in Exercise 5 of Chapter 3, §1). Then the following statements are equivalent:*

(i) *For each i, $1 \leq i \leq n$, there is some $m_i \geq 0$ such that $x_i^{m_i} \in \langle \mathrm{LT}(I) \rangle$.*

(ii) *Let G be a Gröbner basis for I. Then for each i, $1 \leq i \leq n$, there is some $m_i \geq 0$ such that $x_i^{m_i} = \mathrm{LM}(g)$ for some $g \in G$.*

(iii) *The set $\{x^\alpha \mid$ there is $\beta \in \mathbb{Z}_{\geq 0}^m$ such that $x^\alpha y^\beta \notin \langle \mathrm{LT}(I) \rangle\}$ is finite.*

(iv) *The ring $k[x_1, \ldots, x_n, y_1, \ldots, y_m]/I$ is finite over the subring $k[y_1, \ldots, y_m]$.*

Proof. (i) ⟺ (ii) The proof is identical to the proof of (i) ⟺ (ii) in Theorem 6 of §3.

(ii) ⟹ (iii) If some power $x_i^{m_i} \in \langle \mathrm{LT}(I) \rangle$ for $i = 1, \ldots, n$, then any monomial $x^\alpha y^\beta = x_1^{\alpha_1} \cdots x_n^{\alpha_n} y^\beta$ for which some $\alpha_i \geq m_i$ is in $\langle \mathrm{LT}(I) \rangle$. Hence a monomial in the complement of $\langle \mathrm{LT}(I) \rangle$ must have $\alpha_i \leq m_i - 1$ for all $1 \leq i \leq n$. As a result, there are at most $m_1 \cdot m_2 \cdots m_n$ monomials x^α such that $x^\alpha y^\beta \notin \langle \mathrm{LT}(I) \rangle$.

(iii) ⟹ (iv) Take $f \in k[x_1, \ldots, x_n, y_1, \ldots, y_m]$ and divide f by G using the division algorithm. This allows us to write f in the form $f = g + r$, where $g \in I$ and r is a linear combination of monomials $x^\alpha y^\beta \notin \langle \mathrm{LT}(I) \rangle$. By assumption, only finitely many different x^α's appear in these monomials, say $x^{\alpha_1}, \ldots, x^{\alpha_\ell}$. By collecting the terms of r that share the same x^{α_j}, we can write f in the form

$$f = g + B_1 x^{\alpha_1} + \cdots + B_\ell x^{\alpha_\ell}, \quad B_j \in k[y_1, \ldots, y_m].$$

Let $[f]$ denote the equivalence class of f in the quotient ring $k[x_1, \ldots, x_n, y_1, \ldots, y_m]/I$. Since $g \in I$, the above equation implies that

$$[f] = B_1[x^{\alpha_1}] + \cdots + B_\ell[x^{\alpha_\ell}]$$

in $k[x_1, \ldots, x_n, y_1, \ldots, y_m]/I$. Thus $[x^{\alpha_1}], \ldots, [x^{\alpha_\ell}]$ satisfy Definition 2, which proves that $k[x_1, \ldots, x_n, y_1, \ldots, y_m]/I$ is finite over the subring $k[y_1, \ldots, y_m]$.

(iv) $\Rightarrow$ (i) Fix i with $1 \leq i \leq n$. By (iv) and Proposition 3, there is an equation

$$[x_i]^d + A_1[x_i]^{d-1} + \cdots + A_d = 0, \quad A_j \in k[y_1, \ldots, y_m],$$

in $k[x_1, \ldots, x_n, y_1, \ldots, y_m]/I$. Back in $k[x_1, \ldots, x_n, y_1, \ldots, y_m]$, this means that

$$x_i^d + A_1 x_i^{d-1} + \cdots + A_d \in I.$$

Since we are using an n-elimination order, x_i is greater than any monomial in $y_1, \ldots, y_m$. Since $A_j \in k[y_1, \ldots, y_m]$, this implies $x_i^d > \mathrm{LT}(A_j x_i^{d-j})$ for $j = 1, \ldots, d$. It follows that $x_i^d = \mathrm{LT}(x_i^d + A_1 x_i^{d-1} + \cdots + A_d) \in \langle \mathrm{LT}(I) \rangle$, and we are done. $\qquad \square$

The Finiteness Theorem from §3 has parts (i)–(v), where (i)–(iv) were equivalent over any field k. Do you see how the Relative Finiteness Theorem proved here generalizes (i)–(iv) of the Finiteness Theorem?

The final part (v) of the Finiteness Theorem states that $\mathbf{V}(I)$ is finite and is equivalent to the other parts of the theorem when k is algebraically closed. Hence, it is natural to ask if the Relative Finiteness Theorem has a similar geometric meaning in the algebraically closed case.

We begin with the geometric meaning of an ideal $I \subseteq k[x_1, \ldots, x_n, y_1, \ldots, y_m]$. The inclusion of k-algebras $k[y_1, \ldots, y_m] \subseteq k[x_1, \ldots, x_n, y_1, \ldots, y_m]$ corresponds to the projection $k^{n+m} \to k^m$ that sends a point $(\mathbf{a}, \mathbf{b}) = (a_1, \ldots, a_n, b_1, \ldots, b_m)$ in k^{n+m} to its last m coordinates $\mathbf{b} \in k^m$. The ideal I gives a variety $\mathbf{V}(I) \subseteq k^{n+m}$. Composing this inclusion with the projection gives a function

$$(2) \qquad \pi : \mathbf{V}(I) \longrightarrow k^m.$$

Then a point $\mathbf{b} = (b_1, \ldots, b_m) \in k^m$ gives two objects:

- (Algebraic) The ideal $I_\mathbf{b} \subseteq k[x_1, \ldots, x_n]$, which is obtained by setting $y_i = b_i$ in all elements of the ideal I.
- (Geometric) The *fiber* $\pi^{-1}(\mathbf{b}) = \mathbf{V}(I) \cap (k^n \times \{\mathbf{b}\})$, which consists of all points of $\mathbf{V}(I)$ whose last m coordinates are given by $\mathbf{b}$.

The relation between these (which you will establish in the exercises) is that

$$\pi^{-1}(\mathbf{b}) = \mathbf{V}(I_\mathbf{b}) \times \{\mathbf{b}\}.$$

Earlier, we said that $I \subseteq k[x_1, \ldots, x_n, y_1, \ldots, y_m]$ is an ideal depending on parameters $y_1, \ldots, y_m$. We now see what this means: I gives the family of ideals $I_\mathbf{b} \subseteq k[x_1, \ldots, x_n]$, parametrized by $\mathbf{b} \in k^m$, with corresponding varieties $\mathbf{V}(I_\mathbf{b}) \subseteq k^n$. Combining these for all possible parameter values gives the disjoint union

$$(3) \qquad \bigcup_{\mathbf{b} \in k^m} \mathbf{V}(I_\mathbf{b}) \times \{\mathbf{b}\} = \bigcup_{\mathbf{b} \in k^m} \pi^{-1}(\mathbf{b}) = \mathbf{V}(I) \subseteq k^{n+m}.$$

This explains nicely how $\mathbf{V}(I)$ relates to the family of ideals $\{I_\mathbf{b}\}_{\mathbf{b} \in k^m}$.

For an example of this, let us return to our example $k[y] \subseteq k[x,y]/\langle x^2 - y^2 \rangle$, which corresponds to the projection $\mathbf{V}(x^2 - y^2) \to k$ onto the second coordinate. If we set $k = \mathbb{R}$ and pick $\mathbf{b} \in \mathbb{R}$, then for $I = \langle x^2 - y^2 \rangle \subseteq \mathbb{R}[x,y]$, we get the picture:

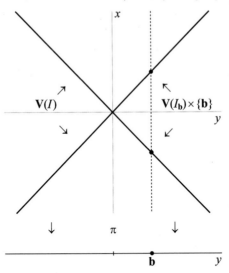

The y-axis is shown horizontally since we want to write the projection π vertically. Note that $\mathbf{V}(I_{\mathbf{b}}) = \mathbf{V}(x^2 - b^2) = \{\pm \mathbf{b}\}$. Also, the union of the fibers $\pi^{-1}(\mathbf{b}) = \mathbf{V}(I_{\mathbf{b}}) \times \{\mathbf{b}\}$ is $\mathbf{V}(I) \subseteq \mathbb{R}^2$, as predicted by (3).

The next step is to interpret our assumption that $I \cap k[y_1, \ldots, y_m] = \{0\}$. The intersection is the n-th elimination ideal I_n of I from Chapter 3. When k is algebraically closed, the Closure Theorem (Theorem 3 of Chapter 3, §2) tells us that $\mathbf{V}(I_n) \subseteq k^m$ the Zariski closure of the image of the variety $\mathbf{V}(I)$ under the projection π. Thus, to say that $I \cap k[y_1, \ldots, y_m] = \{0\}$ is to say that the Zariski closure of the image $\pi(\mathbf{V}(I))$ is equal to the entire affine space k^m. The Closure Theorem also implies that there is a proper variety $W \subsetneq k^m$ such that

$$k^m \setminus W \subseteq \pi(\mathbf{V}(I)).$$

It follows that for "most" $\mathbf{b} \in k^m$ (i.e., outside of W), the fiber $\pi^{-1}(\mathbf{b}) = \mathbf{V}(I_{\mathbf{b}}) \times \{\mathbf{b}\}$ is nonempty. Thus, $I \cap k[y_1, \ldots, y_m] = \{0\}$ means $\mathbf{V}(I_{\mathbf{b}}) \neq \emptyset$ for most $\mathbf{b} \in k^m$.

We now offer the geometric interpretation of the Relative Finiteness Theorem.

Theorem 5 (Geometric Relative Finiteness Theorem). *Let k be algebraically closed and $I \subseteq k[x_1, \ldots, x_n, y_1, \ldots, y_m]$ be an ideal such that $I \cap k[y_1, \ldots, y_m] = \{0\}$. If, in addition, $k[x_1, \ldots, x_n, y_1, \ldots, y_m]/I$ is finite over $k[y]$, then:*

(i) *The projection map $\pi : \mathbf{V}(I) \to k^m$ is onto and has finite fibers.*

(ii) *For each $\mathbf{b} \in k^m$, the variety $\mathbf{V}(I_{\mathbf{b}}) \subseteq k^n$ is finite and nonempty.*

Proof. Note that (i) can be restated as saying that $\pi^{-1}(\mathbf{b})$ is finite and nonempty for all $\mathbf{b} \in k^m$. Since $\pi^{-1}(\mathbf{b}) = \mathbf{V}(I_{\mathbf{b}}) \times \{\mathbf{b}\}$, it follows that (i) and (ii) are equivalent. We will prove the theorem by showing that $\mathbf{V}(I_{\mathbf{b}})$ is finite and $\pi^{-1}(\mathbf{b})$ is nonempty.

Set $\mathbf{x} = (x_1, \ldots, x_n)$ and $\mathbf{y} = (y_1, \ldots, y_m)$, and let $G = \{g_1, \ldots, g_t\}$ be a reduced Gröbner basis for $I \subseteq k[\mathbf{x}, \mathbf{y}]$ for lex order $x_1 > \ldots > x_n > y_1 > \ldots > y_m$. Since G is reduced, the Relative Finiteness Theorem tells us that for each $1 \leq i \leq n$, there exists $g \in G$ such that $\mathrm{LT}(g) = x_i^{N_i}$. Given $\mathbf{b} \in k^m$, we have $g(\mathbf{x}, \mathbf{b}) \in I_{\mathbf{b}}$, and the lex order we are using implies that $\mathrm{LT}(g(\mathbf{x}, \mathbf{b})) = x_i^{N_i}$ in $k[\mathbf{x}]$ for lex order with $x_1 > \cdots > x_n$. Since this holds for all $1 \leq i \leq n$, the Finiteness Theorem from §3 implies that the variety $\mathbf{V}(I_{\mathbf{b}})$ is finite.

It remains to show that the fibers of $\pi : \mathbf{V}(I) \to k^m$ are nonempty. First observe that the n-th elimination ideal of I is $I_n = I \cap k[\mathbf{y}]$. By assumption, this equals $\{0\}$, so that $\mathbf{V}(I_n) = \mathbf{V}(0) = k^m$. Thus, every $\mathbf{b} \in k^m$ is a partial solution. Since $\pi^{-1}(\mathbf{b}) \neq \emptyset$ if and only if there exists $\mathbf{a} \in k^n$ such that $(\mathbf{a}, \mathbf{b}) \in \mathbf{V}(I)$, we need only show that the partial solution $\mathbf{b} \in k^m$ extends to a solution $(\mathbf{a}, \mathbf{b}) \in \mathbf{V}(I) \subseteq k^{n+m}$. We do this by applying the Extension Theorem (Theorem 3 of Chapter 3, §1) n times. To successively eliminate $x_1, \ldots, x_n$, we use the elimination ideals

$$I_0 = I \cap k[x_1, \ldots, x_n, y_1, \ldots, y_m] = I \cap k[\mathbf{x}, \mathbf{y}] = I$$
$$I_1 = I \cap k[x_2 \ldots, x_n, \mathbf{y}]$$
$$\vdots$$
$$I_{n-1} = I \cap k[x_n, \mathbf{y}]$$
$$I_n = I \cap k[\mathbf{y}] = \{0\}.$$

By the Elimination Theorem (Theorem 2 of Chapter 3, §1), the intersection $G_i = G \cap k[x_{i+1}, \ldots, x_n, \mathbf{y}]$ is a Gröbner basis for I_i for $1 \leq i \leq n-1$. Note also that I_i is the first elimination ideal of I_{i-1} for $1 \leq i \leq n-1$. As in the previous paragraph, we have $g \in G$ with $\mathrm{LT}(g) = x_i^{N_i}$. The lex order we are using then implies $g \in G \cap k[x_i, \ldots, x_n, \mathbf{y}] = G_{i-1}$. Now write g as

$$g = x_i^{N_i} + \text{terms in which } x_i \text{ has degree} < N_i.$$

Since k is algebraically closed, the version of the Extension Theorem given in Corollary 4 of Chapter 3, §1 tells us that any partial solution $(a_{i+1}, \ldots, a_n, \mathbf{b}) \in \mathbf{V}(I_i)$ extends to a solution $(a_i, a_{i+1}, \ldots, a_n, \mathbf{b}) \in \mathbf{V}(I_{i-1})$ for each $1 \leq i \leq n$. So, now we are done: start with any $\mathbf{b} \in \mathbf{V}(I_n) = k^m$ and apply Corollary 4 successively with $i = n, n-1, \ldots, 1$ to find $(\mathbf{a}, \mathbf{b}) = (a_1, \ldots, a_n, \mathbf{b}) \in \mathbf{V}(I) \subseteq k^{n+m}$. $\square$

We remark that the converse of the Geometric Relative Finiteness Theorem is not true. There are ideals $I \subseteq k[\mathbf{x}, \mathbf{y}]$ for which the projection of $\mathbf{V}(I) \to k^m$ is onto with finite fibers, but $k[\mathbf{x}, \mathbf{y}]/I$ is not finite over $k[\mathbf{y}]$. In the exercises, you will show that the ideal $\langle x(xy - 1) \rangle \subseteq k[x, y]$ provides an example.

Noether Normalization

When $I \subseteq k[\mathbf{x}, \mathbf{y}]$ satisfies the Relative Finiteness Theorem, the k-algebra $A = k[\mathbf{x}, \mathbf{y}]/I$ is finite over the polynomial ring $k[\mathbf{y}] \subseteq A$. This seems like a very special situation. The surprise that when k is infinite, *any* finitely generated k-algebra is finite over a polynomial subring. This is the *Noether Normalization Theorem*.

Before stating the theorem precisely, we need one bit of terminology: elements $u_1, \ldots, u_m$ in a k-algebra A are *algebraically independent over k* when the only polynomial f with coefficients in k satisfying $f(u_1, \ldots, u_m) = 0$ in A is the zero polynomial. When this condition is satisfied, the subring $k[u_1, \ldots, u_m] \subseteq A$ is isomorphic to a polynomial ring in m variables.

Theorem 6 (Noether Normalization). *Let k be an infinite field and suppose that we are given a finitely generated k-algebra A. Then:*

(i) *There are algebraically independently elements $u_1, \ldots, u_m \in A$ such that A is finite over $k[u_1, \ldots, u_m]$.*

(ii) *If A is generated by $s_1, \ldots, s_\ell$ as a k-algebra, then $m \leq \ell$ and $u_1, \ldots, u_m$ can be chosen to be k-linear combinations of $s_1, \ldots, s_\ell$.*

The proof will involve the following finiteness result.

Lemma 7. *Let $R \subseteq S$ be rings, and assume that there is $s \in S$ such that $S = R[s]$, meaning that every element of S can be written as a polynomial in s with coefficients in R. If, in addition, s satisfies an equation*

$$s^d + r_1 s^{d-1} + \cdots + r_d = 0$$

with $r_1, \ldots, r_d \in R$, then S is finite over R.

Proof. Let $R[x]$ be the ring of polynomials in x with coefficients in R. Also set $f = x^d + r_1 x^{d-1} + \cdots + r_d \in R[x]$. By hypothesis, any element of S can be written as $g(s)$ for some $g \in R[x]$. Dividing g by f gives $a, b \in R[x]$ such that $g = af + b$ and either $\deg(b) < d$ or $b = 0$. The division algorithm presented in §5 of Chapter 1 requires field coefficients, but for more general ring coefficients, the algorithm still works, provided one divides by a monic polynomial such as f. You will check this in the exercises.

Since $f(s) = 0$, we obtain $g(s) = a(s)f(s) + b(s) = a(s) \cdot 0 + b(s) = b(s)$. Since $\deg(b) < d$, we see that any element of S can be expressed as an R-linear combination of $1, s, \ldots, s^{d-1}$. Hence S is finite over R. $\square$

We are now ready to prove the Noether Normalization Theorem.

Proof of Theorem 6. We proceed by induction on the number ℓ of generators. First suppose that $\ell = 1$. Then A is a k-algebra with a single generator s_1, so $A = k[s_1]$. There are two cases to consider. If there is no nonzero polynomial f with $f(s_1) = 0$, then s_1 is algebraically independent over k. So the theorem holds with $m = 1$ and $u_1 = s_1$ since A is finite over $k[s_1] = A$. It remains to consider the case when there is a nonzero polynomial $f \in k[x]$ with $f(s_1) = 0$. We may assume that f is monic. Since f has coefficients in k, Lemma 7 implies A is finite over the subring k. So the theorem holds in this case with $m = 0$.

Now, let $\ell > 1$ and suppose that the theorem is true for finitely generated k-algebras with fewer than ℓ generators. If there is no nonzero polynomial f in ℓ

variables such that $f(s_1, \ldots, s_\ell) = 0$, then $s_1, \ldots, s_\ell$ are algebraically independent and the theorem is true with $m = \ell$ and $u_i = s_i$ for $1 \leq i \leq \ell$ since A is trivially finite over $k[s_1, \ldots, s_\ell] = A$.

Otherwise, choose f nonzero with $f(s_1, \ldots, s_\ell) = 0$. Let $\tilde{s}_1 = s_1 - a_1 s_\ell$, $\tilde{s}_2 = s_2 - a_2 s_\ell, \ldots, \tilde{s}_{\ell-1} = s_{\ell-1} - a_{\ell-1} s_\ell$ for some $a_1, \ldots, a_{\ell-1} \in k$ that we shall choose momentarily. Thus,

$$(4) \qquad s_1 = \tilde{s}_1 + a_1 s_\ell, \ s_2 = \tilde{s}_2 + a_2 s_\ell, \ \ldots, \ s_{\ell-1} = \tilde{s}_{\ell-1} + a_{\ell-1} s_\ell.$$

Replacing s_i with $\tilde{s}_i + a_i s_\ell$ for $1 \leq i \leq \ell - 1$ in f and expanding, we obtain

$$
(5) \qquad
\begin{aligned}
0 = f(s_1, \ldots, s_\ell) &= f(\tilde{s}_1 + a_1 s_\ell, \ldots, \tilde{s}_{\ell-1} + a_{\ell-1} s_\ell, s_\ell) \\
&= c(a_1, \ldots, a_{\ell-1}) s_\ell^d + \text{terms in which } s_\ell \text{ has degree} < d,
\end{aligned}
$$

where d is the total degree of f. We will leave it as an exercise for the reader to show that $c(a_1, \ldots, a_{\ell-1})$ is a nonzero polynomial expression in $a_1, \ldots, a_{\ell-1}$. Since the field k is infinite, we can choose $a_1, \ldots, a_{\ell-1} \in k$ with $c(a_1, \ldots, a_{\ell-1}) \neq 0$ by Proposition 5 of Chapter 1, §1.

For this choice of $a_1, \ldots, a_{\ell-1}$, let $B = k[\tilde{s}_1, \ldots, \tilde{s}_{\ell-1}] \subseteq A$ be the k-algebra generated by $\tilde{s}_1, \ldots, \tilde{s}_{\ell-1}$. We prove that A is finite over B as follows. Dividing each side of (5) by $c(a_1, \ldots, a_{\ell-1}) \neq 0$ gives an equation

$$0 = s_\ell^d + b_1 s_\ell^{d-1} + \cdots + b_d,$$

where $b_1, \ldots, b_d \in B$. Since A is generated as k-algebra by $s_1, \ldots, s_\ell$, (4) implies that the same is true for $\tilde{s}_1, \ldots, \tilde{s}_{\ell-1}, s_\ell$. It follows that $A = B[s_\ell]$. This and the above equation imply that A is finite over B by Lemma 7.

Our inductive hypothesis applies to B since it has $\ell - 1$ algebra generators. Hence there are $m \leq \ell - 1$ algebraically independent elements $u_1, \ldots, u_m \in B$ such that B is finite over $k[u_1, \ldots, u_m]$. Since A is finite over B and "being finite" is transitive (see Exercise 14), it follows that A is finite over $k[u_1, \ldots, u_m]$. Furthermore, since we may assume that $u_1, \ldots, u_m$ are k-linear combinations of $\tilde{s}_1 = s_1 - a_1 s_\ell, \ldots, \tilde{s}_{\ell-1} = s_{\ell-1} - a_{\ell-1} s_\ell$, the u_i are also k-linear combinations of $s_1, \ldots, s_\ell$. This completes the proof of the theorem. □

The inclusion $k[u_1, \ldots, u_m] \subseteq A$ constructed in Theorem 6 is called a *Noether normalization* of A. Part (i) of the theorem still holds when k is finite, though the proof is different—see Theorem 30 in Section 15.3 of DUMMIT and FOOTE (2004).

We next relate Noether normalizations to the Relative Finiteness Theorem. Let $k[u_1, \ldots, u_m] \subseteq A$ be a Noether normalization of A. In the proof of Theorem 8 below, we will see that $u_1, \ldots, u_m$ can be extended to a set of algebra generators $v_1, \ldots, v_n, u_1, \ldots, u_m$ of A over k. This gives an onto k-algebra homomorphism

$$k[x_1, \ldots, x_n, y_1, \ldots, y_m] \longrightarrow A$$

that maps x_i to v_i and y_i to u_i. Let I be the kernel of this map. Then we have a k-algebra isomorphism

(6) $$k[x_1,\ldots,x_n,y_1,\ldots,y_m]/I \cong A.$$

Furthermore, $I \cap k[y_1,\ldots,y_m] = \{0\}$, since a nonzero f in this intersection would satisfy $f(u_1,\ldots,u_m) = 0$, contradicting the algebraic independence of $u_1,\ldots,u_m$. As earlier in the section, $I \cap k[y_1,\ldots,y_m] = \{0\}$ gives an inclusion

$$k[y_1,\ldots,y_m] \subseteq k[x_1,\ldots,x_n,y_1,\ldots,y_m]/I$$

which under the isomorphism (6) corresponds to the inclusion

$$k[u_1,\ldots,u_m] \subseteq A.$$

Since A is finite over $k[u_1,\ldots,u_m]$, it follows that $k[x_1,\ldots,x_n,y_1,\ldots,y_m]/I$ is finite over $k[y_1,\ldots,y_m]$, just as in Theorem 4. In other words, with a suitable choice of algebra generators, the Noether normalization of *any* finitely generated k-algebra becomes an instance of the Relative Finiteness Theorem.

Our final topic is the geometric interpretation of Noether normalization. We will assume that k is algebraically closed, hence infinite by Exercise 4 of Chapter 4, §1. The coordinate ring $k[V]$ of an affine variety $V \subseteq k^\ell$ has a Noether normalization $k[u_1,\ldots,u_m] \subseteq k[V]$ such that the u_i are algebraically independent. If $y_1,\ldots,y_m$ are variables, then mapping y_i to u_i gives a k-algebra homomorphism

$$k[y_1,\ldots,y_m] \longrightarrow k[V],$$

which by Proposition 8 of §4 corresponds to a polynomial map of varieties

$$\pi : V \longrightarrow k^m.$$

The following theorem records the properties of this map.

Theorem 8 (Geometric Noether Normalization). *Let $V \subseteq k^\ell$ be a variety with k algebraically closed. Then a Noether normalization of $k[V]$ can be chosen so that the above map $\pi : V \to k^m$ has the following properties:*

(i) *π is the composition of the inclusion $V \subseteq k^\ell$ with a linear map $k^\ell \to k^m$.*

(ii) *π is onto with finite fibers.*

Proof. Let $\phi_i : V \to k$ be the coordinate function that map $(a_1,\ldots,a_\ell) \in V$ to $a_i \in k$ for $i = 1,\ldots,\ell$. If $t_1,\ldots,t_\ell$ are coordinates on k^ℓ, then the isomorphism

$$k[t_1,\ldots,t_\ell]/\mathbf{I}(V) \cong k[V]$$

from Proposition 3 of §2 takes the equivalence class $[t_i]$ to ϕ_i. Note that $\phi_1,\ldots,\phi_\ell$ are algebra generators of $k[V]$ over k by Exercise 2.

Since each ϕ_i is the restriction to V of a linear map $k^\ell \to k$, the same is true for any k-linear combination of the ϕ_i. By Theorem 6, we can choose $u_1,\ldots,u_m$ to be linear combinations of the ϕ_i. The resulting map $(u_1,\ldots,u_m) : V \to k^m$ is the composition of $V \subseteq k^\ell$ with a linear map $k^\ell \to k^m$. Since $(u_1,\ldots,u_m)$ is the map π described above, part (i) of the theorem follows.

For part (ii), note that $u_1, \ldots, u_m \in k[V]$ are linearly independent over k since they are algebraically independent. When we write $u_j = \sum_{i=1}^{\ell} c_{ij} \phi_i$ for $1 \leq j \leq m$, the resulting $\ell \times m$ matrix $C = (c_{ij})$ has rank m. In the exercises, you will use standard facts from linear algebra to show that there is an invertible $\ell \times \ell$ matrix D whose last m columns are the matrix C. Set $n = \ell - m$, so that $\ell = n + m$.

Using D, we get new coordinates $x_1, \ldots, x_n, y_1, \ldots, y_m$ on k^ℓ, where the x_i (resp. the y_i) are the linear combinations of $t_1, \ldots, t_\ell$ that use the first n (resp. last m) rows of D. With these new coordinates on k^ℓ, we have an isomorphism

$$k[x_1, \ldots, x_n, y_1, \ldots, y_m]/\mathbf{I}(V) \cong k[V]$$

where $[y_i]$ maps to u_i by the construction of D, and $[x_i]$ maps to an element we will denote by v_i. With this setup, we are in the situation of Theorems 4 and 5, and with our new coordinates, the map $\pi : k^\ell = k^{n+m} \to k^m$ is projection onto the last m coordinates. Then Theorem 5 implies that π is onto with finite fibers.   □

This theorem, while nice, does not capture the full geometric meaning of Noether normalization. For this, one needs to study what are called *finite morphisms*, as defined in Section II.3 of HARTSHORNE (1977). We will give a brief hint of the geometry of finite morphisms in Exercise 17.

So, we have two ways to regard the Noether Normalization Theorem. From an algebraic perspective, it is a structure theorem for finitely generated k-algebras over an infinite field, where one finds algebraically independent elements such that the algebra is finite over the subalgebra generated by these elements. From a geometric point of view, the Noether Normalization Theorem asserts that for a variety over an algebraically closed field, there is a number m such that the variety maps *onto* an m-dimensional affine space with finite fibers. As we will see in Chapter 9, the number m is in fact the dimension of the variety.

EXERCISES FOR §6

1. Show that if a ring R is a finitely generated k-algebra, then R is isomorphic to the quotient ring $k[x_1, \ldots, x_n]/I$ for some n and some ideal $I \subseteq k[x_1, \ldots, x_n]$. Hint: If $r_1, \ldots, r_n \in R$ have the property described in part (i) of Definition 1, then map $k[x_1, \ldots, x_n]$ to R by sending x_i to r_i, and consider the ideal in $k[x_1, \ldots, x_n]$ that is the kernel of this map.

2. a. Prove that the coordinate ring $k[V]$ of a variety $V \subseteq k^n$ is a reduced, finitely generated k-algebra. Also explain why the algebra generators can be chosen to be the coordinate functions ϕ_i that map $(a_1, \ldots, a_n) \in V$ to $a_i \in k$ for $i = 1, \ldots, n$.

   b. Conversely, suppose we are given a reduced, finitely generated k-algebra R. Show that R is isomorphic to $k[x_1, \ldots, x_n]/I$ where I is radical. Hint: Combine the previous exercise with Exercise 18 of §2.

   c. Conclude if k is algebraically closed, then every reduced finitely generated k-algebra is isomorphic to the coordinate ring of an affine variety. Hint: The Nullstellensatz.

3. Let $I \subseteq k[x_1, \ldots, x_n, y_1, \ldots, y_m]$ be an ideal satisfying $I \cap k[y_1, \ldots, y_m] = \{0\}$. Prove that the natural map $k[y_1, \ldots, y_m] \to k[x_1, \ldots, x_n, y_1, \ldots, y_m]/I$ is one-to-one.

4. Show that a one-to-one map of k-algebras $k[W] \to k[V]$ corresponds to a *dominant map* $V \to W$ (i.e., a map such that W is the Zariski closure of the image of V). Hint: $\phi : V \to W$ gives $V \to \overline{\phi(V)} \subseteq W$.

5. Let $I \subseteq k[x_1, \ldots, x_n]$ be an ideal, and consider the quotient ring $k[x_1, \ldots, x_n]/I$ with sub-ring $k = k \cdot 1$ consisting of all constant multiples of the identity. Show that $k[x_1, \ldots, x_n]/I$ is finite over k in the sense of Definition 2 if and only if it has finite dimension as a vector space over k. Hint: A vector space has finite dimension if and only if it has a finite spanning set.

6. Show that $k[x, y]/\langle x^a - y^b \rangle$ is finite both over $k[x]$ and $k[y]$, but that $k[x, y]/\langle x^{a+1} - xy^b \rangle$ is finite over $k[y]$ but not $k[x]$. Interpret what this means geometrically.

7. a. Carefully verify that $k[x, y]/\langle xy \rangle$ is not finite over $k[y]$.
   b. However, show that $k[x, y]/\langle x^2 - y^2 \rangle$ (which is finite over $k[y]$) is isomorphic as a k-algebra to $k[x, y]/\langle xy \rangle$. You may assume that $2 \neq 0$ in k. Hint: Use the invertible linear (hence, polynomial) map of k^2 to itself that takes (x, y) to $(x - y, x + y)$.

8. Show that the k-algebras $k[x, y]/\langle xy \rangle$ and $k[x, y]/\langle xy - 1 \rangle$ are not isomorphic.

9. Given rings $R \subseteq S$, we say that S is *finitely generated over* R provided that there exist $s_1, \ldots, s_m \in S$ such that every element of S can be expressed as a polynomial with coefficients in R in $s_1, \ldots, s_m$. Thus a k-algebra is finitely generated in the sense of Definition 1 exactly when it is finitely generated over the subring k.
   a. Given a ring R, explain why the polynomial ring $R[x]$ is finitely generated over R but not finite over R in the sense of Definition 2. Hint: Proposition 3 will be useful.
   b. Now assume that S is finitely generated over R. Strengthen Proposition 3 by showing that S is finite over R if and only if every $s \in S$ satisfies an equation of the form (1), i.e., every $s \in S$ is integral over R as defined in the text. Hint: For the converse, each s_i satisfies an equation of the form $s_i^{\ell_i} + \cdots = 0$. Then consider the finitely many elements of S given by $s_1^{a_1} \cdots s_m^{a_m}$ where $0 \leq a_i \leq \ell_i$ for all i.
   c. Suppose that we have finitely generated k-algebras $R \subseteq S$. Use part (b) to prove that S is finite over R if and only if every element of S is integral over R.

10. Consider the map $\pi : \mathbf{V}(I) \to k^m$ defined in equation (2) in the text. Given a point $\mathbf{b} \in k^m$, prove that $\pi^{-1}(\mathbf{b}) = \mathbf{V}(I_\mathbf{b}) \times \{\mathbf{b}\}$ in k^{n+m}.

11. Suppose that we have a function $f : X \to Y$ for sets X and Y. The *fiber* of $y \in Y$ is the set $f^{-1}(y) = \{x \in X \mid f(x) = y\}$. Prove that we have a disjoint union $X = \bigcup_{y \in Y} f^{-1}(y)$.

12. Consider the ideal $I = \langle x(xy - 1) \rangle = \langle x^2 y - x \rangle \subseteq k[x, y]$, where k is algebraically closed.
    a. Prove that the projection $\pi : \mathbf{V}(I) \to k$ to the y-axis is onto with finite fibers.
    b. Show that the k-algebra $k[x, y]/I$ is not finite over $k[y]$.
    This shows that the converse of the Geometric Relative Finiteness Theorem is not true.

13. We want to show that the polynomial $c(a_1, \ldots, a_{\ell-1})$ in (4) is a nonzero polynomial in $a_1, \ldots, a_{\ell-1}$.
    a. Suppose f has total degree d and write $f = f_d + f_{d-1} + \ldots + f_0$, where f_i is the sum of all terms of f of total degree i, $0 \leq i \leq d$, Show that after the substitution (5), the coefficient $c(a_1, \ldots, a_{\ell-1})$ of s_ℓ^d in is $f_d(a_1, \ldots, a_{\ell-1}, 1)$.
    b. A polynomial $h(z_1, \ldots, z_\ell)$ is *homogeneous of total degree* N if each monomial in h has total degree N. (For example, the polynomial f_i from part (a) is homogeneous of total degree i.) Show that h is the zero polynomial in $k[z_1, \ldots, z_\ell]$ if and only if $h(z_1, \ldots, z_{\ell-1}, 1)$ is the zero polynomial in $k[z_1, \ldots, z_{\ell-1}]$.
    c. Conclude that $c(a_1, \ldots, a_{\ell-1})$ is not the zero polynomial in $a_1, \ldots, a_{\ell-1}$.

14. Let $R \subseteq S \subseteq T$ be rings where S is finite over R via $s_1, \ldots, s_M \in S$ and T is finite over S via $t_1, \ldots, t_N \in T$. Prove that T is finite over R via the products $s_i t_j$, $1 \leq i \leq M, 1 \leq j \leq N$.

15. Let $R[x]$ be the ring of polynomials in x with coefficients in a ring R. Given a monic polynomial $f \in R[x]$, adapt the division algorithm (Proposition 2 of Chapter 1, §5) to show that any $g \in R[x]$ can be written $g = af + b$ where $a, b \in R[x]$ and either $\deg(b) < d$ or $b = 0$.

16. Let C be an $\ell \times m$ matrix of rank m with entries in a field k. Explain why $m \leq \ell$ and prove that there is an invertible $\ell \times \ell$ matrix D whose last m columns are the columns of

C. Hint: The columns of C give m linearly independent vectors in k^ℓ. Extend these to a basis of k^ℓ.

17. In the situation of Theorem 5, we showed that $\pi : \mathbf{V}(I) \to k^m$ is onto with finite fibers. Here we explore some further properties of π.

   a. Let $J \subseteq k[\mathbf{x}, \mathbf{y}]$ be an ideal containing I, so that $\mathbf{V}(J) \subseteq \mathbf{V}(I)$. Prove that $\pi(\mathbf{V}(J)) \subseteq k^m$ is a variety. Hint: Let $J_n = J \cap k[\mathbf{y}]$ and note that $\overline{\pi(\mathbf{V}(J))} = \mathbf{V}(J_n)$ by the Closure Theorem. Then adapt the proof of Theorem 5 to show that all partial solutions in $\mathbf{V}(J_n)$ extend to solutions in $\mathbf{V}(J)$ and conclude that $\phi(\mathbf{V}(J)) = \mathbf{V}(J_n)$.

   b. In general, a polynomial map $\phi : V \to W$ of varieties is *closed* if the image under ϕ of any subvariety of V is a subvariety of W. Thus part (a) implies that $\pi : \mathbf{V}(I) \to k^m$ is closed. Prove more generally that the map $\tilde{\pi} : \mathbf{V}(I) \times k^N \to k^m \times k^N$ defined by $\tilde{\pi}(u, v) = (\pi(u), v)$ is also closed. Hint: All you are doing is adding N more y variables in Theorem 5.

   The text mentioned the notion of a *finite morphism*, which for a polynomial map $\phi : V \to W$ means that $k[V]$ is finite over $\phi^*(k[W])$. Thus the map $\pi : \mathbf{V}(I) \to k^m$ from Theorem 5 is finite. Over an algebraically closed field, a finite map has finite fibers and a property called *universally closed*. We will not define this concept here [see Section II.4 of HARTSHORNE (1977)], but we note that part (b) follows from being universally closed.

18. This exercise will compute the Noether normalization of the k-algebra $A = k[x] \times k[x]$, where we use coordinate-wise multiplication. We assume k is algebraically closed.

   a. A k-algebra contains a copy of the field k. For A, show that $k \cong \{(a, a) \mid a \in k\} \subseteq A$. Also show that A is reduced.

   b. Prove that A is generated by $s_1 = (1, 0)$, $s_2 = (0, 1)$ and $s_3 = (x, x)$ as a k-algebra. Hint: Show that $(f(x), g(x)) = f(s_3)s_1 + g(s_3)s_2$.

   c. Prove that $k[s_3] \subseteq A$ is a Noether normalization of A.

   d. Define a k-algebra homomorphism $k[x_1, x_2, y] \to A$ by $x_1 \mapsto s_1$, $x_2 \mapsto s_2$, and $y \mapsto s_3$. Prove that the kernel of this homomorphism is $I = \langle x_1 + x_2 - 1, x_2^2 - x_2 \rangle$.

   e. In the affine space k^3 with coordinates x_1, x_2, y, show that $V = \mathbf{V}(I)$ is a disjoint union of two lines. Then use this to explain why $k[V]$ is isomorphic to $k[x] \times k[x] = A$.

Chapter 6
Robotics and Automatic Geometric Theorem Proving

In this chapter we will consider two applications of concepts and techniques from algebraic geometry in areas of computer science. First, continuing a theme introduced in several examples in Chapter 1, we will develop a systematic approach that uses algebraic varieties to describe the space of possible configurations of mechanical linkages such as robot "arms." We will use this approach to solve the forward and inverse kinematic problems of robotics for certain types of robots.

Second, we will apply the algorithms developed in earlier chapters to the study of automatic geometric theorem proving, an area that has been of interest to researchers in artificial intelligence. When the hypotheses of a geometric theorem can be expressed as polynomial equations relating the Cartesian coordinates of points in the Euclidean plane, the geometrical propositions deducible from the hypotheses will include all the statements that can be expressed as polynomials in the ideal generated by the hypotheses.

§1 Geometric Description of Robots

To treat the space of configurations of a robot geometrically, we need to make some simplifying assumptions about the components of our robots and their mechanical properties. We will not try to address many important issues in the engineering of actual robots (such as what types of motors and mechanical linkages would be used to achieve what motions, and how those motions would be controlled). Thus, we will restrict ourselves to highly idealized robots. However, within this framework, we will be able to indicate the types of problems that actually arise in robot motion description and planning.

We will always consider robots constructed from rigid links or segments, connected by joints of various types. For simplicity, we will consider only robots in which the segments are connected *in series*, as in a human limb. One end of our robot "arm" will usually be fixed in position. At the other end will be the "hand" or

© Springer International Publishing Switzerland 2015
D.A. Cox et al., *Ideals, Varieties, and Algorithms*, Undergraduate Texts
in Mathematics, DOI 10.1007/978-3-319-16721-3_6

"effector," which will sometimes be considered as a final segment of the robot. In actual robots, this "hand" might be provided with mechanisms for grasping objects or with tools for performing some task. Thus, one of the major goals is to be able to describe and specify the position and orientation of the "hand."

Since the segments of our robots are rigid, the possible motions of the entire robot assembly are determined by the motions of the joints. Many actual robots are constructed using

- planar revolute joints, and
- prismatic joints.

A planar revolute joint permits a *rotation* of one segment relative to another. We will assume that both of the segments in question lie in one plane and all motions of the joint will leave the two segments in that plane. (This is the same as saying that the axis of rotation is perpendicular to the plane in question.)

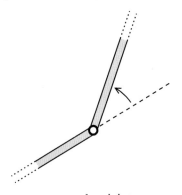

a revolute joint

A prismatic joint permits one segment of a robot to move by sliding or *translation* along an axis. The following sketch shows a schematic view of a prismatic joint between two segments of a robot lying in a plane. Such a joint permits translational motion along a line in the plane.

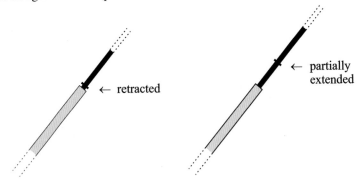

a prismatic joint

If there are several joints in a robot, we will assume for simplicity that the joints all lie in the same plane, that the axes of rotation of all revolute joints are perpendicular to that plane, and, in addition, that the translation axes for the prismatic joints all lie in the plane of the joints. Thus, all motion will take place in one plane. Of course, this leads to a very restricted class of robots. Real robots must usually be capable of 3-dimensional motion. To achieve this, other types and combinations of joints are used. These include "ball" joints allowing rotation about any axis passing through some point in $\mathbb{R}^3$ and helical or "screw" joints combining rotation and translation along the axis of rotation in $\mathbb{R}^3$. It would also be possible to connect several segments of a robot with planar revolute joints, but with *nonparallel* axes of rotation. All of these possible configurations can be treated by methods similar to the ones we will present, but we will not consider them in detail. Our purpose here is to illustrate how affine varieties can be used to describe the geometry of robots, not to present a treatise on practical robotics. The planar robots provide a class of relatively uncomplicated but illustrative examples for us to consider.

Example 1. Consider the following planar robot "arm" with three revolute joints and one prismatic joint. All motions of the robot take place in the plane of the page.

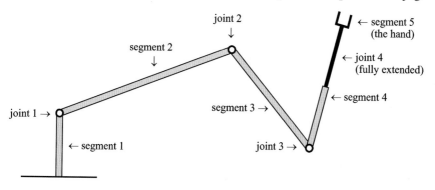

For easy reference, we number the segments and joints of a robot in increasing order out from the fixed end to the hand. Thus, in the above figure, segment 2 connects joints 1 and 2, and so on. Joint 4 is prismatic, and we will regard segment 4 as having variable length, depending on the setting of the prismatic joint. In this robot, the hand of the robot comprises segment 5.

In general, the position or setting of a revolute joint between segments i and $i+1$ can be described by measuring the angle θ (counterclockwise) from segment i to segment $i + 1$. Thus, the totality of settings of such a joint can be parametrized by a *circle* S^1 or by the interval $[0, 2\pi]$ with the endpoints identified. (In some cases, a revolute joint may not be free to rotate through a full circle, and then we would parametrize the possible settings by a subset of S^1.)

Similarly, the setting of a prismatic joint can be specified by giving the distance the joint is extended or, as in Example 1, by the total length of the segment (i.e., the distance between the end of the joint and the previous joint). Either way, the settings of a prismatic joint can be parametrized by a finite interval of real numbers.

If the joint settings of our robot can be specified independently, then the possible settings of the whole collection of joints in a planar robot with r revolute joints and p prismatic joints can be parametrized by the Cartesian product

$$\mathcal{J} = S^1 \times \cdots \times S^1 \times I_1 \times \cdots \times I_p,$$

where there is one S^1 factor for each revolute joint, and I_j gives the settings of the j-th prismatic joint. We will call $\mathcal{J}$ the *joint space* of the robot.

We can describe the space of possible configurations of the "hand" of a planar robot as follows. Fixing a Cartesian coordinate system in the plane, we can represent the possible positions of the "hand" by the points (a, b) of a region $U \subseteq \mathbb{R}^2$. Similarly, we can represent the orientation of the "hand" by giving a unit vector aligned with some specific feature of the hand. Thus, the possible hand orientations are parametrized by vectors $\mathbf{u}$ in $V = S^1$. For example, if the "hand" is attached to a revolute joint, then we have the following picture of the hand configuration:

We will call $C = U \times V$ the *configuration space* or *operational space* of the robot's hand.

Since the robot's segments are assumed to be rigid, each collection of joint settings will place the "hand" in a uniquely determined location, with a uniquely determined orientation. Thus, we have a function or mapping

$$f : \mathcal{J} \longrightarrow C$$

which encodes how the different possible joint settings yield different hand configurations.

The two basic problems we will consider can be described succinctly in terms of the mapping $f : \mathcal{J} \to C$ described above:

- (Forward Kinematic Problem) Can we give an explicit description or formula for f in terms of the joint settings (our coordinates on $\mathcal{J}$) and the dimensions of the segments of the robot "arm"?
- (Inverse Kinematic Problem) Given $c \in C$, can we determine one or all the $j \in \mathcal{J}$ such that $f(j) = c$?

In §2, we will see that the forward problem is relatively easily solved. Determining the position and orientation of the "hand" from the "arm" joint settings is mainly a matter of being systematic in describing the relative positions of the segments on either side of a joint. Thus, the forward problem is of interest mainly as a preliminary to the inverse problem. We will show that the mapping $f : \mathcal{J} \to \mathcal{C}$ giving the "hand" configuration as a function of the joint settings may be written as a polynomial mapping as in Chapter 5, §1.

The inverse problem is somewhat more subtle since our explicit formulas will not be linear if revolute joints are present. Thus, we will need to use the general results on systems of polynomial equations to solve the equation

$$(1) \qquad\qquad\qquad f(j) = c.$$

One feature of nonlinear systems of equations is that there can be several different solutions, even when the entire set of solutions is finite. We will see in §3 that this is true for a planar robot arm with three (or more) revolute joints. As a practical matter, the potential nonuniqueness of the solutions of the systems (1) is sometimes very desirable. For instance, if our real world robot is to work in a space containing physical obstacles or barriers to movement in certain directions, it may be the case that some of the solutions of (1) for a given $c \in \mathcal{C}$ correspond to positions that are not physically reachable:

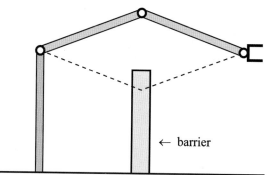

To determine whether it is possible to reach a given position, we might need to determine *all* solutions of (1), then see which one(s) are feasible given the constraints of the environment in which our robot is to work.

EXERCISES FOR §1

1. Give descriptions of the joint space $\mathcal{J}$ and the configuration space $\mathcal{C}$ for the planar robot picture in Example 1 in the text. For your description of $\mathcal{C}$, determine a bounded subset of $U \subseteq \mathbb{R}^2$ containing all possible hand positions. Hint: The description of U will depend on the lengths of the segments.

2. Consider the mapping $f : \mathcal{J} \to \mathcal{C}$ for the robot pictured in Example 1 in the text. On geometric grounds, do you expect f to be a *one-to-one* mapping? Can you find two different ways to put the hand in some particular position with a given orientation? Are there more than two such positions?

The text discussed the joint space $\mathcal{J}$ and the configuration space $\mathcal{C}$ for planar robots. In the following problems, we consider what $\mathcal{J}$ and $\mathcal{C}$ look like for robots capable of motion in three dimensions.

3. What would the configuration space $\mathcal{C}$ look like for a 3-dimensional robot? In particular, how can we describe the possible hand orientations?

4. A "ball" joint at point B allows segment 2 in the robot pictured below to rotate by any angle about any axis in $\mathbb{R}^3$ passing through B. (Note: The motions of this joint are similar to those of the "joystick" found in some computer games.)

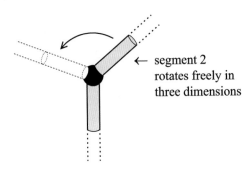

← segment 2
rotates freely in
three dimensions

a ball joint

a. Describe the set of possible joint settings for this joint mathematically. Hint: The distinct joint settings correspond to the possible direction vectors of segment 2.

b. Construct a one-to-one correspondence between your set of joint settings in part (a) and the unit sphere $S^2 \subseteq \mathbb{R}^3$. Hint: One simple way to do this is to use the spherical angular coordinates ϕ, θ on S^2.

5. A helical or "screw" joint at point H allows segment 2 of the robot pictured below to extend out from H along the line L in the direction of segment 1, while rotating about the axis L.

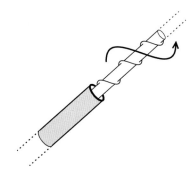

a helical or "screw" joint

The rotation angle θ (measured from the original, unextended position of segment 2) is given by $\theta = l \cdot \alpha$, where $l \in [0, m]$ gives the distance from H to the other end of segment 2 and α is a constant angle. Give a mathematical description of the space of joint settings for this joint.

6. Give a mathematical description of the joint space $\mathcal{J}$ for a 3-dimensional robot with two "ball" joints and one helical joint.

§2   The Forward Kinematic Problem

In this section, we will present a standard method for solving the forward kinematic problem for a given robot "arm." As in §1, we will only consider robots in $\mathbb{R}^2$, which means that the "hand" will be constrained to lie in the plane. Other cases will be studied in the exercises.

All of our robots will have a first segment that is anchored, or fixed in position. In other words, there is no movable joint at the initial endpoint of segment 1. With this convention, we will use a standard rectangular coordinate system in the plane to describe the position and orientation of the "hand." The origin of this coordinate system is placed at joint 1 of the robot arm, which is also fixed in position since all of segment 1 is. For example:

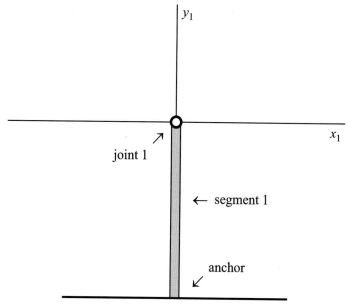

In addition to the global (x_1, y_1) coordinate system, we introduce a local rectangular coordinate system at each of the *revolute joints* to describe the relative positions of the segments meeting at that joint. Naturally, these coordinate systems will *change* as the position of the "arm" varies.

At a revolute joint i, we introduce an (x_{i+1}, y_{i+1}) coordinate system in the following way. The origin is placed at joint i. We take the positive x_{i+1}-axis to lie along the direction of segment $i + 1$ (in the robot's current position). Then the positive y_{i+1}-axis is determined to form a normal right-handed rectangular coordinate system. Note that for each $i \geq 2$, the (x_i, y_i) coordinates of joint i are $(l_i, 0)$, where l_i is the length of segment i.

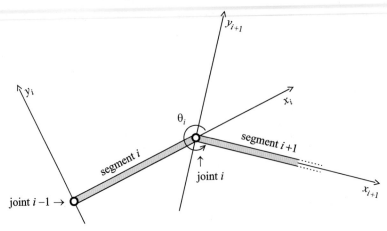

Our first goal is to relate the (x_{i+1}, y_{i+1}) coordinates of a point with the (x_i, y_i) coordinates of that point. Let θ_i be the counterclockwise angle from the x_i-axis to the x_{i+1}-axis. This is the same as the joint setting angle θ_i described in §1. From the diagram above, we see that if a point q has (x_{i+1}, y_{i+1}) coordinates

$$q = (a_{i+1}, b_{i+1}),$$

then to obtain the (x_i, y_i) coordinates of q, say

$$q = (a_i, b_i),$$

we rotate by the angle θ_i (to align the x_i- and x_{i+1}-axes), and then translate by the vector $(l_i, 0)$ (to make the origins of the coordinate systems coincide). In the exercises, you will show that rotation by θ_i is accomplished by multiplying by the rotation matrix

$$\begin{pmatrix} \cos\theta_i & -\sin\theta_i \\ \sin\theta_i & \cos\theta_i \end{pmatrix}.$$

It is also easy to check that translation is accomplished by adding the vector $(l_i, 0)$. Thus, we get the following relation between the (x_i, y_i) and (x_{i+1}, y_{i+1}) coordinates of q:

$$\begin{pmatrix} a_i \\ b_i \end{pmatrix} = \begin{pmatrix} \cos\theta_i & -\sin\theta_i \\ \sin\theta_i & \cos\theta_i \end{pmatrix} \cdot \begin{pmatrix} a_{i+1} \\ b_{i+1} \end{pmatrix} + \begin{pmatrix} l_i \\ 0 \end{pmatrix}.$$

This coordinate transformation is also commonly written in a shorthand form using a 3×3 matrix and 3-component vectors:

$$(1) \qquad \begin{pmatrix} a_i \\ b_i \\ 1 \end{pmatrix} = \begin{pmatrix} \cos\theta_i & -\sin\theta_i & l_i \\ \sin\theta_i & \cos\theta_i & 0 \\ 0 & 0 & 1 \end{pmatrix} \cdot \begin{pmatrix} a_{i+1} \\ b_{i+1} \\ 1 \end{pmatrix} = A_i \cdot \begin{pmatrix} a_{i+1} \\ b_{i+1} \\ 1 \end{pmatrix}.$$

This allows us to combine the rotation by θ_i with the translation along segment i into a single 3×3 matrix A_i.

Example 1. With this notation in hand, let us next consider a general plane robot "arm" with three revolute joints:

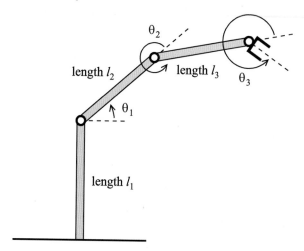

We will think of the hand as segment 4, which is attached via the revolute joint 3 to segment 3. As before, l_i will denote the length of segment i. We have

$$A_1 = \begin{pmatrix} \cos\theta_1 & -\sin\theta_1 & 0 \\ \sin\theta_1 & \cos\theta_1 & 0 \\ 0 & 0 & 1 \end{pmatrix}$$

since the origin of the (x_2, y_2) coordinate system is also placed at joint 1. We also have matrices A_2 and A_3 as in formula (1). The key observation is that the global coordinates of any point can be obtained by starting in the (x_4, y_4) coordinate system and working our way back to the global (x_1, y_1) system one joint at a time. In other words, we multiply the (x_4, y_4) coordinate vector of the point by A_3, A_2, A_1 in turn:

$$\begin{pmatrix} x_1 \\ y_1 \\ 1 \end{pmatrix} = A_1 A_2 A_3 \begin{pmatrix} x_4 \\ y_4 \\ 1 \end{pmatrix}.$$

Using the trigonometric addition formulas, this equation can be written as

$$\begin{pmatrix} x_1 \\ y_1 \\ 1 \end{pmatrix} = \begin{pmatrix} \cos(\theta_1 + \theta_2 + \theta_3) & -\sin(\theta_1 + \theta_2 + \theta_3) & l_3\cos(\theta_1 + \theta_2) + l_2\cos\theta_1 \\ \sin(\theta_1 + \theta_2 + \theta_3) & \cos(\theta_1 + \theta_2 + \theta_3) & l_3\sin(\theta_1 + \theta_2) + l_2\sin\theta_1 \\ 0 & 0 & 1 \end{pmatrix} \begin{pmatrix} x_4 \\ y_4 \\ 1 \end{pmatrix}.$$

Since the (x_4, y_4) coordinates of the hand are $(0,0)$ (because the hand is attached directly to joint 3), we obtain the (x_1, y_1) coordinates of the hand by setting $x_4 = y_4 = 0$ and computing the matrix product above. The result is

(2)
$$\begin{pmatrix} x_1 \\ y_1 \\ 1 \end{pmatrix} = \begin{pmatrix} l_3\cos(\theta_1 + \theta_2) + l_2\cos\theta_1 \\ l_3\sin(\theta_1 + \theta_2) + l_2\sin\theta_1 \\ 1 \end{pmatrix}.$$

The hand orientation is determined if we know the angle between the x_4-axis and the direction of any particular feature of interest to us on the hand. For instance, we might simply want to use the direction of the x_4-axis to specify this orientation. From our computations, we know that the angle between the x_1-axis and the x_4-axis is simply $\theta_1 + \theta_2 + \theta_3$. Knowing the θ_i allows us to also compute this angle.

If we combine this fact about the hand orientation with the formula (2) for the hand position, we get an explicit description of the mapping $f : \mathcal{J} \to \mathcal{C}$ introduced in §1. As a function of the joint angles θ_i, the configuration of the hand is given by

$$
(3) \qquad f(\theta_1 + \theta_2 + \theta_3) = \begin{pmatrix} l_3 \cos(\theta_1 + \theta_2) + l_2 \cos\theta_1 \\ l_3 \sin(\theta_1 + \theta_2) + l_2 \sin\theta_1 \\ \theta_1 + \theta_2 + \theta_3 \end{pmatrix} .
$$

The same ideas will apply when any number of planar revolute joints are present. You will study the explicit form of the function f in these cases in Exercise 7.

Example 2. Prismatic joints can also be handled within this framework. For instance, let us consider a planar robot whose first three segments and joints are the same as those of the robot in Example 1, but which has an additional prismatic joint between segment 4 and the hand. Thus, segment 4 will have variable length and segment 5 will be the hand.

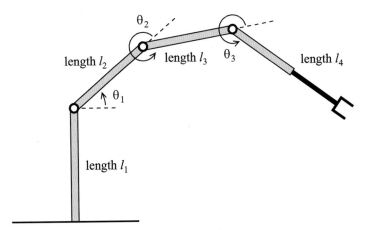

The translation axis of the prismatic joint lies along the direction of segment 4. We can describe such a robot as follows. The three revolute joints allow us exactly the same freedom in placing joint 3 as in the robot studied in Example 1. However, the prismatic joint allows us to change the length of segment 4 to any value between $l_4 = m_1$ (when retracted) and $l_4 = m_2$ (when fully extended). By the reasoning given in Example 1, if the setting l_4 of the prismatic joint is known, then the position of the hand will be given by multiplying the product matrix $A_1 A_2 A_3$ times the (x_4, y_4) coordinate vector of the hand, namely $(l_4, 0)$. It follows that the configuration of the hand is given by

$$\text{(4)} \quad g(\theta_1, \theta_2, \theta_3, l_4) = \begin{pmatrix} l_4\cos(\theta_1 + \theta_2 + \theta_3) + l_3\cos(\theta_1 + \theta_2) + l_2\cos\theta_1 \\ l_4\sin(\theta_1 + \theta_2 + \theta_3) + l_3\sin(\theta_1 + \theta_2) + l_2\sin\theta_1 \\ \theta_1 + \theta_2 + \theta_3 \end{pmatrix}.$$

As before, l_2 and l_3 are constant, but $l_4 \in [m_1, m_2]$ is now another variable. The hand orientation will be given by $\theta_1 + \theta_2 + \theta_3$ as before since the setting of the prismatic joint will not affect the direction of the hand.

We will next discuss how formulas such as (3) and (4) may be converted into representations of f and g as polynomial or rational mappings in suitable variables. The joint variables for revolute and for prismatic joints are handled differently. For the revolute joints, the most direct way of converting to a polynomial set of equations is to use an idea we have seen several times before, for example, in Exercise 8 of Chapter 2, §8. Even though $\cos\theta$ and $\sin\theta$ are transcendental functions, they give a parametrization

$$x = \cos\theta,$$
$$y = \sin\theta$$

of the algebraic variety $\mathbf{V}(x^2 + y^2 - 1)$ in the plane. Thus, we can write the components of the right-hand side of (3) or, equivalently, the entries of the matrix $A_1 A_2 A_3$ in (2) as functions of

$$c_i = \cos\theta_i,$$
$$s_i = \sin\theta_i,$$

subject to the constraints

$$\text{(5)} \quad c_i^2 + s_i^2 - 1 = 0$$

for $i = 1, 2, 3$. Note that the variety defined by these three equations in $\mathbb{R}^6$ is a concrete realization of the joint space $\mathcal{J}$ for this type of robot. Geometrically, this variety is just a Cartesian product of three copies of the circle.

Explicitly, we obtain from (3) an expression for the hand position as a function of the variables $c_1, s_1, c_2, s_2, c_3, s_3$. Using the trigonometric addition formulas, we can write

$$\cos(\theta_1 + \theta_2) = \cos\theta_1 \cos\theta_2 - \sin\theta_1 \sin\theta_2 = c_1 c_2 - s_1 s_2.$$

Similarly,

$$\sin(\theta_1 + \theta_2) = \sin\theta_1 \cos\theta_2 + \sin\theta_2 \cos\theta_1 = s_1 c_2 + s_2 c_1.$$

Thus, the (x_1, y_1) coordinates of the hand position are:

$$\text{(6)} \quad \begin{pmatrix} l_3(c_1 c_2 - s_1 s_2) + l_2 c_1 \\ l_3(s_1 c_2 + s_2 c_1) + l_2 s_1 \end{pmatrix}.$$

In the language of Chapter 5, we have defined a polynomial mapping from the variety $\mathcal{J} = \mathbf{V}(x_1^2 + y_1^2 - 1, x_2^2 + y_2^2 - 1, x_3^2 + y_3^2 - 1)$ to $\mathbb{R}^2$. Note that the hand *position* does not depend on θ_3. That angle enters only in determining the hand orientation.

Since the hand orientation depends directly on the angles θ_i themselves, it is *not* possible to express the orientation itself as a polynomial in $c_i = \cos\theta_i$ and $s_i = \sin\theta_i$. However, we can handle the orientation in a similar way. See Exercise 3.

Similarly, from the mapping g in Example 2, we obtain the polynomial form

$$(7) \quad \begin{pmatrix} l_4(c_1(c_2c_3 - s_2s_3) - s_1(c_2s_3 + c_3s_2)) + l_3(c_1c_2 - s_1s_2) + l_2c_1 \\ l_4(s_1(c_2c_3 - s_2s_3) + c_1(c_2s_3 + c_3s_2)) + l_3(s_1c_2 + s_2c_1) + l_2s_1 \end{pmatrix}$$

for the (x_1, y_1) coordinates of the hand position. Here $\mathcal{J}$ is the subset $V \times [m_1, m_2]$ of the variety $V \times \mathbb{R}$, where $V = \mathbf{V}(x_1^2 + y_1^2 - 1, x_2^2 + y_2^2 - 1, x_3^2 + y_3^2 - 1)$. The length l_4 is treated as another ordinary variable in (7), so our component functions are polynomials in l_4, and the c_i and s_i.

A second way to write formulas (3) and (4) is based on the *rational* parametrization

$$(8) \quad \begin{aligned} x &= \frac{1 - t^2}{1 + t^2}, \\ y &= \frac{2t}{1 + t^2} \end{aligned}$$

of the circle introduced in §3 of Chapter 1. [In terms of the trigonometric parametrization, $t = \tan(\theta/2)$.] This allows us to express the mapping (3) in terms of three variables $t_i = \tan(\theta_i/2)$. We will leave it as an exercise for the reader to work out this alternate explicit form of the mapping $f : \mathcal{J} \to C$ in Example 1. In the language of Chapter 5, the variety $\mathcal{J}$ for the robot in Example 1 is birationally equivalent to $\mathbb{R}^3$. We can construct a rational parametrization $\rho : \mathbb{R}^3 \to \mathcal{J}$ using three copies of the parametrization (8). Hence, we obtain a rational mapping from $\mathbb{R}^3$ to $\mathbb{R}^2$, expressing the hand coordinates of the robot arm as functions of t_1, t_2, t_3 by taking the composition of ρ with the hand coordinate mapping in the form (6).

Both of these forms have certain advantages and disadvantages for practical use. For the robot of Example 1, one immediately visible advantage of the rational mapping obtained from (8) is that it involves only three variables rather than the six variables $s_i, c_i, i = 1, 2, 3$, needed to describe the full mapping f as in Exercise 3. In addition, we do not need the three extra constraint equations (5). However, the t_i values corresponding to joint positions with θ_i close to π are awkwardly large, and there is no t_i value corresponding to $\theta_i = \pi$. We do not obtain every theoretically possible hand position in the image of the mapping f when it is expressed in this form. Of course, this might not actually be a problem if our robot is constructed so that segment $i+1$ is not free to fold back onto segment i (i.e., the joint setting $\theta_i = \pi$ is not possible). The polynomial form (6) is more unwieldy, but since it comes from the trigonometric (unit-speed) parametrization of the circle, it does not suffer from the potential shortcomings of the rational form. It would be somewhat better suited for revolute joints that can freely rotate through a full circle.

EXERCISES FOR §2

1. Consider the plane $\mathbb{R}^2$ with an orthogonal right-handed coordinate system (x_1, y_1). Now introduce a second coordinate system (x_2, y_2) by rotating the first counterclockwise by an angle θ. Suppose that a point q has (x_1, y_1) coordinates (a_1, b_1) and (x_2, y_2) coordinates (a_2, b_2). We claim that

$$\begin{pmatrix} a_1 \\ b_1 \end{pmatrix} = \begin{pmatrix} \cos\theta & -\sin\theta \\ \sin\theta & \cos\theta \end{pmatrix} \cdot \begin{pmatrix} a_2 \\ b_2 \end{pmatrix}.$$

To prove this, first express the (x_2, y_2) coordinates of q in polar form as

$$q = (a_2, b_2) = (r\cos\alpha, \ r\sin\alpha).$$

a. Show that the (x_1, y_1) coordinates of q are given by

$$q = (a_1, b_1) = (r\cos(\alpha + \theta), \ r\sin(\alpha + \theta)).$$

b. Now use trigonometric identities to prove the desired formula.

2. In Examples 1 and 2, we used a 3×3 matrix A to represent each of the changes of coordinates from one local system to another. Those changes of coordinates were rotations, followed by translations. These are special types of *affine transformations*.

a. Show that any affine transformation in the plane

$$x' = ax + by + e,$$
$$y' = cx + dy + f$$

can be represented in a similar way:

$$\begin{pmatrix} x' \\ y' \\ 1 \end{pmatrix} = \begin{pmatrix} a & b & e \\ c & d & f \\ 0 & 0 & 1 \end{pmatrix} \cdot \begin{pmatrix} x \\ y \\ 1 \end{pmatrix}.$$

b. Give a similar representation for affine transformations of $\mathbb{R}^3$ using 4×4 matrices.

3. In this exercise, we will reconsider the hand orientation for the robots in Examples 1 and 2. Namely, let $\alpha = \theta_1 + \theta_2 + \theta_3$ be the angle giving the hand orientation in the (x_1, y_1) coordinate system.

a. Using the trigonometric addition formulas, show that

$$c = \cos\alpha, \ s = \sin\alpha$$

can be expressed as polynomials in $c_i = \cos\theta_i$ and $s_i = \sin\theta_i$. Thus, the whole mapping f can be expressed in polynomial form, at the cost of introducing an extra coordinate function for C.

b. Express c and s using the rational parametrization (8) of the circle.

4. Consider a planar robot with a revolute joint 1, segment 2 of length l_2, a prismatic joint 2 with settings $l_3 \in [0, m_3]$, and a revolute joint 3, with segment 4 being the hand.

a. What are the joint and configuration spaces $\mathcal{J}$ and $\mathcal{C}$ for this robot?

b. Using the method of Examples 1 and 2, construct an explicit formula for the mapping $f : \mathcal{J} \to \mathcal{C}$ in terms of the trigonometric functions of the joint angles.

c. Convert the function f into a polynomial mapping by introducing suitable new coordinates.

5. Rewrite the mappings f and g in Examples 1 and 2, respectively, using the rational parametrization (8) of the circle for each revolute joint. Show that in each case the hand position

and orientation are given by rational mappings on $\mathbb{R}^n$. (The value of n will be different in the two examples.)

6. Rewrite the mapping f for the robot from Exercise 4, using the rational parametrization (8) of the circle for each revolute joint.

7. Consider a planar robot with a fixed segment 1 as in our examples in this section and with n revolute joints linking segments of length $l_2, \ldots, l_n$. The hand is segment $n+1$, attached to segment n by joint n.
   a. What are the joint and configuration spaces for this robot?
   b. Show that the mapping $f : \mathcal{J} \to \mathcal{C}$ for this robot has the form

$$f(\theta_1, \ldots, \theta_n) = \begin{pmatrix} \sum_{i=1}^{n-1} l_{i+1} \cos\left(\sum_{j=1}^{i} \theta_j \right) \\ \sum_{i=1}^{n-1} l_{i+1} \sin\left(\sum_{j=1}^{i} \theta_j \right) \\ \sum_{i=1}^{n} \theta_i \end{pmatrix}.$$

   Hint: Argue by induction on n.

8. Another type of 3-dimensional joint is a "spin" or nonplanar revolute joint that allows one segment to rotate or spin in the plane perpendicular to the other segment. In this exercise, we will study the forward kinematic problem for a 3-dimensional robot containing two "spin" joints. As usual, segment 1 of the robot will be fixed, and we will pick a global coordinate system (x_1, y_1, z_1) with the origin at joint 1 and segment 1 on the z_1-axis. Joint 1 is a "spin" joint with rotation axis along the z_1-axis, so that segment 2 rotates in the (x_1, y_1)-plane. Then segment 2 has length l_2 and joint 2 is a second "spin" joint connecting segment 2 to segment 3. The axis for joint 2 lies along segment 2, so that segment 3 always rotates in the plane perpendicular to segment 2.
   a. Construct a local right-handed orthogonal coordinate system (x_2, y_2, z_2) with origin at joint 1, with the x_2-axis in the direction of segment 2 and the y_2-axis in the (x_1, y_1)-plane. Give an explicit formula for the (x_1, y_1, z_1) coordinates of a general point, in terms of its (x_2, y_2, z_2) coordinates and of the joint angle θ_1.
   b. Express your formula from part (a) in matrix form, using the 4×4 matrix representation for affine space transformations given in part (b) of Exercise 2.
   c. Now, construct a local orthogonal coordinate system (x_3, y_3, z_3) with origin at joint 2, the x_3-axis in the direction of segment 3, and the z_3-axis in the direction of segment 2. Give an explicit formula for the (x_2, y_2, z_2) coordinates of a point in terms of its (x_3, y_3, z_3) coordinates and the joint angle θ_2.
   d. Express your formula from part (c) in matrix form.
   e. Give the transformation relating the (x_3, y_3, z_3) coordinates of a general point to its (x_1, y_1, z_1) coordinates in matrix form. Hint: This will involve suitably multiplying the matrices found in parts (b) and (d).

9. Consider the robot from Exercise 8.
   a. Using the result of part (c) of Exercise 8, give an explicit formula for the mapping $f : \mathcal{J} \to \mathcal{C}$ for this robot.
   b. Express the hand position for this robot as a polynomial function of the variables $c_i = \cos\theta_i$ and $s_i = \sin\theta_i$.
   c. The orientation of the hand (the end of segment 3) of this robot can be expressed by giving a unit vector in the direction of segment 3, expressed in the global coordinate system. Find an expression for the hand orientation.

§3 The Inverse Kinematic Problem and Motion Planning

In this section, we will continue the discussion of the robot kinematic problems introduced in §1. To begin, we will consider the inverse kinematic problem for the planar robot arm with three revolute joints studied in Example 1 of §2. Given a point $(x_1, y_1) = (a, b) \in \mathbb{R}^2$ and an orientation, we wish to determine whether it is possible to place the hand of the robot at that point with that orientation. If it is possible, we wish to find all combinations of joint settings that will accomplish this. In other words, we want to determine the *image* of the mapping $f : \mathcal{J} \to \mathcal{C}$ for this robot; for each c in the image of f, we want to determine the *inverse image* $f^{-1}(c)$.

It is quite easy to see geometrically that if $l_3 = l_2 = l$, the hand of our robot can be placed at any point of the closed disk of radius $2l$ centered at joint 1—the origin of the (x_1, y_1) coordinate system. On the other hand, if $l_3 \neq l_2$, then the hand positions fill out a closed annulus centered at joint 1. (See, for example, the ideas used in Exercise 14 of Chapter 1, §2.) We will also be able to see this using the solution of the forward problem derived in equation (7) of §2. In addition, our solution will give *explicit formulas* for the joint settings necessary to produce a given hand position. Such formulas could be built into a control program for a robot of this kind.

For this robot, it is also easy to control the hand orientation. Since the setting of joint 3 is independent of the settings of joints 1 and 2, we see that, given any θ_1 and θ_2, it is possible to attain any desired orientation $\alpha = \theta_1 + \theta_2 + \theta_3$ by setting $\theta_3 = \alpha - (\theta_1 + \theta_2)$ accordingly.

To simplify our solution of the inverse kinematic problem, we will use the above observation to ignore the hand orientation. Thus, we will concentrate on the position of the hand, which is a function of θ_1 and θ_2 alone. From equation (6) of §2, we see that the possible ways to place the hand at a given point $(x_1, y_1) = (a, b)$ are described by the following system of polynomial equations:

$$
\begin{aligned}
a &= l_3(c_1 c_2 - s_1 s_2) + l_2 c_1, \\
b &= l_3(c_1 s_2 + c_2 s_1) + l_2 s_1, \\
0 &= c_1^2 + s_1^2 - 1, \\
0 &= c_2^2 + s_2^2 - 1
\end{aligned}
$$

(1)

for c_1, s_1, c_2, s_2. To solve these equations, we first compute a grevlex Gröbner basis with

$$c_1 > s_1 > c_2 > s_2.$$

Our solutions will depend on the values of a, b, l_2, l_3, which appear as symbolic parameters in the coefficients of the Gröbner basis:

(2)

$$
\begin{aligned}
&c_1 - \frac{2bl_2 l_3}{2l_2(a^2 + b^2)}s_2 - \frac{a(a^2 + b^2 + l_2^2 - l_3^2)}{2l_2(a^2 + b^2)}, \\
&s_1 + \frac{2al_2 l_3}{2l_2(a^2 + b^2)}s_2 + \frac{b(a^2 + b^2 + l_2^2 - l_3^2)}{2l_2(a^2 + b^2)}, \\
&c_2 - \frac{a^2 + b^2 - l_2^2 - l_3^2}{2l_2 l_3}, \\
&s_2^2 + \frac{(a^2 + b^2)^2 - 2(a^2 + b^2)(l_2^2 + l_3^2) + (l_2^2 - l_3^2)^2}{4l_2^2 l_3^2}.
\end{aligned}
$$

In algebraic terms, this is the reduced Gröbner basis for the ideal I generated by the polynomials in (1) in the ring $\mathbb{R}(a, b, l_2, l_3)[c_1, s_1, c_2, s_2]$. Note that we allow denominators that depend only on the parameters a, b, l_2, l_3.

This is the first time we have computed a Gröbner basis over a field of rational functions, and one has to be a bit careful about how to interpret (2). Working over $\mathbb{R}(a, b, l_2, l_3)$ means that a, b, l_2, l_3 are abstract variables over $\mathbb{R}$, and, in particular, they are algebraically independent [i.e., if p is a polynomial with real coefficients such that $p(a, b, l_2, l_3) = 0$, then p must be the zero polynomial]. Yet, in practice, we want a, b, l_2, l_3 to be certain specific real numbers. When we make such a substitution, the polynomials (1) generate an ideal $\bar{I} \subseteq \mathbb{R}[c_1, s_1, c_2, s_2]$ corresponding to a specific hand position of a robot with specific segment lengths. The key question is whether (2) remains a Gröbner basis for $\bar{I}$ under this substitution. In general, the replacement of variables by specific values in a field is called *specialization*, and the question is how a Gröbner basis behaves under specialization.

A first observation is that we expect problems when a specialization causes any of the denominators in (2) to vanish. This is typical of how specialization works: things usually behave nicely for most (but not all) values of the variables. In the exercises, you will prove that there is a proper subvariety $W \subseteq \mathbb{R}^4$ such that (2) specializes to a Gröbner basis of $\bar{I}$ whenever a, b, l_2, l_3 take values in $\mathbb{R}^4 \setminus W$. We also will see that there is an algorithm for finding W. The subtle point is that, in general, the vanishing of denominators is not the only thing that can go wrong (you will work out some examples in the exercises). Fortunately, in the example we are considering, it can be shown that W is, in fact, defined by the vanishing of the denominators. This means that if we choose values $l_2 \neq 0, l_3 \neq 0$, and $a^2 + b^2 \neq 0$, then (2) still gives a Gröbner basis of (1). The details of the argument will be given in Exercise 9.

Given such a specialization, two observations follow immediately from the leading terms of the Gröbner basis (2). First, any zero s_2 of the last polynomial can be extended uniquely to a full solution of the system. Second, the set of solutions of (1) is a *finite* set for this choice of a, b, l_2, l_3. Indeed, since the last polynomial in (2) is quadratic in s_2, there can be at most *two* distinct solutions. It remains to see which a, b yield *real* values for s_2 (the relevant solutions for the geometry of our robot).

To simplify the formulas somewhat, we will specialize to the case $l_2 = l_3 = 1$. In Exercise 1, you will show that by either substituting $l_2 = l_3 = 1$ directly into (2) or setting $l_2 = l_3 = 1$ in (1) and recomputing a Gröbner basis in $\mathbb{R}(a, b)[c_1, s_1, c_2, s_2]$, we obtain the same result:

(3)
$$
\begin{aligned}
&c_1 - \frac{2b}{2(a^2 + b^2)}s_2 - \frac{a}{2}, \\
&s_1 + \frac{2a}{2(a^2 + b^2)}s_2 + \frac{b}{2}, \\
&c_2 - \frac{a^2 + b^2 - 2}{2}, \\
&s_2^2 + \frac{(a^2 + b^2)(a^2 + b^2 - 4)}{4}.
\end{aligned}
$$

Other choices for l_2 and l_3 will be studied in Exercise 4. [Although (2) remains a Gröbner basis for any nonzero values of l_2 and l_3, the geometry of the situation changes rather dramatically if $l_2 \neq l_3$.]

It follows from our earlier remarks that (3) is a Gröbner basis for (1) for all specializations of a and b where $a^2 + b^2 \neq 0$, which over $\mathbb{R}$ happens whenever the hand is not at the origin. Solving the last equation in (3), we find that

$$s_2 = \pm \frac{1}{2}\sqrt{(a^2 + b^2)(4 - (a^2 + b^2))}.$$

Note that the solution(s) of this equation are real if and only if $a^2 + b^2 \leq 4$, and when $a^2 + b^2 = 4$, we have a double root. From the geometry of the system, that is exactly what we expect. The distance from joint 1 to joint 3 is at most $l_2 + l_3 = 2$, and positions with $l_2 + l_3 = 2$ can be reached in only one way—by setting $\theta_2 = 0$ so that segment 3 and segment 2 point in the same direction.

Given s_2, the other elements of the Gröbner basis (3) give exactly one value for each of c_1, s_1, c_2. Further, since $c_1^2 + s_1^2 - 1$ and $c_2^2 + s_2^2 - 1$ are in the ideal generated by (3), the values we get for c_1, s_1, c_2, s_2 uniquely determine the joint angles θ_1 and θ_2. Thus, the case where $a^2 + b^2 \neq 0$ is fully understood.

It remains to study s_1, c_1, s_2, c_2 when $a = b = 0$. Geometrically, this means that joint 3 is placed at the origin of the (x_1, y_1) system—at the same point as joint 1. The first two polynomials in our basis (3) are undefined when we substitute $a = b = 0$ in their coefficients. This is a case where specialization fails. In fact, setting $l_2 = l_3 = 1$ and $a = b = 0$ into the original system (1) yields the grevlex Gröbner basis

$$
\begin{aligned}
& c_1^2 + s_1^2 - 1, \\
(4) \quad & c_2 + 1, \\
& s_2.
\end{aligned}
$$

With a little thought, the geometric reason for this is visible. There are actually *infinitely many* different possible configurations that will place joint 3 at the origin since segments 2 and 3 have equal lengths. The angle θ_1 can be specified arbitrarily, and then setting $\theta_2 = \pi$ will fold segment 3 back along segment 2, placing joint 3 at $(0, 0)$. These are the only joint settings placing the hand at $(a, b) = (0, 0)$. In Exercise 3 you will verify that this analysis is fully consistent with (4).

Note that the *form* of the specialized Gröbner basis (4) is different from the general form (2). The equation for s_2 now has degree 1, and the equation for c_1 (rather than the equation for s_2) has degree 2. Below we will say more about how Gröbner bases can change under specialization.

This completes the analysis of our robot arm. To summarize, given any (a, b) in (x_1, y_1) coordinates, to place joint 3 at (a, b) when $l_2 = l_3 = 1$, there are
- infinitely many distinct settings of joint 1 when $a^2 + b^2 = 0$,
- two distinct settings of joint 1 when $a^2 + b^2 < 4$,
- one setting of joint 1 when $a^2 + b^2 = 4$,
- no possible settings of joint 1 when $a^2 + b^2 > 4$.

The cases $a^2 + b^2 = 0, 4$ are examples of what are known as *kinematic singularities* for this robot. We will give a precise definition of this concept and discuss some of its meaning below.

In the exercises, you will consider the robot arm with three revolute joints and one prismatic joint introduced in Example 2 of §2. There are more restrictions here for the hand orientation. For example, if l_4 lies in the interval $[0, 1]$, then the hand can be placed in any position in the closed disk of radius 3 centered at $(x_1, y_1) = (0, 0)$. However, an interesting difference is that points on the boundary circle can only be reached with one hand orientation.

Specialization of Gröbner Bases

Before continuing our discussion of robotics, let us make some further comments about specialization. The general version of the approach presented here involves computing the Gröbner basis G of an ideal $I \subseteq k(t_1, \ldots, t_m)[x_1, \ldots, x_n]$, where $t_1, \ldots, t_m$ are the parameters. In Exercises 7 and 8, you will show how to find a proper variety $W \subseteq k^m$ such that G remains a Gröbner basis under all specializations $(t_1, \ldots, t_m) \mapsto (a_1, \ldots, a_m) \in k^m \setminus W$.

Another approach is to turn the parameters into variables. In other words, we work in $k[x_1, \ldots, x_m, t_1, \ldots, t_m]$, which we write as $k[\mathbf{x}, \mathbf{t}]$ for $\mathbf{x} = (x_1, \ldots, x_n)$ and $\mathbf{t} = (t_1, \ldots, t_m)$. Assume as in (1) that our system of equations is $f_1 = \cdots = f_s = 0$, where $f_i \in k[\mathbf{x}, \mathbf{t}]$ for $i = 1, \ldots, s$. Also fix a monomial order on $k[\mathbf{x}, \mathbf{t}]$ with the property that $\mathbf{x}^\alpha > \mathbf{x}^\beta$ implies $\mathbf{x}^\alpha > \mathbf{x}^\beta \mathbf{t}^\gamma$ for all γ. As noted in Chapter 4, §7, examples of such monomial orders include product orders or lex order with $x_1 > \cdots > x_n > t_1 > \cdots > t_m$.

We will also assume that the ideal $\tilde{I} = \langle f_1, \ldots, f_s \rangle \subseteq k[\mathbf{x}, \mathbf{t}]$ satisfies

$$(5) \qquad\qquad\qquad \tilde{I} \cap k[\mathbf{t}] = \{0\}.$$

This tells us that the equations put no constraints on the parameters. Hence we can assign the parameters independently. Also, if the intersection (5) contains a nonzero polynomial $h(\mathbf{t})$, then the f_i generate the unit ideal $\langle 1 \rangle$ in $k(\mathbf{t})[\mathbf{x}]$ since $h(\mathbf{t}) \neq 0$ is invertible in $k(\mathbf{t})$.

With this setup, we have the following result which not only computes a Gröbner basis in $k(\mathbf{t})[\mathbf{x}]$ but also makes it easy to find a variety $W \subseteq k^m$ such that the Gröbner basis specializes nicely away from W.

Proposition 1. *Assume* $\tilde{I} = \langle f_1, \ldots, f_s \rangle \subseteq k[\mathbf{x}, \mathbf{t}]$ *satisfies* (5) *and fix a monomial order as above. If* $G = \{g_1, \ldots, g_t\}$ *is a Gröbner basis for* $\tilde{I}$*, then:*

(i) *G is a Gröbner basis for the ideal of $k(\mathbf{t})[\mathbf{x}]$ generated by the f_i with respect to the induced monomial order.*

(ii) *For $i = 1, \ldots, t$, write $g_i \in G$ in the form*

$$g_i = h_i(\mathbf{t}) \mathbf{x}^{\alpha_i} + \textit{terms} < \mathbf{x}^{\alpha_i},$$

where $h_i(\mathbf{t}) \in k[\mathbf{t}]$ is nonzero. If we set $W = \mathbf{V}(h_1 \cdots h_t) \subseteq k^m$, then for any specialization $\mathbf{t} \mapsto \mathbf{a} \in k^m \setminus W$, the $g_i(\mathbf{x}, \mathbf{a})$ form a Gröbner basis with respect to the induced monomial order for the ideal generated by the $f_i(\mathbf{x}, \mathbf{a})$ in $k[\mathbf{x}]$.

Proof. You will prove part (i) in Exercise 10, and part (ii) follows immediately from Theorem 2 of Chapter 4, §7 since $G \cap k[\mathbf{t}] = \emptyset$ by (5). $\qquad\square$

Proposition 1 can make it easy to find specializations that preserve the Gröbner basis. Unfortunately, the G produced by the proposition may be rather inefficient as a Gröbner basis in $k(\mathbf{t})[\mathbf{x}]$, and the corresponding $W \subseteq k^m$ may be too big.

To illustrate these problems, let us apply Proposition 1 to the main example of this section. Take the ideal generated by (1) in $\mathbb{R}[c_1, s_1, c_2, s_2, a, b, l_2, l_3]$. For the product order built from grevlex with $c_1 > s_1 > c_2 > s_2$ followed by grevlex with $a > b > l_2 > l_3$, the resulting Gröbner basis G has 13 polynomials, as opposed to the Gröbner basis (2) over $\mathbb{R}(a, b, l_2, l_3)[c_1, s_1, c_2, s_2]$, which has four polynomials. Also, you will compute in Exercise 11 that for G, the polynomials $h_i \in \mathbb{R}[a, b, l_2, l_3]$ in Proposition 1 are given by

$$2l_2l_3, \ a^2 + b^2, \ b, \ a, \ a^2 + b^2 - l_2^2 - l_3^2, \ l_2^2 - l_3^2, \ 2l_2l_3, \ 1, \ l_3, \ -l_2, \ l_3, \ l_2, \ 1.$$

Then Proposition 1 implies that G remains a Gröbner basis for any specialization where $a, b, l_2, l_3, a^2 + b^2$ and $a^2 + b^2 - l_2^2 - l_3^2$ are nonzero. In fact, we showed earlier that we only need l_2, l_3 and $a^2 + b^2$ to be nonzero, though that analysis required computations over a function field.

Our discussion of specialization will end with the problem of finding Gröbner bases of *all possible* specializations of a system of equations with parameters. This question led to the idea of a *comprehensive Gröbner basis* [see, for example, the appendix to BECKER and WEISPFENNING (1993)] and has since been refined to concept of a *Gröbner cover*, as defined by MONTES and WIBMER (2010).

Roughly speaking, a Gröbner cover of an ideal in $k[\mathbf{x}, \mathbf{t}]$ consists of pairs (G_i, S_i) for $i = 1, \ldots, N$ with the following properties:

- The G_i are finite subsets of $k[\mathbf{x}, \mathbf{t}]$.
- The *segments* S_i form a constructible partition of the parameter space k^m with coordinates $\mathbf{t} = (t_1, \ldots, t_m)$. More precisely, this means:
  - Each S_i is a constructible subset of k^m, as defined in Chapter 4, §7.
  - $S_i \cap S_j = \emptyset$ for $i \neq j$.
  - $S_1 \cup \cdots \cup S_N = k^m$.
- The specialization $\overline{G}_i$ of G_i for $\mathbf{t} \mapsto \mathbf{a} \in S_i$ is a Gröbner basis of the corresponding specialization of the ideal, and $\overline{G}_i$ has the same leading terms as G_i.

The discussion of Gröbner covers in MONTES and WIBMER (2010) is more precise and in particular addresses the issues of uniqueness and minimality. Here, we are just trying to give the flavor of what it means to be Gröbner cover.

For the robot described by (1), the Gröbner cover package in Singular partitions $\mathbb{R}^4$ into 12 segments. Five segments have nontrivial G_i; the remaining seven have $G_i = \{1\}$. Lumping these together, we get the following table:

	Leading monomials	Segment	Comments
1	c_1, s_1, c_2, s_2^2	$\mathbb{R}^4 \setminus \mathbf{V}(l_2 l_3 (a^2 + b^2))$	See (2)
2	c_1, s_1, c_2^2	$\big(\mathbf{V}(l_2, a^2 + b^2 - l_3^2) \setminus \mathbf{V}(l_3)\big) \cup$ $\big(\mathbf{V}(l_3, a^2 + b^2 - l_2^2) \setminus \mathbf{V}(l_2)\big)$	See Exercise 12
3	c_1^2, c_2^2	$\mathbf{V}(a, b, l_2, l_3)$	See Exercise 13
4	c_1, s_1, c_2, s_2	$\mathbf{V}(a^2 + b^2) \setminus (\cdots \cup \mathbf{V}(a, b))$	Empty over $\mathbb{R}$
5	c_1^2, c_2, s_2	$\big(\mathbf{V}(a, b, l_2 - l_3) \setminus \mathbf{V}(l_3, l_2)\big) \cup$ $\big(\mathbf{V}(a, b, l_2 + l_3) \setminus \mathbf{V}(l_3, l_2)\big)$	See (4) and Exercise 14
6	1	Everything else	No solutions

To save space, we only give the leading terms of each G_i. Segment 1 corresponds to the Gröbner basis (2). By the analysis given earlier in the section, this remains a Gröbner basis under specializations with $l_2 l_3 (a^2 + b^2) \neq 0$, exactly as predicted by Segment 1. Earlier, we also computed the Gröbner basis (4) for the specialization $a = b = 0$, $l_2 = l_3 = 1$. In Exercise 14 you will relate this to Segment 5. Other segments will be studied in the exercises.

Kinematic Singularities

We return to the geometry of robots with a discussion of kinematic singularities and the issues they raise in robot motion planning. Our treatment will use some ideas from multivariable calculus that we have not encountered before.

Let $f : \mathcal{J} \to \mathcal{C}$ be the function expressing the hand configuration as a function of the joint settings. In the explicit parametrizations of the space $\mathcal{J}$ that we have used, each component of f is a differentiable function of the variables θ_i. For example, this is clearly true for the mapping f for a planar robot with three revolute joints:

$$(6) \qquad f(\theta_1, \theta_2, \theta_3) = \begin{pmatrix} l_3 \cos(\theta_1 + \theta_2) + l_2 \cos\theta_1 \\ l_3 \sin(\theta_1 + \theta_2) + l_2 \sin\theta_1 \\ \theta_1 + \theta_2 + \theta_3 \end{pmatrix}.$$

Hence, we can compute the Jacobian matrix (or matrix of partial derivatives) of f with respect to the variables $\theta_1, \theta_2, \theta_3$. We write f_i for the i-th component function of f. Then, by definition, the Jacobian matrix is

$$J_f(\theta_1, \theta_2, \theta_3) = \begin{pmatrix} \dfrac{\partial f_1}{\partial \theta_1} & \dfrac{\partial f_1}{\partial \theta_2} & \dfrac{\partial f_1}{\partial \theta_3} \\[2mm] \dfrac{\partial f_2}{\partial \theta_1} & \dfrac{\partial f_2}{\partial \theta_2} & \dfrac{\partial f_2}{\partial \theta_3} \\[2mm] \dfrac{\partial f_3}{\partial \theta_1} & \dfrac{\partial f_3}{\partial \theta_2} & \dfrac{\partial f_3}{\partial \theta_3} \end{pmatrix}.$$

For example, the mapping f in (6) has the Jacobian matrix

(7) $$J_f(\theta_1, \theta_2, \theta_3) = \begin{pmatrix} -l_3\sin(\theta_1 + \theta_2) - l_2\sin\theta_1 & -l_3\sin(\theta_1 + \theta_2) & 0 \\ l_3\cos(\theta_1 + \theta_2) + l_2\cos\theta_1 & l_3\cos(\theta_1 + \theta_2) & 0 \\ 1 & 1 & 1 \end{pmatrix}.$$

From the matrix of functions J_f, we obtain matrices with constant entries by substituting particular values $j = (\theta_1, \theta_2, \theta_3)$. We will write $J_f(j)$ for the substituted matrix, which plays an important role in multivariable calculus. Its key property is that $J_f(j)$ defines a linear mapping which is the *best linear approximation* of the function f at $j \in \mathcal{J}$. This means that near j, the function f and the linear function given by $J_f(j)$ have roughly the same behavior. In this sense, $J_f(j)$ represents the *derivative* of the mapping f at $j \in \mathcal{J}$.

To define what is meant by a kinematic singularity, we need first to assign *dimensions* to the joint space $\mathcal{J}$ and the configuration space $\mathcal{C}$ for our robot, to be denoted by $\dim(\mathcal{J})$ and $\dim(\mathcal{C})$, respectively. We will do this in a very intuitive way. The dimension of $\mathcal{J}$, for example, will be simply the number of independent "degrees of freedom" we have in setting the joints. Each planar joint (revolute or prismatic) contributes 1 dimension to $\mathcal{J}$. Note that this yields a dimension of 3 for the joint space of the plane robot with three revolute joints. Similarly, $\dim(\mathcal{C})$ will be the number of independent degrees of freedom we have in the configuration (position and orientation) of the hand. For our planar robot, this dimension is also 3.

In general, suppose we have a robot with $\dim(\mathcal{J}) = m$ and $\dim(\mathcal{C}) = n$. Then differentiating f as before, we will obtain an $n \times m$ Jacobian matrix of functions. If we substitute in $j \in \mathcal{J}$, we get the linear map $J_f(j) : \mathbb{R}^m \to \mathbb{R}^n$ that best approximates f near j. An important invariant of a matrix is its *rank*, which is the maximal number of linearly independent columns (or rows). The exercises will review some of the properties of the rank. Since $J_f(j)$ is an $n \times m$ matrix, its rank will always be less than or equal to $\min(m, n)$. For instance, consider our planar robot with three revolute joints and $l_2 = l_3 = 1$. If we let $j = (0, \frac{\pi}{2}, \frac{\pi}{3})$, then formula (7) gives us

$$J_f\left(0, \frac{\pi}{2}, \frac{\pi}{3}\right) = \begin{pmatrix} -1 & -1 & 0 \\ 1 & 0 & 0 \\ 1 & 1 & 1 \end{pmatrix}.$$

This matrix has rank exactly 3 (the largest possible in this case).

We say that $J_f(j)$ has *maximal rank* if its rank is $\min(m, n)$ (the largest possible value), and, otherwise, $J_f(j)$ has *deficient rank*. When a matrix has deficient rank, its kernel is larger and image smaller than one would expect (see Exercise 19). Since $J_f(j)$ closely approximates f, $J_f(j)$ having deficient rank should indicate some special or "singular" behavior of f itself near the point j. Hence, we introduce the following definition.

Definition 2. A **kinematic singularity** for a robot is a point $j \in \mathcal{J}$ such that $J_f(j)$ has rank strictly less than $\min(\dim(\mathcal{J}), \dim(\mathcal{C}))$.

For example, the kinematic singularities of the 3-revolute joint robot occur exactly when the matrix (7) has rank ≤ 2. For square $n \times n$ matrices, having deficient rank is equivalent to the vanishing of the determinant. We have

$$0 = \det(J_f) = \sin(\theta_1 + \theta_2)\cos\theta_1 - \cos(\theta_1 + \theta_2)\sin\theta_1$$
$$= \sin((\theta_1 + \theta_2) - \theta_1) = \sin\theta_2$$

if and only if $\theta_2 = 0$ or $\theta_2 = \pi$. Note that $\theta_2 = 0$ corresponds to a position in which segment 3 extends past segment 2 along the positive x_2-axis, whereas $\theta_2 = \pi$ corresponds to a position in which segment 3 is folded back along segment 2. These are exactly the two special configurations we found earlier in which there are not exactly two joint settings yielding a particular hand configuration.

Kinematic singularities are essentially unavoidable for planar robot arms with three or more revolute joints.

Proposition 3. *Let $f : \mathcal{J} \to \mathcal{C}$ be the configuration mapping for a planar robot with $n \geq 3$ revolute joints. Then there exist kinematic singularities $j \in \mathcal{J}$.*

Proof. By Exercise 7 of §2, we know that f has the form

$$f(\theta_1, \ldots, \theta_n) = \begin{pmatrix} \sum_{i=1}^{n-1} l_{i+1}\cos\left(\sum_{j=1}^{i}\theta_j\right) \\ \sum_{i=1}^{n-1} l_{i+1}\sin\left(\sum_{j=1}^{i}\theta_j\right) \\ \sum_{i=1}^{n}\theta_i \end{pmatrix}.$$

Hence, the Jacobian matrix J_f will be the $3 \times n$ matrix

$$\begin{pmatrix} -\sum_{i=1}^{n-1} l_{i+1}\sin\left(\sum_{j=1}^{i}\theta_j\right) & -\sum_{i=2}^{n-1} l_{i+1}\sin\left(\sum_{j=1}^{i}\theta_j\right) & \cdots & -l_n\sin(\theta_{n-1}) & 0 \\ \sum_{i=1}^{n-1} l_{i+1}\cos\left(\sum_{j=1}^{i}\theta_j\right) & \sum_{i=2}^{n-1} l_{i+1}\cos\left(\sum_{j=1}^{i}\theta_j\right) & \cdots & l_n\cos(\theta_{n-1}) & 0 \\ 1 & 1 & \cdots & 1 & 1 \end{pmatrix}.$$

Since we assume $n \geq 3$, by the definition, a kinematic singularity is a point where the rank of J_f is ≤ 2. If $j \in \mathcal{J}$ is a point where all $\theta_i \in \{0, \pi\}$, then every entry of the first row of $J_f(j)$ is zero. Hence, rank $J_f(j) \leq 2$ for those j. □

Descriptions of the possible motions of robots such as the ones we have developed are used in an essential way in *planning* the motions of the robot needed to accomplish the tasks that are set for it. The methods we have sketched are suitable (at least in theory) for implementation in programs to control robot motion automatically. The main goal of such a program would be to instruct the robot what joint setting changes to make in order to take the hand from one position to another. The basic problems to be solved here would be first, to find a *parametrized path* $c(t) \in \mathcal{C}$ starting at the initial hand configuration and ending at the new desired configuration, and second, to find a corresponding path $j(t) \in \mathcal{J}$ such that $f(j(t)) = c(t)$ for all t. In addition, we might want to impose extra constraints on the paths used such as the following:

1. If the configuration space path $c(t)$ is closed (i.e., if the starting and final configurations are the same), we might also want path $j(t)$ to be a closed path. This would be especially important for robots performing a *repetitive* task such as making a certain weld on an automobile body. Making certain the joint space path is closed means that the whole cycle of joint setting changes can simply be repeated to perform the task again.
2. In any real robot, we would want to limit the *joint speeds* necessary to perform the prescribed motion. Overly fast (or rough) motions could damage the mechanisms.
3. We would want to do as little total joint movement as possible to perform each motion.

Kinematic singularities have an important role to play in motion planning. To see the undesirable behavior that can occur, suppose we have a configuration space path $c(t)$ such that the corresponding joint space path $j(t)$ passes through or near a kinematic singularity. Using the multivariable chain rule, we can differentiate $c(t) = f(j(t))$ with respect to t to obtain

$$(8) \qquad c'(t) = J_f(j(t)) \cdot j'(t).$$

We can interpret $c'(t)$ as the velocity of our configuration space path, whereas $j'(t)$ is the corresponding *joint space velocity*. If at some time t_0 our joint space path passes *through* a kinematic singularity for our robot, then, because $J_f(j(t_0))$ is a matrix of deficient rank, equation (8) may have *no* solution for $j'(t_0)$, which means there may be *no* smooth joint paths $j(t)$ corresponding to configuration paths that move in certain directions. As an example, consider the kinematic singularities with $\theta_2 = \pi$ for our planar robot with three revolute joints. If $\theta_1 = 0$, then segments 2 and 3 point along the x_1-axis:

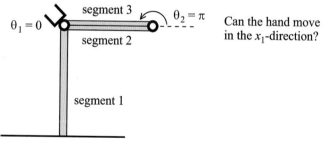

At a kinematic singularity

With segment 3 folded back along segment 2, there is no way to move the hand in the x_1-direction. More precisely, suppose that we have a configuration path such that $c'(t_0)$ is in the direction of the x_1-axis. Then, using formula (7) for J_f, equation (8) becomes

$$c'(t_0) = J_f(t_0) \cdot j'(t_0) = \begin{pmatrix} 0 & 0 & 0 \\ 0 & -1 & 0 \\ 1 & 1 & 1 \end{pmatrix} \cdot j'(t_0).$$

Because the top row of $J_f(t_0)$ is identically zero, this equation has no solution for $j'(t_0)$ since we want the x_1 component of $c'(t_0)$ to be nonzero. Thus, $c(t)$ is a configuration path for which there is no corresponding smooth path in joint space. This is typical of what can go wrong at a kinematic singularity.

For $j(t_0)$ *near* a kinematic singularity, we may still have bad behavior since $J_f(j(t_0))$ may be close to a matrix of deficient rank. Using techniques from numerical linear algebra, it can be shown that in (8), if $J_f(j(t_0))$ is close to a matrix of deficient rank, very large joint space velocities may be needed to achieve a small configuration space velocity. For a simple example of this phenomenon, again consider the kinematic singularities of our planar robot with 3 revolute joints with $\theta_2 = \pi$ (where segment 3 is folded back along segment 2). As the diagram below suggests, in order to move from position A to position B, both near the origin, a large change in θ_1 will be needed to move the hand a short distance.

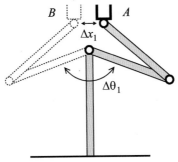

Near a kinematic singularity

To avoid undesirable situations such as this, care must be taken in specifying the desired configuration space path $c(t)$. The study of methods for doing this in a systematic way is an active field of current research in robotics, and unfortunately beyond the scope of this text. For readers who wish to pursue this topic further, a standard basic reference on robotics is the text by PAUL (1981). The survey in BUCHBERGER (1985) contains another discussion of Gröbner basis methods for the inverse kinematic problem. A readable introduction to the inverse kinematic problem and motion control, with references to the original research papers, is given in BAILLIEUL ET AL. (1990).

EXERCISES FOR §3

1. Consider the specialization of the Gröbner basis (2) to the case $l_2 = l_3 = 1$.
   a. First, substitute $l_2 = l_3 = 1$ directly into (2) and simplify to obtain (3).
   b. Now, set $l_2 = l_3 = 1$ in (1) and compute a Gröbner basis for the "specialized" ideal generated by (1), again using grevlex order with $c_1 > s_1 > c_2 > s_2$. The result should be (3), just as in part (a).
2. This exercise studies the geometry of the planar robot with three revolute joints discussed in the text with the dimensions specialized to $l_2 = l_3 = 1$.
   a. Draw a diagram illustrating the two solutions of the inverse kinematic problem for the robot in the general case $0 < a^2 + b^2 < 4$.

b. Explain geometrically why the equations for c_2 and s_2 in (3) do not involve c_1 and s_1. Hint: Use the geometry of the diagram from part (a) to determine how the two values of θ_2 are related.

3. Consider the robot arm discussed in the text. Setting $l_2 = l_3 = 1$ and $a = b = 0$ in (1) gives the Gröbner basis (4). How is this basis different from the basis (3), which only assumes $l_2 = l_3 = 1$? How does this difference explain the properties of the kinematic singularity at $(0, 0)$?

4. In this exercise, you will study the geometry of the robot discussed in the text when $l_2 \neq l_3$.

   a. Set $l_2 = 1, l_3 = 2$ and solve the system (2) for c_1, s_1, c_2, s_2. Interpret your results geometrically, identifying and explaining all special cases. How is this case different from the case $l_2 = l_3 = 1$ done in the text?

   b. Now, set $l_2 = 2, l_3 = 1$ and answer the same questions as in part (a).

As we know from the examples in the text, the form of a Gröbner basis for an ideal can change if symbolic parameters appearing in the coefficients are specialized. In Exercises 5–9, we will study some further examples of this phenomenon and prove some general results.

5. We begin with another example of how denominators in a Gröbner basis can cause problems under specialization. Consider the ideal $I = \langle f, g \rangle$, where $f = x^2 - y, g = (y - tx)(y - t) = -txy + t^2x + y^2 - ty$, and t is a symbolic parameter. We will use lex order with $x > y$.

   a. Compute a reduced Gröbner basis for I in $\mathbb{R}(t)[x, y]$. What polynomials in t appear in the denominators in this basis?

   b. Now set $t = 0$ in f, g and recompute a Gröbner basis. How is this basis different from the one in part (a)? What if we clear denominators in the basis from part (a) and set $t = 0$?

   c. How do the points in the variety $\mathbf{V}(I) \subseteq \mathbb{R}^2$ depend on the choice of $t \in \mathbb{R}$. Is it reasonable that $t = 0$ is a special case?

   d. The first step of Buchberger's algorithm to compute a Gröbner basis for I would be to compute the S-polynomial $S(f, g)$. Compute this S-polynomial by hand in $\mathbb{R}(t)[x, y]$. Note that the special case $t = 0$ is already distinguished at this step.

6. This exercise will explore a more subtle example of what can go wrong during a specialization. Consider the ideal $I = \langle x + ty, x + y \rangle \subseteq \mathbb{R}(t)[x, y]$, where t is a symbolic parameter. We will use lex order with $x > y$.

   a. Show that $\{x, y\}$ is a reduced Gröbner basis of I. Note that neither the original basis nor the Gröbner basis have any denominators.

   b. Let $t = 1$ and show that $\{x+y\}$ is a Gröbner basis for the specialized ideal $\bar{I} \subseteq \mathbb{R}[x, y]$.

   c. To see why $t = 1$ is special, express the Gröbner basis $\{x, y\}$ in terms of the original basis $\{x + ty, x + y\}$. What denominators do you see? In the next problem, we will explore the general case of what is happening here.

7. In this exercise, we will derive a condition under which the form of a Gröbner basis *does not* change under specialization. Consider the ideal

$$I = \langle f_1(\mathbf{x}, \mathbf{t}), \ldots, f_s(\mathbf{x}, \mathbf{t}) \rangle \subseteq k(\mathbf{t})[\mathbf{x}],$$

where $\mathbf{t} = (t_1, \ldots, t_m)$ and $\mathbf{x} = (x_1, \ldots, x_n)$. We think of the t_i as symbolic parameters appearing in the coefficients of $f_1, \ldots, f_s$. Also fix a monomial order. By dividing each f_i by its leading coefficient [which lies in $k(\mathbf{t})$], we may assume that the leading coefficients of the f_i are equal to 1. Then let $\{g_1, \ldots, g_t\}$ be a reduced Gröbner basis for I. Thus the leading coefficients of the g_i are also 1. Finally, let $\mathbf{t} \mapsto \mathbf{a} \in k^m$ be a specialization of the parameters such that none of the denominators of the f_i or g_i vanish at $\mathbf{a}$.

   a. If we use the division algorithm to find $A_{ij} \in k(\mathbf{t})[\mathbf{x}]$ such that

$$f_i = \sum_{j=1}^{t} A_{ij} g_j,$$

then show that none of the denominators of A_{ij} vanish at $\mathbf{a}$.
b. We also know that g_j can be written

$$g_j = \sum_{i=1}^{s} B_{ji} f_i,$$

for some $B_{ij} \in k(\mathbf{t})[\mathbf{x}]$. As Exercise 6 shows, the B_{ji} may introduce new denominators. So assume, in addition, that none of the denominators of the B_{ji} vanish under the specialization $\mathbf{t} \mapsto \mathbf{a}$. Let $\bar{I}$ denote the ideal in $k[\mathbf{x}]$ generated by the specialized f_i. Under these assumptions, prove that the specialized g_j form a basis of $\bar{I}$.
c. Show that the specialized g_j form a Gröbner basis for $\bar{I}$. Hint: The monomial order used to compute $\bar{I}$ only deals with terms in the variables x_j. The parameters t_j are "constants" as far as the ordering is concerned.
d. Let $d_1, \ldots, d_M \in k[\mathbf{t}]$ be all denominators that appear among f_i, g_j, and B_{ji}, and let $W = \mathbf{V}(d_1 \cdot d_2 \cdots d_M) \subseteq k^m$. Conclude that the g_j remain a Gröbner basis for the f_i under all specializations $\mathbf{t} \mapsto \mathbf{a} \in k^m \setminus W$.
8. We next describe an algorithm for finding which specializations preserve a Gröbner basis. We will use the notation of Exercise 7. Thus, we want an algorithm for finding the denominators $d_1, \ldots, d_M$ appearing in the f_i, g_j, and B_{ji}. This is easy to do for the f_i and g_j, but the B_{ji} are more difficult. The problem is that since the f_i are not a Gröbner basis, we cannot use the division algorithm to find the B_{ji}. Fortunately, we only need the denominators. The idea is to work in the ring $k[\mathbf{x}, \mathbf{t}]$. If we multiply the f_i and g_j by suitable polynomials in $k[\mathbf{t}]$, we get

$$\tilde{f}_i, \tilde{g}_j \in k[\mathbf{x}, \mathbf{t}].$$

Let $\tilde{I} \subseteq k(\mathbf{t})[\mathbf{x}]$ be the ideal generated by the $\tilde{f}_i$.
a. Suppose $g_j = \sum_{i=1}^{s} B_{ji} f_i$ in $k(\mathbf{t})[\mathbf{x}]$ and let $d \in k[\mathbf{t}]$ be a polynomial that clears all denominators for the g_j, the f_i, and the B_{ji}. Then prove that

$$d \in (\tilde{I} : \tilde{g}_j) \cap k[\mathbf{t}],$$

where $\tilde{I} : \tilde{g}_j$ is the ideal quotient as defined in §4 of Chapter 4.
b. Give an algorithm for computing $(\tilde{I} : \tilde{g}_j) \cap k[\mathbf{t}]$ and use this to describe an algorithm for finding the subset $W \subseteq k^m$ described in part (d) of Exercise 7.
9. The algorithm described in Exercise 8 can lead to lengthy calculations which may be too much for some computer algebra systems. Fortunately, quicker methods are available in some cases. Let $f_i, g_j \in k(\mathbf{t})[\mathbf{x}]$ be as in Exercises 7 and 8, and suppose we suspect that the g_j will remain a Gröbner basis for the f_i under all specializations where the denominators of the f_i and g_j do not vanish. How can we check this quickly?
a. Let $d \in k[\mathbf{t}]$ be the least common multiple of all denominators in the f_i and g_j and let $\tilde{f}_i, \tilde{g}_j \in k[\mathbf{x}, \mathbf{t}]$ be the polynomials we get by clearing denominators. Finally, let $\tilde{I}$ be the ideal in $k[\mathbf{x}, \mathbf{t}]$ generated by the $\tilde{f}_i$. If $d\tilde{g}_j \in \tilde{I}$ for all j, then prove that specialization works for all $\mathbf{t} \mapsto \mathbf{a} \in k^m \setminus \mathbf{V}(d)$.
b. Describe an algorithm for checking the criterion given in part (a). For efficiency, what monomial order should be used?
c. Apply the algorithm of part (b) to equations (1) in the text. This will prove that (2) remains a Gröbner basis for (1) under all specializations where $l_2 \neq 0, l_3 \neq 0$, and $a^2 + b^2 \neq 0$.

10. This exercise is concerned with part (i) of Proposition 1. Fix a monomial on order $k[\mathbf{x}, \mathbf{t}]$ such that $\mathbf{x}^\alpha > \mathbf{x}^\beta$ implies $\mathbf{x}^\alpha > \mathbf{x}^\beta \mathbf{t}^\gamma$ for all γ. When restricted to monomials involving only the x_i, we get a monomial order on $k(\mathbf{t})[\mathbf{x}]$ called the *induced order*.

   a. A nonzero polynomial $f \in k[\mathbf{x}, \mathbf{t}]$ has leading monomial $\mathrm{LM}(f) = \mathbf{x}^\alpha \mathbf{t}^\beta$. If we regard f as an element of $k(\mathbf{t})[\mathbf{x}]$, prove that its leading monomial with respect to the induced order is $\mathbf{x}^\alpha$.

   b. Use part (a) to prove part (i) of Proposition 1.

11. Consider the ideal generated by (1) in $\mathbb{R}[c_1, s_1, c_2, s_2, a, b, l_2, l_3]$.

   a. Compute a Gröbner basis G using the product order built from grevlex with $c_1 > s_1 > c_2 > s_2$ followed by grevlex with $a > b > l_2 > l_3$.

   b. For each $g_i \in G$, compute the corresponding polynomial $c_i \in \mathbb{R}[a, b, l_2, l_3]$.

12. In the text, we discussed the Gröbner cover for the system (1). In this exercise, we will explore Segment 2, where the set G_2 consists of the polynomials

$$(l_2^2 + l_3^2)c_1 - al_3c_2 - bl_3s_2 - al_2,$$
$$(l_2^2 + l_3^2)s_1 - bl_3c_2 + al_3s_2 - bl_2,$$
$$c_2^2 + s_2^2 - 1.$$

   a. On the first part of Segment 2, we have $l_2 = 0$, $a^2 + b^2 = l_3^2 \neq 0$. Simplify the above polynomials and confirm that the leading monomials are c_1, s_1, c_2^2. Also explain geometrically why the resulting equations have infinitely many solutions.

   b. Do a similar analysis on the second part of Segment 2, where $l_3 = 0$, $a^2 + b^2 = l_2^2 \neq 0$.

13. The most degenerate specialization of (1) is where $a = b = l_2 = l_3 = 0$. What system of equations do we get in this case and how does it relate to Segment 3 in the Gröbner cover of this system?

14. Explain how (4) relates to Segment 5 of the Gröbner cover of (1). Also explain why the second part of the segment is not relevant to this robot arm.

15. Consider the planar robot with two revolute joints and one prismatic joint described in Exercise 4 of §2.

   a. Given a desired hand position and orientation, set up a system of equations as in (1) of this section whose solutions give the possible joint settings to reach that hand configuration. Take the length of segment 2 to be 1.

   b. Using a computer algebra system, solve your equations by computing a Gröbner basis for the ideal generated by the equations from part (a) with respect to a suitable lex order. Note: Some experimentation may be necessary to find a reasonable variable order.

   c. What is the solution of the inverse kinematic problem for this robot. In other words, which hand positions and orientations are possible? How many different joint settings yield a given hand configuration? (Do not forget that the setting of the prismatic joint is limited to a finite interval in $[0, m_3] \subseteq \mathbb{R}$.)

   d. Does this robot have any kinematic singularities according to Definition 2? If so, describe them.

16. Consider the planar robot with three joints and one prismatic joint that we studied in Example 2 of §2. We will assume that segments 2 and 3 have length 1, and that segment 4 varies in length between 1 and 2. The hand configuration consists of a position (a, b) and an orientation θ. We will study this robot using approach suggested by Exercise 3 of §2, which involves parameters $c = \cos\theta$ and $s = \sin\theta$.

   a. Set up a system of equations as in (1) of this section whose solutions give the possible joint settings to reach the hand position (a, b) and orientation (c, s). Hint: If the joint angles are $\theta_1, \theta_2, \theta_3$, then recall that $\theta = \theta_1 + \theta_2 + \theta_3$.

   b. Compute a Gröbner basis over $\mathbb{R}(a, b, c, s)[c_1, s_1, c_2, s_2, c_3, s_3, l_4]$. Your answer should be $\{1\}$.

c. Show that a Gröbner basis over $\mathbb{R}[c_1, s_1, c_2, s_2, c_3, s_3, l_4, a, b, c, s]$ for a suitable monomial order contains $c^2 + s^2 - 1$ and use this to explain the result of part (b).

17. Here we take a different approach to the robot arm of Exercise 16. The idea is to make
c and s variables and write the equations in terms of the cosine and sine of θ_1, θ_2 and
$\theta = \theta_1 + \theta_2 + \theta_3$. Also, by rotating the entire configuration, we can assume that the hand
position is $(a, 0)$.

a. Given the hand position $(a, 0)$, set up a system of equations for the possible joint
settings. The equations will have a as a parameter and variables $c_1, s_1, c_2, s_2, c, s, l_4$.

b. Solve your equations by computing a Gröbner basis for a suitable monomial order.

c. What is the solution of the inverse kinematic problem for this robot? In other words,
which hand positions and orientations are possible? How does the set of possible hand
orientations vary with the position? (Do not forget that the setting l_4 of the prismatic
joint is limited to the finite interval in $[1, 2] \subseteq \mathbb{R}$.)

d. How many different joint settings yield a given hand configuration in general? Are
there special cases?

e. Does this robot have any kinematic singularities according to Definition 2? If so,
describe the corresponding robot configurations and relate them to part (d).

18. Consider the 3-dimensional robot with two "spin" joints from Exercise 8 of §2.

a. Given a desired hand position and orientation, set up a system of equations as in (1)
of this section whose solutions give the possible joint settings to reach that hand
configuration. Take the length of segment 2 to be 4, and the length of segment 3 to be
2, if you like.

b. Solve your equations by computing a Gröbner basis for the ideal generated by your
equations with respect to a suitable lex order. Note: In this case there will be an element of the Gröbner basis that depends only on the hand position coordinates. What
does this mean geometrically? Is your answer reasonable in terms of the geometry of
this robot?

c. What is the solution of the inverse kinematic problem for this robot? That is, which
hand positions and orientations are possible?

d. How many different joint settings yield a given hand configuration in general? Are
there special cases?

e. Does this robot have any kinematic singularities according to Definition 2?

19. Let A be an $m \times n$ matrix with real entries. We will study the rank of A, which is the
maximal number of linearly independent columns (or rows) in A. Multiplication by A
gives a linear map $L_A : \mathbb{R}^n \to \mathbb{R}^m$, and from linear algebra, we know that the rank of
A is the dimension of the image of L_A. As in the text, A has maximal rank if its rank is
$\min(m, n)$. To understand what maximal rank means, there are three cases to consider.

a. If $m = n$, show that A has maximal rank $\Leftrightarrow \det(A) \neq 0 \Leftrightarrow L_A$ is an isomorphism of
vector spaces.

b. If $m < n$, show that A has maximal rank $\Leftrightarrow$ the equation $A \cdot \mathbf{x} = \mathbf{b}$ has a solution for
all $\mathbf{b} \in \mathbb{R}^m \Leftrightarrow L_A$ is an onto mapping.

c. If $m > n$, show that A has maximal rank $\Leftrightarrow$ the equation $A \cdot \mathbf{x} = \mathbf{b}$ has at most one
solution for all $\mathbf{b} \in \mathbb{R}^m \Leftrightarrow L_A$ is a one-to-one mapping.

20. A robot is said to be *kinematically redundant* if the dimension of its joint space $\mathcal{J}$ is
larger than the dimension of its configuration space $\mathcal{C}$.

a. Which of the robots considered in this section (in the text and in Exercises 15–18
above) are kinematically redundant?

b. (This part requires knowledge of the Implicit Function Theorem.) Suppose we have a
kinematically redundant robot and $j \in \mathcal{J}$ is not a kinematic singularity. What can be
said about the inverse image $f^{-1}(f(j))$ in $\mathcal{J}$? In particular, how many different ways
are there to put the robot in the configuration given by $f(j)$?

21. Verify the chain rule formula (8) explicitly for the planar robot with three revolute joints.
Hint: Substitute $\theta_i = \theta_i(t)$ and compute the derivative of the configuration space path
$f(\theta_1(t), \theta_2(t), \theta_3(t))$ with respect to t.

§4 Automatic Geometric Theorem Proving

The geometric descriptions of robots and robot motion we studied in the first three sections of this chapter were designed to be used as tools by a control program to help plan the motions of the robot to accomplish a given task. In the process, the control program could be said to be "reasoning" about the geometric constraints given by the robot's design and its environment and to be "deducing" a feasible solution to the given motion problem. In this section and in the next, we will examine a second subject which has some of the same flavor—automated geometric reasoning in general. We will give two algorithmic methods for determining the validity of general statements in Euclidean geometry. Such methods are of interest to researchers both in artificial intelligence (AI) and in geometric modeling because they have been used in the design of programs that, in effect, can prove or disprove conjectured relationships between, or theorems about, plane geometric objects.

Few people would claim that such programs embody an understanding of the *meaning* of geometric statements comparable to that of a human geometer. Indeed, the whole question of whether a computer is capable of intelligent behavior is one that is still completely unresolved. However, it is interesting to note that a number of new (i.e., apparently previously unknown) theorems have been verified by these methods. In a limited sense, these "theorem provers" are capable of "reasoning" about geometric configurations, an area often considered to be solely the domain of human intelligence.

The basic idea underlying the methods we will consider is that once we introduce Cartesian coordinates in the Euclidean plane, the hypotheses and the conclusions of a large class of geometric theorems can be expressed as *polynomial equations* between the coordinates of collections of points specified in the statements. Here is a simple but representative example.

Example 1. Let A, B, C, D be the vertices of a parallelogram in the plane, as in the figure below.

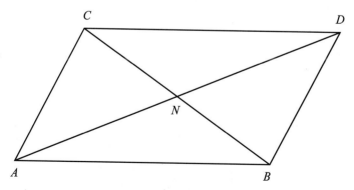

It is a standard geometric theorem that the two diagonals $\overline{AD}$ and $\overline{BC}$ of any parallelogram intersect at a point (N in the figure) which bisects both diagonals. In other words, $AN = DN$ and $BN = CN$, where, as usual, XY denotes the length of the line segment $\overline{XY}$ joining the two points X and Y. The usual proof from geometry is based on showing that the triangles $\triangle ANC$ and $\triangle BND$ are congruent. See Exercise 1.

To relate this theorem to algebraic geometry, we will show how the configuration of the parallelogram and its diagonals (the hypotheses of the theorem) and the statement that the point N bisects the diagonals (the conclusion of the theorem) can be expressed in polynomial form.

The properties of parallelograms are unchanged under translations and rotations in the plane. Hence, we may begin by translating and rotating the parallelogram to place it in any position we like, or equivalently, by choosing our coordinates in any convenient fashion. The simplest way to proceed is as follows. We place the vertex A at the origin and align the side $\overline{AB}$ with the horizontal coordinate axis. In other words, we can take $A = (0,0)$ and $B = (u_1, 0)$ for some $u_1 \neq 0 \in \mathbb{R}$. In what follows we will think of u_1 as an indeterminate or variable whose value can be chosen arbitrarily in $\mathbb{R} \setminus \{0\}$. The vertex C of the parallelogram can be at any point $C = (u_2, u_3)$, where u_2, u_3 are new indeterminates independent of u_1, and $u_3 \neq 0$. The remaining vertex D is now completely determined by the choice of A, B, C.

It will always be true that when constructing the geometric configuration described by a theorem, some of the coordinates of some points will be *arbitrary*, whereas the remaining coordinates of points will be *determined* (possibly up to a finite number of choices) by the arbitrary ones. To indicate arbitrary coordinates, we will consistently use variables u_i, whereas the other coordinates will be denoted x_j. It is important to note that this division of coordinates into two subsets is in no way uniquely specified by the hypotheses of the theorem. Different constructions of a figure, for example, may lead to different sets of arbitrary variables and to different translations of the hypotheses into polynomial equations.

Since D is determined by $A, B,$ and C, we will write $D = (x_1, x_2)$. One hypothesis of our theorem is that the quadrilateral $ABDC$ is a parallelogram or, equivalently, that the opposite pairs of sides are parallel and, hence, have the same slope. Using the slope formula for a line segment, we see that one translation of these statements is as follows:

$$\overline{AB} \parallel \overline{CD} : \ 0 = \frac{x_2 - u_3}{x_1 - u_2},$$

$$\overline{AC} \parallel \overline{BD} : \ \frac{u_3}{u_2} = \frac{x_2}{x_1 - u_1}.$$

Clearing denominators, we obtain the polynomial equations

$$\begin{aligned} h_1 &= x_2 - u_3 = 0, \\ h_2 &= (x_1 - u_1)u_3 - x_2 u_2 = 0. \end{aligned}$$

(1)

(Below, we will discuss another way to get equations for x_1 and x_2.)

Next, we construct the intersection point of the diagonals of the parallelogram. Since the coordinates of the intersection point N are determined by the other data, we write $N = (x_3, x_4)$. Saying that N is the intersection of the diagonals is equivalent to saying that N lies on both of the lines $\overline{AD}$ and $\overline{BC}$, or to saying that the triples

A, N, D and B, N, C are *collinear*. The latter form of the statement leads to the simplest formulation of these hypotheses. Using the slope formula again, we have the following relations:

$$A, N, D \text{ collinear}: \quad \frac{x_4}{x_3} = \frac{x_2}{x_1},$$

$$B, N, C \text{ collinear}: \quad \frac{x_4}{x_3 - u_1} = \frac{u_3}{u_2 - u_1}.$$

Clearing denominators again, we have the polynomial equations

$$(2) \qquad \begin{aligned} h_3 &= x_4 x_1 - x_3 x_2 = 0, \\ h_4 &= x_4(u_2 - u_1) - (x_3 - u_1)u_3 = 0. \end{aligned}$$

The system of four equations formed from (1) and (2) gives one translation of the hypotheses of our theorem.

The conclusions can be written in polynomial form by using the distance formula for two points in the plane (the Pythagorean Theorem) and squaring:

$$\begin{aligned} AN = ND &: \quad x_3^2 + x_4^2 = (x_3 - x_1)^2 + (x_4 - x_2)^2, \\ BN = NC &: \quad (x_3 - u_1)^2 + x_4^2 = (x_3 - u_2)^2 + (x_4 - u_3)^2. \end{aligned}$$

Canceling like terms, the conclusions can be written as

$$(3) \qquad \begin{aligned} g_1 &= x_1^2 - 2x_1 x_3 - 2x_4 x_2 + x_2^2 = 0, \\ g_2 &= 2x_3 u_1 - 2x_3 u_2 - 2x_4 u_3 - u_1^2 + u_2^2 + u_3^2 = 0. \end{aligned}$$

Our translation of the theorem states that the two equations in (3) should hold when the hypotheses in (1) and (2) hold.

As we noted earlier, different translations of the hypotheses and conclusions of a theorem are possible. For instance, see Exercise 2 for a different translation of this theorem based on a different construction of the parallelogram (i.e., a different collection of arbitrary coordinates). There is also a great deal of freedom in the way that hypotheses can be translated. For example, the way we represented the hypothesis that $ABDC$ is a parallelogram in (1) is typical of the way a *computer program* might translate these statements, based on a general method for handling the hypothesis $\overline{AB} \parallel \overline{CD}$. But there is an alternate translation based on the observation that, from the parallelogram law for vector addition, the coordinate vector of the point D should simply be the *vector sum* of the coordinate vectors $B = (u_1, 0)$ and $C = (u_2, u_3)$. Writing $D = (x_1, x_2)$, this alternate translation would be

$$(4) \qquad \begin{aligned} h_1' &= x_1 - u_1 - u_2 = 0, \\ h_2' &= x_2 - u_3 = 0. \end{aligned}$$

These equations are much simpler than the ones in (1). If we wanted to design a geometric theorem-prover that could translate the hypothesis "$ABDC$ is a parallelogram" *directly* (without reducing it to the equivalent form "$\overline{AB} \parallel \overline{CD}$ and $\overline{AC} \parallel \overline{BD}$"), the translation (4) would be preferable to (1).

Further, we could also use h_2' to *eliminate* the variable x_2 from the hypotheses and conclusions, yielding an even simpler system of equations. In fact, with complicated geometric constructions, preparatory simplifications of this kind can sometimes be necessary. They often lead to much more tractable systems of equations.

The following proposition lists some of the most common geometric statements that can be translated into polynomial equations.

Proposition 2. *Let A, B, C, D, E, F be points in the plane. Each of the following geometric statements can be expressed by one or more polynomial equations:*

(i) $\overline{AB}$ *is parallel to* $\overline{CD}$.

(ii) $\overline{AB}$ *is perpendicular to* $\overline{CD}$.

(iii) *A, B, C are collinear.*

(iv) *The distance from A to B is equal to the distance from C to D: $AB = CD$.*

(v) *C lies on the circle with center A and radius AB.*

(vi) *C is the midpoint of $\overline{AB}$.*

(vii) *The acute angle $\angle ABC$ is equal to the acute angle $\angle DEF$*

(viii) *$\overline{BD}$ bisects the angle $\angle ABC$.*

Proof. General methods for translating statements (i), (iii), and (iv) were illustrated in Example 1; the general cases are exactly the same. Statement (v) is equivalent to $AC = AB$. Hence, it is a special case of (iv) and can be treated in the same way. Statement (vi) can be reduced to a conjunction of two statements: A, C, B are collinear, *and $AC = CB$*. We, thus, obtain two equations from (iii) and (iv). Finally, (ii), (vii), and (viii) are left to the reader in Exercise 4.   □

Exercise 3 gives several other types of statements that can be translated into polynomial equations. We will say that a geometric theorem is *admissible* if both its hypotheses and its conclusions admit translations into polynomial equations. There are always many different equivalent formulations of an admissible theorem; the translation will never be unique.

Correctly translating the hypotheses of a theorem into a system of polynomial equations can be accomplished most readily if we think of *constructing* a figure illustrating the configuration in question point by point. This is exactly the process used in Example 1 and in the following example.

Example 3. We will use Proposition 2 to translate the following beautiful result into polynomial equations.

Theorem (The Circle Theorem of Apollonius). Let $\triangle ABC$ be a right triangle in the plane, with right angle at A. The midpoints of the three sides and the foot of the altitude drawn from A to $\overline{BC}$ all lie on one circle.

The theorem is illustrated in the following figure:

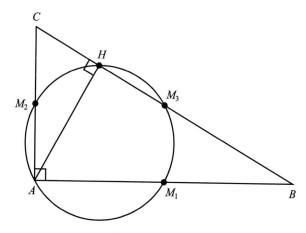

In Exercise 1, you will give a conventional geometric proof of the Circle Theorem. Here we will make the translation to polynomial form, showing that the Circle Theorem is admissible. We begin by constructing the triangle. Placing A at $(0,0)$ and B at $(u_1, 0)$, the hypothesis that $\angle CAB$ is a right angle says $C = (0, u_2)$. (Of course, we are taking a shortcut here; we could also make C a general point and add the hypothesis $CA \perp AB$, but that would lead to more variables and more equations.)

Next, we construct the three midpoints of the sides. These points have coordinates $M_1 = (x_1, 0), M_2 = (0, x_2)$, and $M_3 = (x_3, x_4)$. As in Example 1, we use the convention that u_1, u_2 are to be arbitrary, whereas the x_j are determined by the values of u_1, u_2. Using part (vi) of Proposition 2, we obtain the equations

(5)
$$\begin{aligned} h_1 &= 2x_1 - u_1 = 0, \\ h_2 &= 2x_2 - u_2 = 0, \\ h_3 &= 2x_3 - u_1 = 0, \\ h_4 &= 2x_4 - u_2 = 0. \end{aligned}$$

The next step is to construct the point $H = (x_5, x_6)$, the foot of the altitude drawn from A. We have two hypotheses here:

(6)
$$\begin{aligned} B, H, C \text{ collinear}: \quad h_5 &= u_2 x_5 + u_1 x_6 - u_1 u_2 = 0, \\ AH \perp BC: \quad h_6 &= u_1 x_5 - u_2 x_6 = 0. \end{aligned}$$

Finally, we must consider the statement that M_1, M_2, M_3, H lie on a circle. A general collection of four points in the plane lies on *no* single circle (this is why the statement of the Circle Theorem is interesting). But *three* noncollinear points always do lie on a circle (the *circumscribed circle* of the triangle they form). Thus, our conclusion can be restated as follows: if we construct the circle containing the noncollinear triple M_1, M_2, M_3, then H must lie on *this* circle also. To apply part (v) of Proposition 2, we must know the center of the circle, so this is an additional point that must be constructed. We call the center $O = (x_7, x_8)$ and derive two additional hypotheses:

$$
(7) \quad
\begin{aligned}
M_1 O = M_2 O : \ & h_7 = (x_1 - x_7)^2 + x_8^2 - x_7^2 - (x_8 - x_2)^2 = 0, \\
M_1 O = M_3 O : \ & h_8 = (x_1 - x_7)^2 + (0 - x_8)^2 - (x_3 - x_7)^2 - (x_4 - x_8)^2 = 0.
\end{aligned}
$$

Our conclusion is $HO = M_1 O$, which takes the form

$$
(8) \quad g = (x_5 - x_7)^2 + (x_6 - x_8)^2 - (x_1 - x_7)^2 - x_8^2 = 0.
$$

We remark that both here and in Example 1, the number of hypotheses and the number of *dependent* variables x_j are *the same*. This is typical of properly posed geometric hypotheses. We expect that given values for the u_i, there should be at most finitely many different combinations of x_j satisfying the equations.

We now consider the typical form of an admissible geometric theorem. We will have some number of arbitrary coordinates, or independent variables in our construction, denoted by $u_1, \ldots, u_m$. In addition, there will be some collection of dependent variables $x_1, \ldots, x_n$. The hypotheses of the theorem will be represented by a collection of polynomial equations in the u_i, x_j. As we noted in Example 3, it is typical of a properly posed theorem that the number of hypotheses is equal to the number of dependent variables, so we will write the hypotheses as

$$
(9) \quad
\begin{aligned}
h_1(u_1, \ldots, u_m, x_1, \ldots, x_n) &= 0, \\
&\ \vdots \\
h_n(u_1, \ldots, u_m, x_1, \ldots, x_n) &= 0.
\end{aligned}
$$

The conclusions of the theorem will also be expressed as polynomials in the u_i, x_j. It suffices to consider the case of *one* conclusion since if there are more, we can simply treat them one at a time. Hence, we will write the conclusion as

$$
g(u_1, \ldots, u_m, x_1, \ldots, x_n) = 0.
$$

The question to be addressed is: how can the fact that g follows from $h_1, \ldots, h_n$ be deduced *algebraically*? The basic idea is that we want g to vanish whenever $h_1, \ldots, h_n$ do. We observe that the hypotheses (9) are equations that define a *variety*

$$
V = \mathbf{V}(h_1, \ldots, h_n) \subseteq \mathbb{R}^{m+n}.
$$

This leads to the following definition.

Definition 4. The conclusion g **follows strictly** from the hypotheses $h_1, \ldots, h_n$ if $g \in \mathbf{I}(V) \subseteq \mathbb{R}[u_1, \ldots, u_m, x_1, \ldots, x_n]$, where $V = \mathbf{V}(h_1, \ldots, h_n)$.

Although this definition seems reasonable, we will see later that it is too strict. Most geometric theorems have some "degenerate" cases that Definition 4 does not take into account. But for the time being, we will use the above notion of "follows strictly."

One drawback of Definition 4 is that because we are working over $\mathbb{R}$, we do not have an effective method for determining $\mathbf{I}(V)$. But we still have the following useful criterion.

Proposition 5. *If $g \in \sqrt{\langle h_1, \ldots, h_n \rangle}$, then g follows strictly from $h_1, \ldots, h_n$.*

Proof. The hypothesis $g \in \sqrt{\langle h_1, \ldots, h_n \rangle}$ implies that $g^s \in \langle h_1, \ldots, h_n \rangle$ for some s. Thus, $g^s = \sum_{i=1}^{n} A_i h_i$, where $A_i \in \mathbb{R}[u_1, \ldots, u_m, x_1, \ldots, x_n]$. Then g^s, and, hence, g itself, must vanish whenever $h_1, \ldots, h_n$ do. $\qquad\square$

Note that the converse of this proposition fails whenever $\sqrt{\langle h_1, \ldots, h_n \rangle} \subsetneq \mathbf{I}(V)$, which can easily happen when working over $\mathbb{R}$. Nevertheless, Proposition 5 is still useful because we can test whether $g \in \sqrt{\langle h_1, \ldots, h_n \rangle}$ using the radical membership algorithm from Chapter 4, §2. Let $\tilde{I} = \langle h_1, \ldots, h_n, 1 - yg \rangle$ in the ring $\mathbb{R}[u_1, \ldots, u_m, x_1, \ldots, x_n, y]$. Then Proposition 8 of Chapter 4, §2 implies that

$$g \in \sqrt{\langle h_1, \ldots, h_n \rangle} \iff \{1\} \text{ is the reduced Gröbner basis of } \tilde{I}.$$

If this condition is satisfied, then g follows strictly from $h_1, \ldots, h_n$.

If we work over $\mathbb{C}$, we can get a better sense of what $g \in \sqrt{\langle h_1, \ldots, h_n \rangle}$ means. By allowing solutions in $\mathbb{C}$, the hypotheses $h_1, \ldots, h_n$ define a variety $V_{\mathbb{C}} \subseteq \mathbb{C}^{m+n}$. Then, in Exercise 9, you will use the Strong Nullstellensatz to show that

$$g \in \sqrt{\langle h_1, \ldots, h_n \rangle} \subseteq \mathbb{R}[u_1, \ldots, u_m, x_1, \ldots, x_n]$$
$$\iff g \in \mathbf{I}(V_{\mathbb{C}}) \subseteq \mathbb{C}[u_1, \ldots, u_m, x_1, \ldots, x_n].$$

Thus, $g \in \sqrt{\langle h_1, \ldots, h_n \rangle}$ means that g "follows strictly over $\mathbb{C}$" from $h_1, \ldots, h_n$.

Let us apply these concepts to an example. This will reveal why Definition 4 is too strong.

Example 1 (continued). To see what can go wrong if we proceed as above, consider the theorem on the diagonals of a parallelogram from Example 1, taking as hypotheses the four polynomials from (1) and (2):

$$h_1 = x_2 - u_3,$$
$$h_2 = (x_1 - u_1)u_3 - u_2 x_2,$$
$$h_3 = x_4 x_1 - x_3 x_2,$$
$$h_4 = x_4(u_2 - u_1) - (x_3 - u_1)u_3.$$

We will take as conclusion the *first* polynomial from (3):

$$g = x_1^2 - 2x_1 x_3 - 2x_4 x_2 + x_2^2.$$

To apply Proposition 5, we must compute a Gröbner basis for

$$\tilde{I} = \langle h_1, h_2, h_3, h_4, 1 - yg \rangle \subseteq \mathbb{R}[u_1, u_2, u_3, x_1, x_2, x_3, x_4, y].$$

Surprisingly enough, we do *not* find $\{1\}$. (You will use a computer algebra system in Exercise 10 to verify this.) Since the statement is a true geometric theorem, we must try to understand why our proposed method failed in this case.

The reason can be seen by computing a Gröbner basis for $I = \langle h_1, h_2, h_3, h_4 \rangle$ in $\mathbb{R}[u_1, u_2, u_3, x_1, x_2, x_3, x_4]$, using lex order with $x_1 > x_2 > x_3 > x_4 > u_1 > u_2 > u_3$. The result is

$$f_1 = x_1 x_4 + x_4 u_1 - x_4 u_2 - u_1 u_3,$$
$$f_2 = x_1 u_3 - u_1 u_3 - u_2 u_3,$$
$$f_3 = x_2 - u_3,$$
$$f_4 = x_3 u_3 + x_4 u_1 + x_4 u_2 - u_1 u_3,$$
$$f_5 = x_4 u_1^2 - x_4 u_1 u_2 - \tfrac{1}{2} u_1^2 u_3 + \tfrac{1}{2} u_1 u_2 u_3,$$
$$f_6 = x_4 u_1 u_3 - \tfrac{1}{2} u_1 u_3^2.$$

The variety $V = \mathbf{V}(h_1, h_2, h_3, h_4) = \mathbf{V}(f_1, \ldots, f_6)$ in $\mathbb{R}^7$ defined by the hypotheses is actually *reducible*. To see this, note that f_2 factors as $(x_1 - u_1 - u_2)u_3$, which implies that

$$V = \mathbf{V}(f_1, x_1 - u_1 - u_2, f_3, f_4, f_5, f_6) \cup \mathbf{V}(f_1, u_3, f_3, f_4, f_5, f_6).$$

Since f_5 and f_6 also factor, we can continue this decomposition process. Things simplify dramatically if we recompute the Gröbner basis at each stage, and, in the exercises, you will show that this leads to the decomposition

$$V = V' \cup U_1 \cup U_2 \cup U_3$$

into irreducible varieties, where

$$V' = \mathbf{V}\left(x_1 - u_1 - u_2, x_2 - u_3, x_3 - \frac{u_1 + u_2}{2}, x_4 - \frac{u_3}{2}\right),$$
$$U_1 = \mathbf{V}(x_2, x_4, u_3),$$
$$U_2 = \mathbf{V}(x_1, x_2, u_1 - u_2, u_3),$$
$$U_3 = \mathbf{V}(x_1 - u_2, x_2 - u_3, x_3 u_3 - x_4 u_2, u_1).$$

You will also show that none of these varieties are contained in the others, so that V', U_1, U_2, U_3 are the irreducible components of V.

The problem becomes apparent when we interpret the components U_1, $U_2, U_3 \subseteq V$ in terms of the parallelogram $ABDC$. On U_1 and U_2, we have $u_3 = 0$. This is troubling since u_3 was supposed to be arbitrary. Further, when $u_3 = 0$, the vertex C of our parallelogram lies on $\overline{AB}$ and, hence we do not have a parallelogram at all. This is a *degenerate* case of our configuration, which we intended to *rule out* by the hypothesis that $ABDC$ was an honest parallelogram in the plane. Similarly, we have $u_1 = 0$ on U_3, which again is a degenerate configuration.

You can also check that on $U_1 = \mathbf{V}(x_2, x_4, u_3)$, our conclusion g becomes $g = x_1^2 - 2x_1 x_3$, which is *not* zero since x_1 and x_3 are arbitrary on U_1. This explains why our first attempt failed to prove the theorem. Once we exclude the degenerate cases U_1, U_2, U_3, the above method easily shows that g vanishes on V'. We leave the details as an exercise.

Our goal is to develop a general method that can be used to decide the validity of a theorem, taking into account any degenerate special cases that may need to be excluded. To begin, we use Theorem 2 of Chapter 4, §6 to write $V = \mathbf{V}(h_1, \ldots, h_n) \subseteq \mathbb{R}^{m+n}$ as a finite union of irreducible varieties,

$$(10) \qquad\qquad V = V_1 \cup \cdots \cup V_k.$$

As we saw in the continuation of Example 1, it may be the case that some polynomial equation involving *only* the u_i holds on one or more of these irreducible components of V. Since our intent is that the u_i should be essentially independent, we want to *exclude* these components from consideration if they are present. We introduce the following terminology.

Definition 6. Let W be an irreducible variety in the affine space $\mathbb{R}^{m+n}$ with coordinates $u_1, \ldots, u_m, x_1, \ldots, x_n$. We say that the functions $u_1, \ldots, u_m$ are **algebraically independent on** W if no nonzero polynomial in the u_i alone vanishes identically on W.

Equivalently, Definition 6 states that $u_1, \ldots, u_m$ are algebraically independent on W if $\mathbf{I}(W) \cap \mathbb{R}[u_1, \ldots, u_m] = \{0\}$.

Thus, in the decomposition of the variety V given in (10), we can regroup the irreducible components in the following way:

$$(11) \qquad\qquad V = W_1 \cup \cdots \cup W_p \cup U_1 \cup \cdots \cup U_q,$$

where $u_1, \ldots, u_m$ *are* algebraically independent on the components W_i and *are not* algebraically independent on the components U_j. Thus, the U_j, represent "degenerate" cases of the hypotheses of our theorem. To ensure that the variables u_i are actually arbitrary in the geometric configurations we study, we should consider *only* the subvariety

$$V' = W_1 \cup \cdots \cup W_p \subseteq V.$$

Given a conclusion $g \in \mathbb{R}[u_1, \ldots, u_m, x_1, \ldots, x_n]$ we want to prove, we are not interested in how g behaves on the degenerate cases. This leads to the following definition.

Definition 7. The conclusion g **follows generically** from the hypotheses $h_1, \ldots, h_n$ if $g \in \mathbf{I}(V') \subseteq \mathbb{R}[u_1, \ldots, u_m, x_1, \ldots, x_n]$, where, as above, $V' \subseteq \mathbb{R}^{m+n}$ is the union of the components of the variety $V = \mathbf{V}(h_1, \ldots, h_n)$ on which the u_i are algebraically independent.

Saying a geometric theorem is "true" in the usual sense means precisely that its conclusion(s) follow generically from its hypotheses. The question becomes, given a conclusion g: can we determine when $g \in \mathbf{I}(V')$? In other words, can we develop a criterion that determines whether g vanishes on every component of V on which the u_i *are algebraically independent*, ignoring what happens on the possible "degenerate" components?

Determining the decomposition of a variety into irreducible components is not always easy, so we would like a method to determine whether a conclusion follows generically from a set of hypotheses that *does not* require knowledge of the decomposition (11). Further, even if we could find V', we would still have the problem of computing $\mathbf{I}(V')$.

Fortunately, it is possible to show that g follows generically from $h_1, \ldots, h_n$ without knowing the decomposition of V given in (11). We have the following result.

Proposition 8. *In the above situation, g follows generically from $h_1, \ldots, h_n$ whenever there is some nonzero polynomial $c(u_1, \ldots, u_m) \in \mathbb{R}[u_1, \ldots, u_m]$ such that*

$$c \cdot g \in \sqrt{H},$$

where H is the ideal generated by the hypotheses h_i in $\mathbb{R}[u_1, \ldots, u_m, x_1, \ldots, x_n]$.

Proof. Let V_j be one of the irreducible components of V'. Since $c \cdot g \in \sqrt{H}$, we see that $c \cdot g$ vanishes on V and, hence, on V_j. Thus, the product $c \cdot g$ is in $\mathbf{I}(V_j)$. But V_j is irreducible, so that $\mathbf{I}(V_j)$ is a prime ideal by Proposition 3 of Chapter 4, §5. Thus, $c \cdot g \in \mathbf{I}(V_j)$ implies either c or g is in $\mathbf{I}(V_j)$. We know $c \notin \mathbf{I}(V_j)$ since no nonzero polynomial in the u_i alone vanishes on this component. Hence, $g \in \mathbf{I}(V_j)$, and since this is true for each component of V', it follows that $g \in \mathbf{I}(V')$. $\square$

For Proposition 8 to give a practical way of determining whether a conclusion follows generically from a set of hypotheses, we need a criterion for deciding when there is a nonzero polynomial c with $c \cdot g \in \sqrt{H}$. This is actually quite easy to do. By the definition of the radical, we know that $c \cdot g \in \sqrt{H}$ if and only if

$$(c \cdot g)^s = \sum_{j=1}^{n} A_j h_j$$

for some $A_j \in \mathbb{R}[u_1, \ldots, u_m, x_1, \ldots, x_n]$ and $s \geq 1$. If we divide both sides of this equation by c^s, we obtain

$$g^s = \sum_{j=1}^{n} \frac{A_j}{c^s} h_j,$$

which shows that g is in the radical of the ideal $\widetilde{H}$ generated by $h_1, \ldots, h_n$ over the ring $\mathbb{R}(u_1, \ldots, u_m)[x_1, \ldots, x_n]$ (in which we allow denominators depending only on the u_i). Conversely, if $g \in \sqrt{\widetilde{H}}$, then

$$g^s = \sum_{j=1}^{n} B_j h_j,$$

where the $B_j \in \mathbb{R}(u_1, \ldots, u_m)[x_1, \ldots, x_n]$. If we find a least common denominator c for all terms in all the B_j and multiply both sides by c^s (clearing denominators in the process), we obtain

$$(c \cdot g)^s = \sum_{j=1}^{n} B_j' h_j,$$

where $B_j' \in \mathbb{R}[u_1, \ldots, u_m, x_1, \ldots, x_n]$ and c depends only on the u_i. As a result, $c \cdot g \in \sqrt{H}$. These calculations and the radical membership algorithm from §2 of Chapter 4 establish the following corollary of Proposition 8.

Corollary 9. *In the situation of Proposition 8, the following are equivalent:*

(i) *There is a nonzero polynomial $c \in \mathbb{R}[u_1, \ldots, u_m]$ such that $c \cdot g \in \sqrt{H}$.*

(ii) *$g \in \sqrt{\widetilde{H}}$, where $\widetilde{H}$ is the ideal generated by the h_j in $\mathbb{R}(u_1, \ldots, u_m)[x_1, \ldots, x_n]$.*

(iii) *$\{1\}$ is the reduced Gröbner basis of the ideal*

$$\langle h_1, \ldots, h_n, 1 - yg \rangle \subseteq \mathbb{R}(u_1, \ldots, u_m)[x_1, \ldots, x_n, y].$$

If we combine part (iii) of this corollary with Proposition 8, we get an algorithmic method for proving that a conclusion follows generically from a set of hypotheses. We will call this the Gröbner basis method in geometric theorem proving.

To illustrate the use of this method, we will consider the theorem on parallelograms from Example 1 once more. We first compute a Gröbner basis of the ideal $\langle h_1, h_2, h_3, h_4, 1 - yg \rangle$ in the ring $\mathbb{R}(u_1, u_2, u_3)[x_1, x_2, x_3, x_4, y]$. This computation *does yield* $\{1\}$ as we expect. Making u_1, u_2, u_3 invertible by passing to $\mathbb{R}(u_1, u_2, u_3)$ as our field of coefficients in effect *removes* the degenerate cases encountered above, and the conclusion *does* follow generically from the hypotheses. Moreover, in Exercise 12, you will see that g itself (and not some higher power) actually lies in $\langle h_1, h_2, h_3, h_4 \rangle \subseteq \mathbb{R}(u_1, u_2, u_3)[x_1, x_2, x_3, x_4]$.

Note that the Gröbner basis method does not tell us what the degenerate cases are. The information about these cases is contained in the polynomial $c \in \mathbb{R}[u_1, \ldots, u_m]$, for $c \cdot g \in \sqrt{H}$ tells us that g follows from $h_1, \ldots, h_n$ whenever c does not vanish (this is because $c \cdot g$ vanishes on V). In Exercise 14, we will give an algorithm for finding c.

Over $\mathbb{C}$, we can think of Corollary 9 in terms of the variety $V_{\mathbb{C}} = \mathbf{V}(h_1, \ldots, h_n) \subseteq \mathbb{C}^{m+n}$ as follows. Decomposing $V_{\mathbb{C}}$ as in (11), let $V_{\mathbb{C}}' \subseteq V_{\mathbb{C}}$ be the union of those components where the u_i are algebraically independent. Then Exercise 15 will use the Nullstellensatz to prove that

$$\exists c \neq 0 \text{ in } \mathbb{R}[u_1, \ldots, u_m] \text{ with } c \cdot g \in \sqrt{\langle h_1, \ldots, h_n \rangle} \subseteq \mathbb{R}[u_1, \ldots, u_m, x_1, \ldots, x_u]$$
$$\Longleftrightarrow g \in \mathbf{I}(V_{\mathbb{C}}') \subseteq \mathbb{C}[u_1, \ldots, u_m, x_1, \ldots, x_n].$$

Thus, the conditions of Corollary 9 mean that g "follows generically over $\mathbb{C}$" from the hypotheses $h_1, \ldots, h_n$.

This interpretation points out what is perhaps the main limitation of the Gröbner basis method in geometric theorem proving: it can only prove theorems where the conclusions follow generically over $\mathbb{C}$, even though we are only interested in what happens over $\mathbb{R}$. In particular, there are theorems which are true over $\mathbb{R}$ but not over $\mathbb{C}$ [see STURMFELS (1989) for an example]. Our methods will fail for such theorems.

When using Corollary 9, it is often unnecessary to consider the radical of $\widetilde{H}$. In many cases, the first power of the conclusion is in $\widetilde{H}$ already. So most theorem proving programs in effect use an *ideal* membership algorithm first to test if $g \in \widetilde{H}$, and only go on to the radical membership test if that initial step fails.

To illustrate this, we continue with the Circle Theorem of Apollonius from Example 3. Our hypotheses are the eight polynomials h_i from (5)–(7). We begin by computing a Gröbner basis (using lex order) for the ideal $\widetilde{H}$, which yields

$$
\begin{aligned}
f_1 &= x_1 - u_1/2, \\
f_2 &= x_2 - u_2/2, \\
f_3 &= x_3 - u_1/2, \\
f_4 &= x_4 - u_2/2, \\
f_5 &= x_5 - \frac{u_1 u_2^2}{u_1^2 + u_2^2}, \\
f_6 &= x_6 - \frac{u_1^2 u_2}{u_1^2 + u_2^2}, \\
f_7 &= x_7 - u_1/4, \\
f_8 &= x_8 - u_2/4.
\end{aligned}
$$

(12)

We leave it as an exercise to show that the conclusion (8) reduces to zero on division by this Gröbner basis. Thus, g itself is in $\widetilde{H}$, which shows that g follows generically from $h_1, \ldots, h_8$. Note that we must have either $u_1 \neq 0$ or $u_2 \neq 0$ in order to solve for x_5 and x_6. The equations $u_1 = 0$ and $u_2 = 0$ describe degenerate right "triangles" in which the three vertices are not distinct, so we certainly wish to rule these cases out. It is interesting to note, however, that if either u_1 or u_2 is nonzero, the conclusion is still true. For instance, if $u_1 \neq 0$ but $u_2 = 0$, then the vertices C and A coincide. From (5) and (6), the midpoints M_1 and M_3 coincide, M_2 coincides with A, and H coincides with A as well. As a result, there *is* a circle (infinitely many of them in fact) containing M_1, M_2, M_3, and H in this degenerate case. In Exercise 16, you will study what happens when $u_1 = u_2 = 0$.

We conclude this section by noting that there is one further subtlety that can occur when we use this method to prove or verify a theorem. Namely, there are cases where the given statement of a geometric theorem conceals one or more unstated "extra" hypotheses. These may very well not be included when we make a direct translation to a system of polynomial equations. This often results in a situation where the variety V' is *reducible* or, equivalently, where $p \geq 2$ in (11). In this case, it may be true that the intended conclusion is zero *only* on some of the reducible components of V', so that any method based on Corollary 9 would fail. We will study an example of this type in Exercise 17. If this happens, we may need to reformulate our hypotheses to exclude the extraneous, unwanted components of V'.

In a related development, the Gröbner covers mentioned in §3 have been used to discover the hypotheses under which certain geometric configurations have nice properties. See MONTES and RECIO (2014) for an example.

EXERCISES FOR §4

1. This exercise asks you to give geometric proofs of the theorems that we studied in Examples 1 and 3.
   a. Give a standard Euclidean proof of the theorem of Example 1. Hint: Show $\triangle ANC \cong \triangle BND$.
   b. Give a standard Euclidean proof of the Circle Theorem of Apollonius from Example 3. Hint: First show that $\overline{AB}$ and $\overline{M_2M_3}$ are parallel.
2. This exercise shows that it is possible to give translations of a theorem based on different collections of arbitrary coordinates. Consider the parallelogram $ABDC$ from Example 1 and begin by placing A at the origin.
   a. Explain why it is also possible to consider both of the coordinates of D as arbitrary variables: $D = (u_1, u_2)$.
   b. With this choice, explain why we can specify the coordinates of B as $B = (u_3, x_1)$, i.e., the x-coordinate of B is arbitrary, but the y-coordinate is determined by the choices of u_1, u_2, u_3.
   c. Complete the translation of the theorem based on this choice of coordinates.
3. Let A, B, C, D, E, F, G, H be points in the plane.
   a. Show that the statement $\overline{AB}$ is tangent to the circle through A, C, D can be expressed by polynomial equations. Hint: Construct the center of the circle first. Then, what is true about the tangent and the radius of a circle at a given point?
   b. Show that the statement $AB \cdot CD = EF \cdot GH$ can be expressed by one or more polynomial equations.
   c. Show that the statement $\frac{AB}{CD} = \frac{EF}{GH}$ can be expressed by one or more polynomial equations.
   d. The *cross ratio* of the ordered 4-tuple of distinct *collinear* points (A, B, C, D) is defined to be the real number
$$\frac{AC \cdot BD}{AD \cdot BC}.$$
   Show that the statement "The cross ratio of (A, B, C, D) is equal to $\rho \in \mathbb{R}$" can be expressed by one or more polynomial equations.
4. In this exercise, you will complete the proof of Proposition 2 in the text.
   a. Prove part (ii).
   b. Show that if α, β are acute angles, then $\alpha = \beta$ if and only if $\tan \alpha = \tan \beta$. Use this fact and part (c) of Exercise 3 to prove part (vii) of Proposition 2. Hint: To compute the tangent of an angle, you can construct an appropriate right triangle and compute a ratio of side lengths.
   c. Prove part (viii).
5. Let $\triangle ABC$ be a triangle in the plane. Recall that the *altitude* from A is the line segment from A meeting the opposite side $\overline{BC}$ at a right angle. (We may have to extend $\overline{BC}$ here to find the intersection point.) A standard geometric theorem asserts that the three altitudes of a triangle meet at a single point H, often called the *orthocenter* of the triangle. Give a translation of the hypotheses and conclusion of this theorem as a system of polynomial equations.
6. Let $\triangle ABC$ be a triangle in the plane. It is a standard theorem that if we let M_1 be the midpoint of $\overline{BC}$, M_2 be the midpoint of $\overline{AC}$ and M_3 be the midpoint of $\overline{AB}$, then the segments $\overline{AM_1}$, $\overline{BM_2}$ and $\overline{CM_3}$ meet at a single point M, often called the *centroid* of the triangle. Give a translation of the hypotheses and conclusion of this theorem as a system of polynomial equations.

7. Let $\triangle ABC$ be a triangle in the plane. It is a famous theorem of Euler that the *circumcenter* (the center of the circumscribed circle), the *orthocenter* (from Exercise 5), and the *centroid* (from Exercise 6) are always *collinear*. Translate the hypotheses and conclusion of this theorem into a system of polynomial equations. (The line containing the three "centers" of the triangle is called the *Euler line* of the triangle.)

8. A beautiful theorem ascribed to Pappus concerns two collinear triples of points A, B, C and A', B', C'. Let

$$P = \overline{AB'} \cap \overline{A'B},$$
$$Q = \overline{AC'} \cap \overline{A'C},$$
$$R = \overline{BC'} \cap \overline{B'C}$$

be as in the figure:

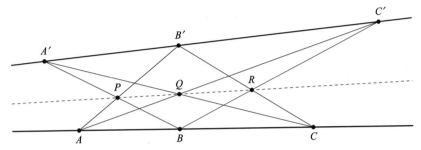

Then it is always the case that P, Q, R are collinear points. Give a translation of the hypotheses and conclusion of this theorem as a system of polynomial equations.

9. Given $h_1, \ldots, h_n \in \mathbb{R}[u_1, \ldots, u_m, x_1, \ldots, x_n]$, let $V_{\mathbb{C}} = \mathbf{V}(h_1, \ldots, h_n) \subseteq \mathbb{C}^{m+n}$. If $g \in \mathbb{R}[u_1, \ldots, u_m, x_1, \ldots, x_n]$, the goal of this exercise is to prove that

$$g \in \sqrt{\langle h_1, \ldots, h_n \rangle} \subseteq \mathbb{R}[u_1, \ldots, u_m, x_1, \ldots, x_n]$$
$$\Longleftrightarrow g \in \mathbf{I}(V_{\mathbb{C}}) \subseteq \mathbb{C}[u_1, \ldots, u_m, x_1, \ldots, x_n].$$

a. Prove the $\Rightarrow$ implication.

b. Use the Strong Nullstellensatz to show that if $g \in \mathbf{I}(V_{\mathbb{C}})$, then there are polynomials $A_j \in \mathbb{C}[u_1, \ldots, u_m, x_1, \ldots, x_n]$ such that $g^s = \sum_{j=1}^{n} A_j h_j$ for some $s \geq 1$.

c. Explain why A_j can be written $A_j = A_j' + iA_j''$, where A_j', A_j'' are polynomials with real coefficients. Use this to conclude that $g^s = \sum_{j=1}^{n} A_j' h_j$, which will complete the proof of the $\Leftarrow$ implication. Hint: g and $h_1, \ldots, h_n$ have real coefficients.

10. Verify the claim made in Example 1 that $\{1\}$ is not the unique reduced Gröbner basis for the ideal $\tilde{I} = \langle h_1, h_2, h_3, h_4, 1 - yg \rangle$.

11. This exercise will study the decomposition into reducible components of the variety defined by the hypotheses of the theorem from Example 1.

a. Verify the claim made in the continuation of Example 1 that

$$V = \mathbf{V}(f_1, x_1 - u_1 - u_2, f_3, \ldots, f_6) \cup \mathbf{V}(f_1, u_3, f_3, \ldots, f_6) = V_1 \cup V_2.$$

b. Compute Gröbner bases for the defining equations of V_1 and V_2. Some of the polynomials should factor and use this to decompose V_1 and V_2.

c. By continuing this process, show that V is the union of the varieties V', U_1, U_2, U_3 defined in the text.

d. Prove that V', U_1, U_2, U_3 are irreducible and that none of them is contained in the union of the others. This shows that V', U_1, U_2, U_3 are the reducible components of V.

    e. On which irreducible component of V is the conclusion of the theorem valid?

    f. Suppose we take as hypotheses the four polynomials in (4) and (2). Is the variety $W = \mathbf{V}(h_1', h_2', h_3, h_4)$ reducible? How many components does it have?

12. Verify the claim made in Example 1 that the conclusion g itself (and not some higher power) is in the ideal generated by h_1, h_2, h_3, h_4 in $\mathbb{R}(u_1, u_2, u_3)[x_1, x_2, x_3, x_4]$.

13. By applying part (iii) of Corollary 9, verify that g follows generically from the h_j for each of the following theorems. What is the lowest power of g which is contained in the ideal $\widetilde{H}$ in each case?

    a. The theorem on the orthocenter of a triangle (Exercise 5).

    b. The theorem on the centroid of a triangle (Exercise 6).

    c. The theorem on the Euler line of a triangle (Exercise 7).

    d. Pappus's Theorem (Exercise 8).

14. In this exercise, we will give an algorithm for finding a nonzero $c \in \mathbb{R}[u_1, \ldots, u_m]$ such that $c \cdot g \in \sqrt{H}$, assuming that such a c exists. We will work with the ideal

$$\overline{H} = \langle h_1, \ldots, h_n, 1 - yg \rangle \subseteq \mathbb{R}[u_1, \ldots, u_m, x_1, \ldots, x_n, y].$$

    a. Show that the conditions of Corollary 9 are equivalent to $\overline{H} \cap \mathbb{R}[u_1, \ldots, u_m] \neq \{0\}$. Hint: Use condition (iii) of the corollary.

    b. If $c \in \overline{H} \cap \mathbb{R}[u_1, \ldots, u_m]$, prove that $c \cdot g \in \sqrt{H}$. Hint: Adapt the argument used in equations (2)–(4) in the proof of Hilbert's Nullstellensatz in Chapter 4, §1.

    c. Describe an algorithm for computing $\overline{H} \cap \mathbb{R}[u_1, \ldots, u_m]$. For maximum efficiency, what monomial order should you use?

    Parts (a)–(c) give an algorithm which decides if there is a nonzero c with $c \cdot g \in \sqrt{H}$ and *simultaneously* produces the required c. Parts (d) and (e) below give some interesting properties of the ideal $\overline{H} \cap \mathbb{R}[u_1, \ldots, u_m]$.

    d. Show that if the conclusion g fails to hold for some choice of $u_1, \ldots, u_m$, then $(u_1, \ldots, u_m) \in W = \mathbf{V}(\overline{H} \cap \mathbb{R}[u_1, \ldots, u_m]) \subseteq \mathbb{R}^m$. Thus, W records the degenerate cases where g fails.

    e. Show that $\sqrt{\overline{H} \cap \mathbb{R}[u_1, \ldots, u_m]}$ gives *all* c's for which $c{\cdot}g \in \sqrt{H}$. Hint: One direction follows from part (a). If $c \cdot g \in \sqrt{H}$, note that $\overline{H}$ contains $(c \cdot g)$'s and $1 - gy$. Now adapt the argument given in Proposition 8 of Chapter 4, §2 to show that $c^s \in \overline{H}$.

15. As in Exercise 9, suppose that we have $h_1, \ldots, h_n \in \mathbb{R}[u_1, \ldots, u_m, x_1, \ldots, x_n]$. Then we get $V_{\mathbb{C}} = \mathbf{V}(h_1, \ldots, h_n) \subseteq \mathbb{C}^{m+n}$. As we did with V, let $V_{\mathbb{C}}'$ be the union of the irreducible components of $V_{\mathbb{C}}$ where $u_1, \ldots, u_n$ are algebraically independent. Given $g \in \mathbb{R}[u_1, \ldots, u_m, x_1, \ldots, x_n]$, we want to show that

$$\exists c \neq 0 \text{ in } \mathbb{R}[u_1, \ldots, u_m] \text{ with } c \cdot g \in \sqrt{\langle h_1, \ldots, h_n \rangle} \subseteq \mathbb{R}[u_1, \ldots, u_m, x_1, \ldots, x_n]$$
$$\Longleftrightarrow g \in \mathbf{I}(V_{\mathbb{C}}') \subseteq \mathbb{C}[u_1, \ldots, u_m, x_1, \ldots, x_n].$$

    a. Prove the $\Rightarrow$ implication. Hint: See the proof of Proposition 8.

    b. Show that if $g \in \mathbf{I}(V_{\mathbb{C}}')$, then there is a nonzero polynomial $c \in \mathbb{C}[u_1, \ldots, u_m]$ such that $c \cdot g \in \mathbf{I}(V_{\mathbb{C}})$. Hint: Write $V_{\mathbb{C}} = V_{\mathbb{C}}' \cup U_1' \cup \cdots \cup U_q'$, where $u_1, \ldots, u_m$ are algebraically dependent on each U_j'. This means there is a nonzero polynomial $c_j \in \mathbb{C}[u_1, \ldots, u_m]$ which vanishes on U_j'.

    c. Show that the polynomial c of part b can be chosen to have real coefficients. Hint: If $\overline{c}$ is the polynomial obtained from c by taking the complex conjugates of the coefficients, show that $\overline{c}$ has real coefficients.

    d. Once we have $c \in \mathbb{R}[u_1, \ldots, u_m]$ with $c \cdot g \in \mathbf{I}(V_{\mathbb{C}})$, use Exercise 9 to complete the proof of the $\Leftarrow$ implication.

16. This exercise deals with the Circle Theorem of Apollonius from Example 3.
    a. Show that the conclusion (8) reduces to 0 on division by the Gröbner basis (12) given in the text.
    b. Discuss the case $u_1 = u_2 = 0$ in the Circle Theorem. Does the conclusion follow in this degenerate case?
    c. Note that in the diagram in the text illustrating the Circle Theorem, the circle is shown passing through the vertex A in addition to the three midpoints and the foot of the altitude drawn from A. Does this conclusion also follow from the hypotheses?

17. In this exercise, we will study a case where a direct translation of the hypotheses of a "true" theorem leads to extraneous components on which the conclusion is actually false. Let $\triangle ABC$ be a triangle in the plane. We construct three new points A', B', C' such that the triangles $\triangle A'BC$, $\triangle AB'C$, $\triangle ABC'$ are *equilateral*. The intended construction is illustrated in the figure below.

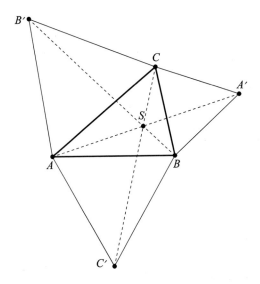

Our theorem is that the three line segments $\overline{AA'}, \overline{BB'}, \overline{CC'}$ all meet in a single point S. (We call S the *Fermat point* of the triangle. If no angle of the original triangle was greater than $\frac{2\pi}{3}$, it can be shown that the three segments $\overline{AS}, \overline{BS}, \overline{CS}$ form a Steiner tree, the network of shortest total length connecting the points A, B, C.)

a. Give a conventional geometric proof of the theorem, assuming the construction is done as in the figure.
b. Now, translate the hypotheses and conclusion of this theorem directly into a set of polynomial equations.
c. Apply the test based on Corollary 9 to determine whether the conclusion follows generically from the hypotheses. The test should fail. Note: This computation may require a *great* deal of ingenuity to push through on some computer algebra systems. This is a complicated system of polynomials.
d. (The key point) Show that there are other ways to construct a figure which is consistent with the hypotheses *as stated*, but which do not agree with the figure above. Hint: Are the points A', B', C' uniquely determined by the hypotheses as stated? Is the statement of the theorem valid for these alternate constructions of the figure? Use this to explain why part (c) did not yield the expected result. (These alternate constructions correspond to points in different components of the variety defined by the hypotheses.)

This example is discussed in detail in pages 69–72 of CHOU (1988). After decomposing the variety defined by the hypotheses, Chou shows that the conclusion follows on a component including the case pictured above and the flipped case where the point A' lies in the half plane determined by BC that contains the triangle, and similarly for B' and C'. These two cases are characterized by the fact that certain rational functions of the coordinates of the points take only positive values. Chou is able to treat this algebraically showing that these cases are where the rational functions in question coincide with certain sums of squares.

§5 Wu's Method

In this section, we will study a second algorithmic method for proving theorems in Euclidean geometry based on systems of polynomial equations. This method, introduced by the Chinese mathematician Wu Wen-Tsün, was developed *before* the Gröbner basis method given in §4. It is also more commonly used than the Gröbner basis method in practice because it is usually more efficient.

Both the elementary version of Wu's method that we will present, and the more refined versions, use an interesting variant of the division algorithm for multivariable polynomials introduced in Chapter 2, §3. The idea here is to follow the one-variable polynomial division algorithm as closely as possible, and we obtain a result known as the *pseudodivision* algorithm. To describe the first step in the process, we consider two polynomials in the ring $k[x_1, \ldots, x_n, y]$, written in the form

$$
\begin{aligned}
f &= c_p y^p + \cdots + c_1 y + c_0, \\
g &= d_m y^m + \cdots + d_1 y + d_0,
\end{aligned}
\tag{1}
$$

where the coefficients c_i, d_j are polynomials in $x_1, \ldots, x_n$. Assume that $m \le p$. Proceeding as in the one-variable division algorithm for polynomials in y, we can attempt to remove the leading term $c_p y^p$ in f by subtracting a multiple of g. However, this is not possible directly unless d_m divides c_p in $k[x_1, \ldots, x_n]$. In pseudodivision, we first *multiply* f by d_m to ensure that the leading coefficient is divisible by d_m, then proceed as in one-variable division. We can state the algorithm formally as follows.

Proposition 1. *Let $f, g \in k[x_1, \ldots, x_n, y]$ be as in (1) and assume $m \le p$ and $d_m \ne 0$.*
(i) *There is an equation*

$$
d_m^s f = qg + r,
$$

where $q, r \in k[x_1, \ldots, x_n, y], s \ge 0$, and either $r = 0$ or the degree of r in y is less than m.
(ii) *$r \in \langle f, g \rangle$ in the ring $k[x_1, \ldots, x_n, y]$.*

Proof. (i) Polynomials q, r satisfying the conditions of the proposition can be constructed by the following algorithm, called *pseudodivision with respect to y*. We use the notation $\deg(h, y)$ for the degree of the polynomial h in the variable y and $LC(h, y)$ for the leading coefficient of h as a polynomial in y—i.e., the coefficient of $y^{\deg(h,y)}$ in h.

Input : f, g
Output : q, r

$m := \deg(g, y); d := LC(g, y)$
$r := f; q := 0$
WHILE $r \neq 0$ AND $\deg(r, y) \geq m$ DO
$\quad r := dr - LC(r, y)gy^{\deg(r,y)-m}$
$\quad q := dq + LC(r, y)y^{\deg(r,y)-m}$
RETURN q, r

Note that if we follow this procedure, the body of the WHILE loop will be executed at most $p - m + 1$ times. Thus, the power s in $d^s f = qg + r$ can be chosen so that $s \leq p - m + 1$. We leave the rest of proof of (i) to the reader as Exercise 1.

From $d^s f = qg + r$, it follows that $r = d^s f - qg \in \langle f, g \rangle$. Since $d = d_m$, we also have $d_m^s f = qg + r$, and the proof of the proposition is complete. $\square$

The polynomials q, r are known as a *pseudoquotient* and a *pseudoremainder* of f on pseudodivision by g, with respect to the variable y. We will use the notation $\text{Rem}(f, g, y)$ for the pseudoremainder produced by the algorithm given in the proof of Proposition 1. For example, if we pseudodivide $f = x^2 y^3 - y$ by $g = x^3 y - 2$ with respect to y by the algorithm above, we obtain the equation

$$(x^3)^3 f = (x^8 y^2 + 2x^5 y + 4x^2 - x^6)g + 8x^2 - 2x^6.$$

In particular, the pseudoremainder is $\text{Rem}(f, g, y) = 8x^2 - 2x^6$.

We note that there is a second, "slicker" way to understand what is happening in this algorithm. The same idea of *allowing denominators* that we exploited in §4 shows that pseudodivision is the same as

- ordinary one-variable polynomial division for polynomials in y, with coefficients in the rational function field $K = k(x_1, \ldots, x_n)$, followed by
- clearing denominators. You will establish this claim in Exercise 2, based on the observation that the only term that needs to be inverted in division of polynomials in $K[y]$ (K any field) is the *leading coefficient* d_m of the divisor g. Thus, the denominators introduced in the process of dividing f by g can all be cleared by multiplying by a suitable power d_m^s, and we get an equation of the form $d_m^s f = qg + r$.

In this second form, or directly, pseudodivision can be readily implemented in most computer algebra systems. Indeed, some systems include pseudodivision as one of the built-in operations on polynomials.

We recall the situation studied in §4, in which the hypotheses and conclusion of a theorem in Euclidean plane geometry are translated into a system of polynomials in variables $u_1, \ldots, u_m, x_1, \ldots, x_n$, with $h_1, \ldots, h_n$ representing the hypotheses and g giving the conclusion. As in equation (11) of §4, we can group the irreducible components of the variety $V = \mathbf{V}(h_1, \ldots, h_n) \subseteq \mathbb{R}^{m+n}$ as

$$V = V' \cup U,$$

where V' is the union of the components on which the u_i are algebraically independent. Our goal is to prove that g vanishes on V'.

The elementary version of Wu's method that we will discuss is tailored for the case where V' is *irreducible*. We note, however, that Wu's method can be extended to the more general reducible case also. The main algebraic tool needed (Ritt's decomposition algorithm based on *characteristic sets* for prime ideals) would lead us too far afield, though, so we will not discuss it. Note that, in practice, we usually do not know in advance whether V' is irreducible or not. Thus, reliable "theorem-provers" based on Wu's method should include these more general techniques too.

Our simplified version of Wu's method uses the pseudodivision algorithm in two ways in the process of determining whether the equation $g = 0$ follows from $h_j = 0$.

- Step 1 of Wu's method uses pseudodivision to reduce the hypotheses to a system of polynomials f_j that are in triangular form in the variables $x_1, \ldots, x_n$. In other words, we seek

(2)
$$\begin{aligned}
f_1 &= f_1(u_1, \ldots, u_m, x_1), \\
f_2 &= f_2(u_1, \ldots, u_m, x_1, x_2), \\
&\vdots \\
f_n &= f_n(u_1, \ldots, u_m, x_1, \ldots, x_n)
\end{aligned}$$

such that $\mathbf{V}(f_1, \ldots, f_n)$ again contains the irreducible variety V', on which the u_i are algebraically independent.

- Step 2 of Wu's method uses *successive* pseudodivision of the conclusion g with respect to each of the variables x_j to determine whether $g \in \mathbf{I}(V')$. We compute

(3)
$$\begin{aligned}
R_{n-1} &= \mathrm{Rem}(g, f_n, x_n), \\
R_{n-2} &= \mathrm{Rem}(R_{n-1}, f_{n-1}, x_{n-1}), \\
&\vdots \\
R_1 &= \mathrm{Rem}(R_2, f_2, x_2), \\
R_0 &= \mathrm{Rem}(R_1, f_1, x_1).
\end{aligned}$$

- Then $R_0 = 0$ implies that g follows from the hypotheses h_j under an additional condition, to be made precise in Theorem 4.

To explain how Wu's method works, we need to explain each of these steps, beginning with the reduction to triangular form.

Step 1. Reduction to Triangular Form

In practice, this reduction can almost always be accomplished using a procedure very similar to Gaussian elimination for systems of linear equations. We will not state any general theorems concerning our procedure, however, because there are some exceptional cases in which it might fail. (See the comments below.) A completely general procedure for accomplishing this kind of reduction may be found in CHOU (1988).

The elementary version is performed as follows. We work one variable at a time, beginning with x_n.

1.1. Among the h_j, find all the polynomials containing the variable x_n. Call the set of such polynomials S. (If there are no such polynomials, the translation of our geometric theorem is most likely incorrect since it would allow x_n to be arbitrary.)

1.2. If there is only one polynomial in S, then we can rename the polynomials, making that one polynomial f'_n, and our system of polynomials will have the form

$$f'_1 = f'_1(u_1, \ldots, u_m, x_1, \ldots, x_{n-1}),$$

(4)
$$\vdots$$
$$f'_{n-1} = f'_{n-1}(u_1, \ldots, u_m, x_1, \ldots, x_{n-1}),$$
$$f'_n = f'_n(u_1, \ldots, u_m, x_1, \ldots, x_n).$$

1.3. If there is more than one polynomial in S, but some element of S has degree 1 in x_n, then we can take f'_n as that polynomial and replace all the other hypotheses in S by their pseudoremainders on division by f'_n with respect to x_n. [One of these pseudoremainders could conceivably be *zero*, but this would mean that f'_n would divide $d^s h$, where h is one of the other hypothesis polynomials and $d = LC(f'_n, x_n)$. This is unlikely since V' is assumed to be irreducible.] We obtain a system in the form (4) again. By part (ii) of Proposition 1, all the f'_j are in the ideal generated by the h_j.

1.4. If there are several polynomials in S, but none has degree 1 in x_n, then pick $a, b \in S$ where $0 < \deg(b, x_n) \le \deg(a, x_n)$ and compute the pseudoremainder $r = \text{Rem}(a, b, x_n)$. Then:

   a. If $\deg(r, x_n) \ge 1$, then replace S by $(S \setminus \{a\}) \cup \{r\}$ (leaving the hypotheses not in S unchanged) and repeat either 1.4 (if $\deg(r, x_n) \ge 2$) or 1.3 (if $\deg(r, x_n) = 1$).

   b. If $\deg(r, x_n) = 0$, then replace S by $S \setminus \{a\}$ (adding r to the hypotheses not in S) and repeat either 1.4 (if the new S has ≥ 2 elements) or 1.2 (if the new S has only one element).

Eventually we are reduced to a system of polynomials of the form (4) again. Since the degree in x_n are reduced each time we compute a pseudoremainder, we will eventually remove the x_n terms from all but one of our polynomials. Moreover, by part (ii) of Proposition 1, each of the resulting polynomials is contained in the ideal generated by the h_j. Again, it is conceivable that we could obtain a zero pseudoremainder at some stage here. This would usually, but not always, imply reducibility, so it is unlikely. We then apply the same process to the polynomials $f'_1, \ldots, f'_{n-1}$ in (4) to remove the x_{n-1} terms from all but one polynomial. Continuing in this way, we will eventually arrive at a system of equations in triangular form as in (2) above.

Once we have the triangular equations, we can relate them to the original hypotheses as follows.

Proposition 2. *Suppose that* $f_1 = \cdots = f_n = 0$ *are the triangular equations obtained from* $h_1 = \cdots = h_n = 0$ *by the above reduction algorithm. Then*

$$V' \subseteq V = \mathbf{V}(h_1, \ldots, h_n) \subseteq \mathbf{V}(f_1, \ldots, f_n).$$

Proof. As we noted above, all the f_j are contained in the ideal generated by the h_j. Thus, $\langle f_1, \ldots, f_n \rangle \subseteq \langle h_1, \ldots, h_n \rangle$ and hence, $V = \mathbf{V}(h_1, \ldots, h_n) \subseteq \mathbf{V}(f_1, \ldots, f_n)$ follows immediately. Since $V' \subseteq V$, we are done. □

Example 3. To illustrate the operation of this triangulation procedure, we will apply it to the hypotheses of the Circle Theorem of Apollonius from §4. Referring back to (5)–(7) of §4, we have

$$h_1 = 2x_1 - u_1,$$
$$h_2 = 2x_2 - u_2,$$
$$h_3 = 2x_3 - u_1,$$
$$h_4 = 2x_4 - u_2,$$
$$h_5 = u_2 x_5 + u_1 x_6 - u_1 u_2,$$
$$h_6 = u_1 x_5 - u_2 x_6,$$
$$h_7 = x_1^2 - x_2^2 - 2x_1 x_7 + 2x_2 x_8,$$
$$h_8 = x_1^2 - 2x_1 x_7 - x_3^2 + 2x_3 x_7 - x_4^2 + 2x_4 x_8.$$

Note that this system is very nearly in triangular form in the x_j. In fact, this is often true, especially in the cases where each step of constructing the geometric configuration involves adding one new point.

In Step 1 of the triangulation procedure, we see that h_7, h_8 are the only polynomials in our set containing x_8. Even better, h_8 has degree 1 in x_8. Hence, we proceed as in 1.3 of the triangulation procedure, making $f_8 = h_8$, and replacing h_7 by

$$f_7 = \mathrm{Rem}(h_7, h_8, x_8)$$
$$= (2x_1 x_2 - 2x_2 x_3 - 2x_1 x_4)x_7 - x_1^2 x_2 + x_2 x_3^2 + x_1^2 x_4 - x_2^2 x_4 + x_2 x_4^2.$$

As this example indicates, we often ignore numerical constants when computing remainders. Only f_7 contains x_7, so nothing further needs to be done there. Both h_6 and h_5 contain x_6, but we are in the situation of 1.3 in the procedure again. We make $f_6 = h_6$ and replace h_5 by

$$f_5 = \mathrm{Rem}(h_5, h_6, x_6) = (u_1^2 + u_2^2)x_5 - u_1 u_2^2.$$

The remaining four polynomials are in triangular form already, so we take $f_i = h_i$ for $i = 1, 2, 3, 4$.

Step 2. Successive Pseudodivision

The key step in Wu's method is the successive pseudodivision operation given in equation (3) computing the *final remainder* R_0. The usefulness of this operation is indicated by the following theorem.

Theorem 4. *Consider the set of hypotheses and the conclusion for a geometric theorem. Let R_0 be the final remainder computed by the successive pseudodivision of g as in (3), using the system of polynomials $f_1, \ldots, f_n$ in triangular form (2). Let d_j be the leading coefficient of f_j as a polynomial in x_j (so d_j is a polynomial in $u_1, \ldots, u_m$ and $x_1, \ldots, x_{j-1}$). Then:*

(i) *There are nonnegative integers $s_1, \ldots, s_n$ and polynomials $A_1, \ldots, A_n$ in the ring $\mathbb{R}[u_1, \ldots, u_m, x_1, \ldots, x_n]$ such that*

$$d_1^{s_1} \cdots d_n^{s_n} g = A_1 f_1 + \cdots + A_n f_n + R_0.$$

(ii) *If R_0 is the zero polynomial, then g is zero at every point of $V' \setminus \mathbf{V}(d_1 d_2 \cdots d_n) \subseteq \mathbb{R}^{m+n}$.*

Proof. Part (i) follows by applying Proposition 1 repeatedly. Pseudodividing g by f_n with respect to x_n, we have

$$R_{n-1} = d_n^{s_n} g - q_n f_n.$$

Hence, when we pseudodivide again with respect to x_{n-1}:

$$
\begin{aligned}
R_{n-2} &= d_{n-1}^{s_{n-1}} (d_n^{s_n} g - q_n f_n) - q_{n-1} f_{n-1} \\
        &= d_{n-1}^{s_{n-1}} d_n^{s_n} g - q_{n-1} f_{n-1} - d_{n-1}^{s_{n-1}} q_n f_n.
\end{aligned}
$$

Continuing in the same way, we will eventually obtain an expression of the form

$$R_0 = d_1^{s_1} \cdots d_n^{s_n} g - (A_1 f_1 + \cdots + A_n f_n),$$

which is what we wanted to show.

(ii) By the result of part (i), if $R_0 = 0$, then at every point of the variety $W = \mathbf{V}(f_1, \ldots, f_n)$, either g or one of the $d_j^{s_j}$ is zero. By Proposition 2, the variety V' is contained in W, so the same is true on V'. The assertion follows. □

Even though they are not always polynomial relations in the u_i alone, the equations $d_j = 0$, where d_j is the leading coefficient of f_j, can often be interpreted as loci defining *degenerate special cases* of our geometric configuration.

Example 3 (continued). For instance, let us complete the application of Wu's method to the Circle Theorem of Apollonius. Our goal is to show that

$$g = (x_5 - x_7)^2 + (x_6 - x_8)^2 - (x_1 - x_7)^2 - x_8^2 = 0$$

is a consequence of the hypotheses $h_1 = \cdots = h_8 = 0$ (see (8) of §4). Using $f_1, \ldots, f_8$ computed above, we set $R_8 = g$ and compute the successive remainders

$$R_{i-1} = \mathrm{Rem}(R_i, f_i, x_i)$$

as i decreases from 8 to 1. When computing these remainders, we always use the minimal exponent s in Proposition 1, and in some cases, we ignore constant factors of the remainder. We obtain the following remainders.

$$R_7 = x_4 x_5^2 - 2x_4 x_5 x_7 + x_4 x_6^2 - x_4 x_1^2 + 2x_4 x_1 x_7 + x_6 x_1^2 - 2x_6 x_1 x_7$$
$$\quad - x_6 x_3^2 + 2x_6 x_3 x_7 - x_6 x_4^2,$$

$$R_6 = x_4^2 x_1 x_5^2 - x_4^2 x_1^2 x_5 - x_4 x_1 x_6 x_3^2 + x_4^2 x_1 x_6^2 - x_4^3 x_1 x_6 + x_4^2 x_2^2 x_5$$
$$\quad - x_4^2 x_2^2 x_1 - x_2 x_4^3 x_5 + x_2 x_4^3 x_1 - x_2 x_1 x_4 x_5^2 - x_2 x_1 x_4 x_6^2$$
$$\quad + x_2 x_3 x_4 x_5^2 + x_2 x_3 x_4 x_6^2 - x_2 x_3 x_4 x_1^2 + x_4 x_1^2 x_6 x_3 + x_4 x_2^2 x_6 x_1$$
$$\quad - x_4 x_2^2 x_6 x_3 + x_2 x_1^2 x_4 x_5 - x_2 x_3^2 x_4 x_5 + x_2 x_3^2 x_4 x_1,$$

$$R_5 = u_2^2 x_4^2 x_1 x_5^2 - u_2^2 x_4^2 x_1^2 x_5 + u_2^2 x_4^2 x_2^2 x_5 - u_2^2 x_4^2 x_2^2 x_1 - u_2^2 x_2 x_4^3 x_5$$
$$\quad + u_2^2 x_2 x_4^3 x_1 - x_4 u_2^2 x_2 x_1 x_5^2 + x_4 u_2^2 x_2 x_3 x_5^2 - x_4 u_2^2 x_2 x_3 x_1^2$$
$$\quad + x_4 u_2^2 x_2 x_1^2 x_5 - x_4 u_2^2 x_2 x_3^2 x_5 + x_4 u_2^2 x_2 x_3^2 x_1 - u_1 x_5 u_2 x_4^2 x_1$$
$$\quad + x_4 u_1 x_5 u_2 x_2^2 x_1 - x_4 u_1 x_5 u_2 x_1 x_3^2 - x_4 u_1 x_5 u_2 x_2^2 x_3$$
$$\quad + x_4 x_1 x_5 u_2 x_1^2 x_3 + u_1^2 x_5^2 x_4^2 x_1 - x_4 u_1^2 x_5^2 x_2 x_1 + x_4 u_1^2 x_5^2 x_2 x_3,$$

$$R_4 = -u_2^4 x_4 x_2 x_3 x_1^2 - u_2^4 x_4^2 x_2^2 x_1 + u_2^4 x_4 x_2 x_3^2 x_1 + u_2^4 x_4^3 x_2 x_1$$
$$\quad - u_2^2 x_4 u_1^2 x_2 x_3 x_1^2 - u_2^2 x_4^2 u_1^2 x_2^2 x_1 + u_2^2 x_4 u_1^2 x_2 x_3^2 x_1$$
$$\quad + u_2^2 x_4^3 u_1^2 x_2 x_1 - u_2^4 x_4^3 u_1 x_2 - u_2^3 x_4^2 u_1^2 x_1 + u_2^4 x_4 u_1 x_2^2$$
$$\quad - u_2^4 x_4^2 u_1 x_1^2 + u_2^3 x_4 u_1^2 x_2^2 x_1 - u_2^3 x_4 u_1^2 x_1 x_3^2 - u_2^4 x_4 u_1 x_2 x_3^2$$
$$\quad + u_2^4 x_4 u_1 x_2 x_1^2 - u_2^3 x_4 u_1^2 x_2^2 x_3 + u_2^3 x_4 u_1^2 x_1^2 x_3 + u_2^4 x_4 u_1^2 x_1$$
$$\quad - u_2^4 x_4 u_1^2 x_2 x_1 + u_2^4 x_4 u_1^2 x_2 x_3,$$

$$R_3 = 4u_2^5 x_2 x_3^2 x_1 - 4u_2^5 u_1 x_2 x_3^2 + 4u_2^5 u_1 x_2 x_1^2 - 4u_2^5 x_2 x_3 x_1^2$$
$$\quad - 3u_2^5 u_1^2 x_2 x_1 + 4u_2^5 u_1^2 x_2 x_3 - 4u_2^4 u_1^2 x_1 x_3^2 - 4u_2^4 u_1^2 x_2^2 x_3$$
$$\quad + 2u_2^4 u_1^2 x_2^2 x_1 + 4u_2^4 u_1^2 x_1^2 x_3 - 4u_2^3 u_1^2 x_2 x_3 x_1^2 + 4u_2^3 u_1^2 x_2 x_3^2 x_1$$
$$\quad - 2u_2^6 x_2^2 x_1 - 2u_2^6 u_1 x_1^2 + 2u_2^6 u_1 x_2^2 + u_2^6 u_1^2 x_1 + u_2^7 x_2 x_1$$
$$\quad - u_2^7 u_1 x_2,$$

$$R_2 = 2u_2^5 u_1 x_2 x_1^2 - 2u_2^5 u_1^2 x_2 x_1 + 2u_2^4 u_1^2 x_2^2 x_1 - 2u_2^6 x_2^2 x_1$$
$$\quad - 2u_2^6 u_1 x_1^2 + 2u_2^6 u_1 x_2^2 + u_2^6 u_1^2 x_1 + u_2^7 x_2 x_1 - u_2^7 u_1 x_2$$
$$\quad + u_2^5 u_1^3 x_2 - 2u_2^4 u_1^3 x_2^2 + 2u_2^4 u_1^3 x_1^2 - 2u_2^3 u_1^3 x_2 x_1^2 + u_2^3 u_1^4 x_2 x_1$$
$$\quad - u_2^4 u_1^4 x_1,$$

$$R_1 = -2u_2^6 u_1 x_1^2 - u_2^4 u_1^4 x_1 + u_2^6 u_1^2 x_1 + 2u_2^4 u_1^3 x_1^2,$$

$$R_0 = 0.$$

By Theorem 4, Wu's method confirms that the Circle Theorem is valid when none of the leading coefficients of the f_j is zero. The nontrivial conditions here are

$$d_5 = u_1^2 + u_2^2 \neq 0,$$
$$d_6 = u_2 \neq 0,$$
$$d_7 = 2x_1 x_2 - 2x_2 x_3 - 2x_1 x_4 \neq 0,$$
$$d_8 = 2x_4 \neq 0.$$

The second condition in this list is $u_2 \neq 0$, which says that the vertices A and C of the right triangle $\triangle ABC$ are distinct [recall we chose coordinates so that $A = (0, 0)$ and $C = (0, u_2)$ in Example 3 of §4]. This also implies the first condition since u_1 and u_2 are real. The condition $2x_4 \neq 0$ is equivalent to $u_2 \neq 0$ by the hypothesis $h_4 = 0$. Finally, $d_7 \neq 0$ says that the vertices of the triangle are distinct (see Exercise 5). From this analysis, we see that the Circle Theorem actually follows generically from its hypotheses as in §4.

The elementary version of Wu's method only gives $g = 0$ under the side conditions $d_j \neq 0$. In particular, note that in a case where V' is reducible, it is entirely conceivable that one of the d_j could vanish on an entire component of V'. If this happened, there would be no conclusion concerning the validity of the theorem for geometric configurations corresponding to points in that component.

Indeed, a much stronger version of Theorem 4 is known when the subvariety V' for a given set of hypotheses is *irreducible*. With the extra algebraic tools we have omitted (Ritt's decomposition algorithm), it can be proved that there are special triangular form sets of f_j (called characteristic sets) with the property that $R_0 = 0$ is a necessary and sufficient condition for g to lie in $\mathbf{I}(V')$. In particular, it is never the case that one of the leading coefficients of the f_j is identically zero on V' so that $R_0 = 0$ implies that g must vanish on all of V'. We refer the interested reader to CHOU (1988) for the details. Other treatments of characteristic sets and the Wu-Ritt algorithm can be found in MISHRA (1993) and WANG (2001).

Finally, we will briefly compare Wu's method with the method based on Gröbner bases introduced in §4. These two methods apply to exactly the same class of geometric theorems and they usually yield equivalent results. Both make essential use of a division algorithm to determine whether a polynomial is in a given ideal or not. However, as we can guess from the triangulation procedure described above, the basic version of Wu's method at least is likely to be much quicker on a given problem. The reason is that simply triangulating a set of polynomials usually requires much less effort than computing a Gröbner basis for the ideal they generate, or for the ideal $\widetilde{H} = \langle h_1, \ldots, h_n, 1 - yg \rangle$. This pattern is especially pronounced when the original polynomials themselves are nearly in triangular form, which is often the case for the hypotheses of a geometric theorem. In a sense, this superiority of Wu's method is only natural since Gröbner bases contain much more information than triangular form sets. Note that we have not claimed anywhere that the triangular form set of polynomials even generates the same ideal as the hypotheses in either $\mathbb{R}[u_1, \ldots, u_m, x_1, \ldots, x_n]$ or $\mathbb{R}(u_1, \ldots, u_m)[x_1, \ldots, x_n]$. In fact, this is not true in general (Exercise 4). Wu's method is an example of a technique tailored to solve a particular problem. Such techniques can often outperform general techniques (such as computing Gröbner bases) that do many other things besides.

Readers interested in pursuing this topic should consult CHOU (1988), the second half of which is an annotated collection of 512 geometric theorems proved by Chou's program implementing Wu's method. WU (1983) is a reprint of the original paper that introduced these ideas.

EXERCISES FOR §5

1. This problem completes the proof of Proposition 1 begun in the text.
   a. Complete the proof of (i) of the proposition.
   b. Show that q, r in the equation $d_m^s f = qg + r$ in the proposition are definitely *not* unique if no condition is placed on the exponent s.

2. Establish the claim stated after Proposition 1 that pseudodivision is equivalent to ordinary polynomial division in the ring $K[y]$, where $K = k(x_1, \ldots, x_n)$.

3. Show that there is a unique minimal $s \le p - m + 1$ in Proposition 1 for which the equation $d_m^s f = qg + r$ exists, and that q and r are unique when s is minimal. Hint: Use the uniqueness of the quotient and remainder for division in $k(x_1, \ldots, x_n)[y]$.

4. Show by example that applying the triangulation procedure described in this section to two polynomials $h_1, h_2 \in k[x_1, x_2]$ can yield polynomials f_1, f_2 that generate an ideal strictly smaller than $\langle h_1, h_2 \rangle$. The same can be true for larger sets of polynomials as well.

5. Show that the nondegeneracy condition $d_7 \ne 0$ for the Circle Theorem is automatically satisfied if u_1 and u_2 are nonzero.

6. Use Wu's method to verify each of the following theorems. In each case, state the conditions $d_j \ne 0$ under which Theorem 4 implies that the conclusion follows from the hypotheses. If you also did the corresponding exercises in §4, try to compare the time and/or effort involved with each method.
   a. The theorem on the diagonals of a parallelogram (Example 1 of §4).
   b. The theorem on the orthocenter of a triangle (Exercise 5 of §4).
   c. The theorem on the centroid of a triangle (Exercise 6 of §4).
   d. The theorem on the Euler line of a triangle (Exercise 7 of §4).
   e. Pappus's Theorem (Exercise 8 of §4).

7. Consider the theorem from Exercise 17 of §4 (for which V' is reducible according to a direct translation of the hypotheses). Apply Wu's method to this problem. (Your final remainder should be nonzero here.)

Chapter 7
Invariant Theory of Finite Groups

Invariant theory has had a profound effect on the development of algebraic geometry. For example, the Hilbert Basis Theorem and Hilbert Nullstellensatz, which play a central role in the earlier chapters in this book, were proved by Hilbert in the course of his investigations of invariant theory.

In this chapter, we will study the invariants of finite groups. The basic goal is to describe *all* polynomials that are unchanged when we change variables according to a given finite group of matrices. Our treatment will be elementary and by no means complete. In particular, we do not presume a prior knowledge of group theory.

§1 Symmetric Polynomials

Symmetric polynomials arise naturally when studying the roots of a polynomial. For example, consider the cubic $f = x^3 + bx^2 + cx + d$ and let its roots be $\alpha_1, \alpha_2, \alpha_3$. Then

$$x^3 + bx^2 + cx + d = (x - \alpha_1)(x - \alpha_2)(x - \alpha_3).$$

If we expand the right-hand side, we obtain

$$x^3 + bx^2 + cx + d = x^3 - (\alpha_1 + \alpha_2 + \alpha_3)x^2 + (\alpha_1\alpha_2 + \alpha_1\alpha_3 + \alpha_2\alpha_3)x - \alpha_1\alpha_2\alpha_3,$$

and thus,

$$
\begin{aligned}
b &= -(\alpha_1 + \alpha_2 + \alpha_3), \\
c &= \alpha_1\alpha_2 + \alpha_1\alpha_3 + \alpha_2\alpha_3, \\
d &= -\alpha_1\alpha_2\alpha_3.
\end{aligned}
$$

(1)

This shows that the coefficients of f are polynomials in its roots. Further, since changing the order of the roots does not affect f, it follows that the polynomials expressing b, c, d in terms of $\alpha_1, \alpha_2, \alpha_3$ are unchanged if we permute $\alpha_1, \alpha_2, \alpha_3$. Such polynomials are said to be *symmetric*. The general concept is defined as follows.

© Springer International Publishing Switzerland 2015
D.A. Cox et al., *Ideals, Varieties, and Algorithms*, Undergraduate Texts in Mathematics, DOI 10.1007/978-3-319-16721-3_7

Definition 1. A polynomial $f \in k[x_1, \ldots, x_n]$ is **symmetric** if

$$f(x_{i_1}, \ldots, x_{i_n}) = f(x_1, \ldots, x_n)$$

for every possible permutation $x_{i_1}, \ldots, x_{i_n}$ of the variables $x_1, \ldots, x_n$.

For example, if the variables are x, y, and z, then $x^2 + y^2 + z^2$ and xyz are obviously symmetric. The following symmetric polynomials will play an important role in our discussion.

Definition 2. Given variables $x_1, \ldots, x_n$, we define the **elementary symmetric polynomials** $\sigma_1, \ldots, \sigma_n \in k[x_1, \ldots, x_n]$ by the formulas

$$\sigma_1 = x_1 + \cdots + x_n,$$

$$\vdots$$

$$\sigma_r = \sum_{i_1 < i_2 < \cdots < i_r} x_{i_1} x_{i_2} \cdots x_{i_r},$$

$$\vdots$$

$$\sigma_n = x_1 x_2 \cdots x_n.$$

Thus, σ_r is the sum of all monomials that are products of r distinct variables. In particular, every term of σ_r has total degree r. To see that these polynomials are indeed symmetric, we will generalize observation (1). Namely, introduce a new variable X and consider the polynomial

(2) $$f(X) = (X - x_1)(X - x_2) \cdots (X - x_n)$$

with roots $x_1, \ldots, x_n$. If we expand the right-hand side, it is straightforward to show that

$$f(X) = X^n - \sigma_1 X^{n-1} + \sigma_2 X^{n-2} + \cdots + (-1)^{n-1} \sigma_{n-1} X + (-1)^n \sigma_n$$

(we leave the details of the proof as an exercise). Now suppose that we rearrange $x_1, \ldots, x_n$. This changes the order of the factors on the right-hand side of (2), but f itself will be unchanged. Thus, the coefficients $(-1)^r \sigma_r$ of f are symmetric polynomials.

One corollary is that for any polynomial with leading coefficient 1, the other coefficients are the elementary symmetric polynomials of its roots (up to a factor of ± 1). The exercises will explore some interesting consequences of this fact.

From the elementary symmetric polynomials, we can construct other symmetric polynomials by taking polynomials in $\sigma_1, \ldots, \sigma_n$. Thus, for example, if $n = 3$,

$$\sigma_2^2 - \sigma_1 \sigma_3 = x^2 y^2 + x^2 yz + x^2 z^2 + xy^2 z + xyz^2 + y^2 z^2$$

is a symmetric polynomial. What is more surprising is that *all* symmetric polynomials can be represented in this way.

Theorem 3 (The Fundamental Theorem of Symmetric Polynomials). *Every symmetric polynomial in* $k[x_1, \ldots, x_n]$ *can be written uniquely as a polynomial in the elementary symmetric polynomials* $\sigma_1, \ldots, \sigma_n$.

Proof. We will use lex order with $x_1 > x_2 > \cdots > x_n$. Given a nonzero symmetric polynomial $f \in k[x_1, \ldots, x_n]$, let $\text{LT}(f) = ax^\alpha$. If $\alpha = (\alpha_1, \ldots, \alpha_n)$, we first claim that $\alpha_1 \geq \alpha_2 \geq \cdots \geq \alpha_n$. To prove this, suppose that $\alpha_i < \alpha_{i+1}$ for some i. Let β be the exponent vector obtained from α by switching α_i and α_{i+1}. We will write this as $\beta = (\ldots, \alpha_{i+1}, \alpha_i, \ldots)$. Since ax^α is a term of f, it follows that ax^β is a term of $f(\ldots, x_{i+1}, x_i, \ldots)$. But f is symmetric, so that $f(\ldots, x_{i+1}, x_i, \ldots) = f$, and thus, ax^β is a term of f. This is impossible since $\beta > \alpha$ under lex order, and our claim is proved.

Now let
$$h = \sigma_1^{\alpha_1 - \alpha_2} \sigma_2^{\alpha_2 - \alpha_3} \cdots \sigma_{n-1}^{\alpha_{n-1} - \alpha_n} \sigma_n^{\alpha_n}.$$

To compute the leading term of h, first note that $\text{LT}(\sigma_r) = x_1 x_2 \cdots x_r$ for $1 \leq r \leq n$. Hence,

(3)
$$
\begin{aligned}
\text{LT}(h) &= \text{LT}(\sigma_1^{\alpha_1 - \alpha_2} \sigma_2^{\alpha_2 - \alpha_3} \cdots \sigma_n^{\alpha_n}) \\
&= \text{LT}(\sigma_1)^{\alpha_1 - \alpha_2} \text{LT}(\sigma_2)^{\alpha_2 - \alpha_3} \cdots \text{LT}(\sigma_n)^{\alpha_n} \\
&= x_1^{\alpha_1 - \alpha_2} (x_1 x_2)^{\alpha_2 - \alpha_3} \cdots (x_1 \ldots x_n)^{\alpha_n} \\
&= x_1^{\alpha_1} x_2^{\alpha_2} \cdots x_n^{\alpha_n} = x^\alpha.
\end{aligned}
$$

It follows that f and ah have the same leading term, and thus,

$$\text{multideg}(f - ah) < \text{multideg}(f)$$

whenever $f - ah \neq 0$.

Now set $f_1 = f - ah$ and note that f_1 is symmetric since f and ah are. Hence, if $f_1 \neq 0$, we can repeat the above process to form $f_2 = f_1 - a_1 h_1$, where a_1 is a constant and h_1 is a product of $\sigma_1, \ldots, \sigma_n$ to various powers. Further, we know that $\text{LT}(f_2) < \text{LT}(f_1)$ when $f_2 \neq 0$. Continuing in this way, we get a sequence of polynomials $f, f_1, f_2, \ldots$ with

$$\text{multideg}(f) > \text{multideg}(f_1) > \text{multideg}(f_2) > \cdots .$$

Since lex order is a well-ordering, the sequence must be finite. But the only way the process terminates is when $f_{t+1} = 0$ for some t. Then it follows easily that

$$f = ah + a_1 h_1 + \cdots + a_t h_t,$$

which shows that f is a polynomial in the elementary symmetric polynomials.

It remains to prove uniqueness. Suppose that we have a symmetric polynomial f which can be written as

$$f = g_1(\sigma_1, \ldots, \sigma_n) = g_2(\sigma_1, \ldots, \sigma_n).$$

Here, g_1 and g_2 are polynomials in n variables, say $y_1, \ldots, y_n$. We need to prove that $g_1 = g_2$ in $k[y_1, \ldots, y_n]$.

If we set $g = g_1 - g_2$, then $g(\sigma_1, \ldots, \sigma_n) = 0$ in $k[x_1, \ldots, x_n]$. Uniqueness will be proved if we can show that $g = 0$ in $k[y_1, \ldots, y_n]$. So suppose that $g \neq 0$. If we write $g = \sum_\beta a_\beta y^\beta$, then $g(\sigma_1, \ldots, \sigma_n)$ is a sum of the polynomials $g_\beta = a_\beta \sigma_1^{\beta_1} \sigma_2^{\beta_2} \cdots \sigma_n^{\beta_n}$, where $\beta = (\beta_1, \ldots, \beta_n)$. Furthermore, the argument used in (3) above shows that

$$\mathrm{LT}(g_\beta) = a_\beta x_1^{\beta_1 + \cdots + \beta_n} x_2^{\beta_2 + \cdots + \beta_n} \cdots x_n^{\beta_n}.$$

It is an easy exercise to show that the map

$$(\beta_1, \ldots \beta_n) \longmapsto (\beta_1 + \cdots + \beta_n, \beta_2 + \cdots + \beta_n, \ldots, \beta_n)$$

is one-to-one. Thus, the g_β's have distinct leading terms. In particular, if we pick β such that $\mathrm{LT}(g_\beta) > \mathrm{LT}(g_\gamma)$ for all $\gamma \neq \beta$, then $\mathrm{LT}(g_\beta)$ will be greater than *all* terms of the g_γ's. It follows that there is nothing to cancel $\mathrm{LT}(g_\beta)$ and, thus, $g(\sigma_1, \ldots, \sigma_n)$ cannot be zero in $k[x_1, \ldots, x_n]$. This contradiction completes the proof of the theorem. $\qquad\qquad\square$

The proof just given is due to Gauss, who needed the properties of symmetric polynomials for his second proof (dated 1816) of the fundamental theorem of algebra. Here is how Gauss states lex order: "Then among the two terms

$$Ma^\alpha b^\beta c^\gamma \cdots \quad \text{and} \quad Ma^{\alpha'} b^{\beta'} c^{\gamma'} \cdots$$

superior order is attributed to the first rather than the second, if

either $\alpha > \alpha'$, or $\alpha = \alpha'$ and $\beta > \beta'$, or $\alpha = \alpha'$, $\beta = \beta'$ and $\gamma > \gamma'$, etc."

[see p. 36 of GAUSS (1876)]. This is the earliest known explicit statement of lex order.

Note that the proof of Theorem 3 gives an algorithm for writing a symmetric polynomial in terms of $\sigma_1, \ldots, \sigma_n$. For an example of how this works, consider

$$f = x^3 y + x^3 z + xy^3 + xz^3 + y^3 z + yz^3 \in k[x, y, z].$$

The leading term of f is $x^3 y = \mathrm{LT}(\sigma_1^2 \sigma)$, which gives

$$f_1 = f - \sigma_1^2 \sigma_2 = -2x^2 y^2 - 5x^2 yz - 2x^2 z^2 - 5xy^2 z - 5xyz^2 - 2y^2 z^2.$$

The leading term is now $-2x^2 y^2 = -2\mathrm{LT}(\sigma_2^2)$, and thus,

$$f_2 = f - \sigma_1^2 \sigma_2 + 2\sigma_2^2 = -x^2 yz - xy^2 z - xyz^2.$$

Then one easily sees that

$$f_3 = f - \sigma_1^2 \sigma_2 + 2\sigma_2^2 + \sigma_1 \sigma_3 = 0$$

and hence,

$$f = \sigma_1^2 \sigma_2 - 2\sigma_2^2 - \sigma_1 \sigma_3$$

is the unique expression of f in terms of the elementary symmetric polynomials.

Surprisingly, we do not need to write a general algorithm for expressing a symmetric polynomial in $\sigma_1, \ldots, \sigma_n$, for we can do this process using the division algorithm from Chapter 2. We can even use the division algorithm to check for symmetry. The precise method is as follows.

Proposition 4. *In the ring $k[x_1, \ldots, x_n, y_1, \ldots, y_n]$, fix a monomial order where any monomial involving one of $x_1, \ldots, x_n$ is greater than all monomials in $k[y_1, \ldots, y_n]$. Let G be a Gröbner basis of $\langle \sigma_1 - y_1, \ldots, \sigma_n - y_n \rangle \subseteq k[x_1, \ldots, x_n, y_1, \ldots, y_n]$. Given $f \in k[x_1, \ldots, x_n]$, let $g = \overline{f}^G$ be the remainder of f on division by G. Then:*

(i) *f is symmetric if and only if $g \in k[y_1, \ldots, y_n]$.*
(ii) *If f is symmetric, then $f = g(\sigma_1, \ldots, \sigma_n)$ is the unique expression of f as a polynomial in the elementary symmetric polynomials $\sigma_1, \ldots, \sigma_n$.*

Proof. As above, we have $f \in k[x_1, \ldots, x_n]$, and $g \in k[x_1, \ldots, x_n, y_1, \ldots, y_n]$ is its remainder on division by $G = \{g_1, \ldots, g_t\}$. This means that

$$f = A_1 g_1 + \cdots + A_t g_t + g,$$

where $A_1, \ldots, A_t \in k[x_1, \ldots x_n, y_1, \ldots, y_n]$. We can assume that $g_i \neq 0$ for all i.

To prove (i), first suppose that $g \in k[y_1, \ldots, y_n]$. Then for each i, substitute σ_i for y_i in the above formula for f. This will not affect f since it involves only $x_1, \ldots, x_n$. The crucial observation is that under this substitution, every polynomial in the ideal $\langle \sigma_1 - y_1, \ldots, \sigma_n - y_n \rangle$ goes to zero. Since $g_1, \ldots, g_t$ lie in this ideal, it follows that

$$f = g(\sigma_1, \ldots, \sigma_n).$$

Hence, f is symmetric.

Conversely, suppose that $f \in k[x_1, \ldots, x_n]$ is symmetric. Then $f = g(\sigma_1, \ldots, \sigma_n)$ for some $g \in k[y_1, \ldots, y_n]$. We want to show that g is the remainder of f on division by G. To prove this, first note that in $k[x_1, \ldots, x_n, y_1, \ldots, y_n]$, a monomial in $\sigma_1, \ldots, \sigma_n$ can be written as follows:

$$\sigma_1^{\alpha_1} \cdots \sigma_n^{\alpha_n} = (y_1 + (\sigma_1 - y_1))^{\alpha_1} \cdots (y_n + (\sigma_n - y_n))^{\alpha_n}$$
$$= y_1^{\alpha_1} \cdots y_n^{\alpha_n} + B_1 \cdot (\sigma_1 - y_1) + \cdots + B_n \cdot (\sigma_n - y_n)$$

for some $B_1, \ldots, B_n \in k[x_1, \ldots, x_n, y_1, \ldots, y_n]$. Multiplying by an appropriate constant and adding over the exponents appearing in g, it follows that

$$g(\sigma_1, \ldots, \sigma_n) = g(y_1, \ldots, y_n) + C_1 \cdot (\sigma_1 - y_1) + \cdots + C_n \cdot (\sigma_n - y_n),$$

where $C_1, \ldots, C_n \in k[x_1, \ldots, x_n, y_1, \ldots, y_n]$. Since $f = g(\sigma_1, \ldots, \sigma_n)$, we can write this as

(4) $$f = C_1 \cdot (\sigma_1 - y_1) + \cdots + C_n \cdot (\sigma_n - y_n) + g(y_1, \ldots, y_n).$$

We want to show that g is the remainder of f on division by G.

The first step is to show that no term of g is divisible by an element of $\mathrm{LT}(G)$. If this were false, then there would be $g_i \in G$ such that $\mathrm{LT}(g_i)$ divides some term of g. Hence, $\mathrm{LT}(g_i)$ would involve only $y_1, \ldots, y_n$ since $g \in k[y_1, \ldots, y_n]$. By our hypothesis on the ordering, it would follow that $g_i \in k[y_1, \ldots, y_n]$. Now replace every y_i with the corresponding σ_i. Since $g_i \in \langle \sigma_1 - y_1, \ldots, \sigma_n - y_n \rangle$, we have already observed that g_i goes to zero under the substitution $y_i \mapsto \sigma_i$. Then $g_i \in k[y_1, \ldots, y_n]$ would mean $g_i(\sigma_1, \ldots, \sigma_n) = 0$. By the uniqueness part of Theorem 3, this would imply $g_i = 0$, which is impossible since $g_i \neq 0$. This proves our claim. It follows that in (4), no term of g is divisible by an element of $\mathrm{LT}(G)$, and since G is a Gröbner basis, Proposition 1 of Chapter 2, §6 tells us that g is the remainder of f on division by G. This proves that the remainder lies in $k[y_1, \ldots, y_n]$ when f is symmetric.

Part (ii) of the proposition follows immediately from the above arguments, and we are done. □

A seeming drawback to the above proposition is the necessity to compute a Gröbner basis for $\langle \sigma_1 - y_1, \ldots, \sigma_n - y_n \rangle$. However, when we use lex order, it is quite simple to write down a Gröbner basis for this ideal. We first need some notation. Given variables $u_1, \ldots, u_s$, let

$$h_j(u_1, \ldots, u_s) = \sum_{|\alpha|=j} u^\alpha$$

be the sum of *all* monomials of total degree j in $u_1, \ldots, u_s$. Then we get the following Gröbner basis.

Proposition 5. *Fix lex order on the polynomial ring* $k[x_1, \ldots, x_n, y_1, \ldots, y_n]$ *with* $x_1 > \cdots > x_n > y_1 > \cdots > y_n$. *Then the polynomials*

$$g_j = h_j(x_j, \ldots, x_n) + \sum_{i=1}^{j}(-1)^i h_{j-i}(x_j, \ldots, x_n)y_i, \quad j = 1, \ldots, n,$$

form a Gröbner basis for the ideal $\langle \sigma_1 - y_1, \ldots, \sigma_n - y_n \rangle$.

Proof. We will sketch the proof, leaving most of the details for the exercises. The first step is to note the polynomial identity

$$(5) \qquad 0 = h_j(x_j, \ldots, x_n) + \sum_{i=1}^{j}(-1)^i h_{j-i}(x_j, \ldots, x_n)\sigma_i.$$

The proof will be covered in Exercises 10 and 11.

The next step is to show that $g_1, \ldots, g_n$ form a basis of $\langle \sigma_1 - y_1, \ldots, \sigma_n - y_n \rangle$. If we subtract the identity (5) from the definition of g_j, we obtain

$$(6) \qquad g_j = \sum_{i=1}^{j}(-1)^i h_{j-i}(x_j, \ldots, x_n)(y_i - \sigma_i),$$

which proves that $\langle g_1, \ldots, g_n \rangle \subseteq \langle \sigma_1 - y_1, \ldots, \sigma_n - y_n \rangle$. To prove the opposite inclusion, note that since $h_0 = 1$, we can write (6) as

$$(7) \qquad g_j = (-1)^j (y_j - \sigma_j) + \sum_{i=1}^{j-1} (-1)^i h_{j-i}(x_j, \ldots, x_n)(y_i - \sigma_i).$$

Then induction on j shows that $\langle \sigma_1 - y_1, \ldots, \sigma_n - y_n \rangle \subseteq \langle g_1, \ldots, g_n \rangle$ (see Exercise 12).

Finally, we need to show that we have a Gröbner basis. In Exercise 12, we will ask you to prove that

$$\mathrm{LT}(g_j) = x_j^j.$$

This is where we use lex order with $x_1 > \cdots > x_n > y_1 > \cdots > y_n$. Thus the leading terms of $g_1, \ldots, g_n$ are relatively prime, and using the theory developed in §§9 and 10 of Chapter 2, it is easy to show that we have a Gröbner basis (see Exercise 12 for the details). This completes the proof. $\qquad\square$

In dealing with symmetric polynomials, it is often convenient to work with ones that are *homogeneous*. Here is the definition.

Definition 6. A polynomial $f \in k[x_1, \ldots, x_n]$ is **homogeneous of total degree** m provided that every term appearing in f has total degree m.

As an example, note that the i-th elementary symmetric polynomial σ_i is homogeneous of total degree i. An important fact is that every polynomial can be written uniquely as a sum of homogeneous polynomials. Namely, given $f \in k[x_1, \ldots, x_n]$, let f_m be the sum of all terms of f of total degree m. Then each f_m is homogeneous and $f = \sum_m f_m$. We call f_m the m-th *homogeneous component of* f.

We can understand symmetric polynomials in terms of their homogeneous components as follows.

Proposition 7. *A polynomial $f \in k[x_1, \ldots, x_n]$ is symmetric if and only if all of its homogeneous components are symmetric.*

Proof. Assume that f is symmetric and let $x_{i_1}, \ldots, x_{i_n}$ be a permutation of $x_1, \ldots, x_n$. This permutation takes a term of f of total degree m to one of the same total degree. Since $f(x_{i_1}, \ldots, x_{i_n}) = f(x_1, \ldots, x_n)$, it follows that the m-th homogeneous component must also be symmetric. The converse is trivial and the proposition follows. $\square$

Proposition 7 tells us that when working with a symmetric polynomial, we can assume that it is homogeneous. In the exercises, we will explore what this implies about how the polynomial is expressed in terms of $\sigma_1, \ldots, \sigma_n$.

The final topic we will explore is a different way of writing symmetric polynomials. Specifically, we will consider the *power sums*

$$s_j = x_1^j + x_2^j + \cdots + x_n^j.$$

Note that s_j is symmetric. Then we can write an arbitrary symmetric polynomial in terms of $s_1, \ldots, s_n$ as follows.

Theorem 8. *If k is a field containing the rational numbers $\mathbb{Q}$, then every symmetric polynomial in $k[x_1, \ldots, x_n]$ can be written as a polynomial in the power sums $s_1, \ldots, s_n$.*

Proof. Since every symmetric polynomial is a polynomial in the elementary symmetric polynomials (by Theorem 3), it suffices to prove that $\sigma_1, \ldots, \sigma_n$ are polynomials in $s_1, \ldots, s_n$. For this purpose, we will use the *Newton identities*, which state that

$$s_j - \sigma_1 s_{j-1} + \cdots + (-1)^{j-1}\sigma_{j-1}s_1 + (-1)^j j\sigma_j = 0, \quad 1 \le j \le n,$$
$$s_j - \sigma_1 s_{j-1} + \cdots + (-1)^{n-1}\sigma_{n-1}s_{j-n+1} + (-1)^n \sigma_n s_{j-n} = 0, \quad j > n.$$

The proof of these identities will be given in the exercises.

We now prove by induction on j that σ_j is a polynomial in $s_1, \ldots, s_n$. This is true for $j = 1$ since $\sigma_1 = s_1$. If the claim is true for $1, 2, \ldots, j - 1$, then the Newton identities imply that

$$\sigma_j = (-1)^{j-1}\frac{1}{j}\left(s_j - \sigma_1 s_{j-1} + \cdots + (-1)^{j-1}\sigma_{j-1}s_1\right).$$

We can divide by the integer j because $\mathbb{Q}$ is contained in the coefficient field (see Exercise 16 for an example of what can go wrong when $\mathbb{Q} \not\subset k$). Then our inductive assumption and the above equation show that σ_j is a polynomial in $s_1, \ldots, s_n$. $\square$

As a consequence of Theorems 3 and 8, every elementary symmetric polynomial can be written in terms of power sums, and vice versa. For example,

$$s_2 = \sigma_1^2 - 2\sigma_2 \longleftrightarrow \sigma_2 = \frac{1}{2}(s_1^2 - s_2),$$
$$s_3 = \sigma_1^3 - 3\sigma_1\sigma_2 + 3\sigma_3 \longleftrightarrow \sigma_3 = \frac{1}{6}(s_1^3 - 3s_1 s_2 + 2s_3).$$

Power sums will be unexpectedly useful in §3 when we give an algorithm for finding the invariant polynomials for a finite group.

EXERCISES FOR §1

1. Prove that $f \in k[x, y, z]$ is symmetric if and only if $f(x, y, z) = f(y, x, z) = f(y, z, x)$.
2. (Requires abstract algebra) Prove that $f \in k[x_1, \ldots, x_n]$ is symmetric if and only if
   $f(x_1, x_2, x_3, \ldots, x_n) = f(x_2, x_1, x_3, \ldots, x_n) = f(x_2, x_3, \ldots x_n, x_1)$. Hint: Show that the cyclic permutations $(1, 2)$ and $(1, 2, \ldots, n)$ generate the symmetric group S_n. See Exercise 4 in Section 3.5 of DUMMIT and FOOTE (2004).
3. Let $\sigma_j^{(i)}$ denote the j-th elementary symmetric polynomial in $x_1, \ldots, x_{i-1}, x_{i+1} \ldots, x_n$ for $j < n$. The superscript "(i)" tells us to omit the variable x_i. Also set $\sigma_n^{(i)} = 0$ and $\sigma_0 = 1$. Prove that $\sigma_j = \sigma_j^{(i)} + x_i\sigma_{j-1}^{(i)}$ for all i, j. This identity is useful in induction arguments involving elementary symmetric polynomials.
4. As in (2), let $f(X) = (X - x_1)(X - x_2)\cdots(X - x_n)$. Prove that $f = X^n - \sigma_1 X^{n-1} + \sigma_2 X^{n-2} + \cdots + (-1)^{n-1}\sigma_{n-1}X + (-1)^n\sigma_n$. Hint: You can give an induction proof using the identities of Exercise 3.

5. Consider the polynomial

$$f = (x^2 + y^2)(x^2 + z^2)(y^2 + z^2) \in k[x, y, z].$$

a. Use the method given in the proof of Theorem 3 to write f as a polynomial in the elementary symmetric polynomials $\sigma_1, \sigma_2, \sigma_3$.

b. Use the method described in Proposition 4 to write f in terms of $\sigma_1, \sigma_2, \sigma_3$.

You can use a computer algebra system for both parts of the exercise. Note that by stripping off the coefficients of powers of X in the polynomial $(X - x)(X - y)(X - z)$, you can get the computer to generate the elementary symmetric polynomials.

6. If the variables are $x_1, \ldots, x_n$, show that $\sum_{i \neq j} x_i^2 x_j = \sigma_1 \sigma_2 - 3\sigma_3$. Hint: If you get stuck, see Exercise 13. Note that a computer algebra system cannot help here!

7. Let $f = x^n + a_1 x^{n-1} + \cdots + a_n \in k[x]$ have roots $\alpha_1, \ldots, \alpha_n$, which lie in some bigger field K containing k.

a. Prove that any symmetric polynomial $g(\alpha_1, \ldots, \alpha_n)$ in the roots of f can be expressed as a polynomial in the coefficients $a_1, \ldots, a_n$ of f.

b. In particular, if the symmetric polynomial g has coefficients in k, conclude that $g(\alpha_1, \ldots, \alpha_n) \in k$.

8. As in Exercise 7, let $f = x^n + a_1 x^{n-1} + \cdots + a_n \in k[x]$ have roots $\alpha_1, \ldots, \alpha_n$, which lie in some bigger field K containing k. The *discriminant* of f is defined to be

$$D(f) = \prod_{i \neq j} (\alpha_i - \alpha_j)$$

a. Use Exercise 7 to show that $D(f)$ is a polynomial in $a_1, \ldots, a_n$.

b. When $n = 2$, express $D(f)$ in terms of a_1 and a_2. Does your result look familiar?

c. When $n = 3$, express $D(f)$ in terms of a_1, a_2, a_3.

d. Explain why a cubic polynomial $x^3 + a_1 x^2 + a_2 x + a_3$ has a multiple root if and only if $-4a_1^3 a_3 + a_1^2 a_2^2 + 18 a_1 a_2 a_3 - 4a_2^3 - 27 a_3^2 = 0$.

9. Given a cubic polynomial $f = x^3 + a_1 x^2 + a_2 x + a_3$, what condition must the coefficients of f satisfy in order for one of its roots to be the average of the other two? Hint: If α_1 is the average of the other two, then $2\alpha_1 - \alpha_2 - \alpha_3 = 0$. But it could happen that α_2 or α_3 is the average of the other two. Hence, you get a condition stating that the product of three expressions similar to $2\alpha_1 - \alpha_2 - \alpha_3$ is equal to zero. Now use Exercise 7.

10. As in Proposition 5, let $h_j(x_1, \ldots, x_n)$ be the sum of all monomials of total degree j in $x_1, \ldots, x_n$. Also, let $\sigma_0 = 1$ and $\sigma_i = 0$ if $i > n$. The goal of this exercise is to show that if $j > 0$, then

$$0 = \sum_{i=0}^{j} (-1)^i h_{j-i}(x_1, \ldots, x_n) \sigma_i(x_1, \ldots, x_n).$$

In Exercise 11, we will use this to prove the closely related identity (5) that appears in the text. To prove the above identity, we will compute the coefficients of the monomials x^α that appear in $h_{j-i} \sigma_i$. Since every term in $h_{j-i} \sigma_i$ has total degree j, we can assume that x^α has total degree j. We will let a denote the number of variables that actually appear in x^α.

a. If x^α appears in $h_{j-i} \sigma_i$, show that $i \leq a$. Hint: How many variables appear in each term of σ_i?

b. If $i \leq a$, show that exactly $\binom{a}{i}$ terms of σ_i involve only variables that appear in x^α. Note that all of these terms have total degree i.

c. If $i \leq a$, show that x^α appears in $h_{j-i} \sigma_i$ with coefficient $\binom{a}{i}$. Hint: This follows from part (b) because h_{j-i} is the sum of all monomials of total degree $j - i$, and each monomial has coefficient 1.

d. Conclude that the coefficient of x^α in $\sum_{i=0}^{j}(-1)^i h_{j-i}\sigma_i$ is $\sum_{i=0}^{a}(-1)^i\binom{a}{i}$. Then use the binomial theorem to show that the coefficient of x^α is zero. This will complete the proof of our identity.

11. In this exercise, we will prove the identity

$$0 = h_j(x_j,\ldots,x_n) + \sum_{i=1}^{j}(-1)^i h_{j-i}(x_k,\ldots,x_n)\sigma_i(x_1,\ldots,x_n)$$

used in the proof of Proposition 5. As in Exercise 10, let $\sigma_0 = 1$, so that the identity can be written more compactly as

$$0 = \sum_{i=0}^{j}(-1)^i h_{j-i}(x_j,\ldots,x_n)\sigma_i(x_1,\ldots,x_n)$$

The idea is to separate out the variables $x_1,\ldots,x_{j-1}$. To this end, if $S \subseteq \{1,\ldots,j-1\}$, let x^S be the product of the corresponding variables and let $|S|$ denote the number of elements in S.

a. Prove that

$$\sigma_i(x_1,\ldots,x_n) = \sum_{S\subseteq\{1,\ldots,j-1\}} x^S \sigma_{i-|S|}(x_j,\ldots,x_n),$$

where we set $\sigma_m = 0$ if $m < 0$.

b. Prove that

$$\sum_{i=0}^{j}(-1)^i h_{j-i}(x_j,\ldots,x_n)\sigma_i(x_1,\ldots,x_n)$$

$$= \sum_{S\subset\{1,\ldots,j-1\}} x^S\left(\sum_{i=|S|}^{j}(-1)^i h_{j-i}(x_j,\ldots,x_n)\sigma_{i-|S|}(x_j,\ldots,x_n)\right).$$

c. Use Exercise 10 to conclude that the sum inside the parentheses is zero for every S. This proves the desired identity. Hint: Let $\ell = i - |S|$.

12. This exercise is concerned with the proof of Proposition 5. Let g_j be as defined in the statement of the proposition.
a. Use equation (7) to prove that $\langle\sigma_1 - y_1,\ldots,\sigma_n - y_n\rangle \subseteq \langle g_1,\ldots,g_n\rangle$.
b. Prove that $\mathrm{LT}(g_j) = x_j^j$.
c. Combine part (b) with Theorem 3 of Chapter 2, §9 and Proposition 1 of Chapter 2, §10 to prove that $g_1,\ldots,g_n$ form a Gröbner basis.

13. Let f be a homogeneous symmetric polynomial of total degree d.
a. Show that f can be written as a linear combination (with coefficients in k) of polynomials of the form $\sigma_1^{i_1}\sigma_2^{i_2}\cdots\sigma_n^{i_n}$ where $d = i_1 + 2i_2 + \cdots + ni_n$.
b. Let m be the maximum degree of x_1 that appears in f. By symmetry, m is the maximum degree in f of any variable. If $\sigma_1^{i_1}\sigma_2^{i_2}\cdots\sigma_n^{i_n}$ appears in the expression of f from part (a), then prove that $i_1 + i_2 + \cdots + i_n \le m$.
c. Show that the symmetric polynomial $\sum_{i\neq j}x_i^2 x_j$ can be written as $a\sigma_1\sigma_2 + b\sigma_3$ for some constants a and b. Then determine a and b. Compare this to what you did in Exercise 6.

14. In this exercise, you will prove the Newton identities used in the proof of Theorem 8. Let the variables be $x_1,\ldots,x_n$.
a. Set $\sigma_0 = 1$ and $\sigma_i = 0$ whenever $i < 0$ or $i > n$. Then show that the Newton identities are equivalent to

$$s_j - \sigma_1 s_{j-1} + \cdots + (-1)^{j-1}\sigma_{j-1}s_1 + (-1)^j j\sigma_j = 0, \quad j \geq 1.$$

b. Prove the identity of part (a) by induction on n. Hint: Similar to Exercise 3, let $s_j^{(n)}$ be the jth power sum of $x_1, \ldots, x_{n-1}$, i.e., all variables except x_n. Then use Exercise 3 and note that $s_j = s_j^{(n)} + x_n^j$.

15. This exercise will use the identity (5) to prove the following *nonsymmetric Newton identities*:

$$x_i^j - \sigma_1 x_i^{j-1} + \cdots + (-1)^{j-1}\sigma_{k-1}x_i + (-1)^j\sigma_k = (-1)^j\sigma_j^{(i)}, \quad 1 \leq j < n,$$

$$x_i^j - \sigma_1 x_i^{j-1} + \cdots + (-1)^{n-1}\sigma_{n-1}x_i^{j-n+1} + (-1)^n\sigma_n x_i^{j-n} = 0, \quad j \geq n,$$

where as in Exercise 3, $\sigma_j^{(i)} = \sigma_j(x_1, \ldots, x_{i-1}, x_{i+1}, \ldots, x_n)$ is the j-th elementary symmetric polynomial of all variables except x_i. We will then give a second proof of the Newton identities.

a. Show that the nonsymmetric Newton identity for $j = n$ follows from (5). Then prove that this implies the nonsymmetric Newton identities for $j \geq n$. Hint: Treat the case $i = n$ first.

b. Show that the nonsymmetric Newton identity for $j = n - 1$ follows from the one for $j = n$. Hint: $\sigma_n = x_i\sigma_{n-1}^{(i)}$.

c. Prove the nonsymmetric Newton identity for $j < n$ by decreasing induction on j. Hint: By Exercise 3, $\sigma_j = \sigma_j^{(i)} + x_i\sigma_{j-1}^{(i)}$.

d. Prove that $\sum_{i=1}^n \sigma_j^{(i)} = (n - j)\sigma_j$. Hint: A term $x_{i_1} \cdots x_{i_j}$, where $1 \leq i_1 < \cdots < i_j \leq n$, appears in how many of the $\sigma_j^{(i)}$'s?

e. Prove the Newton identities.

16. Consider the field $\mathbb{F}_2 = \{0, 1\}$ consisting of two elements. Show that it is impossible to express the symmetric polynomial $xy \in \mathbb{F}_2[x, y]$ as a polynomial in s_1 and s_2 with coefficients $\mathbb{F}_2$. Hint: Show that $s_2 = s_1^2$!

17. Express s_4 as a polynomial in $\sigma_1, \ldots, \sigma_4$ and express σ_4 as a polynomial in $s_1 \ldots, s_4$.

18. We can use the division algorithm to automate the process of writing a polynomial $g(\sigma_1, \ldots, \sigma_n)$ in terms of $s_1, \ldots, s_n$. Namely, regard $\sigma_1, \ldots, \sigma_n, s_1, \ldots, s_n$ as variables and consider the polynomials

$$g_j = s_j = \sigma_1 s_{j-1} + \cdots + (-1)^{j-1}\sigma_{j-1}s_1 + (-1)^j j\sigma_j, \quad 1 \leq j \leq n.$$

Show that if we use the correct lex order, the remainder of $g(\sigma_1, \ldots, \sigma_n)$ on division by $g_1, \ldots, g_n$ will be a polynomial $h(s_1, \ldots, s_n)$ such that $g(\sigma_1, \ldots, \sigma_n) = h(s_1, \ldots, s_n)$. Hint: The lex order you need is *not* $\sigma_1 > \sigma_2 > \cdots > \sigma_n > s_1 > \cdots > s_n$.

§2 Finite Matrix Groups and Rings of Invariants

In this section, we will give some basic definitions for invariants of finite matrix groups and we will compute some examples to illustrate what questions the general theory should address. For the rest of this chapter, we will always assume that our field k contains the rational numbers $\mathbb{Q}$. Such fields are said to be of *characteristic zero*.

Definition 1. Let $GL(n, k)$ be the set of all invertible $n \times n$ matrices with entries in the field k.

If A and B are invertible $n \times n$ matrices, then linear algebra implies that the product AB and inverse A^{-1} are also invertible (see Exercise 1). Also, recall that the $n \times n$ identity matrix I_n has the properties that $A \cdot I_n = I_n \cdot A = A$ and $A \cdot A^{-1} = I_n$ for all $A \in \mathrm{GL}(n, k)$. In the terminology of Appendix A, we say that $\mathrm{GL}(n, k)$ is a *group*.

Note that $A \in \mathrm{GL}(n, k)$ gives an invertible linear map $L_A : k^n \to k^n$ via matrix multiplication. Since every linear map from k^n to itself arises in this way, it is customary to call $\mathrm{GL}(n, k)$ the *general linear group*.

We will be most interested in the following subsets of $\mathrm{GL}(n, k)$.

Definition 2. A finite subset $G \subseteq \mathrm{GL}(n, k)$ is called a **finite matrix group** provided it is nonempty and closed under matrix multiplication. The number of elements of G is called the **order** of G and is denoted $|G|$.

Let us look at some examples of finite matrix groups.

Example 3. Suppose that $A \in \mathrm{GL}(n, k)$ is a matrix such that $A^m = I_n$ for some positive integer m. If m is the smallest such integer, then it is easy to show that

$$C_m = \{I_n, A, \ldots, A^{m-1}\} \subseteq \mathrm{GL}(n, k)$$

is closed under multiplication (see Exercise 2) and, hence, is a finite matrix group. We call C_m a *cyclic group* of order m. An example is given by

$$A = \begin{pmatrix} 0 & -1 \\ 1 & 0 \end{pmatrix} \in \mathrm{GL}(2, k).$$

One can check that $A^4 = I_2$, so that $C_4 = \{I_2, A, A^2, A^3\}$ is a cyclic matrix group of order 4 in $\mathrm{GL}(2, k)$.

Example 4. An important example of a finite matrix group comes from the permutations of variables discussed in §1. Let τ denote a permutation $x_{i_1}, \ldots, x_{i_n}$ of $x_1, \ldots, x_n$. Since τ is determined by what it does to the subscripts, we will set $i_1 = \tau(1), i_2 = \tau(2), \ldots, i_n = \tau(n)$. Then the corresponding permutation of variables is $x_{\tau(1)}, \ldots, x_{\tau(n)}$.

We can create a matrix from τ as follows. Consider the linear map that takes $(x_1, \ldots, x_n)$ to $(x_{\tau(1)}, \ldots, x_{\tau(n)})$. The matrix representing this linear map is denoted M_τ and is called a *permutation matrix*. Thus, M_τ has the property that under matrix multiplication, it permutes the variables according to τ:

$$M_\tau \cdot \begin{pmatrix} x_1 \\ \vdots \\ x_n \end{pmatrix} = \begin{pmatrix} x_{\tau(1)} \\ \vdots \\ x_{\tau(n)} \end{pmatrix}.$$

We leave it as an exercise to show that M_τ is obtained from the identity matrix by permuting its columns according to τ. More precisely, the $\tau(i)$-th column of M_τ is the i-th column of I_n. As an example, consider the permutation τ that takes (x, y, z) to (y, z, x). Here, $\tau(1) = 2, \tau(2) = 3$, and $\tau(3) = 1$, and one can check that

$$
M_\tau \cdot \begin{pmatrix} x \\ y \\ z \end{pmatrix} = \begin{pmatrix} 0 & 1 & 0 \\ 0 & 0 & 1 \\ 1 & 0 & 0 \end{pmatrix} \begin{pmatrix} x \\ y \\ z \end{pmatrix} = \begin{pmatrix} y \\ z \\ x \end{pmatrix}.
$$

Since there are $n!$ ways to permute the variables, we get $n!$ permutation matrices. Furthermore, this set is closed under matrix multiplication, for it is easy to show that

$$
M_\tau \cdot M_\nu = M_{\nu\tau},
$$

where $\nu\tau$ is the permutation that takes i to $\nu(\tau(i))$ (see Exercise 4). Thus, the permutation matrices form a finite matrix group in $\mathrm{GL}(n, k)$. We will denote this matrix group by S_n. (Strictly speaking, the group of permutation matrices is only isomorphic to S_n in the sense of group theory. We will ignore this distinction.)

Example 5. Another important class of finite matrix groups comes from the symmetries of regular polyhedra. For example, consider a cube in $\mathbb{R}^3$ centered at the origin. The set of rotations of $\mathbb{R}^3$ that take the cube to itself is clearly finite and closed under multiplication. Thus, we get a finite matrix group in $\mathrm{GL}(3, \mathbb{R})$. In general, all finite matrix groups in $\mathrm{GL}(3, \mathbb{R})$ have been classified, and there is a rich geometry associated with such groups (see Exercises 5–9 for some examples). To pursue this topic further, the reader should consult BENSON and GROVE (1985), KLEIN (1884), or COXETER (1973).

Finite matrix groups have the following useful properties.

Proposition 6. *Let $G \subseteq \mathrm{GL}(n, k)$ be a finite matrix group. Then:*

(i) $I_n \in G$.

(ii) *If $A \in G$, then $A^m = I_n$ for some positive integer m.*

(iii) *If $A \in G$, then $A^{-1} \in G$.*

Proof. Take $A \in G$. Then $\{A, A^2, A^3, \ldots\} \subseteq G$ since G is closed under multiplication. The finiteness of G then implies that $A^i = A^j$ for some $i > j$, and since A is invertible, we can multiply each side by A^{-j} to conclude that $A^m = I_n$, where $m = i - j > 0$. This proves (ii).

To prove (iii), note that (ii) implies $I_n = A^m = A \cdot A^{m-1} = A^{m-1} \cdot A$. Thus, $A^{-1} = A^{m-1} \in G$ since G is closed under multiplication. As for (i), since $G \neq \emptyset$, we can pick $A \in G$, and then by (ii), $I_n = A^m \in G$. $\qquad \square$

We next observe that elements of $\mathrm{GL}(n, k)$ act on polynomials in $k[x_1, \ldots, x_n]$. To see how this works, let $A = (a_{ij}) \in \mathrm{GL}(n, k)$ and $f \in k[x_1, \ldots, x_n]$. Then

$$
(1) \qquad g(x_1, \ldots, x_n) = f(a_{11}x_1 + \cdots + a_{1n}x_n, \ldots, a_{n1}x_1 + \cdots + a_{nn}x_n)
$$

is again a polynomial in $k[x_1, \ldots, x_n]$. To express this more compactly, let $\mathbf{x}$ denote the column vector of the variables $x_1, \ldots, x_n$. Thus,

$$\mathbf{x} = \begin{pmatrix} x_1 \\ \vdots \\ x_n \end{pmatrix}.$$

Then we can use matrix multiplication to express equation (1) as

$$g(\mathbf{x}) = f(A \cdot \mathbf{x}).$$

If we think of A as a change of basis matrix, then g is simply f viewed using the new coordinates.

For an example of how this works, let $f(x, y) = x^2 + xy + y^2 \in \mathbb{R}[x, y]$ and

$$A = \frac{1}{\sqrt{2}} \begin{pmatrix} 1 & -1 \\ 1 & 1 \end{pmatrix} \in GL(2, \mathbb{R}).$$

Then

$$\begin{aligned}
g(x, y) = f(A \cdot \mathbf{x}) &= f\left(\frac{x - y}{\sqrt{2}}, \frac{x + y}{\sqrt{2}}\right) \\
&= \left(\frac{x - y}{\sqrt{2}}\right)^2 + \frac{x - y}{\sqrt{2}} \cdot \frac{x + y}{\sqrt{2}} + \left(\frac{x + y}{\sqrt{2}}\right)^2 \\
&= \frac{3}{2}x^2 + \frac{1}{2}y^2.
\end{aligned}$$

Geometrically, this shows that we can eliminate the xy term of f by rotating the coordinate axes $45°$.

A remarkable fact is that sometimes this process gives back the same polynomial we started with. For example, if we let $h(x, y) = x^2 + y^2$ and use the above matrix A, then one can check that

$$h(\mathbf{x}) = h(A \cdot \mathbf{x}).$$

In this case, we say that h is *invariant* under A.

This leads to the following fundamental definition.

Definition 7. Let $G \subseteq GL(n, k)$ be a finite matrix group. Then a polynomial $f(\mathbf{x}) \in k[x_1, \ldots, x_n]$ is **invariant** under G if

$$f(\mathbf{x}) = f(A \cdot \mathbf{x})$$

for all $A \in G$. The set of all invariant polynomials is denoted $k[x_1, \ldots, x_n]^G$.

The most basic example of invariants of a finite matrix group is given by the symmetric polynomials.

Example 8. If we consider the group $S_n \subseteq GL(n, k)$ of permutation matrices, then it is obvious that

$$k[x_1, \ldots, x_n]^{S_n} = \{\text{all symmetric polynomials in } k[x_1, \ldots, x_n]\}.$$

By Theorem 3 of §1, we know that symmetric polynomials are polynomials in the elementary symmetric polynomials with coefficients in k. We can write this as

$$k[x_1, \ldots, x_n]^{S_n} = k[\sigma_1, \ldots, \sigma_n].$$

Thus, every invariant can be written as a polynomial in finitely many invariants (the elementary symmetric polynomials). In addition, we know that the representation in terms of the elementary symmetric polynomials is unique. Hence, we have a very explicit knowledge of the invariants of S_n.

One goal of invariant theory is to examine whether all invariants $k[x_1, \ldots, x_n]^G$ are as nice as Example 8. To begin our study of this question, we first show that the set of invariants $k[x_1, \ldots, x_n]^G$ has the following algebraic structure.

Proposition 9. *If $G \subseteq \mathrm{GL}(n, k)$ is a finite matrix group, then the set $k[x_1, \ldots, x_n]^G$ is closed under addition and multiplication and contains the constant polynomials.*

Proof. We leave the easy proof as an exercise.   □

Multiplication and addition in $k[x_1, \ldots, x_n]^G$ automatically satisfy the distributive, associative, etc., properties since these properties are true in $k[x_1, \ldots, x_n]$. In the terminology of Chapter 5, we say that $k[x_1, \ldots, x_n]^G$ is a *commutative ring*. Furthermore, we say that $k[x_1, \ldots, x_n]^G$ is a *subring* of $k[x_1, \ldots, x_n]$.

So far in this book, we have learned three ways to create new rings. In Chapter 5, we saw how to make the quotient ring $k[x_1, \ldots, x_n]/I$ of an ideal $I \subseteq k[x_1, \ldots, x_n]$ and the coordinate ring $k[V]$ of an affine variety $V \subseteq k^n$. Now we can make the ring of invariants $k[x_1, \ldots, x_n]^G$ of a finite matrix group $G \subseteq \mathrm{GL}(n, k)$. In §4, we will see how these constructions are related.

In §1, we saw that the homogeneous components of a symmetric polynomial were also symmetric. We next observe that this holds for the invariants of any finite matrix group.

Proposition 10. *Let $G \subseteq \mathrm{GL}(n, k)$ be a finite matrix group. Then a polynomial $f \in k[x_1, \ldots, x_n]$ is invariant under G if and only if its homogeneous components are invariant.*

Proof. See Exercise 11.   □

Homogeneous invariants play a key role in invariant theory. In §3, we will often use Proposition 10 to reduce to the case of homogeneous invariants.

The following lemma will prove useful in determining whether a given polynomial is invariant under a finite matrix group.

Lemma 11. *Let $G \subseteq \mathrm{GL}(n, k)$ be a finite matrix group and suppose that we have $A_1, \ldots, A_m \in G$ such that every $A \in G$ can be written in the form*

$$A = B_1 B_2 \cdots B_t,$$

where $B_i \in \{A_1, \ldots, A_m\}$ for every i (we say that $A_1, \ldots, A_m$ **generate** G). Then $f \in k[x_1, \ldots, x_n]$ is in $k[x_1, \ldots, x_n]^G$ if and only if

$$f(\mathbf{x}) = f(A_1 \cdot \mathbf{x}) = \cdots = f(A_m \cdot \mathbf{x}).$$

Proof. We first show that if f is invariant under matrices $B_1, \ldots, B_t$, then it is also invariant under their product $B_1 \cdots B_t$. This is clearly true for $t = 1$. If we assume it is true for $t - 1$, then

$$
\begin{aligned}
f((B_1 \cdots B_t) \cdot \mathbf{x}) &= f((B_1 \cdots B_{t-1}) \cdot (B_t \cdot \mathbf{x})) \\
&= f(B_t \cdot \mathbf{x}) \qquad &\text{(by our inductive assumption)} \\
&= f(\mathbf{x}) \qquad &\text{(by the invariance under } B_t\text{).}
\end{aligned}
$$

Now suppose that f is invariant under $A_1, \ldots, A_m$. Since elements $A \in G$ can be written $A = B_1 \cdots B_t$, where every B_i is one of $A_1, \ldots, A_m$, it follows immediately that $f \in k[x_1, \ldots, x_n]^G$. The converse is trivial and the lemma is proved. $\qquad \square$

We can now compute some interesting examples of rings of invariants.

Example 12. Consider the finite matrix group

$$V_4 = \left\{ \begin{pmatrix} \pm 1 & 0 \\ 0 & \pm 1 \end{pmatrix} \right\} \subseteq \mathrm{GL}(2, k).$$

This is sometimes called the *Klein four-group*. In German, "four-group" is written "vierergruppe," which explains the notation V_4. You should check that the matrices

$$\begin{pmatrix} -1 & 0 \\ 0 & 1 \end{pmatrix}, \ \begin{pmatrix} 1 & 0 \\ 0 & -1 \end{pmatrix}$$

generate V_4. Then Lemma 11 implies that a polynomial $f \in k[x, y]$ is invariant under V_4 if and only if

$$f(x, y) = f(-x, y) = f(x, -y).$$

Writing $f = \sum_{ij} a_{ij} x^i y^j$, we can understand the first of these conditions as follows:

$$
\begin{aligned}
f(x, y) = f(-x, y) &\Longleftrightarrow \sum_{ij} a_{ij} x^i y^j = \sum_{ij} a_{ij}(-x)^i y^j \\
&\Longleftrightarrow \sum_{ij} a_{ij} x^i y^j = \sum_{ij} (-1)^i a_{ij} x^i y^j \\
&\Longleftrightarrow a_{ij} = (-1)^i a_{ij} \quad \text{for all } i, j \\
&\Longleftrightarrow a_{ij} = 0 \quad \text{for } i \text{ odd.}
\end{aligned}
$$

It follows that x always appears to an even power. Similarly, the condition $f(x, y) = f(x, -y)$ implies that y appears to even powers. Thus, we can write

$$f(x, y) = g(x^2, y^2).$$

for a unique polynomial $g(x, y) \in k[x, y]$. Conversely, every polynomial f of this form is clearly invariant under V_4. This proves that

$$k[x, y]^{V_4} = k[x^2, y^2].$$

Hence, every invariant of V_4 can be uniquely written as a polynomial in the two homogeneous invariants x^2 and y^2. In particular, the invariants of the Klein four-group behave very much like the symmetric polynomials.

Example 13. For a finite matrix group that is less well-behaved, consider the cyclic group $C_2 = \{\pm I_2\} \subseteq GL(2, k)$ of order 2. In this case, the invariants consist of the polynomials $f \in k[x, y]$ for which $f(x, y) = f(-x, -y)$. We leave it as an exercise to show that this is equivalent to the condition

$$f(x, y) = \sum_{ij} a_{ij} x^i y^j, \quad \text{where } a_{ij} = 0 \text{ whenever } i + j \text{ is odd.}$$

This means that f is invariant under C_2 if and only if the exponents of x and y always have the same parity (i.e., both even or both odd). Hence, we can write a monomial $x^i y^j$ appearing in f in the form

$$x^i y^j = \begin{cases} x^{2k} y^{2l} = (x^2)^k (y^2)^l & \text{if } i, j \text{ are even} \\ x^{2k+1} y^{2l+1} = (x^2)^k (y^2)^l xy & \text{if } i, j \text{ are odd.} \end{cases}$$

This means that every monomial in f, and hence f itself, is a polynomial in the homogeneous invariants x^2, y^2, and xy. We will write this as

$$k[x, y]^{C_2} = k[x^2, y^2, xy].$$

Note also that we need all three invariants to generate $k[x, y]^{C_2}$.

The ring $k[x^2, y^2, xy]$ is fundamentally different from the previous examples because uniqueness breaks down: a given invariant can be written in terms of x^2, y^2, xy in more than one way. For example, $x^4 y^2$ is clearly invariant under C_2, but

$$x^4 y^2 = (x^2)^2 \cdot y^2 = x^2 \cdot (xy)^2.$$

In §4, we will see that the crux of the matter is the algebraic relation $x^2 \cdot y^2 = (xy)^2$ between the basic invariants. In general, a key part of the theory is determining all algebraic relations between invariants. Given this information, one can describe precisely how uniqueness fails.

From these examples, we see that given a finite matrix group G, invariant theory has two basic questions to answer about the ring of invariants $k[x_1, \ldots, x_n]^G$:

- (Finite Generation) Can we find finitely many homogeneous invariants $f_1, \ldots, f_m$ such that every invariant is a polynomial in $f_1, \ldots, f_m$?
- (Uniqueness) In how many ways can an invariant be written in terms of $f_1, \ldots, f_m$? In §4, we will see that this asks for the algebraic relations among $f_1, \ldots, f_m$.

In §§3 and 4, we will give complete answers to both questions. We will also describe algorithms for finding the invariants and the relations between them.

EXERCISES FOR §2

1. If $A, B \in GL(n, k)$ are invertible matrices, show that AB and A^{-1} are also invertible.
2. Suppose that $A \in GL(n, k)$ satisfies $A^m = I_n$ for some positive integer. If m is the smallest such integer, then prove that the set $C_m = \{I_n, A, A^2, \ldots, A^{m-1}\}$ has exactly m elements and is closed under matrix multiplication.
3. Write down the six permutation matrices in $GL(3, k)$.
4. Let M_τ be the matrix of the linear transformation taking $x_1, \ldots, x_n$ to $x_{\tau(1)}, \ldots, x_{\tau(n)}$. This means that if $e_1, \ldots, e_n$ is the standard basis of k^n, then $M_\tau \cdot (\sum_j x_j e_j) = \sum_j x_{\tau(j)} e_j$.
   a. Show that $M_\tau \cdot e_{\tau(i)} = e_i$. Hint: Observe that $\sum_j x_j e_j = \sum_j x_{\tau(j)} e_{\tau(j)}$.
   b. Prove that the $\tau(i)$-th column of M_τ is the i-th column of the identity matrix.
   c. Prove that $M_\tau \cdot M_\nu = M_{\nu\tau}$, where $\nu\tau$ is the permutation taking i to $\nu(\tau(i))$.
5. Consider a cube in $\mathbb{R}^3$ centered at the origin whose edges have length 2 and are parallel to the coordinate axes.
   a. Show that there are finitely many rotations of $\mathbb{R}^3$ about the origin which take the cube to itself and show that these rotations are closed under composition. Taking the matrices representing these rotations, we get a finite matrix group $G \subseteq GL(3, \mathbb{R})$.
   b. Show that G has 24 elements. Hint: Every rotation is a rotation about a line through the origin. So you first need to identify the "lines of symmetry" of the cube.
   c. Write down the matrix of the element of G corresponding to the $120°$ counterclockwise rotation of the cube about the diagonal connecting the vertices $(-1, -1, -1)$ and $(1, 1, 1)$.
   d. Write down the matrix of the element of G corresponding to the $90°$ counterclockwise rotation about the z-axis.
   e. Argue geometrically that G is generated by the two matrices from parts (c) and (d).
6. In this exercise, we will use geometric methods to find some invariants of the rotation group G of the cube (from Exercise 5).
   a. Explain why $x^2 + y^2 + z^2 \in \mathbb{R}[x, y, z]^G$. Hint: Think geometrically in terms of distance to the origin.
   b. Argue geometrically that the union of the three coordinate planes $\mathbf{V}(xyz)$ is invariant under G.
   c. Show that $\mathbf{I}(\mathbf{V}(xyz)) = \langle xyz \rangle$ and conclude that if $f = xyz$, then for each $A \in G$, we have $f(A \cdot \mathbf{x}) = axyz$ for some real number a.
   d. Show that $f = xyz$ satisfies $f(A \cdot \mathbf{x}) = \pm xyz$ for all $A \in G$ and conclude that $x^2 y^2 z^2 \in k[x, y, z]^G$. Hint: Use part (c) and the fact that $A^m = I_3$ for some positive integer m.
   e. Use similar methods to show that the polynomials

   $$((x+y+z)(x+y-z)(x-y+z)(x-y-z))^2, \ ((x^2-y^2)(x^2-z^2)(y^2-z^2))^2$$

   are in $k[x, y, z]^G$. Hint: The plane $x+y+z = 0$ is perpendicular to one of the diagonals of the cube.
7. This exercise will continue our study of the invariants of the rotation group G of the cube begun in Exercise 6.
   a. Show that a polynomial f is in $k[x, y, z]^G$ if and only if $f(x, y, z) = f(y, z, x) = f(-y, x, z)$. Hint: Use parts (c), (d), and (e) of Exercise 5.

b. Let

$$f = xyz,$$
$$g = (x + y + z)(x + y - z)(z - y + z)(x - y - z),$$
$$h = (x^2 - y^2)(x^2 - z^2)(y^2 - z^2).$$

In Exercise 6, we showed that $f^2, g^2, h^2 \in k[x, y, z]^G$. Show that $f, h \notin k[x, y, z]^G$, but $g, fh \in k[x, y, z]^G$. Combining this with the previous exercise, we have invariants $x^2 + y^2 + z^2, g, f^2, fh$, and h^2 of degrees 2, 4, 6, 9, and 12, respectively, in $k[x, y, z]^G$. In §3, we will see that h^2 can be expressed in terms of the others.

8. In this exercise, we will consider an interesting "duality" that occurs among the regular polyhedra.
   a. Consider a cube and an octahedron in $\mathbb{R}^3$, both centered at the origin. Suppose the edges of the cube are parallel to the coordinate axes and the vertices of the octahedron are on the axes. Show that they have the same group of rotations. Hint: Put the vertices of the octahedron at the centers of the faces of the cube.
   b. Show that the dodecahedron and the icosahedron behave the same way. Hint: What do you get if you link up the centers of the 12 faces of the dodecahedron?
   c. Parts (a) and (b) show that in a certain sense, the "dual" of the cube is the octahedron and the "dual" of the dodecahedron is the icosahedron. What is the "dual" of the tetrahedron?

9. (Requires abstract algebra) In this problem, we will consider a tetrahedron centered at the origin of $\mathbb{R}^3$.
   a. Show that the rotations of $\mathbb{R}^3$ about the origin which take the tetrahedron to itself give us a finite matrix group G of order 12 in $\mathrm{GL}(3, \mathbb{R})$.
   b. Since every rotation of the tetrahedron induces a permutation of the four vertices, show that we get a group homomorphism $\rho : G \to S_4$.
   c. Show that ρ is one-to-one and that its image is the alternating group A_4. This shows that the rotation group of the tetrahedron is isomorphic to A_4.

10. Prove Proposition 9.

11. Prove Proposition 10. Hint: If $A = (a_{ij}) \in \mathrm{GL}(n, k)$ and $x_1^{i_1} \cdots x_n^{i_n}$ is a monomial of total degree $m = i_1 + \cdots + i_n$ appearing in f, then show that

$$(a_{11}x_1 + \cdots + a_{1n}x_n)^{i_1} \cdots (a_{n1}x_1 + \cdots + a_{nn}x_n)^{i_n}$$

is homogeneous of total degree m.

12. In Example 13, we studied polynomials $f \in k[x, y]$ with the property that $f(x, y) = f(-x, -y)$. If $f = \sum_{ij} a_{ij} x^i y^j$, show that the above condition is equivalent to $a_{ij} = 0$ whenever $i + j$ is odd.

13. In Example 13, we discovered the algebraic relation $x^2 \cdot y^2 = (xy)^2$ between the invariants x^2, y^2, and xy. We want to show that this is essentially the only relation. More precisely, suppose that we have a polynomial $g(u, v, w) \in k[u, v, w]$ such that $g(x^2, y^2, xy) = 0$. We want to prove that $g(u, v, w)$ is a multiple in $k[u, v, w]$) of $uv - w^2$ (which is the polynomial corresponding to the above relation).
   a. If we divide g by $uv - w^2$ using lex order with $u > v > w$, show that the remainder can be written in the form $uA(u, w) + vB(v, w) + C(w)$.
   b. Show that a polynomial $r = uA(u, w) + vB(v, w) + C(w)$ satisfies $r(x^2, y^2, xy) = 0$ if and only if $r = 0$.

14. Consider the finite matrix group $C_4 \subseteq \mathrm{GL}(2, \mathbb{C})$ generated by

$$A = \begin{pmatrix} i & 0 \\ 0 & -i \end{pmatrix} \in \mathrm{GL}(2, \mathbb{C}).$$

a. Prove that C_4 is cyclic of order 4.
b. Use the method of Example 13 to determine $\mathbb{C}[x, y]^{C_4}$.
c. Is there an algebraic relation between the invariants you found in part (b)? Can you give an example to show how uniqueness fails?
d. Use the method of Exercise 13 to show that the relation found in part (c) is the only relation between the invariants.

15. Consider
$$V_4 = \left\{ \pm \begin{pmatrix} 1 & 0 \\ 0 & 1 \end{pmatrix}, \pm \begin{pmatrix} 0 & 1 \\ 1 & 0 \end{pmatrix} \right\} \subseteq GL(2, k)$$

a. Show that V_4 is a finite matrix group of order 4.
b. Determine $k[x, y]^{V_4}$.
c. Show that any invariant can be written uniquely in terms of the generating invariants you found in part (b).

16. In Example 3, we introduced the finite matrix group C_4 in $GL(2, k)$ generated by
$$A = \begin{pmatrix} 0 & -1 \\ 1 & 0 \end{pmatrix} \in GL(2, k).$$

Try to apply the methods of Examples 12 and 13 to determine $k[x, y]^{C_4}$. Even if you cannot find all of the invariants, you should be able to find some invariants of low total degree. In §3, we will determine $k[x, y]^{C_4}$ completely.

§3 Generators for the Ring of Invariants

The goal of this section is to determine, in an algorithmic fashion, the ring of invariants $k[x_1, \ldots, x_n]^G$ of a finite matrix group $G \subseteq GL(n, k)$. As in §2, we assume that our field k has characteristic zero. We begin by introducing some terminology used implicitly in §2.

Definition 1. Given $f_1, \ldots, f_m \in k[x_1, \ldots, x_n]$, we let $k[f_1, \ldots, f_m]$ denote the subset of $k[x_1, \ldots, x_n]$ consisting of all polynomial expressions in $f_1, \ldots, f_m$ with coefficients in k.

This means that the elements $f \in k[f_1, \ldots, f_m]$ are those polynomials which can be written in the form
$$f = g(f_1, \ldots, f_m),$$
where g is a polynomial in m variables with coefficients in k.

Since $k[f_1, \ldots, f_m]$ is closed under multiplication and addition and contains the constants, it is a subring of $k[x_1, \ldots, x_n]$. We say that $k[f_1, \ldots, f_m]$ is *generated by* $f_1, \ldots, f_m$ over k. One has to be slightly careful about the terminology: the subring $k[f_1, \ldots, f_m]$ and the ideal $\langle f_1, \ldots, f_m \rangle$ are both "generated" by $f_1, \ldots, f_m$, but in each case, we mean something slightly different. In the exercises, we will give some examples to help explain the distinction.

An important tool we will use in our study of $k[x_1, \ldots, x_n]^G$ is the *Reynolds operator*, which is defined as follows.

Definition 2. Given a finite matrix group $G \subseteq \mathrm{GL}(n, k)$, the **Reynolds operator** of G is the map $R_G : k[x_1, \ldots, x_n] \to k[x_1, \ldots, x_n]$ defined by the formula

$$R_G(f)(\mathbf{x}) = \frac{1}{|G|} \sum_{A \in G} f(A \cdot \mathbf{x})$$

for $f(\mathbf{x}) \in k[x_1, \ldots, x_n]$.

One can think of $R_G(f)$ as "averaging" the effect of G on f. Note that division by $|G|$ is allowed since k has characteristic zero. The Reynolds operator has the following crucial properties.

Proposition 3. *Let R_G be the Reynolds operator of the finite matrix group G.*

(i) *R_G is k-linear in f.*
(ii) *If $f \in k[x_1, \ldots, x_n]$, then $R_G(f) \in k[x_1, \ldots, x_n]^G$.*
(iii) *If $f \in k[x_1, \ldots, x_n]^G$, then $R_G(f) = f$.*

Proof. We will leave the proof of (i) as an exercise. To prove (ii), let $B \in G$. Then

$$(1) \qquad R_G(f)(B\mathbf{x}) = \frac{1}{|G|} \sum_{A \in G} f(A \cdot B\mathbf{x}) = \frac{1}{|G|} \sum_{A \in G} f(AB \cdot \mathbf{x}).$$

Writing $G = \{A_1, \ldots, A_{|G|}\}$, note that $A_i B \neq A_j B$ when $i \neq j$ (otherwise, we could multiply each side by B^{-1} to conclude that $A_i = A_j$). Thus the subset $\{A_1 B, \ldots, A_{|G|} B\} \subseteq G$ consists of $|G|$ distinct elements of G and hence must equal G. This shows that

$$G = \{AB \mid A \in G\}.$$

Consequently, in the last sum of (1), the polynomials $f(AB \cdot \mathbf{x})$ are just the $f(A \cdot \mathbf{x})$, possibly in a different order. Hence,

$$\frac{1}{|G|} \sum_{A \in G} f(AB \cdot \mathbf{x}) = \frac{1}{|G|} \sum_{A \in G} f(A \cdot \mathbf{x}) = R_G(f)(\mathbf{x}),$$

and it follows that $R_G(f)(B \cdot \mathbf{x}) = R_G(f)(\mathbf{x})$ for all $B \in G$. This implies $R_G(f) \in k[x_1, \ldots, x_n]^G$.

Finally, to prove (iii), note that if $f \in k[x_1, \ldots, x_n]^G$, then

$$R_G(f)(\mathbf{x}) = \frac{1}{|G|} \sum_{A \in G} f(A \cdot \mathbf{x}) = \frac{1}{|G|} \sum_{A \in G} f(\mathbf{x}) = f(\mathbf{x})$$

since f is invariant. This completes the proof. $\qquad \square$

One nice aspect of this proposition is that it gives us a way of creating invariants. Let us look at an example.

Example 4. Consider the cyclic matrix group $C_4 \subseteq \mathrm{GL}(2, k)$ of order 4 generated by

$$A = \begin{pmatrix} 0 & -1 \\ 1 & 0 \end{pmatrix}.$$

By Lemma 11 of §2, we know that

$$k[x,y]^{C_4} = \{f \in k[x,y] \mid f(x,y) = f(-y,x)\}.$$

One can easily check that the Reynolds operator is given by

$$R_{C_4}(f)(x,y) = \frac{1}{4}(f(x,y) + f(-y,x) + f(-x,-y) + f(y,-x))$$

(see Exercise 3). Using Proposition 3, we can compute some invariants as follows:

$$R_{C_4}(x^2) = \frac{1}{4}(x^2 + (-y)^2 + (-x)^2 + y^2) = \frac{1}{2}(x^2 + y^2),$$

$$R_{C_4}(xy) = \frac{1}{4}(xy + (-y)x + (-x)(-y) + y(-x)) = 0,$$

$$R_{C_4}(x^3y) = \frac{1}{4}(x^3y + (-y)^3x + (-x)^3(-y) + y^3(-x)) = \frac{1}{2}(x^3y - xy^3),$$

$$R_{C_4}(x^2y^2) = \frac{1}{4}(x^2y^2 + (-y)^2x^2 + (-x)^2(-y)^2 + y^2(-x)^2) = x^2y^2.$$

Thus, $x^2 + y^2, x^3y - xy^3, x^2y^2 \in k[x,y]^{C_4}$. We will soon see that these three invariants generate $k[x,y]^{C_4}$.

It is easy to prove that for any monomial x^α, the Reynolds operator gives us a homogeneous invariant $R_G(x^\alpha)$ of total degree $|\alpha|$ whenever it is nonzero. The following wonderful theorem of Emmy Noether shows that we can always find finitely many of these invariants that generate $k[x_1,\dots,x_n]^G$.

Theorem 5. *Given a finite matrix group* $G \subseteq GL(n,k)$, *let* $x^{\beta_1},\dots,x^{\beta_m}$ *be all monomials of total degree* $\leq |G|$. *Then*

$$k[x_1,\dots,x_n]^G = k[R_G(x^{\beta_1}),\dots,R_G(x^{\beta_m})] = k[R_G(x^\beta) \mid |\beta| \leq |G|].$$

In particular, $k[x_1,\dots,x_n]^G$ *is generated by finitely many homogeneous invariants.*

Proof. If $f = \sum_\alpha c_\alpha x^\alpha \in k[x_1,\dots,x_n]^G$, then Proposition 3 implies that

$$f = R_G(f) = R_G\left(\sum_\alpha c_\alpha x^\alpha\right) = \sum_\alpha c_\alpha R_G(x^\alpha).$$

Hence every invariant is a linear combination (over k) of the $R_G(x^\alpha)$. Consequently, it suffices to prove that for all α, $R_G(x^\alpha)$ is a polynomial in the $R_G(x^\beta)$, $|\beta| \leq |G|$.

Noether's clever idea was to fix an integer m and combine *all* $R_G(x^\beta)$ of total degree m into a power sum of the type considered in §1. Using the theory of symmetric polynomials, this can be expressed in terms of finitely many power sums, and the theorem will follow.

The first step in implementing this strategy is to expand $(x_1 + \cdots + x_n)^m$ into a sum of monomials x^α with $|\alpha| = m$:

$$(2) \qquad (x_1 + \cdots + x_n)^m = \sum_{|\alpha|=m} a_\alpha x^\alpha.$$

In Exercise 4, you will prove that a_α is a positive integer for all $|\alpha| = m$.

To exploit this identity, we need some notation. Given $A = (a_{ij}) \in G$, let A_i denote the i-th row of A. Thus, $A_i \cdot \mathbf{x} = a_{i1}x_1 + \cdots + a_{in}x_n$. Then, if $\alpha = (\alpha_1, \ldots, \alpha_n) \in \mathbb{Z}_{\geq 0}^n$, let

$$(A \cdot \mathbf{x})^\alpha = (A_1 \cdot \mathbf{x})^{\alpha_1} \cdots (A_n \cdot \mathbf{x})^{\alpha_n}.$$

In this notation, we have

$$R_G(x^\alpha) = \frac{1}{|G|} \sum_{A \in G} (A \cdot \mathbf{x})^\alpha.$$

Now introduce new variables $u_1, \ldots, u_n$ and substitute $u_i A_i \cdot \mathbf{x}$ for x_i in (2). This gives the identity

$$(u_1 A_1 \cdot \mathbf{x} + \cdots + u_n A_n \cdot \mathbf{x})^m = \sum_{|\alpha|=m} a_\alpha (A \cdot \mathbf{x})^\alpha u^\alpha.$$

If we sum over all $A \in G$, then we obtain

$$(3) \qquad \begin{aligned} S_m = \sum_{A \in G} (u_1 A_1 \cdot \mathbf{x} + \cdots + u_n A_n \cdot \mathbf{x})^m &= \sum_{|\alpha|=m} a_\alpha \left(\sum_{A \in G} (A \cdot \mathbf{x})^\alpha \right) u^\alpha \\ &= \sum_{|\alpha|=m} b_\alpha R_G(x^\alpha) u^\alpha, \end{aligned}$$

where $b_\alpha = |G| a_\alpha$. Note how the sum on the right encodes *all* $R_G(x^\alpha)$ with $|\alpha| = m$. This is why we use the variables $u_1, \ldots, u_n$: they prevent any cancellation from occurring.

The left side of (3) is the m-th power sum S_m of the $|G|$ quantities

$$U_A = u_1 A_1 \cdot \mathbf{x} + \cdots + u_n A_n \cdot \mathbf{x}$$

indexed by $A \in G$. We write this as $S_m = S_m(U_A \mid A \in G)$. By Theorem 8 of §1, every symmetric polynomial in the $|G|$ quantities U_A is a polynomial in $S_1, \ldots, S_{|G|}$. Since S_m is symmetric in the U_A, it follows that

$$S_m = F(S_1, \ldots, S_{|G|})$$

for some polynomial F with coefficients in k. Substituting in (3), we obtain

$$\sum_{|\alpha|=m} b_\alpha R_G(x^\alpha) u^\alpha = F \left(\sum_{|\beta|=1} b_\beta R_G(x^\beta) u^\beta, \ldots, \sum_{|\beta|=|G|} R_G(x^\beta) u^\beta \right).$$

Expanding the right side and equating the coefficients of u^α, it follows that

$$b_\alpha R_G(x^\alpha) = \text{a polynomial in the } R_G(x^\beta), \; |\beta| \le |G|.$$

Since k has characteristic zero, the coefficient $b_\alpha = |G| a_\alpha$ is nonzero in k, and hence $R_G(x^\alpha)$ has the desired form. This completes the proof of the theorem. $\qquad\square$

This theorem solves the *finite generation problem* stated at the end of §2. In the exercises, you will give a second proof of Theorem 5 using the Hilbert Basis Theorem.

To see the power of what we have just proved, let us compute some invariants.

Example 6. We will return to the cyclic group $C_4 \subseteq \mathrm{GL}(2,k)$ of order 4 from Example 4. To find the ring of invariants, we need to compute $R_{C_4}(x^i y^j)$ for all $i+j \le 4$. The following table records the results:

$x^i y^j$	$R_{C_4}(x^i y^j)$	$x^i y^j$	$R_{C_4}(x^i y^j)$
x	0	xy^2	0
y	0	y^3	0
x^2	$\frac{1}{2}(x^2+y^2)$	x^4	$\frac{1}{2}(x^4+y^4)$
xy	0	$x^3 y$	$\frac{1}{2}(x^3 y - xy^3)$
y^2	$\frac{1}{2}(x^2+y^2)$	$x^2 y^2$	$x^2 y^2$
x^3	0	xy^3	$-\frac{1}{2}(x^3 y - xy^3)$
$x^2 y$	0	y^4	$\frac{1}{2}(x^4+y^4)$

By Theorem 5, it follows that $k[x,y]^{C_4}$ is generated by the four invariants $x^2+y^2, x^4+y^4, x^3 y - xy^3$ and $x^2 y^2$. However, we do not need $x^4 + y^4$ since

$$x^4 + y^4 = (x^2+y^2)^2 - 2x^2 y^2.$$

Thus, we have proved that

$$k[x,y]^{C_4} = k[x^2+y^2, x^3 y - xy^3, x^2 y^2].$$

The main drawback of Theorem 5 is that when $|G|$ is large, we need to compute the Reynolds operator for *lots* of monomials. For example, consider the cyclic group $C_8 \subseteq \mathrm{GL}(2,\mathbb{R})$ of order 8 generated by the $45°$ rotation

$$A = \frac{1}{\sqrt{2}} \begin{pmatrix} 1 & -1 \\ 1 & 1 \end{pmatrix} \in \mathrm{GL}(2,\mathbb{R}).$$

In this case, Theorem 5 says that $k[x,y]^{C_8}$ is generated by the 44 invariants $R_{C_8}(x^i y^j)$, $i+j \le 8$. In reality, only 3 are needed. For larger groups, things are even worse, especially if more variables are involved. See Exercise 10 for an example.

Fortunately, there are more efficient methods for finding a generating set of invariants. The main tool is *Molien's Theorem*, which enables one to predict in advance the number of linearly independent homogeneous invariants of given total degree. This theorem can be found in Chapter 2 of STURMFELS (2008) and Chapter 3 of DERKSEN and KEMPER (2002). Both books discuss efficient algorithms for finding invariants that generate $k[x_1, \ldots, x_n]^G$.

Once we know $k[x_1, \ldots, x_n]^G = k[f_1, \ldots, f_m]$, we can ask if there is an algorithm for writing a given invariant $f \in k[x_1, \ldots, x_n]^G$ in terms of $f_1, \ldots, f_m$. For example, it is easy to check that the polynomial

$$(4) \qquad f(x, y) = x^8 + 2x^6y^2 - x^5y^3 + 2x^4y^4 + x^3y^5 + 2x^2y^6 + y^8$$

satisfies $f(x, y) = f(-y, x)$, and hence is invariant under the group C_4 from Example 4. Then Example 6 implies that $f \in k[x, y]^{C_4} = k[x^2 + y^2, x^3y - xy^3, x^2y^2]$. But how do we write f in terms of these three invariants? To answer this question, we will use a method similar to what we did in Proposition 4 of §1.

We will actually prove a bit more, for we will allow $f_1, \ldots, f_m$ to be *arbitrary* elements of $k[x_1, \ldots, x_n]$. The following proposition shows how to test whether a polynomial lies in $k[f_1, \ldots, f_m]$ and, if so, to write it in terms of $f_1, \ldots, f_m$.

Proposition 7. *Suppose that* $f_1, \ldots, f_m \in k[x_1, \ldots, x_n]$ *are given. Fix a monomial order in* $k[x_1, \ldots, x_n, y_1, \ldots, y_m]$ *where any monomial involving one of* $x_1, \ldots, x_n$ *is greater than all monomials in* $k[y_1, \ldots, y_m]$. *Let* G *be a Gröbner basis of the ideal* $\langle f_1 - y_1, \ldots, f_m - y_m \rangle \subseteq k[x_1, \ldots x_n, y_1, \ldots, y_m]$. *Given* $f \in k[x_1, \ldots, x_n]$, *let* $g = \overline{f}^G$ *be the remainder of* f *on division by* G. *Then:*

(i) $f \in k[f_1, \ldots, f_m]$ *if and only if* $g \in k[y_1, \ldots, y_m]$.
(ii) *If* $f \in k[f_1, \ldots, f_m]$, *then* $f = g(f_1, \ldots, f_m)$ *is an expression of* f *as a polynomial in* $f_1, \ldots, f_m$.

Proof. The proof will be similar to the argument given in Proposition 4 of §1 (with one interesting difference). When we divide $f \in k[x_1, \ldots, x_n]$ by $G = \{g_1, \ldots, g_t\}$, we get an expression of the form

$$f = A_1 g_1 + \cdots + A_t g_t + g,$$

with $A_1, \ldots, A_t, g \in k[x_1, \ldots, x_n, y_1, \ldots, y_m]$.

To prove (i), first suppose that $g \in k[y_1, \ldots, y_m]$. Then for each i, substitute f_i for y_i in the above formula for f. This substitution will not affect f since it involves only $x_1, \ldots, x_n$, but it sends every polynomial in $\langle f_1 - y_1, \ldots, f_m - y_m \rangle$ to zero. Since $g_1, \ldots, g_t$ lie in this ideal, it follows that $f = g(f_1, \ldots, f_m)$. Hence, $f \in k[f_1, \ldots, f_m]$.

Conversely, suppose that $f = g(f_1, \ldots, f_m)$ for some $g \in k[y_1, \ldots, y_m]$. Arguing as in §1, one sees that

$$(5) \qquad f = C_1 \cdot (f_1 - y_1) + \cdots + C_m \cdot (f_m - y_m) + g(y_1, \ldots, y_m)$$

[see equation (4) of §1]. Unlike the case of symmetric polynomials, g need not be the remainder of f on division by G—we still need to reduce some more.

Let $G' = G \cap k[y_1, \ldots, y_m]$ consist of those elements of G involving only $y_1, \ldots, y_m$. Renumbering if necessary, we can assume $G' = \{g_1, \ldots, g_s\}$, where $s \leq t$. If we divide g by G', we get an expression of the form

(6) $$g = B_1 g_1 + \cdots + B_s g_s + g',$$

where $B_1, \ldots, B_s, g' \in k[y_1, \ldots, y_m]$. If we combine equations (5) and (6), we can write f in the form

$$f = C_1' \cdot (f_1 - y_1) + \cdots + C_m' \cdot (f_m - y_m) + g'(y_1, \ldots y_m).$$

This follows because, in (6), each g_i lies in $\langle f_1 - y_1, \ldots, f_m - y_m \rangle$. We claim that g' is the remainder of f on division by G. This will prove that the remainder lies in $k[y_1, \ldots, y_m]$.

Since G a Gröbner basis, Proposition 1 of Chapter 2, §6 tells us that g' is the remainder of f on division by G provided that no term of g' is divisible by an element of $\mathrm{LT}(G)$. To prove that g' has this property, suppose that there is $g_i \in G$ where $\mathrm{LT}(g_i)$ divides some term of g'. Then $\mathrm{LT}(g_i)$ involves only $y_1, \ldots, y_m$ since $g' \in k[y_1, \ldots, y_m]$. By our hypothesis on the ordering, it follows that $g_i \in k[y_1, \ldots, y_m]$ and hence, $g_i \in G'$. Since g' is a remainder on division by G', $\mathrm{LT}(g_i)$ cannot divide any term of g'. This contradiction shows that g' is the desired remainder.

Part (ii) of the proposition follows immediately from the above arguments, and we are done. $\square$

In the exercises, you will use this proposition to write the polynomial

$$f(x,y) = x^8 + 2x^6 y^2 - x^5 y^3 + 2x^4 y^4 + x^3 y^5 + 2x^2 y^6 + y^8$$

from (4) in terms of the generating invariants $x^2 + y^2, x^3 y - xy^3, x^2 y^2$ of $k[x, y]^{C_4}$.

The problem of finding generators for the ring of invariants (and the associated problem of finding the relations between them—see §4) played an important role in the development of invariant theory. Originally, the group involved was the group of all invertible matrices over a field. A classic introduction can be found in HILBERT (1993), and STURMFELS (2008) also discusses this case. For more on the invariant theory of finite groups, we recommend DERKSEN and KEMPER (2002), SMITH (1995) and STURMFELS (2008).

EXERCISES FOR §3

1. Given $f_1, \ldots, f_m \in k[x_1, \ldots, x_n]$, we can "generate" the following two objects:
   - The ideal $\langle f_1, \ldots, f_m \rangle \subseteq k[x_1, \ldots, x_n]$ generated by $f_1, \ldots, f_m$. This consists of all expressions $\sum_{i=1}^m h_i f_i$, where $h_1, \ldots, h_m \in k[x_1, \ldots, x_n]$.
   - The subring $k[f_1, \ldots, f_m] \subseteq k[x_1, \ldots, x_n]$ generated by $f_1, \ldots, f_m$ over k. This consists of all expressions $g(f_1, \ldots, f_m)$ where g is a polynomial in m variables with coefficients in k.

   To illustrate the differences between these, we will consider the simple case where $f_1 = x^2 \in k[x]$.
   a. Explain why $1 \in k[x^2]$ but $1 \notin \langle x^2 \rangle$.
   b. Explain why $x^3 \notin k[x^2]$ but $x^3 \in \langle x^2 \rangle$.

2. Let G be a finite matrix group in $\mathrm{GL}(n, k)$. Prove that the Reynolds operator R_G has the following properties:
   a. If $a, b \in k$ and $f, g \in k[x_1, \ldots, x_n]$, then $R_G(af + bg) = aR_G(f) + bR_G(g)$.

b. R_G maps $k[x_1, \ldots, x_n]$ to $k[x_1, \ldots, x_n]^G$ and is onto.

c. $R_G \circ R_G = R_G$.

d. If $f \in k[x_1, \ldots, x_n]^G$ and $g \in k[x_1, \ldots, x_n]$, then $R_G(fg) = f \cdot R_G(g)$.

3. In this exercise, we will work with the cyclic group $C_4 \subseteq GL(2, k)$ from Example 4 in the text.

  a. Prove that the Reynolds operator of C_4 is given by

$$R_{C_4}(f)(x, y) = \frac{1}{4}(f(x, y) + f(-y, x) + f(-x, -y) + f(y, -x)).$$

  b. Compute $R_{C_4}(x^i y^j)$ for all $i + j \leq 4$. Note that some of the computations are done in Example 4. You can check your answers against the table in Example 6.

4. In this exercise, we will study the identity (2) used in the proof of Theorem 5. We will use the *multinomial coefficients*, which are defined as follows. For $\alpha = (\alpha_1, \ldots, \alpha_n) \in \mathbb{Z}_{\geq 0}^n$, let $|\alpha| = m$ and define

$$\binom{m}{\alpha} = \frac{m!}{\alpha_1! \alpha_2! \cdots \alpha_n!}.$$

  a. Prove that $\binom{m}{\alpha}$ is an integer. Hint: Use induction on n and note that when $n = 2$, $\binom{m}{\alpha}$ is a binomial coefficient.

  b. Prove that

$$(x_1 + \cdots + x_n)^m = \sum_{|\alpha| = m} \binom{m}{\alpha} x^\alpha.$$

  In particular, the coefficient a_α in equation (2) is the positive integer $\binom{m}{\alpha}$. Hint: Use induction on n and note that the case $n = 2$ is the binomial theorem.

5. Let $G \subseteq GL(n, k)$ be a finite matrix group. In this exercise, we will give Hilbert's proof that $k[x_1, \ldots, x_n]^G$ is generated by finitely many homogeneous invariants. To begin the argument, let $I \subseteq k[x_1, \ldots, x_n]$ be the *ideal* generated by all homogeneous invariants of positive total degree.

  a. Explain why there are finitely many homogeneous invariants $f_1, \ldots, f_m$ such that $I = \langle f_1, \ldots, f_m \rangle$. The strategy of Hilbert's proof is to show that $k[x_1, \ldots, x_n]^G = k[f_1, \ldots, f_m]$. Since the inclusion $k[f_1, \ldots, f_m] \subseteq k[x_1, \ldots, x_n]^G$ is obvious, we must show that $k[x_1, \ldots, x_n]^G \not\subseteq k[f_1, \ldots, f_m]$ leads to a contradiction.

  b. Prove that $k[x_1, \ldots, x_n]^G \not\subseteq k[f_1, \ldots, f_m]$ implies there is a homogeneous invariant f of positive degree which is not in $k[f_1, \ldots, f_m]$.

  c. For the rest of the proof, pick f as in part (b) with *minimal* total degree d. By definition, $f \in I$, so that $f = \sum_{i=1}^m h_i f_i$ for $h_1, \ldots, h_m \in k[x_1, \ldots, x_n]$. Prove that for each i, we can assume that $h_i f_i$ is either 0 or homogeneous of total degree d.

  d. Use the Reynolds operator to show that $f = \sum_{i=1}^m R_G(h_i) f_i$. Hint: Use Proposition 3 and Exercise 2. Also show that for each i, $R_G(h_i) f_i$ is either 0 or homogeneous of total degree d.

  e. Since f_i has positive total degree, conclude that $R_G(h_i)$ is a homogeneous invariant of total degree $< d$. By the minimality of d, $R_G(h_i) \in k[f_1, \ldots, f_m]$ for all i. Prove that this contradicts $f \notin k[f_1, \ldots, f_m]$.

This proof is a lovely application of the Hilbert Basis Theorem. The one drawback is that it does not tell us how to find the generators—the proof is purely nonconstructive. Thus, for our purposes, Noether's theorem is much more useful.

6. If we have two finite matrix groups G and H such that $G \subseteq H \subseteq GL(n, k)$, prove that $k[x_1, \ldots, x_n]^H \subseteq k[x_1, \ldots, x_n]^G$.

7. Consider the matrix

$$A = \begin{pmatrix} 0 & -1 \\ 1 & -1 \end{pmatrix} \in GL(2, k).$$

a. Show that A generates a cyclic matrix group C_3 of order 3.
b. Use Theorem 5 to find finitely many homogeneous invariants which generate $k[x, y]^{C_3}$.
c. Can you find fewer invariants that generate $k[x, y]^{C_3}$? Hint: If you have invariants $f_1, \ldots, f_m$, you can use Proposition 7 to determine whether $f_1 \in k[f_2, \ldots, f_m]$.

8. Let A be the matrix of Exercise 7.
a. Show that $-A$ generates a cyclic matrix group C_6, of order 6.
b. Show that $-I_2 \in C_6$. Then use Exercise 6 and §2 to show that $k[x, y]^{C_6} \subseteq k[x^2, y^2, xy]$. Conclude that all nonzero homogeneous invariants of C_6 have even total degree.
c. Use part (b) and Theorem 5 to find $k[x, y]^{C_6}$. Hint: There are still a lot of Reynolds operators to compute. You should use a computer algebra program to design a procedure that has i, j as input and $R_{C_6}(x^i y^j)$ as output.

9. Let A be the matrix

$$A = \frac{1}{\sqrt{2}} \begin{pmatrix} 1 & -1 \\ 1 & 1 \end{pmatrix} \in \mathrm{GL}(2, k).$$

a. Show that A generates a cyclic matrix group $C_8 \subseteq \mathrm{GL}(2, k)$ of order 8.
b. Give a geometric argument to explain why $x^2 + y^2 \in k[x, y]^{C_8}$. Hint: A is a rotation matrix.
c. As in Exercise 8, explain why all homogeneous invariants of C_8 have even total degree.
d. Find $k[x, y]^{C_8}$. Hint: Do not do this problem unless you know how to design a procedure (on some computer algebra program) that has i, j as input and $R_{C_8}(x^i y^j)$ as output.

10. Consider the finite matrix group

$$G = \left\{ \begin{pmatrix} \pm 1 & 0 & 0 \\ 0 & \pm 1 & 0 \\ 0 & 0 & \pm 1 \end{pmatrix} \right\} \subseteq \mathrm{GL}(3, k).$$

Note that G has order 8.
a. If we were to use Theorem 5 to determine $k[x, y, z]^G$, for how many monomials would we have to compute the Reynolds operator?
b. Use the method of Example 12 in §2 to determine $k[x, y, z]^G$.

11. Let f be the polynomial (4) in the text.
a. Verify that $f \in k[x, y]^{C_4} = k[x^2 + y^2, x^3 y - xy^3, x^2 y^2]$.
b. Use Proposition 7 to express f as a polynomial in $x^2 + y^2, x^3 y - xy^3, x^2 y^2$.

12. In Exercises 5, 6, and 7 of §2, we studied the rotation group $G \subseteq \mathrm{GL}(3, \mathbb{R})$ of the cube in $\mathbb{R}^3$ and we found that $k[x, y, z]^G$ contained the polynomials

$$\begin{aligned}
f_1 &= x^2 + y^2 + z^2, \\
f_2 &= (x + y + z)(x + y - z)(x - y + z)(x - y - z), \\
f_3 &= x^2 y^2 z^2, \\
f_4 &= xyz(x^2 - y^2)(x^2 - z^2)(y^2 - z^2).
\end{aligned}$$

a. Give an elementary argument using degrees to show that $f_4 \notin k[f_1, f_2, f_3]$.
b. Use Proposition 7 to show that $f_3 \notin k[f_1, f_2]$.
c. In Exercise 6 of §2, we showed that

$$\left((x^2 - y^2)(x^2 - z^2)(y^2 - z^2) \right)^2 \in k[x, y, z]^G.$$

Prove that this polynomial lies in $k[f_1, f_2, f_3]$. Why can we ignore f_4?
Using Molien's Theorem and the methods of STURMFELS (2008), one can prove that $k[x, y, z]^G = k[f_1, f_2, f_3, f_4]$.

§4 Relations Among Generators and the Geometry of Orbits

Given a finite matrix group $G \subseteq \mathrm{GL}(n,k)$, Theorem 5 of §3 guarantees that there are finitely many homogeneous invariants $f_1, \dots, f_m$ such that

$$k[x_1, \dots, x_n]^G = k[f_1, \dots, f_m].$$

In this section, we will describe the *algebraic relations* among the polynomials $f_1, \dots, f_m$. We will also see that these relations have some fascinating algebraic and geometric implications. We continue to assume that k has characteristic zero.

We begin by recalling the *uniqueness problem* stated at the end of §2. For a symmetric polynomial $f \in k[x_1, \dots, x_n]^{S_n} = k[\sigma_1, \dots, \sigma_n]$, we proved that f could be written uniquely as a polynomial in $\sigma_1, \dots \sigma_n$. For a general finite matrix group $G \subseteq \mathrm{GL}(n,k)$, if we know that $k[x_1, \dots, x_n]^G = k[f_1, \dots, f_m]$, then one could similarly ask if $f \in k[x_1, \dots, x_n]^G$ can be uniquely written in terms of $f_1, \dots, f_m$.

To study this question, note that if g_1 and g_2 are polynomials in $k[y_1, \dots, y_m]$, then

$$g_1(f_1, \dots, f_m) = g_2(f_1, \dots, f_m) \iff h(f_1, \dots, f_m) = 0,$$

where $h = g_1 - g_2$. It follows that uniqueness fails if and only if there is a nonzero polynomial $h \in k[y_1, \dots, y_m]$ such that $h(f_1, \dots, f_m) = 0$. Such a polynomial is a *nontrivial algebraic relation* among $f_1, \dots, f_m$.

If we let $F = (f_1, \dots, f_m)$, then the set

$$(1) \qquad I_F = \{ h \in k[y_1, \dots, y_m] \mid h(f_1, \dots, f_m) = 0 \text{ in } k[x_1, \dots, x_n] \}$$

records *all* algebraic relations among $f_1, \dots, f_m$. This set has the following properties.

Proposition 1. *If* $k[x_1, \dots, x_n]^G = k[f_1, \dots, f_m]$, *let* $I_F \subseteq k[y_1, \dots, y_m]$ *be as in* (1). *Then:*

(i) I_F *is a prime ideal of* $k[y_1, \dots, y_m]$.
(ii) *Suppose that* $f \in k[x_1, \dots, x_n]^G$ *and that* $f = g(f_1, \dots, f_m)$ *is one representation of* f *in terms of* $f_1, \dots, f_m$. *Then all such representations are given by*

$$f = g(f_1, \dots, f_m) + h(f_1, \dots, f_m),$$

as h varies over I_F.

Proof. For (i), it is an easy exercise to prove that I_F is an ideal. To show that it is prime, we need to show that $fg \in I_F$ implies that $f \in I_F$ or $g \in I_F$ (see Definition 2 of Chapter 4, §5). But $fg \in I_F$ means that $f(f_1, \dots, f_m)g(f_1, \dots, f_m) = 0$. This is a product of polynomials in $k[x_1, \dots, x_n]$, and hence, $f(f_1, \dots, f_m)$ or $g(f_1, \dots, f_m)$ must be zero. Thus f or g is in I_F.

We leave the proof of (ii) as an exercise. $\qquad \square$

We will call I_F the *ideal of relations* for $F = (f_1, \dots, f_m)$. Another name for I_F used in the literature is the *syzygy ideal*. To see what Proposition 1 tells us about the uniqueness problem, consider $C_2 = \{\pm I_2\} \subseteq \mathrm{GL}(2,k)$. We know from §2 that

$k[x, y]^{C_2} = k[x^2, y^2, xy]$, and, in Example 4, we will see that $I_F = \langle uv - w^2 \rangle \subseteq$ $k[u, v, w]$. Now consider $x^6 + x^3y^3 \in k[x, y]^{C_2}$. Then Proposition 1 implies that *all* possible ways of writing $x^6 + x^3y^3$ in terms of x^2, y^2, xy are given by

$$(x^2)^3 + (xy)^3 + (x^2 \cdot y^2 - (xy)^2) \cdot b(x^2, y^2, xy)$$

since elements of $\langle uv - w^2 \rangle$ are of the form $(uv - w^2) \cdot b(u, v, w)$.

As an example of what the ideal of relations I_F can tell us, let us show how it can be used to reconstruct the ring of invariants.

Proposition 2. *If* $k[x_1, \ldots, x_n]^G = k[f_1, \ldots, f_m]$, *let* $I_F \subseteq k[y_1, \ldots, y_m]$ *be the ideal of relations. Then there is a ring isomorphism*

$$k[y_1, \ldots, y_m]/I_F \cong k[x_1, \ldots, x_n]^G$$

between the quotient ring of $k[y_1, \ldots, y_m]$ *modulo* I_F *(as defined in Chapter 5, §2) and the ring of invariants.*

Proof. Recall from §2 of Chapter 5 that elements of the quotient $k[y_1, \ldots, y_m]/I_F$ are written $[g]$ for $g \in k[y_1, \ldots, y_m]$, where $[g_1] = [g_2]$ if and only if $g_1 - g_2 \in I_F$. Now define $\phi : k[y_1, \ldots, y_m]/I_F \to k[x_1, \ldots, x_n]^G$ by

$$\phi([g]) = g(f_1, \ldots, f_m).$$

We leave it as an exercise to check that ϕ is well-defined and is a ring homomorphism. We need to show that ϕ is one-to-one and onto.

Since $k[x_1, \ldots, x_n]^G = k[f_1, \ldots, f_m]$, it follows immediately that ϕ is onto. To prove that ϕ is one-to-one, suppose that $\phi([g_1]) = \phi([g_2])$. Then $g_1(f_1, \ldots, f_m) = g_2(f_1, \ldots, f_m)$, which implies that $g_1 - g_2 \in I_F$. Thus, $[g_1] = [g_2]$, and hence, ϕ is one-to-one.

It is a general fact that if a ring homomorphism is one-to-one and onto, then its inverse function is a ring homomorphism. Hence, ϕ is a ring isomorphism. $\square$

A more succinct proof of this proposition can be given using the Isomorphism Theorem of Exercise 16 in Chapter 5, §2.

For our purposes, another extremely important property of I_F is that we can compute it explicitly using elimination theory. Namely, consider the system of equations

$$y_1 = f_1(x_1, \ldots, x_n),$$
$$\vdots$$
$$y_m = f_m(x_1, \ldots, x_n).$$

Then I_F can be obtained by eliminating $x_1, \ldots, x_n$ from these equations.

Proposition 3. *If* $k[x_1, \ldots, x_n]^G = k[f_1, \ldots, f_m]$, *consider the ideal*

$$J_F = \langle f_1 - y_1, \ldots, f_m - y_m \rangle \subseteq k[x_1, \ldots, x_n, y_1, \ldots, y_m].$$

(i) I_F is the n-th elimination ideal of J_F. Thus, $I_F = J_F \cap k[y_1, \ldots, y_m]$.

(ii) Fix a monomial order in $k[x_1, \ldots, x_n, y_1, \ldots, y_m]$ where any monomial involving one of $x_1, \ldots, x_n$ is greater than all monomials in $k[y_1, \ldots, y_m]$ and let G be a Gröbner basis of J_F. Then $G \cap k[y_1, \ldots, y_m]$ is a Gröbner basis for I_F in the monomial order induced on $k[y_1, \ldots, y_m]$.

Proof. Note that the ideal J_F appeared earlier in Proposition 7 of §3. To relate J_F to the ideal of relations I_F, we will need the following characterization of J_F: if $p \in k[x_1, \ldots, x_n, y_1, \ldots, y_m]$, then we claim that

$$(2) \qquad p \in J_F \iff p(x_1, \ldots, x_n, f_1, \ldots, f_m) = 0 \text{ in } k[x_1, \ldots, x_n].$$

One implication is obvious since the substitution $y_i \mapsto f_i$ takes all elements of $J_F = \langle f_1 - y_1, \ldots, f_m - y_m \rangle$ to zero. On the other hand, given $p \in k[x_1, \ldots, x_n, y_1, \ldots, y_m]$, if we replace each y_i in p by $f_i - (f_i - y_i)$ and expand, we obtain

$$\begin{aligned} p(x_1, \ldots, x_n, y_1, \ldots, y_m) &= p(x_1, \ldots, x_n, f_1, \ldots, f_m) \\ &\quad + B_1 \cdot (f_1 - y_1) + \cdots + B_m \cdot (f_m - y_m) \end{aligned}$$

for some $B_1, \ldots, B_m \in k[x_1, \ldots, x_n, y_1, \ldots, y_m]$ (see Exercise 4 for the details). In particular, if $p(x_1, \ldots, x_n, f_1, \ldots, f_m) = 0$, then

$$p(x_1, \ldots, x_n, y_1, \ldots, y_m) = B_1 \cdot (f_1 - y_1) + \cdots + B_m \cdot (f_m - y_m) \in J_F.$$

This completes the proof of (2).

Now intersect each side of (2) with $k[y_1, \ldots, y_m]$. For $p \in k[y_1, \ldots, y_m]$, this proves

$$p \in J_F \cap k[y_1, \ldots, y_m] \iff p(f_1, \ldots, f_m) = 0 \text{ in } k[x_1, \ldots, x_n],$$

so that $J_F \cap k[y_1, \ldots, y_m] = I_F$ by the definition of I_F. Thus, (i) is proved, and (ii) is then an immediate consequence of the elimination theory of Chapter 3 (see Theorem 2 and Exercise 5 of Chapter 3, §1). □

We can use this proposition to compute the relations between generators.

Example 4. In §2 we saw that the invariants of $C_2 = \{\pm I_2\} \subseteq \mathrm{GL}(2, k)$ are given by $k[x, y]^{C_2} = k[x^2, y^2, xy]$. Let $F = (x^2, y^2, xy)$ and let the new variables be u, v, w. Then the ideal of relations is obtained by eliminating x, y from the equations

$$u = x^2,$$
$$v = y^2,$$
$$w = xy.$$

If we use lex order with $x > y > u > v > w$, then a Gröbner basis for the ideal $J_F = \langle u - x^2, v - y^2, w - xy \rangle$ consists of the polynomials

$$x^2 - u, \ xy - w, \ xv - yw, \ xw - yu, \ y^2 - v, \ uv - w^2.$$

It follows from Proposition 3 that

$$I_F = \langle uv - w^2 \rangle.$$

This says that all relations between x^2, y^2, and xy are generated by the obvious relation $x^2 \cdot y^2 = (xy)^2$. Then Proposition 2 shows that the ring of invariants can be written as

$$k[x, y]^{C_2} \cong k[u, v, w]/\langle uv - w^2 \rangle.$$

Example 5. In §3, we studied the cyclic matrix group $C_4 \subseteq GL(2, k)$ generated by

$$A = \begin{pmatrix} 0 & -1 \\ 1 & 0 \end{pmatrix}$$

and we saw that

$$k[x, y]^{C_4} = k[x^2 + y^2, x^3 y - xy^3, x^2 y^2].$$

Putting $F = (x^2 + y^2, x^3 y - xy^3, x^2 y^2)$, we leave it as an exercise to show that $I_F \subseteq k[u, v, w]$ is given by $I_F = \langle u^2 w - v^2 - 4w^2 \rangle$. So the one nontrivial relation between the invariants is

$$(x^2 + y^2)^2 \cdot x^2 y^2 = (x^3 y - xy^3)^2 + 4(x^2 y^2)^2.$$

By Proposition 2, we conclude that the ring of invariants can be written as

$$k[x, y]^{C_4} \cong k[u, v, w]/\langle u^2 w - v^2 - 4w^2 \rangle.$$

By combining Propositions 1, 2, and 3 with the theory developed in §3 of Chapter 5, we can solve the *uniqueness problem* stated at the end of §2. Suppose that $k[x_1, \ldots, x_n]^G = k[f_1, \ldots, f_m]$ and let $I_F \subseteq k[y_1, \ldots, y_m]$ be the ideal of relations. If $I_F \neq \{0\}$, we know that a given element $f \in k[x_1, \ldots, x_n]^G$ can be written in more than one way in terms of $f_1, \ldots, f_m$. Is there a consistent choice for how to write f?

To solve this problem, pick a monomial order on $k[y_1, \ldots, y_m]$ and use Proposition 3 to find a Gröbner basis G of I_F. Given $g \in k[y_1, \ldots, y_m]$, let $\bar{g}^G$ be the remainder of g on division by G. In Chapter 5, we showed that the remainders $\bar{g}^G$ uniquely represent elements of the quotient ring $k[y_1, \ldots, y_m]/I_F$ (see Proposition 1 of Chapter 5, §3). Using this together with the isomorphism

$$k[y_1, \ldots, y_m]/I_F \cong k[x_1, \ldots, x_n]^G$$

of Proposition 2, we get a consistent method for writing elements of $k[x_1, \ldots, x_n]^G$ in terms of $f_1, \ldots, f_m$. Thus, Gröbner basis methods help restore the uniqueness lost when $I_F \neq \{0\}$.

So far in this section, we have explored the algebra associated with the ideal of relations I_F. It is now time to turn to the geometry. The basic geometric object associated with an ideal is its variety. Hence, we get the following definition.

Definition 6. If $k[x_1, \ldots, x_n]^G = k[f_1, \ldots, f_m]$, let $I_F \subseteq k[y_1, \ldots, y_m]$ be the ideal of relations for $F = (f_1, \ldots, f_m)$. Then we have the affine variety

$$V_F = \mathbf{V}(I_F) \subseteq k^m.$$

The variety V_F has the following properties.

Proposition 7. *Let I_F and V_F be as in Definition 6. Then:*

(i) *V_F is the smallest variety in k^m containing the parametrization*

$$y_1 = f_1(x_1, \ldots, x_n),$$
$$\vdots$$
$$y_m = f_m(x_1, \ldots, x_n).$$

(ii) *V_F is an irreducible variety.*
(iii) *$I_F = \mathbf{I}(V_F)$, so that I_F is the ideal of all polynomial functions vanishing on V_F.*
(iv) *Let $k[V_F]$ be the coordinate ring of V_F as defined in §4 of Chapter 5. Then there is a ring isomorphism*

$$k[V_F] \cong k[x_1, \ldots, x_n]^G.$$

Proof. Let $J_F = \langle f_1 - y_1, \ldots, f_m - y_m \rangle$. By Proposition 3, I_F is the n-th elimination ideal of J_F. Then part (i) follows immediately from the Polynomial Implicitization Theorem of Chapter 3 (see Theorem 1 of Chapter 3, §3).

For (ii), V_F is irreducible by Proposition 5 from Chapter 4, §5 (k is infinite since it has characteristic zero). Also, in the proof of that proposition, we showed that

$$\mathbf{I}(V_F) = \{g \in k[y_1, \ldots, y_m] \mid g \circ F = 0\}.$$

By (1), this equals I_F, and (iii) follows.

Finally, in Chapter 5, we saw that the coordinate ring $k[V_F]$ could be written as

$$k[V_F] \cong k[y_1, \ldots, y_m]/\mathbf{I}(V_F)$$

(see Theorem 7 of Chapter 5, §2). Since $\mathbf{I}(V_F) = I_F$ by part (iii), we can use the isomorphism of Proposition 2 to obtain

(3) $$k[V_F] \cong k[y_1, \ldots, y_m]/I_F \cong k[x_1, \ldots, x_n]^G.$$

This completes the proof of the proposition. $\qquad\square$

Note how the isomorphisms in (3) link together the three methods (coordinate rings, quotient rings, and rings of invariants) that we have learned for creating new rings.

When we write $k[x_1, \ldots, x_n]^G = k[f_1, \ldots, f_m]$, note that $f_1, \ldots, f_m$ are not uniquely determined. So one might ask how changing to a different set of generators affects the variety V_F. The answer is as follows.

Corollary 8. *Suppose that* $k[x_1, \ldots, x_n]^G = k[f_1, \ldots, f_m] = k[f'_1, \ldots, f'_{m'}]$. *If we set* $F = (f_1, \ldots, f_m)$ *and* $F' = (f'_1, \ldots, f'_{m'})$, *then the varieties* $V_F \subseteq k^m$ *and* $V_{F'} \subseteq k^{m'}$ *are isomorphic (as defined in Chapter 5, §4).*

Proof. Applying Proposition 7 twice, we have isomorphisms $k[V_F] \cong k[x_1, \ldots, x_n]^G$ $\cong k[V_{F'}]$, and it is easy to see that these isomorphisms are the identity on constants. But in Theorem 9 of Chapter 5, §4, we learned that two varieties are isomorphic if and only if there is an isomorphism of their coordinate rings which is the identity on constants. The corollary follows immediately. ☐

One of the lessons we learned in Chapter 4 was that the algebra-geometry correspondence works best over an algebraically closed field k. So for the rest of this section we will assume that k is algebraically closed.

To uncover the geometry of V_F, we need to think about the matrix group $G \subseteq$ $\mathrm{GL}(n, k)$ more geometrically. So far, we have used G to act on polynomials: if $f(\mathbf{x}) \in$ $k[x_1, \ldots, x_n]$, then a matrix $A \in G$ gives us the new polynomial $g(\mathbf{x}) = f(A \cdot \mathbf{x})$. But we can also let G act on the underlying affine space k^n. We will write a point $(a_1, \ldots, a_n) \in k^n$ as a column vector $\mathbf{a}$. Thus,

$$\mathbf{a} = \begin{pmatrix} a_1 \\ \vdots \\ a_n \end{pmatrix}.$$

Then a matrix $A \in G$ gives us the new point $A \cdot \mathbf{a}$ by matrix multiplication.

We can then use G to describe an equivalence relation on k^n: given $\mathbf{a}, \mathbf{b} \in k^n$, we say that $\mathbf{a} \sim_G \mathbf{b}$ if $\mathbf{b} = A \cdot \mathbf{a}$ for some $A \in G$. We leave it as an exercise to verify that $\sim_G$ is indeed an equivalence relation. It is also straightforward to check that the equivalence class of $\mathbf{a} \in k^n$ is given by

$$\{\mathbf{b} \in k^n \mid \mathbf{b} \sim_G \mathbf{a}\} = \{A \cdot \mathbf{a} \mid A \in G\}.$$

These equivalence classes have a special name.

Definition 9. Given a finite matrix group $G \subseteq \mathrm{GL}(n, k)$ and $\mathbf{a} \in k^n$, the G-**orbit** of $\mathbf{a}$ is the set

$$G \cdot \mathbf{a} = \{A \cdot \mathbf{a} \mid A \in G\}.$$

The set of all G-orbits in k^n is denoted k^n/G and is called the **orbit space**.

Note that an orbit $G \cdot \mathbf{a}$ has at most $|G|$ elements. In the exercises, you will show that the number of elements in an orbit is always a divisor of $|G|$.

Since orbits are equivalence classes, it follows that the orbit space k^n/G is the set of equivalence classes of $\sim_G$. Thus, we have constructed k^n/G as a set. But for us, the objects of greatest interest are affine varieties. So it is natural to ask if k^n/G has the structure of a variety in some affine space. The answer is as follows.

Theorem 10. *Let* $G \subseteq \mathrm{GL}(n, k)$ *be a finite matrix group, where* k *is algebraically closed. Suppose that* $k[x_1, \ldots, x_n]^G = k[f_1, \ldots, f_m]$. *Then:*

(i) *The polynomial mapping* $F : k^n \to V_F$ *defined by* $F(\mathbf{a}) = (f_1(\mathbf{a}), \ldots, f_m(\mathbf{a}))$ *is onto. Geometrically, this means that the parametrization* $y_i = f_i(x_1, \ldots, x_n)$ *covers all of* V_F.

(ii) *The map sending the G-orbit* $G \cdot \mathbf{a} \subseteq k^n$ *to the point* $F(\mathbf{a}) \in V_F$ *induces a one-to-one correspondence*

$$k^n/G \cong V_F.$$

Proof. We prove part (i) using elimination theory. Let $J_F = \langle f_1 - y_1, \ldots, f_m - y_m \rangle$ be the ideal defined in Proposition 3. Since $I_F = J_F \cap k[y_1, \ldots, y_m]$ is an elimination ideal of J_F, it follows that a point $(b_1, \ldots, b_m) \in V_F = \mathbf{V}(I_F)$ is a partial solution of the system of equations

$$y_1 = f_1(x_1, \ldots, x_n),$$

$$\vdots$$

$$y_m = f_m(x_1, \ldots, x_n).$$

If we can prove that $(b_1, \ldots, b_m) \in \mathbf{V}(I_F)$ extends to $(a_1, \ldots, a_n, b_1, \ldots, b_m) \in \mathbf{V}(J_F)$, then $F(a_1, \ldots, a_n) = (b_1, \ldots, b_m)$ and the surjectivity of $F : k^n \to V_F$ will follow.

We claim that for each i, there is an element $p_i \in J_F \cap k[x_i, \ldots, x_n, y_1, \ldots, y_m]$ such that

$$(4) \qquad p_i = x_i^N + \text{terms in which } x_i \text{ has degree } < N,$$

where $N = |G|$. For now, we will assume that the claim is true.

Suppose that inductively we have extended $(b_1, \ldots, b_m)$ to a partial solution

$$(a_{i+1}, \ldots, a_n, b_1, \ldots, b_m) \in \mathbf{V}(J_F \cap k[x_{i+1}, \ldots, x_n, y_1, \ldots, y_m]).$$

Since k is algebraically closed, the Extension Theorem of Chapter 3, §1 asserts that we can extend to $(a_i, a_{i+1}, \ldots, a_n, b_1, \ldots, b_m)$, provided the leading coefficient in x_i of one of the generators of $J_F \cap k[x_i, \ldots, x_n, y_1, \ldots, y_m]$ does not vanish at the partial solution. Because of our claim, this ideal contains the above polynomial p_i and we can assume that p_i is a generator (just add it to the generating set). By (4), the leading coefficient is 1, which never vanishes, so that the required a_i exists (see Corollary 4 of Chapter 3, §1).

It remains to prove the existence of p_i. We will need the following lemma.

Lemma 11. *Let* $G \subseteq \mathrm{GL}(n, k)$ *be a finite matrix group and set* $N = |G|$. *Given any* $f \in k[x_1, \ldots, x_n]$, *there are invariants* $g_1, \ldots, g_N \in k[x_1, \ldots, x_n]^G$ *such that*

$$f^N + g_1 f^{N-1} + \cdots + g_N = 0.$$

Proof. Consider the polynomial $\prod_{A \in G}(X - f(A \cdot \mathbf{x}))$. If we multiply it out, we get

$$\prod_{A \in G} (X - f(A \cdot \mathbf{x})) = X^N + g_1(\mathbf{x})X^{N-1} + \cdots + g_N(\mathbf{x}),$$

where the coefficients $g_1, \ldots, g_N$ are in $k[x_1, \ldots, x_n]$. We claim that $g_1, \ldots, g_N$ are invariant under G. To prove this, suppose that $B \in G$. In the proof of Proposition 3 of §3, we saw that the $f(AB \cdot \mathbf{x})$ are just the $f(A \cdot \mathbf{x})$, possibly in a different order. Thus

$$\prod_{A \in G} (X - f(AB \cdot \mathbf{x})) = \prod_{A \in G} (X - f(A \cdot \mathbf{x})),$$

and then multiplying out each side implies that

$$X^N + g_1(B \cdot \mathbf{x})X^{N-1} + \cdots + g_N(B \cdot \mathbf{x}) = X^N + g_1(\mathbf{x})X^{N-1} + \cdots + g_N(\mathbf{x})$$

for each $B \in G$. This proves that $g_1, \ldots, g_N \in k[x_1, \ldots, x_n]^G$.

Since one of the factors is $X - f(I_n \cdot \mathbf{x}) = X - f(\mathbf{x})$, the polynomial vanishes when $X = f$, and the lemma is proved. $\qquad \square$

We can now prove our claim about the polynomial p_i. If we substitute $f = x_i$ in Lemma 11, then we get

(5) $$x_i^N + g_1 x_i^{N-1} + \cdots + g_N = 0$$

for $N = |G|$ and $g_1, \ldots, g_N \in k[x_1, \ldots, x_n]^G$. Since $k[x_1, \ldots, x_n]^G = k[f_1, \ldots, f_m]$, we can write $g_j = h_j(f_1, \ldots, f_m)$ for $j = 1, \ldots, N$. Then let

$$p_i(x_i, y_1, \ldots, y_m) = x_i^N + h_1(y_1, \ldots, y_m)x_i^{N-1} + \cdots + h_N(y_1, \ldots, y_m)$$

in $k[x_i, y_1, \ldots, y_m]$. From (5), it follows that $p_i(x_i, f_1, \ldots, f_m) = 0$ and, hence, by (2), we see that $p_i \in J_F$. Then $p_i \in J_F \cap k[x_i, \ldots, x_n, y_1, \ldots, y_m]$, and our claim is proved.

To prove (ii), first note that the map

$$\tilde{F} : k^n/G \to V_F$$

defined by sending $G \cdot \mathbf{a}$ to $F(\mathbf{a}) = (f_1(\mathbf{a}), \ldots, f_m(\mathbf{a}))$ is well-defined since each f_i is invariant and, hence, takes the same value on all points of a G-orbit $G \cdot \mathbf{a}$. Furthermore, F is onto by part (i) and it follows that $\tilde{F}$ is also onto.

It remains to show that $\tilde{F}$ is one-to-one. Suppose that $G \cdot \mathbf{a}$ and $G \cdot \mathbf{b}$ are distinct orbits. Since $\sim_G$ is an equivalence relation, it follows that the orbits are disjoint. We will construct an invariant $g \in k[x_1, \ldots, x_n]^G$ such that $g(\mathbf{a}) \neq g(\mathbf{b})$. To do this, note that $S = G \cdot \mathbf{b} \cup (G \cdot \mathbf{a} \setminus \{\mathbf{a}\})$ is a finite set of points in k^n and, hence, is an affine variety. Since $\mathbf{a} \notin S$, there must be some defining equation f of S which does not vanish at $\mathbf{a}$. Thus, for $A \in G$, we have

$$f(A \cdot \mathbf{b}) = 0 \quad \text{and} \quad f(A \cdot \mathbf{a}) = \begin{cases} 0 & \text{if } A \cdot \mathbf{a} \neq \mathbf{a} \\ f(\mathbf{a}) \neq 0 & \text{if } A \cdot \mathbf{a} = \mathbf{a}. \end{cases}$$

Then let $g = R_G(f)$. We leave it as an exercise to check that

$$g(\mathbf{b}) = 0 \text{ and } g(\mathbf{a}) = \frac{M}{|G|}f(\mathbf{a}) \neq 0,$$

where M is the number of elements $A \in G$ such that $A \cdot \mathbf{a} = \mathbf{a}$. We have thus found an element $g \in k[x_1, \ldots, x_n]^G$ such that $g(\mathbf{a}) \neq g(\mathbf{b})$.

Now write g as a polynomial $g = h(f_1, \ldots, f_m)$ in our generators. Then $g(\mathbf{a}) \neq g(\mathbf{b})$ implies that $f_i(\mathbf{a}) \neq f_i(\mathbf{b})$ for some i, and it follows that $\tilde{F}$ takes different values on $G \cdot \mathbf{a}$ and $G \cdot \mathbf{b}$. The theorem is now proved. $\qquad\square$

Theorem 10 shows that there is a bijection between the *set* k^n/G and the *variety* V_F. This is what we mean by saying that k^n/G has the structure of an affine variety. Further, whereas I_F depends on the generators chosen for $k[x_1, \ldots, x_n]^G$, we noted in Corollary 8 that V_F is unique up to isomorphism. This implies that the variety structure on k^n/G is unique up to isomorphism.

One nice consequence of Theorem 10 and Proposition 7 is that the "polynomial functions" on the orbit space k^n/G are given by

$$k[V_F] \cong k[x_1, \ldots, x_n]^G.$$

Note how natural this is: an invariant polynomial takes the same value on all points of the G-orbit and, hence, defines a function on the orbit space. Thus, it is reasonable to expect that $k[x_1, \ldots, x_n]^G$ should be the "coordinate ring" of whatever variety structure we put on k^n/G.

Still, the bijection $k^n/G \cong V_F$ is rather remarkable if we look at it slightly differently. Suppose that we start with the geometric action of G on k^n which sends $\mathbf{a}$ to $A \cdot \mathbf{a}$ for $A \in G$. From this, we construct the orbit space k^n/G as the set of orbits. To give this set the structure of an affine variety, look at what we had to do:

- we made the action algebraic by letting G act on polynomials;
- we considered the invariant polynomials and found finitely many generators; and
- we formed the ideal of relations among the generators.

The equations coming from this ideal define the desired variety structure V_F on k^n/G.

In general, an important problem in algebraic geometry is to take a set of interesting objects (G-orbits, lines tangent to a curve, etc.) and give it the structure of an affine (or projective—see Chapter 8) variety. Some simple examples will be given in the exercises.

A final remark is for readers who studied Noether normalization in §6 of Chapter 5. In the terminology of that section, Lemma 11 proved above says that every element of $k[x_1, \ldots, x_n]$ is integral over $k[x_1, \ldots, x_n]^G$. In the exercises, you will use results from Chapter 5, §6 to show that $k[x_1, \ldots, x_n]$ is finite over $k[x_1, \ldots, x_n]^G$ in the sense of Definition 2 of that section.

EXERCISES FOR §4

1. Given $f_1, \ldots, f_m \in k[x_1, \ldots, x_n]$, let $I = \{g \in k[y_1, \ldots, y_m] \mid g(f_1, \ldots, f_m) = 0\}$.
   a. Prove that I is an ideal of $k[y_1, \ldots, y_m]$.
   b. If $f \in k[f_1, \ldots, f_m]$ and $f = g(f_1, \ldots, f_m)$ is one representation of f in terms of $f_1, \ldots, f_m$, prove that all such representations are given by $f = g(f_1, \ldots, f_m) + h(f_1, \ldots, f_m)$ as h varies over I.

2. Let $f_1, \ldots, f_m \in k[x_1, \ldots, x_n]$ and let $I \subseteq k[y_1, \ldots, y_m]$ be the ideal of relations defined in Exercise 1.

   a. Prove that the map sending a coset $[g]$ to $g(f_1, \ldots, f_m)$ defines a well-defined ring homomorphism
   $$\phi : k[y_1, \ldots, y_m]/I \longrightarrow k[f_1, \ldots, f_m].$$

   b. Prove that the map ϕ of part (a) is one-to-one and onto. Thus ϕ is a ring isomorphism.

   c. Use Exercise 13 in Chapter 5, §2 to give an alternate proof that $k[y_1, \ldots, y_m]/I$ and $k[f_1, \ldots, f_m]$ are isomorphic. Hint: Use the ring homomorphism $\Phi : k[y_1, \ldots, y_m] \to k[f_1, \ldots, f_m]$ which sends y_i to f_i.

3. Although Propositions 1 and 2 were stated for $k[x_1, \ldots, x_n]^G$, we saw in Exercises 1 and 2 that these results held for any subring of $k[x_1, \ldots, x_n]$ of the form $k[f_1, \ldots, f_m]$. Give a similar generalization of Proposition 3. Does the proof given in the text need any changes?

4. Given $p \in k[x_1, \ldots, x_n, y_1, \ldots, y_m]$, prove that

   $$
   \begin{aligned}
   p(x_1, \ldots, x_n, y_1, \ldots, y_m) &= p(x_1, \ldots, x_n, f_1, \ldots, f_m) \\
   &\quad + B_1 \cdot (f_1 - y_1) + \cdots + B_m \cdot (f_m - y_m)
   \end{aligned}
   $$

   for some $B_1, \ldots, B_m \in k[x_1, \ldots, x_n, y_1, \ldots, y_m]$. Hint: In p, replace each occurrence of y_i by $f_i - (f_i - y_i)$. The proof is similar to the argument given to prove (4) in §1.

5. Complete Example 5 by showing that $I_F \subseteq k[u, v, w]$ is given by $I_F = \langle u^2 w - v^2 - 4w^2 \rangle$ when $F = (x^2 + y^2, x^3 y - xy^3, x^2 y^2)$.

6. In Exercise 7 of §3, you were asked to compute the invariants of a certain cyclic group $C_3 \subseteq GL(2, k)$ of order 3. Take the generators you found for $k[x, y]^{C_3}$ and find the relations between them.

7. Repeat Exercise 6, this time using the cyclic group $C_6 \subseteq GL(2, k)$ of order 6 from Exercise 8 of §3.

8. In Exercise 12 of §3, we listed four invariants f_1, f_2, f_3, f_4 of the group of rotations of the cube in $\mathbb{R}^3$.

   a. Using $(f_4/xyz)^2$ and part (c) of Exercise 12 of §3, find an algebraic relation between f_1, f_2, f_3, f_4.

   b. Show that there are no nontrivial algebraic relations between f_1, f_2, f_3.

   c. Show that the relation you found in part (a) generates the ideal of all relations between f_1, f_2, f_3, f_4. Hint: If $p(f_1, f_2, f_3, f_4) = 0$ is a relation, use part (a) to reduce to a relation of the form $p_1(f_1, f_2, f_3) + p_2(f_1, f_2, f_3)f_4 = 0$. Then explain how degree arguments imply $p_1(f_1, f_2, f_3) = 0$.

9. Given a finite matrix group $G \subseteq GL(n, k)$, we defined the relation $\sim_G$ on k^n by $\mathbf{a} \sim_G \mathbf{b}$ if $\mathbf{b} = A \cdot \mathbf{a}$ for some $A \in G$.

   a. Verify that $\sim_G$ is an equivalence relation.

   b. Prove that the equivalence class of $\mathbf{a}$ is the set $G \cdot \mathbf{a}$ defined in the text.

10. Consider the group of rotations of the cube in $\mathbb{R}^3$. We studied this group in Exercise 5 of §2, and we know that it has 24 elements.

   a. Draw a picture of the cube which shows orbits consisting of 1, 6, 8, 12 and 24 elements.

   b. Argue geometrically that there is no orbit consisting of four elements.

11. (Requires abstract algebra) Let $G \subseteq GL(n, k)$ be a finite matrix group. In this problem, we will prove that the number of elements in an orbit $G \cdot \mathbf{a}$ divides $|G|$.

   a. Fix $\mathbf{a} \in k^n$ and let $H = \{A \in G \mid A \cdot \mathbf{a} = \mathbf{a}\}$. Prove that H is a subgroup of G. We call H the *isotropy subgroup* or *stabilizer* of $\mathbf{a}$.

   b. Given $A \in G$, we get the *left coset* $AH = \{AB \mid B \in H\}$ of H in G and we let G/H denote the set of all left cosets (note that G/H will not be a group unless H is normal). Prove that the map sending AH to $A \cdot \mathbf{a}$ induces a bijective map $G/H \cong G \cdot \mathbf{a}$.

Hint: You will need to prove that the map is well-defined. Recall that two cosets AH and BH are equal if and only if $B^{-1}A \in H$.

c. Use part (b) to prove that the number of elements in $G \cdot \mathbf{a}$ divides $|G|$.

12. As in the proof of Theorem 10, suppose that we have disjoint orbits $G \cdot \mathbf{a}$ and $G \cdot \mathbf{b}$. Set $S = G \cdot \mathbf{b} \cup G \cdot \mathbf{a} - \{\mathbf{a}\}$, and pick $f \in k[x_1, \ldots, x_n]$ such that $f = 0$ on all points of S but $f(\mathbf{a}) \neq 0$. Let $g = R_G(f)$, where R_G is the Reynolds operator of G.

a. Explain why $g(\mathbf{b}) = 0$.

b. Explain why $g(\mathbf{a}) = \frac{M}{|G|}f(\mathbf{a}) \neq 0$, where M is the number of elements $A \in G$ such that $A \cdot \mathbf{a} = \mathbf{a}$.

13. In this exercise, we will see how Theorem 10 can fail when we work over a field that is not algebraically closed. Consider the group of permutation matrices $S_2 \subseteq \mathrm{GL}(2, \mathbb{R})$.

a. We know that $\mathbb{R}[x, y]^{S_2} = \mathbb{R}[\sigma_1, \sigma_2]$. Show that $I_F = \{0\}$ when $F = (\sigma_1, \sigma_2)$, so that $V_F = \mathbb{R}^2$. Thus, Theorem 10 is concerned with the map $\tilde{F} : \mathbb{R}^2/S_2 \to \mathbb{R}^2$ defined by sending $S_2 \cdot (x, y)$ to $(y_1, y_2) = (x + y, xy)$.

b. Show that the image of $\tilde{F}$ is the set $\{(y_1, y_2) \in \mathbb{R}^2 \mid y_1^2 \geq 4y_2\} \subseteq \mathbb{R}^2$. This is the region lying below the parabola $y_1^2 = 4y_2$. Hint: Interpret y_1 and y_2 as coefficients of the quadratic $X^2 - y_1X + y_2$. When does the quadratic have real roots?

14. There are many places in mathematics where one takes a set of equivalence classes and puts an algebraic structure on them. Show that the construction of a quotient ring $k[x_1, \ldots, x_n]/I$ is an example. Hint: See §2 of Chapter 5.

15. In this exercise, we will give some examples of how something initially defined as a set can turn out to be a variety in disguise. The key observation is that the set of nonvertical lines in the plane k^2 has a natural geometric structure. Namely, such a line L has a unique equation of the form $y = mx + b$, so that L can be identified with the point (m, b) in another 2-dimensional affine space, denoted $k^{2\vee}$. (If we use projective space—to be studied in the next chapter–then we can also include vertical lines.)

Now suppose that we have a curve C in the plane. Then consider all lines which are tangent to C somewhere on the curve. This gives us a subset $C^\vee \subseteq k^{2\vee}$. Let us compute this subset in some simple cases and show that it is an affine variety.

a. Suppose our curve C is the parabola $y = x^2$. Given a point (x_0, y_0) on the parabola, show that the tangent line is given by $y = 2x_0x - x_0^2$ and conclude that $C^\vee$ is the parabola $m^2 + 4b = 0$ in $k^{2\vee}$.

b. Show that $C^\vee$ is an affine variety when C is the cubic curve $y = x^3$.

In general, more work is needed to study $C^\vee$. In particular, the method used in the above examples breaks down when there are vertical tangents or singular points. Nevertheless, one can develop a satisfactory theory of what is called the *dual curve* $C^\vee$ of a curve $C \subseteq k^2$. One can also define the *dual variety* $V^\vee$ of a given irreducible variety $V \subseteq k^n$.

16. (Assumes §6 of Chapter 5) Let $G \subseteq \mathrm{GL}(n, k)$ be a finite matrix group. Prove that $k[x_1, \ldots, x_n]$ is finite over $k[x_1, \ldots, x_n]^G$ as in Definition 1 of Chapter 5, §6. Hint: See Exercise 9 of that section.

Chapter 8
Projective Algebraic Geometry

So far all of the varieties we have studied have been subsets of affine space k^n. In this chapter, we will enlarge k^n by adding certain "points at ∞" to create n-dimensional projective space $\mathbb{P}^n(k)$. We will then define projective varieties in $\mathbb{P}^n(k)$ and study the projective version of the algebra–geometry dictionary. The relation between affine and projective varieties will be considered in §4; in §5, we will study elimination theory from a projective point of view. By working in projective space, we will get a much better understanding of the Extension Theorem in Chapter 3. The chapter will end with a discussion of the geometry of quadric hypersurfaces and an introduction to Bezout's Theorem.

§1 The Projective Plane

This section will study the projective plane $\mathbb{P}^2(\mathbb{R})$ over the real numbers $\mathbb{R}$. We will see that, in a certain sense, the plane $\mathbb{R}^2$ is missing some "points at ∞," and by adding them to $\mathbb{R}^2$, we will get the projective plane $\mathbb{P}^2(\mathbb{R})$. Then we will introduce *homogeneous coordinates*. to give a more systematic treatment of $\mathbb{P}^2(\mathbb{R})$ Our starting point is the observation that two lines in $\mathbb{R}^2$ intersect in a point, *except* when they are parallel. We can take care of this exception if we view parallel lines as meeting at some sort of point at ∞. As indicated by the picture at the top of the following page, there should be different points at ∞, depending on the direction of the lines. To approach this more formally, we introduce an equivalence relation on lines in the plane by setting $L_1 \sim L_2$ if L_1 and L_2 are parallel. Then an equivalence class $[L]$ consists of all lines parallel to a given line L. The above discussion suggests that we should introduce one point at ∞ for each equivalence class $[L]$. We make the following provisional definition.

Definition 1. The **projective plane** over $\mathbb{R}$, denoted $\mathbb{P}^2(\mathbb{R})$, is the set

$$\mathbb{P}^2(\mathbb{R}) = \mathbb{R}^2 \cup \{\text{one point at } \infty \text{ for each equivalence class of parallel lines}\}.$$

© Springer International Publishing Switzerland 2015
D.A. Cox et al., *Ideals, Varieties, and Algorithms*, Undergraduate Texts in Mathematics, DOI 10.1007/978-3-319-16721-3_8

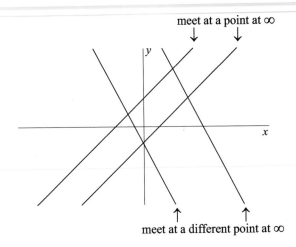

Let $[L]_\infty$ denote the common point at ∞ of all lines parallel to L. Then we call the set $\overline{L} = L \cup [L]_\infty \subseteq \mathbb{P}^2(\mathbb{R})$ the *projective line* corresponding to L. Note that two projective lines always meet at exactly one point: if they are not parallel, they meet at a point in $\mathbb{R}^2$; if they are parallel, they meet at their common point at ∞.

At first sight, one might expect that a line in the plane should have two points at ∞, corresponding to the two ways we can travel along the line. However, the reason why we want only one is contained in the previous paragraph: if there were two points at ∞, then parallel lines would have two points of intersection, not one. So, for example, if we parametrize the line $x = y$ via $(x, y) = (t, t)$, then we can approach its point at ∞ using either $t \to \infty$ or $t \to -\infty$.

A common way to visualize points at ∞ is to make a perspective drawing. Pretend that the earth is flat and consider a painting that shows two roads extending infinitely far in different directions:

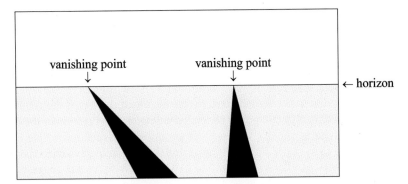

For each road, the two sides (which are parallel, but appear to be converging) meet at the same point on the horizon, which in the theory of perspective is called a *vanishing point*. Furthermore, any line parallel to one of the roads meets at the same vanishing point, which shows that the vanishing point represents the point at ∞ of these lines. The same reasoning applies to any point on the horizon, so that the horizon in the picture represents points at ∞. (Note that the horizon does not contain all of them—it is missing the point at ∞ of lines parallel to the horizon.)

The above picture reveals another interesting property of the projective plane: the points at ∞ form a special projective line, which is called the *line at* ∞. It follows that $\mathbb{P}^2(\mathbb{R})$ has the projective lines $\overline{L} = L \cup [L]_\infty$, where L is a line in $\mathbb{R}^2$, together with the line at ∞. In the exercises, you will prove that two distinct projective lines in $\mathbb{P}^2(\mathbb{R})$ determine a unique point and two distinct points in $\mathbb{P}^2(\mathbb{R})$ determine a unique projective line. Note the symmetry in these statements: when we interchange "point" and "projective line" in one, we get the other. This is an instance of the *principle of duality*, which is one of the fundamental concepts of projective geometry.

For an example of how points at ∞ can occur in other contexts, consider the parametrization of the hyperbola $x^2 - y^2 = 1$ given by the equations

$$x = \frac{1 + t^2}{1 - t^2},$$

$$y = \frac{2t}{1 - t^2}.$$

When $t \neq \pm 1$, it is easy to check that this parametrization covers all of the hyperbola except $(-1, 0)$. But what happens when $t = \pm 1$? Here is a picture of the hyperbola:

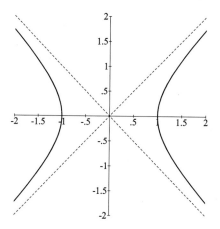

If we let $t \to 1^-$, then the corresponding point (x, y) travels along the first quadrant portion of the hyperbola, getting closer and closer to the asymptote $x = y$. Similarly, if $t \to 1^+$, we approach $x = y$ along the third quadrant portion of the hyperbola. Hence, it becomes clear that $t = 1$ should correspond to the point at ∞ of the asymptote $x = y$. Similarly, one can check that $t = -1$ corresponds to the point at ∞ of $x = -y$. (In the exercises, we will give a different way to see what happens when $t = \pm 1$.)

Thus far, our discussion of the projective plane has introduced some nice ideas, but it is not entirely satisfactory. For example, it is not really clear why the line at ∞ should be called a projective line. A more serious objection is that we have no unified way of naming points in $\mathbb{P}^2(\mathbb{R})$. Points in $\mathbb{R}^2$ are specified by coordinates, but points at ∞ are specified by lines. To avoid this asymmetry, we will introduce *homogeneous coordinates* on $\mathbb{P}^2(\mathbb{R})$.

To get homogeneous coordinates, we will need a new definition of projective space. The first step is to define an equivalence relation on nonzero points of $\mathbb{R}^3$ by setting

$$(x_1, y_1, z_1) \sim (x_2, y_2, z_2)$$

if there is a nonzero real number λ such that $(x_1, y_1, z_1) = \lambda(x_2, y_2, z_2)$. One can easily check that $\sim$ is an equivalence relation on $\mathbb{R}^3 \setminus \{0\}$ (where as usual 0 refers to the origin $(0, 0, 0)$ in $\mathbb{R}^3$). Then we can redefine projective space as follows.

Definition 2. $\mathbb{P}^2(\mathbb{R})$ is the set of equivalence classes of $\sim$ on $\mathbb{R}^3 \setminus \{0\}$. Thus, we can write

$$\mathbb{P}^2(\mathbb{R}) = (\mathbb{R}^3 \setminus \{0\})/\sim .$$

Given a triple $(x, y, z) \in \mathbb{R}^3 \setminus \{0\}$, its equivalence class $p \in \mathbb{P}^2(\mathbb{R})$ will be denoted $p = (x : y : z)$, and we say that $(x : y : z)$ are **homogeneous coordinates** of p. Thus

$$(x_1 : y_1 : z_1) = (x_2 : y_2 : z_2) \Leftrightarrow (x_1, y_1, z_1) = \lambda(x_2, y_2, z_2) \text{ for some } \lambda \in \mathbb{R} \setminus \{0\}.$$

At this point, it is not clear that Definitions 1 and 2 give the same object, although we will see shortly that this is the case.

Homogeneous coordinates are different from the usual notion of coordinates in that they are not unique. For example, the four points $(1 : 1 : 1)$, $(2 : 2 : 2)$, $(\pi : \pi : \pi)$ and $(\sqrt{2} : \sqrt{2} : \sqrt{2})$ are in fact the same point in projective space. But the nonuniqueness of the coordinates is not so bad since they are all multiples of one another.

As an illustration of how we can use homogeneous coordinates, let us define the notion of a projective line.

Definition 3. Given real numbers A, B, C, not all zero, the set

$$\{p \in \mathbb{P}^2(\mathbb{R}) \mid p = (x : y : z) \text{ with } Ax + By + Cz = 0\}$$

is called a **projective line** of $\mathbb{P}^2(\mathbb{R})$.

An important observation is that if the equation $Ax + By + Cz = 0$ holds for one set $(x : y : z)$ of homogeneous coordinates of $p \in \mathbb{P}^2(\mathbb{R})$, then it holds for *all* homogeneous coordinates of p. This is because all of the others can be written as $(\lambda x : \lambda y : \lambda z)$, so that $A \cdot \lambda x + B \cdot \lambda y + C \cdot \lambda z = \lambda(Ax + By + Cz) = 0$. Later in this chapter, we will use the same idea to define varieties in projective space.

To relate our two definitions of projective plane, we will use the map

$$(1) \qquad\qquad\qquad\qquad \mathbb{R}^2 \longrightarrow \mathbb{P}^2(\mathbb{R})$$

defined by sending $(x, y) \in \mathbb{R}^2$ to the point $p \in \mathbb{P}^2(\mathbb{R})$ whose homogeneous coordinates are $(x : y : 1)$. This map has the following properties.

Proposition 4. *The map* (1) *is one-to-one and the complement of its image is the projective line H_∞ defined by $z = 0$.*

Proof. First, suppose that (x, y) and (x', y') map to the same point p in $\mathbb{P}^2(\mathbb{R})$. Then $p = (x : y : 1) = (x' : y' : 1)$, so that $(x, y, 1) = \lambda(x', y', 1)$ for some λ. Looking at the third coordinate, we see that $\lambda = 1$ and it follows that $(x, y) = (x', y')$.

Next, let $p = (x : y : z)$ be a point in $\mathbb{P}^2(\mathbb{R})$. If $z = 0$, then p is on the projective line H_∞. On the other hand, if $z \neq 0$, then we can multiply by $1/z$ to see that $p = (x/z : y/z : 1)$. This shows that p is in the image of map (1). We leave it as an exercise to show that the image of the map is disjoint from H_∞, and the proposition is proved. $\qquad\square$

We will call H_∞ the *line at* ∞. It is customary (though somewhat sloppy) to identify $\mathbb{R}^2$ with its image in $\mathbb{P}^2(\mathbb{R})$, so that we can write projective space as the disjoint union

$$\mathbb{P}^2(\mathbb{R}) = \mathbb{R}^2 \cup H_\infty.$$

This is beginning to look familiar. It remains to show that H_∞ consists of points at ∞ in our earlier sense. Thus, we need to study how lines in $\mathbb{R}^2$ (which we will call *affine lines*) relate to projective lines. The following table tells the story:

affine line		projective line		point at ∞
$L : y = mx + b$	$\to$	$\overline{L} : y = mx + bz$	$\to$	$(1 : m : 0)$
$L : x = c$	$\to$	$\overline{L} : x = cz$	$\to$	$(0 : 1 : 0)$

To understand this table, first consider a nonvertical affine line L defined by $y = mx + b$. Under the map (1), a point (x, y) on L maps to a point $(x : y : 1)$ satisfying the projective equation $y = mx + bz$. Thus, $(x : y : 1)$ lies on the projective line $\overline{L}$ defined by $mx - y + bz = 0$, so that L can be regarded as subset of $\overline{L}$. By Proposition 4, the remaining points of $\overline{L}$ come from where it meets $z = 0$. But the equations $z = 0$ and $y = mx + bz$ clearly imply $y = mx$, so that the solutions are $(x : mx : 0)$. We have $x \neq 0$ since homogeneous coordinates never simultaneously vanish, and dividing by x shows that $(1 : m : 0)$ is the unique point of $\overline{L} \cap H_\infty$. The case of vertical lines is left as an exercise.

The table shows that two lines in $\mathbb{R}^2$ meet at the same point at ∞ if and only if they are parallel. For nonvertical lines, the point at ∞ encodes the slope, and for vertical lines, there is a single (but different) point at ∞. Be sure you understand this. In the exercises, you will check that the points listed in the table exhaust all of H_∞. Consequently, H_∞ consists of a unique point at ∞ for every equivalence class of parallel lines. Then $\mathbb{P}^2(\mathbb{R}) = \mathbb{R}^2 \cup H_\infty$ shows that the projective planes of Definitions 1 and 2 are the same object.

We next introduce a more geometric way of thinking about points in the projective plane. Let $p = (x : y : z)$ be a point in $\mathbb{P}^2(\mathbb{R})$, so that all other homogeneous coordinates for p are given by $(\lambda x : \lambda y : \lambda z)$ for $\lambda \in \mathbb{R} \setminus \{0\}$. The crucial observation is that back in $\mathbb{R}^3$, the points $(\lambda x, \lambda y, \lambda z) = \lambda(x, y, z)$ all lie on the same line through the origin in $\mathbb{R}^3$:

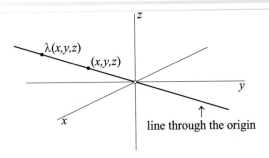

line through the origin

The requirement in Definition 2 that $(x, y, z) \neq (0, 0, 0)$ guarantees that we get a line in $\mathbb{R}^3$. Conversely, given *any* line L through the origin in $\mathbb{R}^3$, a point (x, y, z) on $L \setminus \{0\}$ gives homogeneous coordinates $(x : y : z)$ for a uniquely determined point in $\mathbb{P}^2(\mathbb{R})$ [since any other point on $L \setminus \{0\}$ is a nonzero multiple of (x, y, z)]. This shows that we have a one-to-one correspondence.

(2) $\mathbb{P}^2(\mathbb{R}) \cong \{\text{lines through the origin in } \mathbb{R}^3\}.$

Although it may seem hard to think of a point in $\mathbb{P}^2(\mathbb{R})$ as a line in $\mathbb{R}^3$, there is a strong intuitive basis for this identification. We can see why by studying how to draw a 3-dimensional object on a 2-dimensional canvas. Imagine lines or rays that link our eye to points on the object. Then we draw the object according to where the rays intersect the canvas:

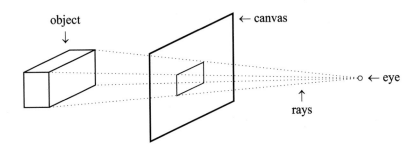

Renaissance texts on perspective would speak of the "pyramid of rays" connecting the artist's eye with the object being painted. For us, the crucial observation is that each ray hits the canvas exactly once, giving a one-to-one correspondence between rays and points on the canvas.

To make this more mathematical, we will let the "eye" be the origin and the "canvas" be the plane $z = 1$ in the coordinate system pictured at the top of the next page. Rather than work with rays (which are half-lines), we will work with lines through the origin. Then, as the picture indicates, every point in the plane $z = 1$ determines a unique line through the origin. This one-to-one correspondence allows us to think of a *point* in the plane as a *line through the origin* in $\mathbb{R}^3$ [which by (2) is a point in $\mathbb{P}^2(\mathbb{R})$]. There are two interesting things to note about this correspondence:

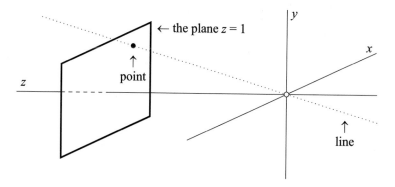

- A point (x, y) in the plane gives the point $(x, y, 1)$ on our "canvas" $z = 1$. The corresponding line through the origin is a point $p \in \mathbb{P}^2(\mathbb{R})$ with homogeneous coordinates $(x : y : 1)$. Hence, the correspondence given above is *exactly* the map $\mathbb{R}^2 \to \mathbb{P}^2(\mathbb{R})$ from Proposition 4.
- The correspondence is not onto since this method will never produce a line in the (x, y)-plane. Do you see how these lines can be thought of as the points at ∞?

In many situations, it is useful to be able to think of $\mathbb{P}^2(\mathbb{R})$ both algebraically (in terms of homogeneous coordinates) and geometrically (in terms of lines through the origin).

As the final topic in this section, we will use homogeneous coordinates to examine the line at ∞ more closely. The basic observation is that although we began with coordinates x and y, once we have homogeneous coordinates, there is nothing special about the extra coordinate z—it is no different from x or y. In particular, if we want, we could regard x and z as the original coordinates and y as the extra one.

To see how this can be useful, consider the parallel lines $L_1 : y = x + 1/2$ and $L_2 : y = x - 1/2$ in the (x, y)-plane:

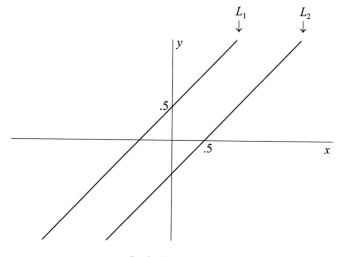

In the (x,y)-plane

We know that these lines intersect at ∞ since they are parallel. But the picture does not show their point of intersection. To view these lines at ∞, consider the projective lines

$$\overline{L}_1 : y = x + (1/2)z,$$
$$\overline{L}_2 : y = x - (1/2)z$$

determined by L_1 and L_2. Now regard x and z as the original variables. Thus, we map the (x, z)-plane $\mathbb{R}^2$ to $\mathbb{P}^2(\mathbb{R})$ via $(x, z) \mapsto (x:1:z)$. As in Proposition 4, this map is one-to-one, and we can recover the (x, z)-plane inside $\mathbb{P}^2(\mathbb{R})$ by setting $y = 1$. If we do this with the equations of the projective lines $\overline{L}_1$ and $\overline{L}_2$, we get the lines $L_1' : z = -2x + 2$ and $L_2' : z = 2x - 2$. This gives the following picture in the (x, z)-plane:

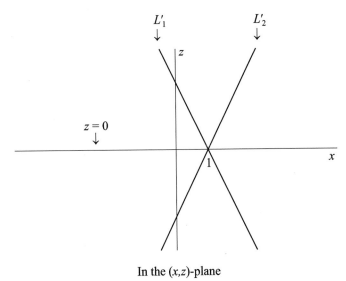

In the (x,z)-plane

In this picture, the x-axis is defined by $z = 0$, which is the line at ∞ as we originally set things up in Proposition 4. Note that L_1' and L_2' meet when $z = 0$, which corresponds to the fact that L_1 and L_2 meet at ∞. Thus, the above picture shows how our two lines behave as they approach the line at ∞. In the exercises, we will study what some other common curves look like at ∞.

It is interesting to compare the above picture with the perspective drawing of two roads given earlier in the section. It is no accident that the horizon in the perspective drawing represents the line at ∞. The exercises will explore this idea in more detail.

Another interesting observation is that the Euclidean notion of distance does not play a prominent role in the geometry of projective space. For example, the lines L_1 and L_2 in the (x, y)-plane are a constant distance apart, whereas L_1' and L_2' get closer and closer in the (x, z)-plane. This explains why the geometry of $\mathbb{P}^2(\mathbb{R})$ is quite different from Euclidean geometry.

EXERCISES FOR §1

1. Using $\mathbb{P}^2(\mathbb{R})$ as given in Definition 1, we saw that the projective lines in $\mathbb{P}^2(\mathbb{R})$ are $\overline{L} = L \cup [L]_\infty$, and the line at ∞.
   a. Prove that *any* two distinct points in $\mathbb{P}^2(\mathbb{R})$ determine a unique projective line. Hint: There are three cases, depending on how many of the points are points at ∞.
   b. Prove that *any* two distinct projective lines in $\mathbb{P}^2(\mathbb{R})$ meet at a unique point. Hint: Do this case-by-case.

2. There are many theorems that initially look like theorems in the plane, but which are really theorems in $\mathbb{P}^2(\mathbb{R})$ in disguise. One classic example is Pappus's Theorem, which goes as follows. Suppose we have two collinear triples of points A, B, C and A', B', C'. Then let

$$P = \overline{AB'} \cap \overline{A'B},$$
$$Q = \overline{AC'} \cap \overline{A'C},$$
$$R = \overline{BC'} \cap \overline{B'C}.$$

Pappus's Theorem states that P, Q, R are always collinear points. In Exercise 8 of Chapter 6, §4, we drew the following picture to illustrate the theorem:

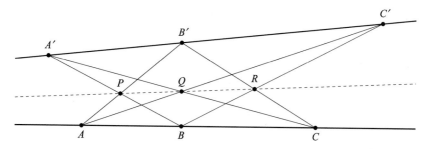

   a. If we let the points on one of the lines go in the opposite order, then we can get the following configuration of points and lines:

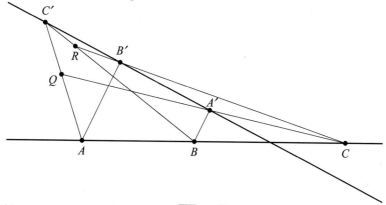

   Note that P is now a point at ∞ when $\overline{AB'}$ and $\overline{A'B}$ are parallel. Is Pappus's Theorem still true [in $\mathbb{P}^2(\mathbb{R})$] for this configuration of points and lines?
   b. By moving the point C in the picture for part (a) show that you can also make Q a point at ∞. Is Pappus's Theorem still true? What line do P, Q, R lie on? Draw a picture to illustrate what happens.

If you made a purely affine version of Pappus's Theorem that took cases (a) and (b) into account, the resulting statement would be rather cumbersome. By working in $\mathbb{P}^2(\mathbb{R})$, we cover these cases simultaneously.

3. We will continue the study of the parametrization $(x, y) = ((1+t^2)/(1-t^2), 2t/(1-t^2))$ of $x^2 - y^2 = 1$ begun in the text.
   a. Given t, show that (x, y) is the point where the hyperbola intersects the line of slope t going through the point $(-1, 0)$. Illustrate your answer with a picture. Hint: Use the parametrization to show that $t = y/(x + 1)$.
   b. Use the answer to part (a) to explain why $t = \pm 1$ maps to the points at ∞ corresponding to the asymptotes of the hyperbola. Illustrate your answer with a drawing.
   c. Using homogeneous coordinates, show that we can write the parametrization as

   $$((1 + t^2)/(1 - t^2) : 2t/(1 - t^2) : 1) = (1 + t^2 : 2t : 1 - t^2),$$

   and use this to explain what happens when $t = \pm 1$. Does this give the same answer as part (b)?
   d. We can also use the technique of part (c) to understand what happens when $t \to \infty$. Namely, in the parametrization $(x : y : z) = (1 + t^2 : 2t : 1 - t^2)$, substitute $t = 1/u$. Then clear denominators (this is legal since we are using homogeneous coordinates) and let $u \to 0$. What point do you get on the hyperbola?

4. This exercise will study what the hyperbola $x^2 - y^2 = 1$ looks like at ∞.
   a. Explain why the equation $x^2 - y^2 = z^2$ gives a well-defined curve C in $\mathbb{P}^2(\mathbb{R})$. Hint: See the discussion following Definition 3.
   b. What are the points at ∞ on C? How does your answer relate to Exercise 3?
   c. In the (x, z) coordinate system obtained by setting $y = 1$, show that C is still a hyperbola.
   d. In the (y, z) coordinate system obtained by setting $x = 1$, show that C is a circle.
   e. Use the parametrization of Exercise 3 to obtain a parametrization of the circle from part (d).

5. Consider the parabola $y = x^2$.
   a. What equation should we use to make the parabola into a curve in $\mathbb{P}^2(\mathbb{R})$?
   b. How many points at ∞ does the parabola have?
   c. By choosing appropriate coordinates (as in Exercise 4), explain why the parabola is tangent to the line at ∞.
   d. Show that the parabola looks like a hyperbola in the (y, z) coordinate system.

6. When we use the (x, y) coordinate system inside $\mathbb{P}^2(\mathbb{R})$, we only view a piece of the projective plane. In particular, we miss the line at ∞. As in the text, we can use (x, z) coordinates to view the line at ∞. Show that there is exactly one point in $\mathbb{P}^2(\mathbb{R})$ that is visible in neither (x, y) nor (x, z) coordinates. How can we view what is happening at this point?

7. In the proof of Proposition 4, show that the image of the map (2) is disjoint from H_∞.

8. As in the text, the line H_∞ is defined by $z = 0$. Thus, points on H_∞ have homogeneous coordinates $(a : b : 0)$, where $(a, b) \neq (0, 0)$.
   a. A vertical affine line $x = c$ gives the projective line $x = cz$. Show that this meets H_∞ at the point $(0 : 1 : 0)$.
   b. Show that a point on H_∞ different from $(0 : 1 : 0)$ can be written uniquely in the from $(1 : m : 0)$ for some real number m.

9. In the text, we viewed parts of $\mathbb{P}^2(\mathbb{R})$ in the (x, y) and (x, z) coordinate systems. In the (x, z) picture, it is natural to ask what happened to y. To see this, we will study how (x, y) coordinates look when viewed in the plane with (x, z) coordinates.
   a. Show that (a, b) in the (x, y)-plane gives the point $(a/b, 1/b)$ in the (x, z)-plane.
   b. Use the formula of part (a) to study what the parabolas $(x, y) = (t, t^2)$ and $(x, y) = (t^2, t)$ look like in the (x, z)-plane. Draw pictures of what happens in both (x, y) and (x, z) coordinates.

10. In this exercise, we will discuss the mathematics behind the perspective drawing given in the text. Suppose we want to draw a picture of a landscape, which we will assume is a horizontal plane. We will make our drawing on a canvas, which will be a vertical plane. Our eye will be a certain distance above the landscape, and to draw, we connect a point on the landscape to our eye with a line, and we put a dot where the line hits the canvas:

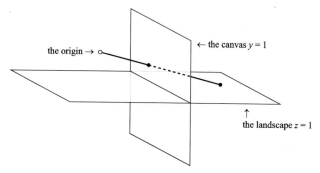

To give formulas for what happens, we will pick coordinates (x, y, z) so that our eye is the origin, the canvas is the plane $y = 1$, and the landscape is the plane $z = 1$ (thus, the positive z-axis points down).

    a. Starting with the point $(a, b, 1)$ on the landscape, what point do we get in the canvas $y = 1$?

    b. Explain how the answer to part (a) relates to Exercise 9. Write a brief paragraph discussing the relation between perspective drawings and the projective plane.

11. As in Definition 3, a projective line in $\mathbb{P}^2(\mathbb{R})$ is defined by an equation of the form $Ax + By + Cz = 0$, where $(A, B, C) \neq (0, 0, 0)$.

    a. Why do we need to make the restriction $(A, B, C) \neq (0, 0, 0)$?

    b. Show that (A, B, C) and (A', B', C') define the same projective line if and only if $(A, B, C) = \lambda(A', B', C')$ for some nonzero real number λ. Hint: One direction is easy. For the other direction, take two distinct points $(a : b : c)$ and $(a' : b' : c')$ on the line $Ax + By + Cz = 0$. Show that the vectors (a, b, c) and (a', b', c') are linearly independent and conclude that the equations $Xa + Yb + Zc = Xa' + Yb' + Zc' = 0$ have a 1-dimensional solution space for the variables X, Y, Z.

    c. Conclude that the set of projective lines in $\mathbb{P}^2(\mathbb{R})$ can be identified with the set $\{(A, B, C) \in \mathbb{R}^3 \mid (A, B, C) \neq (0, 0, 0)\}/\sim$. This set is called the *dual projective plane* and is denoted $\mathbb{P}^2(\mathbb{R})^\vee$.

    d. Describe the subset of $\mathbb{P}^2(\mathbb{R})^\vee$ corresponding to affine lines.

    e. Given a point $p \in \mathbb{P}^2(\mathbb{R})$, consider the set $\tilde{p}$ of all projective lines L containing p. We can regard $\tilde{p}$ as a subset of $\mathbb{P}^2(\mathbb{R})^\vee$. Show that $\tilde{p}$ is a projective line in $\mathbb{P}^2(\mathbb{R})^\vee$. We call $\tilde{p}$ the *pencil of lines* through p.

    f. The Cartesian product $\mathbb{P}^2(\mathbb{R}) \times \mathbb{P}^2(\mathbb{R})^\vee$ has the natural subset

$$I = \{(p, L) \in \mathbb{P}^2(\mathbb{R}) \times \mathbb{P}^2(\mathbb{R})^\vee \mid p \in L\}.$$

Show that I is described by the equation $Ax + By + Cz = 0$, where $(x : y : z)$ are homogeneous coordinates on $\mathbb{P}^2(\mathbb{R})$ and $(A : B : C)$ are homogeneous coordinates on the dual. We will study varieties of this type in §5.

Parts (d), (e), and (f) of Exercise 11 illustrate how collections of naturally defined geometric objects can be given an algebraic structure.

§2 Projective Space and Projective Varieties

The construction of the real projective plane given in Definition 2 of §1 can be generalized to yield projective spaces of any dimension n over any field k. We define an equivalence relation $\sim$ on the nonzero points of k^{n+1} by setting

$$(x_0', \ldots, x_n') \sim (x_0, \ldots, x_n)$$

if there is a nonzero element $\lambda \in k$ such that $(x_0', \ldots, x_n') = \lambda(x_0, \ldots, x_n)$. If we let 0 denote the origin $(0, \ldots, 0)$ in k^{n+1}, then we define projective space as follows.

Definition 1. n-**dimensional projective space** over the field k, denoted $\mathbb{P}^n(k)$, is the set of equivalence classes of $\sim$ on $k^{n+1} \setminus \{0\}$. Thus,

$$\mathbb{P}^n(k) = (k^{n+1} \setminus \{0\})/\sim \, .$$

Given an $(n+1)$-tuple $(x_0, \ldots, x_n) \in k^{n+1} \setminus \{0\}$, its equivalence class $p \in \mathbb{P}^n(k)$ will be denoted $(x_0 : \cdots : x_n)$, and we will say that $(x_0 : \cdots : x_n)$ are **homogeneous coordinates** of p. Thus

$$(x_0' : \cdots : x_n') = (x_0 : \cdots : x_n) \iff (x_0', \ldots, x_n') = \lambda(x_0, \ldots, x_n)$$
$$\text{for some } \lambda \in k \setminus \{0\}.$$

Like $\mathbb{P}^2(\mathbb{R})$, each point $p \in \mathbb{P}^n(k)$ has many sets of homogeneous coordinates. For example, $(0 : \sqrt{2} : 0 : i)$ and $(0 : 2i : 0 : -\sqrt{2})$ are the same point in $\mathbb{P}^3(\mathbb{C})$ since $(0, 2i, 0, -\sqrt{2}) = \sqrt{2}i(0, \sqrt{2}, 0, i)$.

As in §1, we can think of $\mathbb{P}^n(k)$ more geometrically as the set of lines through the origin in k^{n+1}. More precisely, you will show in Exercise 1 that there is a one-to-one correspondence

(1) $\mathbb{P}^n(k) \cong \{\text{lines through the origin in } k^{n+1}\}.$

Just as the real projective plane contains the affine plane $\mathbb{R}^2$ as a subset, $\mathbb{P}^n(k)$ contains the affine space k^n.

Proposition 2. *Let*

$$U_0 = \{(x_0 : \cdots : x_n) \in \mathbb{P}^n(k) \mid x_0 \neq 0\}.$$

Then the map ϕ that sends $(a_1, \ldots, a_n) \in k^n$ to $(1 : a_1 : \cdots : a_n) \in \mathbb{P}^n(k)$ is a one-to-one correspondence between k^n and $U_0 \subseteq \mathbb{P}^n(k)$.

Proof. Since the first component of $\phi(a_1, \ldots, a_n) = (1 : a_1 : \cdots : a_n)$ is nonzero, we get a map $\phi : k^n \to U_0$. We define an inverse map $\psi : U_0 \to k^n$ as follows. Given $p = (x_0 : x_1 : \cdots : x_n) \in U_0$, since $x_0 \neq 0$ we can multiply the homogeneous coordinates by the nonzero scalar $\lambda = \frac{1}{x_0}$ to obtain $p = (1 : \frac{x_1}{x_0} : \cdots : \frac{x_n}{x_0})$. Then set $\psi(p) = (\frac{x_1}{x_0}, \ldots, \frac{x_n}{x_0}) \in k^n$. We leave it as an exercise for the reader to show that ψ

is well-defined and that ϕ and ψ are inverse mappings. This establishes the desired one-to-one correspondence. □

By the definition of U_0, we see that $\mathbb{P}^n(k) = U_0 \cup H$, where

$$(2) \qquad H = \{p \in \mathbb{P}^n(k) \mid p = (0 : x_1 : \cdots : x_n)\}.$$

If we identify U_0 with the affine space k^n, then we can think of H as the *hyperplane at infinity*. It follows from (2) that the points in H are in one-to-one correspondence with nonzero n-tuples $(x_1, \ldots, x_n)$, where two n-tuples represent the same point of H if one is a nonzero scalar multiple of the other (just ignore the first component of points in H). In other words, H is a "copy" of $\mathbb{P}^{n-1}(k)$, the projective space of one smaller dimension. Identifying U_0 with k^n and H with $\mathbb{P}^{n-1}(k)$, we can write

$$(3) \qquad \mathbb{P}^n(k) = k^n \cup \mathbb{P}^{n-1}(k).$$

To see what $H = \mathbb{P}^{n-1}(k)$ means geometrically, note that, by (1), a point $p \in \mathbb{P}^{n-1}(k)$ gives a line $L \subseteq k^n$ going through the origin. Consequently, in the decomposition (3), we should think of p as representing the asymptotic direction of *all* lines in k^n parallel to L. This allows us to regard p as a point at ∞ in the sense of §1, and we recover the intuitive definition of the projective space given there. In the exercises, we will give a more algebraic way of seeing how this works.

A special case worth mentioning is the projective line $\mathbb{P}^1(k)$. Since $\mathbb{P}^0(k)$ consists of a single point (this follows easily from Definition 1), letting $n = 1$ in (3) gives us

$$\mathbb{P}^1(k) = k^1 \cup \mathbb{P}^0(k) = k \cup \{\infty\},$$

where we let ∞ represent the single point of $\mathbb{P}^0(k)$. If we use (1) to think of points in $\mathbb{P}^1(k)$ as lines through the origin in k^2, then the above decomposition reflects the fact these lines are characterized by their slope (where the vertical line has slope ∞). When $k = \mathbb{C}$, it is customary to call

$$\mathbb{P}^1(\mathbb{C}) = \mathbb{C} \cup \{\infty\}$$

the *Riemann sphere*. The reason for this name will be explored in the exercises.

For completeness, we mention that there are many other copies of k^n in $\mathbb{P}^n(k)$ besides U_0. Indeed the proof of Proposition 2 may be adapted to yield the following results.

Corollary 3. *For each* $i = 0, \ldots n$, *let*

$$U_i = \{(x_0 : \cdots : x_n) \in \mathbb{P}^n(k) \mid x_i \neq 0\}.$$

(i) *The points of each* U_i *are in one-to-one correspondence with the points of* k^n.
(ii) *The complement* $\mathbb{P}^n(k) \setminus U_i$ *may be identified with* $\mathbb{P}^{n-1}(k)$.
(iii) *We have* $\mathbb{P}^n(k) = \bigcup_{i=0}^{n} U_i$.

Proof. See Exercise 5. □

Our next goal is to extend the definition of varieties in affine space to projective space. For instance, we can ask whether it makes sense to consider $\mathbf{V}(f)$ for a polynomial $f \in k[x_0, \ldots, x_n]$. A simple example shows that some care must be taken here. For instance, in $\mathbb{P}^2(\mathbb{R})$, we can try to construct $\mathbf{V}(x_1 - x_2^2)$. The point $p = (x_0 : x_1 : x_2) = (1 : 4 : 2)$ appears to be in this set since the components of p satisfy the equation $x_1 - x_2^2 = 0$. However, a problem arises when we note that the same point p can be represented by the homogeneous coordinates $p = (2 : 8 : 4)$. If we substitute these components into our polynomial, we obtain $8 - 4^2 = -8 \neq 0$. We get different results depending on which homogeneous coordinates we choose.

To avoid problems of this type, we use homogeneous polynomials when working in $\mathbb{P}^n(k)$. From Definition 6 of Chapter 7, §1, recall that a polynomial is *homogeneous* of total degree d if every term appearing in f has total degree exactly d. The polynomial $f = x_1 - x_2^2$ in the example is not homogeneous, and this is what caused the inconsistency in the values of f on different homogeneous coordinates representing the same point. For a homogeneous polynomial, this does not happen.

Proposition 4. *Let $f \in k[x_0, \ldots, x_n]$ be a homogeneous polynomial. If f vanishes on any one set of homogeneous coordinates for a point $p \in \mathbb{P}^n(k)$, then f vanishes for all homogeneous coordinates of p. In particular $\mathbf{V}(f) = \{p \in \mathbb{P}^n(k) \mid f(p) = 0\}$ is a well-defined subset of $\mathbb{P}^n(k)$.*

Proof. Let $(a_0 : \cdots : a_n) = (\lambda a_0 : \cdots : \lambda a_n)$ be homogeneous coordinates for a point $p \in \mathbb{P}^n(k)$ and assume that $f(a_0, \ldots, a_n) = 0$. If f is homogeneous of total degree d, then every term in f has the form

$$c x_0^{\alpha_0} \cdots x_n^{\alpha_n},$$

where $\alpha_0 + \cdots + \alpha_n = d$. When we substitute $x_i = \lambda a_i$, this term becomes

$$c(\lambda a_0)^{\alpha_0} \cdots (\lambda a_n)^{\alpha_n} = \lambda^d c a_0^{\alpha_0} \cdots a_n^{\alpha_n}.$$

Summing over the terms in f, we find a common factor of λ^d and, hence,

$$f(\lambda a_0, \ldots, \lambda a_n) = \lambda^d f(a_0, \ldots, a_n) = 0.$$

This proves the proposition.   □

Notice that even if f is homogeneous, the equation $f = a$ does not make sense in $\mathbb{P}^n(k)$ when $0 \neq a \in k$. The equation $f = 0$ is special because it gives a well-defined subset of $\mathbb{P}^n(k)$. We can also consider subsets of $\mathbb{P}^n(k)$ defined by the vanishing of a *system* of homogeneous polynomials (possibly of different total degrees). The correct generalization of the affine varieties introduced in Chapter 1, §2 is as follows.

Definition 5. Let k be a field and let $f_1, \ldots, f_s \in k[x_0, \ldots, x_n]$ be homogeneous polynomials. We set

$$\mathbf{V}(f_1, \ldots, f_s) = \{(a_0 : \cdots : a_n) \in \mathbb{P}^n(k) \mid f_i(a_0, \ldots, a_n) = 0 \text{ for all } 1 \leq i \leq s\}.$$

We call $\mathbf{V}(f_1, \ldots, f_s)$ the **projective variety** defined by $f_1, \ldots, f_s$.

For example, in $\mathbb{P}^n(k)$, any nonzero homogeneous polynomial of degree 1,

$$\ell(x_0,\ldots,x_n) = c_0x_0 + \cdots + c_nx_n,$$

defines a projective variety $\mathbf{V}(\ell)$ called a *hyperplane*. One example we have seen is the hyperplane at infinity, which was defined as $H = \mathbf{V}(x_0)$. When $n = 2$, we call $\mathbf{V}(\ell)$ a projective line, or more simply a *line* in $\mathbb{P}^2(k)$. Similarly, when $n = 3$, we call a hyperplane a *plane* in $\mathbb{P}^3(k)$. Varieties defined by one or more linear polynomials (homogeneous polynomials of degree 1) are called *linear varieties* in $\mathbb{P}^n(k)$. For instance, $\mathbf{V}(x_1,x_2) \subseteq \mathbb{P}^3(k)$ is a linear variety which is a projective line in $\mathbb{P}^3(k)$.

The projective varieties $\mathbf{V}(f)$ defined by a single nonzero homogeneous equation are known collectively as *hypersurfaces*. However, individual hypersurfaces are usually classified according to the total degree of the defining equation. Thus, if f has total degree 2 in $k[x_0,\ldots,x_n]$, we usually call $\mathbf{V}(f)$ a *quadric hypersurface*, or *quadric* for short. For instance, $\mathbf{V}(-x_0^2 + x_1^2 + x_2^2) \subseteq \mathbb{P}^3(\mathbb{R})$ is quadric. Similarly, hypersurfaces defined by equations of total degree 3, 4, and 5 are known as *cubics*, *quartics*, and *quintics*, respectively.

To get a better understanding of projective varieties, we need to discover what the corresponding algebraic objects are. This leads to the notion of *homogeneous ideal*, which will be discussed in §3. We will see that the entire algebra–geometry dictionary of Chapter 4 can be carried over to projective space.

The final topic we will consider in this section is the relation between affine and projective varieties. As we saw in Corollary 3, the subsets $U_i \subseteq \mathbb{P}^n(k)$ are copies of k^n. Thus, we can ask how affine varieties in $U_i \cong k^n$ relate to projective varieties in $\mathbb{P}^n(k)$. First, if we take a projective variety V and intersect it with one of the U_i, it makes sense to ask whether we obtain an affine variety. The answer to this question is always *yes*, and the defining equations of the variety $V \cap U_i$ may be obtained by a process called *dehomogenization*.

We illustrate this by considering $V \cap U_0$. From the proof of Proposition 2, we know that if $p \in U_0$, then p has homogeneous coordinates of the form $(1:x_1:\cdots:x_n)$. If $f \in k[x_0,\ldots,x_n]$ is one of the defining equations of V, then the polynomial $g(x_1,\ldots,x_n) = f(1,x_1,\ldots,x_n) \in k[x_1,\ldots,x_n]$ vanishes at every point of $V \cap U_0$. Setting $x_0 = 1$ in f produces a "dehomogenized" polynomial g which is usually nonhomogeneous. We claim that $V \cap U_0$ is precisely the affine variety obtained by dehomogenizing the equations of V.

Proposition 6. *Let $V = \mathbf{V}(f_1,\ldots,f_s)$ be a projective variety. Then $W = V \cap U_0$ can be identified with the affine variety $\mathbf{V}(g_1,\ldots,g_s) \subseteq k^n$, where $g_i(x_1,\ldots,x_n) = f_i(1,x_1,\ldots,x_n)$ for each $1 \leq i \leq s$.*

Proof. The comments before the statement of the proposition show that using the mapping $\psi : U_0 \to k^n$ from Proposition 2, $\psi(W) \subseteq \mathbf{V}(g_1,\ldots,g_s)$. On the other hand, if $(a_1,\ldots,a_n) \in \mathbf{V}(g_1,\ldots,g_s)$, then the point with homogeneous coordinates $(1:a_1:\cdots:a_n)$ is in U_0 and it satisfies the equations

$$f_i(1,a_1,\ldots,a_n) = g_i(a_1,\ldots,a_n) = 0.$$

Thus, $\phi(\mathbf{V}(g_1, \ldots, g_s)) \subseteq W$. Since the mappings ϕ and ψ are inverses, the points of W are in one-to-one correspondence with the points of $\mathbf{V}(g_1, \ldots, g_s)$.   □

For instance, consider the projective variety

(4) $V = \mathbf{V}(x_1^2 - x_2 x_0, x_1^3 - x_3 x_0^2) \subseteq \mathbb{P}^3(\mathbb{R}).$

To intersect V with U_0, we dehomogenize the defining equations, which gives us the affine variety

$$\mathbf{V}(x_1^2 - x_2, x_1^3 - x_3) \subseteq \mathbb{R}^3.$$

We recognize this as the familiar twisted cubic in $\mathbb{R}^3$.

We can also dehomogenize with respect to other variables. For example, the above proof shows that, for any projective variety $V \subseteq \mathbb{P}^3(\mathbb{R}), V \cap U_1$ can be identified with the affine variety in $\mathbb{R}^3$ defined by the equations obtained by setting $g_i(x_0, x_2, x_3) = f_i(x_0, 1, x_2, x_3)$. When we do this with the projective variety V defined in (4), we see that $V \cap U_1$ is the affine variety $\mathbf{V}(1 - x_2 x_0, 1 - x_3 x_0^2)$. See Exercise 9 for a general statement.

Going in the opposite direction, we can ask whether an affine variety in U_i, can be written as $V \cap U_i$ in some projective variety V. The answer is again *yes*, but there is more than one way to do it, and the results can be somewhat unexpected.

One natural idea is to reverse the dehomogenization process described earlier and "homogenize" the defining equations of the affine variety. For example, consider the affine variety $W = \mathbf{V}(x_2 - x_1^3 + x_1^2)$ in $U_0 = \mathbb{R}^2$. The defining equation is not homogeneous, so we do not get a projective variety in $\mathbb{P}^2(\mathbb{R})$ directly from this equation. But we can use the extra variable x_0 to make $f = x_2 - x_1^3 + x_1^2$ homogeneous. Since f has total degree 3, we modify f so that every term has total degree 3. This leads to the homogeneous polynomial

$$f^h = x_2 x_0^2 - x_1^3 + x_1^2 x_0.$$

Moreover, note that dehomogenizing f^h gives back the original polynomial f in x_1, x_2. The general pattern is the same.

Proposition 7. *Let $g(x_1, \ldots, x_n) \in k[x_1, \ldots, x_n]$ be a polynomial of total degree d.*

(i) *Let $g = \sum_{i=0}^{d} g_i$ be the expansion of g as the sum of its homogeneous components where g_i has total degree i. Then*

$$g^h(x_0, \ldots, x_n) = \sum_{i=0}^{d} g_i(x_1, \ldots, x_n) x_0^{d-i}$$
$$= g_d(x_1, \ldots, x_n) + g_{d-1}(x_1, \ldots x_n) x_0$$
$$+ \cdots + g_0(x_1, \ldots x_n) x_0^d$$

*is a homogeneous polynomial of total degree d in $k[x_0, \ldots, x_n]$. We will call g^h the **homogenization** of g.*

(ii) *The homogenization of g can be computed using the formula*

$$g^h = x_0^d \cdot g\left(\frac{x_1}{x_0}, \ldots, \frac{x_n}{x_0}\right).$$

(iii) *Dehomogenizing g^h yields g, i.e., $g^h(1, x_1, \cdots, x_n) = g(x_1, \ldots, x_n)$.*

(iv) *Let $F(x_0, \ldots, x_n)$ be a homogeneous polynomial and let x_0^e be the highest power of x_0 dividing F. If $f = F(1, x_1, \ldots, x_n)$ is a dehomogenization of F, then $F = x_0^e \cdot f^h$.*

Proof. We leave the proof to the reader as Exercise 10. $\qquad\Box$

As a result of Proposition 7, given any affine variety $W = \mathbf{V}(g_1, \ldots, g_s) \subseteq k^n$, we can homogenize the defining equations of W to obtain a projective variety $V = \mathbf{V}(g_1^h, \ldots, g_s^h) \subseteq \mathbb{P}^n(k)$. Moreover, by part (iii) and Proposition 6, we see that $V \cap U_0 = W$. Thus, our original affine variety W is the *affine portion* of the projective variety V.

As we mentioned before, though, there are some unexpected possibilities.

Example 8. In this example, we will write the homogeneous coordinates of points in $\mathbb{P}^2(k)$ as $(x:y:z)$. Numbering them as $0, 1, 2$, we see that U_2 is the set of points with homogeneous coordinates $(x:y:1)$, and x and y are coordinates on $U_2 \cong k^2$. Now consider the affine variety $W = \mathbf{V}(g) = \mathbf{V}(y - x^3 + x) \subseteq U_2$. We know that W is the affine portion $V \cap U_2$ of the projective variety $V = \mathbf{V}(g^h) = \mathbf{V}(yz^2 - x^3 + xz^2)$.

The variety V consists of W together with the points at infinity $V \cap \mathbf{V}(z)$. The affine portion W is the graph of a cubic polynomial, which is a nonsingular plane curve. The points at infinity, which form the complement of W in V, are given by the solutions of the equations

$$0 = yz^2 - x^3 + xz^2,$$
$$0 = z.$$

It is easy to see that the solutions are $z = x = 0$ and since we are working in $\mathbb{P}^2(k)$, we get the unique point $p = (0:1:0)$ in $V \cap \mathbf{V}(z)$. Thus, $V = W \cup \{p\}$. An unexpected feature of this example is the nature of the extra point p.

To see what V looks like at p, let us dehomogenize the equation of V with respect to y and study the intersection $V \cap U_1$. Since $g^h = yz^2 - x^3 + xz^2$, we find that

$$W' = V \cap U_1 = \mathbf{V}(g^h(x, 1, z)) = \mathbf{V}(z^2 - x^3 + xz^2).$$

From the discussion of singularities in §4 of Chapter 3, one can easily check that p, which becomes the point $(x, z) = (0, 0) \in W'$, is a singular point on W':

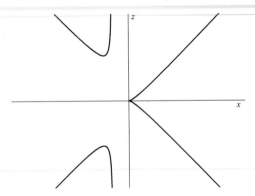

Thus, even if we start from a nonsingular affine variety (that is, one with no singular points), homogenizing the equations and taking the corresponding projective variety may yield a more complicated geometric object. In effect, we are not "seeing the whole picture" in the original affine portion of the variety. In general, given a projective variety $V \subseteq \mathbb{P}^n(k)$, since $\mathbb{P}^n(k) = \bigcup_{i=0}^{n} U_i$, we may need to consider $V \cap U_i$ for each $i = 0, \dots, n$ to see the whole picture of V.

Our next example shows that simply homogenizing the defining equations can lead to the "wrong" projective variety.

Example 9. Consider the affine twisted cubic $W = \mathbf{V}(x_2 - x_1^2, x_3 - x_1^3)$ in $\mathbb{R}^3$. By Proposition 7, $W = V \cap U_0$ for the projective variety $V = \mathbf{V}(x_2 x_0 - x_1^2, x_3 x_0^2 - x_1^3) \subseteq \mathbb{P}^3(\mathbb{R})$. As in Example 8, we can ask what part of V we are "missing" in the affine portion W. The complement of W in V is $V \cap H$, where $H = \mathbf{V}(x_0)$ is the plane at infinity. Thus, $V \cap H = \mathbf{V}(x_2 x_0 - x_1^2, x_3 x_0^2 - x_1^3, x_0)$, and one easily sees that these equations reduce to

$$x_1^2 = 0,$$
$$x_1^3 = 0,$$
$$x_0 = 0.$$

The coordinates x_2 and x_3 are arbitrary here, so $V \cap H$ is the projective line $\mathbf{V}(x_0, x_1) \subseteq \mathbb{P}^3(\mathbb{R})$. Thus we have $V = W \cup \mathbf{V}(x_0, x_1)$.

Since the twisted cubic W is a curve in $\mathbb{R}^3$, our intuition suggests that it should only have a finite number of points at infinity (in the exercises, you will see that this is indeed the case). This indicates that V is probably too big; there should be a smaller projective variety V' containing W. One way to create such a V' is to homogenize other polynomials that vanish on W. For example, the parametrization (t, t^2, t^3) of W shows that $x_1 x_3 - x_2^2 \in \mathbf{I}(W)$. Since $x_1 x_3 - x_2^2$ is already homogeneous, we can add it to the defining equations of V to get

$$V' = \mathbf{V}(x_2 x_0 - x_1^2, x_3 x_0^2 - x_1^3, x_1 x_3 - x_2^2) \subseteq V.$$

Then V' is a projective variety with the property that $V' \cap U_0 = W$, and in the exercises you will show that $V' \cap H$ consists of the single point $p = (0:0:0:1)$.

Thus, $V' = W \cup \{p\}$, so that we have a smaller projective variety that restricts to the twisted cubic. The difference between V and V' is that V has an extra component at infinity. In §4, we will show that V' is the smallest projective variety containing W.

In Example 9, the process by which we obtained V was completely straightforward (we homogenized the defining equations of W), yet it gave us a projective variety that was too big. This indicates that something more subtle is going on. The complete answer will come in §4, where we will learn an algorithm for finding the smallest projective variety containing $W \subseteq k^n \cong U_i$.

EXERCISES FOR §2

1. In this exercise, we will give a more geometric way to describe the construction of $\mathbb{P}^n(k)$. Let $\mathcal{L}$ denote the set of lines through the origin in k^{n+1}.
   a. Show that every element of $\mathcal{L}$ can be represented as the set of scalar multiples of some nonzero vector in k^{n+1}.
   b. Show that two nonzero vectors v' and v in k^{n+1} define the same element of $\mathcal{L}$ if and only if $v' \sim v$ as in Definition 1.
   c. Show that there is a one-to-one correspondence between $\mathbb{P}^n(k)$ and $\mathcal{L}$.

2. Complete the proof of Proposition 2 by showing that the mappings ϕ and ψ defined in the proof are inverses.

3. In this exercise, we will study how lines in $\mathbb{R}^n$ relate to points at infinity in $\mathbb{P}^n(\mathbb{R})$. We will use the decomposition $\mathbb{P}^n(\mathbb{R}) = \mathbb{R}^n \cup \mathbb{P}^{n-1}(\mathbb{R})$ given in (3). Given a line L in $\mathbb{R}^n$, we can parametrize L by the formula $a + bt$, where $a \in L$ and b is a nonzero vector parallel to L. In coordinates, we write this parametrization as $(a_1 + b_1 t, \ldots, a_n + b_n t)$.
   a. We can regard L as lying in $\mathbb{P}^n(\mathbb{R})$ using the homogeneous coordinates
   $$(1 : a_1 + b_1 t : \cdots : a_n + b_n t).$$
   To find out what happens as $t \to \pm\infty$, divide by t to obtain
   $$\left(\frac{1}{t} : \frac{a_1}{t} + b_1 : \cdots : \frac{a_n}{t} + b_n\right).$$
   As $t \to \pm\infty$, what point of $H = \mathbb{P}^{n-1}(\mathbb{R})$ do you get?
   b. The line L will have many parametrizations. Show that the point of $\mathbb{P}^{n-1}(\mathbb{R})$ given by part (a) is the same for all parametrizations of L. Hint: Two nonzero vectors are parallel if and only if one is a scalar multiple of the other.
   c. Parts (a) and (b) show that a line L in $\mathbb{R}^n$ has a well-defined point at infinity in $H = \mathbb{P}^{n-1}(\mathbb{R})$. Show that two lines in $\mathbb{R}^n$ are parallel if and only if they have the same point at infinity.

4. When $k = \mathbb{R}$ or $\mathbb{C}$, the projective line $\mathbb{P}^1(k)$ is easy to visualize.
   a. In the text, we called $\mathbb{P}^1(\mathbb{C}) = \mathbb{C} \cup \{\infty\}$ the Riemann sphere. To see why this name is justified, use the parametrization from Exercise 6 of Chapter 1, §3 to show how the plane corresponds to the sphere minus the north pole. Then explain why we can regard $\mathbb{C} \cup \{\infty\}$ as a sphere.
   b. What common geometric object can we use to represent $\mathbb{P}^1(\mathbb{R})$? Illustrate your reasoning with a picture.

5. Prove Corollary 3.

6. This problem studies the subsets $U_i \subseteq \mathbb{P}^n(k)$ from Corollary 3.
   a. In $\mathbb{P}^4(k)$, identify the points that are in the subsets U_2, $U_2 \cap U_3$, and $U_1 \cap U_3 \cap U_4$.

b. Give an identification of $\mathbb{P}^4(k) \setminus U_2$, $\mathbb{P}^4(k) \setminus (U_2 \cup U_3)$, and $\mathbb{P}^4(k) \setminus (U_1 \cup U_3 \cup U_4)$ as a "copy" of another projective space.

c. In $\mathbb{P}^4(k)$, which points are $\bigcap_{i=0}^{4} U_i$?

d. In general, describe the subset $U_{i_1} \cap \cdots \cap U_{i_s} \subseteq \mathbb{P}^n(k)$, where

$$1 \leq i_1 < i_2 < \cdots < i_s \leq n.$$

7. In this exercise, we will study when a nonhomogeneous polynomial has a well-defined zero set in $\mathbb{P}^n(k)$. Let k be an infinite field. We will show that if $f \in k[x_0, \ldots, x_n]$ is *not* homogeneous, but f vanishes on all homogeneous coordinates of some $p \in \mathbb{P}^n(k)$, then each of the homogeneous components f_i of f (see Definition 6 of Chapter 7, §1) must vanish at p.

a. Write f as a sum of its homogeneous components $f = \sum_i f_i$. If $p = (a_0, \ldots, a_n)$, then show that

$$f(\lambda a_0, \ldots, \lambda a_n) = \sum_i f_i(\lambda a_0, \ldots, \lambda a_n)$$

$$= \sum_i \lambda^i f_i(a_0, \ldots, a_n).$$

b. Deduce that if f vanishes for all $\lambda \neq 0 \in k$, then $f_i(a_0, \ldots, a_n) = 0$ for all i.

8. By dehomogenizing the defining equations of the projective variety V, find equations for the indicated affine varieties.

a. Let $\mathbb{P}^2(\mathbb{R})$ have homogeneous coordinates $(x:y:z)$ and let $V = \mathbf{V}(x^2 + y^2 - z^2) \subseteq \mathbb{P}^2(\mathbb{R})$. Find equations for $V \cap U_0$ and $V \cap U_2$. (Here U_0 is where $x \neq 0$ and U_2 is where $z \neq 0$.) Sketch each of these curves and think about what this says about the projective variety V.

b. $V = \mathbf{V}(x_0 x_2 - x_3 x_4, x_0^2 x_3 - x_1 x_2^2) \subseteq \mathbb{P}^4(k)$ and find equations for the affine variety $V \cap U_0 \subseteq k^4$. Do the same for $V \cap U_3$.

9. Let $V = \mathbf{V}(f_1, \ldots, f_s)$ be a projective variety defined by homogeneous polynomials $f_i \in k[x_0, \ldots, x_n]$. Show that the subset $W = V \cap U_i$, can be identified with the affine variety $\mathbf{V}(g_1, \ldots, g_s) \subseteq k^n$ defined by the dehomogenized polynomials

$$g_j(x_1, \ldots, x_i, x_{i+1}, \ldots, x_n) = f_j(x_1, \ldots, x_i, 1, x_{i+1}, \ldots, x_n), \quad j = 1, \ldots, s.$$

Hint: Follow the proof of Proposition 6, using Corollary 3.

10. Prove Proposition 7.

11. Using part (iv) of Proposition 7, show that if $f \in k[x_1, \ldots, x_n]$ and $F \in k[x_0, \ldots, x_n]$ is any homogeneous polynomial satisfying $F(1, x_1, \ldots, x_n) = f(x_1, \ldots, x_n)$, then $F = x_0^e f^h$ for some $e \geq 0$.

12. What happens if we apply the homogenization process of Proposition 7 to a polynomial g that is itself homogeneous?

13. In Example 8, we were led to consider the variety $W' = \mathbf{V}(z^2 - x^3 + xz^2) \subseteq k^2$. Show that $(x, z) = (0, 0)$ is a singular point of W'. Hint: Use Definition 3 from Chapter 3, §4.

14. For each of the following affine varieties W, apply the homogenization process given in Proposition 7 to write $W = V \cap U_0$, where V is a projective variety. In each case identify $V \setminus W = V \cap H$, where H is the hyperplane at infinity.

a. $W = \mathbf{V}(y^2 - x^3 - ax - b) \subseteq \mathbb{R}^2$, $a, b \in \mathbb{R}$. Is the point $V \cap H$ singular here? Hint: Let the homogeneous coordinates on $\mathbb{P}^2(\mathbb{R})$ be $(z:x:y)$, so that U_0 is where $z \neq 0$.

b. $W = \mathbf{V}(x_1 x_3 - x_2^2, x_1^2 - x_2) \subseteq \mathbb{R}^3$. Is there an extra component at infinity here?

c. $W = \mathbf{V}(x_3^2 - x_1^2 - x_2^2) \subseteq \mathbb{R}^3$.

15. From Example 9, consider the twisted cubic $W = \mathbf{V}(x_2 - x_1^2, x_3 - x_1^3) \subseteq \mathbb{R}^3$.

a. If we parametrize W by (t, t^2, t^3) in $\mathbb{R}^3$, show that as $t \to \pm\infty$, the point $(1 : t : t^2 : t^3)$ in $\mathbb{P}^3(\mathbb{R})$ approaches $(0 : 0 : 0 : 1)$. Thus, we expect W to have one point at infinity.

b. Now consider the projective variety

$$V' = \mathbf{V}(x_2 x_0 - x_1^2, x_3 x_0^2 - x_1^3, x_1 x_3 - x_2^2) \subseteq \mathbb{P}^3(\mathbb{R}).$$

Show that $V' \cap U_0 = W$ and that $V' \cap H = \{(0 : 0 : 0 : 1)\}$.

c. Let $V = \mathbf{V}(x_2 x_0 - x_1^2, x_3 x_0^2 - x_1^3)$ be as in Example 9. Prove that $V = V' \cup \mathbf{V}(x_0, x_1)$. This shows that V is a union of two proper projective varieties.

16. A homogeneous polynomial $f \in k[x_0, \dots, x_n]$ can also be used to define the *affine* variety $C = \mathbf{V}_a(f)$ in k^{n+1}, where the subscript denotes we are working in affine space. We call C the *affine cone* over the projective variety $V = \mathbf{V}(f) \subseteq \mathbb{P}^n(k)$. We will see why this is so in this exercise.

a. Show that if C contains the point $P \neq (0, \dots, 0)$, then C contains the whole line through the origin in k^{n+1} spanned by P.

b. A point $P \in k^{n+1} \setminus \{0\}$ gives homogeneous coordinates for a point $p \in \mathbb{P}^n(k)$. Show that p is in the projective variety V if and only if the line through the origin determined by P is contained in C. Hint: See (1) and Exercise 1.

c. Deduce that C is the union of the collection of lines through the origin in k^{n+1} corresponding to the points in V via (1). This explains the reason for the "cone" terminology since an ordinary cone is also a union of lines through the origin. Such a cone is given by part (c) of Exercise 14.

17. Homogeneous polynomials satisfy an important relation known as *Euler's Formula*. Let $f \in k[x_0, \dots, x_n]$ be homogeneous of total degree d. Then Euler's Formula states that

$$\sum_{i=0}^{n} x_i \cdot \frac{\partial f}{\partial x_i} = d \cdot f.$$

a. Verify Euler's Formula for the homogeneous polynomial $f = x_0^3 - x_1 x_2^2 + 2 x_1 x_3^2$.

b. Prove Euler's Formula (in the case $k = \mathbb{R}$) by considering $f(\lambda x_0, \dots, \lambda x_n)$ as a function of λ and differentiating with respect to λ using the chain rule.

18. In this exercise, we will consider the set of hyperplanes in $\mathbb{P}^n(k)$ in greater detail.

a. Show that two homogeneous linear polynomials,

$$0 = a_0 x_0 + \cdots + a_n x_n,$$
$$0 = b_0 x_0 + \cdots + b_n x_n,$$

define the same hyperplane in $\mathbb{P}^n(k)$ if and only if there is $\lambda \neq 0$ in k such that $b_i = \lambda a_i$ for all $i = 0, \dots, n$. Hint: Generalize the argument given for Exercise 11 of §1.

b. Show that the map sending the hyperplane with equation $a_0 x_0 + \cdots + a_n x_n = 0$ to the vector $(a_0, \dots, a_n)$ gives a one-to-one correspondence

$$\phi : \{\text{hyperplanes in } \mathbb{P}^n(k)\} \to (k^{n+1} \setminus \{0\})/\sim,$$

where $\sim$ is the equivalence relation of Definition 1. The set on the left is called the *dual projective space* and is denoted $\mathbb{P}^n(k)^\vee$. Geometrically, the points of $\mathbb{P}^n(k)^\vee$ are hyperplanes in $\mathbb{P}^n(k)$.

c. Describe the subset of $\mathbb{P}^n(k)^\vee$ corresponding to the hyperplanes containing $p = (1 : 0 : \cdots : 0)$.

19. Let k be an algebraically closed field ($\mathbb{C}$, for example). Show that every homogeneous polynomial $f(x_0, x_1)$ in two variables with coefficients in k can be completely factored into linear homogeneous polynomials in $k[x_0, x_1]$:

$$f(x_0, x_1) = \prod_{i=1}^{d} (a_i x_0 + b_i x_1),$$

where d is the total degree of f. Hint: First dehomogenize f.

20. In §4 of Chapter 5, we introduced the *pencil* defined by two hypersurfaces $V = \mathbf{V}(f)$, $W = \mathbf{V}(g)$. The elements of the pencil were the hypersurfaces $\mathbf{V}(f + cg)$ for $c \in k$. Setting $c = 0$, we obtain V as an element of the pencil. However, W is not (usually) an element of the pencil when it is defined in this way. To include W in the pencil, we can proceed as follows.

   a. Let $(a : b)$ be homogeneous coordinates in $\mathbb{P}^1(k)$. Show that $\mathbf{V}(af + bg)$ is well-defined in the sense that all homogeneous coordinates $(a : b)$ for a given point in $\mathbb{P}^1(k)$ yield the same variety $\mathbf{V}(af + bg)$. Thus, we obtain a family of varieties parametrized by $\mathbb{P}^1(k)$, which is also called the *pencil* of varieties defined by V and W.

   b. Show that both V and W are contained in the pencil $\mathbf{V}(af + bg)$.

   c. Let $k = \mathbb{C}$. Show that every affine curve $\mathbf{V}(f) \subseteq \mathbb{C}^2$ defined by a polynomial f of total degree d is contained in a pencil of curves $\mathbf{V}(aF + bG)$ parametrized by $\mathbb{P}^1(\mathbb{C})$, where $\mathbf{V}(F)$ is a union of lines and G is a polynomial of degree strictly less than d. Hint: Consider the homogeneous components of f. Exercise 19 will be useful.

21. When we have a curve parametrized by $t \in k$, there are many situations where we want to let $t \to \infty$. Since $\mathbb{P}^1(k) = k \cup \{\infty\}$, this suggests that we should let our parameter space be $\mathbb{P}^1(k)$. Here are two examples of how this works.

   a. Consider the parametrization $(x, y) = ((1 + t^2)/(1 - t^2), 2t/(1 - t^2))$ of the hyperbola $x^2 - y^2 = 1$ in $\mathbb{R}^2$. To make this projective, we first work in $\mathbb{P}^2(\mathbb{R})$ and write the parametrization as

   $$((1 + t^2)/(1 - t^2) : 2t/(1 - t^2) : 1) = (1 + t^2 : 2t : 1 - t^2)$$

   (see Exercise 3 of §1). The next step is to make t projective. Given $(a : b) \in \mathbb{P}^1(\mathbb{R})$, we can write it as $(1 : t) = (1 : b/a)$ provided $a \neq 0$. Now substitute $t = b/a$ into the right-hand side and clear denominators. Explain why this gives a well-defined map $\mathbb{P}^1(\mathbb{R}) \to \mathbb{P}^2(\mathbb{R})$.

   b. The twisted cubic in $\mathbb{R}^3$ is parametrized by (t, t^2, t^3). Apply the method of part (a) to obtain a projective parametrization $\mathbb{P}^1(\mathbb{R}) \to \mathbb{P}^3(\mathbb{R})$ and show that the image of this map is precisely the projective variety V' from Example 9.

§3 The Projective Algebra–Geometry Dictionary

In this section, we will study the algebra–geometry dictionary for projective varieties. Our goal is to generalize the theorems from Chapter 4 concerning the $\mathbf{V}$ and $\mathbf{I}$ correspondences to the projective case, and, in particular, we will prove a projective version of the Nullstellensatz.

To begin, we note one difference between the affine and projective cases on the algebraic side of the dictionary. Namely, in Definition 5 of §2, we introduced projective varieties as the common zeros of collections of *homogeneous* polynomials. But being homogeneous is *not* preserved under the sum operation in $k[x_0, \ldots, x_n]$. For example, if we add two homogeneous polynomials of different total degrees, the sum will *never* be homogeneous. Thus, if we form the ideal

$I = \langle f_1, \ldots, f_s \rangle \subseteq k[x_0, \ldots, x_n]$ generated by a collection of homogeneous polynomials, I will contain many *non-homogeneous* polynomials and these would not be candidates for the defining equations of a projective variety.

Nevertheless, each element of I vanishes on all homogeneous coordinates of every point of $V = \mathbf{V}(f_1, \ldots, f_s)$. This follows because each $g \in I$ has the form

$$(1) \qquad g = \sum_{j=1}^{s} A_j f_j$$

for some $A_j \in k[x_0, \ldots, x_n]$. Substituting any homogeneous coordinates of a point in V into g will yield zero since each f_i is zero there.

A more important observation concerns the homogeneous components of g. Suppose we expand each A_j as the sum of its homogeneous components:

$$A_j = \sum_{i=1}^{d} A_{ji}.$$

If we substitute these expressions into (1) and collect terms of the same total degree, it can be shown that the homogeneous components of g also lie in the ideal $I = \langle f_1, \ldots, f_s \rangle$. You will prove this claim in Exercise 2.

Thus, although I contains nonhomogeneous elements g, we see that I also contains the homogeneous components of g. This observation motivates the following definition of a special class of ideals in $k[x_0, \ldots, x_n]$.

Definition 1. An ideal I in $k[x_0, \ldots, x_n]$ is said to be **homogeneous** if for each $f \in I$, the homogeneous components f_i of f are in I as well.

Most ideals *do not* have this property. For instance, let $I = \langle y - x^2 \rangle \subseteq k[x, y]$. The homogeneous components of $f = y - x^2$ are $f_1 = y$ and $f_2 = -x^2$. Neither of these polynomials is in I since neither is a multiple of $y - x^2$. Hence, I is not a homogeneous ideal. However, we have the following useful characterization of when an ideal is homogeneous.

Theorem 2. *Let $I \subseteq k[x_0, \ldots, x_n]$ be an ideal. Then the following are equivalent:*

(i) *I is a homogeneous ideal of $k[x_0, \ldots, x_n]$.*

(ii) *$I = \langle f_1, \ldots, f_s \rangle$, where $f_1, \ldots, f_s$ are homogeneous polynomials.*

(iii) *A reduced Gröbner basis of I (with respect to any monomial ordering) consists of homogeneous polynomials.*

Proof. The proof of (ii) $\Rightarrow$ (i) was sketched above (see also Exercise 2). To prove (i) $\Rightarrow$ (ii), let I be a homogeneous ideal. By the Hilbert Basis Theorem, we have $I = \langle F_1, \ldots, F_t \rangle$ for some polynomials $F_j \in k[x_0, \ldots, x_n]$ (not necessarily homogeneous). If we write F_j as the sum of its homogeneous components, say $F_j = \sum_i F_{ji}$, then each $F_{ji} \in I$ since I is homogeneous. Let I' be the ideal generated by the homogeneous polynomials F_{ji}. Then $I \subseteq I'$ since each F_j is a sum of generators of I'. On the other hand, $I' \subseteq I$ since each of the homogeneous components of F_j is in I. This proves $I = I'$ and it follows that I has a basis of homogeneous polynomials. Finally, the equivalence (ii) $\Leftrightarrow$ (iii) will be covered in Exercise 3. $\square$

As a result of Theorem 2, for any homogeneous ideal $I \subseteq k[x_0, \ldots, x_n]$ we may define

$$\mathbf{V}(I) = \{p \in \mathbb{P}^n(k) \mid f(p) = 0 \text{ for all } f \in I\},$$

as in the affine case. We can prove that $\mathbf{V}(I)$ is a projective variety as follows.

Proposition 3. *Let* $I \subseteq k[x_0, \ldots, x_n]$ *be a homogeneous ideal and suppose that* $I = \langle f_1, \ldots, f_s \rangle$, *where* $f_1, \ldots, f_s$ *are homogeneous. Then*

$$\mathbf{V}(I) = \mathbf{V}(f_1, \ldots, f_s),$$

so that $\mathbf{V}(I)$ *is a projective variety.*

Proof. We leave the easy proof as an exercise. $\square$

One way to create a homogeneous ideal is to consider the ideal generated by the defining equations of a projective variety. But there is another way that a projective variety can give us a homogeneous ideal.

Proposition 4. *Let* $V \subseteq \mathbb{P}^n(k)$ *be a projective variety and let*

$$\mathbf{I}(V) = \{f \in k[x_0, \ldots, x_n] \mid f(a_0, \ldots, a_n) = 0 \text{ for all } (a_0 : \cdots : a_n) \in V\}.$$

(This means that f *must be zero for all homogeneous coordinates of all points in* V.) *If* k *is infinite, then* $\mathbf{I}(V)$ *is a homogeneous ideal in* $k[x_0, \ldots, x_n]$.

Proof. The set $\mathbf{I}(V)$ is closed under sums and closed under products by elements of $k[x_0, \ldots, x_n]$ by an argument exactly parallel to the one for the affine case. Thus, $\mathbf{I}(V)$ is an ideal in $k[x_0, \ldots, x_n]$. Now take $f \in \mathbf{I}(V)$ and a point $p \in V$. By assumption, f vanishes at all homogeneous coordinates $(a_0 : \cdots : a_n)$ of p. Since k is infinite, Exercise 7 of §2 implies that each homogeneous component f_i of f vanishes at $(a_0 : \cdots : a_n)$. This shows that $f_i \in \mathbf{I}(V)$ and, hence, $\mathbf{I}(V)$ is homogeneous. $\square$

Thus, we have all the ingredients of a dictionary relating projective varieties in $\mathbb{P}^n(k)$ and homogeneous ideals in $k[x_0, \ldots, x_n]$. The following theorem is a direct generalization of part (i) of Theorem 7 of Chapter 4, §2 (the affine ideal–variety correspondence).

Theorem 5. *Let* k *be an infinite field. Then the maps*

$$\text{projective varieties} \xrightarrow{\mathbf{I}} \text{homogeneous ideals}$$

and

$$\text{homogeneous ideals} \xrightarrow{\mathbf{V}} \text{projective varieties}$$

are inclusion-reversing. Furthermore, for any projective variety V, *we have*

$$\mathbf{V}(\mathbf{I}(V)) = V.$$

so that **I** *is always one-to-one.*

Proof. The proof is the same as in the affine case.   □

To illustrate the use of this theorem, let us show that every projective variety can be decomposed to irreducible components. As in the affine case, a variety $V \subseteq \mathbb{P}^n(k)$ is *irreducible* if it cannot be written as a union of two strictly smaller projective varieties.

Theorem 6. *Let k be an infinite field.*

(i) *Given a descending chain of projective varieties in* $\mathbb{P}^n(k)$,

$$V_1 \supseteq V_2 \supseteq V_3 \supseteq \cdots,$$

there is an integer N such that $V_N = V_{N+1} = \cdots$.

(ii) *Every projective variety* $V \subseteq \mathbb{P}^n(k)$ *can be written uniquely as a finite union of irreducible projective varieties*

$$V = V_1 \cup \cdots \cup V_m,$$

where $V_i \not\subseteq V_j$ *for* $i \neq j$.

Proof. Since **I** is inclusion-reversing, we get the ascending chain of homogeneous ideals

$$\mathbf{I}(V_1) \subseteq \mathbf{I}(V_2) \subseteq \mathbf{I}(V_3) \subseteq \cdots$$

in $k[x_0, \ldots, x_n]$. Then the ascending chain condition (Theorem 7 of Chapter 2, §5) implies that $\mathbf{I}(V_N) = \mathbf{I}(V_{N+1}) = \cdots$ for some N. By Theorem 5, **I** is one-to-one and (i) follows immediately.

As in the affine case, (ii) is an immediate consequence of (i). See Theorems 2 and 4 of Chapter 4, §6.   □

The relation between operations such as sums, products, and intersections of homogeneous ideals and the corresponding operations on projective varieties is also the same as in affine space. We will consider these topics in more detail in the exercises below.

We define the radical of a homogeneous ideal as usual:

$$\sqrt{I} = \{f \in k[x_0, \ldots, x_n] \mid f^m \in I \text{ for some } m \geq 1\}.$$

As we might hope, the radical of a homogeneous ideal is always itself homogeneous.

Proposition 7. *Let* $I \subseteq k[x_0, \ldots, x_n]$ *be a homogeneous ideal. Then* $\sqrt{I}$ *is also a homogeneous ideal.*

Proof. If $f \in \sqrt{I}$, then $f^m \in I$ for some $m \geq 1$. If $f \neq 0$, decompose f into its homogeneous components

$$f = \sum_i f_i = f_{max} + \sum_{i<max} f_i,$$

where f_{max} is the nonzero homogeneous component of maximal total degree in f. Expanding the power f^m, it is easy to show that

$$(f^m)_{max} = (f_{max})^m.$$

Since I is a homogeneous ideal, $(f^m)_{max} \in I$. Hence, $(f_{max})^m \in I$, which shows that $f_{max} \in \sqrt{I}$.

If we let $g = f - f_{max} \in \sqrt{I}$ and repeat the argument, we get $g_{max} \in \sqrt{I}$. But g_{max} is also one of the homogeneous components of f. Applying this reasoning repeatedly shows that all homogeneous components of f are in $\sqrt{I}$. Since this is true for all $f \in \sqrt{I}$, Definition 1 implies that $\sqrt{I}$ is a homogeneous ideal.   □

The final part of the algebra–geometry dictionary concerns what happens over an algebraically closed field k. Here, we expect an especially close relation between projective varieties and homogeneous ideals. In the affine case, the link was provided by two theorems proved in Chapter 4, the Weak Nullstellensatz and the Strong Nullstellensatz. Let us recall what these theorems tell us about an ideal $I \subseteq k[x_1, \ldots, x_n]$:

- (The Weak Nullstellensatz) $\mathbf{V}_a(I) = \emptyset$ in $k^n \iff I = k[x_1, \ldots, x_n]$.
- (The Strong Nullstellensatz) $\sqrt{I} = \mathbf{I}_a(\mathbf{V}_a(I))$ in $k[x_1, \ldots, x_n]$.

(To prevent confusion, we use $\mathbf{I}_a$ and $\mathbf{V}_a$ for the affine versions of $\mathbf{I}$ and $\mathbf{V}$.) It is natural to ask if these results extend to projective varieties and homogeneous ideals.

The answer, surprisingly, is *no*. In particular, the Weak Nullstellensatz fails for certain homogeneous ideals. To see how this can happen, consider the ideal $I = \langle x_0, \ldots, x_n \rangle \subseteq \mathbb{C}[x_0, \ldots, x_n]$. Then $\mathbf{V}(I) \subseteq \mathbb{P}^n(\mathbb{C})$ is defined by the equations

$$x_0 = \cdots = x_n = 0,$$

which have no solutions in $\mathbb{P}^n(\mathbb{C})$ since we never allow all homogeneous coordinates to vanish simultaneously. It follows that $\mathbf{V}(I) = \emptyset$, yet $I \neq \mathbb{C}[x_0, \ldots, x_n]$.

Fortunately, $I = \langle x_0, \ldots, x_n \rangle$ is one of the few ideals for which $\mathbf{V}(I) = \emptyset$. The following projective version of the Weak Nullstellensatz describes *all* homogeneous ideals with no projective solutions.

Theorem 8 (The Projective Weak Nullstellensatz). *Let k be algebraically closed and let I be a homogeneous ideal in $k[x_0, \ldots, x_n]$. Then the following are equivalent:*

(i) *$\mathbf{V}(I) \subseteq \mathbb{P}^n(k)$ is empty.*

(ii) *Let G be a reduced Gröbner basis for I (with respect to some monomial ordering). Then for each $0 \leq i \leq n$, there is $g \in G$ such that $\mathrm{LT}(g)$ is a nonnegative power of x_i.*

(iii) *For each $0 \leq i \leq n$, there is an integer $m_i \geq 0$ such that $x_i^{m_i} \in I$.*

(iv) *There is some $r \geq 1$ such that $\langle x_0, \ldots, x_n \rangle^r \subseteq I$.*

(v) *$I : \langle x_0, \ldots, x_n \rangle^\infty = k[x_1, \ldots, x_n]$.*

Proof. The ideal I gives us the projective variety $V = \mathbf{V}(I) \subseteq \mathbb{P}^n(k)$. In this proof, we will also work with the *affine* variety $C_V = \mathbf{V}_a(I) \subseteq k^{n+1}$. Note that C_V uses the same ideal I, but now we look for solutions in the affine space k^{n+1}. We call C_V the *affine cone* of V. If we interpret points in $\mathbb{P}^n(k)$ as lines through the origin in k^{n+1}, then C_V is the union of the lines determined by the points of V (see Exercise 16 of §2 for the details of how this works). In particular, C_V contains all homogeneous coordinates of the points in V.

To prove (ii) $\Rightarrow$ (i), first suppose that we have a Gröbner basis where, for each i, there is $g \in G$ with $\mathrm{LT}(g) = x_i^{m_i}$ for some $m_i \geq 0$. Then Theorem 6 of Chapter 5, §3 implies that C_V is a finite set. But suppose there is a point $p \in V$. Then *all* homogeneous coordinates of p lie in C_V. If we write these in the form $\lambda(a_0, \ldots, a_n)$, we see that there are infinitely many since k is algebraically closed and hence infinite. This contradiction shows that $V = \mathbf{V}(I) = \emptyset$.

Turning to (iii) $\Rightarrow$ (ii), let G be a reduced Gröbner basis for I. Then $x_i^{m_i} \in I$ implies that the leading term of some $g \in G$ divides $x_i^{m_i}$, so that $\mathrm{LT}(g)$ must be a power of x_i.

For the remainder of the proof, let $J = \langle x_0, \ldots, x_n \rangle$. The proof of (iv) $\Rightarrow$ (iii) is obvious since $J^r \subseteq I$ implies that $x_i^r \in I$ for all i. For (v) $\Rightarrow$ (iv), we use results about ideal quotients and saturations from Chapter 4 to obtain

$$I : J^{\infty} = k[x_1, \ldots, x_n] \implies J \subseteq \sqrt{I} \implies J^r \subseteq I \text{ for some } r > 0.$$

The first implication uses Proposition 12 of Chapter 4, §4, and the second uses Exercise 12 of Chapter 4, §3.

Finally, since $\mathbf{V}_a(J) = \{0\}$, (i) $\Leftrightarrow$ (v) follows from the equivalences

$$\mathbf{V}(I) = \emptyset \iff \mathbf{V}_a(I) \subseteq \{0\} \iff \mathbf{V}_a(I) \setminus \mathbf{V}_a(J) = \emptyset$$
$$\iff \overline{\mathbf{V}_a(I) \setminus \mathbf{V}_a(J)} = \emptyset$$
$$\iff \mathbf{V}_a(I : J^{\infty}) = \emptyset$$
$$\iff I : J^{\infty} = k[x_1, \ldots, x_n],$$

where the fourth equivalence uses Theorem 10 of Chapter 4, §4 and the fifth uses the Nullstellensatz. This completes the proof of the theorem. □

From part (ii) of the theorem, we get an algorithm for determining if a homogeneous ideal has projective solutions over an algebraically closed field. In Exercise 10, we will discuss other conditions equivalent to $\mathbf{V}(I) = \emptyset$ in $\mathbb{P}^n(k)$.

Once we exclude the ideals described in Theorem 8, we get the following form of the Nullstellensatz for projective varieties.

Theorem 9 (The Projective Strong Nullstellensatz). *Let k be an algebraically closed field and let I be a homogeneous ideal in $k[x_0, \ldots, x_n]$. If $V = \mathbf{V}(I)$ is a nonempty projective variety in $\mathbb{P}^n(k)$, then we have*

$$\mathbf{I}(\mathbf{V}(I)) = \sqrt{I}.$$

Proof. As in the proof of Theorem 8, we will work with the projective variety $V = \mathbf{V}(I) \subseteq \mathbb{P}^n(k)$ and its affine cone $C_V = \mathbf{V}_a(I) \subseteq k^{n+1}$. We first claim that

$$(2) \qquad\qquad\qquad \mathbf{I}_a(C_V) = \mathbf{I}(V)$$

when $V \neq \emptyset$. To see this, suppose that $f \in \mathbf{I}_a(C_V)$. Given $p \in V$, any homogeneous coordinates of p give a point in C_V. Since f vanishes on C_V, it follows that $f(p) = 0$. By definition, this implies $f \in \mathbf{I}(V)$. Conversely, take $f \in \mathbf{I}(V)$. Since any nonzero point of C_V gives homogeneous coordinates for a point in V, it follows that f vanishes on $C_V \setminus \{0\}$. It remains to show that f vanishes at the origin. Since $\mathbf{I}(V)$ is a homogeneous ideal, we know that the homogeneous components f_i of f also vanish on V. In particular, the constant term of f, which is the homogeneous component f_0 of total degree 0, must vanish on V. Since $V \neq \emptyset$, this forces $f_0 = 0$, which means that f vanishes at the origin. Hence, $f \in \mathbf{I}_a(C_V)$ and (2) is proved.

By the affine form of the Strong Nullstellensatz, we know that $\sqrt{I} = \mathbf{I}_a(\mathbf{V}_a(I))$. Then, using (2), we obtain

$$\sqrt{I} = \mathbf{I}_a(\mathbf{V}_a(I)) = \mathbf{I}_a(C_V) = \mathbf{I}(V) = \mathbf{I}(\mathbf{V}(I)).$$

which completes the proof of the theorem. $\qquad\square$

Now that we have the Nullstellensatz, we can complete the projective ideal–variety correspondence begun in Theorem 5. A radical homogeneous ideal in $k[x_0, \ldots, x_n]$ is a homogeneous ideal satisfying $\sqrt{I} = I$. As in the affine case, we have a one-to-one correspondence between projective varieties and radical homogeneous ideals, provided we exclude the cases $\sqrt{I} = \langle x_0, \ldots, x_n \rangle$ and $\sqrt{I} = \langle 1 \rangle$.

Theorem 10. *Let k be an algebraically closed field. If we restrict the correspondences of Theorem 5 to nonempty projective varieties and radical homogeneous ideals properly contained in $\langle x_0, \ldots, x_n \rangle$, then*

$$\{\text{nonempty projective varieties}\} \overset{\mathbf{I}}{\longrightarrow} \left\{ \begin{array}{c} \text{radical homogeneous ideals} \\ \text{properly contained in} \langle x_0, \ldots, x_n \rangle \end{array} \right\}$$

and

$$\left\{ \begin{array}{c} \text{radical homogeneous ideals} \\ \text{properly contained in} \langle x_0, \ldots, x_n \rangle \end{array} \right\} \overset{\mathbf{V}}{\longrightarrow} \{\text{nonempty projective varieties}\}$$

are inclusion-reversing bijections which are inverses of each other.

Proof. First, it is an easy consequence of Theorem 8 that the only radical homogeneous ideals I with $\mathbf{V}(I) = \emptyset$ are $\langle x_0, \ldots, x_n \rangle$ and $k[x_0, \ldots, x_n]$. See Exercise 10 for the details. A second observation is that if I is a homogeneous ideal different from $k[x_0, \ldots, x_n]$, then $I \subseteq \langle x_0, \ldots, x_n \rangle$. This will also be covered in Exercise 9.

These observations show that the radical homogeneous ideals with $\mathbf{V}(I) \neq \emptyset$ are precisely those which satisfy $I \subsetneq \langle x_0, \ldots, x_n \rangle$. Then the rest of the theorem follows as in the affine case, using Theorem 9. $\qquad\square$

We also have a correspondence between *irreducible* projective varieties and homogeneous *prime* ideals, which will be studied in the exercises. In the exercises we will also explore the *field of rational functions* of an irreducible projective variety.

EXERCISES FOR §3

1. In this exercise, you will study the question of determining when a principal ideal $I = \langle f \rangle$ is homogeneous by elementary methods.
   a. Show that $I = \langle x^2 y - x^3 \rangle$ is a homogeneous ideal in $k[x, y]$ without appealing to Theorem 2. Hint: Each element of I has the form $g = A \cdot (x^2 y - x^3)$. Write A as the sum of its homogeneous components and use this to determine the homogeneous components of g.
   b. Show that $\langle f \rangle \subseteq k[x_0, \dots, x_n]$ is a homogeneous ideal if and only if f is a homogeneous polynomial without using Theorem 2.

2. This exercise gives some useful properties of the homogeneous components of polynomials.
   a. Show that if $f = \sum_i f_i$ and $g = \sum_i g_i$ are the expansions of two polynomials as the sums of their homogeneous components, then $f = g$ if and only if $f_i = g_i$ for all i.
   b. Show that if $f = \sum_i f_i$ and $g = \sum_j g_j$ are the expansions of two polynomials as the sums of their homogeneous components, then the homogeneous components h_ℓ of the product $h = f \cdot g$ are given by $h_\ell = \sum_{i+j=\ell} f_i \cdot g_j$.
   c. Use parts (a) and (b) to carry out the proof (sketched in the text) of the implication (ii) $\Rightarrow$ (i) from Theorem 2.

3. This exercise will study how the algorithms of Chapter 2 interact with homogeneous polynomials.
   a. Suppose we use the division algorithm to divide a homogeneous polynomial f by homogeneous polynomials $f_1, \dots, f_s$. This gives an expression of the form $f = a_1 f_1 + \dots + a_s f_s + r$. Prove that the quotients $a_1, \dots, a_s$ and remainder r are homogeneous polynomials (possibly zero). What is the total degree of r?
   b. If f, g are homogeneous polynomials, prove that the S-polynomial $S(f, g)$ is homogeneous.
   c. By analyzing the Buchberger algorithm, show that a homogeneous ideal has a homogeneous Gröbner basis.
   d. Prove the implication (ii) $\Leftrightarrow$ (iii) of Theorem 2.

4. Suppose that an ideal $I \subseteq k[x_0, \dots, x_n]$ has a basis G consisting of homogeneous polynomials.
   a. Prove that G is a Gröbner basis for I with respect to lex order if and only if it is a Gröbner basis for I with respect to grlex (assuming that the variables are ordered the same way).
   b. Conclude that, for a homogeneous ideal, the reduced Gröbner basis for lex and grlex are the same.

5. Prove Proposition 3.

6. In this exercise we study the algebraic operations on ideals introduced in Chapter 4 for homogeneous ideals. Let $I_1, \dots, I_l$ be homogeneous ideals in $k[x_0, \dots, x_n]$.
   a. Show that the ideal sum $I_1 + \dots + I_l$ is also homogeneous. Hint: Use Theorem 2.
   b. Show that the intersection $I_1 \cap \dots \cap I_l$ is also a homogeneous ideal.
   c. Show that the ideal product $I_1 \cdots I_l$ is a homogeneous ideal.

7. The interaction between the algebraic operations on ideals in Exercise 6 and the corresponding operations on projective varieties is the same as in the affine case. Let $I_1, \dots, I_l$ be homogeneous ideals in $k[x_0, \dots, x_n]$ and let $V_i = \mathbf{V}(I_i)$ be the corresponding projective variety in $\mathbb{P}^n(k)$.
   a. Show that $\mathbf{V}(I_1 + \dots + I_l) = \bigcap_{i=1}^l V_i$.

b. Show that $\mathbf{V}(I_1 \cap \cdots \cap I_l) = \mathbf{V}(I_1 \cdots I_l) = \bigcup_{i=1}^l V_i$.

8. Let $f_1, \ldots, f_s$ be homogeneous polynomials of total degrees $d_1 < d_2 \leq \cdots \leq d_s$ and let $I = \langle f_1, \ldots, f_s \rangle \subseteq k[x_0, \ldots, x_n]$.

   a. Show that if g is another homogeneous polynomial of degree d_1 in I, then g must be a constant multiple of f_1. Hint: Use parts (a) and (b) of Exercise 2.

   b. More generally, show that if the total degree of g is d, then g must be an element of the ideal $I_d = \langle f_i \mid \deg(f_i) \leq d \rangle \subseteq I$.

9. This exercise will study some properties of the ideal $I_0 = \langle x_0, \ldots, x_n \rangle \subseteq k[x_0, \ldots, x_n]$.

   a. Show that every proper homogeneous ideal in $k[x_0, \ldots, x_n]$ is contained in I_0.

   b. Show that the r-th power I_0^r is the ideal generated by the collection of monomials in $k[x_0, \ldots, x_n]$ of total degree exactly r and deduce that every homogeneous polynomial of degree $\geq r$ is in I_0^r.

   c. Let $V = \mathbf{V}(I_0) \subseteq \mathbb{P}^n(k)$ and $C_V = \mathbf{V}_a(I_0) \subseteq k^{n+1}$. Show that $\mathbf{I}_a(C_V) \neq \mathbf{I}(V)$, and explain why this does not contradict equation (2) in the text.

10. Given a homogeneous ideal $I \subseteq k[x_0, \ldots, x_n]$, where k is algebraically closed, prove that $\mathbf{V}(I) = \emptyset$ in $\mathbb{P}^n(k)$ is equivalent to either of the following two conditions:

    (i) There is an $r \geq 1$ such that every homogeneous polynomial of total degree $\geq r$ is contained in I.

    (ii) The radical of I is either $\langle x_0, \ldots, x_n \rangle$ or $k[x_0, \ldots, x_n]$.

    Hint: For (i), use Exercise 9, and for (ii), note that the inclusion $\langle x_0, \ldots, x_n \rangle \subseteq \sqrt{I}$ appears in the proof of Theorem 8.

11. A homogeneous ideal is said to be *prime* if it is prime as an ideal in $k[x_0, \ldots, x_n]$.

    a. Show that a homogeneous ideal $I \subseteq k[x_0, \ldots, x_n]$ is prime if and only if whenever the product of two *homogeneous* polynomials F, G satisfies $F \cdot G \in I$, then $F \in I$ or $G \in I$.

    b. Let k be algebraically closed. Let I be a homogeneous ideal. Show that the projective variety $\mathbf{V}(I)$ is irreducible if I is prime. Also, when I is radical, prove that the converse holds, i.e., that I is prime if $\mathbf{V}(I)$ is irreducible. Hint: Consider the proof of the corresponding statement in the affine case (Proposition 3 of Chapter 4, §5).

    c. Let k be algebraically closed. Show that the mappings $\mathbf{V}$ and $\mathbf{I}$ induce one-to-one correspondence between homogeneous prime ideals in $k[x_0, \ldots, x_n]$ properly contained in $\langle x_0, \ldots, x_n \rangle$ and nonempty irreducible projective varieties in $\mathbb{P}^n(k)$.

12. Prove that a homogeneous prime ideal is a radical ideal in $k[x_0, \ldots, x_n]$.

13. In this exercise, we will show how to define the field of rational functions on an irreducible projective variety $V \subseteq \mathbb{P}^n(k)$. If we take a homogeneous polynomial $f \in k[x_0, \ldots, x_n]$, then f does *not* give a well-defined function on V. To see why, let $p = (a_0 : \cdots : a_n) \in V$. Then we also have $p = (\lambda a_0 : \cdots : \lambda a_n)$ for any $\lambda \in k \setminus \{0\}$, and

$$f(\lambda a_0, \ldots, \lambda a_n) = \lambda^d f(a_0, \ldots, a_n),$$

where d is the total degree of f.

    a. Explain why the above equation makes it impossible for us to define $f(p)$ as a single-valued function on V.

    b. If $g \in k[x_0, \ldots, x_n]$ is also homogeneous of total degree d and $g \notin \mathbf{I}(V)$, then show that $\phi = f/g$ is a well-defined function on the nonempty set $V \setminus V \cap \mathbf{V}(g) \subseteq V$.

    c. We say that $\phi = f/g$ and $\phi' = f'/g'$ are *equivalent* on V, written $\phi \sim \phi'$, provided that there is a proper variety $W \subseteq V$ such that ϕ and ϕ' are defined and equal on $V \setminus W$. Prove that $\sim$ is an equivalence relation. An equivalence class for $\sim$ is called a *rational function* on V, and the set of all equivalence classes is denoted $k(V)$. Hint: Your proof will use the fact that V is irreducible.

    d. Show that addition and multiplication of equivalence classes is well-defined and makes $k(V)$ into a field called the *field of rational functions* of the projective variety V.

e. Let U_i be the affine part of $\mathbb{P}^n(k)$ where $x_i = 1$, and assume that $V \cap U_i \neq \emptyset$. Then $V \cap U_i$ is an irreducible affine variety in $U_i \cong k^n$. Show that $k(V)$ is isomorphic to the field $k(V \cap U_i)$ of rational functions on the affine variety $V \cap U_i$. Hint: You can assume $i = 0$. What do you get when you set $x_0 = 1$ in the quotient f/g considered in part (b)?

§4   The Projective Closure of an Affine Variety

In §2, we showed that any affine variety could be regarded as the affine portion of a projective variety. Since this can be done in more than one way (see Example 9 of §2), we would like to find the *smallest* projective variety containing a given affine variety. As we will see, there is an algorithmic way to do this.

Given homogeneous coordinates $x_0, \ldots, x_n$ on $\mathbb{P}^n(k)$, we have the subset $U_0 \subseteq \mathbb{P}^n(k)$ defined by $x_0 \neq 0$. If we identify U_0 with k^n using Proposition 2 of §2, then we get coordinates $x_1, \ldots, x_n$ on k^n. As in §3, we will use $\mathbf{I}_a$ and $\mathbf{V}_a$ for the affine versions of $\mathbf{I}$ and $\mathbf{V}$.

We first discuss how to homogenize an ideal of $k[x_1, \ldots, x_n]$. If we are given an arbitrary ideal $I \subseteq k[x_1, \ldots, x_n]$, the standard way to produce a homogeneous ideal $I^h \subseteq k[x_0, \ldots, x_n]$ is as follows.

Definition 1. Let I be an ideal in $k[x_1, \ldots, x_n]$. We define the **homogenization of I** to be the ideal

$$I^h = \langle f^h \mid f \in I \rangle \subseteq k[x_0, \ldots, x_n],$$

where f^h is the homogenization of f as in Proposition 7 of §2.

Naturally enough, we have the following result.

Proposition 2. *For any ideal $I \subseteq k[x_1, \ldots, x_n]$, the homogenization I^h is a homogeneous ideal in $k[x_0, \ldots, x_n]$.*

Proof. See Exercise 1.   □

Definition 1 is not entirely satisfying as it stands because it does not give us a *finite* generating set for the ideal I^h. There is a subtle point here. Given a particular finite generating set $f_1, \ldots, f_s$ for $I \subseteq k[x_1, \ldots, x_n]$, it is always true that $\langle f_1^h, \ldots, f_s^h \rangle$ is a homogeneous ideal contained in I^h. However, as the following example shows, I^h can be *strictly larger* than $\langle f_1^h, \ldots, f_s^h \rangle$.

Example 3. Consider $I = \langle f_1, f_2 \rangle = \langle x_2 - x_1^2, x_3 - x_1^3 \rangle$, the ideal of the affine twisted cubic in $\mathbb{R}^3$. If we homogenize f_1, f_2, then we get the ideal $J = \langle x_2 x_0 - x_1^2, x_3 x_0^2 - x_1^3 \rangle$ in $\mathbb{R}[x_0, x_1, x_2, x_3]$. We claim that $J \neq I^h$. To prove this, consider the polynomial

$$f_3 = f_2 - x_1 f_1 = x_3 - x_1^3 - x_1(x_2 - x_1^2) = x_3 - x_1 x_2 \in I.$$

Then $f_3^h = x_0 x_3 - x_1 x_2$ is a homogeneous polynomial of degree 2 in I^h. Since the generators of J are also homogeneous, of degrees 2 and 3, respectively, if we had

an equation $f_3^h = A_1 f_1^h + A_2 f_2^h$, then using the expansions of A_1 and A_2 into homogeneous components, we would see that f_3^h was a constant multiple of f_1^h. (See
Exercise 8 of §3 for a general statement along these lines.) Since this is clearly false,
we have $f_3^h \notin J$, and thus, $J \neq I^h$.

Hence, we may ask whether there is some reasonable method for computing a
finite generating set for the ideal I^h. The answer is given in the following theorem.
A *graded* monomial order in $k[x_1, \ldots, x_n]$ is one that orders first by total degree:

$$x^\alpha > x^\beta$$

whenever $|\alpha| > |\beta|$. Note that grlex and grevlex are graded orders, whereas lex is
not.

Theorem 4. *Let I be an ideal in $k[x_1, \ldots, x_n]$ and let $G = \{g_1, \ldots, g_t\}$ be a Gröbner
basis for I with respect to a graded monomial order in $k[x_1, \ldots, x_n]$. Then $G^h =
\{g_1^h, \ldots, g_t^h\}$ is a basis for $I^h \subseteq k[x_0, \ldots, x_n]$.*

Proof. We will prove the theorem by showing the stronger statement that G^h is
actually a Gröbner basis for I^h with respect to an appropriate monomial order in
$k[x_0, \ldots, x_n]$.

Every monomial in $k[x_0, \ldots, x_n]$ can be written

$$x_1^{\alpha_1} \cdots x_n^{\alpha_n} x_0^d = x^\alpha x_0^d,$$

where x^α contains no x_0 factors. Then we can extend the graded order $>$ on monomials in $k[x_1, \ldots, x_n]$ to a monomial order $>_h$ in $k[x_0, \ldots, x_n]$ as follows:

$$x^\alpha x_0^d >_h x^\beta x_0^e \iff x^\alpha > x^\beta \quad \text{or} \quad x^\alpha = x^\beta \text{ and } d > e.$$

In Exercise 2, you will show that this defines a monomial order in $k[x_0, \ldots, x_n]$. Note
that under this ordering, we have $x_i >_h x_0$ for all $i \geq 1$.

For us, the most important property of the order $>_h$ is given in the following
lemma.

Lemma 5. *If $f \in k[x_1, \ldots, x_n]$ and $>$ is a graded order on $k[x_1, \ldots, x_n]$, then*

$$\mathrm{LM}_{>_h}(f^h) = \mathrm{LM}_{>}(f).$$

Proof of Lemma. Since $>$ is a graded order, for any $f \in k[x_1, \ldots, x_n]$, $\mathrm{LM}_{>}(f)$ is
one of the monomials x^α appearing in the homogeneous component of f of *maximal*
total degree. When we homogenize, this term is unchanged. If $x^\beta x_0^e$ is any one of
the other monomials appearing in f^h, then $\alpha > \beta$. By the definition of $>_h$, it follows
that $x^\alpha >_h x^\beta x_0^e$. Hence, $x^\alpha = \mathrm{LM}_{>_h}(f^h)$, and the lemma is proved. □

We will now show that G^h forms a Gröbner basis for the ideal I^h with respect to
the monomial order $>_h$. Each $g_i^h \in I^h$ by definition. Thus, it suffices to show that the
ideal of leading terms $\langle \mathrm{LT}_{>_h}(I^h) \rangle$ is generated by $\mathrm{LT}_{>_h}(G^h)$. To prove this, consider

$F \in I^h$. Since I^h is a homogeneous ideal, each homogeneous component of F is in I^h and, hence, we may assume that F is homogeneous. Because $F \in I^h$, by definition we have

$$(1) \qquad F = \sum_j A_j f_j^h,$$

where $A_j \in k[x_0, \ldots, x_n]$ and $f_j \in I$. We will let $f = F(1, x_1, \ldots, x_n)$ denote the dehomogenization of F. Then setting $x_0 = 1$ in (1) yields

$$f = F(1, x_1, \ldots, x_n) = \sum_j A_j(1, x_1, \ldots, x_n) f_j^h(1, x_1, \ldots, x_n)$$

$$= \sum_j A_j(1, x_1, \ldots, x_n) f_j$$

since $f_j^h(1, x_1, \ldots, x_n) = f_j(x_1, \ldots, x_n)$ by part (iii) of Proposition 7 from §2. This shows that $f \in I \subseteq k[x_1, \ldots, x_n]$. If we homogenize f, then part (iv) of Proposition 7 in §2 implies that

$$F = x_0^e \cdot f^h$$

for some $e \geq 0$. Thus,

$$(2) \qquad \mathrm{LM}_{>_h}(F) = x_0^e \cdot \mathrm{LM}_{>_h}(f^h) = x_0^e \cdot \mathrm{LM}_{>}(f),$$

where the last equality is by Lemma 5. Since G is a Gröbner basis for I, we know that $\mathrm{LM}_{>}(f)$ is divisible by some $\mathrm{LM}_{>}(g_i) = \mathrm{LM}_{>_h}(g_i^h)$ (using Lemma 5 again). Then (2) shows that $\mathrm{LM}_{>_h}(F)$ is divisible by $\mathrm{LM}_{>_h}(g_i^h)$, as desired. This completes the proof of the theorem. □

In Exercise 5, you will see that there is a more elegant formulation of Theorem 4 for the special case of grevlex order.

To illustrate the theorem, consider the ideal $I = \langle x_2 - x_1^2, x_3 - x_1^3 \rangle$ of the affine twisted cubic $W \subseteq \mathbb{R}^3$ once again. Computing a Gröbner basis for I with respect to grevlex order, we find

$$G = \{ x_1^2 - x_2, x_1 x_2 - x_3, x_1 x_3 - x_2^2 \}.$$

By Theorem 4, the homogenizations of these polynomials generate I^h. Thus,

$$(3) \qquad I^h = \langle x_1^2 - x_0 x_2, x_1 x_2 - x_0 x_3, x_1 x_3 - x_2^2 \rangle.$$

Note that this ideal gives us the projective variety $V' = \mathbf{V}(I^h) \subseteq \mathbb{P}^3(\mathbb{R})$ which we discovered in Example 9 of §2.

For the remainder of this section, we will discuss the geometric meaning of the homogenization of an ideal. We will begin by studying what happens when we homogenize the ideal $\mathbf{I}_a(W)$ of all polynomials vanishing on an affine variety W. This leads to the following definition.

Definition 6. Given an affine variety $W \subseteq k^n$, the **projective closure** of W is the projective variety $\overline{W} = \mathbf{V}(\mathbf{I}_a(W)^h) \subseteq \mathbb{P}^n(k)$, where $\mathbf{I}_a(W)^h \subseteq k[x_0, \ldots, x_n]$ is the homogenization of the ideal $\mathbf{I}_a(W) \subseteq k[x_1, \ldots, x_n]$ as in Definition 1.

The projective closure has the following important properties.

Proposition 7. *Let $W \subseteq k^n$ be an affine variety and let $\overline{W} \subseteq \mathbb{P}^n(k)$ be its projective closure. Then:*

(i) $\overline{W} \cap U_0 = \overline{W} \cap k^n = W$.
(ii) *$\overline{W}$ is the smallest projective variety in $\mathbb{P}^n(k)$ containing W.*
(iii) *If W is irreducible, then so is $\overline{W}$.*
(iv) *No irreducible component of $\overline{W}$ lies in the hyperplane at infinity $\mathbf{V}(x_0) \subseteq \mathbb{P}^n(k)$.*

Proof. (i) Let G be a Gröbner basis of $\mathbf{I}_a(W)$ with respect to a graded order on $k[x_1, \ldots, x_n]$. Then Theorem 4 implies that $\mathbf{I}_a(W)^h = \langle g^h \mid g \in G \rangle$. We know that $k^n \cong U_0$ is the subset of $\mathbb{P}^n(k)$, where $x_0 = 1$. Thus, we have

$$\overline{W} \cap U_0 = \mathbf{V}(g^h \mid g \in G) \cap U_0 = \mathbf{V}_a(g^h(1, x_1, \ldots, x_n) \mid g \in G).$$

Since $g^h(1, x_1, \ldots, x_n) = g$ by part (iii) of Proposition 7 of §2, we get $\overline{W} \cap U_0 = W$.

(ii) We need to prove that if V is a projective variety containing W, then $\overline{W} \subseteq V$. Let $V = \mathbf{V}(F_1, \ldots, F_s)$. Then F_i vanishes on V, so that its dehomogenization $f_i = F_i(1, x_1, \ldots, x_n)$ vanishes on W. Thus, $f_i \in \mathbf{I}_a(W)$ and, hence, $f_i^h \in \mathbf{I}_a(W)^h$. This shows that f_i^h vanishes on $\overline{W} = \mathbf{V}(\mathbf{I}_a(W)^h)$. But part (iv) of Proposition 7 from §2 implies that $F_i = x_0^{e_i} f_i^h$ for some integer e_i. Thus, F_i vanishes on $\overline{W}$, and since this is true for all i, it follows that $\overline{W} \subseteq V$.

The proof of (iii) will be left as an exercise. To prove (iv), let $\overline{W} = V_1 \cup \cdots \cup V_m$ be the decomposition of $\overline{W}$ into irreducible components. Suppose that one of them, say V_1, was contained in the hyperplane at infinity $\mathbf{V}(x_0)$. Then

$$\begin{aligned} W = \overline{W} \cap U_0 &= (V_1 \cup \cdots \cup V_m) \cap U_0 \\ &= (V_1 \cap U_0) \cup ((V_2 \cup \cdots \cup V_m) \cap U_0) \\ &= (V_2 \cup \cdots \cup V_m) \cap U_0. \end{aligned}$$

This shows that $V_2 \cup \cdots \cup V_m$ is a projective variety containing W. By the minimality of $\overline{W}$, it follows that $\overline{W} = V_2 \cup \cdots \cup V_m$ and, hence, $V_1 \subseteq V_2 \cup \cdots \cup V_m$. We will leave it as an exercise to show that this is impossible since V_1 is an irreducible component of $\overline{W}$. This contradiction completes the proof. $\qquad \square$

For an example of how the projective closure works, consider the affine twisted cubic $W \subseteq \mathbb{R}^3$. In §4 of Chapter 1, we proved that

$$\mathbf{I}_a(W) = \langle x_2 - x_1^2, x_3 - x_1^3 \rangle.$$

Using Theorem 4, we proved in (3) that

$$\mathbf{I}_a(W)^h = \langle x_1^2 - x_0 x_2, x_1 x_2 - x_0 x_3, x_1 x_3 - x_2^2 \rangle.$$

It follows that the variety $V' = \mathbf{V}(x_1^2 - x_0 x_2, x_1 x_2 - x_0 x_3, x_1 x_3 - x_2^2)$ discussed in Example 9 of §2 is the projective closure of the affine twisted cubic.

The main drawback of the definition of projective closure is that it requires that we know $\mathbf{I}_a(W)$. It would be much more convenient if we could compute the projective closure directly from *any* defining ideal of W. When the field k is algebraically closed, this can always be done.

Theorem 8. *Let k be an algebraically closed field, and let $I \subseteq k[x_1, \ldots, x_n]$ be an ideal. Then $\mathbf{V}(I^h) \subseteq \mathbb{P}^n(k)$ is the projective closure of $\mathbf{V}_a(I) \subseteq k^n$.*

Proof. Let $W = \mathbf{V}_a(I) \subseteq k^n$ and $Z = \mathbf{V}(I^h) \subseteq \mathbb{P}^n(k)$. The proof of part (i) of Proposition 7 shows that Z is a projective variety containing W.

To prove that Z is the smallest such variety, we proceed as in part (ii) of Proposition 7. Thus, let $V = \mathbf{V}(F_1, \ldots, F_s)$ be any projective variety containing W. As in the earlier argument, the dehomogenization $f_i = F_i(1, x_1, \ldots, x_n)$ is in $\mathbf{I}_a(W)$. Since k is algebraically closed, the Nullstellensatz implies that $\mathbf{I}_a(W) = \sqrt{I}$, so that $f_i^m \in I$ for some integer m. This tells us that

$$(f_i^m)^h \in I^h$$

and, consequently, $(f_i^m)^h$ vanishes on Z. In the exercises, you will show that

$$(f_i^m)^h = (f_i^h)^m,$$

and it follows that f_i^h vanishes on Z. Then $F_i = x_0^{e_i} f_i^h$ shows that F_i is also zero on Z. As in Proposition 7, we conclude that $Z \subseteq V$.

This shows that Z is the smallest projective variety containing W. Since the projective closure $\overline{W}$ has the same property by Proposition 7, we see that $Z = \overline{W}$. □

If we combine Theorems 4 and 8, we get **an algorithm for computing the projective closure of an affine variety** over an algebraically closed field k: given $W \subseteq k^n$ defined by $f_1 = \cdots = f_s = 0$, compute a Gröbner basis G of $\langle f_1, \ldots, f_s \rangle$ with respect to a graded order, and then the projective closure in $\mathbb{P}^n(k)$ is defined by $g^h = 0$ for $g \in G$.

Unfortunately, Theorem 8 can fail over fields that are not algebraically closed. Here is an example that shows what can go wrong.

Example 9. Consider $I = \langle x_1^2 + x_2^4 \rangle \subseteq \mathbb{R}[x_1, x_2]$. Then $W = \mathbf{V}_a(I)$ consists of the single point $(0,0)$ in $\mathbb{R}^2$, and hence, the projective closure is the single point $\overline{W} = \{(1:0:0)\} \subseteq \mathbb{P}^2(\mathbb{R})$ (since this is obviously the smallest projective variety containing W). On the other hand, $I^h = \langle x_1^2 x_0^2 + x_2^4 \rangle$, and it is easy to check that

$$\mathbf{V}(I^h) = \{(1:0:0), (0:1:0)\} \subseteq \mathbb{P}^2(\mathbb{R}).$$

This shows that $\mathbf{V}(I^h)$ is strictly larger than the projective closure of $W = \mathbf{V}_a(I)$.

EXERCISES FOR §4

1. Prove Proposition 2.
2. Show that the order $>_h$ defined in the proof of Theorem 4 is a monomial order on $k[x_0, \ldots, x_n]$. Hint: This can be done directly or by using the mixed orders defined in Exercise 9 of Chapter 2, §4.
3. Show by example that the conclusion of Theorem 4 is *not true* if we use an arbitrary monomial order in $k[x_1, \ldots, x_n]$ and homogenize a Gröbner basis with respect to that order. Hint: One example can be obtained using the ideal of the affine twisted cubic and computing a Gröbner basis with respect to a nongraded order.
4. Let $>$ be a graded monomial order on $k[x_1, \ldots, x_n]$ and let $>_h$ be the order defined in the proof of Theorem 4. In the proof of the theorem, we showed that if G is a Gröbner basis for $I \subseteq k[x_1, \ldots, x_n]$ with respect to $>$, then G^h was a Gröbner basis for I^h with respect to $>_h$. In this exercise, we will explore other monomial orders on $k[x_0, \ldots, x_n]$ that have this property.
   a. Define a graded version of $>_h$ by setting

   $$x^\alpha x_0^d >_{gh} x^\beta x_0^e \iff |\alpha| + d > |\beta| + e \quad \text{or} \quad |\alpha| + d = |\beta| + e$$
   $$\text{and } x^\alpha x_0^d >_h x^\beta x_0^e.$$

   Show that G^h is a Gröbner basis with respect to $>_{gh}$.
   b. More generally, let $>'$ be any monomial order on $k[x_0, \ldots, x_n]$ which extends $>$ and which has the property that among monomials of the same total degree, any monomial containing x_0 is smaller than all monomials containing only $x_1, \ldots, x_n$. Show that G^h is a Gröbner basis for $>'$.
5. Let $>$ denote grevlex in the ring $S = k[x_1, \ldots, x_n, x_{n+1}]$. Consider $R = k[x_1, \ldots, x_n] \subseteq S$. For $f \in R$, let f^h denote the homogenization of f with respect to the variable x_{n+1}.
   a. Show that if $f \in R \subseteq S$ (that is, f depends only on $x_1, \ldots, x_n$), then $\mathrm{LT}_>(f) = \mathrm{LT}_>(f^h)$.
   b. Use part (a) to show that if G is a Gröbner basis for an ideal $I \subseteq R$ with respect to grevlex, then G^h is a Gröbner basis for the ideal I^h in S with respect to grevlex.
6. Prove that the homogenization has the following properties for polynomials $f, g \in k[x_1, \ldots, x_n]$:

   $$(fg)^h = f^h g^h,$$
   $$(f^m)^h = (f^h)^m \quad \text{for any integer } m \geq 0.$$

   Hint: Use the formula for homogenization given by part (ii) of Proposition 7 from §2.
7. Show that $I \subseteq k[x_1, \ldots, x_n]$ is a prime ideal if and only if I^h is a prime ideal in $k[x_0, \ldots, x_n]$. Hint: For the $\Rightarrow$ implication, use part (a) of Exercise 11 of §3; for the converse implication, use Exercise 6.
8. Adapt the proof of part (ii) of Proposition 7 to show that $\mathbf{I}(\overline{W}) = \mathbf{I}_a(W)^h$ for any affine variety $W \subseteq k^n$.
9. Prove that an affine variety W is irreducible if and only if its projective closure $\overline{W}$ is irreducible.
10. Let $W = V_1 \cup \cdots \cup V_m$ be the decomposition of a projective variety into its irreducible components such that $V_i \not\subseteq V_j$ for $i \neq j$. Prove that $V_1 \not\subseteq V_2 \cup \cdots \cup V_m$.

In Exercises 11–14, we will explore some interesting varieties in projective space. For ease of notation, we will write $\mathbb{P}^n$ rather than $\mathbb{P}^n(k)$. We will also assume that k is algebraically closed so that we can apply Theorem 8.

11. The twisted cubic that we have used repeatedly for examples is one member of an infinite family of curves known as the *rational normal curves*. The rational normal curve in k^n is the image of the polynomial parametrization $\phi : k \to k^n$ given by

$$\phi(t) = (t, t^2, t^3, \ldots, t^n).$$

By our general results on implicitization from Chapter 3, we know the rational normal curves are affine varieties. Their projective closures in $\mathbb{P}^n$ are also known as rational normal curves.
   a. Find affine equations for the rational normal curves in k^4 and k^5.
   b. Homogenize your equations from part (a) and consider the projective varieties defined by these homogeneous polynomials. Do your equations define the projective closure of the affine curve? Are there any "extra" components at infinity?
   c. Using Theorems 4 and 8, find a set of homogeneous equations defining the projective closures of these rational normal curves in $\mathbb{P}^4$ and $\mathbb{P}^5$, respectively. Do you see a pattern?
   d. Show that the rational normal curve in $\mathbb{P}^n$ is the variety defined by the set of homogeneous quadrics obtained by taking all possible 2×2 subdeterminants of the $2 \times n$ matrix:

$$\begin{pmatrix} x_0 & x_1 & x_2 & \cdots & x_{n-1} \\ x_1 & x_2 & x_3 & \cdots & x_n \end{pmatrix}.$$

12. The affine Veronese surface $S \subseteq k^5$ was introduced in Exercise 6 of Chapter 5, §1. It is the image of the polynomial parametrization $\phi : k^2 \to k^5$ given by

$$\phi(x_1, x_2) = (x_1, x_2, x_1^2, x_1 x_2, x_2^2).$$

The projective closure of S is a projective variety known as the *projective Veronese surface*.
   a. Find a set of homogeneous equations for the projective Veronese surface in $\mathbb{P}^5$.
   b. Show that the parametrization of the affine Veronese surface above can be extended to a mapping from $\Phi : \mathbb{P}^2 \to \mathbb{P}^5$ whose image coincides with the entire projective Veronese surface. Hint: You must show that Φ is well-defined (i.e., that it yields the same point in $\mathbb{P}^5$ for any choice of homogeneous coordinates for a point in $\mathbb{P}^2$).

13. The Cartesian product of two affine spaces is simply another affine space: $k^n \times k^m = k^{m+n}$. If we use the standard inclusions $k^n \subseteq \mathbb{P}^n$, $k^m \subseteq \mathbb{P}^m$, and $k^{n+m} \subseteq \mathbb{P}^{n+m}$ given by Proposition 2 of §2, how is $\mathbb{P}^{n+m}$ different from $\mathbb{P}^n \times \mathbb{P}^m$ (as a set)?

14. In this exercise, we will see that $\mathbb{P}^n \times \mathbb{P}^m$ can be identified with a certain projective variety in $\mathbb{P}^{n+m+nm}$ known as a *Segre variety*. The idea is as follows. Let $p = (x_0 : \cdots : x_n) \in \mathbb{P}^n$ and let $q = (y_0 : \cdots : y_m) \in \mathbb{P}^m$. The Segre mapping $\sigma : \mathbb{P}^n \times \mathbb{P}^m \to \mathbb{P}^{n+m+nm}$ is defined by taking the pair $(p, q) \in \mathbb{P}^n \times \mathbb{P}^m$ to the point in $\mathbb{P}^{n+m+nm}$ with homogeneous coordinates

$$(x_0 y_0 : x_0 y_1 : \cdots : x_0 y_m : x_1 y_0 : \cdots : x_1 y_m : \cdots : x_n y_0 : \cdots : x_n y_m).$$

The components are all the possible products $x_i y_j$ where $0 \le i \le n$ and $0 \le j \le m$. The image is a projective variety called a Segre variety.
   a. Show that σ is a well-defined mapping (i.e., show that we obtain the same point in $\mathbb{P}^{n+m+nm}$ no matter what homogeneous coordinates for p, q we use).
   b. Show that σ is a one-to-one mapping and that the "affine part" $k^n \times k^m$ maps to an affine variety in $k^{n+m+nm} = U_0 \subseteq \mathbb{P}^{n+m+nm}$ that is *isomorphic* to k^{n+m}. (See Chapter 5, §4.)
   c. Taking $n = m = 1$ above, write out $\sigma : \mathbb{P}^1 \times \mathbb{P}^1 \to \mathbb{P}^3$ explicitly and find homogeneous equation(s) for the image. Hint: You should obtain a single quadratic equation. This Segre variety is a quadric surface in $\mathbb{P}^3$.

    d. Now consider the case $n = 2, m = 1$ and find homogeneous equations for the Segre variety in $\mathbb{P}^5$.

    e. What is the intersection of the Segre variety in $\mathbb{P}^5$ and the Veronese surface in $\mathbb{P}^5$? (See Exercise 12.)

§5 Projective Elimination Theory

In Chapter 3, we encountered numerous instances of "missing points" when studying the geometric interpretation of elimination theory. Since our original motivation for projective space was to account for "missing points," it makes sense to look back at elimination theory using what we know about $\mathbb{P}^n(k)$. You may want to review the first two sections of Chapter 3 before reading further.

    We begin with the following example.

Example 1. Consider the variety $V \subseteq \mathbb{C}^2$ defined by the equation

$$xy^2 = x - 1.$$

To eliminate x, we use the elimination ideal $I_1 = \langle xy^2 - x + 1 \rangle \cap \mathbb{C}[y]$, and it is easy to show that $I_1 = \{0\} \subseteq \mathbb{C}[y]$. In Chapter 3, we observed that eliminating x corresponds geometrically to the projection $\pi(V) \subseteq \mathbb{C}$, where $\pi : \mathbb{C}^2 \to \mathbb{C}$ is defined by $\pi(x, y) = y$. We know that $\pi(V) \subseteq \mathbf{V}(I_1) = \mathbb{C}$, but as the following picture shows, $\pi(V)$ does not fill up all of $\mathbf{V}(I_1)$:

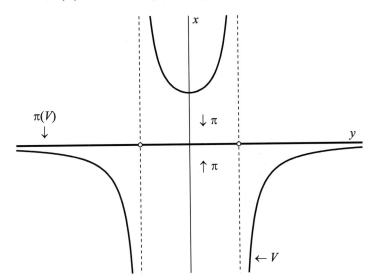

We can control the missing points using the Geometric Extension Theorem (Theorem 2 of Chapter 3, §2). Recall how this works: if we write the defining equation of V as $(y^2 - 1)x + 1 = 0$, then the Extension Theorem guarantees that we can solve for x whenever the leading coefficient of x does not vanish. Thus, $y = \pm 1$ are the only missing points. You can check that these points are missing over $\mathbb{C}$ as well.

To reinterpret the Geometric Extension Theorem in terms of projective space, first observe that the standard projective plane $\mathbb{P}^2(\mathbb{C})$ is not quite what we want. We are really only interested in directions along the projection (i.e., parallel to the x-axis) since all of our missing points lie in this direction. So we do not need all of $\mathbb{P}^2(\mathbb{C})$. A more serious problem is that in $\mathbb{P}^2(\mathbb{C})$, all lines parallel to the x-axis correspond to a *single* point at infinity, yet we are missing *two* points.

To avoid this difficulty, we will use something besides $\mathbb{P}^2(\mathbb{C})$. If we write π as $\pi : \mathbb{C} \times \mathbb{C} \to \mathbb{C}$, the idea is to make the first factor projective rather than the whole thing. This gives us $\mathbb{P}^1(\mathbb{C}) \times \mathbb{C}$, and we will again use π to denote the projection $\pi : \mathbb{P}^1(\mathbb{C}) \times \mathbb{C} \to \mathbb{C}$ onto the second factor.

We will use coordinates $(t:x;y)$ on $\mathbb{P}^1(\mathbb{C}) \times \mathbb{C}$, where the semicolon separates homogeneous coordinates $(t:x)$ on $\mathbb{P}^1(\mathbb{C})$ from usual coordinate y on $\mathbb{C}$. Thus, (in analogy with Proposition 2 of §2) a point $(1:x;y) \in \mathbb{P}^1(\mathbb{C}) \times \mathbb{C}$ corresponds to $(x, y) \in \mathbb{C} \times \mathbb{C}$. We will regard $\mathbb{C} \times \mathbb{C}$ as a subset of $\mathbb{P}^1(\mathbb{C}) \times \mathbb{C}$ and you should check that the complement consists of the "points at infinity" $(0:1;y)$.

We can extend $V \subseteq \mathbb{C} \times \mathbb{C}$ to $\overline{V} \subseteq \mathbb{P}^1(\mathbb{C}) \times \mathbb{C}$ by making the equation of V homogeneous with respect to t and x. Thus, $\overline{V}$ is defined by

$$xy^2 = x - t.$$

In Exercise 1, you will check that this equation is well-defined on $\mathbb{P}^1(\mathbb{C}) \times \mathbb{C}$. To find the solutions of this equation, we first set $t = 1$ to get the affine portion and then we set $t = 0$ to find the points at infinity. This leads to

$$\overline{V} = V \cup \{(0:1;\pm 1)\}$$

(remember that t and x cannot simultaneously vanish since they are homogeneous coordinates). Under the projection $\pi : \mathbb{P}^1(\mathbb{C}) \times \mathbb{C} \to \mathbb{C}$, it follows that $\pi(\overline{V}) = \mathbb{C} = \mathbf{V}(I_1)$ because the two points at infinity map to the "missing points" $y = \pm 1$. As we will soon see, the equality $\pi(\overline{V}) = \mathbf{V}(I_1)$ is a special case of the projective version of the Geometric Extension Theorem.

We will use the following general framework for generalizing the issues raised by Example 1. Suppose we have equations

$$f_1(x_1, \ldots, x_n, y_1, \ldots, y_m) = 0,$$

$$\vdots$$

$$f_s(x_1, \ldots, x_n, y_1, \ldots, y_m) = 0,$$

where $f_1, \ldots, f_s \in k[x_1, \ldots, x_n, y_1, \ldots, y_m]$. Working algebraically, we can eliminate $x_1, \ldots, x_n$ by computing the ideal $I_n = \langle f_1, \ldots, f_s \rangle \cap k[y_1, \ldots, y_m]$ (the Elimination Theorem from Chapter 3, §1 tells us how to do this). If we think geometrically, the above equations define a variety $V \subseteq k^n \times k^m$, and eliminating $x_1, \ldots, x_n$ corresponds to considering $\pi(V)$, where $\pi : k^n \times k^m \to k^m$ is projection onto the last m coordinates. Our goal is to describe the relation between $\pi(V)$ and $\mathbf{V}(I_n)$.

The basic idea is to make the first factor projective. To simplify notation, we will write $\mathbb{P}^n(k)$ as $\mathbb{P}^n$ when there is no confusion about what field we are dealing with. A point in $\mathbb{P}^n \times k^m$ will have coordinates $(x_0 : \cdots : x_n; y_1, \ldots, y_m)$, where $(x_0 : \cdots : x_n)$ are homogeneous coordinates in $\mathbb{P}^n$ and $(y_1, \ldots, y_m)$ are usual coordinates in k^m. Thus, $(1:1; 1, 1)$ and $(2:2; 1, 1)$ are coordinates for the same point in $\mathbb{P}^1 \times k^2$, whereas $(2:2; 2, 2)$ gives a different point. As in Proposition 2 of §2, we will use

$$(x_1, \ldots, x_n, y_1, \ldots, y_m) \longmapsto (1 : x_1 : \cdots : x_n; y_1, \ldots, y_m)$$

to identify $k^n \times k^m$ with the subset of $\mathbb{P}^n \times k^m$ where $x_0 \neq 0$.

We can define varieties in $\mathbb{P}^n \times k^m$ using "partially" homogeneous polynomials as follows.

Definition 2. Let k be a field.

(i) A polynomial $F \in k[x_0, \ldots, x_n, y_1, \ldots, y_m]$ is $(x_0, \ldots, x_n)$-**homogeneous** provided there is an integer $l \geq 0$ such that

$$F = \sum_{|\alpha|=l} h_\alpha(y_1, \ldots, y_m) x^\alpha,$$

where x^α is a monomial in $x_0, \ldots, x_n$ of multidegree α and $h_\alpha \in k[y_1, \ldots, y_m]$.
(ii) The **variety** $\mathbf{V}(F_1, \ldots, F_s) \subseteq \mathbb{P}^n \times k^m$ defined by $(x_0, \ldots, x_n)$-homogeneous polynomials $F_1, \ldots, F_s \in k[x_0, \ldots, x_n, y_1, \ldots, y_m]$ is the set

$$\{(a_0 : \cdots : a_n; b_1, \ldots, b_m) \in \mathbb{P}^n \times k^m \mid F_i(a_0, \ldots, a_n, b_1, \ldots, b_m) = 0$$
$$\text{for } 1 \leq i \leq s\}.$$

In the exercises, you will show that if a $(x_0, \ldots, x_n)$-homogeneous polynomial vanishes at one set of coordinates for a point in $\mathbb{P}^n \times k^m$, then it vanishes for *all* coordinates of the point. This shows that the variety $\mathbf{V}(F_1, \ldots, F_s)$ is a well-defined subset of $\mathbb{P}^n \times k^m$ when $F_1, \ldots, F_s$ are $(x_0, \ldots, x_n)$-homogeneous.

We can now discuss what elimination theory means in this context. Suppose we have $(x_0, \ldots, x_n)$-homogeneous equations

$$F_1(x_0, \ldots, x_n, y_1, \ldots, y_m) = 0,$$

(1)
$$\vdots$$

$$F_s(x_0, \ldots, x_n, y_1, \ldots, y_m) = 0.$$

These define the variety $V = \mathbf{V}(F_1, \ldots, F_s) \subseteq \mathbb{P}^n \times k^m$. We also have the projection map

$$\pi : \mathbb{P}^n \times k^m \to k^m$$

onto the last m coordinates. Then we can interpret $\pi(V) \subseteq k^m$ as the set of all m-tuples $(y_1, \ldots, y_m)$ for which the equations (1) have a *nontrivial* solution in $x_0, \ldots, x_n$ (which means that at least one x_i is nonzero).

To understand what this means algebraically, let us work out an example.

Example 3. In this example, we will use $(u : v; y)$ as coordinates on $\mathbb{P}^1 \times k$. Then consider the equations

$$
(2) \qquad
\begin{aligned}
F_1 &= u + vy = 0, \\
F_2 &= u + uy = 0.
\end{aligned}
$$

Since $(u : v)$ are homogeneous coordinates on $\mathbb{P}^1$, it is straightforward to show that

$$V = \mathbf{V}(F_1, F_2) = \{(0 : 1; 0), (1 : 1; -1)\} \subseteq \mathbb{P}^1 \times k.$$

Under the projection $\pi : \mathbb{P}^1 \times k \to k$, we have $\pi(V) = \{0, -1\}$, so that for a given y, the equations (2) have a nontrivial solution if and only if $y = 0$ or -1. Thus, (2) implies that $y(1 + y) = 0$.

Ideally, there should be a purely algebraic method of "eliminating" u and v from (2) to obtain $y(1 + y) = 0$. Unfortunately, the kind of elimination we did in Chapter 3 does not work. To see why, let $I = \langle F_1, F_2 \rangle \subseteq k[u, v, y]$ be the ideal generated by F_1 and F_2. Since every term of F_1 and F_2 contains u or v, it follows that

$$I \cap k[y] = \{0\}.$$

From the affine point of view, this is the correct answer since the *affine* variety

$$\mathbf{V}_a(F_1, F_2) \subseteq k^2 \times k$$

contains the trivial solutions $(0 : 0; y)$ for *all* $y \in k$. Thus, the affine methods of Chapter 3 will be useful only if we can find an algebraic way of excluding the solutions where $u = v = 0$.

Recall from Chapter 4 that for affine varieties, ideal quotients $I : J$ and saturations $I : J^\infty$ correspond (roughly) to the difference of varieties. Comparing Proposition 7 and Theorem 10 of Chapter 4, §4, we see that the saturation has the best relation to the difference of varieties.

In our situation, the difference

$$\mathbf{V}_a(F_1, F_2) \setminus \mathbf{V}_a(u, v) \subseteq k^2 \times k.$$

consists *exactly* of the nontrivial solutions of (2). Hence, for $I = \langle F_1, F_2 \rangle$, the saturation $I : \langle u, v \rangle^\infty \subseteq k[u, v, y]$ is the algebraic way to remove the trivial solutions.

Since we also want to eliminate u, v, this suggests that we should consider the intersection

$$\hat{I} = (I : \langle u, v \rangle^\infty) \cap k[y].$$

Using the techniques of Chapter 4, it can be shown that $\hat{I} = \langle y(1 + y) \rangle$ in this case. Hence we recover precisely the polynomial we wanted to obtain.

Motivated by this example, we are led to the following definition.

Definition 4. Given an ideal $I \subseteq k[x_0, \ldots, x_n, y_1, \ldots, y_m]$ generated by $(x_0, \ldots, x_n)$-homogeneous polynomials, the **projective elimination ideal** of I is the ideal

$$\hat{I} = (I : \langle x_0, \ldots, x_n \rangle^{\infty}) \cap k[y_1, \ldots, y_m].$$

Exercise 6 will show why saturation is essential—just using the ideal quotient can give the wrong answer. Recall from Definition 8 of Chapter 4, §4 that

$$I : \langle x_0, \ldots, x_n \rangle^{\infty} = \{ f \in k[x_0, \ldots, x_n, y_1, \ldots, y_m] \mid \text{for every } g \in \langle x_0, \ldots, x_n \rangle,$$
$$\text{there is } e \geq 0 \text{ with } fg^e \in I \}.$$

Exercise 7 will give some other ways to represent the projective elimination ideal $\hat{I}$. Here is our first result about $\hat{I}$.

Proposition 5. *Let* $V = \mathbf{V}(F_1, \ldots, F_s) \subseteq \mathbb{P}^n \times k^m$ *be defined by* $(x_0, \ldots, x_n)$-*homogeneous polynomials and let* $\pi : \mathbb{P}^n \times k^m \to k^m$ *be the projection map. Then in* k^m, *we have*

$$\pi(V) \subseteq \mathbf{V}(\hat{I}),$$

where $\hat{I}$ *is the projective elimination ideal of* $I = \langle F_1, \ldots, F_s \rangle$.

Proof. Suppose $(a_0 : \cdots : a_n; b_1, \ldots, b_m) \in V$ and $f \in \hat{I}$. The formula for saturation shows that for every i, $x_i^{e_i} f(y_1, \ldots, y_m) \in I$ for some e_i. Hence this polynomial vanishes on V, so that

$$a_i^{e_i} f(b_1, \ldots, b_m) = 0$$

for all i. Since $(a_0 : \cdots : a_n)$ are homogeneous coordinates, at least one $a_i \neq 0$ and, thus, $f(b_1, \ldots, b_m) = 0$. This proves that f vanishes on $\pi(V)$ and the proposition follows. $\qquad\square$

When the field is algebraically closed, we also have the following projective version of the Extension Theorem.

Theorem 6 (The Projective Extension Theorem). *Let* k *be algebraically closed and assume* $V = \mathbf{V}(F_1, \ldots, F_s) \subseteq \mathbb{P}^n \times k^m$ *is defined by* $(x_0, \ldots, x_n)$-*homogeneous polynomials in* $k[x_0, \ldots, x_n, y_1, \ldots, y_m]$. *Also let* $\hat{I} \subseteq k[y_1, \ldots, y_m]$ *be the projective elimination ideal of* $I = \langle F_1, \ldots, F_s \rangle$. *If*

$$\pi : \mathbb{P}^n \times k^m \to k^m$$

is projection onto the last m *coordinates, then*

$$\pi(V) = \mathbf{V}(\hat{I}).$$

Proof. The inclusion $\pi(V) \subseteq \mathbf{V}(\hat{I})$ follows from Proposition 5. For the opposite inclusion, let $\mathbf{c} = (c_1, \ldots, c_m) \in \mathbf{V}(\hat{I})$ and set

$$F_i(x_0, \ldots, x_n, \mathbf{c}) = F_i(x_0, \ldots, x_n, c_1, \ldots, c_m).$$

This is a homogeneous polynomial in $x_0, \ldots, x_n$, say of total degree $= d_i$ [equal to the total degree of $F_i(x_0, \ldots, x_n, y_1, \ldots, y_m)$ in $x_0, \ldots, x_n$].

We will show that $\mathbf{c} \notin \pi(V)$ leads to a contradiction. To see why, first observe that $\mathbf{c} \notin \pi(V)$ implies that the equations

$$F_1(x_0, \ldots, x_n, \mathbf{c}) = \cdots = F_s(x_0, \ldots, x_n, \mathbf{c}) = 0$$

define the empty variety in $\mathbb{P}^n$. Since the field k is algebraically closed, the Projective Weak Nullstellensatz (Theorem 8 of §3) implies that for some $r \geq 1$, we have

$$\langle x_0, \ldots, x_n \rangle^r \subseteq \langle F_1(x_0, \ldots, x_n, \mathbf{c}), \ldots, F_s(x_0, \ldots, x_n, \mathbf{c}) \rangle.$$

This means that each monomial x^α, $|\alpha| = r$, can be written as a polynomial linear combination of the $F_i(x_0, \ldots, x_n, \mathbf{c})$, say

$$x^\alpha = \sum_{i=1}^{s} H_i(x_0, \ldots, x_n) F_i(x_0, \ldots, x_n, \mathbf{c}).$$

By taking homogeneous components, we can assume that each H_i is homogeneous of total degree $r - d_i$ [since d_i is the total degree of $F_i(x_0, \ldots, x_n, \mathbf{c})$]. Then, writing each H_i as a linear combination of monomials x^{β_i} with $|\beta_i| = r - d_i$, we see that the polynomials

$$x^{\beta_i} F_i(x_0, \ldots, x_n, \mathbf{c}), \quad i = 1, \ldots, s, \; |\beta_i| = r - d_i$$

span the vector space of *all* homogeneous polynomials of total degree r in $x_0, \ldots, x_n$. If the dimension of this space is denoted N_r, then by standard results in linear algebra, we can find N_r of these polynomials which form a basis for this space. We will denote this basis as

$$G_j(x_0, \ldots, x_n, \mathbf{c}), \quad j = 1, \ldots, N_r.$$

This leads to an interesting polynomial in $y_1, \ldots, y_m$ as follows. The polynomial $G_j(x_0, \ldots, x_n, \mathbf{c})$ comes from a polynomial

$$G_j = G_j(x_0, \ldots, x_n, y_1, \ldots, y_m) \in k[x_0, \ldots, x_n, y_1, \ldots, y_m].$$

Each G_j is homogeneous in $x_0, \ldots, x_n$ of total degree r. Thus, we can write

(3) $$G_j = \sum_{|\alpha| = r} a_{j\alpha}(y_1, \ldots, y_m) x^\alpha.$$

Since the x^α with $|\alpha| = r$ form a basis of all homogeneous polynomials of total degree r, there are N_r such monomials. Hence we get a square matrix of polynomials $a_{j\alpha}(y_1, \ldots, y_m)$. Then let

$$D(y_1, \ldots, y_m) = \det(a_{j\alpha}(y_1, \ldots, y_m) \mid 1 \leq j \leq N_r, \; |\alpha| = r)$$

be the corresponding determinant. If we substitute $\mathbf{c}$ into (3), we obtain

$$G_j(x_0, \ldots, x_n, \mathbf{c}) = \sum_{|\alpha|=r} a_{j\alpha}(\mathbf{c}) x^\alpha,$$

and then

$$D(\mathbf{c}) \neq 0$$

since the $G_j(x_0, \ldots, x_n, \mathbf{c})$'s and x^α's are bases of the same vector space.

Our **Main Claim** is that $D(y_1, \ldots, y_m) \in \hat{I}$. This will give the contradiction we seek, since $D(y_1, \ldots, y_m) \in \hat{I}$ and $\mathbf{c} \in \mathbf{V}(\hat{I})$ imply $D(\mathbf{c}) = 0$. The strategy for proving the Main Claim will be to show that for every monomial x^α with $|\alpha| = r$, we have

(4) $$x^\alpha D(y_1, \ldots, y_m) \in I.$$

Once we have (4), it follows easily that $g^r D(y_1, \ldots, y_m) \in I$ for all $g \in \langle x_0, \ldots, x_n \rangle$, which in turn implies that $D \in I : \langle x_0, \ldots, x_n \rangle^\infty$. Since $D \in k[y_1, \ldots, y_m]$, we get $D \in \hat{I}$, proving the Main Claim.

It remains to prove (4). We will use linear algebra over the function field $K = k(x_0, \ldots, x_n, y_1, \ldots, y_m)$ (see Chapter 5, §5). Take N_r variables Y_α for $|\alpha| = r$ and consider the system of linear equations over K given by

$$\sum_{|\alpha|=r} a_{j\alpha}(y_1, \ldots, y_m) Y_\alpha = G_j(x_0, \ldots, x_n, y_1, \ldots, y_m), \quad j = 1, \ldots, N_r.$$

The determinant of this system is $D(y_1, \ldots, y_m)$ and is nonzero since $D(\mathbf{c}) \neq 0$. Hence the system has a unique solution, which by (3) is given by $Y_\alpha = x^\alpha$.

In this situation, Cramer's Rule (Proposition 2 of Appendix A, §4) gives a formula for the solution in terms of the coefficients of the system. More precisely, the solution $Y_\alpha = x^\alpha$ is given by the quotient

$$x^\alpha = \frac{\det(M_\alpha)}{D(y_1, \ldots, y_m)},$$

where M_α is the matrix obtained from $(a_{j\alpha})$ by replacing the column α by the polynomials $G_1, \ldots, G_{N_r}$. If we multiply each side by $D(y_1, \ldots, y_m)$ and expand $\det(M_\alpha)$ along this column, we get an equation of the form

$$x^\alpha D(y_1, \ldots, y_m) = \sum_{j=1}^{N_r} H_{j\alpha}(y_1, \ldots, y_m) G_j(x_0, \ldots, x_n, y_1, \ldots, y_m).$$

However, recall that every G_j is of the form $x^{\beta_i} F_i$, and if we make this substitution and write the sum in terms of the F_i, we obtain

$$x^\alpha D(y_1, \ldots, y_m) \in \langle F_1, \ldots, F_s \rangle = I.$$

This proves (4), and we are done. □

Theorem 6 tells us that when we project a variety $V \subseteq \mathbb{P}^n \times k^m$ into k^m, the result is again a variety. This has the following nice interpretation: if we think of the variables $y_1, \ldots, y_m$ as parameters in the system of equations

$$F_1(x_0, \ldots, x_n, y_1, \ldots, y_m) = \cdots = F_s(x_0, \ldots, x_n, y_1, \ldots, y_m) = 0,$$

then the equations defining $\pi(V) = \mathbf{V}(\hat{I})$ in k^m tell us what conditions the parameters must satisfy in order for the above equations to have a nontrivial solution (i.e., a solution different from $x_0 = \cdots = x_n = 0$).

From a computational perspective, we learned how to compute the saturation $I : \langle x_0, \ldots, x_n \rangle^\infty$ in Chapter 4, §4, and then we can use the elimination theory from Chapter 3 to compute $\hat{I} = (I : \langle x_0, \ldots, x_n \rangle^\infty) \cap k[y_1, \ldots, y_m]$. Hence there is an algorithm for computing projective elimination ideals.

We next relate $\hat{I}$ to the kind of elimination we did in Chapter 3. The basic idea is to reduce to the affine case by dehomogenization. If we fix $0 \le i \le n$, then setting $x_i = 1$ in $F \in k[x_0, \ldots, x_n, y_1, \ldots, y_m]$ gives the polynomial

$$F^{(i)} = F(x_0, \ldots, 1, \ldots, x_n, y_1, \ldots, y_m) \in k[x_0, \ldots, \hat{x}_i, \ldots, x_n, y_1, \ldots, y_m],$$

where $\hat{x}_i$ means that x_i is omitted from the list of variables. Then, given an ideal $I \subseteq k[x_0, \ldots, x_n, y_1, \ldots, y_m]$, we get the *dehomogenization*

$$I^{(i)} = \{F^{(i)} \mid F \in I\} \subseteq k[x_0, \ldots, \hat{x}_i, \ldots, x_n, y_1, \ldots, y_m].$$

It is easy to show that $I^{(i)}$ is an ideal in $k[x_0, \ldots, \hat{x}_i, \ldots, x_n, y_1, \ldots, y_m]$. We also leave it as an exercise to show that if $I = \langle F_1, \ldots, F_s \rangle$, then

$$(5) \qquad\qquad I^{(i)} = \langle F_1^{(i)}, \ldots, F_s^{(i)} \rangle.$$

Let $V \subseteq \mathbb{P}^n \times k^m$ be the variety defined by I. One can think of $I^{(i)}$ as defining the affine portion $V \cap (U_i \times k^m)$, where $U_i \cong k^n$ is the subset of $\mathbb{P}^n$ where $x_i = 1$. Since we are now in a purely affine situation, we can eliminate using the methods of Chapter 3. In particular, we get the n-th elimination ideal

$$I_n^{(i)} = I^{(i)} \cap k[y_1, \ldots, y_m],$$

where the subscript n indicates that the n variables $x_0, \ldots, \hat{x}_i, \ldots, x_n$ have been eliminated. We now compute $\hat{I}$ in terms of its dehomogenizations $I^{(i)}$ as follows.

Proposition 7. *Let $I \subseteq k[x_0, \ldots, x_n, y_1, \ldots, y_m]$ be an ideal that is generated by $(x_0, \ldots, x_n)$-homogeneous polynomials. Then*

$$\hat{I} = I_n^{(0)} \cap I_n^{(1)} \cap \cdots \cap I_n^{(n)}.$$

Proof. First suppose that $f \in \hat{I}$. Then $x_i^{e_i} f(y_1, \ldots, y_m) \in I$ for some $e_i \ge 0$, so that when we set $x_i = 1$, we get $f(y_1, \ldots, y_m) \in I^{(i)}$. This proves $f \in I^{(0)} \cap \cdots \cap I^{(n)}$.

For the other inclusion, we first study the relation between I and $I^{(i)}$. An element $f \in I^{(i)}$ is obtained from some $F \in I$ by setting $x_i = 1$. We claim that F can be assumed to be $(x_0, \ldots, x_n)$-homogeneous. To see why, note that F can be written as a sum $F = \sum_{j=0}^{d} F_j$, where F_j is $(x_0, \ldots, x_n)$-homogeneous of total degree j in $x_0, \ldots, x_n$. Since I is generated by $(x_0, \ldots, x_n)$-homogeneous polynomials, the proof of Theorem 2 of §3 can be adapted to show that $F_j \in I$ for all j (see Exercise 4). This implies that

$$\sum_{j=0}^{d} x_i^{d-j} F_j$$

is a $(x_0, \ldots, x_n)$-homogeneous polynomial in I which dehomogenizes to f when $x_i = 1$. Thus, we can assume that $F \in I$ is $(x_0, \ldots, x_n)$-homogeneous.

As in §2, we can define a homogenization operator which takes a polynomial $f \in k[x_0, \ldots, \hat{x}_i, \ldots, x_n, y_1, \ldots, y_m]$ and uses the extra variable x_i to produce a $(x_0, \ldots, x_n)$-homogeneous polynomial $f^h \in k[x_0, \ldots, x_n, y_1, \ldots, y_m]$. We leave it as an exercise to show that if a $(x_0, \ldots, x_n)$-homogeneous polynomial F dehomogenizes to f using $x_i = 1$, then there is an integer $e_i \geq 0$ such that

(6) $$F = x_i^{e_i} f^h.$$

Now suppose $f \in I^{(i)}$. As we showed above, f comes from $F \in I$ which is $(x_0, \ldots, x_n)$-homogeneous. Since f does not involve $x_0, \ldots, x_n$, we have $f = f^h$, and then (6) implies $x_i^{e_i} f \in I$. This holds for all $0 \leq i \leq n$ if we assume that $f \in I^{(0)} \cap \cdots \cap I^{(n)}$. By Exercise 7, it follows that $f \in \hat{I}$, and we are done. $\qquad\square$

Proposition 7 has a nice interpretation. Namely, $I_n^{(i)}$ can be thought of as eliminating $x_0, \ldots, \hat{x}_i, \ldots, x_n$ on the affine piece of $\mathbb{P}^n \times k^m$ where $x_i = 1$. Then intersecting these affine elimination ideals (which roughly corresponds to the eliminating on the union of the affine pieces) gives the projective elimination ideal.

We can also use Proposition 7 to give another algorithm for finding $\hat{I}$. If $I = \langle F_1, \ldots, F_s \rangle$, we know a basis of $I^{(i)}$ by (5), so that we can compute $I_n^{(i)}$ using the Elimination Theorem of Chapter 3, §1. Then the algorithm for ideal intersections from Chapter 4, §3 tells us how to compute $\hat{I} = I_n^{(0)} \cap \cdots \cap I_n^{(n)}$. Thus we have a second algorithm for computing projective elimination ideals.

To see how this works in practice, consider the equations

$$F_1 = u + vy = 0,$$
$$F_2 = u + uy = 0$$

from Example 3. If we set $I = \langle u + vy, u + uy \rangle \subseteq k[u, v, y]$ and compute suitable Gröbner bases, then we obtain

$$\text{when } u = 1 : \quad I_1^{(u)} = \langle 1 + vy, 1 + y \rangle \cap k[y] = \langle 1 + y \rangle,$$
$$\text{when } v = 1 : \quad I_1^{(v)} = \langle u + y, u + uy \rangle \cap k[y] = \langle y(1 + y) \rangle,$$

and it follows that $\hat{I} = I_1^{(u)} \cap I_1^{(v)} = \langle y(1+y) \rangle$. Can you explain why $I_1^{(u)}$ and $I_1^{(v)}$ are different?

We next return to a question posed earlier concerning the missing points that can occur in the affine case. An ideal $I \subseteq k[x_1, \ldots, x_n, y_1, \ldots, y_m]$ gives a variety $V = \mathbf{V}_a(I) \subseteq k^n \times k^m$, and under the projection $\pi : k^n \times k^m \to k^m$, we know that $\pi(V) \subseteq \mathbf{V}(I_n)$, where I_n is the n-th elimination ideal of I. We want to show that points in $\mathbf{V}(I_n) \setminus \pi(V)$ come from points at infinity in $\mathbb{P}^n \times k^m$.

To decide what variety in $\mathbb{P}^n \times k^m$ to use, we will homogenize with respect to x_0. Recall from the proof of Proposition 7 that $f \in k[x_1, \ldots, x_n, y_1, \ldots, y_m]$ gives us a $(x_0, \ldots, x_n)$-homogeneous polynomial $f^h \in k[x_0, \ldots, x_n, y_1, \ldots, y_m]$. Exercise 9 will study homogenization in more detail. Then the $(x_0, \ldots, x_n)$-*homogenization of* I is defined to be the ideal

$$I^h = \langle f^h \mid f \in I \rangle \subseteq k[x_0, \ldots, x_n, y_1, \ldots, y_m].$$

Using the Hilbert Basis Theorem, it follows easily that I^h is generated by finitely many $(x_0, \ldots, x_n)$-homogeneous polynomials.

The following proposition gives the main properties of I^h.

Proposition 8. *Given* $I \subseteq k[x_1, \ldots, x_n, y_1, \ldots, y_m]$, *let* I^h *be its* $(x_0, \ldots, x_n)$-*homogenization. Then:*

(i) *The projective elimination ideal of* I^h *equals the n-th elimination ideal of I. Thus,*
$$\widehat{I^h} = I_n \subseteq k[y_1, \ldots, y_m].$$
(ii) *If k is algebraically closed, then the variety* $\overline{V} = \mathbf{V}(I^h)$ *is the smallest variety in $\mathbb{P}^n \times k^m$ containing the affine variety* $V = \mathbf{V}_a(I) \subseteq k^n \times k^m$. *We call* $\overline{V}$ *the* **projective closure** *of V in $\mathbb{P}^n \times k^m$.*

Proof. (i) It is straightforward to show that dehomogenizing I^h with respect to x_0 gives $(I^h)^{(0)} = I$. Then the proof of Proposition 7 implies that $\widehat{I^h} \subseteq I_n$. Going the other way, take $f \in I_n$. Since $f \in k[y_1, \ldots, y_m]$, it is already $(x_0, \ldots, x_n)$-homogeneous. Hence, $f = f^h \in I^h$ and it follows that $x_i^0 f \in I^h$ for all i. This shows that $f \in \widehat{I^h}$, and (i) is proved.

Part (ii) is similar to Theorem 8 of §4 and is left as an exercise.   □

Using Theorem 6 and Proposition 8 together, we get the following nice result.

Corollary 9. *Assume that k is algebraically closed and let* $V = \mathbf{V}_a(I) \subseteq k^n \times k^m$, *where* $I \subseteq k[x_1, \ldots, x_n, y_1, \ldots, y_m]$ *is an ideal. Then*

$$\mathbf{V}(I_n) = \pi(\overline{V}),$$

where $\overline{V} \subseteq \mathbb{P}^n \times k^m$ *is the projective closure of V and* $\pi : \mathbb{P}^n \times k^m \to k^m$ *is the projection.*

Proof. Since Proposition 8 tells us that $\overline{V} = \mathbf{V}(I^h)$ and $\widehat{I^h} = I_n$, the corollary follows immediately from Theorem 6.   □

In Chapter 3, points of $\mathbf{V}(I_n)$ were called "partial solutions." Thus, $\mathbf{V}(I_n) \setminus \pi(V)$ consists of those partial solutions which do not extend to solutions in V. The above corollary shows that these points come from points at infinity in the projective closure $\overline{V}$ of V.

To use Corollary 9, we need to be able to compute I^h. As in §4, the difficulty is that $I = \langle f_1, \ldots, f_s \rangle$ need not imply $I^h = \langle f_1^h, \ldots, f_s^h \rangle$. But if we use an appropriate Gröbner basis, we get the desired equality.

Proposition 10. *Let* $>$ *be a monomial order on* $k[x_1, \ldots, x_n, y_1, \ldots, y_m]$ *such that for all monomials* $x^\alpha y^\gamma, x^\beta y^\delta$ *in* $x_1, \ldots, x_n, y_1, \ldots, y_m$, *we have*

$$|\alpha| > |\beta| \Rightarrow x^\alpha y^\gamma > x^\beta y^\delta.$$

If $G = \{g_1, \ldots, g_s\}$ *is a Gröbner basis for* $I \subseteq k[x_1, \ldots, x_n, y_1, \ldots, y_m]$ *with respect to* $>$, *then* $G^h = \{g_1^h, \ldots, g_s^h\}$ *is a basis for* $I^h \subseteq k[x_0, \ldots, x_n, y_1, \ldots, y_m]$.

Proof. This is similar to Theorem 4 of §4 and is left as an exercise.   □

In Example 1, we considered $I = \langle xy^2 - x + 1 \rangle \subseteq \mathbb{C}[x, y]$. This is a principal ideal and, hence, $xy^2 - x + 1$ is a Gröbner basis for any monomial ordering (see Exercise 10 of Chapter 2, §5). If we homogenize with respect to the new variable t, Proposition 10 tells us that I^h is generated by the (t, x)-homogeneous polynomial $xy^2 - x + t$. Now let $\overline{V} = \mathbf{V}(I^h) \subseteq \mathbb{P}^1 \times \mathbb{C}$. Then Corollary 9 shows $\pi(\overline{V}) = \mathbf{V}(I_1) = \mathbb{C}$, which agrees with what we found in Example 1.

Using Corollary 9 and Proposition 10, we can point out a weakness in the Geometric Extension Theorem given in Chapter 3, §2. This theorem stated that if $I = \langle f_1, \ldots, f_s \rangle$, then

(7) $$\mathbf{V}(I_1) = \pi(V) \cup (\mathbf{V}(g_1, \ldots, g_s) \cap \mathbf{V}(I_1)),$$

where $V = \mathbf{V}_a(I)$ and $g_i \in k[x_2, \ldots, x_n]$ is the leading coefficient of f_i with respect to x_1. From the projective point of view, $\{(0:1)\} \times \mathbf{V}(g_1, \ldots, g_s)$ are the points at infinity in $Z = \mathbf{V}(f_1^h, \ldots, f_s^h)$ (this follows from the proof of Theorem 6). Since $f_1, \ldots, f_s$ was an *arbitrary* basis of I, Z may *not* be the projective closure of V and, hence, $\mathbf{V}(g_1, \ldots, g_s)$ may be too large. To make $\mathbf{V}(g_1, \ldots, g_s) \cap \mathbf{V}(I_1)$ as small as possible in (7), we should use a Gröbner basis for I with respect to a monomial ordering of the type described in Proposition 10.

We will end the section with a study of maps between projective spaces. Suppose that $f_0, \ldots, f_m \in k[x_0, \ldots, x_n]$ are homogeneous polynomials of total degree d such that $\mathbf{V}(f_0, \ldots, f_m) = \emptyset$ in $\mathbb{P}^n$. Then we can define a map $F : \mathbb{P}^n \to \mathbb{P}^m$ by the formula

$$F(x_0 : \cdots : x_n) = (f_0(x_0, \ldots, x_n) : \cdots : f_m(x_0, \ldots, x_n)).$$

Since $f_0, \ldots, f_m$ never vanish simultaneously on $\mathbb{P}^n$, $F(x_0 : \cdots : x_n)$ always gives a point in $\mathbb{P}^n$. Furthermore, since the f_i are homogeneous of total degree d, we have

$$F(\lambda x_0 : \cdots : \lambda x_n) = (f_0(\lambda x_0, \ldots, \lambda x_n) : \cdots : f_m(\lambda x_0, \ldots, \lambda x_n))$$
$$= (\lambda^d f_0(x_0, \ldots, x_n) : \cdots : \lambda^d f_m(x_0, \ldots, x_n))$$
$$= (f_0(x_0, \ldots, x_n) : \cdots : f_m(x_0, \ldots, x_n)) = F(x_0 : \cdots : x_n)$$

for all $\lambda \in k \setminus \{0\}$. Thus, F is a well-defined function from $\mathbb{P}^n$ to $\mathbb{P}^m$.

We have already seen examples of such maps between projective spaces. For instance, Exercise 21 of §2 studied the map $F : \mathbb{P}^1 \to \mathbb{P}^2$ defined by

$$F(a:b) = (a^2 + b^2 : 2ab : a^2 - b^2).$$

This is a projective parametrization of $\mathbf{V}(x^2 - y^2 - z^2)$. Also, Exercise 12 of §4 discussed the Veronese map $\phi : \mathbb{P}^2 \to \mathbb{P}^5$ defined by

$$\phi(x_0 : x_1 : x_2) = (x_0^2 : x_0 x_1 : x_0 x_2 : x_1^2 : x_1 x_2 : x_2^2).$$

The image of this map is called the Veronese surface in $\mathbb{P}^5$.

Over an algebraically closed field, we can describe the image of $F : \mathbb{P}^n \to \mathbb{P}^m$ using elimination theory as follows.

Theorem 11. *Let k be algebraically closed and let $F : \mathbb{P}^n \to \mathbb{P}^m$ be defined by homogeneous polynomials $f_0, \ldots, f_m \in k[x_0, \ldots, x_n]$ which have the same total degree > 0 and no common zeros in $\mathbb{P}^n$. In $k[x_0, \ldots, x_n, y_0, \ldots, y_m]$, let I be the ideal $\langle y_0 - f_0, \ldots, y_m - f_m \rangle$ and let $I_{n+1} = I \cap k[y_0, \ldots, y_m]$. Then I_{n+1} is a homogeneous ideal in $k[y_0, \ldots, y_m]$, and*

$$F(\mathbb{P}^n) = \mathbf{V}(I_{n+1}).$$

Proof. The first has three parts. The first part is to show that I_{n+1} is a homogeneous ideal. Suppose that the f_i have total degree d. Since the generators $y_i - f_i$ of I are not homogeneous (unless $d = 1$), we will introduce *weights* on the variables $x_0, \ldots, x_n, y_0, \ldots, y_m$. We say that each x_i has weight 1 and each y_j has weight d. Then a monomial $x^\alpha y^\beta$ has weight $|\alpha| + d|\beta|$, and a polynomial $f \in k[x_0, \ldots, x_n, y_0, \ldots, y_m]$ is *weighted homogeneous* provided every monomial in f has the same weight.

The generators $y_i - f_i$ of I all have weight d, so that I is a weighted homogeneous ideal. If we compute a reduced Gröbner basis G for I with respect to any monomial order, an argument similar to the proof of Theorem 2 of §3 shows that G consists of weighted homogeneous polynomials. For an appropriate lex order, the Elimination Theorem from Chapter 3 shows that $G \cap k[y_0, \ldots, y_m]$ is a basis of the elimination ideal $I_{n+1} = I \cap k[y_0, \ldots, y_m]$. Thus, I_{n+1} has a weighted homogeneous basis. Since the y_i's all have the same weight, a polynomial in $k[y_0, \ldots, y_m]$ is weighted homogeneous if and only if it is homogeneous in the usual sense. This proves that I_{n+1} is a homogeneous ideal.

The second part of the proof is to study the image of F. Here, we will consider varieties in the product $\mathbb{P}^n \times \mathbb{P}^m$. A polynomial $h \in k[x_0, \ldots, x_n, y_0, \ldots, y_m]$ is said to be *bihomogeneous* if it can be written as

$$h = \sum_{|\alpha|=l_1, |\beta|=l_2} a_{\alpha\beta} x^\alpha y^\beta.$$

If $h_1, \ldots, h_s$ are bihomogeneous, we get a well-defined set

$$\mathbf{V}(h_1, \ldots, h_s) \subseteq \mathbb{P}^n \times \mathbb{P}^m$$

which is the *variety* defined by $h_1, \ldots, h_s$. Similarly, if $J \subseteq k[x_0, \ldots, x_n, y_0, \ldots, y_m]$ is generated by bihomogeneous polynomials, then we get a variety $\mathbf{V}(J) \subset \mathbb{P}^n \times \mathbb{P}^m$. (See Exercise 13 for the details.)

Elimination theory applies nicely to a bihomogeneous ideal J. The projective elimination ideal $\hat{J} \subseteq k[y_0, \ldots, y_m]$ is a homogeneous ideal (see Exercise 13). Then, using the projection $\pi : \mathbb{P}^n \times \mathbb{P}^m \to \mathbb{P}^m$, it is an easy corollary of Theorem 6 that

$$(8) \qquad\qquad \pi(\mathbf{V}(J)) = \mathbf{V}(\hat{J})$$

in $\mathbb{P}^m$ (see Exercise 13). As in Theorem 6, this requires that k be algebraically closed.

The particular bihomogeneous ideal we will use is $J = \langle y_i f_j - y_j f_i \rangle$. Note that $y_i f_j - y_j f_i$ has degree d in $x_0, \ldots, x_n$ and degree 1 in $y_0, \ldots, y_m$, so that J is indeed bihomogeneous. Let us first show that $\mathbf{V}(J) \subseteq \mathbb{P}^n \times \mathbb{P}^m$ is the graph of $F : \mathbb{P}^n \to \mathbb{P}^m$. Given $p \in \mathbb{P}^n$, we have $(p, F(p)) \in \mathbf{V}(J)$ since $y_i = f_i(p)$ for all i. Conversely, suppose that $(p, q) \in \mathbf{V}(J)$. Then $q_i f_j(p) = q_j f_i(p)$ for all i, j, where q_i is the i-th homogeneous coordinate of q. We can find j with $q_j \neq 0$, and by our assumption on $f_0, \ldots, f_m$, there is i with $f_i(p) \neq 0$. Then $q_i f_j(p) = q_j f_i(p) \neq 0$ shows that $q_i \neq 0$. Now let $\lambda = f_i(p)/q_i$, which is a nonzero element of k. Then $\lambda q_\ell = (f_i(p)/q_i) q_\ell = f_\ell(p)$ since $q_\ell f_i(p) = q_i f_\ell(p)$. It follows that

$$(p, q) = (p_1 : \cdots : p_n, q_1 : \cdots : q_m) = (p_1 : \cdots : p_n, \lambda q_1 : \cdots : \lambda q_m)$$
$$= (p_1 : \cdots : p_n, f_1(p) : \cdots : f_m(p)) = (p, F(p)).$$

Hence (p, q) is in the graph of F in $\mathbb{P}^n \times \mathbb{P}^m$.

As we saw in §3 of Chapter 3, the projection of the graph is the image of the function. Thus, under $\pi : \mathbb{P}^n \times \mathbb{P}^m \to \mathbb{P}^m$, we have $\pi(\mathbf{V}(J)) = F(\mathbb{P}^n)$. If we combine this with (8), we get $F(\mathbb{P}^n) = \mathbf{V}(\hat{J})$ since k is algebraically closed.

The third and final part of the proof is to show that $\mathbf{V}(\hat{J}) = \mathbf{V}(I_{n+1})$ in $\mathbb{P}^m$. It suffices to work in affine space k^{m+1} and prove that $\mathbf{V}_a(\hat{J}) = \mathbf{V}_a(I_{n+1})$. Observe that the variety $\mathbf{V}_a(I) \subseteq k^{n+1} \times k^{m+1}$ is the graph of the map $k^{n+1} \to k^{m+1}$ defined by $(f_0, \ldots, f_m)$. Under the projection $\pi : k^{n+1} \times k^{m+1} \to k^{m+1}$, we claim that $\pi(\mathbf{V}_a(I)) = \mathbf{V}_a(\hat{J})$. We know that $\mathbf{V}(\hat{J})$ is the image of F in $\mathbb{P}^m$. Once we exclude the origin, this means that $q \in \mathbf{V}_a(\hat{J})$ if and only if there is a some $p \in k^{n+1}$ such that q equals $F(p)$ in $\mathbb{P}^m$. Hence, $q = \lambda F(p)$ in k^{m+1} for some $\lambda \neq 0$. If we set $\lambda' = \sqrt[d]{\lambda}$, then $q = F(\lambda' p)$, which is equivalent to $q \in \pi(\mathbf{V}_a(I))$. The claim now follows easily.

By the Closure Theorem (Theorem 3 of Chapter 3, §2), $\mathbf{V}_a(I_{n+1})$ is the smallest variety containing $\pi(\mathbf{V}_a(I))$. Since this projection equals the variety $\mathbf{V}_a(\hat{J})$, it follows immediately that $\mathbf{V}_a(I_{n+1}) = \mathbf{V}_a(\hat{J})$. This completes the proof of the theorem. $\qquad\square$

EXERCISES FOR §5

1. In Example 1, explain why $xy^2 - x + t = 0$ determines a well-defined subset of $\mathbb{P}^1 \times \mathbb{C}$, where $(t:x)$ are homogeneous coordinates on $\mathbb{P}^1$ and y is a coordinate on $\mathbb{C}$. Hint: See Exercise 2.

2. Suppose $F \in k[x_0, \ldots, x_n, y_1, \ldots, y_m]$ is $(x_0, \ldots, x_n)$-homogeneous. Show that if F vanishes at one set of coordinates for a point in $\mathbb{P}^n \times k^m$, then F vanishes at all coordinates for the point.

3. In Example 3, show that $\mathbf{V}(F_1, F_2) = \{(0:1:0), (1:1;-1)\}$.

4. This exercise will study ideals generated by $(x_0, \ldots, x_n)$-homogeneous polynomials.
   a. Prove that every $F \in k[x_0, \ldots, x_n, y_1, \ldots, y_m]$ can be written uniquely as a sum $\sum_i F_i$ where F_i is a $(x_0, \ldots, x_n)$-homogeneous polynomial of degree i in $x_0, \ldots, x_n$. We call these the $(x_0, \ldots, x_n)$-*homogeneous components* of F.
   b. Prove that an ideal $I \subseteq k[x_0, \ldots, x_n, y_1, \ldots, y_m]$ is generated by $(x_0, \ldots, x_n)$-homogeneous polynomials if and only if I contains the $(x_0, \ldots, x_n)$-homogeneous components of each of its elements.

5. In Example 3, we claimed that $(I : \langle u, v \rangle^\infty) \cap k[y] = \langle y(1+y) \rangle$ when $I = \langle u + vy, u + uy \rangle \subseteq k[u, v, y]$. Prove this using the methods of Chapters 3 and 4.

6. As in Example 3, we will use $(u:v;y)$ as coordinates on $\mathbb{P}^1 \times k$. Let $F_1 = u - vy$ and $F_2 = u^2 - v^2y$ in $k[u, v, y]$.
   a. Compute $\mathbf{V}(F_1, F_2)$ and explain geometrically why eliminating u and v should lead to the equation $y(1 - y) = 0$.
   b. Show that $(I : \langle u, v \rangle) \cap k[y] = \{0\}$ and that $(I : \langle u, v \rangle^\infty) \cap k[y] = \langle y(1 - y) \rangle$. This explains why we need saturations—ideal quotients can give an answer that is too small.

7. Let $I \subseteq k[x_0, \ldots, x_n, y_1, \ldots, y_m]$ be an ideal with projective elimination ideal $\hat{I} = (I : \langle x_0, \ldots, x_n \rangle^\infty) \cap k[y_1, \ldots, y_m]$. By Proposition 9 of Chapter 4, §4, we can write $\hat{I}$ as $\hat{I} = (I : \langle x_0, \ldots, x_n \rangle^e) \cap k[y_1, \ldots, y_m]$ for e sufficiently large. This exercise will explore two other ways to express $\hat{I}$.
   a. Prove that if e is sufficiently large, then
   $$\hat{I} = (I : \langle x_0^e, \ldots, x_n^e \rangle) \cap k[y_1, \ldots, y_m].$$
   b. Prove that
   $$\hat{I} = \{f \in k[y_1, \ldots, y_m] \mid \text{for all } 0 \leq i \leq n, \text{ there is } e_i \geq 0 \text{ with } x_i^{e_i}f \in I\}.$$

8. In this exercise, we will use the dehomogenization operator $F \mapsto F^{(i)}$ defined in the discussion preceding Proposition 7.
   a. Prove that $I^{(i)} = \{F^{(i)} \mid F \in I\}$ is an ideal in $k[x_0, \ldots, \hat{x}_i, \ldots, x_n, y_1, \ldots, y_m]$.
   b. If $I = \langle F_1, \ldots, F_s \rangle$, then show that $I^{(i)} = \langle F_1^{(i)}, \ldots, F_s^{(i)} \rangle$.

9. In the proof of Proposition 7, we needed the homogenization operator, which makes a polynomial $f \in k[x_1, \ldots, x_n, y_1, \ldots, y_m]$ into a $(x_0, \ldots, x_n)$-homogeneous polynomial f^h using the extra variable x_0.
   a. Give a careful definition of f^h.
   b. If we dehomogenize f^h by setting $x_0 = 1$, show that we get $(f^h)^{(0)} = f$.
   c. Let $f = F^{(0)}$ be the dehomogenization of a $(x_0, \ldots, x_n)$-homogeneous polynomial F. Then prove that $F = x_0^e f^h$ for some integer $e \geq 0$.

10. Prove part (ii) of Proposition 8.

11. Prove Proposition 10. Also give an example of a monomial order which satisfies the hypothesis of the proposition. Hint: You can use an appropriate weight order from Exercise 11 of Chapter 2, §4.

12. The proof of Theorem 11 used weighted homogeneous polynomials. The general setup is as follows. Given variables $x_0, \ldots, x_n$, we assume that each variable has a weight q_i, which we assume to be a positive integer. Then the *weight* of a monomial x^α is $\sum_{i=0}^{n} q_i \alpha_i$, where $\alpha = (\alpha_0, \ldots, \alpha_n)$. A polynomial is weighted homogeneous if all of its monomials have the same weight.

    a. Show that every $f \in k[x_0, \ldots, x_n]$ can be written uniquely as a sum of weighted homogeneous polynomials $\sum_i f_i$, where f_i is weighted homogeneous of weight i. These are called the *weighted homogeneous components* of f.

    b. Define what it means for an ideal $I \subseteq k[x_0, \ldots, x_n]$ to be a *weighted homogeneous ideal*. Then formulate and prove a version of Theorem 2 of §3 for weighted homogeneous ideals.

13. This exercise will study the elimination theory of $\mathbb{P}^n \times \mathbb{P}^m$. We will use the polynomial ring $k[x_0, \ldots, x_n, y_0, \ldots, y_m]$, where $(x_0 : \cdots : x_m)$ are homogeneous coordinates on $\mathbb{P}^n$ and $(y_0 : \cdots : y_m)$ are homogeneous coordinates on $\mathbb{P}^m$.

    a. As in the text, $h \in k[x_0, \ldots, x_n, y_0, \ldots, y_m]$ is *bihomogeneous* if it can be written in the form

$$h = \sum_{|\alpha| = l_1, |\beta| = l_2} a_{\alpha\beta} x^\alpha y^\beta.$$

    We say that h has *bidegree* (l_1, l_2). If $h_1, \ldots, h_s$ are bihomogeneous, show that we get a well-defined variety

$$\mathbf{V}(h_1, \ldots, h_s) \subseteq \mathbb{P}^n \times \mathbb{P}^m.$$

    Also, if $J \subseteq k[x_0, \ldots, x_n, y_0, \ldots, y_m]$ is an ideal generated by bihomogeneous polynomials, explain how to define $\mathbf{V}(J) \subseteq \mathbb{P}^n \times \mathbb{P}^m$ and prove that $\mathbf{V}(J)$ is a variety.

    b. If J is generated by bihomogeneous polynomials, we have $V = \mathbf{V}(J) \subseteq \mathbb{P}^n \times \mathbb{P}^m$. Since J is also $(x_0, \ldots, x_n)$-homogeneous, we can form its projective elimination ideal $\hat{J} \subseteq k[y_0, \ldots, y_m]$. Prove that $\hat{J}$ is a homogeneous ideal.

    c. Now assume that k is algebraically closed. Under the projection $\pi : \mathbb{P}^n \times \mathbb{P}^m \to \mathbb{P}^m$, prove that

$$\pi(V) = \mathbf{V}(\hat{J})$$

    in $\mathbb{P}^m$. This is the main result in the elimination theory of varieties in $\mathbb{P}^n \times \mathbb{P}^m$. Hint: J also defines a variety in $\mathbb{P}^n \times k^{m+1}$, so that you can apply Theorem 6 to the projection $\mathbb{P}^n \times k^{m+1} \to k^{m+1}$.

14. For the two examples of maps between projective spaces given in the discussion preceding Theorem 11, compute defining equations for the images of the maps.

15. In Exercise 11 of §1, we considered the projective plane $\mathbb{P}^2$, with coordinates $(x : y : z)$, and the dual projective plane $\mathbb{P}^{2\vee}$, where $(A : B : C) \in \mathbb{P}^{2\vee}$ corresponds to the projective line L defined by $Ax + By + Cz = 0$ in $\mathbb{P}^2$. Show that the subset

$$V = \{(p, L) \in \mathbb{P}^2 \times \mathbb{P}^{2\vee} \mid p \in L\} \subseteq \mathbb{P}^2 \times \mathbb{P}^{2\vee}$$

is the variety defined by a bihomogeneous polynomial in $k[x, y, z, A, B, C]$ of bidegree $(1, 1)$. We call V an *incidence variety*. Hint: See part (f) of Exercise 11 of §1.

§6   The Geometry of Quadric Hypersurfaces

In this section, we will study quadric hypersurfaces in $\mathbb{P}^n(k)$. These varieties generalize conic sections in the plane and their geometry is quite interesting. To simplify notation, we will write $\mathbb{P}^n$ rather than $\mathbb{P}^n(k)$, and we will use $x_0, \ldots, x_n$ as homogeneous coordinates. Throughout this section, we will assume that k is a field not of characteristic 2. This means that $2 = 1 + 1 \neq 0$ in k, so that in particular we can divide by 2.

Before introducing quadric hypersurfaces, we need to define the notion of *projective equivalence*. Let $\mathrm{GL}(n + 1, k)$ be the set of invertible $(n + 1) \times (n + 1)$ matrices with entries in k. We can use elements $A \in \mathrm{GL}(n + 1, k)$ to create transformations of $\mathbb{P}^n$ as follows. Under matrix multiplication, A induces a linear map $A : k^{n+1} \to k^{n+1}$ which is an isomorphism since A is invertible. This map takes subspaces of k^{n+1} to subspaces of the same dimension, and restricting to 1-dimensional subspaces, it follows that A takes a line through the origin to a line through the origin. Thus A induces a map $A : \mathbb{P}^n \to \mathbb{P}^n$ [see (1) from §2]. We call such a map a *projective linear transformation*.

In terms of homogeneous coordinates, we can describe $A : \mathbb{P}^n \to \mathbb{P}^n$ as follows. Suppose that $A = (a_{ij})$, where $0 \leq i, j \leq n$. If $p = (b_0 : \cdots : b_n) \in \mathbb{P}^n$, it follows by matrix multiplication that

$$(1) \quad A(p) = A(b_0 : \cdots : b_n) = (a_{00}b_0 + \cdots + a_{0n}b_n : \cdots : a_{n0}b_0 + \cdots + a_{nn}b_n)$$

are homogeneous coordinates for $A(p)$. This formula makes it easy to work with projective linear transformations. Note that $A : \mathbb{P}^n \to \mathbb{P}^n$ is a bijection, and its inverse is given by the matrix $A^{-1} \in \mathrm{GL}(n + 1, k)$. In Exercise 1, you will study the set of *all* projective linear transformations in more detail.

Given a variety $V \subseteq \mathbb{P}^n$ and an element $A \in \mathrm{GL}(n + 1, k)$, we can apply A to all points of V to get the subset $A(V) = \{A(p) \mid p \in V\} \subseteq \mathbb{P}^n$.

Proposition 1. *If $A \in \mathrm{GL}(n + 1, k)$ and $V \subseteq \mathbb{P}^n$ is a variety, then $A(V) \subseteq \mathbb{P}^n$ is also a variety. We say that V and $A(V)$ are* **projectively equivalent**.

Proof. Suppose that $V = \mathbf{V}(f_1, \ldots, f_s)$, where each f_i is a homogeneous polynomial. Since A is invertible, it has an inverse matrix $B = A^{-1}$. Then for each i, let $g_i = f_i \circ B$. If $B = (b_{ij})$, this means

$$g_i(x_0, \ldots, x_n) = f_i(b_{00}x_0 + \cdots + b_{0n}x_n, \ldots, b_{n0}x_0 + \cdots + b_{nn}x_n).$$

It is easy to see that g_i is homogeneous of the same total degree as f_i, and we leave it as an exercise to show that

$$(2) \qquad\qquad A(\mathbf{V}(f_1, \ldots, f_s)) = \mathbf{V}(g_1, \ldots, g_s).$$

This equality proves the proposition. $\qquad\qquad\qquad\qquad\qquad\qquad\qquad\qquad\square$

We can regard $A = (a_{ij})$ as transforming $x_0, \ldots, x_n$ into new coordinates $X_0, \ldots, X_n$ defined by

$$(3) \qquad\qquad\qquad\qquad X_i = \sum_{j=0}^{n} a_{ij} x_j.$$

These give homogeneous coordinates on $\mathbb{P}^n$ because $A \in \mathrm{GL}(n + 1, k)$. It follows from (1) that we can think of $A(V)$ as the original V viewed using the new homogeneous coordinates $X_0, \ldots, X_n$. An example of how this works will be given in Proposition 2.

In studying $\mathbb{P}^n$, an important goal is to classify varieties up to projective equivalence. In the exercises, you will show that projective equivalence is an equivalence relation. As an example of how this works, let us classify hyperplanes $H \subseteq \mathbb{P}^n$ up to projective equivalence. Recall from §2 that a *hyperplane* is defined by a linear equation of the form

$$a_0 x_0 + \cdots + a_n x_n = 0,$$

where $a_0, \ldots, a_n$ are not all zero.

Proposition 2. *All hyperplanes $H \subseteq \mathbb{P}^n$ are projectively equivalent.*

Proof. We will show that H is projectively equivalent to $\mathbf{V}(x_0)$. Since projective equivalence is an equivalence relation, this will prove the proposition.

Suppose that H is defined by $f = a_0 x_0 + \cdots + a_n x_n$, and assume in addition that $a_0 \neq 0$. Now consider the new homogeneous coordinates

$$
\begin{aligned}
X_0 &= a_0 x_0 + a_1 x_1 + \cdots + a_n x_n, \\
X_1 &= x_1
\end{aligned}
$$

(4)
$$\vdots$$

$$X_n = x_n.$$

Then it is easy to see that $\mathbf{V}(f) = \mathbf{V}(X_0)$.

Thus, in the $X_0, \ldots, X_n$ coordinate system, $\mathbf{V}(f)$ is defined by the vanishing of the first coordinate. As explained in (3), this is the same as saying that $\mathbf{V}(f)$ and $\mathbf{V}(x_0)$ are projectively equivalent via the coefficient matrix

$$
A = \begin{pmatrix}
a_0 & a_1 & \cdots & a_n \\
0 & 1 & \cdots & 0 \\
\vdots & \vdots & \ddots & \vdots \\
0 & 0 & \cdots & 1
\end{pmatrix}
$$

from (4). This is invertible since $a_0 \neq 0$. You should check that $A(\mathbf{V}(f)) = \mathbf{V}(x_0)$, so that we have the desired projective equivalence.

More generally, if $a_i \neq 0$ in f, a similar argument shows that $\mathbf{V}(f)$ is projectively equivalent to $\mathbf{V}(x_i)$. We leave it as an exercise to show that $\mathbf{V}(x_i)$ is projectively equivalent to $\mathbf{V}(x_0)$ for all i, and the proposition is proved. $\qquad\square$

In §2, we observed that $\mathbf{V}(x_0)$ can be regarded as a copy of the projective space $\mathbb{P}^{n-1}$. It follows from Proposition 2 that *all* hyperplanes in $\mathbb{P}^n$ look like $\mathbb{P}^{n-1}$.

Now that we understand hyperplanes, we will study the next simplest case, hypersurfaces defined by a homogeneous polynomial of total degree 2.

Definition 3. A variety $V = \mathbf{V}(f) \subseteq \mathbb{P}^n$, where f is a nonzero homogeneous polynomial of total degree 2, is called a **quadric hypersurface**, or more simply, a **quadric**.

The simplest examples of quadrics come from analytic geometry. Recall that a conic section in $\mathbb{R}^2$ is defined by an equation of the form

$$ax^2 + bxy + cy^2 + dx + ey + f = 0.$$

To get the projective closure in $\mathbb{P}^2(\mathbb{R})$, we homogenize with respect to z to get

$$ax^2 + bxy + cy^2 + dxz + eyz + fz^2 = 0,$$

which is homogeneous of total degree 2. For this reason, quadrics in $\mathbb{P}^2$ are called *conics*.

We can classify quadrics up to projective equivalence as follows.

Theorem 4 (Normal Form for Quadrics). *Let $f = \sum_{i,j=0}^{n} a_{ij}x_ix_j \in k[x_0, \ldots, x_n]$ be a nonzero homogeneous polynomial of total degree 2, and assume that k is a field not of characteristic 2. Then $\mathbf{V}(f)$ is projectively equivalent to a quadric defined by an equation of the form*

$$c_0 x_0^2 + c_1 x_1^2 + \cdots + c_n x_n^2 = 0,$$

where $c_0, \ldots, c_n$ are elements of k, not all zero.

Proof. Our strategy will be to find a change of coordinates $X_i = \sum_{j=0}^{n} b_{ij}x_j$ such that f has the form

$$c_0 X_0^2 + c_1 X_1^2 + \cdots + c_n X_n^2.$$

As in Proposition 2, this will give the desired projective equivalence. Our proof will be an elementary application of completing the square.

We will use induction on the number of variables. For one variable, the theorem is trivial since the only homogeneous polynomials of total degree 2 are of the form $a_{00}x_0^2$. Now assume that the theorem is true when there are n variables.

Given $f = \sum_{i,j=0}^{n} a_{ij}x_ix_j$, we first claim that by a change of coordinates, we can assume $a_{00} \neq 0$. To see this, first suppose that $a_{00} = 0$ and $a_{jj} \neq 0$ for some $1 \leq j \leq n$. In this case, we set

(5) $$X_0 = x_j, \quad X_j = x_0, \quad \text{and} \quad X_i = x_i \text{ for } i \neq 0, j.$$

Then the coefficient of X_0^2 in the expansion of f in terms of $X_0, \ldots, X_n$ is nonzero. On the other hand, if all $a_{ii} = 0$, then since $f \neq 0$, we must have $a_{ij} \neq -a_{ji}$ for some $i \neq j$. Making a change of coordinates as in (5), we may assume that $a_{01} \neq -a_{10}$. Now set

(6) $$X_0 = x_0, \quad X_1 = x_1 - x_0, \quad \text{and} \quad X_i = x_i \text{ for } i \geq 2.$$

We leave it as an easy exercise to show that in terms of $X_0, \ldots, X_n$, the polynomial f has the form $\sum_{i,j=0}^{n} c_{ij}X_iX_j$ where $c_{00} = a_{01} + a_{10} \neq 0$. This establishes the claim.

Now suppose that $f = \sum_{i,j=0}^{n} a_{ij}x_ix_j$ where $a_{00} \neq 0$. Let $b_i = a_{i0} + a_{0i}$ and note that

$$\frac{1}{a_{00}}\left(a_{00}x_0 + \sum_{i=1}^{n}\frac{b_i}{2}x_i\right)^2 = a_{00}x_0^2 + \sum_{i=1}^{n}b_ix_0x_i + \sum_{i,j=1}^{n}\frac{b_ib_j}{4a_{00}}x_ix_j.$$

Since the characteristic of k is not 2, we know that $2 = 1+1 \neq 0$ and, thus, division by 2 is possible in k. Now we introduce new coordinates $X_0, \ldots, X_n$, where

(7) $$X_0 = x_0 + \frac{1}{a_{00}}\sum_{i=1}^{n}\frac{b_i}{2}x_i \quad \text{and} \quad X_i = x_i \text{ for } i \geq 1.$$

Writing f in terms of $X_0, \ldots, X_n$, all of the terms X_0X_i cancel for $1 \leq i \leq n$ and, hence, we get a sum of the form

$$a_{00}X_0^2 + \sum_{i,j=1}^{n}d_{ij}X_iX_j.$$

The sum $\sum_{i,j=1}^{n}d_{ij}X_iX_j$ involves the n variables $X_1, \ldots, X_n$, so that by our inductive assumption, we can find a change of coordinates involving only $X_1, \ldots, X_n$ which transforms $\sum_{i,j=1}^{n}d_{ij}X_iX_j$ into $e_1X_1^2 + \cdots + e_nX_n^2$. We can regard this as a coordinate change for $X_0, X_1, \ldots X_n$ which leaves X_0 fixed. Then we have a coordinate change that transforms $a_{00}X_0^2 + \sum_{i,j=1}^{n}d_{ij}X_iX_j$ into the desired form. This completes the proof of the theorem. $\qquad\square$

In the normal form $c_0x_0^2 + \cdots + c_nx_n^2$ given by Theorem 4, some of the coefficients c_i may be zero. By relabeling coordinates, we may assume that $c_i \neq 0$ if $0 \leq i \leq p$ and $c_i = 0$ for $i > p$. Then the quadric is projectively equivalent to one given by the equation

(8) $$c_0x_0^2 + \cdots + c_px_p^2 = 0, \quad c_0, \ldots, c_p \text{ nonzero}.$$

There is a special name for the number of nonzero coefficients.

Definition 5. Let $V \subseteq \mathbb{P}^n$ be a quadric hypersurface, where the field k is infinite and does not have characteristic 2.

(i) If V is defined by an equation of the form (8), then V has **rank** $p + 1$.
(ii) More generally, if V is an arbitrary quadric, then V has **rank** $p + 1$ if V is projectively equivalent to a quadric defined by an equation of the form (8).

For some examples, suppose we use homogeneous coordinates $(x:y:z)$ in $\mathbb{P}^2(\mathbb{R})$. Then the three conics defined by

$$x^2 + y^2 - z^2 = 0, \ x^2 - z^2 = 0, \ x^2 = 0$$

have ranks 3, 2, and 1, respectively. The first conic is the projective version of the circle, whereas the second is the union of two projective lines $\mathbf{V}(x-z) \cup \mathbf{V}(x+z)$, and the third is the projective line $\mathbf{V}(x)$, which we regard as a degenerate conic of multiplicity two. (In general, we can regard any rank 1 quadric as a hyperplane of multiplicity two.)

In Definition 5, we need to show that the rank of a quadric V is well-defined. This is a somewhat subtle question since k need not be algebraically closed. The exercises at the end of the section will give a careful proof that V has a unique rank. One consequence is that projectively equivalent quadrics have the same rank.

Given a quadric $V = \mathbf{V}(f)$, we next show how to compute the rank directly from the defining polynomial $f = \sum_{i,j=0}^{n} a_{ij}x_ix_j$ of V. We begin with two observations.

A first observation is that we can assume $a_{ij} = a_{ji}$ for all i,j. This follows by setting $b_{ij} = (a_{ij} + a_{ji})/2$ (remember that k has characteristic different from 2). An easy computation shows that $f = \sum_{i,j=0}^{n} b_{ij}x_ix_j$, and our claim follows since $b_{ij} = b_{ji}$.

A second observation is that we can use matrix multiplication to represent f. The coefficients of f form an $(n+1) \times (n+1)$ matrix $Q = (a_{ij})$, which we will assume to be symmetric by our first observation. Let $\mathbf{x}$ be the column vector with entries $x_0, \ldots, x_n$. We leave it as an exercise to show

$$f(\mathbf{x}) = \mathbf{x}^t Q \mathbf{x},$$

where $\mathbf{x}^t$ is the transpose of $\mathbf{x}$.

We can compute the rank of $\mathbf{V}(f)$ in terms of Q as follows.

Proposition 6. *Let $f = \mathbf{x}^t Q \mathbf{x}$, where Q is an $(n+1) \times (n+1)$ symmetric matrix.*
(i) *Given an element $A \in \mathrm{GL}(n+1, k)$, let $B = A^{-1}$. Then*

$$A(\mathbf{V}(f)) = \mathbf{V}(g).$$

where $g(\mathbf{x}) = \mathbf{x}^t B^t Q B \mathbf{x}$.
(ii) *The rank of the quadric hypersurface $\mathbf{V}(f)$ equals the rank of the matrix Q*

Proof. To prove (i), we note from (2) that $A(\mathbf{V}(f)) = \mathbf{V}(g)$, where $g = f \circ B$. We compute g as follows:

$$g(\mathbf{x}) = f(B\mathbf{x}) = (B\mathbf{x})^t Q (B\mathbf{x}) = \mathbf{x}^t B^t Q B \mathbf{x},$$

where we have used the fact that $(UV)^t = V^t U^t$ for all matrices U, V such that UV is defined. This completes the proof of (i).

To prove (ii), first note that Q and $B^t Q B$ have the same rank. This follows since multiplying a matrix on the right or left by an invertible matrix does not change the rank.

Now suppose we have used Theorem 4 to find a matrix $A \in \mathrm{GL}(n+1, k)$ such that $g = c_0 x_0^2 + \cdots + c_p x_p^2$ with $c_0, \ldots, c_p$ nonzero. The matrix of g is a diagonal matrix with $c_0, \ldots, c_p$ on the main diagonal. If we combine this with part (i), we see that

$$B^t Q B = \begin{pmatrix} c_0 & & & & & \\ & \ddots & & & & \\ & & c_p & & & \\ & & & 0 & & \\ & & & & \ddots & \\ & & & & & 0 \end{pmatrix},$$

where $B = A^{-1}$. The rank of a matrix is the maximum number of linearly independent columns and it follows that $B^t Q B$ has rank $p + 1$. The above observation then implies that Q also has rank $p + 1$, as desired. $\square$

When k is an algebraically closed field (such as $k = \mathbb{C}$), Theorem 4 and Proposition 6 show that quadrics are completely classified by their rank.

Proposition 7. *If k is algebraically closed and not of characteristic 2, then a quadric hypersurface of rank $p + 1$ is projectively equivalent to the quadric defined by the equation*

$$\sum_{i=0}^{p} x_i^2 = 0.$$

In particular, two quadrics are projectively equivalent if and only if they have the same rank.

Proof. By Theorem 4, we can assume that we have a quadric defined by a polynomial of the form $c_0 x_0^2 + \cdots + c_p x_p^2 = 0$, where $p + 1$ is the rank. Since k is algebraically closed, the equation $x^2 - c_i = 0$ has a root in k. Pick a root and call it $\sqrt{c_i}$. Note that $\sqrt{c_i} \neq 0$ since c_i is nonzero. Then set

$$X_i = \sqrt{c_i} x_i, \quad 0 \le i \le p,$$
$$X_i = x_i, \quad p < i \le n.$$

This gives the desired form and implies that quadrics of the same rank are projectively equivalent. In the discussion following Definition 5, we noted that projectively equivalent quadrics have the same rank. Hence the proof is complete. $\square$

Over the real numbers, the rank is not the only invariant of a quadric hypersurface. For example, in $\mathbb{P}^2(\mathbb{R})$, the conics $V_1 = \mathbf{V}(x^2 + y^2 + z^2)$ and $V_2 = \mathbf{V}(x^2 + y^2 - z^2)$ have rank 3 but cannot be projectively equivalent since V_1 is empty, yet V_2 is not. In the exercises, you will show given any quadric $\mathbf{V}(f)$ with coefficients in $\mathbb{R}$, there are integers $r \ge -1$ and $s \ge 0$ with $0 \le r + s \le n$ such that $\mathbf{V}(f)$ is projectively equivalent over $\mathbb{R}$ to a quadric of the form

$$x_0^2 + \cdots + x_r^2 - x_{r+1}^2 - \cdots - x_{r+s}^2 = 0.$$

(The case $r = -1$ corresponds to when all of the signs are negative.)

We are most interested in quadrics of maximal rank in $\mathbb{P}^n$.

Definition 8. A quadric hypersurface in $\mathbb{P}^n$ is **nonsingular** if it has rank $n + 1$.

A nonsingular quadric is defined by an equation $f = \mathbf{x}^t Q \mathbf{x} = 0$, where Q has rank $n + 1$. Since Q is an $(n + 1) \times (n + 1)$ matrix, this is equivalent to Q being invertible. An immediate consequence of Proposition 7 is the following.

Corollary 9. *Let k be an algebraically closed field not of characteristic 2. Then all nonsingular quadrics in $\mathbb{P}^n$ are projectively equivalent.*

In the exercises, you will show that a quadric in $\mathbb{P}^n$ of rank $p + 1$ can be represented as the join of a nonsingular quadric in $\mathbb{P}^p$ with a copy of $\mathbb{P}^{n-p-1}$. Thus, we can understand all quadrics once we know the nonsingular ones.

For the remainder of the section, we will discuss some interesting properties of nonsingular quadrics in $\mathbb{P}^2$, $\mathbb{P}^3$, and $\mathbb{P}^5$. For the case of $\mathbb{P}^2$, consider the mapping $F : \mathbb{P}^1 \to \mathbb{P}^2$ defined by

$$F(u:v) = (u^2 : uv : v^2),$$

where $(u:v)$ are homogeneous coordinates on $\mathbb{P}^1$. It is straightforward to show that the image of F is contained in the nonsingular conic $\mathbf{V}(x_0 x_2 - x_1^2)$. In fact, the map $F : \mathbb{P}^1 \to \mathbf{V}(x_0 x_2 - x_1^2)$ is a bijection (see Exercise 11), so that this conic looks like a copy of $\mathbb{P}^1$. When k is algebraically closed, it follows that *all* nonsingular conics in $\mathbb{P}^2$ look like $\mathbb{P}^1$.

When we move to quadrics in $\mathbb{P}^3$, the situation is more interesting. Consider the mapping

$$\sigma : \mathbb{P}^1 \times \mathbb{P}^1 \longrightarrow \mathbb{P}^3$$

which takes $(x_0 : x_1, y_0 : y_1) \in \mathbb{P}^1 \times \mathbb{P}^1$ to the point $(x_0 y_0 : x_0 y_1 : x_1 y_0 : x_1 y_1) \in \mathbb{P}^3$. This map is called a *Segre map* and its properties were studied in Exercise 14 of §4. For us, the important fact is that the image of σ is a nonsingular quadric.

Proposition 10. *The Segre map $\sigma : \mathbb{P}^1 \times \mathbb{P}^1 \to \mathbb{P}^3$ is one-to-one and its image is the nonsingular quadric $\mathbf{V}(z_0 z_3 - z_1 z_2)$.*

Proof. We will use $(z_0 : z_1 : z_2 : z_3)$ as homogeneous coordinates on $\mathbb{P}^3$. An easy calculation shows that

(9) $$\sigma(\mathbb{P}^1 \times \mathbb{P}^1) \subseteq \mathbf{V}(z_0 z_3 - z_1 z_2).$$

To prove equality, σ suppose that $(w_0 : w_1 : w_2 : w_3) \in \mathbf{V}(z_0 z_3 - z_1 z_2)$. If $w_0 \neq 0$, then $(w_0 : w_2, w_0 : w_1) \in \mathbb{P}^1 \times \mathbb{P}^1$ and

$$\sigma(w_0 : w_2, w_0 : w_1) = (w_0^2 : w_0 w_1 : w_0 w_2 : w_1 w_2).$$

However, since $w_0 w_3 - w_1 w_2 = 0$, we can write this as

$$\sigma(w_0 : w_2, w_0 : w_1) = (w_0^2 : w_0 w_1 : w_0 w_2 : w_0 w_3) = (w_0 : w_1 : w_2 : w_3).$$

When a different coordinate is nonzero, the proof is similar, and it follows that (9) is an equality. The above argument can be adapted to show that σ is one-to-one (we leave the details as an exercise), and it is also easy to see that $\mathbf{V}(z_0 z_3 - z_1 z_2)$ is nonsingular. This proves the proposition. $\square$

Proposition 10 has some nice consequences concerning lines on the quadric surface $\mathbf{V}(z_0 z_3 - z_1 z_2) \subseteq \mathbb{P}^3$. But before we can discuss this, we need to learn how to describe projective lines in $\mathbb{P}^3$.

Two points $p \neq q$ in $\mathbb{P}^3$ give linearly independent vectors $\hat{p} = (a_0, a_1, a_2, a_3)$ and $\hat{q} = (b_0, b_1, b_2, b_3)$ in k^4. Now consider the map $F : \mathbb{P}^1 \to \mathbb{P}^3$ given by

$$F(u:v) = (a_0u - b_0v : a_1u - b_1v : a_2u - b_2v : a_3u - b_3v),$$

which for later purposes, we will write as

(10) $$F(u:v) = u(a_0:a_1:a_2:a_3) - v(b_0:b_1:b_2:b_3) = up - vq.$$

Since $\hat{p}$ and $\hat{q}$ are linearly independent, $a_0u - b_0v, \ldots, a_3u - b_3v$ cannot vanish simultaneously when $(u:v) \in \mathbb{P}^1$, so that F is defined on all of $\mathbb{P}^1$. In Exercise 13, you will show that the image of F is a variety $L \subseteq \mathbb{P}^3$ defined by linear equations. We call L the *projective line* (or more simply, the *line*) determined by p and q. Note that L contains both p and q. In the exercises, you will show that all lines in $\mathbb{P}^3$ are projectively equivalent and can be regarded as copies of $\mathbb{P}^1$ sitting inside $\mathbb{P}^3$.

Using the Segre map σ, we can identify the quadric $V = \mathbf{V}(z_0z_3 - z_1z_2) \subseteq \mathbb{P}^3$ with $\mathbb{P}^1 \times \mathbb{P}^1$. If we fix $b = (b_0:b_1) \in \mathbb{P}^1$, the image in V of $\mathbb{P}^1 \times \{b\}$ under σ consists of the points $(ub_0:ub_1:vb_0:vb_1)$ as $(u:v)$ ranges over $\mathbb{P}^1$. By (10), this is the projective line through the points $(b_0:b_1:0:0)$ and $(0:0:b_0:b_1)$. Hence, $b \in \mathbb{P}^1$ determines a line $L_b = \sigma(\mathbb{P}^1 \times \{b\})$ lying on the quadric V. If $b \neq b'$, one can easily show that L_b does not intersect $L_{b'}$ and that every point on V lies on a unique such line. Thus, V is swept out by the family $\{L_b \mid b \in \mathbb{P}^1\}$ of nonintersecting lines. Such a surface is called a *ruled surface*. In the exercises, you will show that $\{\sigma(\{a\} \times \mathbb{P}^1) \mid a \in \mathbb{P}^1\}$ is a second family of lines that sweeps out V. If we look at V in the affine space where $z_0 = 1$, then V is defined by $z_3 = z_1z_2$, and we get the following graph:

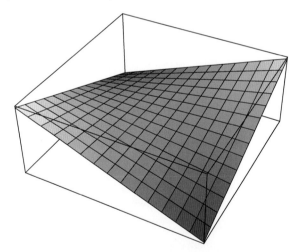

The two families of lines on V are clearly visible in this picture. Over an algebraically closed field, Corollary 9 implies that all nonsingular quadrics in $\mathbb{P}^3$ look like $\mathbf{V}(z_0z_3 - z_1z_2)$ up to projective equivalence. Over the real numbers, however, there are more possibilities—see Exercise 8.

We conclude this section with the problem of describing lines in $\mathbb{P}^3$, which will lead to an interesting quadric in $\mathbb{P}^5$. To motivate what follows, let us first recall the situation of lines in $\mathbb{P}^2$. Here, a line $L \subseteq \mathbb{P}^2$ is defined by a single equation

$A_0 x_0 + A_1 x_1 + A_2 x_2 = 0$. In Exercise 11 of §1, we showed that $(A_0 : A_1 : A_2)$ can be regarded as the "homogeneous coordinates" of L and that the set of all lines forms the dual projective space $\mathbb{P}^{2\vee}$.

It makes sense to ask the same questions for $\mathbb{P}^3$. In particular, can we find "homogeneous coordinates" for lines in $\mathbb{P}^3$? We saw in (10) that a line $L \subseteq \mathbb{P}^3$ can be projectively parametrized using two points $p, q \in L$. This is a good start, but there are infinitely many such pairs on L. How do we get something unique out of this? The idea is the following. Points $p \neq q \in L$ give vectors $\hat{p} = (a_0, a_1, a_2, a_3)$ and $\hat{q} = (b_0, b_1, b_2, b_3)$ in k^4. Then consider the 2×4 matrix whose rows are $\hat{p}$ and $\hat{q}$:

$$\Omega = \begin{pmatrix} a_0 & a_1 & a_2 & a_3 \\ b_0 & b_1 & b_1 & b_3 \end{pmatrix}.$$

We will create coordinates for L using the determinants of 2×2 submatrices of Ω. If we number the columns of Ω using $0, 1, 2, 3$, then the determinant formed using columns i and j will be denoted w_{ij}. We can assume $0 \leq i < j \leq 3$, and we get the six determinants

(11)
$$\begin{aligned} w_{01} &= a_0 b_1 - a_1 b_0, \\ w_{02} &= a_0 b_2 - a_2 b_0, \\ w_{03} &= a_0 b_3 - a_3 b_0, \\ w_{12} &= a_1 b_2 - a_2 b_1, \\ w_{13} &= a_1 b_3 - a_3 b_1, \\ w_{23} &= a_2 b_3 - a_3 b_2. \end{aligned}$$

We will encode them in the 6-tuple

$$(w_{01}, w_{02}, w_{03}, w_{12}, w_{13}, w_{23}) \in k^6.$$

The w_{ij} are called the *Plücker coordinates* of the line L. A first observation is that any line has at least one nonzero Plücker coordinate. To see why, note that Ω has row rank 2 since $\hat{p}$ and $\hat{q}$ are linearly independent. Hence the column rank is also 2, so that there must be two linearly independent columns. These columns give a nonzero Plücker coordinate. From $(w_{01}, w_{02}, w_{03}, w_{12}, w_{13}, w_{23}) \in k^6 \setminus \{0\}$, we get

$$\omega(p, q) = (w_{01} : w_{02} : w_{03} : w_{12} : w_{13} : w_{23}) \in \mathbb{P}^5.$$

To see how the Plücker coordinates depend on the chosen points $p, q \in L$, suppose that we pick a different pair $p' \neq q' \in L$. By (10), L can be described as

$$L = \{up - vq \mid (u, v) \in \mathbb{P}^1\}.$$

In particular, we can write

$$\begin{aligned} p' &= up - vq, \\ q' &= sp - tq \end{aligned}$$

for distinct points $(u:v), (s:t) \in \mathbb{P}^1$. We leave it as an exercise to show that

$$\omega(p', q') = (w'_{01} : w'_{02} : w'_{03} : w'_{12} : w'_{13} : w'_{23}) \in \mathbb{P}^5,$$

where $w'_{ij} = (vs - ut)w_{ij}$ for all $0 \leq i < j \leq 3$. It is easy to see that $vs - ut \neq 0$ since $(u:v) \neq (s:t)$ in $\mathbb{P}^1$. Thus,

$$\omega(p', q') = \omega(p, q)$$

in $\mathbb{P}^5$. This shows that $\omega(p, q)$ gives us a point in $\mathbb{P}^5$ which depends *only* on L. Hence, a line L determines a well-defined point $\omega(L) \in \mathbb{P}^5$.

As we vary L over all lines in $\mathbb{P}^3$, the Plücker coordinates $\omega(L)$ will describe a certain subset of $\mathbb{P}^5$. An straightforward calculation using (11) shows that

$$w_{01}w_{23} - w_{02}w_{13} + w_{03}w_{12} = 0$$

for all sets of Plücker coordinates. If we let z_{ij}, $0 \leq i < j \leq 3$, be homogeneous coordinates on $\mathbb{P}^5$, it follows that the points $\omega(L)$ all lie in the nonsingular quadric $\mathbf{V}(z_{01}z_{23} - z_{02}z_{13} + z_{03}z_{12}) \subseteq \mathbb{P}^5$. Let us prove that this quadric is *exactly* the set of lines in $\mathbb{P}^3$.

Theorem 11. *The map*

$$\{lines\ in\ \mathbb{P}^3\} \longrightarrow \mathbf{V}(z_{01}z_{23} - z_{02}z_{13} + z_{03}z_{12})$$

which sends a line $L \subseteq \mathbb{P}^3$ to its Plücker coordinates $\omega(L) \in \mathbf{V}(z_{01}z_{23} - z_{02}z_{13} + z_{03}z_{12})$ is a bijection.

Proof. The strategy of the proof is to show that a line $L \subseteq \mathbb{P}^3$ can be reconstructed from its Plücker coordinates. Given two points $p = (a_0 : a_1 : a_2 : a_3)$ and $q = (b_0 : b_1 : b_2 : b_3)$ on L, then for the corresponding vectors $\hat{p}, \hat{q} \in k^4$, one can check that (11) implies the following equations in k^4:

$$(12) \qquad \begin{aligned} b_0\hat{p} - a_0\hat{q} &= (0, -w_{01}, -w_{02}, -w_{03}), \\ b_1\hat{p} - a_1\hat{q} &= (w_{01}, 0, -w_{12}, -w_{13}), \\ b_2\hat{p} - a_2\hat{q} &= (w_{02}, w_{12}, 0, -w_{23}), \\ b_3\hat{p} - a_3\hat{q} &= (w_{03}, w_{13}, w_{23}, 0). \end{aligned}$$

It may happen that some of these vectors are 0, but whenever they are nonzero, it follows from (10) that they correspond to points of $L \subseteq \mathbb{P}^3$.

To prove that ω is one-to-one, suppose that we have lines L and L' such that $\omega(L) = \omega(L')$ in $\mathbb{P}^5$. In terms of Plücker coordinates, this means that there is a nonzero λ such that $w_{ij} = \lambda w'_{ij}$ for all $0 \leq i < j \leq 3$. We know that some Plücker coordinate of L is nonzero, and by permuting the coordinates in $\mathbb{P}^3$, we can assume $w_{01} \neq 0$. Then (12) implies that in $\mathbb{P}^3$, the points

$$P = (0 : -w'_{01} : -w'_{02} : -w'_{03}) = (0 : -\lambda w_{01} : -\lambda w_{02} : -\lambda w_{03})$$
$$= (0 : -w_{01} : -w_{02} : -w_{03}),$$
$$Q = (w'_{01} : 0 : -w'_{12} : -w'_{13}) = (\lambda w_{01} : 0 : -\lambda w_{12} : -\lambda w_{13})$$
$$= (w_{01} : 0 : -w_{12} : -w_{13})$$

lie on both L and L'. Since there is a unique line through two points in $\mathbb{P}^3$ (see Exercise 14), it follows that $L = L'$. This proves that our map is one-to-one.

To see that ω is onto, pick a point

$$(w_{01} : w_{02} : w_{03} : w_{12} : w_{13} : w_{23}) \in \mathbf{V}(z_{01}z_{23} - z_{02}z_{13} + z_{03}z_{12}).$$

By changing coordinates in $\mathbb{P}^3$, we can assume $w_{01} \neq 0$. Then the first two vectors in (12) are nonzero and, hence, determine a line $L \subseteq \mathbb{P}^3$. Using the definition of $\omega(L)$ and the relation $w_{01}w_{23} - w_{02}w_{13} + w_{03}w_{12} = 0$, it is straightforward to show that the w_{ij} are the Plücker coordinates of L (see Exercise 16 for the details). This shows that ω is onto and completes the proof of the theorem. □

A nice consequence of Theorem 11 is that the set of lines in $\mathbb{P}^3$ can be given the structure of a projective variety. As we observed at the end of Chapter 7, an important idea in algebraic geometry is that a set of geometrically interesting objects often forms a variety in some natural way.

Theorem 11 can be generalized in many ways. One can study lines in $\mathbb{P}^n$, and it is even possible to define Plücker coordinates for linear subspaces in $\mathbb{P}^n$ of arbitrary dimension. This leads to the study of what are called *Grassmannians*. Using Plücker coordinates, a Grassmannian can be given the structure of a projective variety, although there is usually more than one defining equation. See Exercise 17 for the case of lines in $\mathbb{P}^4$.

We can also think of Theorem 11 from an affine point of view. We already know that there is a natural bijection

$$\{\text{lines through the origin in } k^4\} \cong \{\text{points in } \mathbb{P}^3\},$$

and in the exercises, you will describe a bijection

$$\{\text{planes through the origin in } k^4\} \cong \{\text{lines in } \mathbb{P}^3\}.$$

Thus, Theorem 11 shows that planes through the origin in k^4 have the structure of a quadric hypersurface in $\mathbb{P}^5$. In the exercises, you will see that this has a surprising connection with reduced row echelon matrices. More generally, the Grassmannians mentioned in the previous paragraph can be described in terms of subspaces of a certain dimension in affine space k^{n+1}.

This completes our discussion of quadric hypersurfaces, but by no means exhausts the subject. The classic books by SEMPLE and ROTH (1949) and HODGE and PEDOE (1968) contain a wealth of material on quadric hypersurfaces (and many other interesting projective varieties as well). A more recent reference for quadrics is HARRIS (1995).

EXERCISES FOR §6

1. The set $GL(n + 1, k)$ is closed under inverses and matrix multiplication and is a group in the terminology of Appendix A. In the text, we observed that $A \in GL(n + 1, k)$ induces a projective linear transformation $A : \mathbb{P}^n \to \mathbb{P}^n$. To describe the set of all such transformations, we define a relation on $GL(n + 1, k)$ by

$$A' \sim A \Longleftrightarrow A' = \lambda A \quad \text{for some} \quad \lambda \neq 0.$$

   a. Prove that $\sim$ is an equivalence relation. The set of equivalence classes for $\sim$ is denoted $PGL(n + 1, k)$.
   b. Show that if $A \sim A'$ and $B \sim B'$, then $AB \sim A'B'$. Hence, the matrix product operation is well-defined on the equivalence classes for $\sim$ and, thus, $PGL(n + 1, k)$ has the structure of a group. We call $PGL(n + 1, k)$ the *projective linear group*.
   c. Show that two matrices $A, A' \in GL(n + 1, k)$ define the same mapping $\mathbb{P}^n \to \mathbb{P}^n$ if and only if $A' \sim A$. It follows that we can regard $PGL(n + 1, k)$ as a set of invertible transformations on $\mathbb{P}^n$.

2. Prove equation (2) in the proof of Proposition 1.

3. Prove that projective equivalence is an equivalence relation on the set of projective varieties in $\mathbb{P}^n$.

4. Prove that the hyperplanes $\mathbf{V}(x_i)$ and $\mathbf{V}(x_0)$ are projectively equivalent. Hint: See (5).

5. This exercise is concerned with the proof of Theorem 4.
   a. If $f = \sum_{i,j=0}^{n} a_{ij}x_ix_j$ has $a_{01} \neq -a_{10}$ and $a_{ii} = 0$ for all i, prove that the change of coordinates (6) transforms f into $\sum_{i,j=0}^{n} c_{ij}X_iX_j$ where $c_{00} = a_{01} + a_{10}$.
   b. If $f = \sum_{i,j=0}^{n} a_{ij}x_ix_j$ has $a_{00} \neq 0$, verify that the change of coordinates (7) transforms f into $a_{00}X_0^2 + \sum_{i,j=1}^{n} d_{ij}X_iX_j$.

6. If $f = \sum_{i,j=0}^{n} a_{ij}x_ix_j$, let Q be the $(n + 1) \times (n + 1)$ matrix (a_{ij}).
   a. Show that $f(\mathbf{x}) = \mathbf{x}^tQ\mathbf{x}$.
   b. Suppose that k has characteristic 2 (e.g., $k = \mathbb{F}_2$), and let $f = x_0x_1$. Show that there is *no* symmetric 2×2 matrix Q with entries in k such that $f(\mathbf{x}) = \mathbf{x}^tQ\mathbf{x}$.

7. Use the proofs of Theorem 4 and Proposition 7 to write each of the following as a sum of squares. Assume that $k = \mathbb{C}$.
   a. $x_0x_1 + x_0x_2 + x_2^2$.
   b. $x_0^2 + 4x_1x_3 + 2x_2x_3 + x_4^2$.
   c. $x_0x_1 + x_2x_3 - x_4x_5$.

8. Let $f = \sum_{i,j=0}^{n} a_{ij}x_ix_j \in \mathbb{R}[x_0, \dots, x_n]$ be nonzero.
   a. Show that there are integers $r \geq -1$ and $s \geq 0$ with $0 \leq r + s \leq n$ such that f can be brought to the form

$$x_0^2 + \cdots + x_r^2 - x_{r+1}^2 - \cdots - x_{r+s}^2$$

   by a suitable coordinate change with real coefficients. One can prove that the integers r and s are uniquely determined by f.
   b. Assume $n = 3$ and $f = x_0^2 + \cdots + x_r^2 - x_{r+1}^2 - \cdots - x_3^2$ as in part (a). Of the five possible values $r = -1, \dots, 3$, show that $\mathbf{V}(f)$ is empty in two cases, and in the remaining three, $\mathbf{V}(f) \cap U_0 \subseteq \mathbb{R}^3$ is one of the standard quadric surfaces studied in multivariable calculus.

9. Let $f = \sum_{i,j=0}^{n} a_{ij}x_ix_j \in k[x_0, \dots, x_n]$ be nonzero. In the text, we observed that $\mathbf{V}(f)$ is a nonsingular quadric if and only if $\det(a_{ij}) \neq 0$. We say that $\mathbf{V}(f)$ is *singular* if it is not nonsingular. In this exercise, we will explore a nice way to characterize singular quadrics.

   a. Show that f is singular if and only if there exists a point $a = (a_0 : \cdots : a_n) \in \mathbb{P}^n$ such that

$$\frac{\partial f}{\partial x_0}(a) = \cdots = \frac{\partial f}{\partial x_n}(a) = 0.$$

   b. If $a \in \mathbb{P}^n$ has the property described in part (a), prove that $a \in \mathbf{V}(f)$. In general, a point a of a hypersurface $\mathbf{V}(f)$ (quadric or of higher degree) is called a *singular point* of $\mathbf{V}(f)$ provided that all of the partial derivatives of f vanish at a. Hint: Use Exercise 17 of §2.

10. Let $\mathbf{V}(f) \subseteq \mathbb{P}^n$ be a quadric of rank $p+1$, where $0 < p < n$. Prove that there are $X, Y \subseteq \mathbb{P}^n$ such that (1) $X \simeq \mathbf{V}(g) \subseteq \mathbb{P}^p$ for some nonsingular quadric g, (2) $Y \simeq \mathbb{P}^{n-p-1}$, (3) $X \cap Y = \emptyset$, and (4) $\mathbf{V}(f)$ is the *join* $X * Y$, which is defined to be the set of all lines in $\mathbb{P}^n$ connecting a point of X to a point of Y (and if $X = \emptyset$, we set $X * Y = Y$). Hint: Use Theorem 4.

11. We will study the map $F : \mathbb{P}^1 \to \mathbb{P}^2$ defined by $F(u:v) = (u^2 : uv : v^2)$.

   a. Prove that the image of F lies in $\mathbf{V}(x_0 x_2 - x_1^2)$.

   b. Prove that $F : \mathbb{P}^1 \to \mathbf{V}(x_0 x_2 - x_1^2)$ is a bijection. Hint: Adapt the methods used in the proof of Proposition 10.

12. This exercise will study the Segre map $\sigma : \mathbb{P}^1 \times \mathbb{P}^1 \to \mathbb{P}^3$ defined in the text.

   a. Prove that the image of σ lies in the quadric $\mathbf{V}(z_0 z_3 - z_1 z_2)$.

   b. Use the hint given in the text to prove that σ is one-to-one.

13. In this exercise and the next, we will work out some basic facts about lines in $\mathbb{P}^n$. We start with points $p \neq q \in \mathbb{P}^n$, which correspond to linearly independent vectors $\hat{p}, \hat{q} \in k^{n+1}$.

   a. Adapting the notation used in (10), we can define a map $F : \mathbb{P}^1 \to \mathbb{P}^n$ by $F(u:v) = up - vq$. Show that this map is defined on all of $\mathbb{P}^1$ and is one-to-one.

   b. Let $\ell = a_0 x_0 + \cdots + a_n x_n$ be a linear homogeneous polynomial. Show that ℓ vanishes on the image of F if and only if $p, q \in \mathbf{V}(\ell)$.

   c. Our goal is to show that the image of F is a variety defined by linear equations. Let Ω be the $2 \times (n+1)$ matrix whose rows are $\hat{p}$ and $\hat{q}$. Note that Ω has rank 2. If we multiply column vectors in k^{n+1} by Ω, we get a linear map $\Omega : k^{n+1} \to k^2$. Use results from linear algebra to show that the kernel (or nullspace) of this linear map has dimension $n-1$. Pick a basis $v_1, \ldots, v_{n-1}$ of the kernel, and let ℓ_i be the linear polynomial whose coefficients are the entries of v_i. Then prove that the image of F is $\mathbf{V}(\ell_1 \ldots, \ell_{n-1})$. Hint: Study the subspace of k^{n+1} defined by the equations $\ell_1 = \cdots = \ell_{n-1} = 0$.

14. The exercise will discuss some elementary properties of lines in $\mathbb{P}^n$.

   a. Given points $p \neq q$ in $\mathbb{P}^n$, prove that there is a unique line through p and q.

   b. If L is a line in $\mathbb{P}^n$ and $U_i \cong k^n$ is the affine space where $x_i = 1$, then show that $L \cap U_i$, is either empty or a line in k^n in the usual sense.

   c. Show that all lines in $\mathbb{P}^n$ are projectively equivalent. Hint: In part (c) of Exercise 13, you showed that a line L can be written $L = \mathbf{V}(\ell_1, \ldots, \ell_{n-1})$. Show that you can find ℓ_n and ℓ_{n+1} so that $X_0 = \ell_1, \ldots, X_n = \ell_{n+1}$ is a change of coordinates. What does L look like in the new coordinate system?

15. Let $\sigma : \mathbb{P}^1 \times \mathbb{P}^1 \to \mathbb{P}^3$ be the Segre map.

   a. Show that $L'_a = \sigma(\{a\} \times \mathbb{P}^1)$ is a line in $\mathbb{P}^3$ for all $a \in \mathbb{P}^1$.

   b. Show that every point of $\mathbf{V}(z_0 z_3 - z_1 z_2)$ lies on a unique line L'_a. This proves that the family of lines $\{L'_a \mid a \in \mathbb{P}^1\}$ sweeps out the quadric.

16. This exercise will deal with the proof of Theorem 11.

   a. Show that the Plücker coordinates w'_{ij} of $p' = up - vq$ and $q' = sp - tq$ are related to the Plücker coordinates w_{ij} of p and q via $w'_{ij} = (vs - ut)w_{ij}$.

   b. Use (11) to show that Plücker coordinates satisfy $w_{01}w_{23} - w_{02}w_{13} + w_{03}w_{12} = 0$.

   c. Complete the proof of Theorem 11 by showing that the map ω is onto.

17. In this exercise, we will study Plücker coordinates for lines in $\mathbb{P}^4$

a. Let $L \subseteq \mathbb{P}^4$ be a line. Using the homogeneous coordinates of two points $p, q \in L$, define Plücker coordinates and show that we get a point $\omega(L) \in \mathbb{P}^9$ that depends only on L.

b. Find the relations between the Plücker coordinates and use these to find a variety $V \subseteq \mathbb{P}^4$ such that $\omega(L) \in V$ for all lines L.

c. Show that the map sending a line $L \subseteq \mathbb{P}^4$ to $\omega(L) \in V$ is a bijection.

18. Show that there is a one-to-one correspondence between lines in $\mathbb{P}^3$ and planes through the origin in k^4. This explains why a line in $\mathbb{P}^3$ is different from a line in k^3 or k^4.

19. There is a nice connection between lines in $\mathbb{P}^3$ and 2×4 reduced row echelon matrices of rank 2. Let $V = \mathbf{V}(z_{01}z_{23} - z_{02}z_{13} + z_{03}z_{12})$ be the quadric of Theorem 11.

a. Show that there is a one-to-one correspondence between reduced row echelon matrices of the form
$$\begin{pmatrix} 1 & 0 & a & b \\ 0 & 1 & c & d \end{pmatrix}$$
and points in the affine portion $V \cap U_{01}$, where U_{01} is the affine space in $\mathbb{P}^5$ defined by $z_{01} = 1$. Hint: The rows of the above matrix determine a line in $\mathbb{P}^3$. What are its Plücker coordinates?

b. The matrices given in part (a) do not exhaust all possible 2×4 reduced row echelon matrices of rank 2. For example, we also have the matrices
$$\begin{pmatrix} 1 & a & 0 & b \\ 0 & 0 & 1 & c \end{pmatrix}.$$
Show that there is a one-to-one correspondence between these matrices and points of $V \cap \mathbf{V}(z_{01}) \cap U_{02}$.

c. Show that there are four remaining types of 2×4 reduced row echelon matrices of rank 2 and prove that each of these is in a one-to-one correspondence with a certain portion of V. Hint: The columns containing the leading 1's will correspond to a certain Plücker coordinate being 1.

d. Explain directly (without using V or Plücker coordinates) why 2×4 reduced row echelon matrices of rank 2 should correspond uniquely to lines in $\mathbb{P}^3$. Hint: See Exercise 18.

20. Let k be an algebraically closed field which does not have characteristic 2, and suppose that f, g are nonzero homogeneous polynomials of total degree 2 satisfying $\mathbf{V}(f) = \mathbf{V}(g)$. Use the Nullstellensatz to prove that $f = cg$ for some nonzero $c \in k$. Hint: Proposition 9 of Chapter 4, §2 will be useful. There are three cases to consider: (1) f is irreducible; (2) $f = \ell_1\ell_2$, where ℓ_1, ℓ_2 are linear and neither is a multiple of the other; and (3) $f = \ell^2$.

21. When the field k does not have characteristic 2, Proposition 6 shows that a nonzero homogeneous polynomial f of total degree 2 has a well-defined rank, denoted $\mathrm{rank}(f)$. In order to prove that a quadric hypersurface V has a well-defined rank, we need to show that $V = \mathbf{V}(f) = \mathbf{V}(g)$ implies $\mathrm{rank}(f) = \mathrm{rank}(g)$. If k is algebraically closed, this follows from the previous exercise. Here is a proof that works when k is infinite. The strategy will be to first show that if $\mathrm{rank}(f) = p + 1$, then $\mathrm{rank}(g) \leq p + 1$. This is obvious if $p = n$, so suppose that $p < n$. By a change of coordinates, we may assume $f = c_0x_0^2 + \cdots + c_px_p^2$. Then write
$$g = h_1(x_0, \ldots, x_p) + \sum_{i=0}^{p} x_i\ell_i(x_{p+1}, \ldots, x_n) + h_2(x_{p+1}, \ldots, x_n).$$

a. Show that $h_2 = 0$. Hint: If $(b_{p+1} : \cdots : b_n) \in \mathbb{P}^{n-p-1}$ is arbitrary, then the point
$$(0 : \cdots : 0 : b_{p+1} : \cdots : b_n) \in \mathbb{P}^n$$

lies is $\mathbf{V}(f)$. Then use $\mathbf{V}(f) = \mathbf{V}(g)$ and Proposition 5 of Chapter 1, §1.
b. Show that $\ell_i = 0$ all i. Hint: Suppose there is $1 \leq i \leq p$ such that $\ell_i = \cdots + bx_j + \cdots$ for some $b \neq 0$ and $p + 1 \leq j \leq n$. For any $\lambda \in k$, set

$$p = (0 : \cdots : 0 : 1 : 0 : \cdots : 0 : \lambda : 0 : \cdots : 0) \in \mathbb{P}^n,$$

where the 1 is in position i and the λ is in position j. Show that $f(p) = 1$ and $g(p) = h_1(0, \ldots, 1, \ldots, 0) + \lambda b$. Show that a careful choice of λ makes $g(p) = 0$ and derive a contradiction.
c. Show $g = h_1(x_0, \ldots, x_p)$ implies $\text{rank}(g) \leq p + 1$. Hint: Adapt the proof of Theorem 4.
d. Complete the proof that the rank of V is well-defined.
22. This exercise will study empty quadrics.
    a. Show that $\mathbf{V}(x_0^2 + \cdots + x_n^2) \subseteq \mathbb{P}^n(\mathbb{R})$ is an empty quadric of rank $n + 1$.
    b. Suppose we have a field k and a quadric $V = \mathbf{V}(f) \subseteq \mathbb{P}^n(k)$ which is empty, i.e., $V = \emptyset$. Prove that V has rank $n + 1$.
23. Let $f = x_0^2 + x_1^2$ and $g = x_0^2 + 2x_1^2$ in $\mathbb{Q}[x_0, x_1, x_2]$.
    a. Show that $\mathbf{V}(f) = \mathbf{V}(g) = \{(0 : 0 : 1)\}$ in $\mathbb{P}^2(\mathbb{Q})$.
    b. In contrast to Exercise 20, show that f and g are not multiples of each other and in fact are not projectively equivalent.

§7 Bezout's Theorem

This section will explore what happens when two curves intersect in the plane. We are particularly interested in the *number* of points of intersection. The following examples illustrate why the answer is especially nice when we work with curves in $\mathbb{P}^2(\mathbb{C})$, the projective plane over the complex numbers. We will also see that we need to define the *multiplicity* of a point of intersection. Fortunately, the resultants we learned about in Chapter 3 will make this relatively easy to do.

Example 1. First consider the intersection of a parabola and an ellipse. Suppose that the parabola is $y = x^2$ and the ellipse is $x^2 + 4(y - \lambda)^2 = 4$, where λ is a parameter we can vary. For example, when $\lambda = 2$ or 0, we get the pictures:

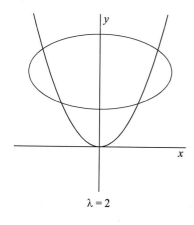

$\lambda = 2$

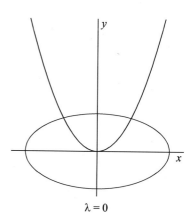

$\lambda = 0$

Over $\mathbb{R}$, we get different numbers of intersections, and it is clear that there are values of λ for which there are *no* points of intersection (see Exercise 1). What is more interesting is that over $\mathbb{C}$, we have four points of intersection in *both* of the above cases. For example, when $\lambda = 0$, we can eliminate x from $y = x^2$ and $x^2 + 4y^2 = 4$ to obtain $y + 4y^2 = 4$. This leads to the solutions

$$(x,y) = \left(\pm\sqrt{\frac{-1 + \sqrt{65}}{8}}, \frac{-1 + \sqrt{65}}{8} \right), \left(\pm\sqrt{\frac{-1 - \sqrt{65}}{8}}, \frac{-1 - \sqrt{65}}{8} \right).$$

The first two are real and the second two are complex (since $-1 - \sqrt{65} < 0$). You can also check that when $\lambda = 2$, working over $\mathbb{C}$ gives no new solutions beyond the four we see in the picture for $\lambda = 2$ (see Exercise 1).

Hence, the number of intersections seems to be more predictable when we work over the complex numbers. As confirmation, you can check that in the cases where there are no points of intersection over $\mathbb{R}$, we still get four points over $\mathbb{C}$ (see Exercise 1).

However, even over $\mathbb{C}$, some unexpected things can happen. For example, suppose we intersect the parabola with the ellipse where $\lambda = 1$:

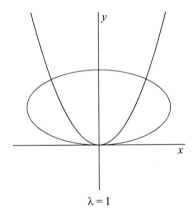

$\lambda = 1$

Here, we see only three points of intersection, and this remains true over $\mathbb{C}$. But the origin is clearly a "special" type of intersection since the two curves are tangent at this point. As we will see later, this intersection has *multiplicity two*, while the other two intersections have *multiplicity one*. If we add up the multiplicities of the points of intersection, we still get four.

Example 2. Now consider the intersection of our parabola $y = x^2$ with a line L. It is easy to see that in most cases, this leads to two points of intersection over $\mathbb{C}$, provided multiplicities are counted properly (see Exercise 2). However, if we intersect with a *vertical* line, then we get the following picture:

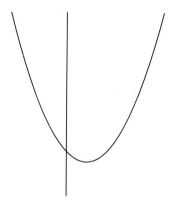

There is just one point of intersection, even over $\mathbb{C}$, and since multiplicities seem to involve tangency, it should be an intersection of multiplicity one. Yet we want the answer to be two, since this is what we get in the other cases. Where is the other point of intersection?

If we change our point of view and work in the projective plane $\mathbb{P}^2(\mathbb{C})$, the above question is easy to answer: the missing point is "at infinity." To see why, let z be the third variable. Then we homogenize $y = x^2$ to get the homogeneous equation $yz = x^2$, and a vertical line $x = c$ gives the projective line $x = cz$. Eliminating x, we get $yz = c^2 z^2$, which is easily solved to obtain $(x:y:z) = (c:c^2:1)$ or $(0:1:0)$ (remember that these are homogeneous coordinates). The first lies in the affine part (where $z = 1$) and is the point we see in the above picture, while the second is on the line at infinity (where $z = 0$).

Example 3. In $\mathbb{P}^2(\mathbb{C})$, consider the two curves given by $C = \mathbf{V}(x^2 - z^2)$ and $D = \mathbf{V}(x^2 y - xz^2 - xyz + z^3)$. It is easy to check that $(1:b:1) \in C \cap D$ for any $b \in \mathbb{C}$, so that the intersection $C \cap D$ is infinite! To see how this could have happened, consider the factorizations

$$x^2 - z^2 = (x - z)(x + z), \quad x^2 y - xz^2 - xyz + z^3 = (x - z)(xy - z^2).$$

Thus, C is a union of two projective lines and D is the union of a line and a conic. In fact, these are the irreducible components of C and D in the sense of §3 (see Proposition 4 below). We now see where the problem occurred: C and D have a common irreducible component $\mathbf{V}(x - z)$, so of course their intersection is infinite.

These examples explain why we want to work in $\mathbb{P}^2(\mathbb{C})$. Hence, for the rest of the section, we will use $\mathbb{C}$ and write $\mathbb{P}^2$ instead of $\mathbb{P}^2(\mathbb{C})$. In this context, a *curve* is a projective variety $\mathbf{V}(f)$ defined by a nonzero homogeneous polynomial $f \in \mathbb{C}[x, y, z]$. Our examples also indicate that we should pay attention to multiplicities of intersections and irreducible components of curves. We begin by studying irreducible components.

Proposition 4. *Let $f \in \mathbb{C}[x, y, z]$ be a nonzero homogeneous polynomial. Then the irreducible factors of f are also homogeneous, and if we factor f into irreducibles:*

$$f = f_1^{a_1} \cdots f_s^{a_s},$$

where f_i is not a constant multiple of f_j for $i \neq j$, then

$$\mathbf{V}(f) = \mathbf{V}(f_1) \cup \cdots \cup \mathbf{V}(f_s)$$

is the minimal decomposition of $\mathbf{V}(f)$ into irreducible components in $\mathbb{P}^2$. Furthermore,

$$\mathbf{I}(\mathbf{V}(f)) = \sqrt{\langle f \rangle} = \langle f_1 \cdots f_s \rangle.$$

Proof. First, suppose that f factors as $f = gh$ for some polynomials $g, h \in \mathbb{C}[x, y, z]$. We claim that g and h must be homogeneous since f is. To prove this, write $g = g_m + \cdots + g_0$, where g_i is homogeneous of total degree i and $g_m \neq 0$. Similarly let $h = h_n + \cdots + h_0$. Then

$$f = gh = (g_m + \cdots + g_0)(h_n + \cdots + h_0)$$
$$= g_m h_n + \text{terms of lower total degree}.$$

Since f is homogeneous, we must have $f = g_m h_n$, and with a little more argument, one can conclude that $g = g_m$ and $h = h_n$ (see Exercise 3). Thus g and h are homogeneous. From here, it follows easily that the irreducible factors f are also homogeneous.

Now suppose f factors as above. Then $\mathbf{V}(f) = \mathbf{V}(f_1) \cup \cdots \cup \mathbf{V}(f_s)$ follows immediately, and this is the minimal decomposition into irreducible components by the projective version of Exercise 10 from Chapter 4, §6. Since $\mathbf{V}(f)$ is nonempty (see Exercise 5), the assertion about $\mathbf{I}(\mathbf{V}(f))$ follows from the Projective Nullstellensatz and Proposition 9 of Chapter 4, §2.   □

A consequence of Proposition 4 is that every curve $C \subseteq \mathbb{P}^2$ has a "best" defining equation. If $C = \mathbf{V}(f)$ for some homogeneous polynomial f, then the proposition implies that $\mathbf{I}(C) = \langle f_1 \cdots f_s \rangle$, where $f_1, \ldots, f_s$ are distinct irreducible factors of f. Thus, any other polynomial defining C is a multiple of $f_1 \cdots f_s$, so that $f_1 \cdots f_s = 0$ is the defining equation of smallest total degree. In the language of Chapter 4, §2, $f_1 \cdots f_s$ is a reduced (or square-free) polynomial. Hence, we call $f_1 \cdots f_s = 0$ the *reduced* equation of C. This equation is unique up to multiplication by a nonzero constant.

When we consider the intersection of two curves C and D in $\mathbb{P}^2$, we will assume that C and D have no common irreducible components. This means that their defining polynomials have no common factors. Our goal is to relate the number of points in $C \cap D$ to the degrees of their reduced equations.

The main tool we will use is the resultant of the defining equations of C and D. Resultants were discussed in §6 of Chapter 3. Readers are advised to review that section up through Corollary 7.

The following lemma will play an important role in our study of this problem.

Lemma 5. *Let $f, g \in \mathbb{C}[x, y, z]$ be homogeneous of total degree m, n respectively. Assume that $f(0, 0, 1)$ and $g(0, 0, 1)$ are nonzero.*

(i) *The resultant* $\mathrm{Res}(f, g, z)$ *is a homogeneous polynomial in x and y of total degree mn if it is nonzero.*

(ii) *If* f, g *have no common factors in* $\mathbb{C}[x, y, z]$*, then* $\mathrm{Res}(f, g, z)$ *is nonzero.*

Proof. First, write f and g as polynomials in z:

$$f = a_0 z^m + \cdots + a_m,$$
$$g = b_0 z^n + \cdots + b_n,$$

and observe that since f is homogeneous of total degree m, each $a_i \in \mathbb{C}[x, y]$ must be homogeneous of degree i. Furthermore, $f(0, 0, 1) \neq 0$ implies that a_0 is a nonzero constant. Similarly, b_i, is homogeneous of degree i and $b_0 \neq 0$.

By Chapter 3, §6, the resultant is given by the $(m + n) \times (m + n)$ determinant

$$\mathrm{Res}(f, g, z) = \det \begin{pmatrix} a_0 & & & b_0 & & \\ \vdots & \ddots & & \vdots & \ddots & \\ a_m & & a_0 & b_n & & b_0 \\ & \ddots & \vdots & & \ddots & \vdots \\ & & a_m & & & b_n \end{pmatrix}$$

$$\underbrace{}_{n \text{ columns}} \underbrace{}_{m \text{ columns}}$$

where the empty spaces are filled by zeros. To show that $\mathrm{Res}(f, g, z)$ is homogeneous of degree mn, let c_{ij} denote the ij-th entry of the matrix. From the pattern of the above matrix, you can check that once we exclude the entries that are obviously zero, we have

$$c_{ij} = \begin{cases} a_{i-j} & \text{if } j \leq n \\ b_{n+i-j} & \text{if } j > n. \end{cases}$$

Thus, a nonzero c_{ij} is homogeneous of total degree $i - j$ (if $j \leq n$) or $n + i - j$ (if $j > n$). By Proposition 1 of Appendix A, §4, the determinant giving $\mathrm{Res}(f, g, z)$ is a sum of products

$$\pm \prod_{i=1}^{m+n} c_{i\sigma(i)},$$

where σ is a permutation of $\{1, \ldots, m + n\}$. We can assume that each $c_{i\sigma(i)}$ in the product is nonzero. If we write the product as

$$\pm \prod_{\sigma(i) \leq n} c_{i\sigma(i)} \prod_{\sigma(i) > n} c_{i\sigma(i)},$$

then, by the above paragraph, this product is a homogeneous polynomial of degree

$$\sum_{\sigma(i) \leq n} (i - \sigma(i)) + \sum_{\sigma(i) > n} (n + i - \sigma(i)).$$

Since σ is a permutation of $\{1, \ldots, m+n\}$, the first sum has n terms and the second has m, and all i's between 1 and $m+n$ appear exactly once. Thus, we can rearrange the sum to obtain

$$mn + \sum_{i=1}^{m+n} i - \sum_{i=1}^{m+n} \sigma(i) = mn.$$

Thus, $\text{Res}(f, g, z)$ is a sum of homogeneous polynomials of degree mn, proving (i).

For (ii), let $k = \mathbb{C}(x, y)$ be the field of rational functions in x, y over $\mathbb{C}$. Let $f = f_1 \cdots f_s$ be the irreducible factorization of f in $\mathbb{C}[x, y, z]$. Since $f = b_0 z^m + \cdots$ for $b_0 \in \mathbb{C} \setminus \{0\}$ and $m > 0$, it follows that every f_i has positive degree in z and hence is irreducible in $k[z]$ by Proposition 3 of Appendix A, §2. Thus, the irreducible factorization of f in $k[z]$ comes from its factorization in $\mathbb{C}[x, y, z]$. The same is clearly true for g.

Now suppose that $\text{Res}(f, g, z)$ is the zero polynomial. If we regard f and g as elements of $k[z]$, then Proposition 3 of Chapter 3, §6 implies that f and g have a common factor in $k[z]$. In particular, they have a common irreducible factor in $k[z]$. By the above paragraph, this implies that they have a common irreducible factor in $\mathbb{C}[x, y, z]$, which contradicts the hypothesis of (ii). $\qquad \square$

This lemma shows that the resultant $\text{Res}(f, g, z)$ is homogeneous in x and y. Homogeneous polynomials in two variables have an especially simple structure.

Lemma 6. *Let $h \in \mathbb{C}[x, y]$ be a nonzero homogeneous polynomial. Then h can be written in the form*

$$h = c(s_1 x - r_1 y)^{m_1} \cdots (s_t x - r_t y)^{m_t},$$

where $c \neq 0$ in $\mathbb{C}$ and $(r_1 : s_1), \ldots, (r_t : s_t)$ are distinct points of $\mathbb{P}^1$. Furthermore,

$$\mathbf{V}(h) = \{(r_1 : s_1), \ldots, (r_t : s_t)\} \subseteq \mathbb{P}^1.$$

Proof. This follows from Exercise 19 of §2. $\qquad \square$

As a first application of these lemmas, we show how to bound the number of points in the intersection of two curves using the degrees of their reduced equations.

Theorem 7. *Let C and D be projective curves in $\mathbb{P}^2$ with no common irreducible components. If the degrees of the reduced equations for C and D are m and n respectively, then $C \cap D$ is finite and has at most mn points.*

Proof. Suppose that $C \cap D$ has more than mn points. Choose $mn + 1$ of them, which we label $p_1, \ldots, p_{mn+1}$, and for $1 \leq i < j \leq mn + 1$, let L_{ij} be the line through p_i and p_j. Then pick a point $q \in \mathbb{P}^2$ such that

$$(1) \qquad\qquad q \notin C \cup D \cup \bigcup_{i<j} L_{ij}$$

(in Exercise 5 you will prove carefully that such points exist).

As in §6, a matrix $A \in \mathrm{GL}(3, \mathbb{C})$ gives a map $A : \mathbb{P}^2 \to \mathbb{P}^2$. It is easy to find an A such that $A(q) = (0:0:1)$ (see Exercise 5). If we regard A as giving new coordinates for $\mathbb{P}^2$ (see (3) in §6), then the point q has coordinates $(0:0:1)$ in the new system. We can thus assume that $q = (0:0:1)$ in (1).

Now suppose that $C = \mathbf{V}(f)$ and $D = \mathbf{V}(g)$, where f and g are reduced of degrees m and n respectively. Then (1) implies $f(0, 0, 1) \neq 0$ since $(0:0:1) \notin C$, and $g(0, 0, 1) \neq 0$ follows similarly. Since f and g have no common factors, Lemma 5 implies that the resultant $\mathrm{Res}(f, g, z)$ is a nonzero homogeneous polynomial of degree mn in x, y.

If we let $p_i = (u_i : v_i : w_i)$, then since the resultant is in the ideal generated by f and g (Proposition 5 of Chapter 3, §6), we have

$$(2) \qquad \mathrm{Res}(f, g, z)(u_i, v_i) = 0.$$

Note that the line connecting $q = (0:0:1)$ to $p_i = (u_i : v_i : w_i)$ intersects $z = 0$ in the point $(u_i : v_i : 0)$ (see Exercise 5). The picture is as follows:

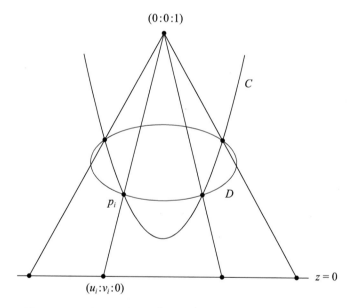

The map taking a point $(u:v:w) \in \mathbb{P}^2 \setminus \{(0:0:1)\}$ to $(u:v:0)$ is an example of a *projection from a point to a line*. Hence, (2) tells us that $\mathrm{Res}(f, g, z)$ vanishes at the points obtained by projecting the $p_i \in C \cap D$ from $(0:0:1)$ to the line $z = 0$.

By (1), $(0:0:1)$ lies on none of the lines connecting p_i and p_j, which implies that the points $(u_i : v_i : 0)$ are distinct for $i = 1, \ldots, mn + 1$. If we regard $z = 0$ as a copy of $\mathbb{P}^1$ with homogeneous coordinates x, y, then we get distinct points $(u_i : v_i) \in \mathbb{P}^1$, and the homogeneous polynomial $\mathrm{Res}(f, g, z)$ vanishes at all $mn + 1$ of them. By Lemmas 5 and 6, this is impossible since $\mathrm{Res}(f, g, z)$ is nonzero of degree mn, and the theorem follows. $\qquad\qquad\qquad\qquad\qquad\qquad\qquad\qquad\qquad\qquad\qquad\square$

Now that we have a criterion for $C \cap D$ to be finite, the next step is to define an *intersection multiplicity* for each point $p \in C \cap D$. There are a variety of ways this can be done, but the simplest involves the resultant.

Thus, we define the intersection multiplicity as follows. Let C and D be curves in $\mathbb{P}^2$ with no common components and reduced equations $f = 0$ and $g = 0$. For each pair of points $p \neq q$ in $C \cap D$, let L_{pq} be the projective line connecting p and q. Pick a matrix $A \in \mathrm{GL}(3, \mathbb{C})$ such that in the new coordinate system given by A, we have

$$(3) \qquad\qquad (0:0:1) \notin C \cup D \cup \bigcup_{p \neq q \text{ in } C \cap D} L_{pq}.$$

(Example 9 below shows how such coordinate changes are done.) As in the proof of Theorem 7, if $p = (u:v:w) \in C \cap D$, then the resultant $\mathrm{Res}(f, g, z)$ vanishes at $(u:v)$, so that by Lemma 6, $vx - uy$ is a factor of $\mathrm{Res}(f, g, z)$.

Definition 8. Let C and D be curves in $\mathbb{P}^2$ with no common components and reduced defining equations $f = 0$ and $g = 0$. Choose coordinates for $\mathbb{P}^2$ so that (3) is satisfied. Then, given $p = (u:v:w) \in C \cap D$, the **intersection multiplicity** $I_p(C, D)$ is defined to be the exponent of $vx - uy$ in the factorization of $\mathrm{Res}(f, g, z)$.

In order for $I_p(C, D)$ to be well-defined, we need to make sure that we get the same answer no matter what coordinate system satisfying (3) we use in Definition 8. For the moment, we will assume this is true and compute some examples of intersection multiplicities.

Example 9. Consider the following polynomials in $\mathbb{C}[x, y, z]$:

$$f = x^3 + y^3 - 2xyz,$$
$$g = 2x^3 - 4x^2y + 3xy^2 + y^3 - 2y^2z.$$

These polynomials [adapted from WALKER (1950)] define cubic curves $C = \mathbf{V}(f)$ and $D = \mathbf{V}(g)$ in $\mathbb{P}^2$. To study their intersection, we first compute the resultant with respect to z:

$$\mathrm{Res}(f, g, z) = -2y(x - y)^3(2x + y).$$

Since the resultant is in the elimination ideal, points in $C \cap D$ satisfy either $y = 0$, $x - y = 0$ or $2x + y = 0$, and from here, it is easy to show that $C \cap D$ consists of the three points

$$p = (0:0:1), \quad q = (1:1:1), \quad r = (4/7:-8/7:1)$$

(see Exercise 6). In particular, this shows that C and D have no common components.

However, the above resultant does *not* give the correct intersection multiplicities since $(0:0:1) \in C$ (in fact, it is a point of intersection). Hence, we must change coordinates. Start with a point such as

$$(0:1:0) \notin C \cup D \cup L_{pq} \cup L_{pr} \cup L_{qr},$$

and find a coordinate change with $A(0:1:0) = (0:0:1)$, say $A(x:y:z) = (z:x:y)$.
Then
$$(0:0:1) \notin A(C) \cup A(D) \cup L_{A(p)A(q)} \cup L_{A(p)A(r)} \cup L_{A(q)A(r)}.$$
To find the defining equation of $A(C)$, note that
$$(u:v:w) \in A(C) \Longleftrightarrow A^{-1}(u:v:w) \in C \Longleftrightarrow f(A^{-1}(u,v,w)) = 0.$$

Thus, $A(C)$ is defined by the equation $f \circ A^{-1}(x,y,z) = f(y,z,x) = 0$, and
similarly, $A(D)$ is given by $g(y,z,x) = 0$. Then, by Definition 8, the resultant
$\mathrm{Res}(f(y,z,x),g(y,z,x),z)$ gives the multiplicities for $A(p) = (1:0:0)$, $A(q) = (1:1:1)$ and $A(r) = (1:4/7:-8/7)$. The resultant is
$$\mathrm{Res}(f(y,z,x),g(y,z,x),\, z) = 8y^5(x-y)^3(4x-7y).$$

so that in terms of p, q and r, the intersection multiplicities are
$$I_p(C,D) = 5, \quad I_q(C,D) = 3, \quad I_r(C,D) = 1.$$

Example 1. [continued] If we let $\lambda = 1$ in Example 1, we get the curves

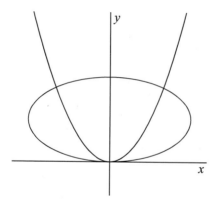

In this picture, the point $(0:0:1)$ is the origin, so we again must change coordinates
before (3) can hold. In the exercises, you will use an appropriate coordinate change
to show that the intersection multiplicity at the origin is in fact equal to 2.

Still assuming that the intersection multiplicities in Definition 8 are well-defined,
we can now prove Bezout's Theorem.

Theorem 10 (Bezout's Theorem). *Let C and D be curves in $\mathbb{P}^2$ with no common
components, and let m and n be the degrees of their reduced defining equations.
Then*
$$\sum_{p \in C \cap D} I_p(C,D) = mn,$$

where $I_p(C,D)$ is the intersection multiplicity at p, as defined in Definition 8.

Proof. Let $f = 0$ and $g = 0$ be the reduced equations of C and D, and assume that coordinates have been chosen so that (3) holds. Write $p \in C \cap D$ as $p = (u_p : v_p : w_p)$. Then we claim that

$$\mathrm{Res}(f, g, z) = c \prod_{p \in C \cap D} (v_p x - u_p y)^{I_p(C,D)},$$

where c is a nonzero constant. For each p, it is clear that $(v_p x - u_p y)^{I_p(C,D)}$ is the exact power of $v_p x - u_p y$ dividing the resultant—this follows by the definition of $I_p(C, D)$. We still need to check that this accounts for *all* roots of the resultant. But if $(u : v) \in \mathbb{P}^1$ satisfies $\mathrm{Res}(f, g, z)(u, v) = 0$, then Corollary 7 of Chapter 3, §6, implies that there is some $w \in \mathbb{C}$ such that f and g vanish at $(u : v : w)$. This is because if we write f and g as in the proof of Lemma 5, a_0 and b_0 are nonzero constants by (3). Thus $(u : v : w) \in C \cap D$, and our claim is proved.

By Lemma 5, $\mathrm{Res}(f, g, z)$ is a nonzero homogeneous polynomial of degree mn. Then Bezout's Theorem follows by comparing the degree of each side in the above equation. $\qquad\square$

Example 9. [continued] In Example 9, we had two cubic curves which intersected in the points $(0 : 0 : 1)$, $(1 : 1 : 1)$ and $(4/7 : -8/7 : 1)$ of multiplicity 5, 3 and 1 respectively. These add up to $9 = 3 \cdot 3$, as desired. If you look back at Example 9, you'll see why we needed to change coordinates in order to compute intersection multiplicities. In the original coordinates, $\mathrm{Res}(f, g, z) = -2y(x - y)^3(2x + y)$, which would give multiplicities 1, 3 and 1. Even without computing the correct multiplicities, we know these cannot be right since they don't add up to 9!

Finally, we show that the intersection multiplicities in Definition 8 are well-defined.

Lemma 11. *In Definition 8, all coordinate change matrices satisfying* (3) *give the same intersection multiplicities* $I_p(C, D)$ *for* $p \in C \cap D$.

Proof. Although this result holds over any algebraically closed field, our proof will use continuity arguments and hence is special to $\mathbb{C}$. We begin by describing carefully the coordinate changes we will use. As in Example 9, pick a point

$$r \notin C \cup D \cup \bigcup_{p \neq q \text{ in } C \cap D} L_{pq}$$

and a matrix $A \in \mathrm{GL}(3, \mathbb{C})$ such that $A(r) = (0 : 0 : 1)$. This means $A^{-1}(0 : 0 : 1) = r$, so that the condition on A is

$$A^{-1}(0 : 0 : 1) \notin C \cup D \cup \bigcup_{p \neq q \text{ in } C \cap D} L_{pq}.$$

Let $l_{pq} = 0$ be the equation of the line L_{pq}, and set

$$h = f \cdot g \cdot \prod_{p \neq q \text{ in } C \cap D} \ell_{pq}.$$

The condition on A is thus $A^{-1}(0:0:1) \notin \mathbf{V}(h)$, i.e., $h(A^{-1}(0,0,1)) \neq 0$.

We can formulate this problem without using matrix inverses as follows. Consider matrices $B \in M_{3\times3}(\mathbb{C})$, where $M_{3\times3}(\mathbb{C})$ is the set of all 3×3 matrices with entries in $\mathbb{C}$, and define the function $H : M_{3\times3}(\mathbb{C}) \to \mathbb{C}$ by

$$H(B) = \det(B) \cdot h(B(0,0,1)).$$

If $B = (b_{ij})$, note that $H(B)$ is a polynomial in the b_{ij}. Since a matrix is invertible if and only if its determinant is nonzero, we have

$$H(B) \neq 0 \iff B \text{ is invertible and } h(B(0,0,1)) \neq 0.$$

Hence the coordinate changes we want are given by $A = B^{-1}$ for matrices B in $M_{3\times3}(\mathbb{C}) \setminus \mathbf{V}(H)$.

Let $C \cap D = \{p_1, \ldots, p_s\}$, and for each $B \in M_{3\times3}(\mathbb{C}) \setminus \mathbf{V}(H)$, let $B^{-1}(p_i) = (u_{i,B} : v_{i,B} : w_{i,B})$. Then, by the argument given in Theorem 10, we can write

$$(4) \qquad \operatorname{Res}(f \circ B, g \circ B, z) = c_B(v_{1,B}x - u_{1,B}y)^{m_{1,B}} \cdots (v_{s,B}x - u_{s,B}y)^{m_{s,B}},$$

where $c_B \neq 0$. This means $I_{p_i}(C, D) = m_{i,B}$ in the coordinate change given by $A = B^{-1}$. Thus, to prove the lemma, we need to show that $m_{i,B}$ takes the same value for *all* $B \in M_{3\times3}(\mathbb{C}) \setminus \mathbf{V}(H)$.

To study the exponents $m_{i,B}$, we consider what happens in general when we have a factorization

$$G(x, y) = (vx - uy)^m H(x, y)$$

where G and H are homogeneous and $(u, v) \neq (0, 0)$. Here, one calculates that

$$(5) \qquad \frac{\partial^{i+j}G}{\partial x^i \partial y^j}(u, v) = \begin{cases} 0 & \text{if } 0 \leq i+j < m \\ m! v^i(-u)^j H(u, v) & \text{if } i+j = m, \end{cases}$$

(see Exercise 8). In particular, if $H(u, v) \neq 0$, then $(u, v) \neq (0, 0)$ implies that some m-th partial of G doesn't vanish at (u, v).

We also need a method for measuring the distance between matrices $B, C \in M_{3\times3}(\mathbb{C})$. If $B = (b_{ij})$ and $C = (c_{ij})$, then the distance between B and C is defined by the formula

$$d(B, C) = \sqrt{\sum_{i,j=1}^{3} |b_{ij} - c_{ij}|^2},$$

where for a complex number $z = a + ib$, $|z| = \sqrt{a^2 + b^2}$. A crucial fact is that any polynomial function $F : M_{3\times3}(\mathbb{C}) \to \mathbb{C}$ is continuous. This means that given $B_0 \in M_{3\times3}(\mathbb{C})$, we can get $F(B)$ arbitrarily close to $F(B_0)$ by taking B sufficiently close to B_0 (as measured by the above distance function). In particular, if $F(B_0) \neq 0$, it follows that $F(B) \neq 0$ for B sufficiently close to B_0.

Now consider the exponent $m = m_{i,B_0}$ for fixed B_0 and i. We claim that $m_{i,B} \leq m$ if B is sufficiently close to B_0. To see this, first note that (4) and (5) imply that some mth partial of $\operatorname{Res}(f \circ B_0, g \circ B_0, z)$ is nonzero at (u_{i,B_0}, v_{i,B_0}). If we write out $(u_{i,B} : v_{i,B})$ and this partial derivative of $\operatorname{Res}(f \circ B, g \circ B, z)$ explicitly, we get formulas which are rational functions with numerators that are polynomials in the entries of B and denominators that are powers of $\det(B)$. Thus this m-th partial of $\text{Res}(f \circ B, g \circ B, z)$, when evaluated at $(u_{i,B} : v_{i,B})$, is a rational function of the same form. Since it is nonzero at B_0, the continuity argument from the previous paragraph shows that this m-th partial of $\text{Res}(f \circ B, g \circ B, z)$ is nonzero at $(u_{i,B}, v_{i,B})$, once B is sufficiently close to B_0. But then, applying (4) and (5) to $\text{Res}(f \circ B, g \circ B, z)$, we conclude that $m_{i,B} \leq m$ [since $m_{i,B} > m$ would imply that all m-th partials would vanish at $(u_{i,B} : v_{i,B})$].

However, if we sum the inequalities $m_{i,B} \leq m = m_{i,B_0}$ for $i = 1, \ldots, s$, we obtain

$$mn = \sum_{i=1}^{s} m_{i,B} \leq \sum_{i=1}^{s} m_{i,B_0} = mn.$$

This implies that we must have term-by-term equalities, so that $m_{i,B} = m_{i,B_0}$ when B is sufficiently close to B_0.

This proves that the function sending B to $m_{i,B}$ is *locally constant*, i.e., its value at a given point is the same as the values at nearby points. In order for us to conclude that the function is actually constant on all of $M_{3\times3}(\mathbb{C}) \setminus \mathbf{V}(H)$, we need to prove that $M_{3\times3}(\mathbb{C}) \setminus \mathbf{V}(H)$ is *path connected*. This will be done in Exercise 9, which also gives a precise definition of path connectedness. Since the Intermediate Value Theorem from calculus implies that a locally constant function on a path connected set is constant (see Exercise 9), we conclude that $m_{i,B}$ takes the same value for all $B \in M_{3\times3}(\mathbb{C}) \setminus \mathbf{V}(H)$. Thus the intersection multiplicities of Definition 8 are well-defined.   □

The intersection multiplicities $I_p(C, D)$ have many properties which make them easier to compute. For example, one can show that $I_p(C, D) = 1$ if and only if p is a nonsingular point of C and D and the curves have distinct tangent lines at p. A discussion of the properties of multiplicities can be found in Chapter 3 of KIRWAN (1992). We should also point out that using resultants to define multiplicities is unsatisfactory in the following sense. Namely, an intersection multiplicity $I_p(C, D)$ is clearly a *local* object—it depends only on the part of the curves C and D near p—while the resultant is a *global* object, since it uses the equations for all of C and D. Local methods for computing multiplicities are available, though they require slightly more sophisticated mathematics. The local point of view is discussed in Chapter 3 of FULTON (1969) and Chapter IV of WALKER (1950). The relation between local methods and resultants is discussed in §2 of Chapter 4 of COX, LITTLE and O'SHEA (2005).

A different approach to the intersection multiplicities $I_p(C, D)$, based on the Euclidean algorithm, can be found in HILMAR and SMYTH (2010).

As an application of what we've done so far in this section, we will prove the following result of Pascal. Suppose we have six distinct points $p_1, \ldots, p_6$ on an irreducible conic in $\mathbb{P}^2$. By Bezout's Theorem, a line meets the conic in at most 2 points (see Exercise 10). Hence, we get six distinct lines by connecting p_1 to p_2, p_2 to p_3, $\ldots$, and p_6 to p_1. If we label these lines $L_1, \ldots, L_6$, then we get the following picture:

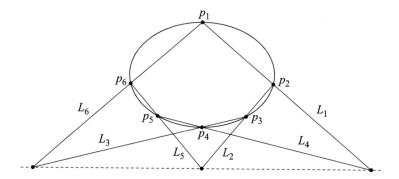

We say that lines L_1, L_4 are *opposite*, and similarly the pairs L_2, L_5 and L_3, L_6 are opposite. The portions of the lines lying inside the conic form a hexagon, and opposite lines correspond to opposite sides of the hexagon.

In the above picture, the intersections of the opposite pairs of lines appear to lie on the same dashed line. The following theorem reveals that this is no accident.

Theorem 12 (Pascal's Mystic Hexagon). *Given six points on an irreducible conic, connected by six lines as above, the points of intersection of the three pairs of opposite lines are collinear.*

Proof. Let the conic be C. As above, we have six points $p_1, \ldots, p_6$ and three pairs of opposite lines $\{L_1, L_4\}$, $\{L_2, L_5\}$, and $\{L_3, L_6\}$. Now consider the two curves $C_1 = L_1 \cup L_3 \cup L_5$ and $C_2 = L_2 \cup L_4 \cup L_6$. These curves are defined by cubic equations, so that by Bezout's Theorem, the number of points in $C_1 \cap C_2$ is 9 (counting multiplicities). However, note that $C_1 \cap C_2$ contains the six original points $p_1, \ldots, p_6$ and the three points of intersection of opposite pairs of lines (you should check this carefully). Thus, these are all of the points of intersection, and all of the multiplicities are one.

Suppose that $C = \mathbf{V}(f)$, $C_1 = \mathbf{V}(g_1)$ and $C_2 = \mathbf{V}(g_2)$, where f has total degree 2 and g_1 and g_2 have total degree 3. Now pick a point $p \in C$ distinct from $p_1, \ldots, p_6$. Thus, $g_1(p)$ and $g_2(p)$ are nonzero (do you see why?), so that $g = g_2(p)g_1 - g_1(p)g_2$ is a cubic polynomial which vanishes at $p, p_1, \ldots, p_6$. Furthermore, g is nonzero since otherwise g_1 would be a multiple of g_2 (or vice versa). Hence, the cubic $\mathbf{V}(g)$ meets the conic C in at least seven points, so that the hypotheses for Bezout's Theorem are not satisfied. Thus, either g is not reduced or $\mathbf{V}(g)$ and C have a common irreducible component. The first of these can't occur, since if g weren't reduced, the curve $\mathbf{V}(g)$ would be defined by an equation of degree at most 2 and $\mathbf{V}(g) \cap C$ would have at most 4 points by Bezout's Theorem. Hence, $\mathbf{V}(g)$ and C must have a common irreducible component. But C is irreducible, which implies that $C = \mathbf{V}(f)$ is a component of $\mathbf{V}(g)$. By Proposition 4, it follows that f must divide g.

Hence, we get a factorization $g = f \cdot l$, where l has total degree 1. Since g vanishes where the opposite lines meet and f does not, it follows that l vanishes at these points. Since $\mathbf{V}(l)$ is a projective line, the theorem is proved. $\qquad\square$

Chapter 8 Projective Algebraic Geometry

Bezout's Theorem serves as a nice introduction to the study of curves in $\mathbb{P}^2$. This part of algebraic geometry is traditionally called *algebraic curves* and includes many interesting topics we have omitted (inflection points, dual curves, elliptic curves, etc.). Fortunately, there are several excellent texts on this subject. In addition to FULTON (1969), KIRWAN (1992) and WALKER (1950) already mentioned, we also warmly recommend CLEMENS (2002), BRIESKORN and KNÖRRER (1986) and FISCHER (2001). For students with a background in complex analysis and topology, we also suggest GRIFFITHS (1989).

EXERCISES FOR §7

1. This exercise is concerned with the parabola $y = x^2$ and the ellipse $x^2 + 4(y - \lambda)^2 = 4$ from Example 1.
   a. Show that these curves have empty intersection over $\mathbb{R}$ when $\lambda < -1$. Illustrate the cases $\lambda < -1$ and $\lambda = -1$ with a picture.
   b. Find the smallest positive real number λ_0 such that the intersection over $\mathbb{R}$ is empty when $\lambda > \lambda_0$. Illustrate the cases $\lambda > \lambda_0$ and $\lambda = \lambda_0$ with a picture.
   c. When $-1 < \lambda < \lambda_0$, describe the possible types of intersections that can occur over $\mathbb{R}$ and illustrate each case with a picture.
   d. In the pictures for parts (a), (b), and (c) use the intuitive idea of multiplicity from Example 1 to determine which ones represent intersections with multiplicity > 1.
   e. Without using Bezout's Theorem, explain why over $\mathbb{C}$, the number of intersections (counted with multiplicity) adds up to 4 when λ is real. Hint: Example 1 gave formulas for the points of intersection when $\lambda = 2$. Do this for general λ.
2. In Example 2, we intersected the parabola $y = x^2$ with a line L in affine space. Assume that L is not vertical.
   a. Over $\mathbb{R}$, show that the number of points of intersection can be 0, 1, or 2. Further, show that you get one point of intersection exactly when L is tangent to $y = x^2$ in the sense of Chapter 3, §4.
   b. Over $\mathbb{C}$, show (without using Bezout's Theorem) that the number of intersections (counted with multiplicity) is exactly 2.
3. In proving Proposition 4, we showed that if $f = gh$ is homogeneous and $g = g_m + \cdots + g_0$, where g_i is homogeneous of total degree i and $g_m \neq 0$, and similarly $h = h_n + \cdots + h_0$, then $f = g_m h_n$. Complete the proof by showing that $g = g_m$ and $h = h_n$. Hint: Let m_0 be the smallest index m_0 such that $g_{m_0} \neq 0$, and define $h_{n_0} \neq 0$ similarly.
4. In this exercise, we sketch an alternate proof of Lemma 5. Given f and g as in the statement of the lemma, let $R(x, y) = \text{Res}(f, g, z)$. It suffices to prove that $R(tx, ty) = t^{mn} R(x, y)$.
   a. Use $a_i(tx, ty) = t^i a_i(x, y)$ and $b_i(tx, ty) = t^i b_i(x, y)$ to show that $R(tx, ty)$ is given by a determinant whose entries are either 0 or $t^i a_i(x, y)$ or $t^i b_i(x, y)$.
   b. In the determinant from part (a), multiply column 2 by t, column 3 by t^2, ..., column n by t^{n-1}, column $n + 2$ by t, column $n + 3$ by t^2, ..., and column $n + m$ by t^{m-1}. Use this to prove that $t^q R(tx, ty)$, where $q = n(n - 1)/2 + m(m - 1)/2$ equals a determinant where in each row, t appears to the *same* power.
   c. By pulling out the powers of t from the rows of the determinant from part (b) prove that $t^q R(tx, ty) = t^r R(x, y)$, where $r = (m + n)(m + n - 1)/2$.
   d. Use part (c) to prove that $R(tx, ty) = t^{mn} R(x, y)$, as desired.
5. This exercise is concerned with the proof of Theorem 7.
   a. Let $f \in \mathbb{C}[x_1, \ldots, x_n]$ be a nonzero polynomial. Prove that $\mathbf{V}(f)$ and $\mathbb{C}^n \setminus \mathbf{V}(f)$ are nonempty. Hint: Use the Nullstellensatz and Proposition 5 of Chapter 1, §1.

b. Use part (a) to prove that you can find $q \notin C \cup D \cup \bigcup_{i<j} L_{ij}$ as claimed in the proof of Theorem 7.

c. Given $q \in \mathbb{P}^2(\mathbb{C})$, find $A \in \mathrm{GL}(3, \mathbb{C})$ such that $A(q) = (0:0:1)$. Hint: The points q and $(0:0:1)$ give nonzero column vectors in $\mathbb{C}^3$. Use linear algebra to find an invertible matrix A taking the first to the second and conclude that $A(q) = (0:0:1)$.

d. Prove that the projective line connecting $(0:0:1)$ to $(u:v:w)$ intersects the line $z = 0$ in the point $(u:v:0)$. Hint: Use equation (10) of §6.

6. In Example 9, we considered the curves $C = \mathbf{V}(f)$ and $D = \mathbf{V}(g)$, where f and g are given in the text.

a. Verify carefully that $p = (0:0:1)$, $q = (1:1:1)$ and $r = (4/7:-8/7:1)$ are the only points of intersection of the curves C and D. Hint: Once you have $\mathrm{Res}(f, g, z)$, you can do the rest by hand.

b. Show that f and g are reduced. Hint: Use a computer.

c. Show that $(0:1:0) \notin C \cup D \cup L_{pq} \cup L_{pr} \cup L_{qr}$.

7. For each of the following pairs of curves, find the points of intersection and compute the intersection multiplicities.

a. $C = \mathbf{V}(yz - x^2)$ and $D = \mathbf{V}(x^2 + 4(y - z)^2 - 4z^2)$. This is the projective version of Example 1 when $\lambda = 1$. Hint: Show that the coordinate change given by $A(x:y:z) = (x:y+z:z)$ has the desired properties.

b. $C = \mathbf{V}(x^2y^3 - 2xy^2z^2 + yz^4 + z^5)$ and $D = \mathbf{V}(x^2y^2 - xz^3 - z^4)$. Hint: There are four solutions, two real and two complex. When finding the complex solutions, computing the gcd of two complex polynomials may help.

8. Prove (5). Hint: Use induction on m, and apply the inductive hypothesis to $\partial G/\partial x$ and $\partial G/\partial y$.

9. (Requires multivariable calculus.) An open set $U \subseteq \mathbb{C}^n$ is *path connected* if for every two points $a, b \in U$, there is a continuous function $\gamma : [0, 1] \to U$ such that $\gamma(0) = a$ and $\gamma(1) = b$.

a. Suppose that $F : U \to \mathbb{Z}$ is locally constant (as in the text, this means that the value of F at a point of U equals its value at all nearby points). Use the Intermediate Value Theorem from calculus to show that F is constant when U is path connected. Hint: If we regard F as a function $F : U \to \mathbb{R}$, explain why F is continuous. Then note that $F \circ \gamma : [0, 1] \to \mathbb{R}$ is also continuous.

b. Let $f \in \mathbb{C}[x]$ be a nonzero polynomial. Prove that $\mathbb{C} \setminus \mathbf{V}(f)$ is path connected.

c. If $f \in \mathbb{C}[x_1, \ldots, x_n]$ is nonzero, prove that $\mathbb{C}^n \setminus \mathbf{V}(f)$ is path connected. Hint: Given $a, b \in \mathbb{C}^n \setminus \mathbf{V}(f)$, consider the complex line $\{ta + (1 - t)b \mid t \in \mathbb{C}\}$ determined by a and b. Explain why $f(ta + (1 - t)b)$ is a nonzero polynomial in t and use part (b).

d. Give an example of $f \in \mathbb{R}[x, y]$ such that $\mathbb{R}^2 \setminus \mathbf{V}(f)$ is not path connected. Further, find a locally constant function $F : \mathbb{R}^2 \setminus \mathbf{V}(f) \to \mathbb{Z}$ which is not constant. Thus, it is essential that we work over $\mathbb{C}$.

10. Let C be an irreducible conic in $\mathbb{P}^2(\mathbb{C})$. Use Bezout's Theorem to explain why a line L meets C in at most two points. What happens when C is reducible? What about when C is a curve defined by an irreducible polynomial of total degree n?

11. In the picture drawn in the text for Pascal's Mystic Hexagon, the six points went clockwise around the conic. If we change the order of the points, we can still form a "hexagon," though opposite lines might intersect inside the conic. For example, the picture could be as follows:

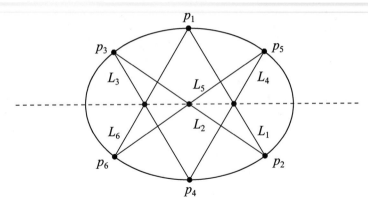

Explain why the theorem remains true in this case.

12. In Pascal's Mystic Hexagon, suppose that the conic is a circle and the six lines come from a regular hexagon inscribed inside the circle. Where do the opposite lines meet and on what line do their intersections lie?

13. Pappus's Theorem from Exercise 8 of Chapter 6, §4, states that if p_3, p_1, p_5 and p_6, p_4, p_2 are two collinear triples of points and we set

$$p = \overline{p_3 p_4} \cap \overline{p_6 p_1}$$
$$q = \overline{p_2 p_3} \cap \overline{p_5 p_6}$$
$$r = \overline{p_4 p_5} \cap \overline{p_1 p_2}.$$

then p, q, r are also collinear. The picture is as follows:

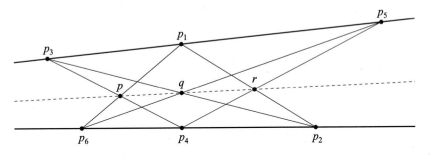

The union of the lines $\overline{p_3 p_1}$ and $\overline{p_6 p_4}$ is a reducible conic C'. Explain why Pappus's Theorem can be regarded as a "degenerate" case of Pascal's Mystic Hexagon. Hint: See Exercise 11. Note that unlike the irreducible case, we can't choose any six points on C': we must avoid the singular point of C', and each component of C' must contain three of the points.

14. The argument used to prove Theorem 12 applies in much more general situations. Suppose that we have curves C and D defined by reduced equations of total degree n such that $C \cap D$ consists of exactly n^2 points. Furthermore, suppose there is an irreducible curve E with a reduced equation of total degree $m < n$ which contains exactly mn of these n^2 points. Then adapt the argument of Theorem 12 to show that there is a curve F with a reduced equation of total degree $n - m$ which contains the remaining $n(n - m)$ points of $C \cap D$.

15. Let C and D be curves in $\mathbb{P}^2(\mathbb{C})$.
   a. Prove that $C \cap D$ must be nonempty.
   b. Suppose that C is nonsingular in the sense of part (a) of Exercise 9 of §6 [if $C = \mathbf{V}(f)$, this means the partial derivatives $\partial f/\partial x$, $\partial f/\partial y$ and $\partial f/\partial z$ don't vanish simultaneously on $\mathbb{P}^2(\mathbb{C})$]. Prove that C is irreducible. Hint: Suppose that $C = C_1 \cup C_2$, which implies $f = f_1 f_2$. How do the partials of f behave at a point of $C_1 \cap C_2$?

16. This exercise will explore an informal proof of Bezout's Theorem. The goal is to give an intuitive explanation of why the number of intersection points is mn.
   a. In $\mathbb{P}^2(\mathbb{C})$, show that a line L meets a curve C of degree n in n points, counting multiplicity. Hint: Choose coordinates so that all of the intersections take place in $\mathbb{C}^2$, and write L parametrically as $x = a + ct$, $y = b + dt$.
   b. If a curve C of degree n meets a union of m lines, use part (a) to predict how many points of intersection there are.
   c. When two curves C and D meet, give an intuitive argument (based on pictures) that the number of intersections (counting multiplicity) doesn't change if one of the curves moves a bit. Your pictures should include instances of tangency and the example of the intersection of the x-axis with the cubic $y = x^3$.
   d. Use the constancy principle from part (c) to argue that if the m lines in part (b) all coincide (giving what is called a line of *multiplicity m*), the number of intersections (counted with multiplicity) is still as predicted.
   e. Using the constancy principle from part (c) argue that Bezout's Theorem holds for general curves C and D by moving D to a line of multiplicity m [as in part (d)]. Hint: If D is defined by $f = 0$, "move" D by letting all but one coefficient of f go to zero.

In technical terms, this is a *degeneration* proof of Bezout's Theorem. A rigorous version of this argument can be found in BRIESKORN and KNÖRRER (1986). Degeneration arguments play an important role in algebraic geometry.

Chapter 9
The Dimension of a Variety

The most important invariant of a linear subspace of affine space is its dimension. For affine varieties, we have seen numerous examples which have a clearly defined dimension, at least from a naive point of view. In this chapter, we will carefully define the dimension of any affine or projective variety and show how to compute it. We will also show that this notion accords well with what we would expect intuitively. In keeping with our general philosophy, we consider the computational side of dimension theory right from the outset.

§1 The Variety of a Monomial Ideal

We begin our study of dimension by considering monomial ideals. In particular, we want to compute the dimension of the variety defined by such an ideal. Suppose, for example, we have the ideal $I = \langle x^2 y, x^3 \rangle$ in $k[x, y]$. Letting H_x denote the line in k^2 defined by $x = 0$ (so $H_x = \mathbf{V}(x)$) and H_y the line $y = 0$, we have

(1)
$$
\begin{aligned}
\mathbf{V}(I) &= \mathbf{V}(x^2 y) \cap \mathbf{V}(x^3) \\
&= (H_x \cup H_y) \cap H_x \\
&= (H_x \cap H_x) \cup (H_y \cap H_x) \\
&= H_x.
\end{aligned}
$$

Thus, $\mathbf{V}(I)$ is the y-axis H_x. Since H_x has dimension 1 as a vector subspace of k^2, it is reasonable to say that it also has dimension 1 as a variety.

As a second example, consider the ideal

$$
I = \langle y^2 z^3, x^5 z^4, x^2 y z^2 \rangle \subseteq k[x, y, z].
$$

Let H_x be the plane defined by $x = 0$ and define H_y and H_z similarly. Also, let H_{xy} be the line $x = y = 0$. Then we have

© Springer International Publishing Switzerland 2015
D.A. Cox et al., *Ideals, Varieties, and Algorithms*, Undergraduate Texts
in Mathematics, DOI 10.1007/978-3-319-16721-3_9

$$\mathbf{V}(I) = \mathbf{V}(y^2 z^3) \cap \mathbf{V}(x^5 z^4) \cap \mathbf{V}(x^2 yz^2)$$
$$= (H_y \cup H_z) \cap (H_x \cup H_z) \cap (H_x \cup H_y \cup H_z)$$
$$= H_z \cup H_{xy}.$$

To verify this, note that the plane H_z belongs to each of the three terms in the second line and, hence, to their intersection. Thus, $\mathbf{V}(I)$ will consist of the plane H_z together, perhaps, with some other subset not contained in H_z. Collecting terms not contained in H_z, we have $H_y \cap H_x \cap (H_x \cup H_y)$, which equals H_{xy}. Thus, $\mathbf{V}(I)$ is the union of the (x, y)-plane H_z and the z-axis H_{xy}. We will say that the dimension of a union of finitely many vector subspaces of k^n is the biggest of the dimensions of the subspaces, and so the dimension of $\mathbf{V}(I)$ is 2 in this example.

The variety of any monomial ideal may be assigned a dimension in much the same fashion. But first we need to describe what a variety of a general monomial ideal looks like. In k^n, a vector subspace defined by setting some subset of the variables $x_1, \dots, x_n$ equal to zero is called a *coordinate subspace*.

Proposition 1. *The variety of a monomial ideal in $k[x_1, \dots, x_n]$ is a finite union of coordinate subspaces of k^n.*

Proof. First, note that if $x_{i_1}^{\alpha_1} \cdots x_{i_r}^{\alpha_r}$ is a monomial in $k[x_1, \dots, x_n]$ with $\alpha_j \geq 1$ for $1 \leq j \leq r$, then

$$\mathbf{V}(x_{i_1}^{\alpha_1} \cdots x_{i_r}^{\alpha_r}) = H_{x_{i_1}} \cup \cdots \cup H_{x_{i_r}},$$

where $H_{x_\ell} = \mathbf{V}(x_\ell)$. Thus, the variety defined by a monomial is a union of coordinate hyperplanes. Note also that there are only n such hyperplanes.

Since a monomial ideal is generated by a finite collection of monomials, the variety corresponding to a monomial ideal is a finite intersection of unions of coordinate hyperplanes. By the distributive property of intersections over unions, any finite intersection of unions of coordinate hyperplanes can be rewritten as a finite union of intersections of coordinate hyperplanes [see (1) for an example of this]. But the intersection of any collection of coordinate hyperplanes is a coordinate subspace. □

When we write the variety of a monomial ideal I as a union of finitely many coordinate subspaces, we can omit a subspace if it is contained in another in the union. Thus, we can write $\mathbf{V}(I)$ as a union of coordinate subspaces.

$$\mathbf{V}(I) = V_1 \cup \cdots \cup V_p,$$

where $V_i \not\subseteq V_j$ for $i \neq j$. In fact, such a decomposition is unique, as you will show in Exercise 8.

Let us make the following provisional definition. We will always assume that k is infinite.

Definition 2. Let V be a variety which is the union of a finite number of linear subspaces of k^n. Then the dimension of V, denoted $\dim V$, is the largest of the dimensions of the subspaces.

Thus, the dimension of the union of two planes and a line is 2, and the dimension of a union of three lines is 1. To compute the dimension of the variety corresponding to a monomial ideal, we merely find the maximum of the dimensions of the coordinate subspaces contained in $\mathbf{V}(I)$.

Although this is easy to do for any given example, it is worth systematizing the computation. Let $I = \langle m_1, \ldots, m_t \rangle$ be a proper ideal generated by the monomials m_j. In trying to compute $\dim \mathbf{V}(I)$, we need to pick out the component of

$$\mathbf{V}(I) = \bigcap_{j=1}^{t} \mathbf{V}(m_j)$$

of largest dimension. If we can find a collection of variables $x_{i_1}, \ldots, x_{i_r}$ such that at least one of these variables appears in each m_j, then the coordinate subspace defined by the equations $x_{i_1} = \cdots = x_{i_r} = 0$ is contained in $\mathbf{V}(I)$. This means we should look for variables which occur in as many of the different m_j as possible. More precisely, for $1 \leq j \leq t$, let

$$M_j = \{\ell \in \{1, \ldots, n\} \mid x_\ell \text{ divides the monomial } m_j\}$$

be the set of subscripts of variables occurring with positive exponent in m_j. (Note that M_j is nonempty by our assumption that $I \neq k[x_1, \ldots, x_n]$.) Then let

$$\mathcal{M} = \{J \subseteq \{1, \ldots, n\} \mid J \cap M_j \neq \emptyset \text{ for all } 1 \leq j \leq t\}$$

consist of all subsets of $\{1, \ldots, n\}$ which have nonempty intersection with *every* set M_j. (Note that $\mathcal{M}$ is not empty because $\{1, \ldots, n\} \in \mathcal{M}$.) If we let $|J|$ denote the number of elements in a set J, then we have the following.

Proposition 3. *With the notation above,*

$$\dim \mathbf{V}(I) = n - \min(|J| \mid J \in \mathcal{M}).$$

Proof. Let $J = \{i_1, \ldots, i_r\}$ be an element of $\mathcal{M}$ such that $|J| = r$ is minimal in $\mathcal{M}$. Since each monomial m_j contains some power of some $x_{i_\ell}, 1 \leq \ell \leq r$, the coordinate subspace $W = \mathbf{V}(x_{i_1}, \ldots, x_{i_r})$ is contained in $\mathbf{V}(I)$. The dimension of W is $n - r = n - |J|$, and hence, by Definition 2, the dimension of $\mathbf{V}(I)$ is at least $n - |J|$.

If $\mathbf{V}(I)$ had dimension larger than $n - r$, then for some $s < r$ there would be a coordinate subspace $W' = \mathbf{V}(x_{\ell_1}, \ldots, x_{\ell_s})$ contained in $\mathbf{V}(I)$. Each monomial m_j would vanish on W' and, in particular, it would vanish at the point $p \in W'$ whose ℓ_i-th coordinate is 0 for $1 \leq i \leq s$ and whose other coordinates are 1. Hence, at least one of the x_{ℓ_i} must divide m_j, and it would follow that $J' = \{\ell_1, \ldots, \ell_s\} \in \mathcal{M}$. Since $|J'| = s < r$, this would contradict the minimality of r. Thus, the dimension of $\mathbf{V}(I)$ must be as claimed. $\qquad\square$

Let us check this on the second example given above. To match the notation of the proposition, we relabel the variables x, y, z as x_1, x_2, x_3, respectively. Then

$$I = \langle x_2^2 x_3^3, x_1^5 x_3^4, x_1^2 x_2 x_3^2 \rangle = \langle m_1, m_2, m_3 \rangle,$$

where
$$m_1 = x_2^2 x_3^3, \quad m_2 = x_1^5 x_3^4, \quad m_3 = x_1^2 x_2 x_3^2.$$

Using the notation of the discussion preceding Proposition 3,
$$M_1 = \{2,3\}, \quad M_2 = \{1,3\}, \quad M_3 = \{1,2,3\},$$

so that
$$\mathcal{M} = \{\{1,2,3\}, \{1,2\}, \{1,3\}, \{2,3\}, \{3\}\}.$$

Then $\min(|J| \mid J \in \mathcal{M}) = 1$, which implies that
$$\dim\mathbf{V}(I) = 3 - \min_{J \in \mathcal{M}} |J| = 3 - 1 = 2.$$

Generalizing this example, note that if some variable, say x_i, appears in every mono-mial in a set of generators for a proper monomial ideal I, then it will be true that $\dim\mathbf{V}(I) = n - 1$ since $J = \{i\} \in \mathcal{M}$. For a converse, see Exercise 4.

It is also interesting to compare a monomial ideal I to its radical $\sqrt{I}$. In the exercises, you will show that $\sqrt{I}$ is a monomial ideal when I is. We also know from Chapter 4 that $\mathbf{V}(I) = \mathbf{V}(\sqrt{I})$ for any ideal I. It follows from Definition 2 that $\mathbf{V}(I)$ and $\mathbf{V}(\sqrt{I})$ have the same dimension (since we defined dimension in terms of the underlying variety). In Exercise 10 you will check that this is consistent with the formula given in Proposition 3.

EXERCISES FOR §1

1. For each of the following monomial ideals I, write $\mathbf{V}(I)$ as a union of coordinate subspaces.
   a. $I = \langle x^5, x^4 yz, x^3 z \rangle \subseteq k[x, y, z]$.
   b. $I = \langle wx^2 y, xyz^3, wz^5 \rangle \subseteq k[w, x, y, z]$.
   c. $I = \langle x_1 x_2, x_3 \cdots x_n \rangle \subseteq k[x_1, \ldots, x_n]$.
2. Find $\dim\mathbf{V}(I)$ for each of the following monomial ideals.
   a. $I = \langle xy, yz, xz \rangle \subseteq k[x, y, z]$.
   b. $I = \langle wx^2 z, w^3 y, wxyz, x^5 z^6 \rangle \subseteq k[w, x, y, z]$.
   c. $I = \langle u^2 vwyz, wx^3 y^3, uxy^7 z, y^3 z, uwx^3 y^3 z^2 \rangle \subseteq k[u, v, w, x, y, z]$.
3. Show that $W \subseteq k^n$ is a coordinate subspace if and only if W can be spanned by a subset of the basis vectors $\{e_i \mid 1 \le i \le n\}$, where e_i is the vector consisting of all zeros except for a 1 in the i-th place.
4. Suppose that $I \subseteq k[x_1, \ldots, x_n]$ is a monomial ideal such that $\dim\mathbf{V}(I) = n - 1$.
   a. Show that the monomials in any generating set for I have a nonconstant common factor.
   b. Write $\mathbf{V}(I) = V_1 \cup \cdots \cup V_p$, where V_i is a coordinate subspace and $V_i \not\subseteq V_j$ for $i \ne j$. Suppose, in addition, that exactly one of the V_i has dimension $n - 1$. What is the maximum that p (the number of components) can be? Give an example in which this maximum is achieved.
5. Let I be a monomial ideal in $k[x_1, \ldots, x_n]$ such that $\dim\mathbf{V}(I) = 0$.
   a. What is $\mathbf{V}(I)$ in this case?
   b. Show that $\dim\mathbf{V}(I) = 0$ if and only if for each $1 \le i \le n$, $x_i^{\ell_i} \in I$ for some $\ell_i \ge 1$. Hint: In Proposition 3, when will it be true that $\mathcal{M}$ contains only $J = \{1, \ldots, n\}$?

6. Let $\langle m_1, \ldots, m_r \rangle \subseteq k[x_1, \ldots, x_n]$ be a monomial ideal generated by $r \leq n$ monomials. Show that $\dim \mathbf{V}(m_1, \ldots, m_r) \geq n - r$.
7. Show that a coordinate subspace is an irreducible variety when the field k is infinite.
8. In this exercise, we will relate the decomposition of the variety of a monomial ideal I as a union of coordinate subspaces given in Proposition 1 with the decomposition of $\mathbf{V}(I)$ into irreducible components. We will assume that the field k is infinite.
   a. If $\mathbf{V}(I) = V_1 \cup \cdots \cup V_p$, where the V_j are coordinate subspaces such that $V_i \not\subseteq V_j$ if $i \neq j$, then show that this union is the minimal decomposition of $\mathbf{V}(I)$ into irreducible varieties given in Theorem 4 of Chapter 4, §6.
   b. Deduce that the V_i in part (a) are unique up to the order in which they are written.
9. Let $I = \langle m_i, \ldots, m_s \rangle$ be a monomial ideal in $k[x_1, \ldots, x_n]$. For each $1 \leq j \leq s$, let $M_j = \{\ell \mid x_\ell \text{ divides } m_j\}$ as in the text, and consider the monomial
$$m_j' = \prod_{\ell \in M_j} x_\ell.$$
Note that m_j' contains exactly the same variables as m_j, but all to the first power.
   a. Show that $m_j' \in \sqrt{I}$ for each $1 \leq j \leq s$.
   b. Show that $\sqrt{I} = \langle m_1', \ldots, m_s' \rangle$. Hint: Use Lemmas 2 and 3 of Chapter 2, §4.
10. Let I be a monomial ideal. Using Exercise 9, show that $\dim \mathbf{V}(I) = \dim \mathbf{V}(\sqrt{I})$ follows from the dimension formula given in Proposition 3.

§2 The Complement of a Monomial Ideal

One of Hilbert's key insights in his famous paper *Über die Theorie der algebraischen Formen* [see HILBERT (1890)] was that the dimension of the variety associated to a monomial ideal could be characterized by the growth of the number of monomials not in the ideal as the total degree increases. We have alluded to this phenomenon in several places in Chapter 5 (notably in Exercise 12 of §3).

In this section, we will make a careful study of the monomials not contained in a monomial ideal $I \subseteq k[x_1, \ldots, x_n]$. Since there may be infinitely many such monomials, our goal will be to find a formula for the number of monomials $x^\alpha \notin I$ which have total degree less than some bound. The results proved here will play a crucial role in §3 when we define the dimension of an arbitrary variety.

Example 1. Consider a proper monomial ideal I in $k[x, y]$. Since I is proper (i.e., $I \neq k[x, y]$), $\mathbf{V}(I)$ is either

a. The origin $\{(0, 0)\}$,
b. the x-axis,
c. the y-axis, or
d. the union of the x-axis and the y-axis.

In case (a), by Exercise 5 of §1, we must have $x^a \in I$ and $y^b \in I$ for some integers $a, b > 0$. Here, the number of monomials not in I will be finite, equal to some constant $C_0 \leq a \cdot b$. If we assume that a and b are as small as possible, we get a picture like the following when we look at exponents:

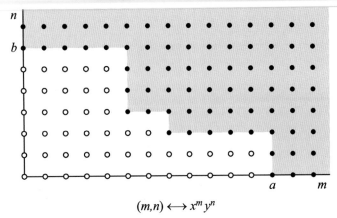

$$(m,n) \longleftrightarrow x^m y^n$$

The monomials in I are indicated by solid dots, while those not in I are open circles.

In case (b), since $\mathbf{V}(I)$ is the x-axis, no power x^a of x can belong to I. On the other hand, since the y-axis does not belong to $\mathbf{V}(I)$, we must have $y^b \in I$ for some minimal integer $b > 0$. The picture would be as follows:

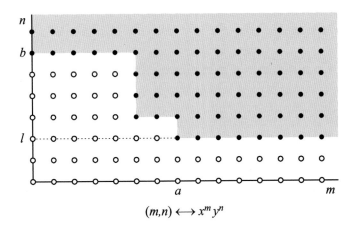

$$(m,n) \longleftrightarrow x^m y^n$$

As the picture indicates, we let l denote the minimum exponent of y that occurs among *all* monomials in I. Note that $l \leq b$, and we also have $l > 0$ since no positive power of x lies in I. Then the monomials in the complement of I are precisely the monomials

$$\{x^i y^j \mid i \in \mathbb{Z}_{\geq 0},\ 0 \leq j \leq l-1\},$$

corresponding to the exponents on l copies of the horizontal axis in $\mathbb{Z}_{\geq 0}^2$, together with a finite number of other monomials. These additional monomials can be characterized as those monomials $m \notin I$ with the property that $x^r m \in I$ for some $r > 0$. In the above picture, they correspond to the open circles on or above the dotted line.

Thus, the monomials in the complement of I consist of l "lines" of monomials together with a finite set of monomials. This description allows us to "count" the number of monomials not in I. More precisely, in Exercise 1, you will show that if $s > l$, the l "lines" contain precisely $l(s+1) - (1 + 2 + \cdots + l - 1)$ monomials of total degree $\leq s$. In particular, if s is large enough (more precisely, we must have $s > a + b$, where a is indicated in the above picture), the number of monomials not in I of total degree $\leq s$ equals $ls + C_0$, where C_0 is some constant depending only on I.

In case (c), the situation is similar to (b), except that the "lines" of monomials are parallel to the vertical axis in the plane $\mathbb{Z}^2_{\geq 0}$ of exponents. In particular, we get a similar formula for the number of monomials not in I of total degree $\leq s$ once s is sufficiently large.

In case (d), let l_1 be the minimum exponent of x that occurs among all monomials of I, and similarly let l_2 be the minimum exponent of y. Note that l_1 and l_2 are positive since xy must divide every monomial in I. Then we have the following picture when we look at exponents:

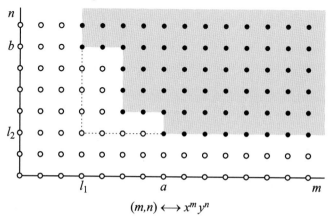

$$(m,n) \longleftrightarrow x^m y^n$$

The monomials in the complement of I consist of the l_1 "lines" of monomials

$$\{x^i y^j \mid 0 \leq i \leq l_1 - 1, \, j \in \mathbb{Z}_{\geq 0}\}$$

parallel to the vertical axis, the l_2 "lines" of monomials

$$\{x^i y^j \mid i \in \mathbb{Z}_{\geq 0}, \, 0 \leq j \leq l_2 - 1\}$$

parallel to the horizontal axis, together with a finite number of other monomials (indicated by open circles inside or on the boundary of the region enclosed by the dotted lines).

Thus, the monomials not in I consist of $l_1 + l_2$ "lines" of monomials together with a finite set of monomials. For s large enough (in fact, for $s > a + b$, where a and b are as in the above picture) the number of monomials not in I of total degree $\leq s$ will be $(l_1 + l_2)s + C_0$, where C_0 is a constant. See Exercise 1 for the details of this claim.

The pattern that appears in Example 1, namely, that the monomials in the complement of a monomial ideal $I \subseteq k[x, y]$ consist of a number of infinite families parallel to the "coordinate subspaces" in $\mathbb{Z}^2_{\geq 0}$, together with a finite collection of monomials, generalizes to arbitrary monomial ideals. In §3, this will be the key to understanding how to define and compute the dimension of an arbitrary variety.

To discuss the general situation, we will introduce some new notation. For each monomial ideal I, we let

$$C(I) = \{\alpha \in \mathbb{Z}_{\geq 0}^n \mid x^\alpha \notin I\}$$

be the set of exponents of monomials *not* in I, i.e., the *complement* of the exponents of monomials in I. This will be our principal object of study. We also set

$$e_1 = (1, 0, \dots, 0),$$
$$e_2 = (0, 1, \dots, 0),$$
$$\vdots$$
$$e_n = (0, 0, \dots, 1).$$

Further, we define the *coordinate subspace of* $\mathbb{Z}_{\geq 0}^n$ determined by $e_{i_1}, \dots, e_{i_r}$, where $i_1 < \dots < i_r$, to be the set

$$[e_{i_1}, \dots, e_{i_r}] = \{a_1 e_{i_1} + \dots + a_r e_{i_r} \mid a_j \in \mathbb{Z}_{\geq 0} \text{ for } 1 \leq j \leq r\}.$$

We say that $[e_{i_1}, \dots, e_{i_r}]$ is an r-dimensional coordinate subspace. Finally, a subset of $\mathbb{Z}_{\geq 0}^n$ is a *translate* of a coordinate subspace $[e_{i_1}, \dots, e_{i_r}]$ if it is of the form

$$\alpha + [e_{i_1}, \dots, e_{i_r}] = \{\alpha + \beta \mid \beta \in [e_{i_1}, \dots, e_{i_r}]\},$$

where $\alpha = \sum_{i \notin \{i_1, \dots, i_r\}} a_i e_i$, $a_i \geq 0$. This restriction on α means that we are translating by a vector perpendicular to $[e_{i_1}, \dots, e_{i_r}]$. For example, $\{(1, l) \mid l \in \mathbb{Z}_{\geq 0}\} = e_1 + [e_2]$ is a translate of the subspace $[e_2]$ in the plane $\mathbb{Z}_{\geq 0}^2$ of exponents.

With these definitions in hand, our discussion of monomial ideals in $k[x, y]$ from Example 1 can be summarized as follows:

a. If $\mathbf{V}(I)$ is the origin, then $C(I)$ consists of a finite number of points.
b. If $\mathbf{V}(I)$ is the x-axis, then $C(I)$ consists of a finite number of translates of $[e_1]$ and, possibly, a finite number of points not on these translates.
c. If $\mathbf{V}(I)$ is the y-axis, then $C(I)$ consists of a finite number of translates of $[e_2]$ and, possibly, a finite number of points not on these translates.
d. If $\mathbf{V}(I)$ is the union of the x-axis and the y-axis, then $C(I)$ consists of a finite number of translates of $[e_1]$, a finite number of translates of $[e_2]$, and, possibly, a finite number of points on neither set of translates.

In the exercises, you will carry out a similar analysis for monomial ideals in the polynomial ring in three variables.

Now let us turn to the general case. We first observe that there is a direct correspondence between the coordinate subspaces in $\mathbf{V}(I)$ and the coordinate subspaces of $\mathbb{Z}_{\geq 0}^n$ contained in $C(I)$.

Proposition 2. *Let $I \subseteq k[x_1, \ldots, x_n]$ be a proper monomial ideal.*

(i) *The coordinate subspace $\mathbf{V}(x_i \mid i \notin \{i_1, \ldots, i_r\})$ is contained in $\mathbf{V}(I)$ if and only if $[e_{i_1}, \ldots, e_{i_r}] \subseteq C(I)$.*

(ii) *The dimension of $\mathbf{V}(I)$ is the dimension of the largest coordinate subspace in $C(I)$.*

Proof. (i) $\Rightarrow$: First observe that $W = \mathbf{V}(x_i \mid i \notin \{i_1, \ldots, i_r\})$ contains the point p whose i_j-th coordinate is 1 for $1 \le j \le r$ and whose other coordinates are 0. For any $\alpha \in [e_{i_1}, \ldots, e_{i_r}]$, the monomial x^α can be written in the form $x^\alpha = x_{i_1}^{\alpha_{i_1}} \cdots x_{i_r}^{\alpha_{i_r}}$. Then $x^\alpha = 1$ at p, so that $x^\alpha \notin I$ since $p \in W \subseteq \mathbf{V}(I)$ by hypothesis. This shows that $\alpha \in C(I)$.

$\Leftarrow$: Suppose that $[e_{i_1}, \ldots e_{i_r}] \subseteq C(I)$. Then every monomial in I contains at least one variable other than $x_{i_1}, \ldots, x_{i_r}$. This means that every monomial in I vanishes on any point $(a_1, \ldots, a_n) \in k^n$ for which $a_i = 0$ when $i \notin \{i_1, \ldots, i_r\}$. So every monomial in I vanishes on the coordinate subspace $\mathbf{V}(x_i \mid i \notin \{i_1, \ldots, i_r\})$, and, hence, the latter is contained in $\mathbf{V}(I)$.

(ii) Note that the coordinate subspace $\mathbf{V}(x_i \mid i \notin \{i_1, \ldots, i_r\})$ has dimension r. It follows from part (i) that the dimensions of the coordinate subspaces of k^n contained in $\mathbf{V}(I)$ and the coordinate subspaces of $\mathbb{Z}_{\ge 0}^n$ contained in $C(I)$ are the same. By Definition 2 of §1, $\dim \mathbf{V}(I)$ is the maximum of the dimensions of the coordinate subspaces of k^n contained in $\mathbf{V}(I)$, so the statement follows. $\square$

We can now characterize the complement of a monomial ideal.

Theorem 3. *If $I \subseteq k[x_1, \ldots, x_n]$ is a proper monomial ideal, then the set $C(I) \subseteq \mathbb{Z}_{\ge 0}^n$ of exponents of monomials not lying in I can be written as a finite (but not necessarily disjoint) union of translates of coordinate subspaces of $\mathbb{Z}_{\ge 0}^n$.*

Before proving the theorem, consider, for example, the ideal $I = \langle x^4 y^3, x^2 y^5 \rangle$.

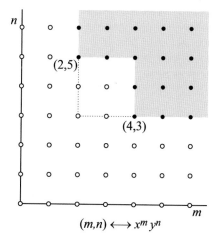

$(m,n) \longleftrightarrow x^m y^n$

Here, it is easy to see that $C(I)$ is the finite union

$$C(I) = [e_1] \cup (e_2 + [e_1]) \cup (2e_2 + [e_1]) \cup [e_2] \cup (e_1 + [e_2])$$
$$\cup \{(3,4)\} \cup \{(3,3)\} \cup \{(2,4)\} \cup \{(2,3)\}.$$

We regard the last four sets in this union as being translates of the 0-dimensional coordinate subspace, which is the origin in $\mathbb{Z}^2_{\geq 0}$.

Proof of Theorem 3. If I is the zero ideal, the theorem is trivially true, so we can assume that $I \neq \{0\}$. The proof is by induction on the number of variables n. If $n = 1$, then $I = \langle x^a \rangle$ for some integer $a > 0$. The only monomials not in I are $1, x, \ldots, x^{a-1}$, and hence $C(I) = \{0, 1, \ldots, a-1\} \subseteq \mathbb{Z}_{\geq 0}$. Thus, the complement consists of a points, all of which are translates of the origin.

Now assume that the result holds for $n - 1$ variables and that we have a monomial ideal $I \subseteq k[x_1, \ldots, x_n]$. For each integer $j \geq 0$, let I_j be the ideal in $k[x_1, \ldots, x_{n-1}]$ generated by monomials m with the property that $m \cdot x_n^j \in I$. Then $C(I_j)$ consists of exponents $\alpha \in \mathbb{Z}^{n-1}_{\geq 0}$ such that $x^\alpha x_n^j \notin I$. Geometrically, this says that $C(I_j) \subseteq \mathbb{Z}^{n-1}_{\geq 0}$ corresponds to the intersection of $C(I)$ and the hyperplane $(0, \ldots, 0, j) + [e_1, \ldots, e_{n-1}]$ in $\mathbb{Z}^n_{\geq 0}$.

Because I is an ideal, we have $I_j \subseteq I_{j'}$ when $j < j'$. By the ascending chain condition for ideals, there is an integer j_0 such that $I_j = I_{j_0}$ for all $j \geq j_0$. For any integer j, we let $C(I_j) \times \{j\}$ denote the set $\{(\alpha, j) \in \mathbb{Z}^n_{\geq 0} \mid \alpha \in C(I_j) \subseteq \mathbb{Z}^{n-1}_{\geq 0}\}$. Then we claim the monomials $C(I)$ not lying in I can be written as

$$(1) \qquad C(I) = (C(I_{j_0}) \times \mathbb{Z}_{\geq 0}) \cup \bigcup_{j=0}^{j_0-1} (C(I_j) \times \{j\}).$$

To prove this claim, first note that $C(I_j) \times \{j\} \subseteq C(I)$ by the definition of $C(I_j)$. To show that $C(I_{j_0}) \times \mathbb{Z}_{\geq 0} \subseteq C(I)$, observe that $I_j = I_{j_0}$ when $j \geq j_0$, so that $C(I_{j_0}) \times \{j\} \subseteq C(I)$ for these j's. When $j < j_0$, we have $x^\alpha x_n^j \notin I$ whenever $x^\alpha x_n^{j_0} \notin I$ since I is an ideal, which shows that $C(I_{j_0}) \times \{j\} \subseteq C(I)$ for $j < j_0$. We conclude that $C(I)$ contains the right-hand side of (1).

To prove the opposite inclusion, take $\alpha = (\alpha_1, \ldots, \alpha_n) \in C(I)$. Then we have $\alpha \in C(I_{\alpha_n}) \times \{\alpha_n\}$ by definition. If $\alpha_n < j_0$, then α obviously lies in the right-hand side of (1). On the other hand, if $\alpha_n \geq j_0$, then $I_{\alpha_n} = I_{j_0}$ shows that $\alpha \in C(I_{j_0}) \times \mathbb{Z}_{\geq 0}$, and our claim is proved.

If we apply our inductive assumption, we can write $C(I_0), \ldots, C(I_{j_0})$ as finite unions of translates of coordinate subspaces of $\mathbb{Z}^{n-1}_{\geq 0}$. Substituting these finite unions into the right-hand side of (1), we immediately see that $C(I)$ is also a finite union of translates of coordinate subspaces of $\mathbb{Z}^n_{\geq 0}$. $\qquad \square$

Our next goal is to find a formula for the number of monomials of total degree $\leq s$ in the complement of a monomial ideal $I \subseteq k[x_1, \ldots, x_n]$. Here is one of the key facts we will need.

Lemma 4. *The number of monomials of total degree $\leq s$ in $k[x_1, \ldots, x_m]$ is the binomial coefficient $\binom{m+s}{s}$.*

Proof. See Exercise 10 of Chapter 5, §3.   □

In what follows, we will refer to $|\alpha| = \alpha_1 + \cdots + \alpha_n$ as the *total degree* of $\alpha \in \mathbb{Z}_{\geq 0}^n$. This is also the total degree of the monomial x^α. Using this terminology, Lemma 4 easily implies that the number of points of total degree $\leq s$ in an m-dimensional coordinate subspace of $\mathbb{Z}_{\geq 0}^n$ is $\binom{m+s}{s}$ (see Exercise 5). Observe that when m is fixed, the expression

$$\binom{m+s}{s} = \binom{m+s}{m} = \frac{1}{m!}(s+m)(s+m-1)\cdots(s+1)$$

is a polynomial of degree m in s. Note that the coefficient of s^m is $1/m!$.

What about the number of monomials of total degree $\leq s$ in a *translate* of an m-dimensional coordinate subspace in $\mathbb{Z}_{\geq 0}^n$? Consider, for instance, the translate $a_{m+1}e_{m+1} + \cdots + a_n e_n + [e_1, \ldots, e_m]$ of the coordinate subspace $[e_1, \ldots, e_m]$. Then, since $a_{m+1}, \ldots, a_n$ are fixed, the number of points in the translate with total degree $\leq s$ is just the number of points in $[e_1, \ldots, e_m]$ of total degree $\leq s - (a_{m+1} + \cdots + a_n)$ provided, of course, that $s > a_{m+1} + \cdots + a_n$. More generally, we have the following.

Lemma 5. *Assume that* $\alpha + [e_{i_1}, \ldots, e_{i_m}]$ *is a translate of the coordinate subspace* $[e_{i_1}, \ldots, e_{i_m}] \subseteq \mathbb{Z}_{\geq 0}^n$, *where as usual* $\alpha = \sum_{i \notin \{i_1, \ldots, i_m\}} a_i e_i$.
(i) *The number of points in* $\alpha + [e_{i_1}, \ldots, e_{i_m}]$ *of total degree* $\leq s$ *is equal to*

$$\binom{m+s-|\alpha|}{s-|\alpha|},$$

provided that $s > |\alpha|$.
(ii) *For* $s > |\alpha|$, *this number of points is a polynomial function of* s *of degree* m, *and the coefficient of* s^m *is* $1/m!$.

Proof. (i) If $s > |\alpha|$, then each point β in $\alpha + [e_{i_1}, \ldots, e_{i_m}]$ of total degree $\leq s$ has the form $\beta = \alpha + \gamma$, where $\gamma \in [e_{i_1}, \ldots, e_{i_m}]$ and $|\gamma| \leq s - |\alpha|$. The formula given in (i) follows using Lemma 4 to count the number of possible γ.
(ii) See Exercise 6.   □

We are now ready to prove a connection between the dimension of $\mathbf{V}(I)$ for a monomial ideal and the degree of the polynomial function which counts the number of points of total degree $\leq s$ in $C(I)$.

Theorem 6. *If* $I \subseteq k[x_1, \ldots, x_n]$ *is a monomial ideal with* $\dim \mathbf{V}(I) = d$, *then for all* s *sufficiently large, the number of monomials not in* I *of total degree* $\leq s$ *is a polynomial of degree* d *in* s. *Furthermore, the coefficient of* s^d *in this polynomial is positive.*

Proof. We need to determine the number of points in $C(I)$ of total degree $\leq s$. By Theorem 3, we know that $C(I)$ can be written as a finite union

$$C(I) = T_1 \cup T_2 \cup \cdots \cup T_t,$$

where each T_i is a translate of a coordinate subspace in $\mathbb{Z}_{\geq 0}^n$. We can assume that $T_i \neq T_j$ for $i \neq j$.

The *dimension* of T_i is the dimension of the associated coordinate subspace. Since I is an ideal, it follows easily that a coordinate subspace $[e_{i_1}, \ldots, e_{i_r}]$ lies in $C(I)$ if and only if some translate does. By hypothesis, $\mathbf{V}(I)$ has dimension d, so that by Proposition 2, each T_i has dimension $\leq d$, with equality occurring for at least one T_i.

We will sketch the remaining steps in the proof, leaving the verification of several details to the reader as exercises. To count the number of points of total degree $\leq s$ in $C(I)$, we must be careful, since $C(I)$ is a union of coordinate subspaces of $\mathbb{Z}_{\geq 0}^n$ that may not be disjoint [for instance, see part (d) of Example 1]. If we use the superscript s to denote the subset consisting of elements of total degree $\leq s$, then it follows that

$$C(I)^s = T_1^s \cup T_2^s \cup \cdots \cup T_t^s.$$

The number of elements in $C(I)^s$ will be denoted $|C(I)^s|$.

In Exercise 7, you will develop a general counting principle (called the Inclusion-Exclusion Principle) that allows us to count the elements in a finite union of finite sets. If the sets in the union have common elements, we cannot simply add to find the total number of elements because that would count some elements in the union more than once. The Inclusion-Exclusion Principle gives "correction terms" that eliminate this multiple counting. Those correction terms are the numbers of elements in double intersections, triple intersections, etc., of the sets in question.

If we apply the Inclusion-Exclusion Principle to the above union for $C(I)^s$, we easily obtain

$$(2) \qquad |C(I)^s| = \sum_i |T_i^s| - \sum_{i<j} |T_i^s \cap T_j^s| + \sum_{i<j<k} |T_i^s \cap T_j^s \cap T_k^s| - \cdots .$$

By Lemma 5, we know that for s sufficiently large, the number of points in T_i^s is a polynomial of degree $m_i = \dim(T_i) \leq d$ in s, and the coefficient of s^{m_i} is $1/m_i!$.

We first note that the first sum in (2) is a polynomial of degree d in s when s is sufficiently large. The degree is exactly d because some of the T_i have dimension d and the coefficients of the leading terms are positive and hence can't cancel. If we can show that the remaining sums in (2) correspond to polynomials of smaller degree, it will follow that $|C(I)^s|$ is given by a polynomial of degree d in s. This will also show that the coefficient of s^d is positive.

You will prove in Exercise 8 that the intersection of two distinct translates of coordinate subspaces of dimensions m and r in $\mathbb{Z}_{\geq 0}^n$ is either empty or a translate of a coordinate subspace of dimension $< \max(m, r)$. Let us see how this applies to a nonzero term $|T_i^s \cap T_j^s|$ in the second sum of (2). Since $T_i \neq T_j$, Exercise 8 implies that $T = T_i \cap T_j$ is the translate of a coordinate subspace of $\mathbb{Z}_{\geq 0}^n$ of dimension $< d$, so that by Lemma 5, the number of points in $T^s = T_i^s \cap T_j^s$ is a polynomial in s of degree $< d$. Adding these up for all $i < j$, we see that the second sum in (2) is a polynomial of degree $< d$ in s for s sufficiently large. The other sums in (2) are handled similarly, and it follows that $|C(I)^s|$ is a polynomial of the desired form when s is sufficiently large. $\qquad \square$

Let us see how this theorem works in the example $I = \langle x^4y^3, x^2y^5 \rangle$ discussed following Theorem 3. Here, we have already seen that $C(I) = C_0 \cup C_1$, where

$$C_1 = [e_1] \cup (e_2 + [e_1]) \cup (2e_2 + [e_1]) \cup [e_2] \cup (e_1 + [e_2]),$$
$$C_0 = \{(3,4), (3,3), (2,4), (2,3)\}.$$

To count the number of points of total degree $\leq s$ in C_1, we count the number in each translate and subtract the number which are counted more than once. (In this case, there are no triple intersections to worry about. Do you see why?) The number of points of total degree $\leq s$ in $[e_2]$ is $\binom{1+s}{s} = \binom{1+s}{1} = s + 1$ and the number in $e_1 + [e_2]$ is $\binom{1+s-1}{s-1} = s$. Similarly, the numbers in $[e_1], e_2 + [e_1]$, and $2e_2 + [e_1]$ are $s+1, s$, and $s-1$, respectively. Of the possible intersections of pairs of these, only six are nonempty and each consists of a single point. You can check that $(1,2)$, $(1,1)$, $(1,0)$, $(0,2)$, $(0,1)$, $(0,0)$ are the six points belonging to more than one translate. Thus, for large s, the number of points of total degree $\leq s$ in C_1 is given by

$$|C_1^s| = (s+1) + s + (s+1) + s + (s-1) - 6 = 5s - 5.$$

Since there are four points in C_0, the number of points of total degree $\leq s$ in $C(I)$ is

$$|C_1^s| + |C_0^s| = (5s - 5) + 4 = 5s - 1,$$

provided that s is sufficiently large. (In Exercise 9 you will show that in this case, s is "sufficiently large" as soon as $s \geq 7$.)

Theorem 6 shows that the dimension of the affine variety defined by a monomial ideal is equal to the degree of the polynomial in s which counts the number of points in $C(I)$ of total degree $\leq s$ for s large. This gives a purely algebraic definition of dimension. In §3, we will extend these ideas to general ideals.

The polynomials that occur in Theorem 6 have the property that they take integer values when the variable s is a sufficiently large integer. For later purposes, it will be useful to characterize this class of polynomials. The first thing to note is that polynomials with this property need *not* have integer coefficients. For example, the polynomial $\frac{1}{2}s(s-1)$ takes integer values whenever s is an integer, but does not have integer coefficients. The reason is that either s or $s-1$ must be even, hence, divisible by 2. Similarly, the polynomial $\frac{1}{3\cdot2}s(s-1)(s-2)$ takes integer values for any integer s: no matter what s is, one of the three consecutive integers $s-2, s-1, s$ must be divisible by 3 and at least one of them divisible by 2. It is easy to generalize this argument and show that

$$\binom{s}{d} = \frac{s(s-1)\cdots(s-(d-1))}{d!} = \frac{1}{d\cdot(d-1)\cdots2\cdot1}s(s-1)\cdots(s-(d-1))$$

takes integer values for any integer s (see Exercise 10). Further, in Exercises 11 and 12, you will show that any polynomial of degree d which takes integer values for sufficiently large integers s can be written uniquely as an integer linear combination of the polynomials

$$\binom{s}{0} = 1, \quad \binom{s}{1} = s, \quad \binom{s}{2} = \frac{s(s-1)}{2}, \quad \ldots, \quad \binom{s}{d} = \frac{s(s-1)\cdots(s-(d-1))}{d!}.$$

Using this fact, we obtain the following sharpening of Theorem 6.

Proposition 7. *If $I \subseteq k[x_1, \ldots, x_n]$ is a monomial ideal with $\dim V(I) = d$, then for all s sufficiently large, the number of points in $C(I)$ of total degree $\leq s$ is a polynomial of degree d in s which can be written in the form*

$$\sum_{i=0}^{d} a_i \binom{s}{d-i},$$

where $a_i \in \mathbb{Z}$ for $0 \leq i \leq d$ and $a_0 > 0$.

In the final part of this section, we will study the projective variety associated with a monomial ideal. This makes sense because every monomial ideal is homogeneous (see Exercise 13). Thus, a monomial ideal $I \subseteq k[x_1, \ldots, x_n]$ determines a projective variety $V_p(I) \subseteq \mathbb{P}^{n-1}(k)$, where we use the subscript p to remind us that we are in projective space. In Exercise 14, you will show that $V_p(I)$ is a finite union of projective linear subspaces which have dimension one less than the dimension of their affine counterparts. As in the affine case, we define the *dimension* of a finite union of projective linear subspaces to be the maximum of the dimensions of the subspaces. Then Theorem 6 shows that the dimension of the projective variety $V_p(I)$ of a monomial ideal I is one less than the degree of the polynomial in s counting the number of monomials not in I of total degree $\leq s$.

In this case it turns out to be more convenient to consider the polynomial in s counting the number of monomials whose total degree is *equal* to s. The reason resides in the following proposition.

Proposition 8. *Let $I \subseteq k[x_1, \ldots, x_n]$ be a monomial ideal and let $V_p(I)$ be the projective variety in $\mathbb{P}^{n-1}(k)$ defined by I. If $\dim V_p(I) = d-1$, then for all s sufficiently large, the number of monomials not in I of total degree s is given by a polynomial of the form*

$$\sum_{i=0}^{d-1} b_i \binom{s}{d-1-i}$$

of degree $d-1$ in s, where $b_i \in \mathbb{Z}$ for $0 \leq i \leq d-1$ and $b_0 > 0$.

Proof. As an affine variety, $V(I) \subseteq k^n$ has dimension d, so that by Theorem 6, the number of monomials not in I of total degree $\leq s$ is a polynomial $p(s)$ of degree d for s sufficiently large. We also know that the coefficient of s^d is positive. It follows that the number of monomials of total degree *equal* to s is given by

$$p(s) - p(s-1)$$

for s large enough. By Exercise 15, this polynomial has degree $d-1$ and the coefficient of s^{d-1} is positive. Since it also takes integer values when s is a sufficiently large integer, it follows from the remarks preceding Proposition 7 that $p(s) - p(s-1)$ has the desired form. $\qquad\square$

In particular, this proposition says that for the projective variety defined by a monomial ideal, the dimension and the degree of the polynomial in the statement are equal. In §3, we will extend these results to the case of arbitrary homogeneous ideals $I \subseteq k[x_1, \ldots, x_n]$.

EXERCISES FOR §2

1. In this exercise, we will verify some of the claims made in Example 1. Remember that $I \subseteq k[x, y]$ is a proper monomial ideal.
   a. In case (b) of Example 1, show that if $s > l$, then the l "lines" of monomials contain $l(s + 1) - (1 + 2 + \cdots + l - 1)$ monomials of total degree $\leq s$.
   b. In case (b), conclude that the number of monomials not in I of total degree $\leq s$ is given by $ls + C_0$ for s sufficiently large. Explain how to compute C_0 and show that $s > a + b$ guarantees that s is sufficiently large. Illustrate your answer with a picture that shows what can go wrong if s is too small.
   c. In case (d) of Example 1, show that the constant C_0 in the polynomial function giving the number of points in $C(I)$ of total degree $\leq s$ is equal to the finite number of monomials not contained in the "lines" of monomials, minus $l_1 \cdot l_2$ for the monomials belonging to both families of lines, minus $1 + 2 + \cdots + (l_1 - 1)$, minus $1 + \cdots + (l_2 - 1)$.
2. Let $I \subseteq k[x_1, \ldots, x_n]$ be a monomial ideal. Suppose that in $\mathbb{Z}_{\geq 0}^n$, the translate $\alpha + [e_{i_1}, \ldots, e_{i_r}]$ is contained in $C(I)$. If $\alpha = \sum_{i \notin \{i_1, \ldots, i_r\}} a_i e_i$, show that $C(I)$ contains all translates $\beta + [e_{i_1}, \ldots, e_{i_r}]$ for all β of the form $\beta = \sum_{i \notin \{i_1, \ldots, i_r\}} b_i e_i$, where $0 \leq b_i \leq a_i$ for all i. In particular, $[e_{i_1}, \ldots, e_{i_r}] \subseteq C(I)$. Hint: I is an ideal.
3. In this exercise, you will find monomial ideals $I \subseteq k[x, y, z]$ with a given $C(I) \subseteq \mathbb{Z}_{\geq 0}^3$.
   a. Suppose that $C(I)$ consists of one translate of $[e_1, e_2]$ and two translates of $[e_2, e_3]$. Use Exercise 2 to show that $C(I) = [e_1, e_2] \cup [e_2, e_3] \cup (e_1 + [e_2, e_3])$.
   b. Find a monomial ideal I so that $C(I)$ is as described in part (a). Hint: Study all monomials of small degree to see whether or not they lie in I.
   c. Suppose now that $C(I)$ consists of one translate of $[e_1, e_2]$, two translates of $[e_2, e_3]$, and one additional translate (not contained in the others) of the line $[e_2]$. Use Exercise 2 to give a precise description of $C(I)$.
   d. Find a monomial ideal I so that $C(I)$ is as in part (c).
4. Let I be a monomial ideal in $k[x, y, z]$. In this exercise, we will study $C(I) \subseteq \mathbb{Z}_{\geq 0}^3$.
   a. Show that $\mathbf{V}(I)$ must be one of the following possibilities: the origin; one, two, or three coordinate lines; one, two, or three coordinate planes; or the union of a coordinate plane and a perpendicular coordinate axis.
   b. Show that if $\mathbf{V}(I)$ contains only the origin, then $C(I)$ has a finite number of points.
   c. Show that if $\mathbf{V}(I)$ is a union of one, two, or three coordinate lines, then $C(I)$ consists of a finite number of translates of $[e_1]$, $[e_2]$, and/or $[e_3]$, together with a finite number of points not on these translates.
   d. Show that if $\mathbf{V}(I)$ is a union of one, two or three coordinate planes, then $C(I)$ consists of a finite number of translates of $[e_1, e_2]$, $[e_1, e_3]$, and/or $[e_2, e_3]$ plus, possibly, a finite number of translates of $[e_1]$, $[e_2]$, and/or $[e_3]$ (where a translate of $[e_i]$ cannot occur unless $[e_i, e_j] \subseteq C(I)$ for some $j \neq i$) plus, possibly, a finite number of points not on these translates.
   e. Finally, show that if $\mathbf{V}(I)$ is the union of a coordinate plane and the perpendicular coordinate axis, then $C(I)$ consists of a finite nonzero number of translates of a single coordinate plane $[e_i, e_j]$, plus a finite nonzero number of translates of $[e_\ell]$, $\ell \neq i, j$, plus, possibly, a finite number of translates of $[e_i]$ and/or $[e_j]$, plus a finite number of points not on any of these translates.
5. Show that the number of points in any m-dimensional coordinate subspace of $\mathbb{Z}_{\geq 0}^n$ of total degree $\leq s$ is given by $\binom{m+s}{s}$.

6. Prove part (ii) of Lemma 5.

7. In this exercise, you will develop a counting principle, called the Inclusion-Exclusion Principle. The idea is to give a general method for counting the number of elements in a union of finite sets. We will use the notation $|A|$ for the number of elements in the finite set A.

   a. Show that for any two finite sets A and B,

   $$|A \cup B| = |A| + |B| - |A \cap B|.$$

   b. Show that for any three finite sets A, B, C,

   $$|A \cup B \cup C| = |A| + |B| + |C| - |A \cap B| - |A \cap C| - |B \cap C| + |A \cap B \cap C|.$$

   c. Using induction on the number of sets, show that the number of elements in a union of n finite sets $A_1 \cup \cdots \cup A_n$ is equal to the sum of the $|A_i|$, minus the sum of all double intersections $|A_i \cap A_j|$, $i < j$, plus the sum of all the threefold intersections $|A_i \cap A_j \cap A_k|$, $i < j < k$, minus the sum of the fourfold intersections, etc. This can be written as the following formula:

   $$|A_1 \cup \cdots \cup A_n| = \sum_{r=1}^{n} (-1)^{r-1} \left(\sum_{1 \le i_1 < \cdots < i_r \le n} |A_{i_1} \cap \cdots \cap A_{i_r}| \right).$$

8. In this exercise, you will show that the intersection of two translates of different coordinate subspaces of $\mathbb{Z}_{\geq 0}^n$ is a translate of a lower dimensional coordinate subspace.

   a. Let $A = \alpha + [e_{i_1} \ldots, e_{i_m}]$, where $\alpha = \sum_{i \notin \{i_1,\ldots,i_m\}} a_i e_i$, and let $B = \beta + [e_{j_1}, \ldots, e_{j_r}]$, where $\beta = \sum_{i \notin \{j_1,\ldots,j_r\}} b_i e_i$. If $A \neq B$ and $A \cap B \neq \emptyset$, then show that

   $$[e_{i_1}, \ldots, e_{i_m}] \neq [e_{j_1}, \ldots, e_{j_r}]$$

   and that $A \cap B$ is a translate of

   $$[e_{i_1}, \ldots, e_{i_m}] \cap [e_{j_1}, \ldots, e_{j_r}].$$

   b. Deduce that $\dim A \cap B < \max(m, r)$.

9. Show that if $s \geq 7$, then the number of elements in $C(I)$ of total degree $\leq s$ for the monomial ideal I in the example following Theorem 6 is given by the polynomial $5s - 1$.

10. Show that the polynomial

   $$p(s) = \binom{s}{d} = \frac{s(s-1) \cdots (s - (d-1))}{d!}$$

   takes integer values for all integers s. Note that p is a polynomial of degree d in s.

11. In this exercise, we will show that every polynomial $p(s)$ of degree $\leq d$ which takes integer values for every $s \in \mathbb{Z}_{\geq 0}$ can be written as a unique linear combination with integer coefficients of the polynomials $\binom{s}{0}, \binom{s}{1}, \binom{s}{2}, \ldots, \binom{s}{d}$.

   a. Show that the polynomials

   $$\binom{s}{0}, \binom{s}{1}, \binom{s}{2}, \ldots, \binom{s}{d}$$

   are linearly independent in the sense that

   $$a_0 \binom{s}{0} + a_1 \binom{s}{1} + \cdots + a_d \binom{s}{d} = 0$$

   for all s implies that $a_0 = a_1 = \cdots = a_d = 0$.

b. Show that any two polynomials $p(s)$ and $q(s)$ of degree $\leq d$ which take the same values at the $d+1$ points $s = 0, 1, \ldots, d$ must be identical. Hint: How many roots does the polynomial $p(s) - q(s)$ have?

c. Suppose we want to construct a polynomial $p(s)$ that satisfies

$$
\begin{aligned}
p(0) &= c_0, \\
p(1) &= c_1, \\
&\vdots \\
p(d) &= c_d,
\end{aligned}
$$

where the c_i are given values in $\mathbb{Z}$. Show that if we set

$$
\begin{aligned}
\Delta_0 &= c_0, \\
\Delta_1 &= c_1 - c_0, \\
\Delta_2 &= c_2 - 2c_1 + c_0, \\
&\vdots \\
\Delta_d &= \sum_{n=0}^{d} (-1)^n \binom{d}{n} c_{d-n},
\end{aligned}
$$

then the polynomial

$$
p(s) = \Delta_0 \binom{s}{0} + \Delta_1 \binom{s}{1} + \cdots + \Delta_d \binom{s}{d}
$$

satisfies $p(i) = c_i$ for $i = 0, \ldots, d$. Hint: Argue by induction on d. [The polynomial $p(s)$ is called a Newton–Gregory interpolating polynomial.]

d. Explain why the Newton–Gregory polynomial takes integer values for all integer s. Hint: Recall that the c_i are integers. See also Exercise 10.

e. Deduce from parts (a)–(d) that every polynomial of degree d which takes integer values for all integer $s \geq 0$ can be written as a unique integer linear combination of $\binom{s}{0}, \ldots, \binom{s}{d}$.

12. Suppose that $p(s)$ is a polynomial of degree d which takes integer values when s is a sufficiently large integer, say $s \geq a$. We want to prove that $p(s)$ is an integer linear combination of the polynomials $\binom{s}{0}, \ldots, \binom{s}{d}$ studied in Exercises 10 and 11. We can assume that a is a positive integer.

a. Show that the polynomial $p(s + a)$ can be expressed in terms of $\binom{s}{0}, \ldots, \binom{s}{d}$ and conclude that $p(s)$ is an integer linear combination of $\binom{s-a}{0}, \ldots, \binom{s-a}{d}$.

b. Use Exercise 10 to show that $p(s)$ takes integer values for all $s \in \mathbb{Z}$ and conclude that $p(s)$ is an integer linear combination of $\binom{s}{0}, \ldots, \binom{s}{d}$.

13. Show that every monomial ideal is a homogeneous ideal.

14. Let $I \subseteq k[x_1, \ldots, x_n]$ be a monomial ideal.

a. In k^n, let $\mathbf{V}(x_{i_1}, \ldots, x_{i_r})$ be a coordinate subspace of dimension $n - r$ contained in $\mathbf{V}(I)$. Prove that $\mathbf{V}_p(x_{i_1}, \ldots, x_{i_r}) \subseteq \mathbf{V}_p(I)$ in $\mathbb{P}^{n-1}(k)$. Also show that $\mathbf{V}_p(x_{i_1}, \ldots, x_{i_r})$ looks like a copy of $\mathbb{P}^{n-r-1}$ sitting inside $\mathbb{P}^{n-1}$. Thus, we say that $\mathbf{V}_p(x_{i_1}, \ldots, x_{i_r})$ is a projective linear subspace of dimension $n - r - 1$.

b. Prove the claim made in the text that $\mathbf{V}_p(I)$ is a finite union of projective linear subspaces of dimension one less than their affine counterparts.

15. Verify the statement in the proof of Proposition 8 that if $p(s)$ is a polynomial of degree d in s with a positive coefficient of s^d, then $p(s) - p(s-1)$ is a polynomial of degree $d - 1$ with a positive coefficient of s^{d-1}.

§3 The Hilbert Function and the Dimension of a Variety

In this section, we will define the Hilbert function of an ideal I and use it to define the dimension of a variety V. We will give the definitions in both the affine and projective cases. The basic idea will be to use the experience gained in the last section and define dimension in terms of the number of monomials not contained in the ideal I. In the affine case, we will use the number of monomials not in I of total degree $\leq s$, whereas in the projective case, we consider those of total degree *equal* to s.

However, we need to note that the results from §2 do not apply directly because when I is not a monomial ideal, different monomials not in I can be dependent on one another. For instance, if $I = \langle x^2 - y^2 \rangle$, neither the monomial x^2 nor y^2 belongs to I, but their difference does. So we should not regard x^2 and y^2 as two monomials not in I. Rather, to generalize §2, we will need to consider the number of monomials of total degree $\leq s$ which are "linearly independent modulo" I.

In Chapter 5, we defined the quotient of a ring modulo an ideal. There is an analogous operation on vector spaces which we will use to make the above ideas precise. Given a vector space V and a subspace $W \subseteq V$, it is not difficult to show that the relation on V defined by $v \sim v'$ if $v - v' \in W$ is an equivalence relation (see Exercise 1). The set of equivalence classes of $\sim$ is denoted V/W, so that

$$V/W = \{[v] \mid v \in V\}.$$

In the exercises, you will check that the operations $[v]+[v'] = [v+v']$ and $a[v] = [av]$, where $a \in k$ and $v, v' \in V$, are well-defined and make V/W into a k-vector space, called the *quotient space* of V modulo W.

When V is finite-dimensional, we can compute the dimension of V/W as follows.

Proposition 1. *Let W be a subspace of a finite-dimensional vector space V. Then W and V/W are also finite-dimensional vector spaces, and*

$$\dim V = \dim W + \dim V/W.$$

Proof. If V is finite-dimensional, it is a standard fact from linear algebra that W is also finite-dimensional. Let $v_1, \ldots, v_m$ be a basis of W, so that $\dim W = m$. In V, the vectors $v_1, \ldots, v_m$ are linearly independent and, hence, can be extended to a basis $v_1, \ldots, v_m, v_{m+1}, \ldots v_{m+n}$ of V. Thus, $\dim V = m + n$. We claim that $[v_{m+1}], \ldots, [v_{m+n}]$ form a basis of V/W.

To see that they span, take $[v] \in V/W$. If we write $v = \sum_{i=1}^{m+n} a_i v_i$, then $v \sim a_{m+1} v_{m+1} + \cdots + a_{m+n} v_{m+n}$ since their difference is $a_1 v_1 + \cdots + a_m v_m \in W$. It follows that in V/W, we have

$$[v] = [a_{m+1} v_{m+1} + \cdots + a_{m+n} v_{m+n}] = a_{m+1}[v_{m+1}] + \cdots + a_{m+n}[v_{m+n}].$$

The proof that $[v_{m+1}], \ldots, [v_{m+n}]$ are linearly independent is left to the reader (see Exercise 2). This proves the claim, and the proposition follows immediately. $\square$

The Dimension of an Affine Variety

Considered as a vector space over k, the polynomial ring $R = k[x_1, \ldots, x_n]$ has infinite dimension, and the same is true for any nonzero ideal (see Exercise 3). To get something finite-dimensional, we will restrict ourselves to polynomials of total degree $\leq s$. Hence, we let

$$R_{\leq s} = k[x_1, \ldots, x_n]_{\leq s}$$

denote the set of polynomials of total degree $\leq s$ in R. By Lemma 4 of §2, it follows that $R_{\leq s}$ is a vector space of dimension $\binom{n+s}{s}$. Then, given an ideal $I \subseteq R$, we let

$$I_{\leq s} = I \cap R_{\leq s}$$

denote the set of polynomials in I of total degree $\leq s$. Note that $I_{\leq s}$ is a vector subspace of $R_{\leq s}$. We are now ready to define the affine Hilbert function of I.

Definition 2. Let I be an ideal in $R = k[x_1, \ldots, x_n]$. The **affine Hilbert function** of I is the function on the nonnegative integers s defined by

$$^a\!HF_{R/I}(s) = \dim R_{\leq s}/I_{\leq s} = \dim R_{\leq s} - \dim I_{\leq s},$$

where the second equality is by Proposition 1.

Strictly speaking, $^a\!HF_{R/I}$ is the affine Hilbert function of R/I, but we prefer to call it the affine Hilbert function of I. With this terminology, the results of §2 for monomial ideals can be restated as follows.

Proposition 3. *Let I be a proper monomial ideal in $R = k[x_1, \ldots, x_n]$.*

(i) *For $s \geq 0$, $^a\!HF_{R/I}(s)$ is the number of monomials not in I of total degree $\leq s$.*

(ii) *For s sufficiently large, the affine Hilbert function of I is given by a polynomial function*

$$^a\!HF_{R/I}(s) = \sum_{i=0}^{d} b_i \binom{s}{d-i},$$

where $b_i \in \mathbb{Z}$ and b_0 is positive.

(iii) *The degree of the polynomial in part (ii) is the maximum of the dimensions of the coordinate subspaces contained in $\mathbf{V}(I)$.*

Proof. To prove (i), first note that $\{x^\alpha \mid |\alpha| \leq s\}$ is a basis of $R_{\leq s}$ as a vector space over k. Further, Lemma 3 of Chapter 2, §4 shows that $\{x^\alpha \mid |\alpha| \leq s, \ x^\alpha \in I\}$ is a basis of $I_{\leq s}$. Consequently, the monomials in $\{x^\alpha \mid |\alpha| \leq s, \ x^\alpha \notin I\}$ are exactly what we add to a basis of $I_{\leq s}$ to get a basis of $R_{\leq s}$. It follows from the proof of Proposition 1 that $\{[x^\alpha] \mid |\alpha| \leq s, \ x^\alpha \notin I\}$ is a basis of the quotient space $R_{\leq s}/I_{\leq s}$, which completes the proof of (i).

Parts (ii) and (iii) follow easily from (i) and Proposition 7 of §2. $\qquad\square$

We are now ready to link the ideals of §2 to arbitrary ideals in $R = k[x_1, \ldots, x_n]$. The key ingredient is the following observation due to Macaulay. As in Chapter 8, §4, we say that a monomial order $>$ on $k[x_1, \ldots, x_n]$ is a *graded order* if $x^\alpha > x^\beta$ whenever $|\alpha| > |\beta|$.

Proposition 4. *Let $I \subseteq R = k[x_1, \ldots, x_n]$ be an ideal and let $>$ be a graded order on R. Then the monomial ideal $\langle \mathrm{LT}(I) \rangle$ has the same affine Hilbert function as I.*

Proof. Since $\langle \mathrm{LT}(I) \rangle = \{0\}$ when $I = \{0\}$, we may assume $I \neq \{0\}$. Fix s and consider the leading monomials $\mathrm{LM}(f)$ of all elements $f \in I_{\leq s}$. There are only finitely many such monomials, so that

(1) $\{\mathrm{LM}(f) \mid f \in I_{\leq s}\} = \{\mathrm{LM}(f_1), \ldots, \mathrm{LM}(f_m)\}$

for some polynomials $f_1, \ldots, f_m \in I_{\leq s}$. By rearranging and deleting duplicates, we can assume that $\mathrm{LM}(f_1) > \mathrm{LM}(f_2) > \cdots > \mathrm{LM}(f_m)$. We claim that $f_1, \ldots, f_m$ form a basis of $I_{\leq s}$ as a vector space over k.

To prove this, consider a nontrivial linear combination $a_1 f_1 + \cdots + a_m f_m$ and choose the smallest i such that $a_i \neq 0$. Given how we ordered the leading monomials, there is nothing to cancel $a_i \mathrm{LT}(f_i)$, so the linear combination is nonzero. Hence, $f_1, \ldots, f_m$ are linearly independent. Next, let $W = [f_1, \ldots, f_m] \subseteq I_{\leq s}$ be the subspace spanned by $f_1, \ldots, f_m$. If $W \neq I_{\leq s}$, pick $f \in I_{\leq s} \setminus W$ with $\mathrm{LM}(f)$ minimal. By (1), $\mathrm{LM}(f) = \mathrm{LM}(f_i)$ for some i, and hence, $\mathrm{LT}(f) = \lambda \mathrm{LT}(f_i)$ for some $\lambda \in k$. Then $f - \lambda f_i \in I_{\leq s}$ has a smaller leading monomial, so that $f - \lambda f_i \in W$ by the minimality of $\mathrm{LM}(f)$. This implies $f \in W$, which is a contradiction. It follows that $W = [f_1, \ldots, f_m] = I_{\leq s}$, and we conclude that $f_1, \ldots, f_m$ form a basis.

The monomial ideal $\langle \mathrm{LT}(I) \rangle$ is generated by the leading terms (or leading monomials) of elements of I. Thus, $\mathrm{LM}(f_i) \in \langle \mathrm{LT}(I) \rangle_{\leq s}$ since $f_i \in I_{\leq s}$. We claim that $\mathrm{LM}(f_1), \ldots, \mathrm{LM}(f_m)$ form a vector space basis of $\langle \mathrm{LT}(I) \rangle_{\leq s}$. Arguing as above, it is easy to see that they are linearly independent. It remains to show that they span, i.e., that $[\mathrm{LM}(f_1), \ldots, \mathrm{LM}(f_m)] = \langle \mathrm{LT}(I) \rangle_{\leq s}$. By Lemma 3 of Chapter 2, §4, it suffices to show that

(2) $\{\mathrm{LM}(f_1), \ldots, \mathrm{LM}(f_m)\} = \{\mathrm{LM}(f) \mid f \in I, \mathrm{LM}(f) \text{ has total degree } \leq s\}$.

To relate this to (1), note that $>$ is a graded order, which implies that for *any* nonzero polynomial $f \in k[x_1, \ldots, x_n]$, $\mathrm{LM}(f)$ has the same total degree as f. In particular, if $\mathrm{LM}(f)$ has total degree $\leq s$, then so does f, which means that (2) follows immediately from (1).

Thus, $I_{\leq s}$ and $\langle \mathrm{LT}(I) \rangle_{\leq s}$ have the same dimension (since they both have bases consisting of m elements), and then the dimension formula of Proposition 1 implies

$$^a HF_{R/I}(s) = \dim R_{\leq s}/I_{\leq s} = \dim R_{\leq s}/\langle \mathrm{LT}(I) \rangle_{\leq s} = {}^a HF_{R/\langle \mathrm{LT}(I) \rangle}(s).$$

This proves the proposition. □

If we combine Propositions 3 and 4, it follows immediately that if I is *any* ideal in $k[x_1, \ldots, x_n]$ and s is sufficiently large, the affine Hilbert function of I can be written

$$^{a}HF_{R/I}(s) = \sum_{i=0}^{d} b_i \binom{s}{d-i},$$

where the b_i are integers and b_0 is positive. This leads to the following definition.

Definition 5. The polynomial which equals $^{a}HF_{R/I}(s)$ for sufficiently large s is called the **affine Hilbert polynomial** of I and is denoted $^{a}HP_{R/I}(s)$.

As with the affine Hilbert function, $^{a}HP_{R/I}(s)$ should be called the affine Hilbert polynomial of R/I, but we will use the terminology of Definition 5.

As an example, consider the ideal $I = \langle x^3y^2 + 3x^2y^2 + y^3 + 1 \rangle \subseteq R = k[x,y]$. If we use grlex, then $\langle \mathrm{LT}(I) \rangle = \langle x^3y^2 \rangle$, and using the methods of §2, one can show that the number of monomials not in $\langle \mathrm{LT}(I) \rangle$ of total degree $\leq s$ equals $5s - 5$ when $s \geq 3$. From Propositions 3 and 4, we obtain

$$^{a}HF_{R/I}(s) = {}^{a}HF_{R/\langle \mathrm{LT}(I) \rangle}(s) = 5s - 5$$

when $s \geq 3$. It follows that the affine Hilbert polynomial of I is

$$^{a}HP_{R/I}(s) = 5s - 5.$$

By definition, the affine Hilbert function of an ideal I coincides with the affine Hilbert polynomial of I when s is sufficiently large. The smallest integer s_0 such that $^{a}HP_{R/I}(s) = {}^{a}HF_{R/I}(s)$ for all $s \geq s_0$ is called the *index of regularity* of I. Determining the index of regularity is of considerable interest and importance in many computations with ideals, but we will not pursue this topic in detail here.

We next compare the degrees of the affine Hilbert polynomials of I and $\sqrt{I}$.

Proposition 6. *If $I \subseteq R = k[x_1, \ldots, x_n]$ is an ideal, then the affine Hilbert polynomials of I and $\sqrt{I}$ have the same degree.*

Proof. For a monomial ideal I, we know that the degree of the affine Hilbert polynomial is the dimension of the largest coordinate subspace of k^n contained in $\mathbf{V}(I)$. Since $\sqrt{I}$ is monomial by Exercise 9 of §1 and $\mathbf{V}(I) = \mathbf{V}(\sqrt{I})$, it follows immediately that $^{a}HP_{R/I}$ and $^{a}HP_{R/\sqrt{I}}$ have the same degree.

Now let I be an arbitrary ideal in $R = k[x_1, \ldots, x_n]$ and pick any graded order $>$ in R. We claim that

$$(3) \qquad \langle \mathrm{LT}(I) \rangle \subseteq \langle \mathrm{LT}(\sqrt{I}) \rangle \subseteq \sqrt{\langle \mathrm{LT}(I) \rangle}.$$

The first containment is immediate from $I \subseteq \sqrt{I}$. To establish the second, let x^α be a monomial in $\mathrm{LT}(\sqrt{I})$. This means that there is a polynomial $f \in \sqrt{I}$ such that $\mathrm{LT}(f) = x^\alpha$. We know that $f^r \in I$ for some $r \geq 0$, and it follows that $(x^\alpha)^r = \mathrm{LT}(f)^r = \mathrm{LT}(f^r) \in \langle \mathrm{LT}(I) \rangle$. Thus, $x^\alpha \in \sqrt{\langle \mathrm{LT}(I) \rangle}$.

If we set $J = \langle \mathrm{LT}(I) \rangle$, then we can write (3) as

$$J \subseteq \langle \mathrm{LT}(\sqrt{I}) \rangle \subseteq \sqrt{J}.$$

In Exercise 8, we will prove that if $I_1 \subseteq I_2$ are any ideals of $R = k[x_1, \ldots, x_n]$, then $\deg(^aHP_{R/I_1}) \geq \deg(^aHP_{R/I_2})$. If we apply this fact to the above inclusions, we obtain the inequalities

$$\deg(^aHP_{R/J}) \geq \deg(^aHP_{R/\langle \mathrm{LT}(\sqrt{I}) \rangle}) \geq \deg(^aHP_{R/\sqrt{J}}).$$

By the result for monomial ideals, the two outer terms here are equal, so that all three degrees are equal. Since $J = \langle \mathrm{LT}(I) \rangle$, we obtain

$$\deg(^aHP_{R/\langle \mathrm{LT}(I) \rangle}) = \deg(^aHP_{R/\langle \mathrm{LT}(\sqrt{I}) \rangle}).$$

By Proposition 4, the same is true for $^aHP_{R/I}$ and $^aHP_{R/\sqrt{I}}$, and we are done.   □

This proposition is evidence of something that is not at all obvious, namely, that the degree of the affine Hilbert polynomial has *geometric* meaning in addition to its *algebraic* significance in indicating how far $I_{\leq s}$ is from being all of $k[x_1, \ldots, x_n]_{\leq s}$. Recall that $\mathbf{V}(I) = \mathbf{V}(\sqrt{I})$ for all ideals. Thus, the degree of the affine Hilbert polynomial is the same for a large collection of ideals defining the same variety. Moreover, we know from §2 that the degree of the affine Hilbert polynomial is the same as our intuitive notion of the dimension of the variety of a monomial ideal. So it should be no surprise that in the general case, we define dimension in terms of the degree of the affine Hilbert function. We will always assume that the field k is infinite.

Definition 7. The **dimension** of a nonempty affine variety $V \subseteq k^n$, denoted $\dim V$, is the degree of the affine Hilbert polynomial of the corresponding ideal $I = \mathbf{I}(V) \subseteq R = k[x_1, \ldots, x_n]$.

When $V = \emptyset$, we have $1 \in \mathbf{I}(V)$, which implies $R_{\leq s} = \mathbf{I}(V)_{\leq s}$ for all s. Hence, $^aHP_{R/\mathbf{I}(V)} = 0$. Since the zero polynomial does not have a degree, we do not assign a dimension to the empty variety.

As an example, consider the twisted cubic $V = \mathbf{V}(y - x^2, z - x^3) \subseteq \mathbb{R}^3$. In Chapter 1, we showed that $I = \mathbf{I}(V) = \langle y - x^2, z - x^3 \rangle \subseteq R = \mathbb{R}[x, y, z]$. Using grlex, a Gröbner basis for I is $\{y^3 - z^2, x^2 - y, xy - z, xz - y^2\}$, so that $\langle \mathrm{LT}(I) \rangle = \langle y^3, x^2, xy, xz \rangle$. Then

$$\dim V = \deg(^aHP_{R/I})$$
$$= \deg(^aHP_{R/\langle \mathrm{LT}(I) \rangle})$$
$$= \text{maximum dimension of a coordinate subspace in } \mathbf{V}(\langle \mathrm{LT}(I) \rangle)$$

by Propositions 3 and 4. Since

$$\mathbf{V}(\langle \mathrm{LT}(I) \rangle) = \mathbf{V}(y^3, x^2, xy, xz) = \mathbf{V}(x, y) \subseteq \mathbb{R}^3,$$

we conclude that $\dim V = 1$. This agrees with our intuition that the twisted cubic should be 1-dimensional since it is a curve in $\mathbb{R}^3$.

For another example, let us compute the dimension of the variety of a monomial ideal. In Exercise 10, you will show that $\mathbf{I}(\mathbf{V}(I)) = \sqrt{I}$ when I is a monomial ideal and k is infinite. Then Proposition 6 implies that

$$\dim \mathbf{V}(I) = \deg(^aHP_{R/\mathbf{I}(\mathbf{V}(I))}) = \deg(^aHP_{R/\sqrt{I}}) = \deg(^aHP_{R/I}),$$

and it follows from part (iii) of Proposition 3 that $\dim \mathbf{V}(I)$ is the maximum dimension of a coordinate subspace contained in $\mathbf{V}(I)$. This agrees with the provisional definition of dimension given in §2. In Exercise 10, you will see that this can fail when k is a finite field.

One drawback of Definition 7 is that to find the dimension of a variety V, we need to know $\mathbf{I}(V)$, which, in general, is difficult to compute. It would be much nicer if $\dim V$ were the degree of $^aHP_{R/I}$, where I is an arbitrary ideal defining V. Unfortunately, this is not true in general. For example, if $I = \langle x^2 + y^2 \rangle \subseteq R = \mathbb{R}[x, y]$, it is easy to check that $^aHP_{R/I}$ has degree 1. Yet $V = \mathbf{V}(I) = \{(0,0)\} \subseteq \mathbb{R}^2$ is easily seen to have dimension 0. Thus, $\dim \mathbf{V}(I) \neq \deg(^aHP_{R/I})$ in this case (see Exercise 11 for the details).

When the field k is algebraically closed, these difficulties go away. More precisely, we have the following theorem that tells us how to compute the dimension in terms of any defining ideal.

Theorem 8 (The Dimension Theorem). *Let $V = \mathbf{V}(I)$ be a nonempty affine variety, where $I \subseteq R = k[x_1, \ldots, x_n]$ is an ideal. If k is algebraically closed, then*

$$\dim V = \deg(^aHP_{R/I}).$$

Furthermore, if $>$ is a graded order on R, then

$$\dim V = \deg(^aHP_{R/\langle \mathrm{LT}(I) \rangle})$$
$$= \text{maximum dimension of a coordinate subspace in } \mathbf{V}(\langle \mathrm{LT}(I) \rangle).$$

Finally, the last two equalities hold over any field k when $I = \mathbf{I}(V)$.

Proof. Since k is algebraically closed, the Nullstellensatz implies that $\mathbf{I}(V) = \mathbf{I}(\mathbf{V}(I)) = \sqrt{I}$. Then

$$\dim V = \deg(^aHP_{R/\mathbf{I}(V)}) = \deg(^aHP_{R/\sqrt{I}}) = \deg(^aHP_{R/I}),$$

where the last equality is by Proposition 6. The second part of the theorem now follows immediately using Propositions 3 and 4. $\qquad\square$

In other words, over an algebraically closed field, to compute the dimension of a variety $V = \mathbf{V}(I)$, one can proceed as follows:

- Compute a Gröbner basis for I using a graded order such as grlex or grevlex.
- Compute the maximal dimension d of a coordinate subspace that is contained in $\mathbf{V}(\langle \mathrm{LT}(I) \rangle)$. Note that Proposition 3 of §1 gives an algorithm for doing this.

Then $\dim V = d$ follows from Theorem 8.

The Dimension of a Projective Variety

Our discussion of the dimension of a projective variety $V \subseteq \mathbb{P}^n(k)$ will parallel what we did in the affine case and, in particular, many of the arguments are the same. We start by defining the Hilbert function and the Hilbert polynomial for an arbitrary homogeneous ideal $I \subseteq S = k[x_0, \ldots, x_n]$. As above, we assume that k is infinite.

As we saw in §2, the projective case uses total degree equal to s rather than $\leq s$. Since polynomials of total degree s do not form a vector space (see Exercise 13), we will work with *homogeneous* polynomials of total degree s. Let

$$S_s = k[x_0, \ldots, x_n]_s$$

denote the set of homogeneous polynomials of total degree s in S, together with the zero polynomial. In Exercise 13, you will show that S_s is a vector space of dimension $\binom{n+s}{s}$. If $I \subseteq S$ is a homogeneous ideal, we let

$$I_s = I \cap S_s$$

denote the set of homogeneous polynomials in I of total degree s (and the zero polynomial). Note that I_s is a vector subspace of S_s. Then the *Hilbert function* of I is defined by

$$HF_{S/I}(s) = \dim S_s/I_s.$$

Strictly speaking, we should call this the projective Hilbert function of S/I, but the above terminology is what we will use in this book.

When $I \subseteq S$ is a monomial ideal, the argument of Proposition 3 adapts easily to show that $HF_{S/I}(s)$ is the number of monomials not in I of total degree s. It follows from Proposition 8 of §2 that for s sufficiently large, we can express the Hilbert function of a monomial ideal in the form

$$(4) \qquad\qquad HF_{S/I}(s) = \sum_{i=0}^{d} b_i \binom{s}{d-i},$$

where $b_i \in \mathbb{Z}$ and b_0 is positive. We also know that d is the largest dimension of a projective coordinate subspace contained in $\mathbf{V}(I) \subseteq \mathbb{P}^n(k)$.

As in the affine case, we can use a monomial order to link the Hilbert function of a homogeneous ideal to the Hilbert function of a monomial ideal.

Proposition 9. *Let* $I \subseteq S = k[x_0, \ldots, x_n]$ *be a homogeneous ideal and let* $>$ *be a monomial order on* S. *Then the monomial ideal* $\langle \mathrm{LT}(I) \rangle$ *has the same Hilbert function as* I.

Proof. The argument is similar to the proof of Proposition 4. However, since we do *not* require that $>$ be a graded order, some changes are needed.

For a fixed s, we can find $f_1, \ldots, f_m \in I_s$ such that

$$(5) \qquad\qquad \{\mathrm{LM}(f) \mid f \in I_s\} = \{\mathrm{LM}(f_1), \ldots, \mathrm{LM}(f_m)\}$$

and we can assume that $\text{LM}(f_1) > \text{LM}(f_2) > \cdots > \text{LM}(f_m)$. As in the proof of Proposition 4, $f_1, \ldots, f_m$ form a basis of I_s as a vector space over k.

Now consider $\langle \text{LT}(I) \rangle_s$. We know $\text{LM}(f_i) \in \langle \text{LT}(I) \rangle_s$ since $f_i \in I_s$, and we need to show that $\text{LM}(f_1), \ldots, \text{LM}(f_m)$ form a vector space basis of $\langle \text{LT}(I) \rangle_s$. The leading terms are distinct, so as above, they are linearly independent. It remains to prove that they span. By Lemma 3 of Chapter 2, §4, it suffices to show that

(6) $\{\text{LM}(f_1), \ldots, \text{LM}(f_m)\} = \{\text{LM}(f) \mid f \in I, \ \text{LM}(f) \text{ has total degree } s\}.$

To relate this to (5), suppose that $\text{LM}(f)$ has total degree s for some $f \in I$. If we write f as a sum of homogeneous polynomials $f = \sum_i h_i$, where h_i has total degree i, it follows that $\text{LM}(f) = \text{LM}(h_s)$ since $\text{LM}(f)$ has total degree s. Since I is a homogeneous ideal, we have $h_s \in I$. Thus, $\text{LM}(f) = \text{LM}(h_s)$ where $h_s \in I_s$, and, consequently, (6) follows from (5). From here, the argument is identical to what we did in Proposition 4, and we are done.   □

If we combine Proposition 9 with the description of the Hilbert function for a monomial ideal given by (4), we see that for any homogeneous ideal $I \subseteq S = k[x_0, \ldots, x_n]$, the Hilbert function can be written

$$HF_{S/I}(s) = \sum_{i=0}^{d} b_i \binom{s}{d-i}$$

for s sufficiently large. The polynomial on the right of this equation is called the *Hilbert polynomial* of I and is denoted $HP_{S/I}(s)$.

We then define the dimension of a projective variety in terms of the Hilbert polynomial as follows.

Definition 10. The **dimension** of a nonempty projective variety $V \subset \mathbb{P}^n(k)$, denoted $\dim V$, is the degree of the Hilbert polynomial of the corresponding homogeneous ideal $I = \mathbf{I}(V) \subseteq S = k[x_0, \ldots, x_n]$. (Note that I is homogeneous by Proposition 4 of Chapter 8, §3.)

As in the affine case, the dimension of the empty variety is not defined. Over an algebraically closed field, we can compute the dimension as follows.

Theorem 11 (The Dimension Theorem). *Let $V = \mathbf{V}(I) \subseteq \mathbb{P}^n(k)$ be a projective variety, where $I \subseteq S = k[x_0, \ldots, x_n]$ is a homogeneous ideal. If V is nonempty and k is algebraically closed, then*

$$\dim V = \deg(HP_{S/I}).$$

Furthermore, for any monomial order on S, we have

$\dim V = \deg(HP_{S/\langle \text{LT}(I) \rangle})$
$\quad\quad = $ *maximum dimension of a projective coordinate subspace in $\mathbf{V}(\langle \text{LT}(I) \rangle)$.*

Finally, the last two equalities hold over any field k when $I = \mathbf{I}(V)$.

Proof. The first step is to show that I and $\sqrt{I}$ have Hilbert polynomials of the same degree. The proof is similar to what we did in Proposition 6 and is left as an exercise.

By the projective Nullstellensatz, we know that $\mathbf{I}(V) = \mathbf{I}(\mathbf{V}(I)) = \sqrt{I}$, and, from here, the proof is identical to what we did in the affine case (see Theorem 8). $\quad\square$

For our final result, we compare the dimension of affine and projective varieties.

Theorem 12. (i) *Let $I \subseteq S = k[x_0, \ldots, x_n]$ be a homogeneous ideal. Then for $s \geq 1$, we have*

$$HF_{S/I}(s) = {}^aHF_{S/I}(s) - {}^aHF_{S/I}(s-1).$$

There is a similar relation between Hilbert polynomials. Consequently, if $V \subseteq \mathbb{P}^n(k)$ is a nonempty projective variety and $C_V \subseteq k^{n+1}$ is its affine cone (see Chapter 8, §3), then

$$\dim C_V = \dim V + 1.$$

(ii) *Let $I \subseteq R = k[x_1, \ldots, x_n]$ be an ideal and let $I^h \subseteq S = k[x_0, \ldots, x_n]$ be its homogenization with respect to x_0 (see Chapter 8, §4). Then for $s \geq 0$, we have*

$$^aHF_{R/I}(s) = HF_{S/I^h}(s).$$

There is a similar relation between Hilbert polynomials. Consequently, if $V \subseteq k^n$ is a nonempty affine variety and $\overline{V} \subset \mathbb{P}^n(k)$ is its projective closure (see Chapter 8, §4), then

$$\dim V = \dim \overline{V}.$$

Proof. We will use the subscripts a and p to indicate the affine and projective cases respectively. The first part of (i) follows easily by reducing to the case of a monomial ideal and using the results of §2. We leave the details as an exercise. For the second part of (i), note that the affine cone C_V is simply the *affine* variety in k^{n+1} defined by $\mathbf{I}_p(V)$. Further, it is easy to see that $\mathbf{I}_a(C_V) = \mathbf{I}_p(V)$ (see Exercise 19). Thus, the dimensions of V and C_V are the degrees of $HP_{S/\mathbf{I}_p(V)}$ and $^aHP_{S/\mathbf{I}_p(V)}$, respectively. Then $\dim C_V = \dim V + 1$ follows from Exercise 15 of §2 and the relation just proved between the Hilbert polynomials.

To prove the first part of (ii), consider the maps

$$\phi : R_{\leq s} = k[x_1, \ldots, x_n]_{\leq s} \longrightarrow S_s = k[x_0, \ldots, x_n]_s,$$
$$\psi : S_s = k[x_0, \ldots, x_n]_s \longrightarrow R_{\leq s} = k[x_1, \ldots, x_n]_{\leq s}$$

defined by the formulas

(7)
$$\phi(f) = x_0^s f\left(\frac{x_1}{x_0}, \ldots, \frac{x_n}{x_0}\right) \quad \text{for } f \in R_{\leq s},$$
$$\psi(F) = F(1, x_1, \ldots, x_n) \quad \text{for } F \in S_s.$$

We leave it as an exercise to check that these are linear maps that are inverses of each other, and hence, $R_{\leq s}$ and S_s are isomorphic vector spaces. You should also check that if $f \in R_{\leq s}$ has total degree $d \leq s$, then

$$\phi(f) = x_0^{s-d} f^h,$$

where f^h is the homogenization of f as defined in Proposition 7 of Chapter 8, §2.

Under these linear maps, you will check in the exercises that

(8)
$$\phi(I_{\leq s}) \subseteq I_s^h,$$
$$\psi(I_s^h) \subseteq I_{\leq s},$$

and it follows easily that the above inclusions are equalities. Thus, $I_{\leq s}$ and I_s^h are also isomorphic vector spaces.

This shows that $R_{\leq s}$ and S_s have the same dimension, and the same holds for $I_{\leq s}$ and I_s^h. By the dimension formula of Proposition 1, we conclude that

$$^aHF_{R/I}(s) = \dim R_{\leq s}/I_{\leq s} = \dim S_s/I_s^h = HF_{I^h}(s),$$

which then implies that $^aHP_{R/I} = HP_{I^h}$.

For the second part of (ii), suppose $V \subseteq k^n$. Let $I = \mathbf{I}_a(V) \subseteq k[x_1, \ldots, x_n]$ and let $I^h \subseteq k[x_0, \ldots, x_n]$ be the homogenization of I with respect to x_0. Then $\overline{V}$ is defined to be $\mathbf{V}_p(I^h) \subset \mathbb{P}^n(k)$. Furthermore, we know from Exercise 8 of Chapter 8, §4 that $I^h = \mathbf{I}_p(\overline{V})$. Then

$$\dim V = \deg(^aHP_{R/I}) = \deg(HP_{S/I^h}) = \dim \overline{V}$$

follows immediately from $^aHP_{R/I} = HP_{I^h}$, and the theorem is proved.   □

The computer algebra systems Maple and Sage can compute the Hilbert polynomial of a homogeneous ideal, and the same is true for the more specialized programs CoCoA, Macaulay2 and Singular.

EXERCISES FOR §3

1. In this exercise, you will verify that if V is a vector space and W is a subspace of V, then V/W is a vector space.
   a. Show that the relation on V defined by $v \sim v'$ if $v - v' \in W$ is an equivalence relation.
   b. Show that the addition and scalar multiplication operations on the equivalence classes defined in the text are well-defined. Thus, if $v, v', w, w' \in V$ are such that $[v] = [v']$ and $[w] = [w']$, then show that $[v + w] = [v' + w']$ and $[av] = [av']$ for all $a \in k$.
   c. Verify that V/W is a vector space under the operations given in part (b).
2. Let V be a finite-dimensional vector space and let W be a vector subspace of V. If $\{v_1, \ldots, v_m, v_{m+1}, \ldots, v_{m+n}\}$ is a basis of V such that $\{v_1, \ldots, v_m\}$ is a basis for W, then show that $[v_{m+1}], \ldots, [v_{m+n}]$ are linearly independent in V/W.
3. Show that a nonzero ideal $I \subseteq k[x_1, \ldots, x_n]$ has infinite dimension as a vector space over k. Hint: Pick $f \neq 0$ in I and consider $x^\alpha f$.

4. The proofs of Propositions 4 and 9 involve finding vector space bases of $k[x_1,\ldots,x_n]_{\leq s}$ and $k[x_0,\ldots,x_n]_s$ where the elements in the bases have distinct leading terms. We showed that such bases exist, but our proof was nonconstructive. In this exercise, we will illustrate a method for actually finding such a basis. We will only discuss the homogeneous case, but the method applies equally well to the affine case.

The basic idea is to start with any basis of I, and order the elements according to their leading terms. If two of the basis elements have the same leading monomial, we can replace one of them with a k-linear combination that has a smaller leading monomial. Continuing in this way, we will get the desired basis.

To see how this works in practice, let I be a homogeneous ideal in $k[x,y,z]$, and suppose that $\{x^3 - xy^2, x^3 + x^2y - z^3, x^2y - y^3\}$ is a basis for I_3. We will use grlex order with $x > y > z$.

   a. Show that if we subtract the first polynomial from the second, leaving the third polynomial unchanged, then we get a new basis for I_3.
   b. The second and third polynomials in this new basis now have the same leading monomial. Show that if we change the third polynomial by subtracting the second polynomial from it and multiplying the result by -1, we end up with a basis $\{x^3 - xy^2, x^2y + xy^2 - z^3, xy^2 + y^3 - z^3\}$ for I_3 in which all three leading monomials are distinct.

5. Let $I = \langle x^3 - xyz, y^4 - xyz^2, xy - z^2\rangle$. Using grlex with $x > y > z$, find bases of I_3 and I_4 where the elements in the bases have distinct leading monomials. Hint: Use the method of Exercise 4.

6. Use the methods of §2 to compute the affine Hilbert polynomials for each of the following ideals.
   a. $I = \langle x^3y, xy^2\rangle \subseteq k[x,y]$.
   b. $I = \langle x^3y^2 + 3x^2y^2 + y^3 + 1\rangle \subseteq k[x,y]$.
   c. $I = \langle x^3yz^5, xy^3z^2\rangle \subseteq k[x,y,z]$.
   d. $I = \langle x^3 - yz^2, y^4 - x^2yz\rangle \subseteq k[x,y,z]$.

7. Find the index of regularity [that is, the smallest s_0 such that $^aHF_I(s) = {}^aHP_I(s)$ for all $s \geq s_0$] for each of the ideals in Exercise 6.

8. In this exercise, we will show that if $I_1 \subseteq I_2$ are ideals in $R = k[x_1,\ldots,x_n]$, then

$$\deg(^aHP_{R/I_1}) \geq \deg(^aHP_{R/I_2}).$$

   a. Show that $I_1 \subseteq I_2$ implies $C(\langle \mathrm{LT}(I_1)\rangle) \supseteq C(\langle \mathrm{LT}(I_2)\rangle)$ in $\mathbb{Z}^n_{\geq 0}$.
   b. Show that for $s \geq 0$, the affine Hilbert *functions* satisfy the inequality

$$^aHF_{R/I_1}(s) \geq {}^aHF_{R/I_2}(s).$$

   c. From part (b), deduce the desired statement about the degrees of the affine Hilbert *polynomials*. Hint: Argue by contradiction and consider the values of the polynomials as $s \to \infty$.
   d. If $I_1 \subseteq I_2$ are homogeneous ideals in $S = k[x_0,\ldots,x_n]$, prove an analogous inequality for the degrees of the Hilbert polynomials of I_1 and I_2.

9. Use Definition 7 to show that a point $p = (a_1,\ldots,a_n) \in k^n$ gives a variety of dimension zero. Hint: Use Exercise 7 of Chapter 4, §5 to describe $\mathbf{I}(\{p\})$.

10. Let $I \subseteq k[x_1,\ldots,x_n]$ be a monomial ideal, and assume that k is an infinite field. In this exercise, we will study $\mathbf{I}(\mathbf{V}(I))$.
   a. Show that $\mathbf{I}(\mathbf{V}(x_{i_1},\ldots,x_{i_r})) = \langle x_{i_1},\ldots,x_{i_r}\rangle$. Hint: Use Proposition 5 of Chapter 1, §1.
   b. Show that an intersection of monomial ideals is a monomial ideal. Hint: Use Lemma 3 of Chapter 2, §4.
   c. Show that $\mathbf{I}(\mathbf{V}(I))$ is a monomial ideal. Hint: Use parts(a) and (b) together with Theorem 15 of Chapter 4, §3.

d. The final step is to show that $I(V(I)) = \sqrt{I}$. We know that $\sqrt{I} \subseteq I(V(I))$, and since $I(V(I))$ is a monomial ideal, you need only prove that $x^\alpha \in I(V(I))$ implies that $x^{r\alpha} \in I$ for some $r > 0$. Hint: If $I = \langle m_1, \ldots, m_t \rangle$ and $x^{r\alpha} \notin I$ for $r > 0$, show that for every j, there is x_{i_j} such that x_{i_j} divides m_j but not x^α. Use $x_{i_1}, \ldots, x_{i_t}$ to obtain a contradiction.

e. Let $\mathbb{F}_2$ be a field with of two elements and let $I = \langle x \rangle \subseteq \mathbb{F}_2[x, y]$. Show that $I(V(I)) = \langle x, y^2 - y \rangle$. This is bigger than $\sqrt{I}$ and is not a monomial ideal.

11. Let $I = \langle x^2 + y^2 \rangle \subseteq R = \mathbb{R}[x, y]$.
    a. Show carefully that $\deg(^a HP_{R/I}) = 1$.
    b. Use Exercise 9 to show that $\dim V(I) = 0$.

12. Compute the dimension of the affine varieties defined by the following ideals. You may assume that k is algebraically closed.
    a. $I = \langle xz, xy - 1 \rangle \subseteq k[x, y, z]$.
    b. $I = \langle zw - y^2, xy - z^3 \rangle \subseteq k[x, y, z, w]$.

13. Consider the polynomial ring $S = k[x_0, \ldots, x_n]$.
    a. Give an example to show that the set of polynomials of total degree s is not closed under addition and, hence, does not form a vector space.
    b. Show that the set of homogeneous polynomials of total degree s (together with the zero polynomial) is a vector space over k.
    c. Use Lemma 5 of §2 to show that this vector space has dimension $\binom{n+s}{s}$. Hint: Consider the number of polynomials of total degree $\leq s$ and $\leq s - 1$.
    d. Give a second proof of the dimension formula of part (c) using the isomorphism of Exercise 20 below.

14. If $I \subseteq S = k[x_0, \ldots, x_n]$ is a homogeneous ideal, show that the Hilbert polynomials $HP_{S/I}$ and $HP_{S/\sqrt{I}}$ have the same degree. Hint: The quickest way is to use Theorem 12.

15. We will study when the Hilbert polynomial is zero.
    a. If $I \subseteq S = k[x_0, \ldots, x_n]$ is a homogeneous ideal, prove that $\langle x_0, \ldots, x_n \rangle^r \subseteq I$ for some $r \geq 0$ if and only if the Hilbert polynomial of I is the zero polynomial.
    b. Conclude that if $V \subseteq \mathbb{P}^n(k)$ is a variety, then $V = \emptyset$ if and only if its Hilbert polynomial is the zero polynomial. Thus, the empty variety in $\mathbb{P}^n(k)$ does not have a dimension.

16. Compute the dimension of projective varieties defined by the following ideals. Assume that k is algebraically closed.
    a. $I = \langle x^2 - y^2, x^3 - x^2 y + y^3 \rangle \subseteq k[x, y, z]$.
    b. $I = \langle y^2 - xz, x^2 y - z^2 w, x^3 - yzw \rangle \subseteq k[x, y, z, w]$.

17. In this exercise, we will see that in general, the relation between the number of variables n, the number r of polynomials in a basis of I, and the dimension of $V = V(I)$ is subtle. Let $V \subseteq \mathbb{P}^3(k)$ be the curve parametrized by $x = t^3 u^2, y = t^4 u, z = t^5, w = u^5$ for $(t:u) \in \mathbb{P}^1(k)$. Since this is a curve in 3-dimensional space, our intuition would lead us to believe that V should be defined by two equations. Assume that k is algebraically closed.

    a. Use Theorem 11 of Chapter 8, §5 to find an ideal $I \subseteq k[x, y, z, w]$ such that $V = V(I)$ in $\mathbb{P}^3(k)$. If you use grevlex for a certain ordering of the variables, you will get a basis of I containing three elements.
    b. Show that I_2 is 1-dimensional and I_3 is 6-dimensional.
    c. Show that I cannot be generated by two elements. Hint: Suppose that $I = \langle A, B \rangle$, where A and B are homogeneous. By considering I_2, show that A or B must be a multiple of $y^2 - xz$, and then derive a contradiction by looking at I_3.

    A much more difficult question would be to prove that there are no two homogeneous polynomials A, B such that $V = V(A, B)$.

18. This exercise is concerned with the proof of part (i) of Theorem 12.
    a. Use the methods of §2 to show that $HF_{S/I}(s) = {}^aHF_{S/I}(s) - {}^aHF_{S/I}(s-1)$ whenever I is a monomial ideal in $S = k[x_0, \ldots, x_n]$.
    b. Prove that $HF_I(s) = {}^aHF_I(s) - {}^aHF_I(s-1)$ for all homogeneous ideals $I \subseteq S$.
19. If $V \subseteq \mathbb{P}^n(k)$ is a nonempty projective variety and $C_V \subseteq k^{n+1}$ is its affine cone, then prove that $\mathbf{I}_p(V) = \mathbf{I}_a(C_V)$ in $k[x_0, \ldots, x_n]$.
20. This exercise is concerned with the proof of part (ii) of Theorem 12.
    a. Show that the maps ϕ and ψ defined in (7) are linear maps and verify that they are inverses of each other.
    b. Prove (8) and conclude that $\phi : I_{\leq s} \to I_s^h$ is an isomorphism whose inverse is ψ.

§4 Elementary Properties of Dimension

Using the definition of the dimension of a variety from §3, we can now state several basic properties of dimension. As in §3, we assume that the field k is infinite.

Proposition 1. *Let V_1 and V_2 be projective or affine varieties. If $V_1 \subseteq V_2$, then $\dim V_1 \leq \dim V_2$.*

Proof. We leave the proof to the reader as Exercise 1. $\qquad\square$

We next will study the relation between the dimension of a variety and the number of defining equations. We begin with the case where V is defined by a single equation.

Proposition 2. *Let k be an algebraically closed field and let $f \in k[x_0, \ldots, x_n]$ be a nonconstant homogeneous polynomial. Then the dimension of the projective variety in $\mathbb{P}^n(k)$ defined by f is*

$$\dim \mathbf{V}(f) = n - 1.$$

Proof. Fix a monomial order $>$ on $k[x_0, \ldots, x_n]$. Since k is algebraically closed, Theorem 11 of §3 says the dimension of $\mathbf{V}(f)$ is the maximum dimension of a projective coordinate subspace contained in $\mathbf{V}(\langle \mathrm{LT}(I) \rangle)$, where $I = \langle f \rangle$. One can check that $\langle \mathrm{LT}(I) \rangle = \langle \mathrm{LT}(f) \rangle$, and since $\mathrm{LT}(f)$ is a nonconstant monomial, the projective variety $\mathbf{V}(\mathrm{LT}(f))$ is a union of subspaces of $\mathbb{P}^n(k)$ of dimension $n - 1$. It follows that $\dim \mathbf{V}(I) = n - 1$. $\qquad\square$

Thus, when k is algebraically closed, a hypersurface $\mathbf{V}(f)$ in $\mathbb{P}^n$ always has dimension $n - 1$. We leave it as an exercise for the reader to prove the analogous statement for affine hypersurfaces.

It is important to note that these results are *not valid* if k is not algebraically closed. For instance, let $I = \langle x^2 + y^2 \rangle$ in $\mathbb{R}[x, y]$. In §3, we saw that $\mathbf{V}(f) = \{(0,0)\} \subseteq \mathbb{R}^2$ has dimension 0, yet Proposition 2 would predict that the dimension was 1. In fact, over a field that is not algebraically closed, the variety in k^n or $\mathbb{P}^n$ defined by a single polynomial can have any dimension between 0 and $n - 1$.

The following theorem establishes the analogue of Proposition 2 when the ambient space $\mathbb{P}^n(k)$ is replaced by an arbitrary variety V. Note that if I is an ideal and f is a polynomial, then $\mathbf{V}(I + \langle f \rangle) = \mathbf{V}(I) \cap \mathbf{V}(f)$.

Theorem 3. *Let k be an algebraically closed and let I be a homogeneous ideal in $S = k[x_0, \ldots, x_n]$. If $\dim \mathbf{V}(I) > 0$ and f is any nonconstant homogeneous polynomial, then*

$$\dim \mathbf{V}(I) \geq \dim \mathbf{V}(I + \langle f \rangle) \geq \dim \mathbf{V}(I) - 1.$$

Proof. To compute the dimension of $\mathbf{V}(I + \langle f \rangle)$, we will need to compare the Hilbert polynomials $HP_{S/I}$ and $HP_{S/(I + \langle f \rangle)}$. We first note that since $I \subseteq I + \langle f \rangle$, Exercise 8 of §3 implies that

$$\deg(HP_{S/I}) \geq \deg(HP_{S/(I + \langle f \rangle)}),$$

from which we conclude that $\dim \mathbf{V}(I) \geq \dim \mathbf{V}(I + \langle f \rangle)$ by Theorem 11 of §3.

To obtain the other inequality, suppose that $f \in S = k[x_0, \ldots, x_n]$ has total degree $r > 0$. Fix a total degree $s \geq r$ and consider the map

$$\pi : S_s/I_s \longrightarrow S_s/(I + \langle f \rangle)_s$$

which sends $[g] \in S_s/I_s$ to $\pi([g]) = [g] \in S_s/(I + \langle f \rangle)_s$. In Exercise 4, you will check that π is a well-defined linear map. It is easy to see that π is onto, and to investigate its kernel, we will use the map

$$\alpha_f : S_{s-r}/I_{s-r} \longrightarrow S_s/I_s$$

defined by sending $[h] \in S_{s-r}/I_{s-r}$ to $\alpha_f([h]) = [fh] \in S_s/I_s$. In Exercise 5, you will show that α_f is also a well-defined linear map.

We claim that the kernel of π is exactly the image of α_f, i.e., that

$$(1) \qquad \alpha_f(S_{s-r}/I_{s-r}) = \{[g] \mid \pi([g]) = [0] \text{ in } S_s/(I + \langle f \rangle)_s\}.$$

To prove this, note that if $h \in S_{s-r}$, then $fh \in (I + \langle f \rangle)_s$ and, hence, $\pi([fh]) = [0]$ in $S_s/(I + \langle f \rangle)_s$. Conversely, if $g \in S_s$ and $\pi([g]) = [0]$, then $g \in (I + \langle f \rangle)_s$. This means $g = g' + fh$ for some $g' \in I$. If we write $g' = \sum_i g_i'$ and $h = \sum_i h_i$ as sums of homogeneous polynomials, where g_i' and h_i have total degree i, it follows that $g = g_s' + f h_{s-r}$ since g and f are homogeneous. Since I is a homogeneous ideal, we have $g_s' \in I_s$, and it follows that $[g] = [f h_{s-r}] = \alpha_f([h_{s-r}])$ in S_s/I_s. This shows that $[g]$ is in the image of α_f and completes the proof of (1).

Since π is onto and we know its kernel by (1), the dimension theorem for linear mappings shows that

$$\dim S_s/I_s = \dim \alpha_f(S_{s-r}/I_{s-r}) + \dim S_s/(I + \langle f \rangle)_s.$$

Now certainly,

$$(2) \qquad \dim \alpha_f(S_{s-r}/I_{s-r}) \leq \dim S_{s-r}/I_{s-r},$$

with equality if and only if α_f is one-to-one. Hence,

$$\dim S_s/(I + \langle f \rangle)_s \geq \dim S_s/I_s - \dim S_{s-r}/I_{s-r}.$$

In terms of Hilbert functions, this tells us that

$$HF_{S/(I+\langle f\rangle)}(s) \geq HF_{S/I}(s) - HF_{S/I}(s-r)$$

whenever $s \geq r$. Thus, if s is sufficiently large, we obtain the inequality

$$(3) \qquad\qquad HP_{S/(I+\langle f\rangle)}(s) \geq HP_{S/I}(s) - HP_{S/I}(s-r)$$

for the Hilbert polynomials.

Suppose that $HP_{S/I}$ has degree d. Then it is easy to see that the polynomial on the right-hand side of (3) has degree $d - 1$ (the argument is the same as used in Exercise 15 of §2). Thus, (3) shows that $HP_{S/(I+\langle f\rangle)}$ is $\geq$ a polynomial of degree $d - 1$ for s sufficiently large, which implies $\deg(HP_{S/(I+\langle f\rangle)}) \geq d - 1$ [see, for example, part (c) of Exercise 8 of §3]. Since k is algebraically closed, we conclude that $\dim\mathbf{V}(I + \langle f\rangle) \geq \dim\mathbf{V}(I) - 1$ by Theorem 11 of §3. □

By carefully analyzing the proof of Theorem 3, we can give a condition that ensures that $\dim\mathbf{V}(I + \langle f\rangle) = \dim\mathbf{V}(I) - 1$. We need some new terminology for this. A *zero divisor* in a commutative ring R is a nonzero element a such that $a \cdot b = 0$ for some nonzero element b in R.

Corollary 4. *Let k be an algebraically closed field and let $I \subseteq S = k[x_0, \ldots, x_n]$ be a homogeneous ideal. Let f be a nonconstant homogeneous polynomial whose class in the quotient ring S/I is not a zero divisor. Then*

$$\dim\mathbf{V}(I + \langle f\rangle) = \dim\mathbf{V}(I) - 1$$

when $\dim\mathbf{V}(I) > 0$. Furthermore, $\mathbf{V}(I + \langle f\rangle) = \emptyset$ when $\dim\mathbf{V}(I) = 0$.

Proof. As we observed in the proof of Theorem 3, the inequality (2) is an equality if the multiplication map α_f is one-to-one. We claim that the latter is true if $[f] \in S/I$ is not a zero divisor. Namely, suppose that $[h] \in S_{s-r}/I_{s-r}$ is nonzero. This implies that $h \notin I_{s-r}$ and, hence, $h \notin I$ since $I_{s-r} = I \cap S_{s-r}$. Thus, $[h] \in S/I$ is nonzero, so that $[f][h] = [fh]$ is nonzero in S/I by our assumption on f. Hence, $fh \notin I$ and, hence, $\alpha_f([h]) = [fh]$ is nonzero in S_s/I_s. This shows that α_f is one-to-one.

Since (2) is an equality, the proof of Theorem 3 shows that we also get the equality

$$\dim S_s/(I + \langle f\rangle)_s = \dim S_s/I_s - \dim S_{s-r}/I_{s-r}$$

when $s \geq r$. In terms of Hilbert polynomials, this says $HP_{S/(I+\langle f\rangle)}(s) = HP_{S/I}(s) - HP_{S/I}(s - r)$, and it follows immediately that $\dim\mathbf{V}(I + \langle f\rangle) = \dim\mathbf{V}(I) - 1$. □

We remark that Theorem 3 can fail for affine varieties, even when k is algebraically closed. For example, consider the ideal $I = \langle xz, yz\rangle \subseteq \mathbb{C}[x, y, z]$. One easily sees that in $\mathbb{C}^3$, we have $\mathbf{V}(I) = \mathbf{V}(z) \cup V(x, y)$, so that $\mathbf{V}(I)$ is the union of the (x, y)-plane and the z-axis. In particular, $\mathbf{V}(I)$ has dimension 2 (do you see why?). Now, let $f = z - 1 \in \mathbb{C}[x, y, z]$. Then $\mathbf{V}(f)$ is the plane $z = 1$, and it follows that $\mathbf{V}(I + \langle f\rangle) = \mathbf{V}(I) \cap \mathbf{V}(f)$ consists of the single point $(0, 0, 1)$ (you will check this carefully in Exercise 7). By Exercise 9 of §3, we know that a point has dimension 0. Yet Theorem 3 would predict that $\mathbf{V}(I + \langle f \rangle)$ had dimension at least 1.

What goes "wrong" here is that the planes $z = 0$ and $z = 1$ are parallel and, hence, do not meet in affine space. We are missing a component of dimension 1 at infinity. This is an example of the way dimension theory works more satisfactorily for homogeneous ideals and projective varieties. It is possible to formulate a version of Theorem 3 that is valid for affine varieties, but we will not pursue that question here.

Our next result extends Theorem 3 to the case of several polynomials $f_1, \ldots, f_r$.

Proposition 5. *Let k be an algebraically closed field and let I be a homogeneous ideal in $k[x_0, \ldots, x_n]$. Let $f_1, \ldots, f_r$ be nonconstant homogeneous polynomials in $k[x_0, \ldots, x_n]$ such that $r \leq \dim \mathbf{V}(I)$. Then*

$$\dim \mathbf{V}(I + \langle f_1, \ldots, f_r \rangle) \geq \dim \mathbf{V}(I) - r.$$

Proof. The result follows immediately from Theorem 3 by induction on r.   □

In the exercises, we will ask you to derive a condition on the polynomials $f_1, \ldots, f_r$ which guarantees that the dimension of $\mathbf{V}(f_1, \ldots, f_r)$ is exactly equal to $n - r$.

Our next result concerns varieties of dimension zero.

Proposition 6. *Let V be a nonempty affine or projective variety. Then V consists of finitely many points if and only if $\dim V = 0$.*

Proof. We will give the proof only in the affine case. Let $>$ be a graded order on $k[x_1, \ldots, x_n]$. If V is finite, then let $a_j, j = 1, \ldots, m_i$, be the distinct elements of k appearing as i-th coordinates of points of V. Then

$$f = \prod_{j=1}^{m_i} (x_i - a_j) \in \mathbf{I}(V)$$

and we conclude that $\mathrm{LT}(f) = x_i^{m_i} \in \langle \mathrm{LT}(\mathbf{I}(V)) \rangle$. This implies that $\mathbf{V}(\langle \mathrm{LT}(\mathbf{I}(V)) \rangle) = \{0\}$ and then Theorem 8 of §3 implies that $\dim V = 0$.

Now suppose that $\dim V = 0$. Then the affine Hilbert polynomial of $\mathbf{I}(V)$ is a constant C, so that

$$\dim k[x_1, \ldots, x_n]_{\leq s} / \mathbf{I}(V)_{\leq s} = C$$

for s sufficiently large. If we also have $s \geq C$, then the classes $[1], [x_i], [x_i^2], \ldots, [x_i^s] \in k[x_1, \ldots, x_n]_{\leq s} / \mathbf{I}(V)_{\leq s}$ are $s + 1$ vectors in a vector space of dimension $C \leq s$ and, hence, they must be linearly dependent. But a nontrivial linear relation

$$[0] = \sum_{j=0}^{s} a_j [x_i^j] = \left[\sum_{j=0}^{s} a_j x_i^j \right]$$

means that $\sum_{j=0}^{s} a_j x_i^j$ is a nonzero polynomial in $\mathbf{I}(V)_{\leq s}$. This polynomial vanishes on V, which implies that there are only finitely many distinct i-th coordinates among the points of V. Since this is true for all $1 \leq i \leq n$, we see that V must be finite. $\square$

If, in addition, k is algebraically closed, then the conditions of Theorem 6 of Chapter 5, §3 are equivalent to $\dim V = 0$. In particular, given any defining ideal I of V, we get a simple criterion for detecting when a variety has dimension 0.

Now that we understand varieties of dimension 0, let us record some interesting properties of positive dimensional varieties.

Proposition 7. *Let k be algebraically closed.*

(i) *Let $V \subseteq \mathbb{P}^n(k)$ be a projective variety of dimension > 0. Then $V \cap \mathbf{V}(f) \neq \emptyset$ for every nonconstant homogeneous polynomial $f \in k[x_0, \ldots, x_n]$. Thus, a positive dimensional projective variety meets every hypersurface in $\mathbb{P}^n(k)$.*

(ii) *Let $W \subseteq k^n$ be an affine variety of dimension > 0. If $\overline{W}$ is the projective closure of W in $\mathbb{P}^n(k)$, then $W \neq \overline{W}$. Thus, a positive dimensional affine variety always has points at infinity.*

Proof. (i) Let $V = \mathbf{V}(I)$. Since $\dim V > 0$, Theorem 3 shows that $\dim V \cap \mathbf{V}(f) \geq \dim V - 1 \geq 0$. Let us check carefully that this guarantees $V \cap \mathbf{V}(f) \neq \emptyset$.

If $V \cap \mathbf{V}(f) = \emptyset$, then the projective Nullstellensatz implies that $\langle x_0, \ldots, x_n \rangle^r \subseteq I + \langle f \rangle$ for some $r \geq 0$. By Exercise 15 of §3, it follows that $HP_{S/(I+\langle f \rangle)}$ is the zero polynomial, where $S = k[x_0, \ldots, x_n]$. Yet if you examine the proof of Theorem 3, the inequality given for $HP_{S/(I+\langle f \rangle)}$ shows that this polynomial cannot be zero when $\dim V > 0$. We leave the details as an exercise.

(ii) The points at infinity of W are $\overline{W} \cap \mathbf{V}(x_0)$, where $\mathbf{V}(x_0)$ is the hyperplane at infinity. By Theorem 12 of §3, we have $\dim \overline{W} = \dim W > 0$, and then (i) implies that $\overline{W} \cap \mathbf{V}(x_0) \neq \emptyset$. $\square$

We next study the dimension of the union of two varieties.

Proposition 8. *If V and W are nonempty varieties either both in k^n or both in $\mathbb{P}^n(k)$, then*
$$\dim(V \cup W) = \max(\dim V, \dim W).$$

Proof. The proofs for the affine and projective cases are nearly identical, so we will give only the affine proof. Let $R = k[x_1, \ldots, x_n]$.

Let $I = \mathbf{I}(V)$ and $J = \mathbf{I}(W)$, so that $\dim V = \deg(^aHP_{R/I})$ and $\dim W = \deg(^aHP_{R/J})$. It is easy to show that $\mathbf{I}(V \cup W) = \mathbf{I}(V) \cap \mathbf{I}(W) = I \cap J$. It is more convenient to work with the product ideal IJ, and we note that
$$IJ \subseteq I \cap J \subset \sqrt{IJ}$$

(see Exercise 15). By Exercise 8 of §3, we conclude that
$$\deg(^aHP_{R/IJ}) \geq \deg(^aHP_{R/I\cap J}) \geq \deg(^aHP_{R/\sqrt{IJ}}).$$

by Proposition 6 of §3, the outer terms are equal. We conclude that $\dim(V \cup W) = \deg(^aHP_{IJ})$.

Now fix a graded order $>$ on $R = k[x_1, \ldots, x_n]$. By Propositions 3 and 4 of §3, it follows that $\dim V$, $\dim W$, and $\dim(V \cup W)$ are given by the maximal dimension of a coordinate subspace contained in $\mathbf{V}(\langle \mathrm{LT}(I) \rangle)$, $\mathbf{V}(\langle \mathrm{LT}(J) \rangle)$ and $\mathbf{V}(\langle \mathrm{LT}(IJ) \rangle)$ respectively. In Exercise 16, you will prove that

$$\langle \mathrm{LT}(IJ) \rangle \supseteq \langle \mathrm{LT}(I) \rangle \cdot \langle \mathrm{LT}(J) \rangle.$$

This implies

$$\mathbf{V}(\langle \mathrm{LT}(IJ) \rangle) \subseteq \mathbf{V}(\langle \mathrm{LT}(I) \rangle) \cup \mathbf{V}(\langle \mathrm{LT}(J) \rangle).$$

Since k is infinite, every coordinate subspace is irreducible (see Exercise 7 of §1), and as a result, a coordinate subspace contained in $\mathbf{V}(\langle \mathrm{LT}(IJ) \rangle)$ lies in either $\mathbf{V}(\langle \mathrm{LT}(I) \rangle)$ or $\mathbf{V}(\langle \mathrm{LT}(J) \rangle)$. This implies $\dim(V \cup W) \leq \max(\dim V, \dim W)$. The opposite inequality follows from Proposition 1, and the proposition is proved. □

This proposition has the following useful corollary.

Corollary 9. *The dimension of a variety is the largest of the dimensions of its irreducible components.*

Proof. If $V = V_1 \cup \cdots \cup V_r$ is the decomposition of V into irreducible components, then Proposition 8 and an induction on r shows that

$$\dim V = \max(\dim V_1, \ldots, \dim V_r),$$

as claimed. □

This corollary allows us to reduce to the case of an irreducible variety when computing dimensions. The following result shows that for irreducible varieties, the notion of dimension is especially well-behaved.

Proposition 10. *Let k be an algebraically closed field and let $V \subseteq \mathbb{P}^n(k)$ be an irreducible variety.*

(i) *If $f \in k[x_0, \ldots, x_n]$ is a homogeneous polynomial which does not vanish identically on V, then $\dim(V \cap \mathbf{V}(f)) = \dim V - 1$ when $\dim V > 0$. Furthermore, $V \cap \mathbf{V}(f) = \emptyset$ when $\dim V = 0$.*

(ii) *If $W \subseteq V$ is a variety such that $W \neq V$, then $\dim W < \dim V$.*

Proof. (i) By Proposition 4 of Chapter 5, §1, we know that $\mathbf{I}(V)$ is a prime ideal and $k[V] \cong k[x_0, \ldots, x_n]/\mathbf{I}(V)$ is an integral domain. Since $f \notin \mathbf{I}(V)$, the class of f is nonzero in $k[x_0, \ldots, x_n]/\mathbf{I}(V)$ and, hence, is not a zero divisor. The desired conclusion then follows from Corollary 4.

(ii) If W is a proper subvariety of V, then we can find $f \in \mathbf{I}(W) \setminus \mathbf{I}(V)$. Thus, $W \subseteq V \cap \mathbf{V}(f)$, and it follows from (i) and Proposition 1 that

$$\dim W \leq \dim(V \cap \mathbf{V}(f)) = \dim V - 1 < \dim V.$$

This completes the proof of the proposition. □

Part (i) of Proposition 10 asserts that when V is irreducible and f does not vanish on V, then some component of $V \cap \mathbf{V}(f)$ has dimension $\dim V - 1$. With some more work, it can be shown that *every* component of $V \cap \mathbf{V}(f)$ has dimension $\dim V - 1$. See, for example, Theorem 3.8 in Chapter IV of KENDIG (2015) or Theorem 5 of Chapter 1, §6 of SHAFAREVICH (2013).

In the next section, we will see that there is a way to understand the meaning of the dimension of an *irreducible* variety V in terms of the coordinate ring $k[V]$ and the field of rational functions $k(V)$ of V that we introduced in Chapter 5.

EXERCISES FOR §4

1. Prove Proposition 1. Hint: Use Exercise 8 of the previous section.
2. Let k be an algebraically closed field. If $f \in k[x_1, \ldots, x_n]$ is a nonconstant polynomial, show that the affine hypersurface $\mathbf{V}(f) \subseteq k^n$ has dimension $n - 1$.
3. In $\mathbb{R}^4$, give examples of four different affine varieties, each defined by a single equation, that have dimensions $0, 1, 2, 3$, respectively.
4. Let $S = k[x_0, \ldots, x_n]$. In this exercise, we study the mapping

$$\pi : S_s/I_s \longrightarrow S_s/(I + \langle f \rangle)_s$$

   defined by $\pi([g]) = [g]$ for all $g \in S_s$.
   a. Show that π is well-defined. This means showing that the image of the class $[g]$ does not depend on which representative g in the class that we choose. We call π the *natural projection* from S_s/I_s to $S_s/(I + \langle f \rangle)_s$.
   b. Show that π is a linear mapping of vector spaces.
   c. Show that the natural projection π is onto.
5. Show that if f is a homogeneous polynomial of degree r and $I \subseteq S = k[x_0, \ldots, x_n]$ is a homogeneous ideal, then the map

$$\alpha_f : S_{s-r}/I_{s-r} \longrightarrow S_s/I_s$$

   defined by $\alpha_f([h]) = [f \cdot h]$ is a well-defined linear mapping. Hence, you need to show that $\alpha_f([h])$ does not depend on the representative h for the class and that α preserves the vector space operations.
6. Let $f \in S = k[x_0, \ldots, x_n]$ be a homogeneous polynomial of total degree $r > 0$.
   a. Find a formula for the Hilbert polynomial of $\langle f \rangle$. Your formula should depend only on n and r (and, of course, s). In particular, all such polynomials f have the same Hilbert polynomial. Hint: Examine the proofs of Theorem 3 and Corollary 4 in the case when $I = \{0\}$.
   b. More generally, suppose that $V = \mathbf{V}(I)$ and that the class of f is not a zero divisor in S/I. Then show that the Hilbert polynomial of $I + \langle f \rangle$ depends only on I and r.
   If we vary f, we can regard the varieties $\mathbf{V}(f) \subseteq \mathbb{P}^n(k)$ as an algebraic family of hypersurfaces. Similarly, varying f gives the family of varieties $V \cap \mathbf{V}(f)$. By parts (a) and (b), the Hilbert polynomials are constant as we vary f. In general, if a technical condition called "flatness" is satisfied, the Hilbert polynomials are constant on an algebraic family of varieties.
7. Let $I = \langle xz, yz \rangle$. Show that $\mathbf{V}(I + \langle z - 1 \rangle) = \{(0,0,1)\}$.
8. Let $S = k[x_0, \ldots, x_n]$. A sequence $f_1, \ldots, f_r$ of $r \leq n+1$ nonconstant homogeneous polynomials is called an S-sequence if the class $[f_{j+1}]$ is not a zero divisor in $S/\langle f_1, \ldots, f_j \rangle$ for each $1 \leq j < r$.
   a. Show for example that for $r \leq n$, $x_0, \ldots, x_r$ is an S-sequence.

b. Show that if k is algebraically closed and $f_1, \ldots, f_r$ is an S-sequence, then

$$\dim \mathbf{V}(f_1, \ldots, f_r) = n - r.$$

Hint: Use Corollary 4 and induction on r. Work with the ideals $I_j = \langle f_1, \ldots, f_j \rangle$ for $1 \le j \le r$.

9. Let $S = k[x_0, \ldots, x_n]$. A homogeneous ideal I is said to be a *complete intersection* if it can be generated by an S-sequence. A projective variety V is called a *complete intersection* if $\mathbf{I}(V)$ is a complete intersection.
   a. Show that every irreducible linear subspace of $\mathbb{P}^n(k)$ is a complete intersection.
   b. Show that hypersurfaces are complete intersections when k is algebraically closed.
   c. Show that projective closure of the union of the (x, y)- and (z, w)-planes in k^4 is not a complete intersection.
   d. Let V be the affine twisted cubic $\mathbf{V}(y - x^2, z - x^3)$ in k^3. Is the projective closure of V a complete intersection?
   Hint for parts (c) and (d): Use the technique of Exercise 17 of §3.

10. Suppose that $I \subseteq R = k[x_1, \ldots, x_n]$ is an ideal. In this exercise, we will prove that the affine Hilbert polynomial of I is a constant if and only if the quotient ring R/I is finite-dimensional as a vector space over k. Furthermore, when this happens, we will show that the constant is the dimension of R/I as a vector space over k.
    a. Let $\alpha_s : R_{\le s}/I_{\le s} \to R/I$ be the map defined by $\alpha_s([f]) = [f]$. Show that α_s is well-defined and one-to-one.
    b. If R/I is finite-dimensional, show that α_s is an isomorphism for s sufficiently large and conclude that the affine Hilbert polynomial is a constant (and equals the dimension of R/I). Hint: Pick a basis $[f_1], \ldots, [f_m]$ of R/I and let s be bigger than the total degrees of $f_1, \ldots, f_m$.
    c. Now suppose the affine Hilbert polynomial is a constant. Show that if $s \le t$, the image of α_t contains the image of α_s. If s and t are large enough, conclude that the images are equal. Use this to show that α_s is an isomorphism for s sufficiently large and conclude that R/I is finite-dimensional.

11. Let $V \subseteq k^n$ be finite. In this exercise, we will prove that $k[x_1, \ldots, x_n]/\mathbf{I}(V)$ is finite-dimensional and that its dimension is $|V|$, the number of points in V. If we combine this with the previous exercise, we see that the affine Hilbert polynomial of $\mathbf{I}(V)$ is the constant $|V|$. Suppose that $V = \{p_1, \ldots, p_m\}$, where $m = |V|$.
    a. Define a map $\phi : k[x_1, \ldots, x_n]/\mathbf{I}(V) \to k^m$ by $\phi([f]) = (f(p_1), \ldots, f(p_m))$. Show that ϕ is a well-defined linear map and show that it is one-to-one.
    b. Fix i and let $W_i = \{p_j \mid j \ne i\}$. Show that $1 \in \mathbf{I}(W_i) + \mathbf{I}(\{p_i\})$. Hint: Show that $\mathbf{I}(\{p_i\})$ is a maximal ideal.
    c. By part (b), we can find $f_i \in \mathbf{I}(W_i)$ and $g_i \in \mathbf{I}(\{p_i\})$ such that $f_i + g_i = 1$. Show that $\phi(f_i)$ is the vector in k^m which has a 1 in the i-th coordinate and 0's elsewhere.
    d. Conclude that ϕ is an isomorphism and that $\dim k[x_1, \ldots, x_n]/\mathbf{I}(V) = |V|$.

12. Let $I \subseteq S = k[x_0, \ldots, x_n]$ be a homogeneous ideal. In this exercise we will study the geometric significance of the coefficient b_0 of the Hilbert polynomial

$$HP_{S/I}(s) = \sum_{i=0}^{d} b_i \binom{s}{d - i}.$$

We will call b_0 the *degree* of I. The *degree* of a projective variety V is defined to be the degree of $\mathbf{I}(V)$ and, as we will see, the degree is in a sense a generalization of the total degree of the defining equation for a hypersurface. Note also that we can regard Exercises 10 and 11 as studying the degrees of ideals and varieties with constant affine Hilbert polynomial.

a. Show that the degree of the ideal $\langle f \rangle$ is the same as the total degree of f. Also, if k is algebraically closed, show that the degree of the hypersurface $\mathbf{V}(f)$ is the same as the total degree of f_{red}, the reduction of f defined in Chapter 4, §2. Hint: Use Exercise 6.

b. Show that if $I \subseteq S = k[x_0, \ldots, x_n]$ is a complete intersection (Exercise 9) generated by the elements of an S-sequence $f_1, \ldots, f_r$, then the degree of I is the product

$$\deg(f_1) \cdot \deg(f_2) \cdots \deg(f_r)$$

of the total degrees of the f_i. Hint: Look carefully at the proof of Theorem 3. The hint for Exercise 8 may be useful.

c. Determine the degree of the projective closure of the standard twisted cubic.

13. Verify carefully the claim made in the proof of Proposition 7 that $HP_{S/(I+\langle f \rangle)}$ cannot be the zero polynomial when $\dim V > 0$. Hint: Look at the inequality (3) from the proof of Theorem 3.

14. This exercise will explore what happens if we weaken the hypotheses of Proposition 7.

a. Let $V = \mathbf{V}(x) \subseteq k^2$. Show that $V \cap \mathbf{V}(x - 1) = \emptyset$ and explain why this does not contradict part (a) of the proposition.

b. Let $W = \mathbf{V}(x^2 + y^2 - 1) \subseteq \mathbb{R}^2$. Show that $W = \overline{W}$ in $\mathbb{P}^2(\mathbb{R})$ and explain why this does not contradict part (b) of the proposition.

15. If $I, J \subseteq k[x_1, \ldots, x_n]$ are ideals, prove that $IJ \subseteq I \cap J \subseteq \sqrt{IJ}$.

16. Show that if I and J are any ideals and $>$ is any monomial ordering, then

$$\langle \mathrm{LT}(I) \rangle \cdot \langle \mathrm{LT}(J) \rangle \subseteq \langle \mathrm{LT}(IJ) \rangle.$$

17. Using Proposition 10, we can get an alternative definition of the dimension of an irreducible variety. We will assume that the field k is algebraically closed and that $V \subseteq \mathbb{P}^n(k)$ is irreducible.

a. If $\dim V > 0$, prove that there is an irreducible variety $W \subseteq V$ such that $\dim W = \dim V - 1$. Hint: Use Proposition 10 and look at the irreducible components of $V \cap \mathbf{V}(f)$.

b. If $\dim V = m$, prove that one can find a chain of $m + 1$ irreducible varieties

$$V_0 \subseteq V_1 \subseteq \cdots \subseteq V_m = V$$

such that $V_i \neq V_{i+1}$ for $0 \leq i \leq m - 1$.

c. Show that it is impossible to find a similar chain of length greater than $m + 1$ and conclude that the dimension of an irreducible variety is one less than the length of the longest strictly increasing chain of irreducible varieties contained in V.

18. Prove an affine version of part (ii) of Proposition 10.

§5 Dimension and Algebraic Independence

In §3, we defined the dimension of an affine variety as the degree of the affine Hilbert polynomial. This was useful for proving the properties of dimension in §4, but Hilbert polynomials do not give the full story. In algebraic geometry, there are many ways to formulate the concept of dimension and we will explore two of the more interesting approaches in this section and the next.

Algebraic Independence

If $V \subseteq k^n$ is an affine variety, recall from Chapter 5 that the *coordinate ring* $k[V]$ consists of all polynomial functions on V. This is related to the ideal $\mathbf{I}(V) \subseteq R = k[x_1, \ldots, x_n]$ by the natural ring isomorphism $k[V] \cong R/\mathbf{I}(V)$ (which is the identity on k) discussed in Theorem 7 of Chapter 5, §2. To see what $k[V]$ has to do with dimension, note that for any $s \geq 0$, there is a well-defined linear map

$$(1) \qquad R_{\leq s}/\mathbf{I}(V)_{\leq s} \longrightarrow R/\mathbf{I}(V) \cong k[V]$$

which is one-to-one (see Exercise 10 of §4). Thus, we can regard $R_{\leq s}/\mathbf{I}(V)_{\leq s}$ as a finite-dimensional "piece" of $k[V]$ that approximates $k[V]$ more and more closely as s gets larger. Since the degree of $^a HP_{R/\mathbf{I}(V)}$ measures how fast these finite-dimensional approximations are growing, we see that $\dim V$ tells us something about the "size" of $k[V]$.

This discussion suggests that we should be able to formulate the dimension of V directly in terms of the ring $k[V]$. To do this, we will use the notion of algebraically independent elements.

Definition 1. We say that elements $\phi_1, \ldots, \phi_r \in k[V]$ are **algebraically independent over** k if there is no nonzero polynomial F of r variables with coefficients in k such that $F(\phi_1, \ldots, \phi_r) = 0$ in $k[V]$.

Note that if $\phi_1, \ldots, \phi_r \in k[V]$ are algebraically independent over k, then the ϕ_i's are distinct and nonzero. It is also easy to see that any subset of $\{\phi_1, \ldots, \phi_r\}$ is also algebraically independent over k (see Exercise 1 for the details).

The simplest example of algebraically independent elements occurs when $V = k^n$. If k is an infinite field, we have $\mathbf{I}(V) = \{0\}$ and, hence, the coordinate ring is $k[V] = R = k[x_1, \ldots, x_n]$. Here, the elements $x_1, \ldots, x_n$ are algebraically independent over k since $F(x_1, \ldots, x_n) = 0$ means that F is the zero polynomial.

For another example, let V be the twisted cubic in $\mathbb{R}^3$ with $\mathbf{I}(V) = \langle y - x^2, z - x^3 \rangle$. Let us show that $[x] \in \mathbb{R}[V]$ is algebraically independent over $\mathbb{R}$. Suppose F is a polynomial with coefficients in $\mathbb{R}$ such that $F([x]) = [0]$ in $\mathbb{R}[V]$. By the way we defined the ring operations in $\mathbb{R}[V]$, this means $[F(x)] = [0]$, so that $F(x) \in \mathbf{I}(V)$. It is easy to show that $\mathbb{R}[x] \cap \langle y - x^2, z - x^3 \rangle = \{0\}$, which proves that F is the zero polynomial. On the other hand, we leave it to the reader to verify that $[x], [y] \in \mathbb{R}[V]$ are not algebraically independent over $\mathbb{R}$ since $[y] - [x]^2 = [0]$ in $\mathbb{R}[V]$.

We can relate the dimension of V to the number of algebraically independent elements in the coordinate ring $k[V]$ as follows.

Theorem 2. *Let $V \subseteq k^n$ be an affine variety. Then the dimension of V equals the maximal number of elements of $k[V]$ which are algebraically independent over k.*

Proof. We will first show that if $d = \dim V$, then we can find d elements of $k[V]$ which are algebraically independent over k. To do this, let $I = \mathbf{I}(V)$ and consider the ideal of leading terms $\langle \mathrm{LT}(I) \rangle$ for a graded order on $R = k[x_1, \ldots, x_n]$. By Theorem 8 of §3, we know that d is the maximum dimension of a coordinate subspace contained in $\mathbf{V}(\langle \mathrm{LT}(I) \rangle)$. A coordinate subspace $W \subseteq \mathbf{V}(\langle \mathrm{LT}(I) \rangle)$ of dimension d is defined by the vanishing of $n - d$ coordinates, so that we can write $W = \mathbf{V}(x_j \mid j \notin \{i_1, \ldots, i_d\})$ for some $1 \leq i_1 < \cdots < i_d \leq n$. We will show that $[x_{i_1}], \ldots, [x_{i_d}] \in k[V]$ are algebraically independent over k.

The first step is to prove

(2) $I \cap k[x_{i_1}, \ldots, x_{i_d}] = \{0\}.$

Let $p \in k^n$ be the point whose i_j-th coordinate is 1 for $1 \leq j \leq d$ and whose other coordinates are 0, and note that $p \in W \subseteq \mathbf{V}(\langle \mathrm{LT}(I) \rangle)$. Then every monomial in $\langle \mathrm{LT}(I) \rangle$ vanishes at p and, hence, no monomial in $\langle \mathrm{LT}(I) \rangle$ can involve only $x_{i_1}, \ldots, x_{i_d}$ (this is closely related to the proof of Proposition 2 of §2). Since $\langle \mathrm{LT}(I) \rangle$ is a monomial ideal, this implies that $\langle \mathrm{LT}(I) \rangle \cap k[x_{i_1}, \ldots, x_{i_d}] = \{0\}$. Then (2) follows since a nonzero element $f \in I \cap k[x_{i_1}, \ldots, x_{i_d}]$ would give the nonzero element $\mathrm{LT}(f) \in \langle \mathrm{LT}(I) \rangle \cap k[x_{i_1}, \ldots, x_{i_d}]$.

From (2), one easily sees that the map

$$k[x_{i_1}, \ldots, x_{i_d}] \longrightarrow R/I$$

sending $f \in k[x_{i_1}, \ldots, x_{i_d}]$ to $[f] \in R/I$ is a one-to-one ring homomorphism that is the identity on k (see Exercise 3 of Chapter 5, §6). The $x_{i_j} \in k[x_{i_1}, \ldots, x_{i_d}]$ are algebraically independent, so the same is true for their images $[x_{i_j}] \in R/I$ since the map is one-to-one. This shows that R/I contains $d = \dim V$ algebraically independent elements.

The final step in the proof is to show that if r elements of $k[V]$ are algebraically independent over k, then $r \leq \dim V$. So assume that $[f_1], \ldots, [f_r] \in k[V]$ are algebraically independent. Let N be the largest of the total degrees of $f_1, \ldots, f_r$ and let $y_1, \ldots, y_r$ be new variables. If $F \in k[y_1, \ldots, y_r]$ is a polynomial of total degree $\leq s$, then the polynomial $F(f_1, \ldots, f_r) \in R = k[x_1, \ldots, x_n]$ has total degree $\leq Ns$ (see Exercise 2). Then consider the map

(3) $\alpha : k[y_1, \ldots, y_r]_{\leq s} \longrightarrow R_{\leq Ns}/I_{\leq Ns}$

which sends $F(y_1, \ldots, y_r) \in k[y_1, \ldots, y_r]_{\leq s}$ to $[F(f_1, \ldots, f_r)] \in R_{\leq Ns}/I_{\leq Ns}$. We leave it as an exercise to show that α is a well-defined linear map.

We claim that α is one-to-one. To see why, suppose that $F \in k[y_1, \ldots, y_r]_{\leq s}$ and $[F(f_1, \ldots, f_r)] = [0]$ in $R_{\leq Ns}/I_{\leq Ns}$. Using the map (1), it follows that

$$[F(f_1, \ldots, f_r)] = F([f_1], \ldots, [f_r]) = [0] \quad \text{in } R/I \cong k[V].$$

Since $[f_1], \ldots, [f_r]$ are algebraically independent and F has coefficients in k, it follows that F must be the zero polynomial. Hence, α is one-to-one.

Comparing dimensions in (3), we see that

(4) ${}^a HF_{R/I}(Ns) = \dim R_{\leq Ns}/I_{\leq Ns} \geq \dim k[y_1, \ldots, y_r]_{\leq s}.$

Since the y_i are variables, Lemma 4 of §2 shows that the dimension of $k[y_1, \ldots, y_r]_{\leq s}$ is $\binom{r+s}{s} = \binom{r+s}{r}$, which is a polynomial of degree r in s. In terms of the affine Hilbert polynomial, this implies

$$^aHP_{R/I}(Ns) \geq \binom{r+s}{r} = \text{a polynomial of degree } r \text{ in } s$$

for s sufficiently large. It follows that $^aHP_{R/I}(Ns)$ and, hence, $^aHP_{R/I}(s)$ must have degree at least r. Thus, $r \leq \dim V$, which completes the proof of the theorem. □

As an application, we show that isomorphic varieties have the same dimension.

Corollary 3. *Let V and V' be affine varieties which are isomorphic (as defined in Chapter 5, §4). Then $\dim V = \dim V'$.*

Proof. By Theorem 9 of Chapter 5, §4, we know V and V' are isomorphic if and only if there is a ring isomorphism $\alpha : k[V] \to k[V']$ which is the identity on k. Then elements $\phi_1, \ldots, \phi_r \in k[V]$ are algebraically independent over k if and only if $\alpha(\phi_1), \ldots, \alpha(\phi_r) \in k[V']$ are. We leave the easy proof of this assertion as an exercise. From here, the corollary follows immediately from Theorem 2. □

In the proof of Theorem 2, note that the $d = \dim V$ algebraically independent elements we found in $k[V]$ came from the coordinates. We can use this to give another formulation of dimension.

Corollary 4. *Let $V \subseteq k^n$ be an affine variety. Then the dimension of V is equal to the largest integer r for which there exist r variables $x_{i_1}, \ldots, x_{i_r}$ such that $\mathbf{I}(V) \cap k[x_{i_1}, \ldots, x_{i_r}] = \{0\}$ [i.e., such that $\mathbf{I}(V)$ contain no nonzero polynomials in only these variables].*

Proof. From (2), it follows that we can find $d = \dim V$ such variables. Suppose we could find $d+1$ variables, $x_{j_1}, \ldots, x_{j_{d+1}}$ such that $I \cap k[x_{j_1}, \ldots, x_{j_{d+1}}] = \{0\}$. Then the argument following (2) would imply that $[x_{j_1}], \ldots, [x_{j_{d+1}}] \in k[V]$ were algebraically independent over k. This is impossible by Theorem 2 since $d = \dim V$. □

In the exercises, you will show that if k is algebraically closed, then Corollary 4 remains true if we replace $\mathbf{I}(V)$ with *any* defining ideal I of V. Since we know how to compute $I \cap k[x_{i_1}, \ldots, x_{i_r}]$ by elimination theory, Corollary 4 then gives us an alternative method (though not an efficient one) for computing the dimension of a variety.

Projections and Noether Normalization

We can also interpret Corollary 4 in terms of projections. If we choose r variables $x_{i_1}, \ldots, x_{i_r}$, then we get the projection map $\pi : k^n \to k^r$ defined by $\pi(a_1, \ldots, a_n) = (a_{i_1}, \ldots, a_{i_r})$. Also, let $\tilde{I} = \mathbf{I}(V) \cap k[x_{i_1}, \ldots, x_{i_r}]$ be the appropriate elimination ideal. If k is algebraically closed, then the Closure Theorem from §2 of Chapter 3 shows that $\mathbf{V}(\tilde{I}) \cap k^r$ is the smallest variety containing the projection $\pi(V)$. It follows that

$$\tilde{I} = \{0\} \iff \mathbf{V}(\tilde{I}) = k^r$$
$$\iff \text{the smallest variety containing } \pi(V) \text{ is } k^r$$
$$\iff \pi(V) \text{ is Zariski dense in } k^r.$$

Thus, Corollary 4 shows that the dimension of V is the largest dimension of a coordinate subspace for which the projection of V is Zariski dense in the subspace.

We can regard the above map π as a linear map from k^n to itself which leaves the i_j-th coordinate unchanged for $1 \le j \le r$ and sends the other coordinates to 0. It is then easy to show that $\pi \circ \pi = \pi$ and that the image of π is $k^r \subseteq k^n$ (see Exercise 8). More generally, a linear map $\pi : k^n \to k^n$ is called a *projection* if $\pi \circ \pi = \pi$. If π has rank r, then the image of π is an r-dimensional subspace H of k^n, and we say that π is *a projection onto H*.

Now let π be a projection onto a subspace $H \subseteq k^n$. Under π, any variety $V \subseteq k^n$ gives a subset $\pi(V) \subseteq H$. Then we can interpret the dimension of V in terms of its projections $\pi(V)$ as follows.

Proposition 5. *Let k be an algebraically closed field and let $V \subseteq k^n$ be an affine variety. Then the dimension of V is the largest dimension of a subspace $H \subseteq k^n$ for which a projection of V onto H is Zariski dense.*

Proof. If V has dimension d, then by the above remarks, we can find a projection of V onto a d-dimensional coordinate subspace which has Zariski dense image.

Now let $\pi : k^n \to k^n$ be an arbitrary projection onto an r-dimensional subspace H of k^n. We need to show that $r \le \dim V$ whenever $\pi(V)$ is Zariski dense in H. From linear algebra, we can find a basis of k^n so that in the new coordinate system, $\pi(a_1, \ldots, a_n) = (a_1, \ldots, a_r)$. Since changing coordinates does not affect the dimension (this follows from Corollary 3 since a coordinate change gives isomorphic varieties), we are reduced to the case of a projection onto a coordinate subspace, and then the proposition follows from the above remarks. $\square$

Let π be a projection of k^n onto a subspace H of dimension r. By the Closure Theorem from Chapter 3, §2 we know that if $\pi(V)$ is Zariski dense in H, then we can find a proper variety $W \subseteq H$ such that $H \setminus W \subset \pi(V)$. Thus, $\pi(V)$ "fills up" most of the r-dimensional subspace H, and hence, it makes sense that this should force V to have dimension at least r. So Proposition 5 gives a very geometric way of thinking about the dimension of a variety.

The idea of projection is also closely related to the Noether normalization of $k[V] \cong k[x_1, \ldots, x_n]/\mathbf{I}(V)$ studied in Chapter 5, §6. The proof of Theorem 8 of that section implies that we can find linear combinations $u_1, \ldots, u_m$ of the coordinate functions $\phi_i = [x_i] \in k[V]$ such that the inclusion

$$k[u_1, \ldots, u_m] \subseteq k[V]$$

is a *Noether normalization*, meaning that $u_1, \ldots, u_m$ are algebraically independent over k and $k[V]$ is finite over $k[u_1, \ldots, u_m]$ as in Definition 2 of Chapter 5, §6. The above inclusion corresponds to a projection map $\pi : V \to k^m$ and relates to Proposition 5 as follows.

Theorem 6. *In the above situation, m is the dimension of V. Also, the projection $\pi : V \to k^m$ is onto with finite fibers.*

Proof. The second assertion follows from Theorem 8 of Chapter 5, §6. Turning to $\dim V$, note that $m \leq \dim V$ by Theorem 2 since $u_1, \ldots, u_m$ are algebraically independent over k.

It remains to show $m \geq \dim V$. By a suitable change of coordinates in k^n, we can assume that $u_i = [x_i]$ for $i = 1, \ldots, m$. Let $I = \mathbf{I}(V)$ and consider lex order on $k[x_1, \ldots, x_n]$ with $x_{m+1} > \cdots > x_n > x_1 > \cdots > x_m$.

Arguing as in the proof of Theorem 2, we see that $k[x_1, \ldots, x_m] \cap I = \{0\}$, which allows us to regard $k[x_1, \ldots, x_m]$ as a subring of $k[x_1, \ldots, x_n]/I \cong k[V]$. Since $k[V]$ is finite over $k[x_1, \ldots, x_m]$, the Relative Finiteness Theorem (Theorem 4 from Chapter 5, §6) implies that

$$(5) \qquad\qquad x_{m+1}^{a_{m+1}}, \ldots, x_n^{a_n} \in \langle \mathrm{LT}(I) \rangle.$$

for some $a_{m+1}, \ldots, a_n \geq 0$.

By the Dimension Theorem (Theorem 8 of §3), $\dim V$ is the maximum dimension of a coordinate subspace contained in $\mathbf{V}(\langle \mathrm{LT}(I) \rangle)$, which by Proposition 2 of §2 equals the maximum dimension of a coordinate subspace $[e_{i_1}, \ldots, e_{i_r}]$ contained in the complement $C(\langle \mathrm{LT}(I) \rangle) \subseteq \mathbb{Z}_{\geq 0}^n$. However,

$$[e_{i_1}, \ldots, e_{i_r}] \subseteq C(\langle \mathrm{LT}(I) \rangle) \implies x_{i_j}^a \notin \langle \mathrm{LT}(I) \rangle \text{ for all } a \geq 0.$$

Comparing this with (5), we must have $\{i_1, \ldots, i_r\} \cap \{m+1, \ldots, n\} = \emptyset$. In other words, $\{i_1, \ldots, i_r\} \subseteq \{1, \ldots, m\}$, so that $r \leq m$. This implies $\dim V \leq m$, and the theorem follows. $\qquad\qquad\qquad\qquad\qquad\qquad\qquad\qquad\qquad\qquad\qquad\quad\square$

Notice how Theorem 6 gives a sharper version of Proposition 5, since the projection is now onto and the fibers are finite. The theorem also justifies the claim made in Chapter 5, §6 that a Noether normalization determines the dimension.

Irreducible Varieties and Transcendence Degrees

For the final part of the section, we will assume that V is an *irreducible* variety. By Proposition 4 of Chapter 5, §1, we know that $\mathbf{I}(V)$ is a prime ideal and that $k[V]$ is an integral domain. As in §5 of Chapter 5, we can then form the field of fractions of $k[V]$, which is the *field of rational functions* on V and is denoted $k(V)$. For elements of $k(V)$, the definition of algebraic independence over k is the same as that given for elements of $k[V]$ in Definition 1. We can relate the dimension of V to $k(V)$ as follows.

Theorem 7. *Let $V \subseteq k^n$ be an irreducible affine variety. Then the dimension of V equals the maximal number of elements of $k(V)$ which are algebraically independent over k.*

Proof. Let $d = \dim V$. Since $k[V] \subseteq k(V)$, any d elements of $k[V]$ which are algebraically independent over k will have the same property when regarded as elements of $k(V)$. So it remains to show that if $\phi_1, \ldots, \phi_r \in k(V)$ are algebraically independent, then $r \leq \dim V$. Each ϕ_i is a quotient of elements of $k[V]$, and if we pick a common denominator f, then we can write $\phi_i = [f_i]/[f]$ for $1 \leq i \leq r$. Note also that $[f] \neq [0]$ in $k[V]$. We need to modify the proof of Theorem 2 to take the denominator f into account. As usual, we set $R = k[x_1, \ldots, x_n]$.

Let N be the largest of the total degrees of $f, f_1, \ldots, f_r$, and let $y_1, \ldots, y_r$ be new variables. If $F \in k[y_1, \ldots, y_r]$ is a polynomial of total degree $\leq s$, then we leave it as an exercise to show that

$$f^s F(f_1/f, \ldots, f_r/f)$$

is a polynomial in R of total degree $\leq Ns$ (see Exercise 10). Then consider the map

(6) $$\beta : k[y_1, \ldots, y_r]_{\leq s} \longrightarrow R_{\leq Ns}/I_{\leq Ns}$$

sending a polynomial $F(y_1, \ldots, y_r) \in k[y_1, \ldots, y_r]_{\leq s}$ to $[f^s F(f_1/f, \ldots, f_r/f)] \in R_{\leq Ns}/I_{\leq Ns}$. We leave it as an exercise to show that β is a well-defined linear map.

To show that β is one-to-one, suppose that we have $F \in k[y_1, \ldots, y_r]_{\leq s}$ such that $[f^s F(f_1/f, \ldots, f_r/f)] = [0]$ in $R_{\leq Ns}/I_{\leq Ns}$. Using the map (1), it follows that

$$[f^s F(f_1/f, \ldots, f_r/f)] = [0] \quad \text{in } R/I \cong k[V].$$

However, if we work in $k(V)$ and use $\phi_i = [f_i]/[f]$, then we can write this as

$$[f]^s F([f_1]/[f], \ldots, [f_r]/[f]) = [f]^s F(\phi_1, \ldots, \phi_r) = [0] \text{ in } k(V).$$

Since $k(V)$ is a field and $[f] \neq [0]$, it follows that $F(\phi_1, \ldots, \phi_r) = [0]$. Then F must be the zero polynomial since $\phi_1, \ldots, \phi_r$ are algebraically independent and F has coefficients in k. Thus, β is one-to-one.

Once we know that β is one-to-one in (6), we get the inequality (4), and from here, the proof of Theorem 2 shows that $\dim V \geq r$. This proves the theorem. $\square$

As a corollary of this theorem, we can prove that birationally equivalent varieties have the same dimension.

Corollary 8. *Let V and V' be irreducible affine varieties which are birationally equivalent (as defined in Chapter 5, §5). Then $\dim V = \dim V'$.*

Proof. In Theorem 10 of Chapter 5, §5, we showed that two irreducible affine varieties V and V' are birationally equivalent if and only if there is an isomorphism $k(V) \cong k(V')$ of their function fields which is the identity on k. The remainder of the proof is identical to what we did in Corollary 3. $\square$

In field theory, there is a concept of *transcendence degree* which is closely related to what we have been studying. In general, when we have a field K containing k, we have the following definition.

Definition 9. Let K be a field containing k. Then we say that K has **transcendence degree** d **over** k provided that d is the largest number of elements of K which are algebraically independent over k.

If we combine this definition with Theorem 7, then for any irreducible affine variety V, we have

$$\dim V = \text{the transcendence degree of } k(V) \text{ over } k.$$

Many books on algebraic geometry use this as the definition of the dimension of an irreducible variety. The dimension of an arbitrary variety is then defined to be the maximum of the dimensions of its irreducible components.

For an example of transcendence degree, suppose that k is infinite, so that $k(V) = k(x_1, \ldots, x_n)$ when $V = k^n$. Since k^n has dimension n, we conclude that the field $k(x_1, \ldots, x_n)$ has transcendence degree n over k. It is clear that the transcendence degree is at least n, but it is less obvious that no $n + 1$ elements of $k(x_1, \ldots, x_n)$ can be algebraically independent over k. So our study of dimension yields some insights into the structure of fields.

To fully understand transcendence degree, one needs to study more about algebraic and transcendental field extensions. See Section 14.9 of DUMMIT and FOOTE (2004).

EXERCISES FOR §5

1. Let $\phi_1, \ldots, \phi_r \in k[V]$ be algebraically independent over k.
    a. Prove that the ϕ_i are distinct and nonzero.
    b. Prove that any nonempty subset of $\{\phi_1, \ldots, \phi_r\}$ consists of algebraically independent elements over k.
    c. Let $y_1, \ldots, y_r$ be variables and consider the map $\alpha : k[y_1, \ldots, y_r] \to k[V]$ defined by $\alpha(F) = F(\phi_1, \ldots, \phi_r)$. Show that α is a one-to-one ring homomorphism.
2. This exercise is concerned with the proof of Theorem 2.
    a. If $f_1, \ldots, f_r \in k[x_1, \ldots, x_n]$ have total degree $\leq N$ and $F \in k[x_1, \ldots, x_n]$ has total degree $\leq s$, show that $F(f_1, \ldots, f_r)$ has total degree $\leq Ns$.
    b. Show that the map α defined in the proof of Theorem 2 is a well-defined linear map.
3. Complete the proof of Corollary 3.
4. Let k be an algebraically closed field and let $I \subseteq R = k[x_1, \ldots, x_n]$ be an ideal. Show that the dimension of $\mathbf{V}(I)$ is equal to the largest integer r for which there exist r variables $x_{i_1}, \ldots, x_{i_r}$ such that $I \cap k[x_{i_1}, \ldots, x_{i_r}] = \{0\}$. Hint: Use I rather than $\mathbf{I}(V)$ in the proof of Theorem 2. Be sure to explain why $\dim V = \deg(^a HP_{R/I})$.
5. Let $I = \langle xy - 1 \rangle \subseteq k[x, y]$. What is the projection of $\mathbf{V}(I)$ to the x-axis and to the y-axis? Note that $\mathbf{V}(I)$ projects densely to both axes, but in neither case is the projection the whole axis.
6. Let k be infinite and let $I = \langle xy, xz \rangle \subseteq k[x, y, z]$.
    a. Show that $I \cap k[x] = 0$, but that $I \cap k[x, y]$ and $I \cap k[x, z]$ are not equal to 0.
    b. Show that $I \cap k[y, z] = 0$, but that $I \cap k[x, y, z] \neq 0$.
    c. What do you conclude about the dimension of $\mathbf{V}(I)$?
7. Here is a more complicated example of the phenomenon exhibited in Exercise 6. Again, assume that k is infinite and let $I = \langle zx - x^2, zy - xy \rangle \subseteq k[x, y, z]$.
    a. Show that $I \cap k[z] = 0$. Is either $I \cap k[x, z]$ or $I \cap k[y, z]$ equal to 0?
    b. Show that $I \cap k[x, y] = 0$, but that $I \cap k[x, y, z] \neq 0$.
    c. What does part (b) say about $\dim \mathbf{V}(I)$?

8. Given $1 \le i_1 < \cdots < i_r \le n$, define a linear map $\pi : k^n \to k^n$ by letting $\pi(a_1, \ldots, a_n)$ be the vector whose i_j-th coordinate is a_{i_j} for $1 \le j \le r$ and whose other coordinates are 0. Show that $\pi \circ \pi = \pi$ and determine the image of π.

9. In this exercise, we will show that there can be more than one projection onto a given subspace $H \subseteq k^n$.

   a. Show that the matrices
   $$\begin{pmatrix} 1 & 0 \\ 0 & 0 \end{pmatrix}, \begin{pmatrix} 1 & 1 \\ 0 & 0 \end{pmatrix}$$
   both define projections from $\mathbb{R}^2$ onto the x-axis. Draw a picture that illustrates what happens to a typical point of $\mathbb{R}^2$ under each projection.

   b. Show that there is a one-to-one correspondence between projections of $\mathbb{R}^2$ onto the x-axis and nonhorizontal lines in $\mathbb{R}^2$ through the origin.

   c. More generally, fix an r-dimensional subspace $H \subseteq k^n$. Show that there is a one-to-one correspondence between projections of k^n onto H and $(n-r)$-dimensional subspaces $H' \subseteq k^n$ which satisfy $H \cap H' = \{0\}$. Hint: Consider the kernel of the projection.

10. This exercise is concerned with the proof of Theorem 7.

    a. If $f_1, \ldots, f_r \in k[x_1, \ldots, x_n]$ have total degree $\le N$ and $F \in k[x_1, \ldots, x_n]$ has total degree $\le s$, show that $f^s F(f_1/f, \ldots, f_r/f)$ is a polynomial in $k[x_1, \ldots, x_n]$.

    b. Show that the polynomial of part (a) has total degree $\le Ns$.

    c. Show that the map β defined in the proof of Theorem 7 is a well-defined linear map.

11. Complete the proof of Corollary 8.

12. Suppose that $\phi : V \to W$ is a polynomial map between affine varieties (see Chapter 5, §1). We proved in §4 of Chapter 5 that ϕ induces a ring homomorphism $\phi^* : k[W] \to k[V]$ which is the identity on k. From ϕ, we get the subset $\phi(V) \subseteq W$. We say that ϕ is *dominating* if the smallest variety of W containing $\phi(V)$ is W itself. Thus, ϕ is dominating if its image is Zariski dense in W.

    a. Show that ϕ is dominating if and only if the homomorphism $\phi^* : k[W] \to k[V]$ is one-to-one. Hint: Show that $W' \subseteq W$ is a proper subvariety if and only if there is nonzero element $[f] \in k[W]$ such that $W' \subseteq W \cap \mathbf{V}(f)$.

    b. If ϕ is dominating, show that $\dim V \ge \dim W$. Hint: Use Theorem 2 and part (a).

13. This exercise will study the relation between parametrizations and dimension. Assume that k is an infinite field.

    a. Suppose that $F : k^m \to V$ is a polynomial parametrization of a variety V (as defined in Chapter 3, §3). Thus, m is the number of parameters and V is the smallest variety containing $F(k^m)$. Prove that $m \ge \dim V$.

    b. Give an example of a polynomial parametrization $F : k^m \to V$ where $m > \dim V$.

    c. Now suppose that $F : k^m \setminus W \to V$ is a rational parametrization of V (as defined in Chapter 3, §3). We know that V is irreducible by Proposition 6 of Chapter 4, §5. Show that we can define a field homomorphism $F^* : k(V) \to k(t_1, \ldots, t_m)$ which is one-to-one. Hint: See the proof of Theorem 10 of Chapter 5, §5.

    d. If $F : k^m \setminus W \to V$ is a rational parametrization, show that $m \ge \dim V$.

14. Suppose that $V \subseteq \mathbb{P}^n(k)$ is an irreducible projective variety and let $k(V)$ be its field of rational functions as defined in Exercise 13 of Chapter 8, §3.

    a. Prove that $\dim V$ is the transcendence degree of $k(V)$ over k. Hint: Reduce to the affine case.

    b. We say that two irreducible projective varieties V and V' (lying possibly in different projective spaces) are *birationally equivalent* if any of their nonempty affine portions $V \cap U_i$ and $V' \cap U_j$ are birationally equivalent in the sense of Chapter 5, §5. Prove that V and V' are birationally equivalent if and only if there is a field isomorphism $k(V) \cong k(V')$ which is the identity on k. Hint: Use Exercise 13 of Chapter 8, §3 and Theorem 10 of Chapter 5, §5.

    c. Prove that birationally equivalent projective varieties have the same dimension.

§6 Dimension and Nonsingularity

This section will explore how dimension is related to the geometric properties of a variety V. The discussion will be rather different from §5, where the algebraic properties of $k[V]$ and $k(V)$ played a dominant role. We will introduce some rather sophisticated concepts, and some of the theorems will be proved only in special cases. For convenience, we will assume that V is always an affine variety.

When we look at a surface $V \subseteq \mathbb{R}^3$, one intuitive reason for saying that it is 2-dimensional is that at a point p on V, a small portion of the surface looks like a small portion of the plane. This is reflected by the way the tangent plane approximates V at p:

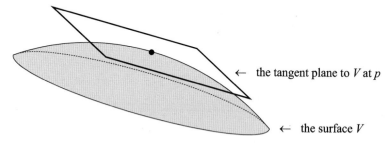

← the tangent plane to V at p

← the surface V

Of course, we have to be careful because the surface may have points where there does not seem to be a tangent plane. For example, consider the cone $\mathbf{V}(x^2 + y^2 - z^2)$. There seems to be a nice tangent plane everywhere except at the origin:

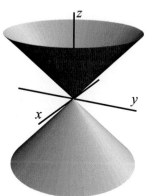

In this section, we will introduce the concept of a *nonsingular point* p of a variety V, and we will give a careful definition of the *tangent space* $T_p(V)$ of V at p. Our discussion will generalize what we did for curves in §4 of Chapter 3. The tangent space gives useful information about how the variety V behaves near the point p. This is the so-called "local viewpoint." Although we have not discussed this topic previously, it plays an important role in algebraic geometry. In general, properties which reflect the behavior of a variety near a given point are called *local properties*.

We begin with a discussion of the tangent space. For a curve V defined by an equation $f(x, y) = 0$ in $\mathbb{R}^2$, we saw in Chapter 3 that the line tangent to the curve at a point $(a, b) \in V$ is defined by the equation

$$\frac{\partial f}{\partial x}(a, b) \cdot (x - a) + \frac{\partial f}{\partial y}(a, b) \cdot (y - b) = 0,$$

provided that the two partial derivatives do not both vanish (see Exercise 4 of Chapter 3, §4). We can generalize this to an arbitrary variety as follows.

Definition 1. Let $V \subseteq k^n$ be an affine variety and let $p = (p_1, \ldots, p_n) \in V$ be a point.

(i) If $f \in k[x_1, \ldots, x_n]$ is a polynomial, the **linear part** of f at p, denoted $d_p(f)$, is defined to be the polynomial

$$d_p(f) = \frac{\partial f}{\partial x_1}(p)(x_1 - p_1) + \cdots + \frac{\partial f}{\partial x_n}(p)(x_n - p_n).$$

Note that $d_p(f)$ has total degree ≤ 1.

(ii) The **tangent space** of V at p, denoted $T_p(V)$, is the variety

$$T_p(V) = \mathbf{V}(d_p(f) \mid f \in \mathbf{I}(V)).$$

If we are working over $\mathbb{R}$, then the partial derivative $\frac{\partial f}{\partial x_i}$ has the usual meaning. For other fields, we use the *formal partial derivative*, which is defined by

$$\frac{\partial}{\partial x_i} \left(\sum_{\alpha_1, \ldots, \alpha_n} c_{\alpha_1 \ldots \alpha_n} x_1^{\alpha_1} \cdots x_i^{\alpha_i} \cdots x_n^{\alpha_n} \right) = \sum_{\alpha_1, \ldots, \alpha_n} c_{\alpha_1 \ldots \alpha_n} \alpha_i x_1^{\alpha_1} \cdots x_i^{\alpha_i - 1} \cdots x_n^{\alpha_n}.$$

In Exercise 1, you will show that the usual rules of differentiation apply to $\frac{\partial}{\partial x_i}$. We first prove some simple properties of $T_p(V)$.

Proposition 2. *Let $p \in V \subseteq k^n$.*

(i) *If $\mathbf{I}(V) = \langle f_1, \ldots, f_s \rangle$, then $T_p(V) = \mathbf{V}(d_p(f_1), \ldots, d_p(f_s))$.*

(ii) *$T_p(V)$ is a translate of a linear subspace of k^n.*

Proof. (i) By the product rule, it is easy to show that

$$d_p(hf) = h(p) \cdot d_p(f) + d_p(h) \cdot f(p)$$

(see Exercise 2). This implies $d_p(hf) = h(p) \cdot d_p(f)$ when $f(p) = 0$, and it follows that if $g = \sum_{i=1}^{s} h_i f_i \in \mathbf{I}(V) = \langle f_1, \ldots, f_s \rangle$, then

$$d_p(g) = \sum_{i=1}^{s} d_p(h_i f_i) = \sum_{i=1}^{s} h_i(p) \cdot d_p(f_i) \in \langle d_p(f_1), \ldots, d_p(f_s) \rangle.$$

This shows that $T_p(V)$ is defined by the vanishing of the $d_p(f_i)$.

(ii) Introduce new coordinates on k^n by setting $X_i = x_i - p_i$ for $1 \leq i \leq n$. This coordinate system is obtained by translating p to the origin. By part (i), we know that $T_p(V)$ is given by $d_p(f_1) = \cdots = d_p(f_s) = 0$. Since each $d_p(f_i)$ is linear in $X_1, \ldots, X_n$, we see that $T_p(V)$ is a linear subspace with respect to the X_i. In terms of the original coordinates, this means that $T_p(V)$ is a translate of a subspace of k^n. $\square$

We can get an intuitive idea of what the tangent space means by thinking about *Taylor's formula* for a polynomial of several variables. For a polynomial of one variable, one has the standard formula

$$f(x) = f(a) + f'(a)(x - a) + \text{terms involving higher powers of } x - a.$$

For $f \in k[x_1, \ldots, x_n]$, you will show in Exercise 3 that if $p = (p_1, \ldots, p_n)$, then

$$f = f(p) + \frac{\partial f}{\partial x_1}(p)(x_1 - p_1) + \cdots + \frac{\partial f}{\partial x_n}(p)(x_n - p_n)$$
$$+ \text{ terms of total degree } \geq 2 \text{ in } x_1 - p_1, \ldots, x_n - p_n.$$

This is part of Taylor's formula for f at p. When $p \in V$ and $f \in \mathbf{I}(V)$, we have $f(p) = 0$, so that

$$f = d_p(f) + \text{terms of total degree } \geq 2 \text{ in } x_1 - p_1, \ldots, x_n - p_n.$$

Thus $d_p(f)$ is the best linear approximation of f near p. Now suppose that $\mathbf{I}(V) = \langle f_1, \ldots, f_s \rangle$. Then V is defined by the vanishing of the f_i, so that the best linear approximation to V near p should be defined by the vanishing of the $d_p(f_i)$. By Proposition 2, this is exactly the tangent space $T_p(V)$.

We can also think about $T_p(V)$ in terms of lines that meet V with "higher multiplicity" at p. In Chapter 3, this was how we defined the tangent line for curves in the plane. In the higher dimensional case, suppose that we have $p \in V$ and let L be a line through p. We can parametrize L by $F(t) = p + tv$, where $v \in k^n$ is a vector parallel to L. If $f \in k[x_1, \ldots, x_n]$, then $f \circ F(t)$ is a polynomial in the variable t, and note that $f \circ F(0) = f(p)$. Thus, 0 is a root of this polynomial whenever $f \in \mathbf{I}(V)$. We can use the multiplicity of this root to decide when L is contained in $T_p(V)$.

Proposition 3. *If L is a line through p parametrized by $F(t) = p + tv$, then $L \subseteq T_p(V)$ if and only if 0 is a root of multiplicity ≥ 2 of $f \circ F(t)$ for all $f \in \mathbf{I}(V)$.*

Proof. If we write the parametrization of L as $x_i = p_i + tv_i$ for $1 \leq i \leq n$, where $p = (p_1, \ldots, p_n)$ and $v = (v_1, \ldots, v_n)$, then, for any $f \in \mathbf{I}(V)$, we have

$$g(t) = f \circ F(t) = f(p_1 + tv_1, \ldots, p_n + tv_n).$$

As we noted above, $g(0) = 0$ because $p \in V$, so that $t = 0$ is a root of $g(t)$. In Exercise 5 of Chapter 3, §4, we showed that $t = 0$ is a root of multiplicity ≥ 2 if and only if we also have $g'(0) = 0$. Using the chain rule for functions of several variables, we obtain

$$\frac{dg}{dt} = \frac{\partial f}{\partial x_1}\frac{dx_1}{dt} + \cdots + \frac{\partial f}{\partial x_n}\frac{dx_n}{dt} = \frac{\partial f}{\partial x_1}v_1 + \cdots + \frac{\partial f}{\partial x_n}v_n.$$

If follows that $g'(0) = 0$ if and only if

$$0 = \sum_{i=1}^{n}\frac{\partial f}{\partial x_i}(p)v_i = \sum_{i=1}^{n}\frac{\partial f}{\partial x_i}(p)((p_i + v_i) - p_i).$$

The expression on the right in this equation is $d_p(f)$ evaluated at the point $p+v \in L$, and it follows that $p + v \in T_p(V)$ if and only if $g'(0) = 0$ for all $f \in \mathbf{I}(V)$. Since $p \in L$, we know that $L \subseteq T_p(V)$ is equivalent to $p + v \in T_p(V)$, and the proposition is proved. $\square$

It is time to look at some examples.

Example 4. Let $V \subseteq \mathbb{C}^n$ be the hypersurface defined by $f = 0$, where $f \in k[x_1, \ldots, x_n]$ is a nonconstant polynomial. By Proposition 9 of Chapter 4, §2, we have

$$\mathbf{I}(V) = \mathbf{I}(\mathbf{V}(f)) = \sqrt{\langle f \rangle} = \langle f_{\mathrm{red}} \rangle,$$

where $f_{\mathrm{red}} = f_1 \cdots f_r$ is the product of the distinct irreducible factors of f. We will assume that $f = f_{\mathrm{red}}$. This implies that

$$V = \mathbf{V}(f) = \mathbf{V}(f_1 \cdots f_r) = \mathbf{V}(f_1) \cup \cdots \cup \mathbf{V}(f_r)$$

is the decomposition of V into irreducible components (see Exercise 10 of Chapter 4, §6). In particular, every component of V has dimension $n - 1$ by the affine version of Proposition 2 of §4.

Since $\mathbf{I}(V) = \langle f \rangle$, it follows from Proposition 2 that for any $p \in V$, $T_p(V)$ is the linear space defined by the single equation

$$\frac{\partial f}{\partial x_1}(p)(x_1 - p_1) + \cdots + \frac{\partial f}{\partial x_n}(p)(x_n - p_n) = 0.$$

This implies that

(1) $$\dim T_p(V) = \begin{cases} n - 1 & \text{at least one } \frac{\partial f}{\partial x_i}(p) \neq 0 \\ n & \text{all } \frac{\partial f}{\partial x_i}(p) = 0. \end{cases}$$

You should check how this generalizes Proposition 2 of Chapter 3, §4.

For a specific example, consider $V = \mathbf{V}(x^2 - y^2z^2 + z^3)$. In Exercise 4, you will show that $f = x^2 - y^2z^2 + z^3 \in \mathbb{C}[x, y, z]$ is irreducible, so that $\mathbf{I}(V) = \langle f \rangle$. The partial derivatives of f are

$$\frac{\partial f}{\partial x} = 2x, \quad \frac{\partial f}{\partial y} = -2yz^2, \quad \frac{\partial f}{\partial z} = -2y^2z + 3z^2.$$

We leave it as an exercise to show that on V, the partials vanish simultaneously only on the y-axis, which lies in V. Thus, the tangent spaces $T_p(V)$ are all 2-dimensional, except along the y-axis, where they are all of $\mathbb{C}^3$. Over $\mathbb{R}$, we get the following picture of V (which appeared earlier in §2 of Chapter 1):

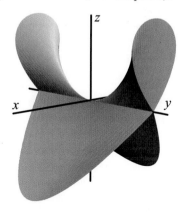

When we give the definition of nonsingular point later in this section, we will see that the points of V on the y-axis are the singular points, whereas other points of V are nonsingular.

Example 5. Now consider the curve $C \subseteq \mathbb{C}^3$ obtained by intersecting the surface V of Example 4 with the plane $x + y + z = 0$. Thus, $C = \mathbf{V}(x + y + z, x^2 - y^2 z^2 + z^3)$. Using the techniques of §3, you can verify that $\dim C = 1$.

In the exercises, you will also show that $\langle f_1, f_2 \rangle = \langle x + y + z, x^2 - y^2 z^2 + z^3 \rangle$ is a prime ideal, so that C is an irreducible curve. Since a prime ideal is radical, the Nullstellensatz implies that $\mathbf{I}(C) = \langle f_1, f_2 \rangle$. Thus, for $p = (a, b, c) \in C$, it follows that $T_p(C)$ is defined by the linear equations

$$d_p(f_1) = 1 \cdot (x - a) + 1 \cdot (y - b) + 1 \cdot (z - c) = 0,$$
$$d_p(f_2) = 2a \cdot (x - a) + (-2bc^2) \cdot (y - b) + (-2b^2 c + 3c^2) \cdot (z - c) = 0.$$

This is a system of linear equations in $x - a, y - b, z - c$ with coefficient matrix

$$J_p(f_1, f_2) = \begin{pmatrix} 1 & 1 & 1 \\ 2a & -2bc^2 & -2b^2 c + 3c^2 \end{pmatrix}.$$

Let $\mathrm{rank}(J_p(f_1, f_2))$ denote the rank of this matrix. Since $T_p(C)$ is a translate of the kernel of $J_p(f_1, f_2)$, it follows that

$$\dim T_p(C) = 3 - \mathrm{rank}(J_p(f_1, f_2)).$$

In the exercises, you will show that $T_p(C)$ is 1-dimensional at all points of C except for the origin, where $T_0(C)$ is the 2-dimensional plane $x + y + z = 0$.

In these examples, we were careful to always compute $\mathbf{I}(V)$. It would be much nicer if we could use any set of defining equations of V. Unfortunately, this does not always work: if $V = \mathbf{V}(f_1, \ldots, f_s)$, then $T_p(V)$ need *not* be defined by $d_p(f_1) = \cdots = d_p(f_s) = 0$. For example, let V be the y-axis in k^2. Then V is defined by $x^2 = 0$, but you can easily check that $T_p(V) \neq \mathbf{V}(d_p(x^2))$ for all $p \in V$. However, in Theorem 9, we will find a nice condition on $f_1, \ldots, f_s$ which, when satisfied, will allow us to compute $T_p(V)$ using the $d_p(f_i)$'s.

Examples 4 and 5 indicate that the nicest points on V are the ones where $T_p(V)$ has the same dimension as V. But this principle does not apply when V has irreducible components of different dimensions. For example, let $V = \mathbf{V}(xz, yz) \subseteq \mathbb{R}^3$. This is the union of the (x, y)-plane and the z-axis, and it is easy to check that

$$\dim T_p(V) = \begin{cases} 2 & p \text{ is on the } (x, y)\text{-plane minus the origin} \\ 1 & p \text{ is on the } z\text{-axis minus the origin} \\ 3 & p \text{ is the origin.} \end{cases}$$

Excluding the origin, points on the z-axis have a 1-dimensional tangent space, which seems intuitively correct. Yet at such a point, we have $\dim T_p(V) < \dim V$. The problem, of course, is that we are on a component of the wrong dimension.

To avoid this difficulty, we need to define the dimension of a variety at a point.

Definition 6. Let V be an affine variety. For $p \in V$, the **dimension of V at p**, denoted $\dim_p V$, is the maximum dimension of an irreducible component of V containing p.

By Corollary 9 of §4, we know that $\dim V$ is the maximum of $\dim_p V$ as p varies over all points of V. If V is a hypersurface in $\mathbb{C}^n$, it is easy to compute $\dim_p V$, for in Example 4, we showed that every irreducible component of V has dimension $n - 1$. It follows that $\dim_p V = n - 1$ for all $p \in V$. On the other hand, if $V \subseteq k^n$ is an arbitrary variety, the theory developed in §§3 and 4 enables us to compute $\dim V$, but unless we know how to decompose V into irreducible components, more subtle tools are needed to compute $\dim_p V$. This will be discussed in §7 when we study the properties of the tangent cone.

We can now define what it means for a point $p \in V$ to be nonsingular.

Definition 7. Let p be a point on an affine variety V. Then p is **nonsingular** (or **smooth**) provided $\dim T_p(V) = \dim_p V$. Otherwise, p is a **singular** point of V.

If we look back at our previous examples, it is easy to identify which points are nonsingular and which are singular. In Example 5, the curve C is irreducible, so that $\dim_p C = 1$ for all $p \in C$ and, hence, the singular points are where $\dim T_p(C) \neq 1$ (only one in this case). For the hypersurfaces $V = \mathbf{V}(f)$ considered in Example 4, we know that $\dim_p V = n - 1$ for all $p \in V$, and it follows from (1) so that p is singular if and only if all of the partial derivatives of f vanish at p. This means that the singular points of V form the variety

$$(2) \qquad \qquad \Sigma = \mathbf{V}\left(f, \frac{\partial f}{\partial x_1}, \ldots, \frac{\partial f}{\partial x_n}\right).$$

In general, the singular points of a variety V have the following properties.

Theorem 8. *Let $V \subseteq k^n$ be an affine variety and let*

$$\Sigma = \{p \in V \mid p \text{ is a singular point of } V\}.$$

*We call Σ the **singular locus** of V. Then:*

(i) *Σ is an affine variety contained in V.*
(ii) *If $p \in \Sigma$, then $\dim T_p(V) > \dim_p V$.*
(iii) *Σ contains no irreducible component of V.*
(iv) *If V_i and V_j are distinct irreducible components of V, then $V_i \cap V_j \subseteq \Sigma$.*

Proof. A complete proof of this theorem is beyond the scope of the book. Instead, we will assume that V is a hypersurface in $\mathbb{C}^n$ and show that the theorem holds in this case. As we discuss each part of the theorem, we will give references for the general case.

Let $V = \mathbf{V}(f) \subseteq \mathbb{C}^n$ be a hypersurface such that $\mathbf{I}(V) = \langle f \rangle$. We noted earlier that $\dim_p V = n - 1$ and that Σ consists of those points of V where all of the partial derivatives of f vanish simultaneously. Then (2) shows that Σ is an affine variety, which proves (i) for hypersurfaces. A proof in the general case is given in the Corollary to Theorem 6 in Chapter II, §2 of SHAFAREVICH (2013).

Part (ii) of the theorem says that at a singular point of V, the tangent space is too big. When V is a hypersurface in $\mathbb{C}^n$, we know from (1) that if p is a singular point, then $\dim T_p(V) = n > n - 1 = \dim_p V$. This proves (ii) for hypersurfaces, and the general case follows from Theorem 3 in Chapter II, §1 of SHAFAREVICH (2013).

Part (iii) says that on each irreducible component of V, the singular locus consists of a proper subvariety. Hence, most points of a variety are nonsingular. To prove this for a hypersurface, let $V = \mathbf{V}(f) = \mathbf{V}(f_1) \cup \cdots \cup \mathbf{V}(f_r)$ be the decomposition of V into irreducible components, as discussed in Example 4. Suppose that Σ contains one of the components, say $\mathbf{V}(f_1)$. Then every $\frac{\partial f}{\partial x_i}$ vanishes on $\mathbf{V}(f_1)$. If we write $f = f_1 g$, where $g = f_2 \cdots f_r$, then

$$\frac{\partial f}{\partial x_i} = f_1 \frac{\partial g}{\partial x_i} + \frac{\partial f_1}{\partial x_i} g$$

by the product rule. Since f_1 certainly vanishes on $\mathbf{V}(f_1)$, it follows that $\frac{\partial f_1}{\partial x_i} g$ also vanishes on $\mathbf{V}(f_1)$. By assumption, f_1 is irreducible, so that

$$\frac{\partial f_1}{\partial x_i} g \in \mathbf{I}(\mathbf{V}(f_1)) = \langle f_1 \rangle.$$

This says that f_1 divides $\frac{\partial f_1}{\partial x_i} g$ and, hence, f_1 divides $\frac{\partial f_1}{\partial x_i}$ or g. The latter is impossible since g is a product of irreducible polynomials distinct from f_1 (meaning that none of them is a constant multiple of f_1). Thus, f_1 must divide $\frac{\partial f_1}{\partial x_i}$ for all i. Since $\frac{\partial f_1}{\partial x_i}$ has smaller total degree than f_1, we must have $\frac{\partial f_1}{\partial x_i} = 0$ for all i, and it follows that f_1 is constant (see Exercise 9). This contradiction proves that Σ contains no component of V.

When V is an arbitrary irreducible variety, a proof that Σ is a proper subvariety can be found in the corollary to Theorems 4.1 and 4.3 in Chapter IV of KENDIG (2015). See also the discussion preceding the definition of singular point in Chapter II, §1 of SHAFAREVICH (2013). If V has two or more irreducible components, the claim follows from the irreducible case and part (iv) below. See Exercise 11 for the details.

Finally, part (iv) of the theorem says that a nonsingular point of a variety lies on a unique irreducible component. In the hypersurface case, suppose that $V = \mathbf{V}(f) = \mathbf{V}(f_1) \cup \cdots \cup \mathbf{V}(f_r)$ and that $p \in \mathbf{V}(f_i) \cap \mathbf{V}(f_j)$ for $i \neq j$. Then we can write $f = gh$, where f_i divides g and f_j divides h. Hence, $g(p) = h(p) = 0$, and then an easy argument using the product rule shows that $\frac{\partial f}{\partial x_i}(p) = 0$ for all i. This proves that $\mathbf{V}(f_i) \cap \mathbf{V}(f_j) \subseteq \Sigma$, so that (iv) is true for hypersurfaces. When V is an arbitrary variety, see Theorem 6 in Chapter II, §2 of SHAFAREVICH (2013). $\square$

In some cases, it is also possible to show that a point of a variety V is nonsingular without having to compute $\mathbf{I}(V)$. To formulate a precise result, we will need some notation. Given $f_1, \ldots, f_r \in k[x_1, \ldots, x_n]$, as in Chapter 6 let $J(f_1, \ldots, f_r)$ be the *Jacobian matrix*, the $r \times n$ matrix of partial derivatives:

$$J(f_1, \ldots, f_r) = \begin{pmatrix} \frac{\partial f_1}{\partial x_1} & \cdots & \frac{\partial f_1}{\partial x_n} \\ \vdots & & \vdots \\ \frac{\partial f_r}{\partial x_1} & \cdots & \frac{\partial f_r}{\partial x_n} \end{pmatrix}.$$

Given $p \in k^n$, evaluating this matrix at p gives an $r \times n$ matrix of numbers denoted $J_p(f_1, \ldots, f_r)$. Then we have the following result.

Theorem 9. *Let $V = \mathbf{V}(f_1, \ldots, f_r) \subseteq \mathbb{C}^n$ be an arbitrary variety and suppose that $p \in V$ is a point where $J_p(f_1, \ldots, f_r)$ has rank r. Then p is a nonsingular point of V and lies on a unique irreducible component of V of dimension $n - r$.*

Proof. As with Theorem 8, we will only prove this for a hypersurface $V = \mathbf{V}(f) \subseteq \mathbb{C}^n$, which is the case $r = 1$ of the theorem. Here, note that f is now *any* defining equation of V, and, in particular, it could happen that $\mathbf{I}(V) \neq \langle f \rangle$. But we still know that f vanishes on V, and it follows from the definition of tangent space that

$$(3) \qquad\qquad T_p(V) \subseteq \mathbf{V}(d_p(f)).$$

Since $r = 1$, $J_p(f)$ is the row vector whose entries are $\frac{\partial f}{\partial x_i}(p)$, and our assumption that $J_p(f)$ has rank 1 implies that at least one of the partials is nonzero at p. Thus, $d_p(f)$ is a nonzero linear function of $x_i - p_i$, and it follows from (3) that $\dim T_p(V) \leq n - 1$. If we compare this to (1), we see that p is a nonsingular point of V, and by part (iv) of Theorem 8, it lies on a unique irreducible component of V. Since the component has the predicted dimension $n - r = n - 1$, we are done. For the general case, see Theorem (1.16) of MUMFORD (1981). $\square$

Theorem 9 is important for several reasons. First of all, it is very useful for determining the nonsingular points and dimension of a variety. For instance, it is now possible to redo Examples 4 and 5 without having to compute $\mathbf{I}(V)$ and $\mathbf{I}(C)$. Another aspect of Theorem 9 is that it relates nicely to our intuition that the dimension should drop by one for each equation defining V. This is what happens in the theorem, and in fact we can sharpen our intuition as follows. Namely, the dimension should drop by one for each defining equation, provided the defining equations are sufficiently independent [which means that rank$(J_p(f_1,\ldots,f_r)) = r$]. In Exercise 16, we will give a more precise way to state this. Furthermore, note that our intuition applies to the nonsingular part of V.

Theorem 9 is also related to some important ideas in the calculus of several variables. In particular, the *Implicit Function Theorem* has the same hypothesis concerning $J_p(f_1,\ldots,f_r)$ as Theorem 9. When $V = \mathbf{V}(f_1,\ldots,f_r)$ satisfies this hypothesis, the complex variable version of the Implicit Function Theorem asserts that near p, the variety V looks like the graph of a nice function of $n - r$ variables, and we get a vivid picture of why V has dimension $n - r$ at p. To understand the full meaning of Theorem 9, one needs to study the notion of a *manifold*. A nice discussion of this topic and its relation to nonsingularity and dimension can be found in KENDIG (2015).

EXERCISES FOR §6

1. We will discuss the properties of the formal derivative defined in the text.
   a. Show that $\frac{\partial}{\partial x_i}$ is k-linear and satisfies the product rule.
   b. Show that $\frac{\partial}{\partial x_i}\left(\frac{\partial}{\partial x_j}f\right) = \frac{\partial}{\partial x_j}\left(\frac{\partial}{\partial x_i}f\right)$ for all i and j.
   c. If $f_1,\ldots,f_r \in k[x_1,\ldots,x_n]$, compute $\frac{\partial}{\partial x_i}(f_1^{\alpha_1}\cdots f_r^{\alpha_r})$.
   d. Formulate and prove a version of the chain rule for computing the partial derivatives of a polynomial of the form $F(f_1,\ldots,f_r)$. Hint: Use part (c).
2. Prove that $d_p(hf) = h(p) \cdot d_p(f) + d_p(h) \cdot f(p)$.
3. Let $p = (p_1,\ldots,p_n) \in k^n$ and let $f \in k[x_1,\ldots,x_n]$.
   a. Show that f can be written as a polynomial in $x_i - p_i$. Hint: $x_i^m = ((x_i - p_i) + p_i)^m$.
   b. Suppose that when we write f as a polynomial in $x_i - p_i$, every term has total degree at least 2. Show that $\frac{\partial f}{\partial x_i}(p) = 0$ for all i.
   c. If we write f as a polynomial in $x_i - p_i$, show that the constant term is $f(p)$ and the linear term is $d_p(f)$. Hint: Use part (b).
4. As in Example 4, let $f = x^2 - y^2 z^2 + z^3 \in \mathbb{C}[x, y, z]$ and let $V = \mathbf{V}(f) \subseteq \mathbb{C}^3$.
   a. Show carefully that f is irreducible in $\mathbb{C}[x, y, z]$.
   b. Show that V contains the y-axis.
   c. Let $p \in V$. Show that the partial derivatives of f all vanish at p if and only if p lies on the y-axis.
5. Let A be an $m \times n$ matrix, where $n \geq m$. If $r \leq m$, we say that a matrix B is an $r \times r$ *submatrix* of A provided that B is the matrix obtained by first selecting r columns of A, and then selecting r rows from those columns.
   a. Pick a 3×4 matrix of numbers and write down all of its 3×3 and 2×2 submatrices.
   b. Show that A has rank $< r$ if and only if all $t \times t$ submatrices of A have determinant zero for all $r \leq t \leq m$. Hint: The rank of a matrix is the maximum number of linearly independent columns. If A has rank s, it follows that you can find an $m \times s$ submatrix of rank s. Now use the fact that the rank is also the maximum number of linearly independent rows. What is the criterion for an $r \times r$ matrix to have rank $< r$?
6. As in Example 5, let $C = \mathbf{V}(x + y + z, x^2 - y^2 z^2 + z^3) \subseteq \mathbb{C}^3$ and let I be the ideal $I = \langle x + y + z, x^2 - y^2 z^2 + z^3 \rangle \subseteq \mathbb{C}[x, y, z]$.

a. Show that I is a prime ideal. Hint: Introduce new coordinates $X = x + y + z, Y = y$, and $Z = z$. Show that $I = \langle X, F(Y,Z) \rangle$ for some polynomial in Y, Z. Prove that $\mathbb{C}[X, Y, Z]/I \cong \mathbb{C}[Y, Z]/\langle F \rangle$ and show that $F \in \mathbb{C}[Y, Z]$ is irreducible.

b. Conclude that C is an irreducible variety and that $\mathbf{I}(C) = I$.

c. Compute the dimension of C.

d. Determine all points $(a, b, c) \in C$ such that the 2×3 matrix

$$J_p(f_1, f_2) = \begin{pmatrix} 1 & 1 & 1 \\ 2a & -2bc^2 & -2b^2c + 3c^2 \end{pmatrix}$$

has rank < 2. Hint: Use Exercise 5.

7. Let $f = x^2 \in k[x, y]$. In k^2, show that $T_p(\mathbf{V}(f)) \neq \mathbf{V}(d_p(f))$ for all $p \in V$.

8. Let $V = \mathbf{V}(xy, xz) \subseteq k^3$ and assume that k is an infinite field.
   a. Compute $\mathbf{I}(V)$.
   b. Verify the formula for $\dim T_p(V)$ given in the text.

9. Suppose that $f \in k[x_1, \ldots, x_n]$ is a polynomial such that $\frac{\partial}{\partial x_i} f = 0$ for all i. If the field k has characteristic 0 (which means that k contains a field isomorphic to $\mathbb{Q}$), then show that f must be the constant.

10. The result of Exercise 9 may be false if k does not have characteristic 0.
    a. Let $f = x^2 + y^2 \in \mathbb{F}_2[x, y]$, where $\mathbb{F}_2$ is a field with two elements. What are the partial derivatives of f?
    b. To analyze the case when k does not have characteristic 0, we need to define the *characteristic* of k. Given any field k, show that there is a ring homomorphism $\phi :$ $\mathbb{Z} \to k$ which sends $n > 0$ in $\mathbb{Z}$ to $1 \in k$ added to itself n times. If ϕ is one-to-one, argue that k contains a copy of $\mathbb{Q}$ and hence has characteristic 0.
    c. If k does not have characteristic 0, it follows that the map ϕ of part (b) cannot be one-to-one. Show that the kernel must be the ideal $\langle p \rangle \subseteq \mathbb{Z}$ for some prime number p. We say that k has *characteristic p* in this case. Hint: Use the Isomorphism Theorem from Exercise 16 of Chapter 5, §2 and remember that k is an integral domain.
    d. If k has characteristic p, show that $(a + b)^p = a^p + b^p$ for every $a, b \in k$.
    e. Suppose that k has characteristic p and let $f \in k[x_1, \ldots, x_n]$. Show that all partial derivatives of f vanish identically if and only if every exponent of every monomial appearing in f is divisible by p.
    f. Suppose that k is algebraically closed and has characteristic p. If $f \in k[x_1, \ldots, x_n]$ is irreducible, then show that some partial derivative of f must be nonzero. This shows that Theorem 8 is true for hypersurfaces over *any* algebraically closed field. Hint: If all partial derivatives vanish, use parts (d) and (e) to write f as a p-th power. Why do you need k to be algebraically closed?

11. Let $V = V_1 \cup \cdots \cup V_r$ be a decomposition of a variety into its irreducible components.
    a. Suppose that $p \in V$ lies in a unique irreducible component V_i. Show that $T_p(V) = T_p(V_i)$. This reflects the local nature of the tangent space. Hint: One inclusion follows easily from $V_i \subseteq V$. For the other inclusion, pick a function $f \in \mathbf{I}(W) \backslash \mathbf{I}(V_i)$, where W is the union of the other irreducible components. Then $g \in \mathbf{I}(V_i)$ implies $fg \in \mathbf{I}(V)$.
    b. With the same hypothesis as part (a), show that p is nonsingular in V if and only if it is nonsingular in V_i.
    c. Let Σ be the singular locus of V and let Σ_i be the singular locus of V_i. Prove that

$$\Sigma = \bigcup_{i \neq j}(V_i \cap V_j) \cup \bigcup_i \Sigma_i.$$

Hint: Use part (b) and part (iv) of Theorem 8.

d. If each Σ_i is a proper subset of V_i, then show that Σ contains no irreducible components of V. This shows that part (iii) of Theorem 8 follows from the irreducible case.

12. Find all singular points of the following curves in k^2. Assume that k is algebraically closed.
    a. $y^2 = x^3 - 3$.
    b. $y^2 = x^3 - 6x^2 + 9x$.
    c. $x^2y^2 + x^2 + y^2 + 2xy(x + y + 1) = 0$.
    d. $x^2 = x^4 + y^4$.
    e. $xy = x^6 + y^6$.
    f. $x^2y + xy^2 = x^4 + y^4$.
    g. $x^3 = y^2 + x^4 + y^4$.
13. Find all singular points of the following surfaces in k^3. Assume that k is algebraically closed.
    a. $xy^2 = z^2$.
    b. $x^2 + y^2 = z^2$.
    c. $x^2y + x^3 + y^3 = 0$.
    d. $x^3 - zxy + y^3 = 0$.
14. Show that $\mathbf{V}(y - x^2 + z^2, 4x - y^2 + w^3) \subseteq \mathbb{C}^4$ is a nonempty smooth surface.
15. Let $V \subseteq k^n$ be a hypersurface with $\mathbf{I}(V) = \langle f \rangle$. Show that if V is not a hyperplane and $p \in V$ is nonsingular, then either the variety $V \cap T_p(V)$ has a singular point at p or the restriction of f to $T_p(V)$ has an irreducible factor of multiplicity ≥ 2. Hint: Pick coordinates so that $p = 0$ and $T_p(V)$ is defined by $x_1 = 0$. Thus, we can regard $T_p(V)$ as a copy of k^{n-1}, then $V \cap T_p(V)$ is a hypersurface in k^{n-1}. Then the restriction of f to $T_p(V)$ is the polynomial $f(0, x_2, \ldots, x_n)$. See also Example 4.
16. Let $V \subseteq \mathbb{C}^n$ be irreducible and let $p \in V$ be a nonsingular point. Suppose that V has dimension d.
    a. Show that we can always find polynomials $f_1, \ldots, f_{n-d} \in \mathbf{I}(V)$ with the property that $T_p(V) = \mathbf{V}(d_p(f_1), \ldots, d_p(f_{n-d}))$.
    b. If $f_1, \ldots, f_{n-d}$ are as in part (a) show that $J_p(f_1, \ldots, f_{n-d})$ has rank $n - d$ and conclude that V is an irreducible component of $\mathbf{V}(f_1, \ldots, f_{n-d})$. This shows that although V itself may not be defined by $n - d$ equations, it is a component of a variety that is. Hint: Use Theorem 9.
17. Suppose that $V \subseteq \mathbb{C}^n$ is irreducible of dimension d and suppose that $\mathbf{I}(V) = \langle f_1, \ldots, f_s \rangle$.
    a. Show that $p \in V$ is nonsingular if and only if $J_p(f_1, \ldots, f_s)$ has rank $n - d$. Hint: Use Proposition 2.
    b. By Theorem 8, we know that V has nonsingular points. Use this and part (a) to prove that $d \geq n - s$. How does this relate to Proposition 5 of §4?
    c. Let $\mathcal{D}$ be the set of determinants of all $(n - d) \times (n - d)$ submatrices of $J(f_1, \ldots, f_s)$. Prove that the singular locus of V is $\Sigma = V \cap \mathbf{V}(g \mid g \in \mathcal{D})$. Hint: Use part (a) and Exercise 5. Also, what does part (ii) of Theorem 8 tell you about the rank of $J_p(f_1, \ldots, f_s)$?

§7 The Tangent Cone

In this final section of the chapter, we will study the *tangent cone* of a variety V at a point p. When p is nonsingular, we know that, near p, V is nicely approximated by its tangent space $T_p(V)$. This clearly fails when p is singular, for as we saw in Theorem 8 of §6, the tangent space has the wrong dimension (it is too big). To approximate V near a singular point, we need something different.

We begin with an example.

Example 1. Consider the curve $y^2 = x^2(x+1)$, which has the following picture in the plane $\mathbb{R}^2$:

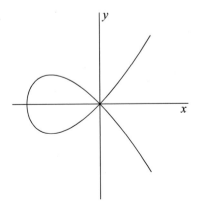

We see that the origin is a singular point. Near this point, the curve is approximated by the lines $x = \pm y$. These lines are defined by $x^2 - y^2 = 0$, and if we write the defining equation of the curve as $f(x, y) = x^2 - y^2 + x^3 = 0$, we see that $x^2 - y^2$ is the nonzero homogeneous component of f of smallest total degree.

Similarly, consider the curve $y^2 - x^3 = 0$:

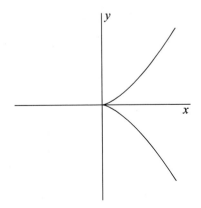

The origin is again a singular point, and the nonzero homogeneous component of $y^2 - x^3$ of smallest total degree is y^2. Here, $\mathbf{V}(y^2)$ is the x-axis and gives a nice approximation of the curve near $(0,0)$.

In both of the above curves, we approximated the curve near the singular point using the smallest nonzero homogeneous component of the defining equation. To generalize this idea, suppose that $p = (p_1, \ldots, p_n) \in k^n$. If $\alpha = (\alpha_1, \ldots, \alpha_n) \in \mathbb{Z}^n_{\geq 0}$, let

$$(x - p)^\alpha = (x_1 - p_1)^{\alpha_1} \cdots (x_n - p_n)^{\alpha_n},$$

and note that $(x - p)^\alpha$ has total degree $|\alpha| = \alpha_1 + \cdots + \alpha_n$. Now, given any polynomial $f \in k[x_1, \ldots, x_n]$ of total degree d, we can write f as a polynomial in

$x_i - p_i$, so that f is a k-linear combination of $(x - p)^\alpha$ for $|\alpha| \leq d$. If we group according to total degree, we can write

$$(1) \qquad\qquad f = f_{p,0} + f_{p,1} + \cdots + f_{p,d},$$

where $f_{p,j}$ is a k-linear combination of $(x-p)^\alpha$ for $|\alpha| = j$. Note that $f_{p,0} = f(p)$ and $f_{p,1} = d_p(f)$ (as defined in Definition 1 of the previous section). In the exercises, you will discuss Taylor's formula, which shows how to express $f_{p,j}$ in terms of the partial derivatives of f at p. In many situations, it is convenient to translate p to the origin so that we can use homogeneous components. We can now define the tangent cone.

Definition 2. Let $V \subseteq k^n$ be an affine variety and let $p = (p_1, \ldots, p_n) \in V$.

(i) If $f \in k[x_1, \ldots, x_n]$ is a nonzero polynomial, then $f_{p,\min}$ is defined to be $f_{p,j}$, where j is the smallest integer such that $f_{p,j} \neq 0$ in (1).

(ii) The **tangent cone** of V at p, denoted $C_p(V)$, is the variety

$$C_p(V) = \mathbf{V}(f_{p,\min} \mid f \in \mathbf{I}(V)).$$

The tangent cone gets its name from the following proposition.

Proposition 3. *Let $p \in V \subseteq k^n$. Then $C_p(V)$ is the translate of the affine cone of a variety in $\mathbb{P}^{n-1}(k)$.*

Proof. Introduce new coordinates on k^n by letting $X_i = x_i - p_i$. Relative to this coordinate system, we can assume that p is the origin 0. Then $f_{0,\min}$ is a homogeneous polynomial in $X_1, \ldots, X_n$, and as f varies over $\mathbf{I}(V)$, the $f_{0,\min}$ generate a homogeneous ideal $J \subseteq k[X_1, \ldots, X_n]$. Then $C_p(V) = \mathbf{V}_a(J) \subseteq k^n$ by definition. Since J is homogeneous, we also get a projective variety $W = \mathbf{V}_p(J) \subseteq \mathbb{P}^{n-1}(k)$, and as we saw in Chapter 8, this means that $C_p(V)$ is an affine cone $C_W \subseteq k^n$ of W. This proves the proposition. $\qquad\square$

The tangent cone of a hypersurface $V \subseteq k^n$ is especially easy to compute. In Exercise 2 you will show that if $\mathbf{I}(V) = \langle f \rangle$, then $C_p(V)$ is defined by the single equation $f_{p,\min} = 0$. This is exactly what we did in Example 1. However, when $\mathbf{I}(V) = \langle f_1, \ldots, f_s \rangle$ has more generators, it need *not* follow that $C_p(V) = \mathbf{V}((f_1)_{p,\min}, \ldots, (f_s)_{p,\min})$. For example, suppose that V is defined by $xy = xz + z(y^2 - z^2) = 0$. In Exercise 3, you will show that $\mathbf{I}(V) = \langle xy, xz + z(y^2 - z^2) \rangle$. To see that $C_0(V) \neq \mathbf{V}(xy, xz)$, note that $f = yz(y^2 - z^2) = y(xz + z(y^2 - z^2)) - z(xy) \in \mathbf{I}(V)$. Then $f_{0,\min} = yz(y^2 - z^2)$ vanishes on $C_0(V)$, yet does not vanish on all of $\mathbf{V}(xy, xz)$.

We can overcome this difficulty by using an appropriate Gröbner basis. The result is stated most efficiently when the point p is the origin.

Proposition 4. *Assume that the origin 0 is a point of $V \subseteq k^n$. Let x_0 be a new variable and pick a monomial order on $k[x_0, x_1, \ldots, x_n]$ such that among monomials of the same total degree, any monomial involving x_0 is greater than any monomial involving only $x_1, \ldots, x_n$ (lex and grlex with $x_0 > \cdots > x_n$ satisfy this condition). Then:*

(i) *Let* $\mathbf{I}(V)^h \subseteq k[x_0, x_1, \ldots, x_n]$ *be the homogenization of* $\mathbf{I}(V)$ *and let* $G_1, \ldots, G_s$ *be a homogeneous Gröbner basis of* $\mathbf{I}(V)^h$ *with respect to the above monomial order. Then*

$$C_0(V) = \mathbf{V}((g_1)_{0,\min}, \ldots, (g_s)_{0,\min}),$$

where $g_i = G_i(1, x_1, \ldots, x_n)$ *is the dehomogenization of* G_i.

(ii) *Suppose that* k *is algebraically closed, and let* I *be any ideal such that* $V = \mathbf{V}(I)$. *If* $G_1, \ldots, G_s$ *are a homogeneous Gröbner basis of* I^h, *then*

$$C_0(V) = \mathbf{V}((g_1)_{0,\min}, \ldots, (g_s)_{0,\min}),$$

where $g_i = G_i(1, x_1, \ldots, x_n)$ *is the dehomogenization of* G_i.

Proof. In this proof, we will write f_j and $f_{\min}$ rather than $f_{0,j}$ and $f_{0,\min}$.

(i) Let $I = \mathbf{I}(V)$. It suffices to show that $f_{\min} \in \langle (g_1)_{\min}, \ldots, (g_s)_{\min} \rangle$ for all $f \in I$. If this fails to hold, then we can find $f \in I$ with $f_{\min} \notin \langle (g_1)_{\min}, \ldots, (g_s)_{\min} \rangle$ such that $\mathrm{LT}(f_{\min})$ is minimal [note that we can regard $f_{\min}$ as a polynomial in $k[x_0, x_1, \ldots, x_n]$, so that $\mathrm{LT}(f_{\min})$ is defined]. If we write f as a sum of homogeneous components

$$f = f_{\min} + \cdots + f_d,$$

where d is the total degree of f, then

$$f^h = f_{\min} \cdot x_0^a + \cdots + f_d \in I^h$$

for some a. By the way we chose the monomial order on $k[x_0, x_1, \ldots, x_n]$, it follows that $\mathrm{LT}(f^h) = \mathrm{LT}(f_{\min})x_0^a$. Since $G_1, \ldots, G_s$ form a Gröbner basis, we know that some $\mathrm{LT}(G_i)$ divides $\mathrm{LT}(f_{\min})x_0^a$.

If g_i is the dehomogenization of G_i, then $g_i \in I$ follows easily. Since G_i is homogeneous, we have

$$\mathrm{LT}(G_i) = \mathrm{LT}((g_i)_{\min})x_0^b$$

for some b (see Exercise 4). This implies that $\mathrm{LT}(f_{\min}) = cx^\alpha \mathrm{LT}((g_i)_{\min})$ for some nonzero $c \in k$ and some monomial x^α in $x_1, \ldots, x_n$. Now let $\tilde{f} = f - cx^\alpha g_i \in I$. Since $f_{\min} \notin \langle (g_1)_{\min}, \ldots, (g_s)_{\min} \rangle$, we know that $f_{\min} - cx^\alpha (g_i)_{\min} \neq 0$, and it follows that

$$\tilde{f}_{\min} = f_{\min} - cx^\alpha (g_i)_{\min}.$$

Then $\mathrm{LT}(\tilde{f}_{\min}) < \mathrm{LT}(f_{\min})$ since the leading terms of $f_{\min}$ and $cx^\alpha (g_i)_{\min}$ are equal. This contradicts the minimality of $\mathrm{LT}(f_{\min})$, and (i) is proved. In the exercises, you will show that $g_1, \ldots, g_s$ are a basis of I, though not necessarily a Gröbner basis.

(ii) Let W denote the variety $\mathbf{V}(f_{\min} \mid f \in I)$. If we apply the argument of part (i) to the ideal I, we see immediately that

$$W = \mathbf{V}((g_1)_{\min}, \ldots, (g_s)_{\min}).$$

It remains to show that W is the tangent cone at the origin. Since $I \subseteq \mathbf{I}(V)$, the inclusion $C_0(V) \subseteq W$ is obvious by the definition of tangent cone. Going the other way, suppose that $g \in \mathbf{I}(V)$. We need to show that $g_{\min}$ vanishes on W. By the Nullstellensatz, we know that $g^m \in I$ for some m and, hence, $(g^m)_{\min} = 0$ on W. In the exercises, you will check that $(g^m)_{\min} = (g_{\min})^m$ and it follows that $g_{\min}$ vanishes on W. This completes the proof of the proposition.   □

In practice, this proposition is usually used over an algebraically closed field, for here, part (ii) says that we can compute the tangent cone using *any* set of defining equations of the variety.

For an example of how to use Proposition 4, suppose $V = \mathbf{V}(xy, xz + z(y^2 - z^2))$. If we set $I = \langle xy, xz + z(y^2 - z^2) \rangle$, the first step is to determine $I^h \subseteq k[w, x, y, z]$, where w is the homogenizing variable. Using grlex order on $k[x, y, z]$, a Gröbner basis for I is $\{xy, xz + z(y^2 - z^2), x^2 z - xz^3\}$. By the theory developed in §4 of Chapter 8, $\{xy, xzw + z(y^2 - z^2), x^2 zw - xz^3\}$ is a basis of I^h. In fact, it is a Gröbner basis for grlex order, with the variables ordered $x > y > z > w$ (see Exercise 5). However, this monomial order does not satisfy the hypothesis of Proposition 4, but if we use grlex with $w > x > y > z$, then a Gröbner basis is

$$\{xy, xzw + z(y^2 - z^2), yz(y^2 - z^2)\}.$$

Proposition 4 shows that if we dehomogenize and take minimal homogeneous components, then the tangent cone at the origin is given by

$$C_0(V) = \mathbf{V}(xy, xz, yz(y^2 - z^2)).$$

In the exercises, you will show that this tangent cone is the union of five lines through the origin in k^3.

We will next study how the tangent cone approximates the variety V near the point p. Recall from Proposition 3 that $C_p(V)$ is the translate of an affine cone, which means that $C_p(V)$ is made up of lines through p. So to understand the tangent cone, we need to describe which lines through p lie in $C_p(V)$.

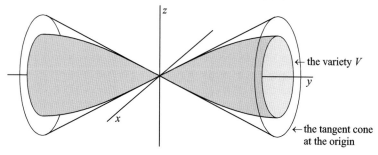

We will do this using secant lines. More precisely, let L be a line in k^n through p. Then L is a *secant line* of V if it meets V in a point distinct from p. Here is the crucial idea: if we take secant lines determined by points of V getting closer and closer to p, then the "limit" of the secant lines should lie on the tangent cone. You can see this in the preceding picture.

To make this idea precise, we will work over the complex numbers $\mathbb{C}$. Here, it is possible to define what it means for a sequence of points $q_i \in \mathbb{C}^n$ to converge to $q \in \mathbb{C}^n$. For instance, if we think of $\mathbb{C}^n$ as $\mathbb{R}^{2n}$, this means that the coordinates of q_i converge to the coordinates of q. We will assume that the reader has had some experience with sequences of this sort.

We will treat lines through their parametrizations. Suppose we have parametrized L via $p + tv$, where $v \in \mathbb{C}^n$ is a nonzero vector parallel to L and $t \in \mathbb{C}$. Then we define a limit of lines as follows.

Definition 5. We say that a line $L \subseteq \mathbb{C}^n$ through a point $p \in \mathbb{C}^n$ is a **limit of lines** $\{L_i\}_{i=1}^{\infty}$ through p if given a parametrization $p + tv$ of L, there exist parametrizations $p + tv_i$ of L_i such that $\lim_{i \to \infty} v_i = v$ in $\mathbb{C}^n$.

This notion of convergence corresponds to the following picture:

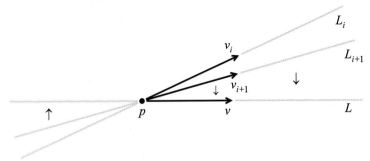

Now we can state a precise version of how the tangent cone approximates a complex variety near a point.

Theorem 6. *Let $V \subseteq \mathbb{C}^n$ be an affine variety. Then a line L through $p \in V$ lies in the tangent cone $C_p(V)$ if and only if there exists a sequence $\{q_i\}_{i=1}^{\infty}$ of points in $V \setminus \{p\}$ converging to p such that if L_i is the secant line containing p and q_i, then the lines L_i converge to the given line L.*

Proof. By translating p to the origin, we may assume that $p = 0$. Let $\{q_i\}_{i=1}^{\infty}$ be a sequence of points in $V \setminus \{0\}$ converging to the origin and suppose the lines L_i through 0 and q_i converge (in the sense of Definition 5) to some line L through the origin. We want to show that $L \subseteq C_0(V)$.

By the definition of L_i converging to L, we can find parametrizations tv_i of L_i (remember that $p = 0$) such that the v_i converge to v as $i \to \infty$. Since $q_i \in L_i$, we can write $q_i = t_i v_i$ for some complex number t_i. Note that $t_i \neq 0$ since $q_i \neq 0$. We claim that the t_i converge to 0. This follows because as $i \to \infty$, we have $v_i \to v \neq 0$ and $t_i v_i = q_i \to 0$. (A more detailed argument will be given in Exercise 8.)

Now suppose that f is any polynomial that vanishes on V. As in the proof of Proposition 4, we write $f_{\min}$ and f_j rather than $f_{0,\min}$ and $f_{0,j}$. If f has total degree d, then we can write $f = f_\ell + f_{\ell+1} + \cdots + f_d$, where $f_\ell = f_{\min}$. Since $q_i = t_i v_i \in V$, we have

$$(2) \qquad 0 = f(t_i v_i) = f_\ell(t_i v_i) + \cdots + f_d(t_i v_i).$$

Each f_j is homogeneous of degree j, so that $f_j(t_i v_i) = t_i^j f_j(v_i)$. Thus,

$$(3) \qquad 0 = t_i^\ell f_\ell(v_i) + \cdots + t_i^d f_d(v_i).$$

Since $t_i \neq 0$, we can divide through by t_i^ℓ to obtain

$$(4) \qquad 0 = f_\ell(v_i) + t_i f_{\ell+1}(v_i) + \cdots + t_i^{d-\ell} f_d(v_i).$$

Letting $i \to \infty$, the right-hand side in (4) tends to $f_\ell(v)$ since $v_i \to v$ and $t_i \to 0$. We conclude that $f_\ell(v) = 0$, and since $f_\ell(tv) = t^\ell f_\ell(v) = 0$ for all t, it follows that $L \subseteq C_0(V)$. This shows that $C_0(V)$ contains all limits of secant lines determined by sequences of points in V converging to 0.

To prove the converse, we will first study the set

$$(5) \qquad \mathcal{V} = \{(v,t) \in \mathbb{C}^n \times \mathbb{C} \mid tv \in V, t \neq 0\} \subseteq \mathbb{C}^{n+1}.$$

If $(v,t) \in \mathcal{V}$, note that the L determined by 0 and $tv \in V$ is a secant line. Thus, we want to know what happens to $\mathcal{V}$ as $t \to 0$. For this purpose, we will study the Zariski closure $\overline{\mathcal{V}}$ of $\mathcal{V}$, which is the smallest variety in $\mathbb{C}^{n+1}$ containing $\mathcal{V}$. We claim that

$$(6) \qquad \overline{\mathcal{V}} = \mathcal{V} \cup (C_0(V) \times \{0\}).$$

From §4 of Chapter 4, we know that $\overline{\mathcal{V}} = \mathbf{V}(\mathbf{I}(\mathcal{V}))$. So we need to calculate the functions that vanish on $\mathcal{V}$. If $f \in \mathbf{I}(V)$, write $f = f_\ell + \cdots + f_d$ where $f_\ell = f_{\min}$, and set

$$\tilde{f} = f_\ell + t f_{\ell+1} + \cdots + t^{d-\ell} f_d \in \mathbb{C}[t, x_1, \ldots, x_n].$$

We will show that

$$(7) \qquad \mathbf{I}(\mathcal{V}) = \langle \tilde{f} \mid f \in \mathbf{I}(V) \rangle \subseteq \mathbb{C}[t, x_1, \ldots, x_n].$$

One direction of the proof is easy, for $f \in \mathbf{I}(V)$ and $(v,t) \in \mathcal{V}$ imply $f(tv) = 0$, and then equations (2), (3), and (4) show that $\tilde{f}(v,t) = 0$. Conversely, suppose that $g \in \mathbb{C}[t, x_1, \ldots, x_n]$ vanishes on $\mathcal{V}$. Write $g = \sum_i g_i t^i$, where $g_i \in \mathbb{C}[x_1, \ldots, x_n]$, and let $g_i = \sum_j g_{ij}$ be the decomposition of g_i into the sum of its homogeneous components. If $(v,t) \in \mathcal{V}$, then for every $\lambda \in \mathbb{C} \setminus \{0\}$, we have $(\lambda v, \lambda^{-1} t) \in \mathcal{V}$ since $(\lambda^{-1} t) \cdot (\lambda v) = tv \in V$. Thus,

$$0 = g(\lambda v, \lambda^{-1}t) = \sum_{i,j} g_{ij}(\lambda v)(\lambda^{-1}t)^i = \sum_{i,j} \lambda^i g_{ij}(v)\lambda^{-i}t^i = \sum_{i,j} \lambda^{j-i} g_{ij}(v)t^i$$

for all $\lambda \neq 0$. Letting $m = j - i$, we can organize this sum according to powers of λ:

$$0 = \sum_m \left(\sum_i g_{i,m+i}(v)t^i \right) \lambda^m.$$

Since this holds for all $\lambda \neq 0$, it follows that $\sum_i g_{i,m+i}(v)t^i = 0$ for all m and, hence, $\sum_i g_{i,m+i}t^i \in \mathbf{I}(\mathcal{V})$. Let $f_m = \sum_i g_{i,m+i} \in \mathbb{C}[x_1,\ldots,x_n]$. Since $(v,1) \in \mathcal{V}$ for all $v \in V$, it follows that $f_m \in \mathbf{I}(V)$. Let i_0 be the smallest i such that $g_{i,m+i} \neq 0$. Then

$$\tilde{f}_m = g_{i_0,m+i_0} + g_{i_0+1,m+i_0+1}t + \cdots,$$

so that $\sum_i g_{i,m+i}t^i = t^{i_0}\tilde{f}_m$. From this, it follows easily that $g \in \langle \tilde{f} \mid f \in \mathbf{I}(V) \rangle$, and (7) is proved.

From (7), we have $\overline{\mathcal{V}} = \mathbf{V}(\tilde{f} \mid f \in \mathbf{I}(V))$. To compute this variety, let $(v,t) \in \mathbb{C}^{n+1}$, and first suppose that $t \neq 0$. Using (2), (3), and (4), it is straightforward to show that $\tilde{f}(v,t) = 0$ if and only if $f(tv) = 0$. Thus,

$$\overline{\mathcal{V}} \cap \{(v,t) \mid t \neq 0\} = \mathcal{V}.$$

Now suppose $t = 0$. If $f = f_{\min} + \cdots + f_d$, it follows from the definition of $\tilde{f}$ that $\tilde{f}(v,0) = 0$ if and only if $f_{\min}(v) = 0$. Hence,

$$\overline{\mathcal{V}} \cap \{(v,t) \mid t = 0\} = C_0(V) \times \{0\},$$

and (6) is proved.

To complete the proof of Theorem 6, we will need the following fact about Zariski closure.

Proposition 7. *Let $Z \subseteq W \subseteq \mathbb{C}^n$ be affine varieties and assume that W is the Zariski closure of $W \setminus Z$. If $z \in Z$ is any point, then there is a sequence of points $\{w_i\}_{i=1}^{\infty}$ in $W \setminus Z$ which converges to z.*

Proof. The proof of this is beyond the scope of the book. In Theorem (2.33) of MUMFORD (1981), this result is proved for irreducible varieties in $\mathbb{P}^n(\mathbb{C})$. Exercise 9 will show how to deduce Proposition 7 from Mumford's theorem. $\square$

To apply this proposition to our situation, let $Z = C_0(V) \times \{0\} \subseteq W = \overline{\mathcal{V}}$. By (6), we see that $W \setminus Z = \overline{\mathcal{V}} \setminus (C_0(V) \times \{0\}) = \mathcal{V}$ and, hence, $W = \overline{\mathcal{V}}$ is the Zariski closure of $W \setminus Z$. Then the proposition implies that any point in $Z = C_0(V) \times \{0\}$ is a limit of points in $W \setminus Z = \mathcal{V}$.

We can now finish the proof of Theorem 6. Suppose a line L parametrized by tv is contained in $C_0(V)$. Then $v \in C_0(V)$, which implies that $(v,0) \in C_0(V) \times \{0\}$. By the above paragraph, we can find points $(v_i,t_i) \in \mathcal{V}$ which converge to $(v,0)$. If we let L_i be the line parametrized by tv_i, then $v_i \to v$ shows that $L_i \to L$. Furthermore, since $q_i = t_i v_i \in V$ and $t_i \neq 0$, we see that L_i is the secant line determined by $q_i \in V$.

Finally, as $k \to \infty$, we have $q_i = t_i \cdot v_i \to 0 \cdot v = 0$, which shows that L is a limit of secant lines of points $q_i \in V$ converging to 0. This completes the proof of the theorem. □

If we are working over an infinite field k, we may not be able to define what it means for secant lines to converge to a line in the tangent cone. So it is not clear what the analogue of Theorem 6 should be. But if $p = 0$ is in V over k, we can still form the set $\mathcal{V}$ as in (5), and every secant line still gives a point $(v, t) \in \mathcal{V}$ with $t \neq 0$. A purely algebraic way to discuss limits of secant lines as $t \to 0$ would be to take the smallest variety containing $\mathcal{V}$ and see what happens when $t = 0$. This means looking at $\overline{\mathcal{V}} \cap (k^n \times \{0\})$, which by (6) is exactly $C_0(V) \times \{0\}$. You should check that the proof of (6) is valid over k, so that the decomposition

$$\overline{\mathcal{V}} = \mathcal{V} \cup (C_0(V) \times \{0\})$$

can be regarded as the extension of Theorem 6 to the infinite field k. In Exercise 10, we will explore some other interesting aspects of the variety $\overline{\mathcal{V}}$.

Another way in which the tangent cone approximates the variety is in terms of dimension. Recall from §6 that $\dim_p V$ is the maximum dimension of an irreducible component of V containing p.

Theorem 8. *Let p be a point on an affine variety $V \subseteq k^n$. Then $\dim_p V = \dim C_p(V)$.*

Proof. This is a standard result in advanced courses in commutative algebra [see, for example, Theorem 13.9 in MATSUMURA (1989)]. As in §6, we will only prove this for the case of a hypersurface in $\mathbb{C}^n$. If $V = \mathbf{V}(f)$, we know that $C_p(V) = \mathbf{V}(f_{p,\min})$ by Exercise 2. Thus, both V and $C_p(V)$ are hypersurfaces, and, hence, both have dimension $n - 1$ at all points. This shows that $\dim_p V = \dim C_p(V)$. □

This is a nice result because it enables us to compute $\dim_p V$ without having to decompose V into its irreducible components.

The final topic of this section will be the relation between the tangent cone and the tangent space. In the exercises, you will show that for any point p of a variety V, we have

$$C_p(V) \subseteq T_p(V).$$

In terms of dimensions, this implies that

$$\dim C_p(V) \leq \dim T_p(V).$$

Then the following corollary of Theorem 8 tells us when these coincide.

Corollary 9. *Assume that k is algebraically closed and let p be a point of a variety $V \subseteq k^n$. Then the following are equivalent:*

(i) *p is a nonsingular point of V.*
(ii) *$\dim C_p(V) = \dim T_p(V)$.*
(iii) *$C_p(V) = T_p(V)$.*

Proof. Since $\dim C_p(V) = \dim_p V$ by Theorem 8, the equivalence of (i) and (ii) is immediate from the definition of a nonsingular point. The implication (iii) $\Rightarrow$ (ii) is trivial, so that it remains to prove (ii) $\Rightarrow$ (iii).

Since k is algebraically closed, we know that k is infinite, which implies that the linear space $T_p(V)$ is an irreducible variety in k^n. [When $T_p(V)$ is a coordinate subspace, this follows from Exercise 7 of §1. See Exercise 12 below for the general case.] Thus, if $C_p(V)$ has the same dimension $T_p(V)$, the equality $C_p(V) = T_p(V)$ follows immediately from the affine version of Proposition 10 of §4 (see Exercise 18 of §4). $\qquad\square$

If we combine Theorem 6 and Corollary 9, it follows that at a nonsingular point p of a variety $V \subseteq \mathbb{C}^n$, the tangent space at p is the union of all limits of secant lines determined by sequences of points in V converging to p. This is a powerful generalization of the idea from elementary calculus that the tangent line to a curve is a limit of secant lines.

EXERCISES FOR §7

1. Suppose that k is a field of characteristic 0. Given $p \in k^n$ and $f \in k[x_1, \ldots, x_n]$, we know that f can be written in the form $f = \sum_\alpha c_\alpha(x - p)^\alpha$, where $c_\alpha \in k$ and $(x - p)^\alpha$ is as in the text. Given α, define

$$\frac{\partial^\alpha}{\partial x^\alpha} = \frac{\partial^{\alpha_1}}{\partial x_1^{\alpha_1}} \cdots \frac{\partial^{\alpha_n}}{\partial x_n^{\alpha_n}},$$

where $\frac{\partial^{\alpha_i}}{\partial^{\alpha_i} x_i}$ means differentiation α_i times with respect to x_i. Finally, set

$$\alpha! = \alpha_1! \cdot \alpha_2! \cdots \alpha_n!.$$

a. Show that

$$\frac{\partial^\alpha (x - p)^\beta}{\partial x^\alpha}(p) = \begin{cases} \alpha! & \text{if } \alpha = \beta \\ 0 & \text{otherwise.} \end{cases}$$

   Hint: There are two cases to consider: when $\beta_i < \alpha_i$ for some i, and when $\beta_i \geq \alpha_i$ for all i.

b. If $f = \sum_\alpha c_\alpha(x - p)^\alpha$, then show that

$$c_\alpha = \frac{1}{\alpha!} \frac{\partial^\alpha f}{\partial x^\alpha}(p),$$

   and conclude that

$$f = \sum_\alpha \frac{1}{\alpha!} \frac{\partial^\alpha f}{\partial x^\alpha}(p)(x - p)^\alpha.$$

   This is *Taylor's formula* for f at p. Hint: Be sure to explain where you use the characteristic 0 assumption.

c. Write out the formula of part (b) explicitly when $f \in k[x, y]$ has total degree 3.

d. What formula do we get for $f_{p,j}$ in terms of the partial derivatives of f?

e. Give an example to show that over a finite field, it may be impossible to express f in terms of its partial derivatives. Hint: See Exercise 10 of §6.

2. Let $V \subseteq k^n$ be a hypersurface.

a. If $\mathbf{I}(V) = \langle f \rangle$, prove that $C_p(V) = \mathbf{V}(f_{p,\min})$.

b. If k is algebraically closed and $V = \mathbf{V}(f)$, prove that the conclusion of part (a) is still true. Hint: See the proof of part (ii) of Proposition 4.

3. In this exercise, we will show that the ideal $I = \langle xy, xz + z(y^2 - z^2) \rangle \subseteq k[x, y, z]$ is a radical ideal when k has characteristic 0.

a. Show that
$$\langle x, z(y^2 - z^2) \rangle = \langle x, z \rangle \cap \langle x, y - z \rangle \cap \langle x, y + z \rangle.$$
Furthermore, show that the three ideals on the right-hand side of the equation are prime. Hint: Work in $k[x, y, z]/\langle x \rangle \cong k[y, z]$ and use the fact that $k[y, z]$ has unique factorization. Also explain why this result fails if k is the field $\mathbb{F}_2$ consisting of two elements.

b. Show that
$$\langle y, xz - z^3 \rangle = \langle y, z \rangle \cap \langle y, x - z^2 \rangle,$$
and show that the two ideals on the right-hand side of the equation are prime.

c. Prove that $I = \langle x, z(y^2 - z^2) \rangle \cap \langle y, xz - z^3 \rangle$. Hint: One way is to use the ideal intersection algorithm from Chapter 4, §3. There is also an elementary argument.

d. By parts (a), (b) and (c) we see that I is an intersection of five prime ideals. Show that I is a radical ideal. Also, use this decomposition of I to describe $V = \mathbf{V}(I) \subseteq k^3$.

e. If k is algebraically closed, what is $\mathbf{I}(V)$?

4. This exercise is concerned with the proof of Proposition 4. Fix a monomial order $>$ on $k[x_0, \dots, x_n]$ with the properties described in the statement of the proposition.

a. If $g \in k[x_1, \dots, x_n]$ is the dehomogenization of a homogeneous polynomial $G \in k[x_0, \dots, x_n]$, prove that $\mathrm{LT}(G) = \mathrm{LT}(g_{\min})x_0^b$ for some b.

b. If $G_1, \dots, G_s$ is a basis of I^h, prove that the dehomogenizations $g_1, \dots, g_s$ form a basis of I. In Exercise 5, you will show that if the G_i's are a Gröbner basis for $>$, the g_i's may fail to be a Gröbner basis for I with respect to the induced monomial order on $k[x_1, \dots, x_n]$.

c. If $f, g \in k[x_1, \dots, x_n]$, show that $(f \cdot g)_{\min} = f_{\min} \cdot g_{\min}$ and $(f^m)_{\min} = (f_{\min})^m$.

5. We will continue our study of the variety $V = \mathbf{V}(xy, xz + z(y^2 - z^2))$ begun in the text.

a. If we use grlex with $w > x > y > z$, show that a Gröbner basis for $I^h \subseteq k[w, x, y, z]$ is $\{xy, xzw + z(y^2 - z^2), yz(y^2 - z^2)\}$.

b. If we dehomogenize the Gröbner basis of part (a), we get a basis of I. Show that this basis is *not* a Gröbner basis of I for grlex with $x > y > z$.

c. Use Proposition 4 to show that the tangent cone $C_0(V)$ is a union of five lines through the origin in k^3 and compare your answer to part (e) of Exercise 3.

6. Compute the dimensions of the tangent cone and the tangent space at the origin of the varieties defined by the following ideals:

a. $\langle xz, xy \rangle \subseteq k[x, y, z]$.

b. $\langle x - y^2, x - z^3 \rangle \subseteq k[x, y, z]$.

7. In §3 of Chapter 3, we used elimination theory to show that the tangent surface of the twisted cubic $\mathbf{V}(y - x^2, z - x^3) \subseteq \mathbb{R}^3$ is defined by the equation
$$x^3 z - (3/4)x^2 y^2 - (3/2)xyz + y^3 + (1/4)z^2 = 0.$$

a. Show that the singular locus of the tangent surface S is exactly the twisted cubic. Hint: Two different ideals may define the same variety. For an example of how to deal with this, see equation (14) in Chapter 3, §4.

b. Compute the tangent space and tangent cone of the surface S at the origin.

8. Suppose that in $\mathbb{C}^n$ we have two sequences of vectors v_i and $t_i v_i$, where $t_i \in \mathbb{C}$, such that $v_i \to v \neq 0$ and $t_i v_i \to 0$. We claim that $t_i \to 0$ in $\mathbb{C}$. To prove this, define the length of a complex number $z = x + y\sqrt{-1}$ to be $|z| = \sqrt{x^2 + y^2}$ and define the length of $v = (z_1, \dots, z_n) \in \mathbb{C}^n$ to be $|v| = \sqrt{|z_1|^2 + \dots + |z_n|^2}$. Recall that $v_i \to v$ means that for every $\epsilon > 0$, there is N such that $|v_i - v| < \epsilon$ for $k \geq N$.

    a. If we write $v = (z_1, \ldots, z_n)$ and $v_i = (z_{i1}, \ldots, z_{in})$, then show that $v_i \to v$ implies $z_{ij} \to z_j$ for all j. Hint: Observe that $|z_j| \le |v|$.

    b. Pick a nonzero component z_j of v. Show that $z_{ij} \to z_j \ne 0$ and $t_i z_{ij} \to 0$. Then divide by z_j and conclude that $t_i \to 0$.

9. Theorem (2.33) of MUMFORD (1981) states that if $W \subset \mathbb{P}^n(\mathbb{C})$ is an irreducible projective variety and $Z \subseteq W$ is a projective variety not equal to W, then any point in Z is a limit of points in $W \setminus Z$. Our goal is to apply this to prove Proposition 7.

    a. Let $Z \subseteq W \subseteq \mathbb{C}^n$ be affine varieties such that W is the Zariski closure of $W \setminus Z$. Show that Z contains no irreducible component of W.

    b. Show that it suffices to prove Proposition 7 in the case when W is irreducible. Hint: If p lies in Z, then it lies in some component W_1 of W. What does part (a) tell you about $W_1 \cap Z \subseteq W_1$?

    c. Let $Z \subseteq W \subseteq \mathbb{C}^n$, where W is irreducible and $Z \ne W$, and let $\overline{Z}$ and $\overline{W}$ be their projective closures in $\mathbb{P}^n(\mathbb{C})$. Show that the irreducible case of Proposition 7 follows from Mumford's Theorem (2.33). Hint: Use $\overline{Z} \cup (\overline{W} \setminus W) \subseteq \overline{W}$.

    d. Show that the converse of the proposition is true in the following sense. Let $p \in \mathbb{C}^n$. If $p \notin V \setminus W$ and p is a limit of points in $V \setminus W$, then show that $p \in W$. Hint: Show that $p \in V$ and recall that polynomials are continuous.

10. Let $V \subseteq k^n$ be a variety containing the origin and let $\mathcal{V} \subseteq k^{n+1}$ be the set described in (5). Given $\lambda \in k$, consider the "slice" $(k^n \times \{\lambda\}) \cap \mathcal{V}$. Assume that k is infinite.

    a. When $\lambda \ne 0$, show that this slice equals $V_\lambda \times \{\lambda\}$, where $V_\lambda = \{v \in k^n \mid \lambda v \in V\}$. Also show that V_λ is an affine variety.

    b. Show that $V_1 = V$, and, more generally, for $\lambda \ne 0$, show that V_λ is isomorphic to V. Hint: Consider the polynomial map defined by sending $(x_1, \ldots, x_n)$ to $(\lambda x_1, \ldots, \lambda x_n)$.

    c. Suppose that $k = \mathbb{R}$ or $\mathbb{C}$ and that $\lambda \ne 0$ is close to the origin. Explain why V_λ gives a picture of V where we have expanded the scale by a factor of $1/\lambda$. Conclude that as $\lambda \to 0$, V_λ shows what V looks like as we "zoom in" at the origin.

    d. Use (6) to show that $V_0 = C_0(V)$. Explain what this means in terms of the "zooming in" described in part (c).

11. If $p \in V \subseteq k^n$, show that $C_p(V) \subseteq T_p(V)$.

12. If k is an infinite field and $V \subseteq k^n$ is a subspace (in the sense of linear algebra), then prove that V is irreducible. Hint: In Exercise 7 of §1, you showed that this was true when V was a coordinate subspace. Now pick an appropriate basis of k^n.

13. Let $W \subset \mathbb{P}^{n-1}(\mathbb{C})$ be a nonempty projective variety and let $C_W \subseteq \mathbb{C}^n$ be its affine cone.

    a. Prove that the tangent cone of C_W at the origin is C_W.

    b. Prove that the origin is a smooth point of C_W if and only if W is a projective linear subspace of $\mathbb{P}^{n-1}(\mathbb{C})$. Hint: Use Corollary 9.

In Exercises 14–17, we will study the "blow-up" of a variety V at a point $p \in V$. The blowing-up process gives us a map of varieties $\pi : \mathrm{Bl}_p V \to V$ such that away from p, the two varieties look the same, but at p, $\mathrm{Bl}_p V$ can be much larger than V, depending on what the tangent cone $C_p(V)$ looks like.

14. Let k be an arbitrary field. In §5 of Chapter 8, we studied varieties in $\mathbb{P}^{n-1} \times k^n$, where $\mathbb{P}^{n-1} = \mathbb{P}^{n-1}(k)$. Let $y_1, \ldots, y_n$ be homogeneous coordinates in $\mathbb{P}^{n-1}$ and let $x_1, \ldots, x_n$ be coordinates in k^n. Then the $(y_1, \ldots, y_n)$-homogeneous polynomials $x_i y_j - x_j y_i$ (this is the terminology of Chapter 8, §5) define a variety $\Gamma \subseteq \mathbb{P}^{n-1} \times k^n$. This variety has some interesting properties.

    a. Fix $(p, q) \in \mathbb{P}^{n-1} \times k^n$. Picking homogeneous coordinates of p gives a nonzero vector $\hat{p} \in k^n \setminus \{0\}$. Show that $(p, q) \in \Gamma$ if and only if $q = t\hat{p}$ for some $t \in k$ (which might be zero).

    b. If $q \ne 0$ is in k^n, show that $(\mathbb{P}^{n-1} \times \{q\}) \cap \Gamma$ consists of a single point (p, q) where q gives homogeneous coordinates of $p \in \mathbb{P}^{n-1}$. On the other hand, when $q = 0$, show that $(\mathbb{P}^{n-1} \times \{0\}) \cap \Gamma = \mathbb{P}^{n-1} \times \{0\}$.

c. Let $\pi : \Gamma \to k^n$ be the projection map. Show that $\pi^{-1}(q)$ consists of a single point, except when $q = 0$, in which case $\pi^{-1}(0)$ is a copy of $\mathbb{P}^{n-1}$. This shows that we can regard Γ as the variety obtained by removing the origin from k^n and replacing it by a copy of $\mathbb{P}^{n-1}$.

d. To see what the $\mathbb{P}^{n-1} \times \{0\} \subseteq \Gamma$ means, consider a line L through the origin parametrized by tv, where $v \in k^n \setminus \{0\}$. Although there are many choices for v, they all give the same point $w \in \mathbb{P}^{n-1}$. Show that the points $(w, tv) \in \mathbb{P}^{n-1} \times k^n$ lie in Γ and, hence, describe a curve $L \subseteq \Gamma$. Investigate where this curve meets $\mathbb{P}^{n-1} \times \{0\}$ and conclude that distinct lines through the origin in k^n give distinct points in $\pi^{-1}(0)$. Thus, the difference between Γ and k^n is that Γ separates tangent directions at the origin. We call $\pi : \Gamma \to k^n$ the *blow-up* of k^n at the origin.

15. This exercise is a continuation of Exercise 14. Let $V \subseteq k^n$ be a variety containing the origin and assume that the origin is not an irreducible component of V. Our goal here is to define the blow-up of V at the origin. Let $\Gamma \subseteq \mathbb{P}^{n-1} \times k^n$ be as in the previous exercise. Then *blow-up* of V at 0, denoted $\mathrm{Bl}_0 V$, is defined to be the Zariski closure of $(\mathbb{P}^{n-1} \times (V \setminus \{0\})) \cap \Gamma$ in $\mathbb{P}^{n-1} \times k^n$.

    a. Prove that $\mathrm{Bl}_0 V \subseteq \Gamma$ and $\mathrm{Bl}_0 k^n = \Gamma$.
    b. If $\pi : \Gamma \to k^n$ is as in Exercise 14, then prove that $\pi(\mathrm{Bl}_0 V) \subseteq V$, so $\pi : \mathrm{Bl}_0 V \to V$. Hint: First show that $\mathrm{BL}_0 V \subseteq \mathbb{P}^{n-1} \times V$.
    c. Use Exercise 14 to show that $\pi^{-1}(q)$ consists of a single point for $q \neq 0$ in V.

    In Exercise 16, you will describe $\pi^{-1}(0)$ in terms of the tangent cone of V at the origin.

16. Let $V \subseteq k^n$ be a variety containing the origin and assume that the origin is not an irreducible component of V. We know that tangent cone $C_0(V)$ is the affine cone C_W over some projective variety $W \subset \mathbb{P}^{n-1}$. We call W the *projectivized tangent cone* of V at 0. The goal of this exercise is to show that if $\pi : \mathrm{Bl}_0 V \to V$ is the blow-up of V at 0 as defined in Exercise 15, then $\pi^{-1}(0) = W \times \{0\}$. Assume that k is infinite.

    a. Show that our assumption that $\{0\}$ is not an irreducible component of V implies that V is the Zariski closure of $V \setminus \{0\}$.
    b. Show that $g \in k[y_1, \ldots, y_n, x_1, \ldots, x_n]$ lies in $\mathbf{I}(\mathrm{Bl}_0 V)$ if and only if $g(q, tq) = 0$ for all $q \in V \setminus \{0\}$ and all $t \in k \setminus \{0\}$. Hint: Use part (a) of Exercise 14.
    c. Then show that $g \in \mathbf{I}(\mathrm{Bl}_0 V)$ if and only if $g(q, tq) = 0$ for all $q \in V$ and all $t \in k$. Hint: Use parts (a) and (b).
    d. Explain why $\mathbf{I}(\mathrm{Bl}_0 V)$ is generated by $(y_1, \ldots, y_n)$-homogeneous polynomials.
    e. Assume that $g = \sum_\alpha g_\alpha(y_1, \ldots, y_n) x^\alpha \in \mathbf{I}(\mathrm{Bl}_0 V)$. By part (d), we may assume that the g_α are all homogeneous of the same total degree d. Let

    $$f(x_1, \ldots, x_n) = \sum_\alpha g_\alpha(x_1, \ldots, x_n) x^\alpha.$$

    Prove that $f \in \mathbf{I}(V)$. Hint: Show that $g(x_1, \ldots, x_n, tx_1, \ldots, tx_n) = f(x_1, \ldots, x_n) t^d$, and then use part (c).
    f. Prove that $W \times \{0\} \subseteq \mathrm{Bl}_0 V \cap (\mathbb{P}^{n-1} \times \{0\})$. Hint: It suffices to show that $g(v, 0) = 0$ for $g \in \mathbf{I}(\mathrm{Bl}_0 V)$ and $v \in C_0(V)$. In the notation of part (e) note that $g(v, 0) = g_0(v)$. If $g_0 \neq 0$, show that $g_0 = f_{\min}$, where f is the polynomial defined in part (e).
    g. Prove that $\mathrm{Bl}_0 V \cap (\mathbb{P}^{n-1} \times \{0\}) \subseteq W \times \{0\}$. Hint: Recall that W is the projective variety defined by the polynomials $f_{\min}$ for $f \in \mathbf{I}(V)$. Write $f_\ell = f_{\min}$ and let g be the remainder of $t^\ell f$ on division by $tx_1 - y_1, \ldots, tx_n - y_n$ for a monomial order with $\mathrm{LT}(tx_i - y_i) = tx_i$ for all i. Show that t does not appear in g and that $g_0 = f_\ell$ when we write $g = \sum_\alpha g_\alpha(y_1, \ldots, y_n) x^\alpha$. Then use part (c) to prove that $g \in \mathbf{I}(\mathrm{Bl}_0 V)$. Now complete the proof using $f_\ell(v) = g_0(v) = g(v, 0)$.

A line in the tangent cone can be regarded as a way of approaching the origin through points of V. So we can think of the projectivized tangent cone W as describing all possible ways of approaching the origin within V. Then $\pi^{-1}(0) = W \times \{0\}$ means that each of these different ways gives a distinct point in the blow-up. Note how this generalizes Exercise 14.

17. Assume that k is an algebraically closed field and suppose that $V = \mathbf{V}(f_1, \ldots, f_s) \subseteq k^n$ contains the origin.

   a. By analyzing what you did in part (g) of Exercise 16, explain how to find defining equations for the blow-up $\mathrm{Bl}_0 V$.

   b. Compute the blow-up at the origin of $\mathbf{V}(y^2 - x^2 - x^3)$ and describe how your answer relates to the first picture in Example 1.

   c. Compute the blow-up at the origin of $\mathbf{V}(y^2 - x^3)$.

   Note that in parts (b) and (c), the blow-up is a smooth curve. In general, blowing-up is an important tool in what is called *desingularizing* a variety with singular points.

Chapter 10
Additional Gröbner Basis Algorithms

In §10 of Chapter 2 we discussed some criteria designed to identify situations where it is possible to see in advance that an S-polynomial remainder will be zero in Buchberger's algorithm. Those unnecessary S-polynomial remainder calculations are in fact the main computational bottleneck for the basic form of the algorithm. Finding ways to avoid them, or alternatively to replace them with less expensive computations, is the key to improving the efficiency of Gröbner basis calculation. The algorithms we discuss in this chapter apply several different approaches to achieve greater efficiency. Some of them use Gröbner bases of homogeneous ideals or ideas inspired by the special properties of Gröbner bases in that case. So we begin in §1 by showing that the computation of a homogeneous Gröbner basis can be organized to proceed degree by degree. This gives the framework for Traverso's Hilbert driven Buchberger algorithm, discussed in §2, which uses the Hilbert function of a homogeneous ideal to control the computation and bypass many unnecessary S-polynomial remainder calculations. We also show in §1 that the information generated by several S-polynomial remainder computations can be obtained simultaneously via row operations on a suitable matrix. This connection with linear algebra is the basis for Faugère's F_4 algorithm presented in §3. Finally, we introduce the main ideas behind signature-based Gröbner basis algorithms, including Faugère's F_5 algorithm, in §4.

§1 Preliminaries

From now on in this chapter, when we refer to Buchberger's algorithm, we will mean a version of the algorithm similar to the one from Theorem 9 of Chapter 2, §10. Readers may wish to review that section before reading farther. In particular, the algorithm maintains a list B of pairs of indices (i,j) in the current partial Gröbner basis $G = (f_1, \ldots, f_t)$ for which the S-polynomials $S(f_i, f_j)$ remain to be considered. Moreover, the algorithm only inserts additional polynomials into the initial list G and updates the set of pairs accordingly. Finding the corresponding reduced

© Springer International Publishing Switzerland 2015
D.A. Cox et al., *Ideals, Varieties, and Algorithms*, Undergraduate Texts in Mathematics, DOI 10.1007/978-3-319-16721-3_10

Gröbner basis will be treated as a separate operation. Several criteria for bypassing unnecessary S-polynomial remainders that are more powerful than those considered in Chapter 2 will be introduced in later sections.

Homogeneous Gröbner Bases

There are a number of special features of the computation of Gröbner bases of homogeneous ideals that present opportunities for simplifications and shortcuts. Indeed, to take advantage of these shortcuts some of the first-generation computer algebra systems for these computations, including the original Macaulay program developed by Bayer and Stillman, accepted only homogeneous inputs.

We studied homogeneous polynomials and ideals in Chapter 8 where Theorem 2 of §3 established the equivalence of these statements:

- $I \subseteq k[x_1, \ldots, x_n]$ is a homogeneous ideal.
- $I = \langle f_1, \ldots, f_s \rangle$, where all the f_i are homogeneous polynomials.
- The unique reduced Gröbner basis of I with respect to any monomial ordering consists of homogeneous polynomials.

In this chapter, the *degree* of a nonzero polynomial f, denoted $\deg(f)$, will always refer to its total degree. We write $k[x_1, \ldots, x_n]_m$ (respectively $k[x_1, \ldots, x_n]_{\leq m}$) for the vector space of polynomials of degree m (respectively $\leq m$) in $k[x_1, \ldots, x_n]$. By definition the zero polynomial is included in both vector spaces as the additive identity element.

From Exercise 3 of Chapter 8, §3, we know that in applying Buchberger's algorithm to compute a Gröbner basis from homogeneous input polynomials,

- all the nonzero S-polynomials generated by the algorithm are homogeneous, and
- all nonzero remainders on division of S-polynomials by the current partial Gröbner basis are homogeneous.

We will next study how homogeneity can be used to organize the computations involved in Buchberger's algorithm. First, there is a natural notion of the degree of a pair (i, j) once we fix a monomial order on $k[x_1, \ldots, x_n]$.

Definition 1. Let $G = (f_1, \ldots, f_t)$ be an ordered list of homogeneous polynomials. The **degree** of the pair (i, j) relative to the list G is $\deg(\text{lcm}(\text{LM}(f_i), \text{LM}(f_j)))$.

For instance, if we use grevlex order with $x > y > z$ on $\mathbb{Q}[x, y, z]$, and

$$f_1 = x^2 - 2y^2 - 2yz - z^2,$$
$$f_2 = -xy + 2yz + z^2,$$
$$f_3 = -x^2 + xy + xz + z^2,$$

then the pair $(1, 2)$ has degree 3 since $\text{lcm}(\text{LM}(f_1), \text{LM}(f_2)) = x^2 y$. However, the pair $(1, 3)$ has degree 2 since $\text{lcm}(\text{LM}(f_1), \text{LM}(f_3)) = x^2$.

Now if the f_i are homogeneous, it is easy to see that the degree of (i, j) coincides with the degree of $S(f_i, f_j)$ and with the degree of the remainder $\overline{S(f_i, f_j)}^G$ if the

S-polynomial and the remainder are nonzero (see Exercise 1). Next, in the course of Buchberger's algorithm, each time a nonzero S-polynomial remainder is found, recall that new pairs are created.

Proposition 2. *In Buchberger's algorithm with homogeneous input polynomials, when a new polynomial f_t is included in the partial Gröbner basis G, all the pairs (i, t) with $i < t$ have degree strictly greater than $\deg(f_t)$.*

Proof. Say $\deg(f_t)$ is equal to m. Since f_t is homogeneous, this is also the degree of $\mathrm{LM}(f_t)$. Consider a pair (i, t) with $i < t$. Then the degree of (i, t) is $\deg(\mathrm{lcm}(\mathrm{LM}(f_i), \mathrm{LM}(f_t)))$ which is clearly greater than or equal to m. To establish the claim, suppose the pair (i, t) has degree equal to m. Since $\mathrm{LM}(f_t)$ has degree m, this implies $\mathrm{lcm}(\mathrm{LM}(f_i), \mathrm{LM}(f_t)) = \mathrm{LM}(f_t)$ and hence $\mathrm{LM}(f_i)$ divides $\mathrm{LM}(f_t)$. But this is a contradiction since f_t is a remainder on division by a list of polynomials containing f_i. Hence the degree of (i, t) must be strictly greater than m. $\square$

Our next proposition shows that if pairs of lower degree are always processed before pairs of higher degree (this would hold, for instance, if a graded monomial ordering and the *normal selection strategy* discussed in Chapter 2, §10 is used), then the degrees of elements of a homogeneous Gröbner basis are very orderly and predictable. In particular, when all pairs of degree $\leq m$ have been processed, all elements of a Gröbner basis in degrees $\leq m$ have been found—lower-degree elements never appear later, as they easily can with nonhomogeneous inputs.

Proposition 3. *Assume that the input $F = (f_1, \ldots, f_s)$ in Buchberger's algorithm consists of homogeneous polynomials and the algorithm is terminated at a time when B contains no pairs of degree $\leq m$ for some $m \in \mathbb{Z}_{\geq 0}$. Let G_m be the set of elements in G of degree $\leq m$ at that time, i.e., $G_m = G \cap k[x_1, \ldots, x_n]_{\leq m}$. Then:*

(i) *There is a Gröbner basis $\widetilde{G}$ for $I = \langle f_1, \ldots, f_s \rangle$ such that the set of elements of degree $\leq m$ in $\widetilde{G}$ coincides with G_m, i.e., $\widetilde{G} \cap k[x_1, \ldots, x_n]_{\leq m} = G_m$.*

(ii) *If $f \in I$ is homogeneous of degree $\leq m$, then $\mathrm{LT}(f)$ is divisible by $\mathrm{LT}(g)$ for some $g \in G_m$.*

Proof. To prove part (i), let $\widetilde{G}$ be the Gröbner basis of I obtained by letting the algorithm run to completion. Then $G_m \subseteq \widetilde{G}$ since the algorithm only inserts new polynomials into G. Moreover, if $f \in \widetilde{G} \setminus G_m$, then f is either an input polynomial of degree $> m$ or a remainder on division of an S-polynomial for a pair of degree $> m$. So by the remark following Definition 1 above in the second case, $\deg(f) > m$. It follows that $\widetilde{G} \cap k[x_1, \ldots, x_n]_{\leq m} = G_m$ as claimed. Part (ii) follows since $\mathrm{LT}(f)$ is divisible by $\mathrm{LT}(g)$ for some $g \in \widetilde{G}$ by the definition of a Gröbner basis. But then $g \in G_m$ must hold since $\deg(f) \leq m$. $\square$

A set of polynomials in I that satisfies the property given for G_m in part (ii) of Proposition 3 is called a *Gröbner basis of I up to degree m*. You will show in Exercise 9 that an equivalent statement is:

(1) $$\overline{S(f_i, f_j)}^G = 0 \text{ for all pairs } (i, j) \text{ of degree } \leq m.$$

To summarize, here is a version of Buchberger's algorithm tailored for the homogeneous case.

Theorem 4. *Let $f_1, \ldots, f_s$ be homogeneous and let $I = \langle f_1, \ldots, f_s \rangle$.*

(i) *The algorithm below terminates and correctly computes a Gröbner basis for I.*

(ii) *The values of m in successive passes through the outer WHILE loop are strictly increasing.*

(iii) *When the pass through the outer WHILE loop with a given m is complete, $G_m = G \cap k[x_1, \ldots, x_n]_{\leq m}$ is a Gröbner basis for I up to degree m.*

Input : $F = (f_1, \ldots, f_s)$ with all f_i homogeneous

Output : G, a Gröbner basis for I

$B := \{(i,j) \mid 1 \leq i < j \leq s\}$

$G := F$

$l := s$

WHILE $B \neq \emptyset$ DO

    $m :=$ minimal degree of pairs remaining in B

    $B' := \{(i,j) \in B \mid \deg(i,j) = m\}$

    $B := B \setminus B'$

    WHILE $B' \neq \emptyset$ DO

        $(i,j) :=$ first element in B'

        $B' := B' \setminus \{(i,j)\}$

        $S := \overline{S(f_i,f_j)}^G$

        IF $S \neq 0$ THEN

            $l := l + 1; f_l := S$

            $G := G \cup \{f_l\}$

            $B := B \cup \{(i,l) \mid 1 \leq i \leq l - 1\}$

RETURN G

Proof. Part (i) follows from the termination and correctness proofs for the usual Buchberger algorithm. You will complete the details in Exercise 10. When the pass through the outer WHILE loop with a given m is complete, all pairs of degree m in B at the start of that pass have been removed and processed. By Proposition 2 no new pairs of degree $\leq m$ have been added. So B contains only pairs of degrees strictly larger than m. This proves the claim in part (ii). Part (iii) then follows from Proposition 3.   □

In what follows, we will refer to this as the *degree by degree* version of Buchberger's algorithm. We are not claiming that this, by itself, is any improvement over the algorithms studied in Chapter 2. Indeed, this version merely singles out one particular way of choosing the next pairs to be processed at each stage and exploits the good consequences when the input polynomials are homogeneous. However, it does give a framework for developing improved algorithms. Note that in some cases all elements of the set G_m in part (iii) of the theorem might be found before all the pairs of degree m in B' have been processed. If that happened, no further S-polynomial remainders for pairs in degree m would be necessary and the inner loop could be terminated early. In the process, some S-polynomial remainder calculations could be bypassed. This is an opportunity for exactly the sort of improvement in efficiency we mentioned in the introduction to this chapter. The Hilbert driven algorithm to be presented in §2 uses the Hilbert function of the ideal I to recognize when this happens.

Homogeneous Gröbner bases and algorithms for computing them are treated in greater generality and in more detail in BECKER and WEISPFENNING (1993) and KREUZER and ROBBIANO (2005).

Homogenization and Dehomogenization

Since many applications involve nonhomogeneous ideals, if we want to use facts about homogeneous Gröbner basis algorithms, we need a way to relate the two cases. We will use the following notation from Proposition 7 of Chapter 8, §2. If $f(x_1, \ldots, x_n)$ is a polynomial of degree m and x_0 is a new *homogenizing variable*, then the homogenization of f is

$$f^h(x_0, x_1, \ldots, x_n) = x_0^m f\left(\frac{x_1}{x_0}, \ldots, \frac{x_n}{x_0}\right).$$

Similarly, if $F = (f_1, \ldots, f_s)$ is an ordered list of polynomials, then we will write $F^h = (f_1^h, \ldots, f_s^h)$. Going the other way, if $g = g(x_0, x_1, \ldots, x_n)$ is a homogeneous polynomial, then its dehomogenization is

$$g^d(x_1, \ldots, x_n) = g(1, x_1, \ldots, x_n).$$

Similarly, if $G = (g_1, \ldots, g_t)$ is an ordered list of homogeneous polynomials, then $G^d = (g_1^d, \ldots, g_t^d)$. In this chapter, the superscripts h and d always refer to homogenization and dehomogenization; they are never exponents. By Proposition 7 of Chapter 8, §2, $(f^h)^d = f$ for all f in $k[x_1, \ldots, x_n]$. It is not necessarily the case that $(g^d)^h = g$, though, since $(g^d)^h$ could differ from g by a power of the homogenizing variable x_0.

We next discuss the behavior of Gröbner bases under homogenization. Part of the following has appeared in Chapter 8.

- Let I be a nonhomogeneous ideal in $k[x_1, \ldots, x_n]$ and let $>$ be a *graded* monomial order. In this case, the relation between a Gröbner basis G for I and a particular Gröbner basis for the homogeneous ideal $I^h = \langle f^h \mid f \in I \rangle$ is pleasantly simple. From Theorem 4 of Chapter 8, §4, G^h is a Gröbner basis for I^h with respect to a product order $>_h$. If x^α and x^β are monomials in $x_1, \ldots, x_n$, then

(2) $x^\alpha x_0^a >_h x^\beta x_0^b \Leftrightarrow x^\alpha > x^\beta$, or $x^\alpha = x^\beta$ and $a > b$.

- Let $I = \langle f_1, \ldots, f_s \rangle$, where the f_i are an arbitrary ideal basis and consider $J = \langle f_1^h, \ldots, f_s^h \rangle$. This ideal can be strictly smaller than I^h, as in Example 3 from Chapter 8, §4. Moreover, the corresponding projective variety, $\mathbf{V}(J) \subseteq \mathbb{P}^{n-1}$, can be strictly larger than $\mathbf{V}(I^h)$. In particular, $\mathbf{V}(J)$ can have additional irreducible components in the hyperplane at infinity $\mathbf{V}(x_0)$ not contained in $\mathbf{V}(I^h)$.

Even though the ideal $J = \langle f_1^h, \ldots, f_s^h \rangle$ can differ from I^h, it is a homogeneous ideal. So we could apply any improved Gröbner basis algorithm tailored for the homogeneous case to these polynomials. Moreover, the order $>$ used to define the product order $>_h$ can be chosen arbitrarily. But there is a remaining question: How does a Gröbner basis for J relate to Gröbner bases for I? The following statement is a new observation.

Theorem 5. *Let G be the reduced Gröbner basis for $J = \langle f_1^h, \ldots, f_s^h \rangle$ with respect to the monomial order $>_h$ from (2). Then G^d is a Gröbner basis for $I = \langle f_1, \ldots, f_s \rangle$ with respect to the order $>$.*

Proof. We write $G = \{g_1, \ldots, g_t\}$. The proof is a variation on the proof of Theorem 4 of Chapter 8, §4. We claim first that each of the dehomogenized polynomials g_j^d is in I, so the ideal they generate is contained in I. In "fancy" terms, the underlying reason is that setting $x_0 = 1$ defines a ring homomorphism from $k[x_0, x_1, \ldots, x_n]$ to $k[x_1, \ldots, x_n]$, as you will see in Exercise 2. The g_j are in the ideal generated by the f_i^h, so for each j we have an equation

$$g_j(x_0, x_1, \ldots, x_n) = \sum_{i=1}^{s} B_i(x_0, \ldots, x_n) f_i^h(x_0, x_1, \ldots x_n)$$

for some polynomials B_i. Setting $x_0 = 1$ and using the homomorphism property to get the second line below, we have

$$\begin{aligned}
g_j^d(x_1, \ldots, x_n) &= g_j(1, x_1, \ldots, x_n) \\
&= \sum_{i=1}^{s} B_i(1, x_1, \ldots, x_n) f_i^h(1, x_1, \ldots, x_n) \\
&= \sum_{i=1}^{s} B_i(1, x_1, \ldots, x_n) f_i(x_1, \ldots, x_n),
\end{aligned}$$

since $(f^h)^d = f$ for all $f \in k[x_1, \ldots, x_n]$. This establishes the claim.

Next, we show that the opposite inclusion also holds. Since G is a Gröbner basis for the ideal J, for each i, $1 \leq i \leq s$, we have

$$f_i^h(x_0, x_1, \ldots, x_n) = \sum_{j=1}^{t} A_j(x_0, x_1, \ldots, x_n) g_j(x_0, x_1, \ldots x_n)$$

for some polynomials A_j. So setting $x_0 = 1$ again we have

$$f_i(x_1, \ldots, x_n) = (f_i^h(x_0, x_1 \ldots, x_n))^d$$

$$= \sum_{j=1}^t A_j(1, x_1, \ldots, x_n) g_j(1, x_1, \ldots, x_n)$$

$$= \sum_{j=1}^t A_j(1, x_1, \ldots, x_n) g_j^d(x_1, \ldots, x_n).$$

Together with the result from the first paragraph, this shows G^d is a basis for I.

It remains to show that G^d is a Gröbner basis. For this we need first to understand where the leading monomials of the g_j^d come from. The g_j are homogeneous since they are elements of the reduced Gröbner basis for the homogeneous ideal J with respect to $>_h$. So let $\mathrm{LT}_{>_h}(g_j) = x^\alpha x_0^a$ and consider any other monomial $x^\beta x_0^b$ in g_j. By definition, $x^\alpha x_0^a >_h x^\beta x_0^b$. However the case $\alpha = \beta$ and $a > b$ from (2) never occurs here since g_j is homogeneous: $|\alpha| + a = |\beta| + b$, so if $\alpha = \beta$, then $a = b$ as well. In other words, we have a result parallel to (2) from Chapter 8, §4:

(3) $$\mathrm{LM}_{>_h}(g_j) = x_0^a \mathrm{LM}_>(g_j^d).$$

Now, since G is a Gröbner basis for J, by one implication in Theorem 3 of Chapter 2, §9,

$$S(g_i, g_j) \to_G 0$$

for all $i \neq j$. Because of (3) you will show in detail in Exercise 3 that after dehomogenization we have

$$S(g_i^d, g_j^d) \to_{G^d} 0.$$

This shows that G^d is a Gröbner basis, using the other implication in Theorem 3 of Chapter 2, §9 and the proof is complete. □

Hence Gröbner basis algorithms designed for the homogeneous case can be applied to *nonhomogeneous* ideals in $k[x_1, \ldots, x_n]$ as well by homogenizing the original polynomials, applying the algorithm to the homogenized forms and then dehomogenizing. We should mention that even though G is a reduced Gröbner basis for J the dehomogenized basis G^d can fail to be reduced. Here is a simple example.

Example 6. Let $I = \langle x^2 + y^2 - 1, x + y^2 - 2 \rangle$ in $\mathbb{Q}[x, y]$, using lex order with $x > y$. If we homogenize with a new variable z (rather than x_0 as in the general notation), we obtain $J = \langle x^2 + y^2 - z^2, xz + y^2 - 2z^2 \rangle$. Theorem 5 is stated using the order $>_h$ defined in (2). But note that $>_h$ is the same as lex order with $x > y > z$ in this case, so $\mathrm{LM}(f_2^h) = xz$ contains the homogenizing variable. Computing a Gröbner basis for J we find

$$G = \{y^4 - 3y^2z^2 + 3z^4, xz + y^2 - 2z^2, xy^2 + y^2z - 3z^3, x^2 + y^2 - z^2\},$$

(where the leading term in each polynomial is listed first). Thus

$$G^d = \{y^4 - 3y^2 + 3, x + y^2 - 2, xy^2 + y^2 - 3, x^2 + y^2 - 1\}.$$

We can see immediately that this is not a reduced Gröbner basis since the leading terms of the third and fourth polynomials are divisible by $\mathrm{LT}(x + y^2 - 2) = x$. Indeed, the unique reduced lex Gröbner basis for I is $\{y^4 - 3y^2 + 3, x + y^2 - 2\}$.

As in this example, further work may be necessary to produce a reduced Gröbner basis for I if we proceed by homogenizing, applying an algorithm tailored for homogeneous polynomials, and then dehomogenizing.

Gröbner Basis Algorithms and Linear Algebra

We will see next that S-polynomial remainder computations can, in a sense, be replaced by computations that produce the same information in a different way. The new idea involved is to make a translation from polynomial algebra into linear algebra [see LAZARD (1983) for a fuller exposition of the connections]. This is especially clear in the homogeneous case so we will again assume I is a homogeneous ideal. As usual, we write $I_m = I \cap k[x_1, \ldots, x_n]_m$ for the vector space of homogeneous polynomials of degree m in I (together with the zero polynomial). We record the following easy property of homogeneous ideals.

Lemma 7. *Let $I = \langle f_1, \ldots, f_s \rangle$ where the f_i are homogeneous. Let $m \in \mathbb{Z}_{\geq 0}$. Then every element of I_m is a linear combination with coefficients in k of the polynomials $x^\alpha f_i$ where $|\alpha| + \deg(f_i) = m$.*

Proof. The proof is left to the reader as Exercise 5. □

Let S_m be the set of all pairs (α, i) where $\alpha \in \mathbb{Z}_{\geq 0}^n$ and $|\alpha| + \deg(f_i) = m$, listed in any particular order. We fix a particular monomial ordering and let T_m be the set of monomials x^β with $|\beta| = m$, listed in *decreasing order*. Construct an $|S_m| \times |T_m|$ matrix M_m with entries in k whose rows are the vectors of coefficients of the polynomials $x^\alpha f_i$. If

$$x^\alpha f_i = \sum_\beta c_\beta x^\beta,$$

then the entries on the row of the matrix M_m for (α, i) are the c_β.

Example 8. In $\mathbb{Q}[x, y, z]$, use grevlex order with $x > y > z$ and let I be generated by the homogeneous polynomials

$$\begin{aligned} f_1 &= x^2 - 2y^2 - 2yz - z^2, \\ f_2 &= -xy + 2yz + z^2, \\ f_3 &= -x^2 + xy + xz + z^2. \end{aligned}$$

For this ordering $x^2 > xy > y^2 > xz > yz > z^2$ so the matrix M_2 is:

$$(4) \qquad M_2 = \begin{pmatrix} 1 & 0 & -2 & 0 & -2 & -1 \\ 0 & -1 & 0 & 0 & 2 & 1 \\ -1 & 1 & 0 & 1 & 0 & 1 \end{pmatrix}.$$

Similarly, the monomials of degree 3 are ordered

$$x^3 > x^2y > xy^2 > y^3 > x^2z > xyz > y^2z > xz^2 > yz^2 > z^3.$$

Ordering the rows following the list $xf_1, yf_1, zf_1, xf_2, yf_2, zf_2, xf_3, yf_3, zf_3$, we obtain

$$(5) \qquad M_3 = \begin{pmatrix} 1 & 0 & -2 & 0 & 0 & -2 & 0 & -1 & 0 & 0 \\ 0 & 1 & 0 & -2 & 0 & 0 & -2 & 0 & -1 & 0 \\ 0 & 0 & 0 & 0 & 1 & 0 & -2 & 0 & -2 & -1 \\ 0 & -1 & 0 & 0 & 0 & 2 & 0 & 1 & 0 & 0 \\ 0 & 0 & -1 & 0 & 0 & 0 & 2 & 0 & 1 & 0 \\ 0 & 0 & 0 & 0 & 0 & -1 & 0 & 0 & 2 & 1 \\ -1 & 1 & 0 & 0 & 1 & 0 & 0 & 1 & 0 & 0 \\ 0 & -1 & 1 & 0 & 0 & 1 & 0 & 0 & 1 & 0 \\ 0 & 0 & 0 & 0 & -1 & 1 & 0 & 1 & 0 & 1 \end{pmatrix}.$$

Using row operations (including row interchanges) the Gauss-Jordan algorithm takes the matrix M_m to the (unique) row reduced echelon form matrix N_m (see Chapter 2, §1). Each nonzero row in N_m is a linear combination with coefficients $a_{(\alpha,i)}$ in k of the rows of M_m. Hence each such row represents an element

$$g = \sum_{(\alpha,i)\in S_m} a_{(\alpha,i)} x^\alpha f_i \in I_m.$$

You will show in Exercise 6 that the nonzero rows in the matrix N_m represent a vector space basis for I_m, so each nonzero S-polynomial remainder of degree m corresponds to a linear combination of these rows.

Example 9. Continuing from Example 8, the row reduced echelon form matrix computed from M_2 in (4) is

$$N_2 = \begin{pmatrix} 1 & 0 & 0 & -1 & -2 & -2 \\ 0 & 1 & 0 & 0 & -2 & -1 \\ 0 & 0 & 1 & -\frac{1}{2} & 0 & -\frac{1}{2} \end{pmatrix}.$$

The rows in N_2 correspond to polynomials g_1, g_2, g_3 that form a new basis for the vector space I_2 spanned by the original $F = (f_1, f_2, f_3)$. Note that the leading term of g_3 is $\mathrm{LT}(g_3) = y^2$, which is not in the monomial ideal $\langle \mathrm{LT}(f_1), \mathrm{LT}(f_2), \mathrm{LT}(f_3) \rangle = \langle x^2, xy \rangle$. In other words, the computation of N_2 has accomplished the same sort of uncovering of new leading terms that S-polynomial remainders are designed to produce. Moreover, it is easy to check that $g_3 = y^2 - \frac{1}{2}xz - \frac{1}{2}z^2$ is a constant multiple of the remainder $\overline{S(f_1, f_3)}^F$ (from the pair of degree 2 for these polynomials). So the row reduction has emulated a part of the computation of a Gröbner basis by

Buchberger's algorithm. In Exercise 7, you will compute the row reduced echelon form matrix N_3 for M_3 given in (5) and see an interesting interpretation of the results. We will also use this example in §2.

The results seen in Example 9 are consequences of the way the columns in M_2 and N_2 are ordered using the monomial ordering. Similar patterns hold for all collections of homogeneous polynomials. The following proposition shows how row reducing the matrix M_m produces information equivalent to some of the S-polynomial remainder computations in Buchberger's algorithm. The row reduction essentially does *all of* the S-polynomial remainders in degree m *at once*.

Proposition 10. *Let $f_1, \ldots, f_s$ be any collection of homogeneous polynomials, and let I be the ideal they generate. Let $g_1, \ldots, g_t$ be the polynomials corresponding to the nonzero rows in the row reduced echelon form matrix N_m. If $g \in I_m$ is any nonzero polynomial, then $\mathrm{LM}(g)$ is equal to $\mathrm{LM}(g_i)$ for some i, $1 \le i \le t$.*

Proof. Because of the way the set T_m of monomials is ordered, the leading 1 in each nonzero row of the row reduced echelon form matrix N_m is the leading coefficient of the corresponding polynomial g_i in I_m. Moreover, because of the properties of matrices in row reduced echelon form, the leading 1's appear in distinct columns, so the leading monomials of the g_i are all distinct. Finally, as noted before, the g_i are a vector space basis for I_m. So all $g \in I_m$, including all the nonzero S-polynomial remainders for pairs of degree m, are equal to linear combinations $g = \sum_{i=1}^{t} c_i g_i$ with $c_i \in k$. Hence $\mathrm{LM}(g)$ is equal to $\mathrm{LM}(g_i)$ for the smallest i, $1 \le i \le t$, such that $c_i \ne 0$. $\square$

The proposition implies that polynomials in $I = \langle f_1, \ldots, f_s \rangle$ with *every possible* leading monomial could be found by computing the row reduced echelon form matrices N_m for each $m \in \mathbb{Z}_{\ge 0}$ in turn. Hence, stopping after degree m, the resulting set of polynomials from the N_j, $0 \le j \le m$, will form a Gröbner basis up to degree m for I as in Proposition 3. You will show this in Exercise 8.

However, this idea by itself does not give an alternative to Buchberger's algorithm because we have not said how to determine when the process can be stopped with a full Gröbner basis for I. In addition, we have not said how to use the fact that S-polynomials $S(f_i, f_j)$ for pairs (i, j) of degree m give elements of I_m. The F_4 algorithm to be presented in §3 takes some of these ideas and combines them to produce a very efficient alternative to Buchberger's algorithm that also applies when homogeneity is not present.

EXERCISES FOR §1

1. Let $F = (f_1, \ldots, f_s)$ where the f_i are homogeneous. Show that the degree of (i, j) coincides with the degree of $S(f_i, f_j)$ and with the degree of the remainder on division of $S(f_i, f_j)$ by F if the S-polynomial and the remainder are not zero.

2. Show that the mapping

$$k[x_0, x_1, \ldots, x_n] \longrightarrow k[x_1, \ldots, x_n]$$
$$g(x_0, \ldots, x_n) \longmapsto g^d(x_1, \ldots, x_n) = g(1, x_1, \ldots, x_n)$$

is a ring homomorphism and is onto.

3. Consider the situation in the last part of the proof of Theorem 5, where we have homogeneous g_i, g_j satisfying $S(g_i, g_j) \to_G 0$ and for all g, we have an equation $\mathrm{LM}_{>_h}(g) = x_0^a \mathrm{LM}_>(g^d)$ for some $a \in \mathbb{Z}_{\geq 0}$.

   a. Use Exercise 2 to show that $S(g_i, g_j)^d = S(g_i^d, g_j^d)$ for all $i \neq j$.

   b. Suppose that $S(g_i, g_j) = A_1 g_1 + \cdots + A_t g_t$ is a standard representation. Show that we obtain another standard representation if we dehomogenize both sides of this equation.

   c. Deduce that $S(g_i^d, g_j^d) \to_{G^d} 0$.

4. Let $>$ be lex order on $k[x_1, \ldots, x_n]$ with $x_1 > \cdots > x_n$. Show that the $>_h$ monomial order on $k[x_0, x_1, \ldots, x_n]$ defined in (4) is the same as lex order with $x_1 > \cdots > x_n > x_0$ (i.e., with the homogenizing variable ordered last).

5. Prove Lemma 7. Hint: Exercise 8 of Chapter 8, §3 will be useful.

6. Show that the polynomials corresponding to the nonzero rows of the row reduced echelon form matrix N_m form a vector space basis for I_m for each $m \in \mathbb{Z}_{\geq 0}$.

7. a. Compute the row reduced echelon form matrix N_3 from the matrix M_3 in (3). You should obtain entire rows of zeroes in N_3.

   b. By keeping track of the polynomials $x^\alpha f_i$ corresponding to the rows in M_3 as the row operations are performed, show that the result of part (a) means there are relations

   $$\ell_1 f_1 + \ell_2 f_2 + \ell_3 f_3 = 0,$$

   where the ℓ_i are homogeneous polynomials of degree 1 in x, y, z. Find two different explicit relations of this form (i.e., relations where the corresponding ℓ_i are not just constant multiples of each other).

   c. Check that the polynomials f_1, f_2, f_3 in this example are the determinants of the 2×2 submatrices of the 2×3 matrix

   $$\begin{pmatrix} x+y & y+z & x-z \\ x+2y+z & x+z & x \end{pmatrix}.$$

   d. Show that the result of part (c) gives an explanation for the relations found in part (b). [Hint: There are two "obvious" ways of appending an additional row to the matrix in part (c) to obtain a 3×3 matrix of homogeneous polynomials of degree 1 whose determinant is identically zero. Expanding the determinant by cofactors along the new third row will give relations as in part (b).]

   e. Determine a Gröbner basis for $I = \langle f_1, f_2, f_3 \rangle$ with respect to grevlex order with $x > y > z$.

8. a. Let $f_1, \ldots, f_s$ be homogeneous. Show that if all the matrices M_j for $j \leq m$ are put in row reduced echelon form, then the resulting polynomials form a Gröbner basis up to degree m for $I = \langle f_1, \ldots, f_s \rangle$.

   b. Show by example, though, that in the procedure from part (a) it is possible to get a set of polynomials that is strictly larger than the Gröbner basis up to degree m obtained in the proof of Proposition 3.

9. In this exercise you will prove another characterization of Gröbner bases up to degree m for a homogeneous ideal I. Let $G = (f_1, \ldots, f_t)$ be a list of homogeneous generators of I. By adapting the proof of Buchberger's Criterion (Theorem 6 of Chapter 2, §6) to the homogeneous case, prove that the following are equivalent:

   i. For all $f \in I_{\leq m}$, $\mathrm{LT}(f)$ is divisible by $\mathrm{LT}(f_i)$ for some i, $1 \leq i \leq t$.

   ii. For all pairs (i, j) of degree $\leq m$ relative to G, $\overline{S(f_i, f_j)}^G = 0$.

10. a. Show that G is a Gröbner basis for a homogeneous ideal I if and only if it is a Gröbner basis for I up to degree m for all $m \in \mathbb{Z}_{\geq 0}$.

    b. Use this to finish a proof of Theorem 4.

§2   Hilbert Driven Buchberger Algorithms

In this section we will write $S = k[x_1, \ldots, x_n]$. Recall from Chapter 9, §3 that if I is a homogeneous ideal and $m \in \mathbb{Z}_{\geq 0}$, the value of the Hilbert function $HF_{S/I}$ at m is defined by

$$HF_{S/I}(m) = \dim S_m/I_m = \dim S_m - \dim I_m,$$

where, as in the previous section I_m is the vector space of homogeneous elements of degree m in I (together with the zero polynomial), and similarly S_m is the vector space of all homogeneous polynomials of degree m (again together with the zero polynomial). The notation dim here means the dimension of the indicated vector space over k.

For example, with $S = \mathbb{Q}[x, y, z]$ and the ideal

$$I = \langle x^2 - 2y^2 - 2yz - z^2, -xy + 2yz + z^2, -x^2 + xy + xz + z^2 \rangle$$

from Examples 8 and 9 of §1 we have $\dim \mathbb{Q}[x, y, z]_2 = 6$ and $\dim I_2 = 3$, so $HF_{S/I}(2) = 6 - 3 = 3$. Similarly, $\dim \mathbb{Q}[x, y, z]_3 = 10$ and $\dim I_3 = 7$. So $HF_{S/I}(3) = 10 - 7 = 3$.

Hilbert functions for general I are computable because of the result from Proposition 9 of Chapter 9, §3:

- If I is a homogeneous ideal and $>$ is any monomial order, then the Hilbert functions $HF_{S/\langle \mathrm{LT}(I) \rangle}$ and $HF_{S/I}$ are equal.

A first connection with Gröbner bases is an easy corollary of this statement. If G is a set of polynomials, then as usual we will write

$$\langle \mathrm{LT}(G) \rangle = \langle \mathrm{LT}(g) \mid g \in G \rangle$$

for the monomial ideal generated by the leading terms of the elements of G.

Proposition 1. *Let $G \subseteq I$ be a finite subset. Then*

$$HF_{S/\langle \mathrm{LT}(I) \rangle}(m) \leq HF_{S/\langle \mathrm{LT}(G) \rangle}(m)$$

for all $m \in \mathbb{Z}_{\geq 0}$. Equality holds for all m if and only if G is Gröbner basis for I.

Proof. The first statement holds since $\langle \mathrm{LT}(G) \rangle_m \subseteq \langle \mathrm{LT}(I) \rangle_m$ for all m. The second statement follows then since $\langle \mathrm{LT}(G) \rangle_m = \langle \mathrm{LT}(I) \rangle_m$ for all m is equivalent to $\langle \mathrm{LT}(G) \rangle = \langle \mathrm{LT}(I) \rangle$, which is equivalent to saying G is a Gröbner basis for I.   □

In this section we will show how information from $HF_{S/I}$ can be used to control the computation of a Gröbner basis for I using the degree by degree version of Buchberger's algorithm presented in Theorem 4 from §1. The goal, as indicated in the comments following that theorem, is to recognize when some S-polynomial remainder calculations are unnecessary and can be bypassed. The algorithm we will present was first described in TRAVERSO (1997).

Since the way we usually determine the Hilbert function $HF_{S/I}$ is to compute a Gröbner basis for I and then find the Hilbert function $HF_{S/I} = HF_{S/\langle \text{LT}(I) \rangle}$ using the monomial ideal $\langle \text{LT}(I) \rangle$, this might seem somewhat circular at first glance and some explanation is required. The idea is that the algorithm we will develop in this section is designed for certain special circumstances such as the following.

- Suppose we want to compute a Gröbner basis for a not necessarily homogeneous ideal I with respect to a lex order. These computations can be extremely complex; both the space required to store the intermediate polynomials generated by Buchberger's algorithm and the time required can be large. However, the calculation of a Gröbner basis G for the same ideal with respect to a graded order such as grevlex is often much easier. By the discussion in §1, if G is a Gröbner basis for I with respect to a graded order, then G^h is a Gröbner basis for the homogeneous ideal I^h. From this information we can compute the Hilbert function of I^h. Hence we are led to the problem of *converting* the basis G^h into a Gröbner basis for I^h with respect to a different monomial order, say a lex order with the homogenizing variable as the smallest variable. It is for this *Gröbner basis conversion* step that the Hilbert function and the algorithm to be presented in this section will be of use. We can then derive a lex Gröbner basis for I by dehomogenization. If a reduced lex Gröbner basis is required, then further remainder computations as in the proof of Theorem 5 of Chapter 2, §7 can be used. Experience has shown that this seemingly roundabout approach is often more efficient than the straight lex Gröbner basis computation.

- In some cases, the computation of an initial Gröbner basis for I may even be unnecessary because the given ideal generators are themselves automatically a Gröbner basis with respect to *some* monomial order. This happens, for instance, in the implicitization problems for polynomial parametrizations studied in §2 of Chapter 2, where we considered ideals of the form

$$I = \langle x_1 - f_1(t_1, \ldots, t_r), \ldots, x_n - f_n(t_1, \ldots, t_r) \rangle.$$

Let $>$ be a monomial order with the property that for each i, x_i is greater than any monomial containing only the t_j. This would be true, for instance, for lex order with $x_1 > \cdots > x_n > t_1 > \cdots > t_r$. The leading terms of the generators of I are the x_i, which are pairwise relatively prime. Hence by Proposition 1 of Chapter 2, §10 and Theorem 3 of Chapter 2, §9, the given generators for I are a Gröbner basis. Some additional care is required in this case to determine an appropriate way to homogenize and apply the information from the Hilbert function. We will return to this situation later.

Now assume I is a homogeneous ideal and that we already know $HF_{S/I}(m)$ for all m. In order to make use of Proposition 1 in the computation of a Gröbner basis for I, we must be able to compare the known $HF_{S/I} = HF_{S/\langle \text{LT}(I) \rangle}$ with the $HF_{S/\langle \text{LT}(G) \rangle}$ for the intermediate partial Gröbner bases G generated as the algorithm proceeds. So another necessary ingredient is an efficient algorithm that computes the Hilbert functions of monomial ideals "on the fly."

Hilbert Functions of Monomial Ideals

Since we are discussing algorithms for computing Gröbner bases for an ideal I, in this subsection, we use J to denote a monomial ideal. Eventually, the J here will correspond to one of the ideals $\langle \mathrm{LT}(G) \rangle$ for a partial Gröbner basis G of I.

The first thing to understand here is that algorithms for computing the Hilbert function do not compute $HF_{S/J}(m)$ one m at a time, but rather produce information that packages the values $HF_{S/J}(m)$ for all m simultaneously. This is done by means of a standard trick, namely use of the *generating function* for the Hilbert function, a formal power series in an auxiliary variable, say t, whose coefficients encode the Hilbert function values. This power series is known as the *Hilbert-Poincaré series* (or sometimes just the *Hilbert series*) of S/J and is defined as:

$$P_{S/J}(t) = \sum_{m=0}^{\infty} HF_{S/J}(m) t^m.$$

Example 2. Some easy first cases are the following.

a. The simplest case of all is $J = \langle 1 \rangle$, for which $S/J = \{[0]\}$. Hence $HF_{S/\langle 1 \rangle}(m) = 0$ for all $m \geq 0$. The Hilbert-Poincaré series is

$$P_{S/\langle 1 \rangle} = 0.$$

b. If $J = \langle x_1, \ldots, x_n \rangle$, then $S/J \cong k$. Hence $HF_{S/\langle x_1, \ldots, x_n \rangle}(0) = 1$, while for all $m \geq 1$, $HF_{S/\langle x_1, \ldots, x_n \rangle}(m) = 0$. The Hilbert-Poincaré series has just one nonzero term:

$$P_{S/\langle x_1, \ldots, x_n \rangle} = 1.$$

c. The next simplest case is $J = \langle 0 \rangle$, so $S/J \cong S$. In this case the Hilbert function will just count the number of monomials in each degree. From Exercise 13 of Chapter 9, §3, we know that the number of monomials of degree m in $k[x_1, \ldots, x_n]$ is equal to the binomial coefficient $\binom{n+m-1}{m}$. Hence

$$HF_{S/\langle 0 \rangle}(m) = \binom{n+m-1}{m},$$

and the Hilbert-Poincaré series is

$$P_{S/\langle 0 \rangle}(t) = \sum_{m=0}^{\infty} \binom{n+m-1}{m} t^m.$$

In Exercise 1, you will show that this is the same as the Taylor series (at $t = 0$) of the rational function $\frac{1}{(1-t)^n}$. Hence we will also write

(1) $$P_{S/\langle 0 \rangle}(t) = \frac{1}{(1-t)^n}.$$

d. Generalizing parts (b) and (c), let $J = \langle x_1, \ldots, x_p \rangle$ for $p \le n$. Then $S/J \cong S'/\langle 0 \rangle$ where S' is a polynomial ring in $n - p$ variables. Moreover degrees are preserved for monomials not in J under this isomorphism. Then dim S_m/J_m will be the same as the number of distinct monomials of degree m in the polynomial ring in $n - p$ variables and hence

$$P_{S/\langle x_1, \ldots, x_p \rangle}(t) = \frac{1}{(1-t)^{n-p}} = \frac{(1-t)^p}{(1-t)^n}.$$

Note that the rational functions in all parts of this example can be written with the denominator $(1-t)^n$.

Here is a somewhat more interesting example.

Example 3. Let $J = \langle x^2, xy, xz \rangle \subseteq S = k[x, y, z]$. We have $HF_{S/J}(0) = 1$ and $HF_{S/J}(1) = 3$, since all the generators of J have degree 2. The monomials of degree 2 that are not in J are y^2, yz, z^2. Similarly for all $m \ge 0$, the set of all monomials that contain no factor of x will lie outside J and give linearly independent equivalence classes in S_m/J_m that span the quotient as a vector space over k. There are exactly $m + 1$ of those monomials of degree m:

$$y^m, y^{m-1}z, \ldots, yz^{m-1}, z^m$$

as in part (c) of Example 2 with $n = 2$. Apart from these, the only other contribution to the Hilbert function comes from the monomial x of degree 1. Therefore the Hilbert-Poincaré series of S/J will have the form:

$$\begin{aligned} P_{S/J}(t) &= 1 + (1 + 2)t + 3t^2 + 4t^3 + 5t^4 + \cdots \\ &= 1 + 3t + 3t^2 + 4t^3 + 5t^4 + \cdots. \end{aligned}$$

From part (c) of Example 2, or by computing a Taylor expansion for the rational function $\frac{1}{(1-t)^2}$, we recognize that this can be rewritten as

$$\begin{aligned} t + \frac{1}{(1-t)^2} &= \frac{t^3 - 2t^2 + t + 1}{(1-t)^2} \\ &= \frac{-t^4 + 3t^3 - 3t^2 + 1}{(1-t)^3}. \end{aligned}$$

In the last line we have multiplied the numerator and denominator by $1 - t$ to match the denominator for $J = \langle 0 \rangle$ and $n = 3$ variables as in (1) above. The reason for doing this will become clear in the theorem below.

Theorem 4. *Let J be a monomial ideal in $S = k[x_1, \ldots, x_n]$. The Hilbert-Poincaré series for S/J can be written in the form*

$$P_{S/J}(t) = \frac{N_{S/J}(t)}{(1-t)^n},$$

where $N_{S/J}(t) \in \mathbb{Z}[t]$ is a polynomial with integer coefficients.

We include the following proof because it serves as the basis for a first algorithm for computing the Hilbert-Poincaré series of S/J for a monomial ideal but this will be admittedly something of a digression from our main topic. Readers who are willing to accept the theorem without proof may wish to proceed directly to Theorem 6 and the example at the end of this subsection.

Proof. Let $T = \{x^{\alpha(1)}, \ldots, x^{\alpha(s)}\}$ be the *minimal* set of generators for J. We include the case $s = 0$, where $T = \emptyset$ and $J = \{0\}$. The proof will use a slightly tricky induction on the sum of the degrees of the generators, the integer quantity

$$\Sigma = \sum_{j=1}^{s} |\alpha(j)|.$$

Since $s \geq 0$ and $|\alpha(j)| \geq 0$ for all j when $s > 0$, the base cases will have either
- $\Sigma = 0$, $s = 0$, in which case $J = \{0\}$, or
- $\Sigma = 0$ but $s \geq 1$, which implies $J = \langle 1 \rangle$, or
- $\Sigma \leq n$ with $|\alpha(j)| = 1$ for all j.

All of these cases are covered in Example 2 above. If $J = \{0\}$, then $S/J \cong S$, so $P_{S/J}(t) = \frac{1}{(1-t)^n}$. If $J = \langle 1 \rangle$, then $S/J = \{[0]\}$, so $P_{S/J}(t) = 0 = \frac{0}{(1-t)^n}$. If all the generators of J have degree 1, then $P_{S/J}(t) = \frac{1}{(1-t)^{n-p}} = \frac{(1-t)^p}{(1-t)^n}$ for some $p \geq 1$, which has the required form. So the statement holds in all of these cases.

For the induction step, we will need to make use of the following lemma.

Lemma 5. *Let J be a monomial ideal and let $h \in S_r$ be a monomial of degree r. For each $m \geq r$, consider the linear mapping*

$$\alpha_h : S_{m-r}/J_{m-r} \longrightarrow S_m/J_m$$

defined by multiplication by h. Then
  (i) *The kernel of α_h is equal to $(J:h)_{m-r}/J_{m-r}$, where $J:h$ is the quotient, or colon, ideal defined in Chapter 4.*
 (ii) *The cokernel of α_h (i.e., the target space S_m/J_m modulo the image of α_h), is equal to $S_m/(J + \langle h \rangle)_m$.*
(iii) *We have*

$$HF_{S/J}(m) - HF_{S/J}(m - r) = \dim \operatorname{coker}(\alpha_h) - \dim \ker(\alpha_h).$$

 (iv) *The Hilbert-Poincaré series of S/J, $S/(J + \langle h \rangle)$ and $S/(J:h)$ satisfy*

$$P_{S/J}(t) = P_{S/(J+\langle h\rangle)}(t) + t^r \cdot P_{S/(J:h)}(t).$$

Proof of Lemma 5. The main content of parts (i)–(iii) of the lemma has already appeared in the proof of Theorem 3 from Chapter 9, §4 (although what we did there was slightly more general in that we considered an arbitrary homogeneous ideal I instead of a monomial ideal J and an arbitrary homogeneous polynomial f rather than a monomial h). Part (i) is also more precise in that we did not identify the kernel of α_h in this way before. However, the statement is immediate by the definition of the quotient ideal and you will verify this and the rest in Exercise 2.

We now turn to the proof of part (iv). Since $HF_{S/J}(m) = \dim S_m/J_m$, parts (i)–(iii) imply

$$
\begin{aligned}
(2) \quad \dim S_m/J_m &= \dim \operatorname{coker}(\alpha_h) - \dim \ker(\alpha_h) + \dim S_{m-r}/J_{m-r} \\
&= \dim S_m/(J + \langle h \rangle)_m - \dim (J:h)_{m-r}/J_{m-r} + \dim S_{m-r}/J_{m-r}
\end{aligned}
$$

By Proposition 1 of Chapter 9, §3, the last two terms of the second line simplify to

$$
-\dim (J:h)_{m-r}/J_{m-r} + \dim S_{m-r}/J_{m-r} = \dim S_{m-r}/(J:h)_{m-r}
$$

since the $\dim J_{m-r}$ terms cancel. Combining this with (2), we obtain

$$
\dim S_m/J_m = \dim S_m/(J + \langle h \rangle)_m + \dim S_{m-r}/(J:h)_{m-r}.
$$

This implies the equality on the Hilbert-Poincaré series since the coefficient of t^m in the product $t^r \cdot P_{S/(J:h)}(t)$ is $\dim S_{m-r}/(J:h)_{m-r}$ whenever $m \geq r$. $\qquad \square$

Using the lemma, we will now complete the induction step in the proof of the theorem. Assume that we have proved the statement for all S and J with $\Sigma \leq \ell$ and consider a monomial ideal J with $\Sigma = \ell + 1$ and minimal generating set T as before. Since $\Sigma > 0$, it follows that $s > 0$. Moreover, since we have covered the cases where all the generators have degree equal to 1 in a base case, we may assume that some $x^{\alpha(j_0)}$ in T has $|\alpha(j_0)| \geq 2$ and let x_i be a variable appearing in this monomial.

By part (iv) of the lemma with $h = x_i$ and $r = 1$ we have an equation

$$
(3) \qquad P_{S/J}(t) = P_{S/(J+\langle x_i \rangle)}(t) + t \cdot P_{S/(J:x_i)}(t).
$$

In $J + \langle x_i \rangle$, the monomial $x^{\alpha(j_0)}$ in T cannot appear in a minimal generating set since x_i divides it. Hence the sum of the degrees of the minimal generators must drop. The induction hypothesis applies and we have

$$
P_{S/(J+\langle x_i \rangle)}(t) = \frac{N_1(t)}{(1-t)^n}
$$

for some $N_1(t) \in \mathbb{Z}[t]$.

It remains to analyze the second term on the right in (3). Let us change our notation for the minimal generating set T so

$$
x^{\alpha(j)} = x_i^{a_j} \hat{x}^{\hat{\alpha}(j)},
$$

where $\hat{x}^{\hat{\alpha}(j)}$ is a monomial in the variables other than x_i and $\hat{\alpha}(j) \in \mathbb{Z}_{\geq 0}^{n-1}$ contains the other exponents. By Theorem 14 of Chapter 4, §4 we get a basis for $J:x_i$ by taking a basis $\{g_1, \ldots, g_t\}$ for $J \cap \langle x_i \rangle$ and dividing each of these polynomials by x_i. In Exercise 3 you will show that this amounts to replacing $x^{\alpha(j)} = x_i^{a_j} \hat{x}^{\hat{\alpha}(j)}$ by $x_i^{a_j-1} \hat{x}^{\hat{\alpha}(j)}$ if $a_j > 0$ and leaving it unchanged otherwise. Since we assumed $a_{j_0} > 0$

in $x^{\alpha(j_0)}$, this means that the sum of the degrees of the minimal generators of $J : x_i$ also decreases and by induction

$$P_{S/(J:x_i)}(t) = \frac{N_2(t)}{(1-t)^n}$$

where $N_2(t)$ is some polynomial in $\mathbb{Z}[t]$. The proof is complete when we put the two terms on the right of (3) over a common denominator. □

The equality in (3) and the proof of the theorem also serve as the bases for the following (somewhat rudimentary) recursive algorithm for computing $P_{S/J}(t)$.

Theorem 6. *The following function* HPS *terminates and correctly computes the Hilbert-Poincaré series* $P_{S/J}(t)$ *for a monomial ideal J in $k[x_1, \ldots, x_n]$.*

> Input : $J \subseteq S = k[x_1, \ldots, x_n]$, a monomial ideal
> Output : HPS$(J) = $ the Hilbert-Poincaré series $P_{S/J}(t)$ of S/J
>
> $T := $ minimal set of generators for J
> IF $T = \emptyset$ THEN
> $\quad$ HPS$(J) := \dfrac{1}{(1-t)^n}$
> ELSE IF $T = \{1\}$ THEN
> $\quad$ HPS$(J) := 0$
> ELSE IF T consists of p monomials of degree 1 THEN
> $\quad$ HPS$(J) := \dfrac{1}{(1-t)^{n-p}}$
> ELSE
> $\quad$ Select x_i appearing in a monomial of degree > 1 in T
> $\quad$ HPS$(J) := $ HPS$(J + \langle x_i \rangle) + t \cdot$ HPS$(J : x_i)$
> RETURN HPS(J)

Proof. If we are not in one of the base cases, the same function is applied recursively to the monomial ideals $J + \langle x_i \rangle$ and $J : x_i$ and the results are combined using (3). As shown in the proof of Theorem 4, both of the ideals $J + \langle x_i \rangle$ and $J : x_i$ are closer to the base cases in the sense that the sum of the degrees of the minimal generators decreases. Hence all chains of recursive calls will reach the base cases eventually. In other words, the function always terminates. The correctness follows from (3), a special case of part (iv) of Lemma 5. □

Example 7. We reconsider Example 3 using the function HPS described in Theorem 6. The variable x_i is sometimes called the *pivot* variable in the process. If we apply HPS to $J = \langle x^2, xy, xz \rangle$ in $\mathbb{Q}[x, y, z]$, we are not in a base case to start. Choosing x as the pivot variable, we see that

$$J + \langle x \rangle = \langle x^2, xy, xz, x \rangle = \langle x \rangle$$

and

$$J : x = \langle x, y, z \rangle.$$

(Note how the sum of the degrees of the generators has dropped in both cases.) Applying HPS recursively to the first ideal, we see all generators have degree 1, so

$$P_{S/(J+\langle x \rangle)}(t) = \frac{1}{(1-t)^2} = \frac{1-t}{(1-t)^3}.$$

Moreover, $J : x$ also has all generators of degree 1, so we are in a base case on this recursive call too and

$$P_{S/(J:x)}(t) = 1 = \frac{(1-t)^3}{(1-t)^3}.$$

Combining these values as in (3) yields the result we saw before in Example 3 for $P_{S/J}(t)$:

$$\frac{1}{(1-t)^2} + t \cdot 1 = \frac{-t^4 + 3t^3 - 3t^2 + 1}{(1-t)^3}.$$

You will investigate what happens if a different pivot variable is chosen in Exercise 5.

More refined recursive algorithms for computing Hilbert-Poincaré series are discussed in BIGATTI (1997). These still make use of part (iv) of Lemma 5, but include other improvements; see Exercise 4 for the idea behind one possible additional "divide and conquer" strategy. By Theorem 4, it is actually only necessary to compute the numerator polynomial $N_{S/J}(t)$, so some versions of these algorithms are set up to compute just that polynomial.

Hilbert Functions and Buchberger's Algorithm

We are now ready to discuss the improved version of Buchberger's algorithm making use of the information from the Hilbert function. We begin by noting the following corollary of Proposition 1 stated using the Hilbert-Poincaré series.

Proposition 8. *Let G be a finite subset of a homogeneous ideal I.*

(i) *The series expansions of $P_{S/I}(t)$ and $P_{S/\langle \mathrm{LT}(G) \rangle}(t)$ agree up to and including the t^m terms if and only if G is a Gröbner basis of I up to degree m.*

(ii) *$P_{S/I}(t) = P_{S/\langle \mathrm{LT}(G) \rangle}(t)$ if and only if G is a Gröbner basis for I.*

Proof. This is left to the reader as Exercise 6.   □

The following algorithm uses information from the Hilbert function to control when S-polynomial remainder calculations can stop in each degree and when the algorithm can terminate.

Theorem 9. *The following version of Buchberger's algorithm terminates and correctly computes a Gröbner basis for $I = \langle f_1, \ldots, f_s \rangle$ with f_i homogeneous, using the function* HPS *for monomial ideals from Theorem 6 and knowledge of $P_{S/I}(t)$.*

Input : $F = (f_1, \ldots, f_s)$ homogeneous, $P_{S/I}(t)$
Output : G, a Gröbner basis for I

$B := \{(i,j) \mid 1 \leq i < j \leq s\}$
$G := F$
$\Delta := \textit{infinity}$
$m' := 0$
$l := s$
WHILE $B \neq \emptyset$ DO
$\qquad B' := \{(i,j) \in B \mid \deg(i,j) = m'\}$
$\qquad B := B \setminus B'$
$\qquad$ WHILE $B' \neq \emptyset$ AND $\Delta > 0$ DO
$\qquad\qquad (i,j) :=$ first element in B'
$\qquad\qquad B' := B' \setminus \{(i,j)\}$
$\qquad\qquad S := \overline{S(f_i,f_j)}^{\,G}$
$\qquad\qquad$ IF $S \neq 0$ THEN
$\qquad\qquad\qquad l := l + 1;\ f_l := S$
$\qquad\qquad\qquad G := G \cup \{f_l\}$
$\qquad\qquad\qquad B := B \cup \{(i,l) \mid 1 \leq i \leq l-1\}$
$\qquad\qquad\qquad \Delta := \Delta - 1$
$\qquad P_{S/\langle \mathrm{LT}(G)\rangle}(t) := \mathrm{HPS}(\langle \mathrm{LT}(G)\rangle)$
$\qquad$ IF $P_{S/\langle \mathrm{LT}(G)\rangle}(t) = P_{S/I}(t)$ THEN
$\qquad\qquad$ RETURN G
$\qquad$ ELSE
$\qquad\qquad m' := \min(m \mid HF_{S/\langle \mathrm{LT}(G)\rangle}(m) \neq HF_{S/I}(m))$
$\qquad\qquad \Delta := HF_{S/\langle \mathrm{LT}(G)\rangle}(m') - HF_{S/I}(m')$
$\qquad\qquad B' := \{(i,j) \in B \mid \deg(i,j) < m'\}$
$\qquad\qquad B := B \setminus B'$

In the literature this is often called the *Hilbert driven* Buchberger algorithm, and we will follow this practice. See Exercise 7 below for additional information about

how the tests based on the Hilbert function might be implemented using information
from the numerator polynomial $N_{S/J}(t)$ in the rational function form of $P_{S/J}(t)$.

Proof. The algorithm terminates and is correct for the same reasons that the degree
by degree version of the standard Buchberger algorithm from Theorem 4 of §1 terminates and is correct. To prove this, we will show that exactly the same pairs are
inserted into B and exactly the same polynomials are inserted into G by this algorithm as in the degree by degree Buchberger algorithm (under a reasonable assumption about the order in which the pairs from the sets called B' in both algorithms are
processed). The important difference between the two algorithms is that the Hilbert
driven algorithm discards pairs without computing the S-polynomial remainders in
some cases. But in fact that only happens when the remainders would automatically
be zero; so no new polynomials would be inserted into G and no new pairs would
be generated and inserted into B by the degree by degree Buchberger algorithm.

To prove this claim, for simplicity, assume there are no pairs of degree 0 in
B at the start. (If there are, then $I = k[x_1,\ldots,x_n]$ and it is easy to see that the
Hilbert driven algorithm performs correctly in that case.) Then $B' = \emptyset$ on the
first pass through the outer WHILE loop in the Hilbert driven algorithm. If the
Hilbert-Poincaré series of S/I and $S/\langle \mathrm{LT}(G)\rangle$ agree at the start, then by part (ii)
of Proposition 8 we see that G is already a Gröbner basis and the algorithm terminates immediately. Otherwise, this pass simply finds an integer $m' \geq 1$ such that
$HF_{S/\langle \mathrm{LT}(G)\rangle}(m) = HF_{S/I}(m)$ for $m < m'$ but $HF_{S/\langle \mathrm{LT}(G)\rangle}(m') < HF_{S/I}(m')$, i.e., m'
is the smallest power of t for which the coefficients in the Hilbert-Poincaré series
of $S/\langle \mathrm{LT}(G)\rangle$ and S/I differ. It follows from part (i) of Proposition 8 that G is a
Gröbner basis up to degree $m' - 1 \geq 0$ at the end of that pass. If there were pairs in
any degrees $\leq m' - 1$ in B to start, this implies that the degree by degree Buchberger
algorithm would compute all those S-polynomials but all the remainders would be
zero by (1) in §1. So we have reached the end of passes through the outer loops in
the two algorithms with the same B and G in both. Moreover, both algorithms will
next consider the same set B' of pairs of degree m'. We will now use this first part
of the argument as the base case for a proof by induction.

Assume that we have reached the end of some passes through the outer loops in
the two algorithms with the same B and G in both, that both algorithms will next
consider the same set B' of pairs with the same degree m', and that elements in B'
are processed in the same order. Before this point the Hilbert driven algorithm also
computed

$$\Delta = HF_{S/\langle \mathrm{LT}(G)\rangle}(m') - HF_{S/I}(m').$$

This is the *number of new linearly independent basis elements needed in degree m'*
to obtain a Gröbner basis up to degree m'. Both algorithms now do the same computations up to a point. Each nonzero S-polynomial remainder gives an new element
of G and additional pairs in B. Note that each of these new pairs has degree $> m'$ by
Proposition 2 from §1. In the Hilbert driven algorithm, each nonzero S-polynomial
remainder found decreases Δ by 1. So the correctness of the degree by degree
Buchberger algorithm tells us that the value of Δ must eventually reach 0 before
the current pass through the outer WHILE loop is completed. When that happens

$HF_{S/\langle \mathrm{LT}(G)\rangle}(m')$ will equal $HF_{S/I}(m')$. The Hilbert-Poincaré series now agree up to and including the $t^{m'}$ terms, hence G is a Gröbner basis for I up to degree m' by part (i) of Proposition 8. The Hilbert driven algorithm's inner WHILE loop will terminate at this point. Any remaining pairs of degree m' at this time are discarded since the S-polynomial remainders will necessarily be zero (see (1) and Exercise 9 in §1). Hence the degree by degree Buchberger algorithm will find exactly the same elements in degree m' in G and form the same pairs in B using those new elements.

Moreover, with the new value of G, the Hilbert driven algorithm recomputes $P_{S/\langle \mathrm{LT}(G)\rangle}(t)$ using the function HPS and compares the result with $P_{S/I}(t)$. If the series are equal, then G is a Gröbner basis for I by part (ii) of Proposition 8. Any pairs remaining to be processed at that point *may also be discarded* for the same reason as above. Otherwise a new, strictly larger, m' is found such that $P_{S/\langle \mathrm{LT}(G)\rangle}(t)$ and $P_{S/I}(t)$ agree up to the $t^{m'-1}$ terms and G is a Gröbner basis up to degree $m'-1$. As in discussion of the base case, the degree by degree algorithm finds only zero remainders for any pairs of degree between the previous m' and the new m'. Thus both algorithms will start processing the pairs of the new m' with the same G and B and our claim is proved by induction.

Since we know the degree by degree algorithm terminates and computes a Gröbner basis for I, the same must be true for the Hilbert driven algorithm.   □

Here is a first example to illustrate the Hilbert driven algorithm in action.

Example 10. Consider $S = \mathbb{Q}[x, y, z]$ and the homogeneous ideal

$$I = \langle x^2 - 2y^2 - 2yz - z^2, -xy + 2yz + z^2, -x^2 + xy + xz + z^2\rangle$$

from Examples 8 and 9 of §1. The Hilbert-Poincaré series for S/I has the form:

$$P_{S/I}(t) = \frac{2t^3 - 3t^2 + 1}{(1 - t)^3} = \frac{2t + 1}{1 - t}.$$

The Taylor expansion of this rational function starts out as follows:

(4) $P_{S/I}(t) = 1 + 3t + 3t^2 + 3t^3 + \cdots,$

with all coefficients after the first equal to 3. Hence $HF_{S/I}(0) = 1$ and $HF_{S/I}(m) = 3$ for all $m \geq 1$. Several of the computer algebra systems discussed in Appendix C can compute Hilbert-Poincaré series in this form; we used Maple and grevlex order with $x > y > z$ on $\mathbb{Q}[x, y, z]$ to find this.

We note that a grevlex Gröbner basis for I was computed in the process of deriving the formula in (4), but we will not set up this computation as a Gröbner basis conversion. Instead, we will simply compute a Gröbner basis for I for lex order with $x > y > z$, using the Hilbert driven algorithm with the three given generators for I and the Hilbert-Poincaré series as the input.

The initial B is $\{(1, 3), (1, 2), (2, 3)\}$, where the first pair has degree 2 and the remaining two pairs have degree 3. The first pass through the outer WHILE loop serves to determine the value $m' = 2$ where the S-polynomial remainder

computations actually begin. In more detail, note that before any computation of
S-polynomials occurs, $\langle \text{LT}(G) \rangle = \langle x^2, xy \rangle$. The Hilbert-Poincaré series for
$S/\langle \text{LT}(G) \rangle$ is

$$\frac{t^3 - 2t^2 + 1}{(1-t)^3} = 1 + 3t + 4t^2 + 5t^3 + 6t^4 + \cdots .$$

Hence from (4), the smallest m' for which we are "missing" leading terms for I is
$m' = 2$, and $\Delta = 4 - 3 = 1$.

In the next pass through the outer WHILE loop, the algorithm makes $B' = \{(1,3)\}$ and $B = \{(1,2),(2,3)\}$. Then $S(f_1, f_3)$ is

$$(x^2 - 2y^2 - 2yz - z^2) + (-x^2 + xy + xz + z^2) = xy + xz - 2y^2 - 2yz$$

and dividing by G yields the remainder

$$\overline{S(f_1,f_3)}^G = xz - 2y^2 + z^2.$$

This is not zero, so we include this as a new element f_4 in G and B is updated to

$$B = \{(1,2),(2,3),(1,4),(2,4),(3,4)\}.$$

All these pairs have degree 3. So the inner loop terminates with $B' = \emptyset$ and $\Delta = 0$.

Now the algorithm again compares the Hilbert-Poincaré series $P_{S/I}(t)$ from (4)
and $P_{S/\langle \text{LT}(G)\rangle}(t) = P_{S/\langle x^2, xy, xz\rangle}(t)$. From Example 3 or Example 7, the Hilbert-Poincaré series for this monomial ideal is

$$\frac{-t^4 + 3t^3 - 3t^2 + 1}{(1-t)^3} = 1 + 3t + 3t^2 + 4t^3 + 5t^4 + \cdots .$$

Hence the new m' is $m' = 3$ and $\Delta = 4 - 3 = 1$. We seek one additional polynomial
of degree 3.

For $m' = 3$, we have $B' = \{(1,2),(2,3),(1,4),(2,4),(3,4)\}$, and B is updated
to $\emptyset$. The first pair of degree 3 to be processed is $(1,2)$. For this pair,

$$\overline{S(f_1,f_2)}^G = -2y^3 + 3yz^2 + z^3,$$

where $G = \{f_1, f_2, f_3, f_4\}$. Hence this nonzero polynomial becomes f_5, and this is
inserted into G. B is updated to $\{(2,5),(1,5),(3,5),(4,5)\}$, where the pair $(2,5)$
has degree 4 and the other new pairs have degree 5. Δ is reduced to 0 again and
the inner WHILE loop is exited. In the process, the other pairs of degree 3 in B'
above are *discarded*. Since we have found the one new polynomial of degree 3 immediately, those pairs are unnecessary. This illustrates one advantage of the Hilbert
driven approach.

Now another very interesting thing happens. For the monomial ideal $J = \langle x^2, xy, xz, y^3 \rangle$ the Hilbert-Poincaré series $P_{S/J}(t)$ is equal to $P_{S/I}(t)$ from (4). Hence

$$HF_{S/I}(m) = HF_{S/\langle \mathrm{LT}(G)\rangle}(m)$$

for all $m \in \mathbb{Z}_{\geq 0}$ and the algorithm terminates. It is not necessary to compute and reduce S-polynomials for any of the remaining pairs of degrees 4 or 5 in B. The procedure returns the final G containing the three original polynomials and the f_4, f_5 found above.

Application to Implicitization

We now return to the second proposed use for the Hilbert driven Buchberger algorithm mentioned at the start of the section. Recall that we noted that in polynomial implicitization problems, the generators of an ideal of the form

(5) $$I = \langle x_1 - f_1(t_1, \ldots, t_r), \ldots, x_n - f_n(t_1, \ldots, t_r)\rangle$$

in $k[x_1, \ldots, x_n, t_1, \ldots, t_r]$ are already a Gröbner basis for any monomial order that makes the x_i the leading terms, since those monomials are pairwise relatively prime. Assume for simplicity that each $f_i(t_1, \ldots, t_r)$ is homogeneous, with $\deg f_i = d_i \geq 1$ for each i. (If this were not true, we could introduce a homogenizing variable t_0 to attain this without changing the x_i terms.)

The goal here would be to eliminate the $t_1, \ldots, t_r$ to obtain a basis for the elimination ideal

$$I \cap k[x_1, \ldots, x_n],$$

for instance by converting the basis from (5) into a Gröbner basis with respect to an elimination order such as lex order with $t_1 > \cdots > t_r > x_1 > \cdots > x_n$.

Now we run into a small problem because the Hilbert driven Buchberger algorithm requires homogeneous inputs. That would be true for the given generators only in the relatively uninteresting case that $d_i = 1$ for all i. The implicitization operation can be done most easily in that case by standard linear algebra techniques (see Exercise 8) so computing a Gröbner basis is overkill in a sense.

We could homogenize the generators for I by introducing a new variable as in §1. But there is a somewhat better alternative that arrives at a homogeneous ideal in a different, seemingly rather underhanded, way. Namely, we can introduce *new variables* in order to write x_i as a d_i-th power. Let $x_i = \xi_i^{d_i}$, and consider the ideal

$$\tilde{I} = \langle \xi_1^{d_1} - f_1(t_1, \ldots, t_r), \ldots, \xi_n^{d_n} - f_n(t_1, \ldots, t_r)\rangle$$

in $S = k[\xi_1, \ldots, \xi_n, t_1, \ldots, t_r]$. These generators are a Gröbner basis for $\tilde{I}$ with respect to any monomial order that makes each ξ_i greater than any monomial containing only the t_j, for the same reason we saw before.

Since this new ideal is homogeneous we can apply the Hilbert driven algorithm to $\tilde{I}$. Moreover, applying the result of Exercise 4, the Hilbert-Poincaré series for an ideal of this form can be written down directly—with no computations—from the form of the generators:

$$(6) \qquad P_{S/\tilde{I}}(t) = \frac{\prod_{i=1}^{n}(1-t^{d_i})}{(1-t)^{n+r}}.$$

In the computation of a Gröbner basis for $\tilde{I}$, it is not difficult to see that every occurrence of the variable ξ_i will be as $(\xi_i^{d_i})^r = \xi_i^{d_i r}$ for some $r \in \mathbb{Z}_{\geq 0}$. As a result we will be able to convert back to the original variables x_i immediately once we have found the generators for the elimination ideal

$$\tilde{I} \cap k[\xi_1, \ldots, \xi_n].$$

Here is a calculation illustrating this idea.

Example 11. Let us see how the Hilbert driven Buchberger algorithm could be applied to convert the given basis for

$$I = \langle x - u^2 - v^2, y - uv, z - u^2 \rangle$$

to a lex Gröbner basis with $u > v > x > y > z$ and hence find the elimination ideal $I \cap \mathbb{Q}[x, y, z]$. This will be generated by a single polynomial giving the implicit equation for the variety parametrized by

$$x = u^2 + v^2,$$
$$y = uv,$$
$$z = u^2.$$

As in the general discussion above, we introduce new variables ξ, η, ζ with $\xi^2 = x$, $\eta^2 = y$, and $\zeta^2 = z$ to make the generators of

$$\tilde{I} = \langle \xi^2 - u^2 - v^2, \eta^2 - uv, \zeta^2 - u^2 \rangle$$

homogeneous, and let $S = \mathbb{Q}[\xi, \eta, \zeta, u, v]$. These polynomials also constitute a Gröbner basis already for any monomial order that makes the ξ^2, η^2, ζ^2 the leading terms, for instance, lex with $\xi > \eta > \zeta > u > v$. The Hilbert-Poincaré series $P_{S/\tilde{I}}(t)$ can be written down immediately using (6):

$$(7) \qquad P_{S/\tilde{I}}(t) = \frac{(1-t^2)^3}{(1-t)^5}$$
$$= 1 + 5t + 12t^2 + 20t^3 + 28t^4 + \cdots.$$

Now, with respect to lex order with $u > v > \xi > \eta > \zeta$, the leading monomials of the three ideal generators in the initial $G = (f_1, f_2, f_3)$ are $\mathrm{LM}(f_1) = u^2$, $\mathrm{LM}(f_2) = uv$, $\mathrm{LM}(f_3) = u^2$. the initial terms in the corresponding Hilbert-Poincaré series are

$$P_{S/\langle u^2, uv \rangle}(t) = 1 + 5t + 13t^2 + \cdots.$$

So the initial $m' = 2$, $\Delta = 4 - 3 = 1$ and we are "missing" one generator in degree 2. There is also exactly one pair of degree 2, namely $(1,3)$. We find

$$\overline{S(f_1,f_3)}^G = -v^2 + \xi^2 - \zeta^2.$$

This becomes a new element f_4 in G and we update the set of pairs to be processed to $B = \{(1,2),(2,3),(1,4),(2,4),(3,4)\}$.

Now, we find

$$P_{S/\langle u^2, uv, v^2 \rangle}(t) = 1 + 5t + 12t^2 + 22t^3 + \cdots,$$

and comparing with (7), $m' = 3$ and $\Delta = 22 - 20 = 2$. This means we need to find 2 additional elements in degree 3. There are three pairs in degree 3: $(1,2),(2,3),(2,4)$. With the first of these, $\overline{S(f_1,f_2)}^G = -u\eta^2 + v\zeta^2$ and we have found one new element f_5 in degree 3. With f_5 included in G, you will check that $\overline{S(f_2,f_3)}^G = 0$, but $\overline{S(f_2,f_4)}^G = -u\xi^2 + u\zeta^2 + v\eta^2 \neq 0$, so this polynomial is included in G as f_6.

At this point

$$P_{S/\langle u^2, uv, v^2, u\eta^2, u\xi^2 \rangle}(t) = 1 + 5t + 12t^2 + 20t^3 + 29t^4 + \cdots$$

so $m' = 4$, $\Delta = 29 - 28 = 1$ and we must find 1 new element in degree 4. The S-polynomial remainder

$$f_7 = \overline{S(f_2,f_5)}^G = -\xi^2\zeta^2 + \eta^4 + \zeta^4$$

is such an element. Now the Hilbert-Poincaré series agree:

$$P_{S/\langle u^2, uv, v^2, u\eta^2, u\xi^2, \xi^2\zeta^2 \rangle}(t) = P_{S/I}(t)$$

and the algorithm terminates. As in the previous example, there are a number of remaining pairs that are unnecessary and hence discarded at termination. Substituting back to the original variables, the polynomial

$$-xz + y^2 + z^2$$

(from f_7 above) generates the elimination ideal $I \cap \mathbb{Q}[x,y,z]$.

A different (but equally underhanded) method for recovering homogeneity will be explored in the exercises. You will see that basis conversions as in this example can also be carried out in an alternative way.

Before leaving this topic, we should mention that there are other Gröbner basis conversion algorithms. The FGLM algorithm [see FAUGÈRE, GIANNI, LAZARD, and MORA (1993)] applies for ideals satisfying the conditions of Theorem 6 in Chapter 5, §3, which are known as *zero-dimensional ideals* because of the result of Proposition 6 from §4 of Chapter 9. A second conversion algorithm called the Gröbner walk was introduced in COLLART, KALKBRENER and MALL (1998) and applies to general ideals. Both algorithms are also discussed in COX, LITTLE and O'SHEA (2005), in Chapters 2 and 8, respectively.

EXERCISES FOR §2

1. Show that, as claimed in Example 2,

$$\sum_{m=0}^{\infty} \binom{n+m-1}{m} t^m = \frac{1}{(1-t)^n}.$$

   Hint: The slickest proof is to use induction on n and to note that the induction step is closely related to the result of differentiating both sides of the equality above with respect to t.

2. Carefully verify the statements in parts (i)–(iii) of Lemma 5. Use the sketch of a proof given in the text as an outline for the steps and as a source of hints.

3. Let $J = \langle x^{\alpha(j)} \mid j = 1, \ldots, t \rangle$ be a monomial ideal. Let x_i be one of the variables that appears in the generators and write $x^{\alpha(j)} = x_i^{a_j} \hat{x}^{\hat{\alpha}(j)}$, where the components of $\hat{\alpha}(j) \in \mathbb{Z}_{\geq 0}^{n-1}$ contain the other exponents.

   a. Show that $J \cap \langle x_i \rangle$ is generated by the $x^{\alpha(j)}$ such that $a_j > 0$ and the monomials $x_i x^{\alpha(j)}$ for the j such that $a_j = 0$.

   b. Using Theorem 14 of Chapter 4, §4, show that $J : x_i$ is generated by the $x_i^{a_j-1} \hat{x}^{\hat{\alpha}(j)}$ for the j such that $a_j > 0$ and the $x^{\alpha(j)}$ for the j such that $a_j = 0$.

4. This exercise explores another "divide and conquer" strategy for computing Hilbert-Poincaré series. Procedures for computing these series often make use of part (a).

   a. Assume that the monomial ideal J has the form $J = J_1 + J_2$, where J_1, J_2 are monomial ideals such that the minimal generating set for J_1 contains only the variables $x_1, \ldots, x_p$ and the minimal generating set for J_2 contains only the variables $x_{p+1}, \ldots, x_n$. Let $S_1 = k[x_1, \ldots, x_p]$ and $S_2 = k[x_{p+1}, \ldots, x_n]$ and write $\bar{J}_i = J_i \cap S_i$ for $i = 1, 2$. Show that the Hilbert-Poincaré series for S/J factors as

   $$P_{S/J}(t) = P_{S_1/\bar{J}_1}(t) \cdot P_{S_2/\bar{J}_2}(t).$$

   b. Use the result from part (a) to give another proof of the result from Exercise 1 above.

   c. Use the result from part (a) to prove (6) in the text. Hint: Show first that $P_{k[x]/\langle x^d \rangle}(t) = (1-t^d)/(1-t)$.

5. a. What happens in Example 7 if y or z is chosen as the initial pivot variable?

   b. Apply the recursive algorithm discussed in the text to compute $P_{S/J}(t)$ for $J = \langle x^4 y, u^2 \rangle$ in $S = k[x, y, u, v]$. Check your result using part (a) of Exercise 4.

   c. Apply the recursive algorithm discussed in the text to compute $P_{S/J}(t)$ for $J = \langle x^2 yz, xy^2 z, xyz^2 \rangle$ in $S = k[x, y, z]$.

6. Prove Proposition 8. Hint: Use Proposition 1.

7. In this problem we will consider some additional aspects of the relation between values of the Hilbert function $HF_{S/J}(m)$ and the form of the Hilbert-Poincaré series $P_{S/J}(t) = N_{S/J}(t)/(1-t)^n$, where $N_{S/J}(t) \in \mathbb{Z}[t]$ is the polynomial from Theorem 4.

   a. Let $J_2 \subseteq J_1$ be two monomial ideals in S with $N_{S/J_1}(t) = 1 + a_1 t + a_2 t^2 + \cdots$ and $N_{S/J_2}(t) = 1 + b_1 t + b_2 t^2 + \cdots$. Show that

   $$HF_{S/J_1}(s) = HF_{S/J_2}(s) \text{ for } s = 1, \ldots, m-1, \text{ and } HF_{S/J_1}(m) < HF_{S/J_2}(m)$$

   if and only if

   $$a_s = b_s \text{ for } s = 1, \ldots, m-1, \text{ and } a_m < b_m.$$

   b. Deduce that if the equivalent conditions of part (a) are satisfied, then the following holds: $\dim (J_1)_m - \dim (J_2)_m = b_m - a_m$.

This shows that all the information needed for the Hilbert driven Buchberger algorithm is accessible in the numerator polynomial from the rational function form of the Hilbert-Poincaré series.

8. In the ideal I from (5) suppose f_i is homogeneous of degree 1 for all i.

   a. Show that the image of the corresponding parametrization mapping from $\mathbb{R}^r \to \mathbb{R}^n$ is a vector subspace of $\mathbb{R}^n$ of dimension at most r.

   b. Show how to obtain implicit equations for that subspace by row-reducing a suitable matrix. Hint: If $n > r$, then the $f_1, \ldots, f_n$ are linearly dependent. Write (5) in matrix form with the coefficients of the t_j to left of those of the x_i on each row. Row-reduce to find rows with zeroes in the columns corresponding to the t_j.

9. In this exercise, you will consider some of the calculations in Example 11.

   a. Check the computations in degree 3 in the example. In particular, explain why $HF_{S/\langle u^2, uv, v^2 \rangle}(3) = 22$, so we are looking for $22 - 20 = 2$ generators in degree 3.

   b. Check the computations in degree 4 in the example. In particular, explain why $HF_{S/\langle u^2, uv, v^2, u\eta^2, u\xi^2 \rangle}(4) = 29$, and $29 - 28 = 1$ generator in degree 4 must be found.

10. In this exercise, we will define a Hilbert function and a Hilbert-Poincaré series for a homogeneous ideal and relate them to the objects studied in the text. Let I be a homogeneous ideal and $m \in \mathbb{Z}_{\geq 0}$. Then we can define $HF_I(m) = \dim I_m$, where dim is the dimension as a vector space over k. Using $HF_I(m)$, let $P_I(t)$ be the generating function

$$P_I(t) = \sum_{m=0}^{\infty} HF_I(m) t^m.$$

We also write $HF_S(m) = HF_{S/\langle 0 \rangle}(m)$ and $P_S(t) = P_{S/\langle 0 \rangle}(t)$.

   a. Show that with these definitions $HF_I(m) = HF_S(m) - HF_{S/I}(m)$ for all $m \geq 0$.

   b. Deduce that the Hilbert-Poincaré series satisfy $P_I(t) = P_S(t) - P_{S/I}(t)$.

   c. Show that $P_I(t) = N_I(t)/(1-t)^n$ for some $N_I(t) \in \mathbb{Z}[t]$.

   In other words, the Hilbert-Poincaré series $P_I(t)$ has the same form given in Theorem 4 of this section. Hilbert functions and Hilbert-Poincaré series can be defined for any finitely generated positively graded module over S and the Hilbert-Poincaré series of any such module has the same form as in the statement of the theorem. See Chapter 5 of COX, LITTLE and O'SHEA (2005).

We next develop another way to apply the Hilbert driven Buchberger algorithm in implicitization. Instead of introducing new variables, we can redefine the way degrees are calculated. To start, suppose $S = k[x_1, \ldots, x_n]$, where $\deg x_i = d_i$ is strictly positive for each i. Let $d = (d_1, \ldots, d_n)$. The d-degree of a monomial x^α is the dot product $\alpha \cdot d$. A polynomial all of whose terms have the same d-degree is *weighted homogeneous*.

11. In this exercise we will see that most basic properties of homogeneous polynomials also hold with a general weight vector $d = (d_1, \ldots, d_n)$.

   a. Show that each polynomial $f \in S$ can be written uniquely as a sum $f = \sum_m f_m$ where the f_m are weighted homogeneous polynomials of d-degree m.

   b. An ideal $I \subseteq S$ is said to be *weighted homogeneous* if $f \in I$ implies $f_m \in I$ for all m. Show that ideal I is weighted homogeneous if and only if it has a set of generators that are weighted homogeneous polynomials.

   c. If f, g are weighted homogeneous, then using any monomial order, $S(f,g)$ is also weighted homogeneous, and we get a notion of a weighted homogeneous degree for each pair in a Gröbner basis computation.

   d. If G consists of weighted homogeneous polynomials and h is weighted homogeneous, then a nonzero remainder $\overline{h}^G$ is weighted homogeneous of the same d-degree as h.

   e. Show that an ideal I is weighted homogeneous if and only if a reduced Gröbner basis consists of weighted homogeneous polynomials.

12. This exercise studies how the Hilbert driven Buchberger algorithm extends to the weighted homogeneous case. If I is a weighted homogeneous ideal, the Hilbert function $HF_{S/I}(m)$ and the Hilbert-Poincaré series $P_{S/I}(t)$ can be defined in the same way as in the text, except that now, in the formula

$$HF_{S/I}(m) = \dim S_m/I_m = \dim S_m - \dim I_m,$$

the notation S_m refers to the vector space of weighted homogeneous polynomials of d-degree equal to m (together with the zero polynomial) and $I_m = I \cap S_m$.

a. If I is a weighted homogeneous ideal and $>$ is any monomial order, show that the Hilbert functions $HF_{S/\langle \text{LT}_>(I)\rangle}$ and $HF_{S/I}$ are equal.

b. Show that the Hilbert-Poincaré series for $J = \langle 0 \rangle$ has a form similar to that seen in Example 2, but incorporating the weights d_i:

$$P_{S/\langle 0 \rangle}(t) = \frac{1}{\left(1 - t^{d_1}\right) \cdots \left(1 - t^{d_n}\right)},$$

so the factors in the denominator are in correspondence with the variables x_i. Hint: Show first that if x has degree d, then $P_{k[x]/\langle 0 \rangle}(t) = 1/(1 - t^d)$. Then show there is a factorization as in Exercise 4.

c. Generalize Theorem 4 for a general weight vector d. The rational function denominator will be the same as in part (b).

d. The Hilbert-Poincaré series $P_{S/J}(t)$ for a monomial ideal J can be computed by a recursive algorithm as in the case studied in the text. But the degrees of the variables are needed in order to get the proper generalization of part (iv) of Lemma 5 and to include the proper factors in the denominators in the base cases. Develop a recursive algorithm for computing the Hilbert-Poincaré series $P_{S/J}(t)$.

e. Exactly what modifications (if any) are needed in the Hilbert driven Buchberger algorithm for weighted homogeneous inputs?

13. Carry out the computations in Example 11 using the approach from the previous exercises. Let $S = \mathbb{Q}[x, y, z, u, v]$ with weight vector $d = (2, 2, 2, 1, 1)$, which makes $x - u^2 - v^2, y - uv, z - u^2$ into weighted homogeneous polynomials. You will see that the results are completely equivalent to the way we presented the calculation in the text, although the Hilbert-Poincaré series will look different because they are defined using d.

§3 The F_4 Algorithm

The F_4 algorithm was introduced in FAUGÈRE (1999). We will use this common terminology although it is perhaps more accurate to regard F_4 as a *family* of algorithms since there are many variants and alternatives. Nevertheless, the algorithms in the F_4 family have the following common features:

• F_4 algorithms do not compute S-polynomial remainders one at a time. Instead, they use techniques from *linear algebra*—row-reduction of matrices similar to those described in §1—to accomplish what amounts to simultaneous computation of several S-polynomial remainders.

• The idea of the F_4 algorithms is, in a sense, inspired by the homogeneous case and F_4 algorithms behave especially nicely then. But the algorithms are *not* restricted to homogeneous input polynomials and, as we will see, significantly

greater care is required to find the correct matrices for the algorithm to work
with when homogeneity is not present. Moreover, F_4 algorithms do not necessarily proceed strictly degree by degree.

- The matrix-based Gröbner basis strategy discussed at the end of §1 had no way
  to prove termination. In contrast, F_4 algorithms use termination criteria derived
  from Buchberger's Criterion.

In this section we will present a basic algorithm showing the common framework
of the F_4 family, but not incorporating any of the variants and alternatives that produce major gains in efficiency. Even so, a lot of this will require explanation, so we
will present the F_4 algorithm in pseudocode form and give a line-by-line overview
with special attention to the trickiest features involved before we state and prove a
theorem concerning its behavior.

Here is the pseudocode for the *basic F_4 algorithm* we will discuss:

> Input : $F = (f_1, \ldots, f_s)$
> Output : G, a Gröbner basis for $I = \langle f_1, \ldots, f_s \rangle$
>
> $G := F$
> $t := s$
> $B := \{\{i, j\} \mid 1 \le i < j \le s\}$
> WHILE $B \ne \emptyset$ DO
>   Select $B' \ne \emptyset, B' \subseteq B$
> $B := B \setminus B'$
> $L := \left\{ \dfrac{\mathrm{lcm}(\mathrm{LM}(f_i), \mathrm{LM}(f_j))}{\mathrm{LT}(f_i)} \cdot f_i \;\middle|\; \{i, j\} \in B' \right\}$
> $M := \mathrm{ComputeM}(L, G)$
> $N := $ row reduced echelon form of M
> $N^+ := \{n \in \mathrm{rows}(N) \mid \mathrm{LM}(n) \notin \langle \mathrm{LM}(\mathrm{rows}(M)) \rangle\}$
>   FOR $n \in N^+$ DO
> $t := t + 1$
> $f_t := $ polynomial form of n
> $G := G \cup \{f_t\}$
> $B := B \cup \{\{i, t\} \mid 1 \le i < t\}$
> RETURN G

As in the other algorithms we have considered, G starts out as the set of input polynomials. Any new polynomials generated are simply included in the set G and the
value of t records the cardinality of G. The algorithm maintains a list B of pairs
for which the corresponding S-polynomials are not known to reduce to zero. But

note that B is now a set of *unordered pairs*. Parallel to the discussion in §1 in the homogeneous case, we can define the *degree* of a pair $\{i,j\}$ to be the integer

$$\deg(\mathrm{lcm}(\mathrm{LM}(f_i), \mathrm{LM}(f_j))).$$

This is the same as the total degree of the leading monomial of both "halves"

(1) $$\frac{\mathrm{lcm}(\mathrm{LM}(f_i), \mathrm{LM}(f_j))}{\mathrm{LT}(f_i)} \cdot f_i \quad \text{and} \quad \frac{\mathrm{lcm}(\mathrm{LM}(f_i), \mathrm{LM}(f_j))}{\mathrm{LT}(f_j)} \cdot f_j$$

of the S-polynomial $S(f_i, f_j)$. However, this is not necessarily the same as the total degree of the S-polynomial in the nonhomogeneous case. The total degree of the S-polynomial can be the same, smaller, or larger depending on the other terms in f_i and f_j and on the monomial order.

The body of the WHILE loop is repeated until $B = \emptyset$, i.e., until all pairs have been processed. The selection of $B' \subseteq B$ in the line following the WHILE statement is a first place that variants and alternatives come into play. If F did consist of homogeneous polynomials, and we wanted to proceed degree by degree as in Theorem 4 in §1, we could take B' to be the set of *all pairs of the minimal degree remaining in B* in each pass. This is also the *normal selection strategy* suggested by Faugère in the nonhomogeneous case. But other strategies are possible too.

In the standard approach to computing a Gröbner basis, the next step would be to compute the remainders $\overline{S(f_i, f_j)}^G$ for $\{i,j\} \in B'$. However, as we learned in §9 of Chapter 2, we are not limited to remainders produced by the division algorithm. We can also use any equation of the form

(2) $$S(f_i, f_j) - c_1 x^{\alpha(1)} f_{\ell_1} - c_2 x^{\alpha(2)} f_{\ell_2} - \cdots = \overline{S(f_i, f_j)}$$

that leads to a standard representation of $S(f_i, f_j)$ when the resulting $\overline{S(f_i, f_j)}$ is included in G. Equation (2) uses a *linear combination* of $x^{\alpha(1)} f_{\ell_1}$, $x^{\alpha(2)} f_{\ell_2}$, etc.. This suggests that linear algebra has a role to play and F_4 will make crucial use of this observation, but in a very clever way.

The F_4 approach encodes the raw material for reductions as in (2) for all the pairs in B' simultaneously. By "raw material," we mean a set H of polynomials and a corresponding matrix M of coefficients. H is built up in stages, starting from the polynomials in L, namely the pair of polynomials from (1) for each pair $\{i,j\} \in B'$. Both are included in L since $\{i,j\}$ is unordered. The difference of these two polynomials is the S-polynomial $S(f_i, f_j)$ but F_4 does not compute the difference—the two "halves" are separate elements of the set L.

Then enough additional polynomials $x^\alpha f_\ell$ are inserted into H so that it has the information needed to create the new elements to be added to the Gröbner basis. In more detail, whenever x^β is a monomial of an element of H that is divisible by some $\mathrm{LM}(f_\ell)$, then a unique element $x^\alpha f_\ell$ with $x^\beta = \mathrm{LM}(x^\alpha f_\ell)$ is included in H. The intuition behind this comes from (2), where such a monomial x^β cannot appear on the right-hand side and hence must be canceled by something on the left-hand side.

Including $x^\alpha f_\ell$ in H guarantees that we can achieve this. The ComputeM procedure to be described below generates H and the corresponding matrix of coefficients M.

The matrix M is reduced to the echelon form N in the next step and this linear algebra calculation effectively produces the equations of the form (2). The "new" polynomials from N^+ have leading terms not divisible by the leading terms of any of the polynomials corresponding to the rows of M, so they correspond to the $\overline{S(f_i, f_j)}$ in (2). These new polynomials are included in G and the set of pairs is updated. The claims are that the main loop terminates, and when it does G is a Gröbner basis for $I = \langle f_1, \ldots, f_s \rangle$.

The Procedure ComputeM

We next describe in detail how the F_4 procedure ComputeM generates the set H discussed above and the corresponding matrix of coefficients. Given the sets of polynomials L and G, as indicated above, the goal is to produce a set of polynomials H such that

i. $L \subseteq H$, and

ii. whenever x^β is a monomial appearing in some $f \in H$ and there exists some $f_\ell \in G$ with $\mathrm{LM}(f_\ell) \mid x^\beta$, then H contains a product $x^\alpha f_\ell$ whose leading monomial equals x^β.

The idea is that H can be constructed with a loop that continues as long as there are more monomials x^β to be considered.

- First, H is initialized as the set of polynomials in L, namely the

$$\frac{\mathrm{lcm}(\mathrm{LM}(f_i), \mathrm{LM}(f_j))}{\mathrm{LT}(f_j)} \cdot f_i \quad \text{and} \quad \frac{\mathrm{lcm}(\mathrm{LM}(f_i), \mathrm{LM}(f_j))}{\mathrm{LT}(f_j)} \cdot f_j$$

for $\{i, j\} \in B'$. We see that condition (ii) essentially holds when x^β is the leading monomial of one of these pairs of polynomials, since H contains both polynomials in the pair. The leading coefficient of f_j is inverted too in that element of L so the second polynomial might differ from the desired form $x^\alpha f_j$ by a constant multiple. But that is actually irrelevant for the row reduction that comes next; those leading monomials x^β are "done" at this point and they are not considered again.

- Now let x^β be the largest monomial not yet considered appearing in some polynomial in H. If there is some f_ℓ in G such that the leading monomial of f_ℓ divides x^β, then we include *exactly one* product $x^\alpha f_\ell$ with leading monomial x^β in H. If there is no such element in G, then we do nothing. The monomial x^β is now also "done"; so we do not consider it again.

- We claim that continuing in this way as long as there are monomials not yet considered appearing in polynomials in H eventually gives a set H with the desired properties.

But of course we must still show that this process eventually terminates.

We will need a notation for the set of monomials (not including the coefficients) contained in a polynomial f and we write $\text{Mon}(f)$ for this. Similarly, if K is a set of polynomials, then

$$\text{Mon}(K) = \bigcup_{f \in K} \text{Mon}(f)$$

is the set of all monomials contained in some $f \in K$. In addition, $\text{LM}(K)$ will denote the set of leading monomials of the $f \in K$. To understand the algorithm below, readers should be aware that we are working under the convention that when new polynomials are inserted in H, the set $\text{Mon}(H)$ updates immediately (even though there is no explicit command that does that operation). Assuming termination for the moment, the procedure returns an $|H| \times |\text{Mon}(H)|$ matrix M containing the coefficients of all the polynomials in H. As in §1, the monomials in $\text{Mon}(H)$ and the columns of M should be arranged in decreasing order according the monomial order in question.

Proposition 1. *The following procedure ComputeM terminates and computes a set of polynomials H satisfying (i) and (ii) above. It returns the corresponding matrix of coefficients M.*

> Input : $L, G = (f_1, \ldots, f_t)$
> Output : M
>
>
> $H := L$
> $done := \text{LM}(H)$
> WHILE $done \neq \text{Mon}(H)$ DO
> > Select largest $x^\beta \in (\text{Mon}(H) \setminus done)$ with respect to $>$
> > $done := done \cup \{x^\beta\}$
> > IF there exists $f_\ell \in G$ such that $\text{LM}(f_\ell) | x^\beta$ THEN
> > > Select any one $f_\ell \in G$ such that $\text{LM}(f_\ell) | x^\beta$
> > >
> > > $$H := H \cup \left\{ \frac{x^\beta}{\text{LM}(f_\ell)} \cdot f_\ell \right\}$$
>
> $M :=$ matrix of coefficients of H with respect to $\text{Mon}(H)$,
> > columns in decreasing order according to $>$
> RETURN M

Proof. To show that the algorithm terminates, note first that after x^β is selected at the start of a pass through the WHILE loop, the monomial x^β is included in the set *done* so it is not considered again. Any new monomials added to $\text{Mon}(H) \setminus done$ in that pass must come from a polynomial $\frac{x^\beta}{\text{LM}(f_\ell)} \cdot f_\ell$ where $\text{LM}(f_\ell) | x^\beta$. This has leading monomial equal to x^β, hence any monomials added to $\text{Mon}(H) \setminus done$ are smaller than x^β in the specified monomial order $>$. It follows that the x^β considered in successive passes through the loop form a strictly decreasing sequence in the monomial order. In Exercise 1 you will show that this implies no new monomials will be included in $\text{Mon}(H)$ after some point, and ComputeM will terminate because it is eventually true that $done = \text{Mon}(H)$. The final set H then has property (i) by the initialization. Property (ii) holds since all the x^β appearing in H have been considered at some point. $\square$

We note that this procedure is called SymbolicPreprocessing in FAUGÈRE (1999) and in many discussions of F_4 algorithms. An example of ComputeM will be given later in this section.

The Main Theorem

We are now ready to state and prove the main theorem of this section.

Theorem 2. *The basic F_4 algorithm given above, using the procedure ComputeM and any selection strategy for the set of pairs B', terminates and computes a Gröbner basis for $I = \langle f_1, \ldots, f_s \rangle$.*

Proof. We will prove that the algorithm terminates first. In each pass through the WHILE loop the pairs in the set B' are removed from B and processed. Note that one of two things happens—either no new polynomials are included in the set G and B does not change after B' is removed, or else new polynomials are included in G (by way of a nonempty N^+) and B is updated to include more pairs from those new polynomials. We will show that in any pass for which $N^+ \neq \emptyset$, writing G_{old} for the value of G at the start of the body of the loop and G_{new} for the value of G at the end, there is a strict containment

(3) $\langle \text{LM}(G_{\text{old}}) \rangle \subsetneq \langle \text{LM}(G_{\text{new}}) \rangle.$

By the ACC, such a strict containment can happen only finitely many times before the chain of monomial ideals stabilizes. From that point on, pairs are only removed from B, so the main loop eventually terminates with $B = \emptyset$.

So we need to show that (3) is valid when $N^+ \neq \emptyset$. In fact, in case $N^+ \neq \emptyset$, we claim that *none* of the leading monomials of the polynomial forms of elements of N^+ are contained in the monomial ideal $\langle \text{LT}(G_{\text{old}}) \rangle$. We show this next with an argument by contradiction.

Suppose that n is a row of the echelon form matrix N corresponding to a polynomial f such that $\text{LM}(f) \notin \langle \text{LM}(\text{rows}(M)) \rangle$, but $\text{LM}(f) \in \langle \text{LM}(G_{\text{old}}) \rangle$. The second statement implies that there exists $f_\ell \in G_{\text{old}}$ satisfying $\text{LM}(f_\ell) | \text{LM}(f)$. The monomial $x^\beta = \text{LM}(f)$ must be one of the monomials corresponding to the columns in M (and N). That means it must be one of the monomials appearing in the final set $\text{Mon}(H)$ built up in the ComputeM procedure. However, in that procedure, note that every time there exists an element $f_\ell \in G = G_{\text{old}}$ and $x^\beta \in \text{Mon}(H)$ such that $\text{LM}(f_\ell) | x^\beta$, then H is enlarged by including the polynomial $\frac{x^\beta}{\text{LM}(f_\ell)} \cdot f_\ell$. Hence a row was included in the matrix M which corresponds to a polynomial with leading monomial x^β. This contradicts what we said above. Hence the claim is established.

It remains to show that the output of the algorithm is a Gröbner basis for the ideal I generated by the input polynomials. First note that for each pair $\{i,j\}$ with $1 \leq i,j \leq t$ for the final value of t, the pair $\{i,j\}$ will have been included in B at some point and then removed. In the pass through the main WHILE loop in which it was removed, the set L contained both

$$\frac{\mathrm{lcm}(\mathrm{LM}(f_i), \mathrm{LM}(f_j))}{\mathrm{LT}(f_i)} \cdot f_i \quad \text{and} \quad \frac{\mathrm{lcm}(\mathrm{LM}(f_i), \mathrm{LM}(f_j))}{\mathrm{LT}(f_j)} \cdot f_j$$

and hence the matrix M contained rows corresponding to each of these polynomials. The S-polynomial $S(f_i, f_j)$ is the difference and hence it corresponds to a linear combination of the rows of M. However, the rows of the echelon form matrix N form a basis for the vector space spanned by the rows of M. Hence $S(f_i, f_j)$ is equal to a linear combination of the polynomials corresponding to the rows of N. In Exercise 2 you will show that this yields an *standard representation* for $S(f_i, f_j)$ using the set G_{new} at the end of that pass through the main loop. Hence the final G is a Gröbner basis by Theorem 3 of Chapter 2, §9. $\qquad\square$

Here is a small example of the algorithm in action.

Example 3. Let $f_1 = x^2 + xy - 1$, $f_2 = x^2 - z^2$ and $f_3 = xy + 1$. Let us compute a Gröbner basis for $I = \langle f_1, f_2, f_3 \rangle$ with respect to grevlex order with $x > y > z$ using the algorithm given before, leaving some checks to the reader in Exercise 3.

The initial value set of pairs is $B = \{\{1,2\}, \{1,3\}, \{2,3\}\}$, where the first pair has degree 2 and the remaining pairs have degree 3. If we use the normal selection strategy suggested by Faugère, we consider the pair of degree 2 first, take $B' = \{\{1,2\}\}$, and remove that pair from B. We have $S(f_1, f_2) = f_1 - f_2$, so the set L and the initial H in the procedure ComputeM equals $\{f_1, f_2\}$. We note that $\mathrm{LM}(f_3) = xy$ divides one of the terms in $f_1 \in L$ and hence we update H to $\{f_1, f_2, f_3\}$, making $\mathrm{Mon}(H) = \{x^2, xy, z^2, 1\}$. There are no other monomials in $\mathrm{Mon}(H)$ whose leading terms are divisible by leading monomials from G, hence the matrix M produced by ComputeM in the first pass through the main loop is

$$M = \begin{pmatrix} 1 & 1 & 0 & -1 \\ 1 & 0 & -1 & 0 \\ 0 & 1 & 0 & 1 \end{pmatrix}.$$

The corresponding row reduced echelon form matrix is

$$N = \begin{pmatrix} 1 & 0 & 0 & -2 \\ 0 & 1 & 0 & 1 \\ 0 & 0 & 1 & -2 \end{pmatrix}$$

and the set N^+ consists of row 3, which corresponds to the polynomial

$$f_4 = z^2 - 2.$$

This is inserted into G and the set of pairs B is updated to

$$B = \{\{1,3\}, \{2,3\}, \{1,4\}, \{2,4\}, \{3,4\}\}.$$

There are two pairs of degree 3 and three of degree 4 now. We select $B' = \{\{1,3\}, \{2,3\}\}$. Then $S(f_1,f_3) = yf_1 - xf_3$ and $S(f_2,f_3) = yf_2 - xf_3$, so $L = \{yf_1, yf_2, xf_3\}$. The procedure ComputeM starts with $H = L$ and appends the new polynomial yf_3 to H since the leading term of that polynomial equals the xy^2 contained in yf_1, Moreover, the yz^2 in yf_2 is divisible by $\mathrm{LT}(f_4)$, so yf_4 is also appended to H. At this point no further terms in the polynomials in H are divisible by leading terms from G, so $H = \{yf_1, yf_2, xf_3, yf_3, yf_4\}$, $\mathrm{Mon}(H) = \{x^2y, xy^2, yz^2, x, y\}$, and

$$M = \begin{pmatrix} 1 & 1 & 0 & 0 & -1 \\ 1 & 0 & -1 & 0 & 0 \\ 1 & 0 & 0 & 1 & 0 \\ 0 & 1 & 0 & 0 & 1 \\ 0 & 0 & 1 & 0 & -2 \end{pmatrix}.$$

When we row-reduce M, we find that N^+ contains one new entry corresponding to

$$f_5 = x + 2y.$$

Note one feature of F_4 in the nonhomogeneous case—new polynomials found and inserted into G can have lower degrees than any previously known polynomials, and the computation does not necessarily proceed degree by degree as in the homogeneous case from §1.

In fact at this point three of the remaining pairs ($\{1,5\}$, $\{2,5\}$, and $\{3,5\}$) have degree 2 and you will check in Exercise 3 that the next M is a 6×5 matrix of rank 4 corresponding to $H = \{f_1, f_2, f_3, xf_5, yf_5, f_4\}$ with $\mathrm{Mon}(H) = \{x^2, xy, y^2, z^2, 1\}$. There is one new leading term that we find here from the row reduced echelon matrix N, namely the y^2 in

$$f_6 = y^2 - \frac{1}{2}.$$

At this point the remaining pairs do not introduce any new leading terms, so you will check in Exercise 3 that the algorithm terminates with $G = \{f_1, \ldots, f_6\}$, a nonreduced Gröbner basis for I.

Comments and Extensions

In FAUGÈRE (1999) it is suggested that the F_4-type algorithms are intended primarily for use with graded orders such as grevlex. If a lex Gröbner basis is desired, then a grevlex computation, followed by a Gröbner basis conversion using one of the algorithms mentioned in §2 might in fact be the best approach.

Our basic F_4 algorithm is intended to illustrate the key features of the F_4 family, but it does not incorporate some of the improvements discussed in FAUGÈRE (1999). The improved F_4 algorithms discussed there are a definite advance over the improved Buchberger algorithm from Chapter 2 for nonhomogeneous Gröbner bases.

In fact F_4 was the first algorithm to succeed on several notoriously difficult bench-mark problems. However, the advantages of this approach are probably difficult to appreciate in small examples such as Example 3. Those advantages come to the fore only in larger problems because they stem mainly from the possibility of ap-plying well-developed and efficient algorithms from linear algebra to the task of row-reducing the matrices M. For ideals in larger numbers of variables generated by polynomials of higher degrees, the matrices involved can be quite large (e.g., hun-dreds or thousands of rows and columns). Moreover, those matrices are typically quite *sparse* (i.e., a large fraction of their entries are zero) so the data structures used to represent them efficiently in an actual implementation might involve lists rather than large arrays. There are quite efficient algorithms known for this sort of computation and they can easily be incorporated into the F_4 framework. It is also unnecessary to compute the full row-reduced echelon form of M—any "triangu-lar" form where the leading entries in the nonzero rows are in distinct columns will accomplish the same thing [see part (e) of Exercise 2].

As indicated before, there are many ways to modify the basic F_4 algorithm to im-prove its efficiency or to take advantage of special features of a particular problem:

- Different selection strategies for the set B' are possible.
- If a reduced Gröbner basis is desired, each time a new element is adjoined to G, it can be used to reduce all the other elements in G.
- Additional criteria can be used to discard unnecessary pairs as we did in the improved Buchberger algorithm from §10 of Chapter 2. (This would be possible for several of the pairs in Example 3, for instance, since the leading monomials are relatively prime.)
- In some cases there will be several different f_ℓ whose leading monomials divide the monomial $x^\beta \in \text{Mon}(H)$ in the course of carrying out the ComputeM proce-dure. Different strategies for choosing which $x^\alpha f_\ell$ should be inserted into H are possible and the matrices M will depend on those choices.

We will not pursue these variations and other modifications farther here—interested readers should consult FAUGÈRE (1999).

EXERCISES FOR §3

1. Complete the proof of Proposition 1 and show that the main loop in ComputeM always terminates. Hint: Monomial orders are well-orderings.
2. In this exercise, you will prove the last claim in the proof of Theorem 2.
   a. Let G be any finite set of polynomials and let H be a finite subset of
   $$\{x^\alpha g \mid g \in G, \ x^\alpha \text{ a monomial}\}.$$
   Let M be the matrix of coefficients of H with respect to $\text{Mon}(H)$ and let N be the row reduced echelon form of M. Show that if f is the polynomial corresponding to any linear combination of the rows of N, then either $f = 0$, or else $\text{LM}(f)$ is equal to $\text{LM}(h)$ for the polynomial h corresponding to one of the rows of N. Hint: The idea is the same as in Proposition 10 from §1.
   b. Now let N^+ be the set of rows in N whose leading entries are not in the set of leading entries from M. In the situation of part (a), show that every f in the linear span of H

has a standard representation with respect to the set $G \cup \widetilde{N}^+$, where $\widetilde{N}^+$ is the set of polynomials corresponding to the rows N^+.

c. Let G_{old} be the set G at the start of a pass through the main WHILE loop of the basic F_4 procedure. At the end of that pass, G has been enlarged to $G_{new} = G_{old} \cup \widetilde{N}^+$, where $\widetilde{N}^+$ is defined as in part (b). Let f be the polynomial form of any linear combination of the rows of that N. Show that f has a standard representation with respect to G_{new}.

d. Deduce that if $\{i, j\}$ is any pair in the set B' on a pass through the main WHILE loop, then the S-polynomial $S(f_i, f_j)$ has a standard representation in terms of G_{new} at the end of that pass.

e. Show that parts (a), (b), (c), and (d) remain true if instead of the row reduced echelon form N, we use any N' obtained from M by row operations in which the leading entries in the nonzero rows appear in distinct columns, and a submatrix $(N')^+$ is selected in the same way used to produce N^+ above.

3. Check the claims made in the computation in Example 3.

4. Apply the basic F_4 algorithm to compute a grevlex Gröbner basis for the ideal

$$I = \langle x + y + z + w, xy + yz + zw + wx, xyz + yzw + zwx + wxy, xyzw - 1 \rangle.$$

This is called the *cyclic 4* problem and is discussed in Example 2.6 in FAUGÈRE (1999). The analogous cyclic 9 problem in 9 variables was one of the notoriously difficult benchmark problems first solved with the F_4 algorithm. See FAUGÈRE (2001).

5. Suppose $I = \langle f_1, \ldots, f_s \rangle$ with f_i homogeneous.

a. If the normal selection strategy is used to produce B' and all pairs of some degree m are considered, exactly how is the matrix M generated by the F_4 ComputeM procedure for that m related to the matrix M_m constructed in §1? Hint: They will not usually be the same.

b. Which approach is more efficient in this case, the F_4 approach or the ideas from §1? Why?

§4   Signature-based Algorithms and F_5

In this section we will present a brief overview of another family of Gröbner basis algorithms, the *signature-based* algorithms stemming from FAUGÈRE (2002), including the F_5 algorithm presented there. These algorithms have been the subject of a large number of articles by many different authors proposing different modifications and refinements. This family is large and replete with complicated interconnections. (It overlaps the F_4 family as well, in a sense that we will describe later, so the situation is even more complicated.) This tangled story is due to two circumstances.

- F_5 and related algorithms have been extremely successful in solving previously intractable problems, especially in cryptographic applications. So they have attracted a tremendous amount of interest in the computational algebra research community.

- However, Faugère's presentation of the original F_5 algorithm did not contain a complete termination and correctness proof. So the termination of F_5 in every case was only conjectural until about 2012. Many of the articles mentioned above presented versions of the F_5 algorithm in special cases or with other additional or modified features designed to facilitate termination proofs.

There are now complete proofs of termination and correctness for the original F_5, but these are still sufficiently complicated that we will not present them here. We will not trace the history of the different attempts to prove termination and correctness either. For that, we direct interested readers to the excellent and exhaustive survey of this area in EDER and FAUGÈRE (2014). Our goal is to present a simple algorithm in the signature-based family that represents a synthesis of much of this work, following parts of Eder and Faugère's survey. Two disclaimers are in order, though. First, this family makes use of a number of new ideas and data structures, some of which require the mathematical theory of modules over a ring for a full development. Since that is beyond the scope of this text, we will need to develop some of this in an *ad hoc* fashion. Second, we will omit some proofs to streamline the presentation.

Motivating the Signature-Based Approach

One of the principal new features of the signature-based family of Gröbner basis algorithms is the systematic use of information indicating how the polynomials generated in the course of the computation depend on the original input polynomials $f_1, \ldots, f_s$. The goal is to eliminate unnecessary S-polynomial remainder calculations as much as possible by exploiting relations between the f_i. Some of those relations can be written down immediately—see (3) below for an example—while others are found as consequences of steps in the calculation. We will use the following examples to motivate the approach used in these algorithms.

Example 1. Let $R = \mathbb{Q}[x,y,z]$ and use grevlex order with $x > y > z$. Consider $f_1 = x^2 + z$ and $f_2 = xy - z$. In a Gröbner basis computation by Buchberger's algorithm we would immediately find

$$S(f_1, f_2) = yf_1 - xf_2 = xz + yz.$$

This is also the remainder on division by $\{f_1, f_2\}$ so we have a new basis element f_3 with $\mathrm{LT}(f_3) = xz$. But now

$$(1) \qquad S(f_1, f_3) = zf_1 - xf_3 = z(x^2 + z) - x(xz + yz) = -xyz + z^2 = -zf_2.$$

Hence $\overline{S(f_1, f_3)}^{\{f_1, f_2, f_3\}} = 0$ and we see this second remainder calculation turned out to be unnecessary—it produced no useful information.

The criterion from Proposition 8 of Chapter 2, §10 *does not* capture what is going on in this example. After computing $S(f_2, f_3) = -y^2z - z^2 = f_4$, neither of the pairs $(1,2)$ and $(2,3)$ is in the set B in the improved Buchberger algorithm. But $\mathrm{LT}(f_2) = xy$ does not divide $\mathrm{lcm}(\mathrm{LT}(f_1), \mathrm{LT}(f_3)) = x^2z$. So we would not be able to rule out computing and reducing $S(f_1, f_3)$ on that basis.

The S-polynomials and remainders in this computation each have the form

$$(a_1, a_2) \cdot (f_1, f_2) = a_1 f_1 + a_2 f_2$$

for some vector of polynomials $(a_1, a_2) \in \mathbb{Q}[x, y, z]^2$. Let us keep track of the vector (a_1, a_2) separately rather than immediately going to the combination $a_1 f_1 + a_2 f_2$.

When we do this, we see the beginnings of a systematic way to see why $S(f_1, f_3) = -z f_2$. Rewriting everything in terms of f_1, f_2, we have

$$\begin{aligned} S(f_1, f_3) &= z f_1 - x f_3 \\ &= z f_1 - x(y f_1 - x f_2) \\ &= (xy + z, x^2) \cdot (f_1, f_2). \end{aligned}$$

The first component of the vector $(-xy + z, x^2)$ is nothing other than $-f_2$ and the second component is $f_1 - z$. Hence

(2) $S(f_1, f_3) = (-f_2, f_1 - z) \cdot (f_1, f_2)$

and the vector $(-f_2, f_1 - z)$ is one way of producing $S(f_1, f_3)$ from the input polynomials f_1, f_2. Note that we also have the relation

(3) $(-f_2, f_1) \cdot (f_1, f_2) = -f_2 f_1 + f_1 f_2 = 0.$

Subtracting (3) from (2), we obtain

$$\begin{aligned} S(f_1, f_3) &= S(f_1, f_3) - 0 \\ &= (-f_2, f_1 - z) \cdot (f_1, f_2) - (-f_2, f_1) \cdot (f_1, f_2) \\ &= (0, -z) \cdot (f_1, f_2) = -z f_2. \end{aligned}$$

Hence we recover the equation $S(f_1, f_3) = -z f_2$ found earlier.

To say more about how signature-based algorithms would make use of such calculations, we need a brief digression about vectors of polynomials. A nonzero vector of polynomials can be written uniquely as a sum of vectors with just one nonzero component, and where that nonzero component is a constant in k times a monomial. Moreover, each of those special vectors can be written using the usual standard basis vectors $e_1, \ldots, e_n$. We will call those the *terms* appearing in the vector. For instance, the vector $(a_1, a_2) = (x^2 + 3yz, -xz)$ is the sum of the terms

$$(x^2 + 3yz, -xz) = (x^2, 0) + (3yz, 0) + (0, -xz) = x^2 e_1 + 3yz e_1 - xz e_2.$$

In each vector of polynomials, we will see below in (5) that it is possible to order the terms. For instance, here we might first decree that any term containing e_2 is larger than any term containing e_1, and then use grevlex order on $\mathbb{Q}[x, y, z]$ on the monomials in terms containing the same e_i. This will single out a largest term in a vector of polynomials in a fashion analogous to identifying the leading term in a single polynomial. For instance, in Example 1, we saw that $S(f_1, f_3)$ uses the vector

$$(xy + z, x^2) = (xy, 0) + (z, 0) + (0, x^2) = -xy e_1 + z e_1 + x^2 e_2.$$

Here the largest term is $x^2 \mathbf{e}_2 = (0, x^2)$.

For Example 1, a signature-based algorithm would proceed as above to compute the vector $(xy + z, x^2)$ giving $S(f_1, f_3)$ in terms of (f_1, f_2). Then, it would recognize that this has the *same* leading term as the vector $(-f_2, f_1)$ from (3), which is known from the start. Since the right-hand side of (3) is *zero*, the algorithm would know that computing $S(f_1, f_3)$ is unnecessary—Proposition 14 below guarantees that $S(f_1, f_3)$ reduces to zero in this situation.

Example 2. Now consider $f_1 = x^2 + xy$ and $f_2 = x^2 + y$ in $\mathbb{Q}[x, y]$ and use grevlex order with $x > y$. Using some of the ideas from Example 1 and the following discussion, note that the S-polynomial $S(f_1, f_2)$ can be written as

$$S(f_1, f_2) = (1, -1) \cdot (f_1, f_2) = xy - y.$$

Since that does not reduce to zero under $\{f_1, f_2\}$, we would include f_3 as a new Gröbner basis element. Let us write $\mathbf{a} = (1, -1)$ for the vector producing $S(f_1, f_2)$. The largest term in $\mathbf{a}$ according to the order proposed in Example 1 is the $-\mathbf{e}_2$.

Now consider what happens when we compute $S(f_1, f_3)$. We have

$$S(f_1, f_3) = yf_1 - xf_3 = yf_1 - x(f_1 - f_2) = (y - x)f_1 + xf_2.$$

Hence this S-polynomial corresponds to the vector $\mathbf{b} = (y - x, x)$. The S-polynomial itself is $S(f_1, f_3) = xy^2 + xy$, and

$$\overline{S(f_1, f_3)}^{\{f_1, f_2, f_3\}} = y^2 + y.$$

This gives another Gröbner basis element. Similarly,

$$S(f_2, f_3) = yf_2 - xf_3 = yf_2 - x(f_1 - f_2) = -xf_1 + (x + y)f_2.$$

Hence this S-polynomial corresponds to the vector $\mathbf{c} = (-x, x + y)$. The S-polynomial itself is $S(f_2, f_3) = -xy + y^2$, and we find

$$\overline{S(f_2, f_3)}^{\{f_1, f_2, f_3\}} = y^2 + y.$$

Note that these two remainder calculations have led to precisely the same result! So if we had included $f_4 = y^2 + y$ in the set of divisors, then the second remainder would be zero. This means that one of these computations is unnecessary and it is natural to ask whether we could have predicted this from the form of the vectors $\mathbf{b}$ and $\mathbf{c}$. We will see in Proposition 15 that the answer is yes, *because* the largest terms in the order proposed above are the same—in both vectors, the largest term is the $x\mathbf{e}_2$. We will see that once we have done one computation, the other becomes unnecessary.

It might still be unclear exactly why we are claiming that we have identified something of interest in these examples, but this should become more transparent later in the section. For now, it will have to suffice to say that if $I = \langle f_1, \ldots, f_s \rangle$ is any collection of polynomials, then the S-polynomials and remainders produced in the course of a Gröbner basis computation can all be written as

$$(a_1, \ldots, a_s) \cdot (f_1, \ldots, f_s) = a_1 f_1 + \cdots + a_s f_s$$

for certain $\mathbf{a} = (a_1, \ldots, a_s)$ in $k[x_1, \ldots, x_n]^s$. The orderings on terms in vectors described before generalize immediately. We will see that there are key features of the vectors $\mathbf{a}$ corresponding to some S-polynomials that make computing the S-polynomial remainder unnecessary. Moreover as in the examples above, those key features can be recognized *directly from the largest term in the vector* and other information similar to (3) known to the algorithm. In particular, it is not necessary to compute the combination $a_1 f_1 + \cdots + a_s f_s$ to recognize that a key feature is present. This can lead to the sort of increase in efficiency mentioned in the introduction to this chapter. This approach is the principal new feature of F_5 and other signature-based Gröbner basis algorithms. The *signature* of a vector is another name for the largest term. We will give a precise definition of "largest term" in (4) below.

Vectors of Polynomials and Syzygies

Now we indicate some of the theoretical background needed to make this all more precise. Let $R = k[x_1, \ldots, x_n]$. In R^s we have vector addition (using component-wise sums in R) and scalar multiplication by elements of R (using component-wise products in R). It is these operations that make R^s an example of a *module* over the ring R, but we will not use that language.

Since each element f of the ideal $I = \langle f_1, \ldots, f_s \rangle$ has the form $f = \sum_{i=1}^{s} a_i f_i$ for some $a_i \in R$, we have an onto mapping

$$\phi : R^s \longrightarrow I$$
$$\mathbf{a} = (a_1, \ldots, a_s) \longmapsto \sum_{i=1}^{s} a_i f_i.$$

As in the examples above, we will work with the vectors $\mathbf{a} \in R^s$, *and* with the polynomials $\phi(\mathbf{a}) \in I$. If $s > 1$, the mapping ϕ is *never* one-to-one. In particular, ϕ will map nonzero vectors in R^s to 0 in R. We introduce the following terminology.

Definition 3. We say $\mathbf{a} \in R^s$ is a **syzygy** on the polynomials in $(f_1, \ldots, f_s)$ if

$$\phi(\mathbf{a}) = \sum_{i=1}^{s} a_i f_i = 0 \in R.$$

The definition of a syzygy given here generalizes the definition of a syzygy on the leading terms of a set of polynomials from Chapter 2, §10. The difference is that here, the sum involves the whole polynomials $f_1, \ldots, f_s$, not just their leading terms.

For example, the vector $(-f_2, f_1) = -f_2 \mathbf{e}_2 + f_1 \mathbf{e}_2$ from Example 1 is a syzygy on (f_1, f_2). More generally if $(f_1, \ldots, f_s)$ is any list of polynomials, then for each pair

(i,j) with $1 \leq i < j \leq s$ we have an analogous syzygy

$$\mathbf{k}_{ij} = -f_j \mathbf{e}_i + f_i \mathbf{e}_j,$$

known as a *Koszul syzygy*. These are often considered to be "trivial" (i.e., not especially interesting) as syzygies. But we will see that they play a key role in F_5 and other signature-based algorithms because they are known directly from the f_i. A more interesting example comes from the relations $\ell_1 f_1 + \ell_2 f_2 + \ell_3 f_3 = 0$ found in Exercise 7 from §1. These equations correspond to syzygies $(\ell_1, \ell_2, \ell_3) \in \mathbb{Q}[x,y,z]^3$ on the (f_1, f_2, f_3).

Signatures and $\mathfrak{s}$-Reduction

The s-tuples $\mathbf{g} = (g_1, \ldots, g_s)$ in R^s can be expressed as sums of terms $cx^\alpha \mathbf{e}_i$, where the $\mathbf{e}_i$ are the standard basis vectors in R^s and $c \in k$. A monomial order $>$ on R can be extended to an analogous order on the terms $cx^\alpha \mathbf{e}_i$ in several ways. This is discussed in detail, for instance, in Chapter 5 of COX, LITTLE and O'SHEA (2005). In this overview, we will use just one, the POT or *position-over-term* order extending the order $>$ on R. As usual, we ignore the coefficient in k and then the POT order is defined by

(4) $$x^\alpha \mathbf{e}_i >_{POT} x^\beta \mathbf{e}_j \iff i > j, \text{ or } i = j \text{ and } x^\alpha > x^\beta.$$

For example, if $R = \mathbb{Q}[x,y,z]$ and $>$ is lex order with $x > y > z$, then $x^3 y \mathbf{e}_2 >_{POT} x^4 \mathbf{e}_1$ since the index 2 from the $\mathbf{e}_2$ in the first term is greater than the index 1 from the $\mathbf{e}_1$ in the second. On the other hand, $x^3 z \mathbf{e}_1 >_{POT} xy^5 \mathbf{e}_1$ since both terms include $\mathbf{e}_1$ and $x^3 z > xy^5$ in lex order.

See Exercise 1 for some of the general properties of the $>_{POT}$ order, which are parallel to properties of monomial orders in $R = k[x_1, \ldots, x_n]$. It would be possible to write LT($\mathbf{g}$) for the largest term in the $>_{POT}$ order. However, to be consistent with the literature, we will use the following terminology.

Definition 4. Let $\mathbf{g} = (g_1, \ldots, g_s) \in R^s$. Then the **signature** of $\mathbf{g}$, denoted $\mathfrak{s}(\mathbf{g})$, is the term appearing in $\mathbf{g}$ that is largest in the $>_{POT}$ order.

For example, if $\mathbf{g} = (x^3, y, x + z^2)$ in $\mathbb{Q}[x,y,z]^3$, with $>_{POT}$ extending the grevlex order with $x > y > z$, then $\mathfrak{s}(\mathbf{g}) = z^2 \mathbf{e}_3$.

We next consider what is known as $\mathfrak{s}$-reduction (or signature reduction) of vectors. Given two vectors $\mathbf{g}, \mathbf{h}$ in R^s, an $\mathfrak{s}$-reduction produces a new vector, a linear combination of the form $\mathbf{g} - cx^\alpha \mathbf{h}$. This uses the vector sum and the component-wise multiplication of the vector $\mathbf{h}$ by $cx^\alpha \in R$ mentioned before.

Definition 5. Let $\mathbf{g}, \mathbf{h} \in R^s$. Let $x^\alpha \in R$ be a monomial and let $c \in k$. We say that $\mathbf{g} - cx^\alpha \mathbf{h} \in R^s$ is the result of an $\mathfrak{s}$-**reduction** of $\mathbf{g}$ by $\mathbf{h}$, or that $\mathbf{h}$ $\mathfrak{s}$-reduces $\mathbf{g}$ to $\mathbf{k} = \mathbf{g} - cx^\alpha \mathbf{h}$, if

(i) There is a term bx^β in the polynomial $\phi(\mathbf{g}) \in R$ such that

$$\text{LT}(cx^\alpha \phi(\mathbf{h})) = cx^\alpha \text{LT}(\phi(\mathbf{h})) = bx^\beta, \quad \text{and}$$

(ii) $\mathfrak{s}(\mathbf{g}) \geq_{POT} \mathfrak{s}(x^\alpha \mathbf{h})$.

In the case of equality in part (ii), we say the reduction is a **singular $\mathfrak{s}$-reduction**; otherwise the reduction is a **regular $\mathfrak{s}$-reduction**.

There are two important aspects of this definition. On the level of the polynomial $\phi(\mathbf{g} - cx^\alpha \mathbf{h})$, the $\mathfrak{s}$-reduction performs an operation analogous to one step in the division algorithm, in which one term of $\phi(\mathbf{g})$ is canceled by the leading term of $cx^\alpha \phi(\mathbf{h})$ [see part (i)]. But then there is an additional condition coming from the signatures of the corresponding vectors in R^s [see part (ii)] that puts restrictions on which of these reduction steps are allowed. On the level of the vectors, the idea of $\mathfrak{s}$-reductions is that they either leave the signature of $\mathbf{g}$ unchanged (the regular case), or else cancel a term in $\mathbf{g}$ and yield a new signature that is smaller in the $>_{POT}$ order (the singular case).

Example 6. Let $R = \mathbb{Q}[x, y, z]$ with grevlex order with $x > y > z$ and let $f_1 = x^2y + xy, f_2 = xy^2 - xy$ and $f_3 = xy - xz$. Let $\mathbf{g} = (y, -x, 0)$, so the largest term for the $>_{POT}$ order is $\mathfrak{s}(\mathbf{g}) = -x\mathbf{e}_2$. We have

$$\phi(\mathbf{g}) = y(x^2y + xy) - x(xy^2 - xy) = -x^2y + xy^2.$$

Consider the term $-x^2y$ in this polynomial. This is equal to $\text{LT}(f_1)$, so we can cancel that term by adding f_1. Moreover, the vector $\mathbf{h} = (1, 0, 0) = \mathbf{e}_1$ corresponding to f_1 satisfies

$$\mathfrak{s}(\mathbf{g}) = x\mathbf{e}_2 >_{POT} \mathfrak{s}(\mathbf{h}) = \mathbf{e}_1.$$

Hence we have a regular $\mathfrak{s}$-reduction, and the result is

$$\mathbf{g} + \mathbf{h} = (y, -x, 0) + (1, 0, 0) = (y + 1, -x, 0),$$

for which

$$\phi(\mathbf{g} + \mathbf{h}) = xy^2 + xy.$$

Note that this *does not change the signature*: $\mathfrak{s}(\mathbf{g} + \mathbf{h}) = \mathfrak{s}(\mathbf{g}) = -x\mathbf{e}_2$.

The leading term in f_3 could also be used to cancel the term $-x^2y$ in $\phi(\mathbf{g})$. But note that the vector corresponding to f_3 is $\mathbf{e}_3$ and

$$\mathfrak{s}(\mathbf{g} + x\mathbf{e}_3) = \mathfrak{s}(y\mathbf{e}_1 - x\mathbf{e}_2 + x\mathbf{e}_3) = x\mathbf{e}_3,$$

which is different from the signature of $\mathbf{g}$ and larger in the $>_{POT}$ order. Therefore, this reduction of $\phi(\mathbf{g})$ does not correspond to an $\mathfrak{s}$-reduction of $\mathbf{g}$.

Allowing several $\mathfrak{s}$-reductions in sequence yields the following notions.

Definition 7. Let H be a set of elements of R^s. We say $\mathbf{g}$ is **$\mathfrak{s}$-reduced to $\mathbf{k}$ by H** if there is a finite sequence of $\mathfrak{s}$-reductions that takes $\mathbf{g}$ to $\mathbf{k}$:

$$\mathbf{g} - c_1 x^{\alpha(1)} \mathbf{h}_1 - \cdots - c_\ell x^{\alpha(\ell)} \mathbf{h}_\ell = \mathbf{k}$$

with $\mathbf{h}_i \in \mathbf{H}$ and $c_i \in k$.

When no regular $\mathfrak{s}$-reductions are possible on $\mathbf{g}$ using $\mathbf{H}$, then we have a vector analogous to a remainder on division.

Definition 8. We say $\mathbf{g}$ is **regular $\mathfrak{s}$-reduced with respect to H** if $\mathbf{g}$ has no regular $\mathfrak{s}$-reductions by any element of $\mathbf{H}$.

It is possible to develop a signature-based version of the division algorithm in R to produce regular $\mathfrak{s}$-reduced vectors with respect to any $\mathbf{H}$. Since the idea is essentially the same as in the usual division algorithm, we will pursue that in Exercise 5. Such a process will be used in the signature-based algorithm presented later.

Signature Gröbner Bases

For convenience, we will assume from now on that the given generators of I and any additional polynomials produced in the course of a Gröbner basis computation have leading coefficient equal to 1. Such polynomials are said to be *monic*. This will simplify the process of forming S-polynomials and the corresponding vectors. We begin with the notions corresponding to the S-polynomials in the usual case.

Definition 9. Let $\mathbf{g}, \mathbf{h} \in R^s$ correspond to monic $\phi(\mathbf{g})$ and $\phi(\mathbf{h})$ in R. Then the corresponding **S-vector** is the element of R^s defined by:

$$S(\mathbf{g}, \mathbf{h}) = \frac{\mathrm{lcm}(\mathrm{LM}(\phi(\mathbf{g})), \mathrm{LM}(\phi(\mathbf{h})))}{\mathrm{LM}(\phi(\mathbf{g}))} \cdot \mathbf{g} - \frac{\mathrm{lcm}(\mathrm{LM}(\phi(\mathbf{g})), \mathrm{LM}(\phi(\mathbf{h})))}{\mathrm{LM}(\phi(\mathbf{h}))} \cdot \mathbf{h}.$$

It is not difficult to see that

$$\phi(S(\mathbf{g}, \mathbf{h})) = S(\phi(\mathbf{g}), \phi(\mathbf{h})), \tag{5}$$

where the right-hand side is the usual S-polynomial (see Exercise 3). Hence this definition is compatible with the usual non-signature-based definition.

The formula in Definition 9 expresses $S(\mathbf{g}, \mathbf{h})$ as a difference of two vectors. The S-vector is *singular* if these vectors have the same leading term and is *regular* otherwise. In the signature-based approach, the regular S-vectors play a much larger role than the singular ones. The important cancellations of leading terms happen when we apply ϕ to the S-vector, not by reason of cancellations between leading terms in the S-vector (as happens in the singular case).

In the basic Buchberger algorithm, we check to see whether S-polynomial remainders are equal to 0. There is also a notion of "reduction to zero" in signature-based algorithms, but this term has a somewhat different meaning here. The idea behind this is the same as the observation in Example 1.

Definition 10. We say that $\mathbf{g} \in R^s$ **$\mathfrak{s}$-reduces to zero** by some set of vectors $\mathbf{H}$ if there exists a syzygy $\mathbf{k}$ such that $\mathbf{g}$ can be $\mathfrak{s}$-reduced to $\mathbf{k}$ using vectors from $\mathbf{H}$.

In particular, for a vector $\mathbf{g}$ that $\mathfrak{s}$-reduces to zero, the syzygy $\mathbf{k}$ will typically be a nonzero vector of R^s, but by the definition of syzygy $\phi(\mathbf{k}) = 0$ in R. See Exercise 2 for an example.

Definition 11. Let I be an ideal in R and let $\mathbf{G} = \{\mathbf{g}_1, \dots, \mathbf{g}_t\} \subseteq R^s$ where all $\phi(\mathbf{g}_i)$ are monic. Then $\mathbf{G}$ is said to be a **signature Gröbner basis** for I if every element of R^s $\mathfrak{s}$-reduces to zero using $\mathbf{G}$. Similarly, if $M = x^\alpha \mathbf{e}_i$ is a term in R^s, we say $\mathbf{G}$ is a **signature Gröbner basis below** M if all $\mathbf{h} \in R^s$ with $\mathfrak{s}(\mathbf{h}) <_{POT} M$ $\mathfrak{s}$-reduce to zero using $\mathbf{G}$.

There is an analog of the usual Buchberger Criterion as well.

Proposition 12. Let I be an ideal in R and $\mathbf{G} = \{\mathbf{g}_1, \dots, \mathbf{g}_t\} \subseteq R^s$ and assume $\phi(\mathbf{g}_i)$ is monic for all i.

(i) *Then $\mathbf{G}$ is a signature Gröbner basis for I if and only if all S-vectors $S(\mathbf{g}_i, \mathbf{g}_j)$ with $1 \leq i < j \leq t$ and all $\mathbf{e}_i$ with $1 \leq i \leq t$ $\mathfrak{s}$-reduce to zero using $\mathbf{G}$.*

(ii) *Similarly, if $M = x^\alpha \mathbf{e}_i$ is a term in R^s, $\mathbf{G}$ is a signature Gröbner basis below M for I if and only if all S-vectors $S(\mathbf{g}_i, \mathbf{g}_j)$ and all $\mathbf{e}_i$ with signature less than M in the $>_{POT}$ order $\mathfrak{s}$-reduce to zero using $\mathbf{G}$.*

The requirement that all the $\mathbf{e}_i$ $\mathfrak{s}$-reduce to zero in part (i) ensures that $\phi(\mathbf{G})$ is a Gröbner basis for I and not for some ideal strictly contained in I. The connection between signature Gröbner bases and Gröbner bases in the usual sense is given by the following statement.

Proposition 13. *If $\mathbf{G} = \{\mathbf{g}_1, \dots, \mathbf{g}_t\} \subseteq R^s$ is a signature Gröbner basis for I, then*

$$\phi(\mathbf{G}) = \{\phi(\mathbf{g}_1), \dots, \phi(\mathbf{g}_t)\}$$

is a Gröbner basis for I.

We omit the proofs of Propositions 12 and 13. The idea for Proposition 13 is that Definition 11 and (5) imply that every element of $I = \phi(R^s)$, including each of the usual S-polynomials $S(\phi(\mathbf{g}_i), \phi(\mathbf{g}_j))$, satisfies

$$S(\phi(\mathbf{g}_i), \phi(\mathbf{g}_j)) \rightarrow_{\phi(\mathbf{G})} 0.$$

We now want to identify situations where it can be seen immediately that it is not necessary to reduce an S-vector. Our first result requires some terminology. If $\mathbf{g}$ and $\mathbf{h}$ are in R^s, we say that $\mathfrak{s}(\mathbf{g})$ *divides* $\mathfrak{s}(\mathbf{h})$, or $\mathfrak{s}(\mathbf{h})$ *is divisible by* $\mathfrak{s}(\mathbf{g})$, if

$$\mathfrak{s}(\mathbf{h}) = cx^\gamma \mathfrak{s}(\mathbf{g})$$

for some monomial x^γ in R and some $c \in k$. This implies the terms $\mathfrak{s}(\mathbf{g})$ and $\mathfrak{s}(\mathbf{h})$ contain the same standard basis vector $\mathbf{e}_\ell$. We then have the following proposition that covers the situation encountered in Example 1.

Proposition 14. *Let $\mathbf{G} = \{\mathbf{g}_1, \dots, \mathbf{g}_t\}$ and $\mathbf{h} = S(\mathbf{g}_i, \mathbf{g}_j)$. If $\mathbf{G}$ is a signature Gröbner basis below $\mathfrak{s}(\mathbf{h})$ for I and there is a syzygy $\mathbf{k}$ such that $\mathfrak{s}(\mathbf{k})$ divides $\mathfrak{s}(\mathbf{h})$, then $\mathbf{h}$ $\mathfrak{s}$-reduces to zero using $\mathbf{G}$.*

Proof. If **k** is a syzygy as in the statement of the proposition, then you will show that x^γ **k** is also a syzygy in Exercise 4. This new syzygy has signature $\mathfrak{s}(x^\gamma$ **k**$) = \mathfrak{s}(\mathbf{h})$. Therefore, $\mathfrak{s}(\mathbf{h} - x^\gamma$ **k**$) <_{POT} \mathfrak{s}(\mathbf{h})$. Since we assume **G** is a Gröbner basis below $\mathfrak{s}(\mathbf{h})$, the vector $\mathbf{h} - x^\gamma$ **k** $\mathfrak{s}$-reduces to zero by **G**. But then the same is true for **h** by Definition 10 and the fact that the collection of syzygies is closed under sums in R^s (again, see Exercise 4). □

In Example 1, for instance, we had $f_1 = x^2 + z$ and $f_2 = xy - z$. and the first step in a Gröbner basis computation found

$$f_3 = \overline{S(f_1, f_2)} = xz + yz.$$

The Koszul syzygy $\mathbf{k} = -f_2 \mathbf{e}_1 + f_1 \mathbf{e}_2$ has $\mathfrak{s}(\mathbf{k}) = x^2 \mathbf{e}_2$. Moreover, knowing the vector corresponding to f_3, namely $y\mathbf{e}_1 - x\mathbf{e}_1$, the vector **h** for $S(f_1, f_3)$ is

$$S(f_1, f_3) = zf_1 - xf_3 \Rightarrow \mathbf{h} = z\mathbf{e}_1 - x(y\mathbf{e}_1 - x\mathbf{e}_2) = (-xy + z)\mathbf{e}_1 + x^2 \mathbf{e}_2.$$

Note that $\mathfrak{s}(\mathbf{h}) = x^2 \mathbf{e}_2$ divides $\mathfrak{s}(\mathbf{k})$. Provided we have already computed a a signature Gröbner basis in signatures below $x^2 \mathbf{e}_2$, then Proposition 14 applies and **h** $\mathfrak{s}$-reduces to zero (in the sense of Definition 10). As we will see, our signature Gröbner basis algorithm will process S-vectors in increasing order of their signatures, which will ensure that by the time the computation corresponding to $S(f_1, f_3)$ is reached, Proposition 14 will apply and the unnecessary reduction will be detected.

The second result will cover the situation seen in Example 2.

Proposition 15. *Let* $\mathbf{g}, \mathbf{h} \in R^s$ *with* $\mathfrak{s}(\mathbf{g}) = \mathfrak{s}(\mathbf{h})$ *and let* **G** *be a signature Gröbner basis below this signature. If both* **g** *and* **h** *are regular* $\mathfrak{s}$*-reduced with respect to* **G**, *then* $\phi(\mathbf{g}) = \phi(\mathbf{h})$.

Proof. Aiming for a contradiction, suppose that $\phi(\mathbf{g}) \neq \phi(\mathbf{h})$. Then by assumption $\mathfrak{s}(\mathbf{g} - \mathbf{h})$ is smaller in the $>_{POT}$ order. But this implies that $\mathbf{g} - \mathbf{h}$ $\mathfrak{s}$-reduces to zero under **G**. Interchanging the roles of **g** and **h** if necessary, we can assume that the $>_{POT}$ leading term of $\mathbf{g} - \mathbf{h}$ appears in **g**. But this contradicts the assumption that both **g** and **h** were regular $\mathfrak{s}$-reduced. □

The main consequence of this proposition is that if S-vectors are processed in increasing order by signature, *at most one S-vector with any given signature* need be processed. For instance, in Example 2, we found that the S-vectors corresponding to $S(f_1, f_3)$ and $S(f_2, f_3)$ had the same signature and the remainders were equal.

In some earlier presentations of F_5-type algorithms, Proposition 14 and Proposition 15 were used to develop two separate criteria for eliminating unnecessary S-vectors (a *syzygy criterion* and a *rewriting criterion*). But in fact, Eder and Faugère note in §7 of EDER and FAUGÈRE (2014) that those two criteria can be combined into one, since both amount to checking whether the signature of an S-vector is divisible by the signature of a known vector—either a syzygy or a previously computed element of the intermediate signature Gröbner basis.

A Signature-Based Algorithm

We will present a signature-based algorithm following an outline very much like that of Buchberger's algorithm to make connections with other approaches we have studied more apparent. Faugère's presentation of the original F_5 algorithm looked quite different.

> Input: $F = (f_1, \dots, f_s)$, $f_i \in R$
> Output: $\phi(\mathbf{G})$, a Gröbner basis for $I = \langle f_1, \dots, f_s \rangle$
>
> $\mathbf{G} := \emptyset$
> $\mathbf{P} := \{\mathbf{e}_1, \dots, \mathbf{e}_s\}$
> $\mathbf{S} := \{-f_j\mathbf{e}_i + f_i\mathbf{e}_j \mid 1 \le i < j \le s\}$
> WHILE $\mathbf{P} \ne \emptyset$ DO
> $\quad$ $\mathbf{g} :=$ the element of smallest signature in $\mathbf{P}$
> $\quad$ $\mathbf{P} := \mathbf{P} \setminus \{\mathbf{g}\}$
> $\quad$ IF Criterion$(\mathbf{g}, \mathbf{G} \cup \mathbf{S}) = $ false THEN
> $\quad\quad$ $\mathbf{h} :=$ a regular $\mathfrak{s}$-reduction of $\mathbf{g}$ by $\mathbf{G}$
> $\quad\quad$ IF $\phi(\mathbf{h}) = 0$ THEN
> $\quad\quad\quad$ $\mathbf{S} := \mathbf{S} \cup \{\mathbf{h}\}$
> $\quad\quad$ ELSE
> $\quad\quad\quad$ $\mathbf{h} := \dfrac{1}{\mathrm{LC}(\phi(\mathbf{h}))}\mathbf{h}$
> $\quad\quad\quad$ $\mathbf{P} := \mathbf{P} \cup \{S(\mathbf{k}, \mathbf{h}) \mid \mathbf{k} \in \mathbf{G} \text{ and } S(\mathbf{k}, \mathbf{h}) \text{ is regular}\}$
> $\quad\quad\quad$ $\mathbf{G} := \mathbf{G} \cup \{\mathbf{h}\}$
> RETURN $\phi(\mathbf{G})$

In this algorithm, $\mathbf{G}$ represents the current intermediate signature Gröbner basis, and $\mathbf{S}$ represents a set of known syzygies on the input polynomials $f_1, \dots, f_s$. The initial value of $\mathbf{S}$ is the set of Koszul syzygies—syzygies of the form encountered in Example 1—namely, the vectors

$$\mathbf{k}_{ij} = -f_j\mathbf{e}_i + f_i\mathbf{e}_j,$$

for all pairs of indices $1 \le i < j \le s$. Note that the choice of signs here and the definition of the $>_{POT}$ order makes $\mathfrak{s}(\mathbf{k}_{ij}) = \mathrm{LM}(f_i)\mathbf{e}_j$ (recall that we assume all polynomials occurring are monic). The initial value of $\mathbf{G}$ is $\emptyset$ and the set of standard basis vectors in R^s [with $\phi(\mathbf{e}_i) = f_i$] are placed in a set $\mathbf{P}$ that will also contain S-vectors of pairs later in the computation. Each of the $\mathbf{e}_i$ will be considered as the algorithm proceeds and each of them will either $\mathfrak{s}$-reduce to zero immediately, or else an element will be inserted in $\mathbf{G}$ that will imply $\mathbf{e}_i$ is $\mathfrak{s}$-reduced to zero by $\mathbf{G}$. The condition on the $\mathbf{e}_i$ from Proposition 12 will hold because of this.

We may assume that the set **P** is always sorted in increasing order according to the signature. The computation will proceed by increasing signatures and we will have intermediate values of **G** that are signature Gröbner bases in signatures below some term M at all times. This means that, by Proposition 15, only *regular S*-vectors need to be saved and processed.

The algorithm is organized as a loop that continues as long as **P** is not empty. In each pass, the **g** remaining in **P** with the smallest signature is selected and removed. The algorithm applies a Boolean function Criterion based on Propositions 14 and 15 to discard that **g** immediately, if possible. This criterion tests $\mathfrak{s}(\mathbf{g})$ for divisibility by the signatures of the elements of $\mathbf{G} \cup \mathbf{S}$ and returns the value true if $\mathfrak{s}(\mathbf{g})$ is divisible by one of those signatures. If the function Criterion returns false, then the algorithm $\mathfrak{s}$-reduces **g** by **G**. The signature division algorithm from Exercise 5 would be used for this. If the remainder **h** is a syzygy, then **h** is included in **S**. If not, then a constant multiple of **h** [making $\phi(\mathbf{h})$ monic] is included in **G** and the set **P** is updated with additional S-vectors. When $\mathbf{P} = \emptyset$, the hypotheses of Proposition 12 are satisfied, so **G** will be a signature Gröbner basis and by Proposition 13 we have a Gröbner basis of I in the usual sense as well.

A termination and correctness proof for the Buchberger-style signature-based algorithm can be constructed using EDER and FAUGÈRE (2014) and the references in that survey article. The proof for this form is not as hard as that for the original version of F_5, which does things somewhat differently.

Example 16. As we have done with the other algorithms presented in this chapter, we will trace a part of the computation of a Gröbner basis using the signature-based algorithm described above. Let us take $R = \mathbb{Q}[x, y, z, t]$ with grevlex order and $x > y > z > t$. Let

$$I = \langle xy - yt, x^2 - zt, z^3 - t^3 \rangle$$

and call the generators f_1, f_2, f_3 respectively. Note that the grevlex leading terms are the first terms in each case and that these are listed in increasing order. At the start of the first pass through the main loop we have

$$\mathbf{G} = \emptyset,$$
$$\mathbf{S} = \{-(x^2 - zt)\mathbf{e}_1 + (xy - yt)\mathbf{e}_2, -(z^3 - t^3)\mathbf{e}_1 + (xy - yt)\mathbf{e}_3,$$
$$-(z^3 - t^3)\mathbf{e}_2 + (x^2 - zt)\mathbf{e}_3\},$$
$$\mathbf{P} = \{\mathbf{e}_1, \mathbf{e}_2, \mathbf{e}_3\}.$$

The first and second passes through the main loop only remove $\mathbf{e}_1$ and then $\mathbf{e}_2$ from **P**, insert them it into **G**, and then update **P**. On the second pass, $S(\mathbf{e}_1, \mathbf{e}_2) = x\mathbf{e}_1 - y\mathbf{e}_2$ is a regular S-vector, and **P** is updated to

$$\mathbf{P} = \{x\mathbf{e}_1 - y\mathbf{e}_2, \mathbf{e}_3\}.$$

(Note how this is a bit different from the set-up in the traditional Buchberger algorithm, but accomplishes the generation of S-vectors in another way.)

In the third pass, we are now ready to consider the S-vector from f_1, f_2: $\mathbf{g} = x\mathbf{e}_1 - y\mathbf{e}_2$. We have $\phi(\mathbf{g}) = yzt - xyt$, where the grevlex leading term is the second term. That term is reducible using $\mathrm{LT}(f_1)$, and this is a valid $\mathfrak{s}$-reduction since

$$\mathfrak{s}(\mathbf{g}) = -y\mathbf{e}_2 >_{POT} t\,\mathbf{e}_1.$$

The signature does not change if we add $t\mathbf{e}_1$. We compute

$$\mathbf{h} = \mathbf{g} + t\mathbf{e}_1 = (x + t)\mathbf{e}_1 - y\mathbf{e}_2$$

for which $\phi(\mathbf{h}) = yzt - yt^2$ and $\mathrm{LT}(yzt - yt^2) = yzt$. No further $\mathfrak{s}$-reduction is possible so $\mathbf{P}$ is updated to include

(6) $S(\mathbf{e}_1, (x + t)\mathbf{e}_1 - y\mathbf{e}_2) = (-x^2 - xt + zt)\mathbf{e}_1 + xy\,\mathbf{e}_2,$

(7) $S(\mathbf{e}_2, (x + t)\mathbf{e}_1 - y\mathbf{e}_2) = -(x^3 + x^2t)\mathbf{e}_1 + (yzt + x^2y)\mathbf{e}_2,$

and $\mathbf{h} = (x+t)\mathbf{e}_1 - y\,\mathbf{e}_2$ becomes a new element in $\mathbf{G}$. You will check the computation of the S-vectors in Exercise 6.

In the next two passes through the main loop, the algorithm processes the S-vectors from (6) and then (7). In Exercise 6, you will show that both S-vectors are discarded without any calculation because the function Criterion returns **true** in both cases. Hence after those two passes (the fourth and fifth, if you are counting), $\mathbf{P}$ has been reduced to the singleton set $\{\mathbf{e}_3\}$. If f_3 were not present, the algorithm would terminate at this point. Moreover, all further vectors to be considered will contain $\mathbf{e}_3$ terms. It follows that $\phi(\mathbf{G})$ at this point gives a Gröbner basis for the ideal $\langle f_1, f_2 \rangle \subseteq I$.

The sixth pass through the main loop first sets up the S-vectors for the pairs involving $\mathbf{e}_3$ and the previously found elements of $\mathbf{G}$. After that pass we have, with signatures listed in increasing $>_{POT}$ order and only the leading (signature) term shown:

(8) $S((x + t)\mathbf{e}_1 - y\mathbf{e}_2, \mathbf{e}_3) = -yt\,\mathbf{e}_3 + \cdots,$

(9) $S(\mathbf{e}_1, \mathbf{e}_3) = -xy\mathbf{e}_3 + \cdots,$

(10) $S(\mathbf{e}_2, \mathbf{e}_3) = -x^2\mathbf{e}_3 + \cdots.$

The next passes through the main loop process the S-vectors from (8), (9), and (10) one at a time. In Exercise 6, you will show that the vector in (8) reduces to zero and a new element is added to $\mathbf{S}$ with signature $yt\mathbf{e}_3$. The remaining S-vectors in (9) and (10) can be discarded since the function Criterion returns **true** for both.

At this point the algorithm terminates and returns the Gröbner basis

$$\{xy - yt, x^2 - zt, yzt - yt^2, z^3 - t^3\}$$

for I. Note how the signature criteria were able to identify several unnecessary reductions.

Comments

We have only scratched the surface concerning the family of algorithms including F_5 but we hope this is enough to give the reader an idea of what is involved.

- An interesting observation, noted in Example 16, is that with our choices (in particular the $>_{POT}$ order and the ordering of the set **P**), the computation of the Gröbner basis by this algorithm is *incremental* in the sense that Gröbner bases for $\langle f_1, f_2 \rangle$, then $\langle f_1, f_2, f_3 \rangle$, then $\langle f_1, f_2, f_3, f_4 \rangle$ and so forth are computed in turn. The order in which the f_i are listed can thus affect the course of the computation rather drastically. Moreover, there are a number of other optimizations that could be introduced based on this. For instance, only Koszul syzygies involving the initial segment of the list of f_i are needed at any time in this case, so they could be generated along the way.

- Different strategies for selecting **g** from **P** are also possible. For instance, a number of authors have considered *Matrix F_5* algorithms incorporating ideas from the F_4 family. These would select several elements of **P** and reduce them simultaneously. This is the sense in which the F_4 and F_5 families overlap that we mentioned at the start of this section.

- There are many different ways to set up more powerful criteria for eliminating unnecessary S-vectors. In particular, our function Criterion looks only at the signature of the S-vector. In EDER and FAUGÈRE (2014), more general rewriting criteria are considered that actually consider terms appearing in both "halves" of the S-vector.

The next comments are intended for readers with more commutative algebra background.

- Since the main goal of the F_5-type algorithms is to identify and ignore unnecessary S-polynomial remainder computations, we may ask whether there are theoretical results implying that this strategy is actually successful. One of the most impressive is stated as Corollary 7.16 in EDER and FAUGÈRE (2014). If the input polynomials form a *regular sequence*, then the algorithm actually does no explicit S-vector reductions to 0; any S-vector that would reduce to 0 is "caught" and discarded by the Criterion. This is the reason for the title of FAUGÈRE (2002).

- Finally, we mention that the final contents of **S** might also be of great interest in some situations. The vectors in that set form a basis for the module of all syzygies on the $f_1, \ldots, f_s$.

Modern Gröbner Basis Software

We conclude this chapter with some general comments regarding the current state of understanding about computing Gröbner bases. With the accumulated experience of many people over the course of 30 years or so, it has become clear that this is one area where "one size fits all" is a poor strategy. As a result, modern Gröbner basis software often combines several of the approaches we have described. When several methods are applicable, heuristics for selecting which method will be used in a particular computation become especially important.

The `Basis` command of the recent versions of Maple's `Groebner` package, for instance, incorporates five different procedures for computing Gröbner bases:

- A compiled implementation of an improved F_4 algorithm (from Faugère's `FGb` library). This is often the fastest option, at least for grevlex orders and larger problems. But this is not completely general in that it does not support coefficients in a rational function field, provides only some monomial orders, and does not allow for coefficients mod p for large primes p.
- An interpreted Maple implementation of F_4 which is completely general but usually a lot slower than the `FGb` version of F_4.
- The traditional Buchberger algorithm, with the normal pair selection strategy and the criteria for discarding unnecessary pairs discussed in §10 of Chapter 2 (which can actually be superior in some special cases, e.g., for the case where a single polynomial is adjoined to a known Gröbner basis).
- The FGLM basis conversion algorithm.
- The Gröbner walk basis conversion algorithm.

For instance, for many problems in more than 5 or 6 variables using grevlex or other graded monomial orders, the `FGb` F_4 algorithm (or the Maple F_4) will be the default choice. For lex Gröbner bases for zero-dimensional ideals, a grevlex Gröbner basis computation, followed by FGLM basis conversion is the default method, and the Gröbner walk might be employed in other cases.

Magma has a very similar suite of Gröbner basis routines.

For lex Gröbner bases, Singular makes another heuristic choice and typically computes a grevlex Gröbner basis, homogenizes, converts to a lex basis via the Hilbert driven Buchberger algorithm from §2, then dehomogenizes and does further remainder calculations to produce a reduced Gröbner basis.

This is one area where further developments can be expected and the state of the art may look quite different in the future.

EXERCISES FOR §4

1. Show that the $>_{POT}$ order on the terms $x^\alpha e_i$ satisfies the following properties
   a. $>_{POT}$ is a well-ordering on the set of terms in R^s.
   b. If $x^\alpha e_i >_{POT} x^\beta e_j$ then for all $x^\gamma \in R$, $x^\gamma \cdot x^\alpha e_i >_{POT} x^\gamma \cdot x^\beta e_j$.
   c. The $>_{POT}$ order on R^s is compatible with the $>$ order on S in the sense that $x^\alpha > x^\beta \Leftrightarrow x^\alpha e_i > x^\beta e_i$ for all $i = 1, \ldots, s$.
2. In this problem, you will reconsider the computations from Example 1 in the light of the general language introduced following that example.
   a. Show that the vector $\mathbf{g} = (-f_2, f_1 - z)$ s-reduces to zero (as in Definition 10) using $\mathbf{h}$ from the set $\{\mathbf{e}_1, \mathbf{e}_2\}$ of vectors corresponding to (f_1, f_2).
   b. If we also allow reduction by the syzygy $\mathbf{h} = (-f_2, f_1)$, show that we can s-reduce $\mathbf{g}$ to $(0, 0)$.
   c. Explain why the computation in part (b) is actually unnecessary, though.
3. Show that if $S(\mathbf{g}, \mathbf{h})$ is the S-vector of $\mathbf{g}, \mathbf{h}$ from Definition 10, then

$$\phi(S(\mathbf{g}, \mathbf{h})) = S(\phi(\mathbf{g}), \phi(\mathbf{h})),$$

where the right side is the S-polynomial of two polynomials.

4. If $f_1, \ldots, f_s$ are polynomials in R, show that the collection of syzygies of the f_i in R^s is closed under sums and also closed under scalar multiplication by arbitrary elements of R. (This says that the syzygies form a submodule of the module R^s.)

5. In this problem, you will show that the algorithm below performs regular s-reduction of $\mathbf{g}$ by the set $\mathbf{H}$ and returns a regular s-reduced "remainder" $\mathbf{u}$.

Input: $\mathbf{g} \in R^s, \mathbf{H} = \{\mathbf{h}_1, \ldots, \mathbf{h}_\ell\} \subseteq R^s$
Output: $\mathbf{u} \in R^s$

$\mathbf{u} := \mathbf{g}$
$r := 0$
WHILE $\phi(\mathbf{u}) \neq r$ DO
    $m := \mathrm{LT}(\phi(\mathbf{u}) - r)$
    $i := 1$
    reductionoccurred := false
    WHILE $i \leq \ell$ AND *reductionoccurred* = false DO
        IF $\mathrm{LT}(\phi(\mathbf{h}_i)) | m$ AND $\mathfrak{s}\left(\dfrac{m}{\mathrm{LT}(\phi(\mathbf{h}_i))} \mathbf{h}_i \right) < \mathfrak{s}(\mathbf{u})$ THEN
            $\mathbf{d} := \dfrac{m}{\mathrm{LT}(\phi(\mathbf{h}_i))} \mathbf{h}_i$
            $\mathbf{u} := \mathbf{u} - \mathbf{d}$
            reductionoccurred := true
        ELSE
            $i := i + 1$
    IF *reductionoccurred* = false THEN
        $r := r + m$
RETURN $\mathbf{u}$

  a. Show that the algorithm always terminates.
  b. Show that only regular s-reductions are performed on $\mathbf{u}$ in the course of the algorithm and the output $\mathbf{u}$ is regular s-reduced.
  c. Modify the algorithm so that it also performs singular s-reductions whenever possible.

6. In this exercise you will check a number of the steps of the computation in Example 16, then complete the calculation.
  a. Check the computations of the S-vectors in (6) and (7).
  b. In the next passes through the main loop, check that the Criterion function returns true both times. Hint: Look at the signatures of the elements of $\mathbf{S}$ and the signatures of the S-vectors.
  c. If we did not use the function Criterion, we would have to do s-reductions on these S-vectors. Do that explicitly, and show that both reduce to multiples of syzygies in $\mathbf{S}$.
  d. Check the computations of the S-vectors in (8), (9), and (10).
  e. Show that the S-vector in (8) reduces to zero, and determine the syzygy included in $\mathbf{S}$ at this step.
  f. Verify that the S-vectors in (9) and (10) can be discarded because the function Criterion returns true for each of them.

Appendix A
Some Concepts from Algebra

This appendix contains precise statements of various algebraic facts and definitions used in the text. For students who have had a course in abstract algebra, much of this material will be familiar. For students seeing these terms for the first time, keep in mind that the abstract concepts defined here are used in the text in very concrete situations.

§1 Fields and Rings

We first give a precise definition of a field.

Definition 1. *A* **field** *consists of a set k and two binary operations "+" and "·" defined on k for which the following conditions are satisfied:*

(i) $(a+b)+c = a+(b+c)$ and $(a{\cdot}b){\cdot}c = a{\cdot}(b{\cdot}c)$ for all $a,b,c \in k$ (associativity).
(ii) $a + b = b + a$ and $a \cdot b = b \cdot a$ for all $a,b \in k$ (commutativity).
(iii) $a \cdot (b + c) = a \cdot b + a \cdot c$ for all $a,b,c \in k$ (distributivity).
(iv) There are $0,1 \in k$ such that $a + 0 = a \cdot 1 = a$ for all $a \in k$ (identities).
(v) Given $a \in k$, there is $b \in k$ such that $a + b = 0$ (additive inverses).
(vi) Given $a \in k$, $a \neq 0$, there is $c \in k$ such that $a \cdot c = 1$ (multiplicative inverses).

The fields most commonly used in the text are $\mathbb{Q}$, $\mathbb{R}$, and $\mathbb{C}$. In the exercises to §1 of Chapter 1, we mention the field $\mathbb{F}_2$ which consists of the two elements 0 and 1. Some more complicated fields are discussed in the text. For example, in §3 of Chapter 1, we define the field $k(t_1, \ldots, t_m)$ of rational functions in $t_1, \ldots, t_m$ with coefficients in k. Also, in §5 of Chapter 5, we introduce the field $k(V)$ of rational functions on an irreducible variety V.

If we do not require multiplicative inverses, then we get a commutative ring.

Definition 2. *A* **commutative ring** *consists of a set R and two binary operations "+" and "·" defined on R for which the following conditions are satisfied:*

© Springer International Publishing Switzerland 2015
D.A. Cox et al., *Ideals, Varieties, and Algorithms*, Undergraduate Texts
in Mathematics, DOI 10.1007/978-3-319-16721-3

(i) $(a + b) + c = a + (b + c)$ and $(a \cdot b) \cdot c = a \cdot (b \cdot c)$ for all $a, b, c \in R$ (associativity).

(ii) $a + b = b + a$ and $a \cdot b = b \cdot a$ for all $a, b \in R$ (commutativity).

(iii) $a \cdot (b + c) = a \cdot b + a \cdot c$ for all $a, b, c \in R$ (distributivity).

(iv) There are $0, 1 \in R$ such that $a + 0 = a \cdot 1 = a$ for all $a \in R$ (identities).

(v) Given $a \in R$, there is $b \in R$ such that $a + b = 0$ (additive inverses).

Note that any field is obviously a commutative ring. Other examples of commutative rings are the integers $\mathbb{Z}$ and the polynomial ring $k[x_1, \ldots, x_n]$. The latter is the most commonly used ring in the book. In Chapter 5, we construct two other commutative rings: the coordinate ring $k[V]$ of polynomial functions on an affine variety V and the quotient ring $k[x_1, \ldots, x_n]/I$, where I is an ideal of $k[x_1, \ldots, x_n]$.

A special case of commutative rings are the *integral domains*.

Definition 3. A commutative ring R is an **integral domain** if whenever $a, b \in R$ and $a \cdot b = 0$, then either $a = 0$ or $b = 0$.

A *zero divisor* in a commutative ring R is a nonzero element $a \in R$ such that $a \cdot b = 0$ for some nonzero $b \in R$. Hence integral domains have no zero divisors. Any field is an integral domain, and the polynomial ring $k[x_1, \ldots, x_n]$ is an integral domain. In Chapter 5, we prove that the coordinate ring $k[V]$ of a variety V is an integral domain if and only if V is irreducible.

Finally, we note that the concept of ideal can be defined for any ring.

Definition 4. Let R be a commutative ring. A subset $I \subseteq R$ is an **ideal** if it satisfies:

(i) $0 \in I$.

(ii) If $a, b \in I$, then $a + b \in I$.

(iii) If $a \in I$ and $b \in R$, then $b \cdot a \in I$.

Note how this generalizes the definition of ideal given in §4 of Chapter 1.

§2 Unique Factorization

Definition 1. Let k be a field. A polynomial $f \in k[x_1, \ldots, x_n]$ is **irreducible over** k if f is nonconstant and is not the product of two nonconstant polynomials in $k[x_1, \ldots, x_n]$.

This definition says that if a nonconstant polynomial f is irreducible over k, then up to a constant multiple, its only nonconstant factor is f itself. Also note that the concept of irreducibility depends on the field. For example, $x^2 + 1$ is irreducible over $\mathbb{Q}$ and $\mathbb{R}$, but over $\mathbb{C}$ we have $x^2 + 1 = (x - i)(x + i)$.

Every nonconstant polynomial is a product of irreducible polynomials as follows.

Theorem 2. *Every nonconstant $f \in k[x_1, \ldots, x_n]$ can be written as a product $f = f_1 \cdot f_2 \cdots f_r$ of irreducibles over k. Further, if $f = g_1 \cdot g_2 \cdots g_s$ is another factorization into irreducibles over k, then $r = s$ and the g_i's can be permuted so that each f_i is a nonzero constant multiple of g_i.*

The final assertion of the theorem says that *unique factorization* holds in the polynomial ring $k[x_1, \ldots, x_n]$.

Proof. The proof of Theorem 2 is by induction on the number of variables. The base case $k[x_1]$ is covered in §5 of Chapter 1. Now suppose that $k[x_1, \ldots, x_{n-1}]$ has unique factorization. The key tool for proving unique factorization in $k[x_1, \ldots, x_n]$ is *Gauss's Lemma*, which in our situation can be stated as follows.

Proposition 3. *Let $k(x_1, \ldots, x_{n-1})$ be the field of rational functions in $x_1, \ldots, x_{n-1}$. If $f \in k[x_1, \ldots, x_n]$ is irreducible and has positive degree in x_n, then f is irreducible in $k(x_1, \ldots, x_{n-1})[x_n]$.*

This follows from Proposition 5 of Section 9.3 of DUMMIT and FOOTE (2004) since $k[x_1, \ldots, x_{n-1}]$ has unique factorization.

Combining Proposition 3 with unique factorization in the rings $k[x_1, \ldots, x_{n-1}]$ and $k(x_1, \ldots, x_{n-1})[x_n]$, it is straightforward to prove that $k[x_1, \ldots, x_n]$ has unique factorization. See Theorem 7 of Section 9.3 of DUMMIT and FOOTE (2004) for the details.   □

For polynomials in $\mathbb{Q}[x_1, \ldots, x_n]$, there are algorithms for factoring into irreducibles over $\mathbb{Q}$. A classical algorithm due to Kronecker is discussed in Theorem 4.8 of MINES, RICHMAN, and RUITENBERG (1988), and a more efficient method is given in Section 16.6 of VON ZUR GATHEN and GERHARD (2013).

Most computer algebra systems have a command for factoring polynomials in $\mathbb{Q}[x_1, \ldots, x_n]$. Factoring polynomials in $\mathbb{R}[x_1, \ldots, x_n]$ or $\mathbb{C}[x_1, \ldots, x_n]$ is *much* more difficult.

§3   Groups

A group can be defined as follows.

Definition 1. *A* **group** *consists of a set G and a binary operation "$\cdot$" defined on G for which the following conditions are satisfied:*

(i) $(a \cdot b) \cdot c = a \cdot (b \cdot c)$ *for all $a, b, c \in G$ (associativity).*
(ii) *There is $1 \in G$ such that $1 \cdot a = a \cdot 1 = a$ for all $a \in G$ (identity).*
(iii) *Given $a \in G$, there is $b \in G$ such that $a \cdot b = b \cdot a = 1$ (inverses).*

A simple example of a group is given by the integers $\mathbb{Z}$ under addition. Note $\mathbb{Z}$ is not a group under multiplication. A more interesting example comes from linear algebra. Let k be a field and define

$$GL(n, k) = \{A \mid A \text{ is an invertible } n \times n \text{ matrix with entries in } k\}.$$

From linear algebra, we know that the product AB of two invertible matrices A and B is again invertible. Thus, matrix multiplication defines a binary operation on $GL(n, k)$, and it is easy to verify that all of the group axioms are satisfied.

In Chapter 7, we will need the notion of a subgroup.

Definition 2. Let G be a group. A nonempty subset $H \subseteq G$ is called *a* **subgroup** if it satisfies:

(i) $1 \in H$.
(ii) If $a, b \in H$, then $a \cdot b \in H$.
(iii) If $a \in H$, then $a^{-1} \in H$, where a^{-1} is the inverse of a in G.

One important group is the *symmetric group* S_n. Let n be a positive integer and consider the set

$$S_n = \{\sigma : \{1, \dots, n\} \to \{1, \dots, n\} \mid \sigma \text{ is one-to-one and onto}\}.$$

Then composition of functions turns S_n into a group. Since an element $\sigma \in S_n$ permutes the numbers 1 through n, we call σ a *permutations*. Note that S_n has $n!$ elements.

A *transposition* is an element of S_n that interchanges two numbers in $\{1, \dots, n\}$ and leaves all other numbers unchanged. Every permutation is a product of transpositions, though not in a unique way.

The *sign* of a permutation is defined to be

$$\text{sgn}(\sigma) = \begin{cases} +1 & \text{if } \sigma \text{ is a product of an even number of transpositions,} \\ -1 & \text{if } \sigma \text{ is a product of an odd number of transpositions.} \end{cases}$$

One can show that $\text{sgn}(\sigma)$ is well-defined. Proofs of these assertions about S_n can be found in Section 3.5 of DUMMIT and FOOTE (2004).

§4   Determinants

In linear algebra, one usually encounters the determinant $\det(A)$ of an $n \times n$ matrix A with entries in a field such as $\mathbb{R}$ or $\mathbb{C}$. Typical formulas are

$$\det \begin{pmatrix} a_{11} & a_{12} \\ a_{21} & a_{22} \end{pmatrix} = a_{11}a_{22} - a_{12}a_{21}$$

and

$$\det \begin{pmatrix} a_{11} & a_{12} & a_{13} \\ a_{21} & a_{22} & a_{23} \\ a_{31} & a_{32} & a_{33} \end{pmatrix} = a_{11} \det \begin{pmatrix} a_{22} & a_{23} \\ a_{32} & a_{33} \end{pmatrix} - a_{12} \det \begin{pmatrix} a_{21} & a_{23} \\ a_{31} & a_{33} \end{pmatrix} + a_{13} \det \begin{pmatrix} a_{21} & a_{22} \\ a_{31} & a_{32} \end{pmatrix},$$

which simplifies to

$$a_{11}a_{22}a_{33} - a_{11}a_{23}a_{32} - a_{12}a_{21}a_{33} + a_{12}a_{23}a_{31} + a_{13}a_{21}a_{32} + a_{13}a_{22}a_{31}.$$

In Chapters 3 and 8, we will use determinants whose entries are polynomials. Fortunately, the theory of determinants works for $n \times n$ matrices with entries in a commutative ring, such as a polynomial ring. In this generality, the above formulas can be extended to express the determinant of an $n \times n$ matrix as a sum of $n!$ terms indexed by permutations $\sigma \in S_n$, where the sign of the term is $\text{sgn}(\sigma)$ from §3. More precisely, we have the following result.

Proposition 1. *If $A = (a_{ij})$ is an $n \times n$ matrix with entries in a commutative ring, then*

$$\det(A) = \sum_{\sigma \in S_n} \text{sgn}(\sigma) a_{1\sigma(1)} \cdots a_{n\sigma(n)}.$$

Proofs of all properties of determinants stated here can be found in Section 11.4 of DUMMIT and FOOTE (2004).

A second fact about determinants we will need concerns the solution of a linear system of n equations in n unknowns. In matrix form, the system is written

$$AX = b,$$

where $A = (a_{ij})$ is the $n \times n$ coefficient matrix, b is an $n \times 1$ column vector, and X is the column vector whose entries are the unknowns $x_1, \ldots, x_n$.

Assume that we are working over a field and A is invertible. Then $\det(A) \neq 0$ and A^{-1} exists. Furthermore, the system $AX = b$ has the unique solution given by $X = A^{-1}b$. However, rather than finding the solution by Gaussian elimination as done in most linear algebra courses, we need a *formula* for the solution.

Proposition 2 (Cramer's Rule). *Suppose we have a system of equations $AX = b$ over a field. If A is invertible, then the unique solution is given by*

$$x_i = \frac{\det(M_i)}{\det(A)},$$

where M_i is the matrix obtained from A by replacing its i-th column with b.

We use Propositions 1 and 2 to prove properties of resultants in Chapter 3, §6 and Chapter 8, §7. We also use Proposition 2 in the proof of the Projective Extension Theorem given in Chapter 8, §5. When we apply Cramer's Rule in these proofs, the entries of A and b will typically be polynomials and the field will be the associated field of rational functions.

We end with a fact about cofactors. The (i,j)-*cofactor* of an $n \times n$ matrix A is $c_{ij} = (-1)^{i+j}\det(A_{ij})$, where A_{ij} is the $(n-1) \times (n-1)$ matrix obtained from A by deleting row i and column j. Also let I_n be the $n \times n$ identity matrix.

Proposition 3. *Let A be an $n \times n$ matrix with entries in a commutative ring, and let B be the transpose of the matrix of cofactors (c_{ij}). Then*

$$BA = AB = \det(A)I_n.$$

Proposition 3 is used in our treatment of Noether normalization in Chapter 5, §6.

Appendix B
Pseudocode

Pseudocode is commonly used in mathematics and computer science to present algorithms. In this appendix, we will describe the pseudocode used in the text. If you have studied a programming language, you may see a similarity between our pseudocode and the language you studied. This is no accident, since programming languages are also designed to express algorithms. The syntax, or "grammatical rules," of our pseudocode will not be as rigid as that of a programming language since we do not require that it run on a computer. However, pseudocode serves much the same purpose as a programming language.

As indicated in the text, an algorithm is a specific set of instructions for performing a particular calculation with numerical or symbolic information. Algorithms have *inputs* (the information the algorithm will work with) and *outputs* (the information that the algorithm produces). At each step of an algorithm, the next operation to be performed must be completely determined by the current state of the algorithm. Finally, an algorithm must always terminate after a finite number of steps.

Whereas a simple algorithm may consist of a sequence of instructions to be performed one after the other, most algorithms also use the following special structures:

- Repetition structures, which allow a sequence of instructions to be repeated. These structures are also known as *loops*. The decision whether to repeat a group of instructions can be made in several ways, and our pseudocode includes different types of repetition structures adapted to different circumstances.
- Branching structures, which allow the possibility of performing different sequences of instructions under different circumstances that may arise as the algorithm is executed.

These structures, as well as the rest of the pseudocode, will be described in more detail in the following sections.

© Springer International Publishing Switzerland 2015
D.A. Cox et al., *Ideals, Varieties, and Algorithms*, Undergraduate Texts in Mathematics, DOI 10.1007/978-3-319-16721-3

§1   Inputs, Outputs, Variables, and Constants

We always specify the inputs and outputs of our algorithms on two lines before the start of the algorithm proper. The inputs and outputs are given by symbolic names in usual mathematical notation. Sometimes, we do not identify what *type* of information is represented by the inputs and outputs. In this case, their meaning should be clear from the context of the discussion preceding the algorithm. Variables (information stored for use during execution of the algorithm) are also identified by symbolic names. We freely introduce new variables in the course of an algorithm. Their types are determined by the context. For example, if a new variable called a appears in an instruction, and we set a equal to a polynomial, then a should be treated as a polynomial from that point on. Numerical constants are specified in usual mathematical notation. The two words true and false are used to represent the two possible truth values of an assertion.

§2   Assignment Statements

Since our algorithms are designed to describe mathematical operations, by far the most common type of instruction is the *assignment* instruction. The syntax is

$$<\text{variable}> := <\text{expression}>.$$

The symbol := is the assignment operator in many computer languages. The meaning of this instruction is as follows. First, we *evaluate* the expression of the right of the assignment operator, using the currently stored values for any variables that appear. Then the result is stored in the variable on the left-hand side. If there was a previously stored value in the variable on the left-hand side, the assignment *erases* it and *replaces* it with the computed value from the right-hand side. For example, if a variable called i has the numerical value 3, and we execute the instruction

$$i := i + 1,$$

the value $3 + 1 = 4$ is computed and stored in i. After the instruction is executed, i will contain the value 4.

§3   Looping Structures

Three different types of repetition structures are used in the algorithms given in the text. They are similar to the ones used in many languages. The most general and most frequently used repetition structure in our algorithms is the WHILE structure. The syntax is

$$\text{WHILE} <\text{condition}> \text{DO} <\text{action}>.$$

Here, <action> is a sequence of instructions. In a WHILE structure, the action is the group of statements to be repeated. We always indent this sequence of instructions. The *end* of the action is signaled by a return to the level of indentation used for the WHILE statement itself.

The <condition> after the WHILE is an assertion about the values of variables, etc., that is either true or false at each step of the algorithm. For instance, the condition

$$i \leq s \text{ AND } \textit{divisionoccurred} = \mathsf{false}$$

appears in a WHILE loop in the division algorithm from Chapter 2, §3.

When we reach a WHILE structure in the execution of an algorithm, we determine whether the condition is true or false. If it is true, then the action is performed once, and we go back and test the condition again. If it is still true, we repeat the action once again. Continuing in the same way, the action will be repeated as long as the condition remains true. When the condition becomes false (at some point during the execution of the action), that iteration of the action will be completed, and then the loop will terminate. To summarize, in a WHILE loop, the condition is tested *before* each repetition, and that condition must be true for the repetition to continue.

A second repetition structure that we use on occasion is the REPEAT structure. A REPEAT loop has the syntax

$$\text{REPEAT} <\text{action}> \text{UNTIL} <\text{condition}>.$$

Reading this as an English sentence indicates its meaning. Unlike the condition in a WHILE, the condition in a REPEAT loop tells us when to *stop*. In other words, the action will be repeated as long as the condition is false. In addition, the action of a REPEAT loop is always performed at least once since we only test the condition *after* doing the sequence of instructions representing the action. As with a WHILE structure, the instructions in the action are indented.

The final repetition structure that we use is the FOR structure. We use the syntax

$$\text{FOR each } s \text{ in } S \text{ DO} <\text{action}>$$

to represent the instruction: "perform the indicated action for each element $s \in S$." Here S is a finite set of objects and the action to be performed will usually depend on which s we are considering. The order in which the elements of S are considered is not important. Unlike the previous repetition structures, the FOR structure will necessarily cause the action to be performed a fixed number of times (namely, the number of elements in S).

§4   Branching Structures

We use only one type of branching structure, which is general enough for our purposes. The syntax is

IF <condition> THEN <action1> ELSE <action2>.

The meaning is as follows. If the condition is true at the time the IF is reached, action1 is performed (once only). Otherwise (that is, if the condition was false), action2 is performed (again, once only). The instructions in action1 and action2 are indented, and the ELSE separates the two sequences of instructions. The end of action2 is signaled by a return to the level of indentation used for the IF and ELSE statements.

In this branching structure, the truth or falsity of the condition selects which action to perform. In some cases, we *omit* the ELSE and action2, i.e.,

IF <condition> THEN <action1>.

This form is equivalent to

IF <condition> THEN <action1> ELSE <do nothing>.

§5   Output Statements

As already mentioned, the first two lines of our algorithms give its input and output. We also always include a RETURN statements with the syntax

RETURN <output of the algorithm>

to indicate precisely where the final output of the algorithm is returned. Most of the time, the RETURN statement is the last line of the algorithm.

Appendix C
Computer Algebra Systems

This appendix will discuss several computer algebra systems that can be used in conjunction with this book. Our comments here are addressed both to new users, including students, and to instructors considering which system might be most appropriate for use in a course. We will consider Maple, *Mathematica*, Sage, and the specialized systems CoCoA, Macaulay2, and Singular in detail. In their different ways, these are all amazingly powerful programs, and our brief discussions will not do justice to their true capabilities. Readers should not expect a complete, general introduction to any of these systems, instructions for downloading them, details about how those systems might be installed in computer labs, or what other software might be used in conjunction with them. Instead, we will assume that you already know or can access local or online documentation from the web sites indicated below concerning:

- How to enter and exit the program, save work and continue it in a later session.
- How to enter commands, execute them, and refer to the results generated by previous commands in an interactive session.
- If applicable, how to insert comments or other annotations in the worksheet or notebook interfaces provided in general-purpose systems such as Maple, *Mathematica*, and Sage.
- How to work with lists. For example, in the Gröbner basis command, the input contains a list of polynomials, and the output is another list which a Gröbner basis for the ideal generated by the polynomials in the input list. You should be able to find the length of a list and extract polynomials from a list.
- How to save results to an external file as text. This can be important, especially when output fills more than one computer screen. You should be able to save output in a file and examine it or print it out for further study.
- More advanced users will probably also want to know how to create and read in external files containing sequences of input commands or code for procedures.

For courses taught from this book with a laboratory component, we suggest that instructors may find that an efficient way to get students up to speed is to use a first lab meeting to cover aspects of the particular computer algebra being used.

© Springer International Publishing Switzerland 2015
D.A. Cox et al., *Ideals, Varieties, and Algorithms*, Undergraduate Texts
in Mathematics, DOI 10.1007/978-3-319-16721-3

§1   General Purpose Systems: Maple, *Mathematica*, Sage

The systems discussed in this section—Maple, *Mathematica*, and Sage—include components for computations of Gröbner bases and other operations on polynomial ideals. However, this is only a very small part of their functionality. They also include extensive facilities for numerical computation, for generating high quality graphics, and for many other types of symbolic computation. They also incorporate programming languages to automate multistep calculations and create new procedures. In addition, they have sophisticated notebook or worksheet user interfaces that can be used to generate interactive documents containing input commands, output and text annotations.

Maple

Maple is one of the leading commercial systems of this type. The Maplesoft web site `http://www.maplesoft.com` includes, among many other things, information about the different versions available and the full electronic documentation for the system. For us, the most important part of Maple is the `Groebner` package. Our discussion applies to the versions of this package distributed starting with Maple 11. As of summer 2014, the current version is MAPLE18 (2014).

To have access to the commands in the `Groebner` package, the command
`with(Groebner)`
must be executed before any of the following commands. (Note: In the current version of the worksheet interface, no input prompt is generated and commands need not be terminated with semicolons. The user can toggle back and forth between inert text for comments and input commands using "buttons" in the Maple window. If desired, a previous version of the interface can also be used. There each input line is marked with an input prompt [> and commands generating visible output are terminated with semicolons.) In any case, once the `Groebner` package is loaded, you can perform the division algorithm, compute Gröbner bases, and carry out a variety of other commands described below.

The definition of a monomial order in Maple always involves an explicit list of variables. All of the monomial orders discussed in Chapter 2 and Chapter 3 are provided, together with general mechanisms for specifying others. With $x > y > z$, for instance,

- lex order is specified by `plex(x,y,z)`,
- grlex order is specified by `grlex(x,y,z)` and
- grevlex order is specified by `tdeg(x,y,z)`.

The `lexdeg` command provides a mechanism for specifying elimination orders. For example, `lexdeg([x_1,...,x_k],[x_{k+1},...,x_n])` specifies an order that eliminates $x_1, \ldots, x_k$ leaving polynomials in $x_{k+1}, \ldots, x_n$. The idea is similar to (but not exactly the same as) the elimination orders discussed in Exercise 6 of Chapter 3, §1. Weight orders as in Exercise 10 of Chapter 2, §4, and general matrix orders are also available. The documentation for the `Groebner` package gives full information about the syntax of these declarations.

The basic commands in Maple's `Groebner` package are `NormalForm` for doing the division algorithm and `Basis`, for computing Gröbner bases. The syntax for the `NormalForm` command is

`NormalForm(f,polylist,order,options)`

where `f` is the dividend polynomial and `polylist` is the list of divisors. No options need be specified. The output is the remainder on division. If the quotients on division are required, then an option as follows:

`NormalForm(f,polylist,order,'Q')`

instructs Maple to return the list of quotients as the value of the variable `Q`. That list is not shown automatically, but can be seen by entering the variable name `Q` as a command on a subsequent line.

The syntax of the `Basis` command is similar:

`Basis(polylist,order,options)`

The output will be a reduced Gröbner basis (in the sense of Chapter 2, §7) except for clearing denominators. Optional inputs can be used to specify the algorithm used to compute the Gröbner basis (see the discussion at the end of Chapter 10, §4), to compute a transformation matrix giving the Gröbner basis polynomials as combinations of the inputs, or to specify the characteristic of the coefficient field. If no characteristic is specified, it is taken to be zero by default, and computations are done over $\mathbb{Q}$ with no limitation on the sizes of coefficients.

As an example of how this all works, consider the command

`gb := Basis([x^2 + y,2*x*y + y^2],plex(x,y))`

This computes a list which is a Gröbner basis for the ideal $\langle x^2 + y, 2xy + y^2 \rangle$ in $\mathbb{Q}[x, y]$ using lex order with $x > y$ and assigns it the symbolic name `gb`. With an additional option as follows:

`gb := Basis([x^2 + y,2*x*y + y^2],plex(x,y),characteristic=p)`

where p is a specific prime number, the computation is done in the ring of polynomials in x, y with coefficients in the finite field of integers modulo p. The same option works in `NormalForm` as well.

To tell Maple that a certain variable is in the base field (a "parameter"), simply omit it from the variable list in the monomial order specification. Thus

`Basis([v*x^2 + y,u*x*y + y^2],plex(x,y))`

will compute a Gröbner basis for $\langle vx^2 + y, uxy + y^2 \rangle$ in $\mathbb{Q}(u, v)[x, y]$ using lex order with $x > y$. In each case, the answer is reduced up to clearing denominators (so the leading coefficients of the Gröbner basis are polynomials in u and v).

The symbol `I` is a predefined constant in Maple, equal to the imaginary unit $i = \sqrt{-1}$. Computations of Gröbner bases over $\mathbb{Q}(i)$ can be done simply by including `I` at the appropriate places in the coefficients of the input polynomials. (This also explains why trying to use the name `I` for the list of generators of an ideal will cause an error.) Coefficients in other algebraic extensions of $\mathbb{Q}$ (or other base fields) can be included either by means of radical expressions or `RootOf` expressions. Thus, for instance, to include a $\sqrt{2}$ in a coefficient of a polynomial, we could simply enter the polynomial using `2^(1/2)` or `RootOf(u^2 - 2)` in the appropriate location. We refer the reader to Maple's documentation for the details.

Other useful commands in the `Groebner` package include

- `LeadingTerm, LeadingMonomial, LeadingCoefficient` which take as input a polynomial and a monomial order and return the indicated information. The names of these commands follow the terminology used in this text.
- `SPolynomial`, which computes the S-polynomial of two polynomials with respect to a monomial order. Note: the results here can differ from those in this text by constant factors since Maple does not divide by the leading coefficients.
- `IsProper`, which uses the consistency algorithm from Chapter 4, §1 to determine if a set of polynomial equations has a solution over an algebraically closed field.
- `IsZeroDimensional` which uses the finiteness algorithm from Chapter 5, §3 to determine if a system of polynomial equations has only a finite number of solutions over an algebraically closed field.
- `UnivariatePolynomial`, which given a variable and a set of generators for an ideal computes the polynomial of lowest degree in the variable which lies in the ideal (the generator of the corresponding elimination ideal).
- `HilbertPolynomial`, which computes the Hilbert polynomial of a homogeneous ideal as defined in Chapter 9. A related command `HilbertSeries` computes the Hilbert-Poincaré series for a homogeneous ideal defined in Chapter 10, §2. These commands are also defined for nonhomogeneous ideals and compute the corresponding asymptotic polynomial and generating function for the first difference of the affine Hilbert function.

We should mention that Maple also includes a `PolynomialIdeals` package containing a number of commands closely related to the content of this text and with functionality overlapping that of the `Groebner` package to some extent. The basic data structure for the `PolynomialIdeals` package is a polynomial ideal, defined as follows:

```
with(PolynomialIdeals)
J := <x^2 + y,2*x*y + y^2>
K := <x^3 + x*y - 1>
```

(The < and > from the keyboard act as $\langle$ and $\rangle$.) `PolynomialIdeals` contains implementations of the algorithms developed in Chapters 2–4 for ideal membership, radical membership, ideal containment, operations such as sums, products, intersections, ideal quotients and saturation, and primary decomposition. Much of this is based on the routines in the `Groebner` package, but the `PolynomialIdeals` package is set up so that users need only understand the higher-level descriptions of the ideal operations and not the underlying algorithms. For example,

```
Intersect(J,K)
```

computes the intersection of the two ideals defined above.

Mathematica

Mathematica is the other leading commercial system in this area. The web site http://www.wolfram.com/mathematica contains information about the different versions available and the full electronic documentation for the system. As of summer 2014, the current version is MATHEMATICA 10 (2014).

There is no special package to load in order to compute Gröbner bases: the basic commands are part of the *Mathematica* kernel. *Mathematica* knows most of the basic monomial orderings considered in Chapter 2. Lex order is called Lexicographic, grlex is called DegreeLexicographic and grevlex is called DegreeReverseLexicographic. The monomial order is specified by including a MonomialOrder option within the *Mathematica* commands described below. If you omit the MonomialOrder option, *Mathematica* will use lex as the default. *Mathematica* can also use weight orders as described in the comments at the end of the exercises to Chapter 2, §4.

Since a monomial order also depends on how the variables are ordered, *Mathematica* also needs to know a list of variables in order to specify the monomial order you want. For example, to tell *Mathematica* to use lex order with variables $x > y > z$, you would input {x,y,z} (*Mathematica* uses braces {...} for lists and square brackets [...] to delimit the inputs to a command or function).

For our purposes, important *Mathematica* commands are PolynomialReduce and GroebnerBasis. PolynomialReduce implements a variant of the division algorithm from Chapter 2 that does not necessarily respect the ordering of the list of divisors. This means that the quotients and the remainder may differ from the ones computed by our algorithm, though the quotients and remainders still satisfy the conditions of Theorem 3 of Chapter 2, §3. The syntax is as follows:

In[1]:= PolynomialReduce[f,polylist,varlist,options]

(The input prompt In[1]:= is generated automatically by *Mathematica*.) This computes quotients and a remainder of the polynomial f by the polynomials in polylist, using the monomial order specified by varlist and the optional MonomialOrder declaration. For example, to divide $x^3 + 3y^2$ by $x^2 + y$ and $2xy + y^2$ using grlex order with $x > y$, one would enter:

In[2]:= PolynomialReduce[x^3 + 3y^2,{x^2 + y,2xy + y^2},
        {x,y},MonomialOrder -> DegreeLexicographic]

The output is a list with two entries: the first is a list of the quotients and the second is the remainder.

The command for computing Gröbner bases has the following syntax:

In[3]:= GroebnerBasis[polylist,varlist,options]

This computes a Gröbner basis for the ideal generated by the polynomials in polylist with respect to the monomial order given by the MonomialOrder option with the variables ordered according to varlist. The answer is a reduced Gröbner basis (in the sense of Chapter 2, §7), except for clearing denominators. As an example of how GroebnerBasis works, consider

In[4]:= gb = GroebnerBasis[{x^2 +y,2xy + y^2},{x,y}]

The output is a list (assigned to the symbolic name gb) which is a Gröbner basis for the ideal $\langle x^2 + y, 2xy + y^2 \rangle \subseteq \mathbb{Q}[x, y]$ using lex order with $x > y$. We omitted the `MonomialOrder` option since lex is the default.

If you use polynomials with integer or rational coefficients in `GroebnerBasis` or `PolynomialReduce`, *Mathematica* will assume that you are working over the field $\mathbb{Q}$. There is no limitation on the size of the coefficients. Another possible coefficient field is the Gaussian rational numbers $\mathbb{Q}(i) = \{a + bi \mid a, b \in \mathbb{Q}\}$, where $i = \sqrt{-1}$ (*Mathematica* uses I to denote the imaginary unit). To compute a Gröbner basis over a finite field with p elements (where p is a prime number), you need to include the option `Modulus -> p` in the `GroebnerBasis` command. This option also works in `PolynomialReduce`.

Mathematica can also work with coefficients that lie in a rational function field. The strategy is that the variables in the base field (the "parameters") should be omitted from the variable list in the input, and then one sets the `CoefficientDomain` option to `RationalFunctions`. For example, the command:
```
In[5]:= GroebnerBasis[{v x^2 + y,u x y + y^2},{x,y},
        CoefficientDomain -> RationalFunctions]
```
will compute a Gröbner basis for $\langle vx^2 + y, uxy + y^2 \rangle \subseteq \mathbb{Q}(u, v)[x, y]$ using lex order with $x > y$. The `CoefficientDomain` option is also available for remainders using `PolynomialReduce`.

Here are some other useful *Mathematica* commands:

- `MonomialList`, which lists the terms of a polynomial according to the monomial order. Using this, `MonomialList[f,vars,monomialorder][[1]]` can be used to pick out the leading term.
- `Eliminate`, which uses the Elimination Theorem from Chapter 3, §1 to eliminate variables from a system of equations.
- `Solve`, which attempts to find all solutions of a system of equations.

Mathematica also allows some control over the algorithm used to produce a Gröbner basis. For instance, it is possible to specify that the Gröbner walk basis conversion algorithm mentioned in Chapter 10 will be used via the `Method` option in the `GroebnerBasis` command. Further descriptions and examples can be found in the electronic documentation at `http://www.wolfram.com/mathematica`.

Sage

Sage is a free and open-source computer algebra system under continual development since its initial release in 2005. As of summer 2014, the latest version is STEIN ET AL. (2014). The leader of the Sage project is William Stein of the University of Washington; hundreds of other mathematicians have contributed code and packages. The source code, as well as Linux and Mac OS X binary executables are available for download from the web site `http://sagemath.org`; a version of Sage that runs on Windows systems in combination with the VirtualBox operating system virtualization software is also available there.

The design of Sage is rather different from that of other packages we discuss in this appendix in that Sage has been built to provide a common front end for pre-existing open-source packages, including in particular the Singular system described in §2 below. Sage provides a browser-based notebook interface allowing users to create interactive documents. It also incorporates command constructs based on features of the Python programming language. It is intended to be comparable in power and scope to the commercial packages Maple and *Mathematica*. But it also has particular strengths in computational algebra and number theory because of its development history and the other packages it subsumes.

To do Gröbner basis computations in Sage, one must first define a polynomial ring that contains all polynomials involved and that specifies the monomial order to be used. For instance, to compute a Gröbner basis for the ideal $I = \langle x^2 + y, 2xy + y^2 \rangle \subseteq \mathbb{Q}[x, y]$ with respect to the lex order with $x > y$, we could proceed as follows. First define the ring with an input command like this:

```
R.<x,y> = PolynomialRing(QQ,order='lex')
```

(the QQ is Sage's built-in field of rational numbers and the order of the variables is determined by their ordering in the ring definition). There are many other equivalent ways to do this too; Sage's syntax is very flexible. Then the ideal I can be defined with the command

```
I = Ideal(x^2 + y,2*x*y + y^2)
```

and the Gröbner basis can be computed and displayed via the command

```
I.groebner_basis()
```

The syntax here is the same as that of the object-oriented features of Python. The last part of this, the `.groebner_basis()`, indicates that we are applying a function, or "method" defined for all objects that are ideals in polynomial rings that have been defined, and requiring no other input. In the notebook interface, pressing the TAB key on an input line with the partial command `I.` will generate a listing of all methods that can be applied to an object of the type of the object `I`. This can be helpful if you are unsure of the correct syntax for the command you want or what operations are permitted on an object.

Sage's polynomial ring definition mechanism is very general. The field of co-efficients can be any finitely generated extension of $\mathbb{Q}$ or a finite field. The finite field with p elements for a prime p is denoted `GF(p)`—this would replace the `QQ` in the definition of the polynomial ring if we wanted to work over that finite field instead. Algebraic elements, including the imaginary unit $i = \sqrt{-1}$, may be defined via the `NumberField` command, while fields of rational functions may be defined via the `FractionField` command. For example, suppose we wanted to work with polynomials in variables x, y with coefficients in the field $\mathbb{Q}(\sqrt{2})$. Here is one way to construct the ring we want. The first step, or something equivalent, is necessary to define the variable in the polynomial used in the definition of the number field; every object used in a Sage session must be defined in the context of some structure previously defined. The rt2 is the symbolic name for $\sqrt{2}$ in the field F:

```
R.<z> = PolynomialRing(QQ)
F.<rt2> = NumberField(z^2 - 2)
R.<x,y> = PolynomialRing(F)
```

Then we could define ideals in the ring R, compute Gröbner bases, etc. Computations over $\mathbb{Q}(i)$ can be done in the same way by using `NumberField` with the polynomial $z^2 + 1$ satisfied by $i = \sqrt{-1}$.

To define a rational function field $\mathbb{Q}(a)$ as the coefficient field, we may proceed like this:

```
P.<a> = PolynomialRing(F)
K = FractionField(P)
R.<x,y> = PolynomialRing(K)
```

These constructions can even be combined to produce fields such as $\mathbb{Q}(\sqrt{2})(a)$. However, it is important to realize that Sage makes use of the program Singular, whose coefficient fields are not quite this general. For ideals in polynomial rings with coefficients in a field like $\mathbb{Q}(\sqrt{2})(a)$, Sage falls back to its own slower Gröbner basis routines.

General monomial orders, including all of those discussed in the text, can be specified in Sage. We have seen how to specify lex orders above. Graded lex orders are obtained with `order = 'deglex'`, while graded reverse lex orders use `order = 'degrevlex'`. Various weight orders as in Exercise 10 of Chapter 2, §4, and general matrix orders are also available.

Most of the other basic operations we have discussed are provided in Sage as methods that are defined on ideals or individual polynomials. For instance, the leading term, leading monomial, and leading coefficient of a polynomial f are computed by `f.lt()`, `f.lm()`, and `f.lc()`, respectively. If G is a Gröbner basis for an ideal I and f is a polynomial in the ring containing I, then the remainder on division by G is computed by

```
f.reduce(G)
```

If desired, the list of quotients in the division can be recovered using

```
(f - f.reduce(G)).lift(G)
```

The computation of S-polynomials can be done by using an "educational" implementation of the basic Buchberger algorithm that is made accessible through the command

```
from sage.rings.polynomial.toy_buchberger import spol
```

Then to compute the S-polynomial of f and g, use `spol(f,g)`.

Other useful methods defined on ideals include:

- `.gens()`, which returns the list of generators of the ideal. One caution: As in Python, all lists are indexed starting from 0 rather than 1.
- `.elimination_ideal(varlist)`, where `varlist` is the list of variables to be eliminated,
- `.hilbert_polynomial()`, defined only for homogeneous ideals,
- `.hilbert_series()`, defined only for homogeneous ideals,
- `.primary_decomposition()`

All of the "real work" in these computations in Sage is being done by Singular, but comparing the form of the commands here with those in the section on Singular below shows that the Sage syntax is different. In effect, the Sage input is being translated for presentation to Singular, and results are passed back to Sage. In particular, the output of these commands are Sage objects and can be used by parts of the Sage system outside of Singular. It is also possible to communicate directly with Singular if that is desired. In our experience, though, the Python-esque Sage syntax is somewhat easier to work with for many purposes.

§2   Special Purpose Programs: CoCoA, Macaulay2, Singular

Unlike the systems discussed in §1, the programs discussed in this section were
developed primarily for use by researchers in algebraic geometry and commutative algebra. With some guidance, though, beginners can also make effective use
of them. These systems offer minimal numerical computation and graphics (at best)
and the current versions feature text-based user interfaces. They also tend to provide
direct access to less of the infrastructure of Gröbner basis computations such as S-
polynomials, remainders on division by general sets of divisors, and so on. But like
the general purpose programs, they incorporate complete programming languages
so it is possible to extend their basic functionality by creating new procedures. These
programs tend to be much more powerful within their limited application domain
since they include a number of more advanced algorithms and more sophisticated
higher-level constructs.

CoCoA

CoCoA (for "Computations in Commutative Algebra") is a free, open-source computer algebra system for polynomial computations. Versions for Unix systems, Mac
OS X, and Windows are available from `http://cocoa.dima.unige.it`. All
documentation is also posted there. As of summer 2014, the current standard version
is ABBOTT ET AL. (2014). CoCoA-5 involved a major redesign of many features of
the system and a new C++ library of underlying functions. This means that some aspects of the new system and its documentation are still under development and may
change in the future. Previous versions of CoCoA-4 are also available for a number of operating systems. The development of CoCoA is currently led by Lorenzo
Robbiano, John Abbott, Anna Bigatti, and Giovanni Lagorio, of the University of
Genoa in Italy. Many other current and former members of the CoCoA team have
participated in this effort. A large number of advanced algorithms and contributed
packages are provided.

To do Gröbner basis computations in CoCoA, one must first define a polynomial
ring that contains all polynomials involved and that specifies the monomial order to
be used. For instance, to compute a Gröbner basis for $I = \langle x^2 + y, 2xy + y^2 \rangle \subseteq \mathbb{Q}[x, y]$
with respect to the lex order with $x > y$, we could proceed as follows. First define
the ring with an input command like this:
`use R ::= QQ[x,y], Lex;`
(all CoCoA commands end with a semicolon and the double colon is correct for `use`
ring specifications). This sets the current ring (called R) to the polynomial ring with
coefficient field equal to the predefined field QQ (that is, the field of rational numbers
$\mathbb{Q}$). The ordering of the variables is determined by the order within the list.

Then the ideal I can be defined with the command
`I := ideal(x^2 + y,2*x*y + y^2);`
Unlike older versions of CoCoA, the new version does require explicit asterisks for
multiplication. The Gröbner basis can be computed and displayed via the command

```
GBasis(I);
```
If the results are to be saved and used later under a different name, an assignment
```
GB := GBasis(I);
```
could be used. A separate command `ReducedGBasis` is provided for reduced Gröbner bases.

   CoCoA's polynomial ring definition mechanism is quite flexible. Coefficients in
a field of rational functions over $\mathbb{Q}$ or over a finite field may also be defined as follows. For example, suppose we wanted to work with polynomials in variables x, y
with coefficients in the field $\mathbb{Q}(a)$. Here is one way to construct the ring we want:
```
K := NewFractionField(NewPolynomialRing(QQ,["a"]));
use R ::= K[x,y];
```
Note that this `use` did not specify a monomial order. The default is grevlex, which
can also be specified explicitly as `DegRevLex`. The grlex order is `DegLex`. Elimination orders are specified as `Elim(vars)`, where `vars` are the variables to be
eliminated (either a single variable or a range of variables in the list defining the
current ring), indicated like this:
```
use R ::= QQ[x,y,z,w], Elim(x..y);
```
This would define one of the elimination orders considered in Exercise 11 in Chapter 2, §4 eliminating x, y in favor of z, w. General monomial orders, including all of
the others discussed in the text, can be specified in CoCoA by means of matrices.

   Many of the other basic operations we have discussed are provided in CoCoA as
functions that are defined on ideals or individual polynomials. For instance, what we
call the leading term (including the coefficient) is `LM(f)` in CoCoA, while `LT(f)`
or `LPP(f)` compute the what we call the leading monomial. `LC(f)` is the leading
coefficient. The command `NF(f,I)` computes the remainder on division of a polynomial f with respect to a Gröbner basis for I. A Gröbner basis for the ideal will be
computed in order to carry this out if that has not been done previously.

   Other useful commands defined on ideals include:

- `intersect(I,J)` computes the intersection of the ideals I and J.
- `colon(I,J)` computes the quotient ideal $I:J$.
- `saturate(I,J)` computes the saturation $I:J^\infty$ of I with respect to J.
- `elim(X,I)` computes an elimination ideal, where X is a variable, range of variables, or list of variables to be eliminated.
- `HilbertFunction(R/I)` computes a representation of the Hilbert function of
  R/I. This is intended for homogeneous ideals. If I is not homogeneous, then the
  output is the Hilbert function of $\langle \text{LT}(I) \rangle$.
- `HilbertPoly(R/I)` computes a representation of the Hilbert polynomial of
  R/I. This behaves the same way as `HilbertPoly` if I is not a homogeneous
  ideal.
- `HilbertSeries(R/I)` gives a representation of the Hilbert-Poincaré series of
  R/I as defined in Chapter 10, §2. This is defined only for homogeneous ideals.
- `PrimaryDecomposition(I)` is available but is implemented only for square-
  free monomial ideals at the current time.

Macaulay2

Macaulay2 is a free, open-source computer algebra system for computations in algebraic geometry and commutative algebra. Versions for most Unix systems, Mac OS X, and Windows (the last running under the Cygwin operating system virtualization software) may be downloaded from the Macaulay2 web site http://www. math.uiuc.edu/Macaulay2. Complete electronic documentation is also available there. As of summer 2014, the current version is GRAYSON and STILLMAN (2013). Macaulay2 has been developed by Daniel Grayson of the University of Illinois and Michael Stillman of Cornell University; a number of packages extending the basic functionality of the system have been contributed by other mathematicians.

Macaulay2 has a special emphasis on computations in algebraic geometry, specifically computations of syzygies, free resolutions of modules over polynomial rings, and information about varieties that can be derived from those computations. Much of this is beyond the scope of this book, but all of it is based on the Gröbner basis computations we discuss and enough of that infrastructure is accessible to make Macaulay2 useful for courses based on this text.

To do Gröbner basis computations in Macaulay2, one must first define a polynomial ring that contains all the polynomials involved and the monomial order to be used. For instance, to compute a Gröbner basis for the ideal $I = \langle x^2 + y, 2xy + y^2 \rangle \subseteq \mathbb{Q}[x, y]$ with respect to the lex order with $x > y$, we could proceed as follows. First define the ring with an input command like this:

```
i1 : R = QQ[x,y,MonomialOrder=>Lex]
```
(the i1 represents the input prompt; executing this command will generate two lines of output labeled o1 showing the name R and its type PolynomialRing). The QQ is Macaulay2's notation for the field $\mathbb{Q}$. The order on the ring variables is specified by the ordering of the list in the square brackets.

The ideal I can be defined with the command
```
i2 : I = ideal(x^2 + y,2*x*y + y^2)
```
Then
```
i3 : gens gb I
```
computes the required Gröbner basis and presents the result as a matrix of polynomials with one row. There are many options that can be specified to control how the computation is performed and what algorithms are used.

The remainder on division of a polynomial f by a Gröbner basis for an ideal I is computed by
```
i4 : f % I
```
in Macaulay2. If a Gröbner basis for I has already been computed it is used, otherwise it is computed in order to find the unique remainder.

General monomial orders, including all of those discussed in the text, can be specified in Macaulay2. Grevlex orders are the default. We have seen how to specify lex orders above, while grlex orders are obtained with MonomialOrder=>GLex. The elimination orders from Exercise 11 in Chapter 2, §4 are specified like this: MonomialOrder=>Elimination n, where *n* is the number of variables to eliminate (from the start of the list). Various weight orders as in Exercise 11 of Chapter 2,

§4, and product orders are also available. These can be combined in very flexible ways giving orders equivalent to any matrix order.

Macaulay2 allows very general coefficient fields in polynomial rings. For instance, to define the ring of polynomials $\mathbb{Q}(u,v)[x,y]$ with coefficients in the field $\mathbb{Q}(u,v)$ and the grlex order with $x > y$ we could proceed as follows:

```
i5 : R = QQ[u,v]
i6 : K = frac(R)
```

(this computes the field of fractions of the ring R, that is the field $\mathbb{Q}(u,v)$ of rational functions in u,v). Then the ring we want is

```
i7 : S = K[x,y,MonomialOrder=>GLex]
```

The coefficient field of a polynomial ring can also be a finite field. Use ZZ/p for the field of integers modulo the prime p, for instance. Finite extensions of known fields can be specified like this. For instance suppose we wanted to use polynomials with coefficients in the field $\mathbb{Q}(\sqrt{2})$. We could use:

```
i8 : A = QQ[rt2]/(rt2^2 - 2)
```

(as the notation seems to indicate, this is a quotient ring of the polynomial ring in one variable modulo the ideal generated by the polynomial rt2^2 - 2). Then

```
i9 : L = toField(A)
```

"converts" this to a field that can be used as the coefficient field for a new polynomial ring. The field $\mathbb{Q}(i)$ could be defined in a similar way.

The leading term, leading monomial, and leading coefficient of a polynomial with respect to the current monomial order are computed by commands leadTerm, leadMonomial, and leadCoefficient. The leadTerm function can also be applied to an ideal I and the output will be a set of generators for the monomial ideal $\langle \mathrm{LT}(I) \rangle$.

If we have ideals I,J in the current ring, Macaulay2 allows us to compute the sum as I + J, the ideal product as I*J, the ideal quotient as quotient(I,J), the saturation as saturate(I,J), and the intersection as intersect(I,J). Other useful commands include:

- radical for the radical of an ideal.
- primaryDecomposition for primary decomposition.
- hilbertFunction(m,I) for a value of the Hilbert function.
- hilbertPolynomial(I,Projective=>false) gives the Hilbert polynomial in the form we have discussed.
- hilbertSeries computes the Hilbert-Poincaré series from Chapter 10, §2.

Singular

Singular is a free, open-source computer algebra system for polynomial computations. Versions for most Unix systems, Mac OS X, and Windows may be downloaded from http://www.singular.uni-kl.de. Complete documentation can also be found there. The version of Singular that runs on Windows systems uses the Cygwin operating system virtualization software. As of summer 2014, the current standard version is DECKER ET AL. (2012). The development of Singular has been directed by Wolfram Decker, Gert-Martin Greuel, Gerhard Pfister, and Hans Schönemann at the University of Kaiserslautern in Germany. Many other current and former members of the Singular team have also participated in this effort.

Singular has a special emphasis on commutative algebra, algebraic geometry, and singularity theory. It provides usable features for certain numerical computations, but not the same level of support for those areas or for graphics found in general-purpose packages. Singular's major strength is that it provides highly efficient implementations of its central algorithms (especially Gröbner basis computations in polynomial rings and standard basis computations in localizations, free resolutions, resultants, and so forth). A large number of advanced algorithms and contributed packages in the fields mentioned above, plus a procedural programming language with syntax similar to C are also provided. Interfaces with third-party software for convex geometry, tropical geometry and visualization, plus a comprehensive online manual and help resource are available.

Assignment statements in Singular generally indicate the type of the result (if that has not been previously specified), then a name for the result, an equals sign, and the command specifying the procedure used to compute the result.

To do Gröbner basis computations in Singular, one must first define a polynomial ring that contains all polynomials involved and that specifies the monomial order to be used. For instance, to compute a Gröbner basis of $I = \langle x^2 + y, 2xy + y^2 \rangle \subseteq \mathbb{Q}[x, y]$ with respect to lex order with $x > y$, we could proceed as follows. First define the ring with an input command like this:

```
> ring r = 0, (x,y), lp;
```

(the > represents the input prompt; all Singular commands end with a semicolon). This defines a ring called r with a coefficient field of characteristic zero (that is, the field $\mathbb{Q}$). The (x,y) is the list of ring variables; the lp indicates the lexicographic order (with variables ordered as in the list). For a polynomial ring over a finite field with p elements (p a prime), just change the 0 in the ring definition to the p desired. General monomial orders, including all of those discussed in the text, can be specified in Singular. We have seen how to specify lex orders. Graded lex orders are obtained with Dp, while grevlex orders use dp. Various weight orders as in Exercise 10 of Chapter 2, §4, and general matrix orders are also available.

Then the ideal I can be defined with the command

```
> ideal i = x2 + y,2xy + y2;
```

No special symbol is necessary to indicate the exponents, although the long form

```
> ideal i = x^2 + y,2*x*y + y^2;
```

is also recognized.

The Gröbner basis can be computed and displayed via the command

```
> groebner(i);
```

If the results are to be saved and used later under a different name, an assignment like

```
> ideal gi = groebner(i);
```

could be used. A command

```
> i = groebner(i);
```

could also be used if we wanted to overwrite the name i. No type specification is needed there since i has already been defined as an ideal in the ring r.

Singular's polynomial ring definition mechanism is quite flexible. The field of coefficients can be any finite extension of $\mathbb{Q}$ or a finite field. Coefficients in a field of rational functions over $\mathbb{Q}$ or over a prime field may also be defined in the `ring` command. For example, suppose we wanted to work with polynomials in variables x, y with coefficients in the field $\mathbb{Q}(\sqrt{2})$. Here is one way to construct the ring we want:

```
> ring r = (0,a), (x,y), lp;
> minpoly = a2 - 2;
```

Then the name a represents $\sqrt{2}$ in the coefficient field. By changing the `minpoly` declaration we could also use $\mathbb{Q}(i)$ as coefficient field. The `minpoly` declaration should come immediately after the ring definition; also Singular does not check for irreducibility so this should be done manually before using a field definition of this type. Without the `minpoly` declaration, we would have the rational function field $\mathbb{Q}(a)$ as the field of coefficients. Any number of such symbolic parameters can be defined by placing their names in the list with the characteristic of the coefficient field. One limitation here is that it is not possible to define a polynomial ring whose coefficient field is a rational function field over a finite extension of $\mathbb{Q}$.

Most of the other basic operations we have discussed are provided in Singular as functions that are defined on ideals or individual polynomials. For instance, the leading term, leading monomial, and leading coefficient of a polynomial f are computed by `lead(f)`, `leadmonom(f)`, and `leadcoef(f)` respectively. If G is a Gröbner basis for an ideal I and f is a polynomial in the ring containing I, then the remainder on division by G is computed by

```
> reduce(f,G);
```

If G is not a Gröbner basis, then a warning is generated since the remainder is not uniquely determined.

Other useful commands defined on ideals include:

- `intersect(I,J)` computes the intersection of the ideals I and J.
- `sat(I,J)` computes the saturation $I : J^\infty$ of I with respect to J.
- `eliminate(I,m)` computes an elimination ideal, where m is a monomial containing the variables to be eliminated.
- `finduni(I)` computes univariate polynomials in I for all the variables appearing, provided that I is zero-dimensional.
- `hilb(I)` gives a representation of the Hilbert-Poincaré series defined in Chapter 10, §2. This is defined only for homogeneous ideals.
- `hilbPoly(I)`, computes a representation of the Hilbert polynomial. This is part of an external package, so the command LIB "poly.lib"; must be entered before it is accessible. This is defined only for homogeneous ideals.
- Primary decompositions can be computed in several different ways using the functions in the `primdec.lib` library. We refer the interested reader to the Singular documentation or to the book GREUEL and PFISTER (2008).

§3 Other Systems and Packages

In addition to the computer algebra systems described in the previous sections, the following software may also be used for some or all the computations we have discussed.

- The REDUCE system with the Groebner and Cali packages described in previous editions of this book is still available at `http://reduce-algebra.com`.
- The computer algebra system Magma is designed for computations in commutative algebra, group theory, number theory, and combinatorics. It has a very efficient implementation of Gröbner basis algorithms and functionality comparable to that of CoCoA, Macaulay2, and Singular. More information, documentation, and a web-based Magma "calculator" for small computations can be found at `http://magma.maths.usyd.edu.au/magma`.
- For MATLAB users, the Symbolic Math Toolbox contains the Gröbner basis implementation from the MuPAD system. More information can be found at `http://www.mathworks.com/products/symbolic`.
- The `FGb` package incorporating the F_4 algorithm discussed in Chapter 10, §3 is now standard in Maple, and is available in a standalone version callable from C programs at `http://www-polsys.lip6.fr/~jcf/Software/FGb`.

Appendix D
Independent Projects

Unlike the rest of the book, this appendix is addressed to the instructor. We will discuss several ideas for research papers or projects supplementing topics introduced in the text.

§1 General Comments

Independent projects in a course based on this text can be valuable in several ways:

- They can help students to develop a deeper understanding of the ideas presented in the text by applying what they have learned.
- They can expose students to further developments in subjects beyond what is discussed in the text.
- They can give students more experience and sophistication as users of computer algebra systems.
- Projects can be excellent opportunities for small groups of two or three students to work together and learn collaboratively.
- More extensive and open-ended projects can even give students a taste of doing mathematical research.

There is much more material in our book than can be covered in a single semester. So a project could simply be to learn a part of the text that was not covered in class. In this appendix, though, we will concentrate on additional topics beyond what is in this book. In most cases students would need to start by reading from additional sources and learn the mathematics involved. In some cases, the focus might then be on implementing algorithms and computing examples. In others, the primary goal might be to write an expository paper and/or give an oral presentation about what they have learned.

The descriptions we give for each project area are rather brief. Although a few references are provided, most of the descriptions would need to be narrowed down and fleshed out before being given to students as assignments. The list is in no way

© Springer International Publishing Switzerland 2015
D.A. Cox et al., *Ideals, Varieties, and Algorithms*, Undergraduate Texts
in Mathematics, DOI 10.1007/978-3-319-16721-3

definitive or exhaustive, and users of the text are encouraged to contact the authors with comments or suggestions concerning these or other projects they have used.

§2   Suggested Projects

We discuss some ideas for more theoretical projects first, then indicate project topics where implementing algorithms in a computer algebra system might form a part of the project. Finally, we indicate some ideas for additional areas where the techniques we have discussed have been applied and that might form the basis for other projects.

1. **The Complexity of the Ideal Membership Problem**. In §10 of Chapter 2, we briefly discussed some of the worst-case complexity results concerning the computation of Gröbner bases and solving the ideal membership problem. The main purposes of this project would be to have students learn about the Mayr and Meyer examples, understand the *double exponential* growth of degree bounds for the ideal membership problem, and appreciate the implications for computational algebra. A suggested first reference here is BAYER and STILLMAN (1988). For larger projects taking the ideas in different directions toward the frontier of research, the following sources may be useful. The article KOH (1998) shows that similar double exponential behavior can be obtained even with ideals generated by polynomials of total degree 2. The article SWANSON (2004) studies the algebraic structure of the Mayr-Meyer ideals in much greater detail and includes a list of questions aimed at identifying the precise features producing their behavior. This involves quite a few topics not discussed in our text (embedded primes, etc.) but studying this in a more extensive project might be an interesting way to motivate learning those additional topics in commutative algebra. Finally, ASCHENBRENNER (2004) discusses these complexity questions for ideals in polynomial rings with coefficients in $\mathbb{Z}$ rather than a field.

2. **Symbolic Recipes for Solving Polynomial Systems**. One of the applications of computing Gröbner bases with respect to lex and other elimination orders discussed in the text is finding the points in $\mathbf{V}(I)$ for zero-dimensional ideals I. However, for larger and more realistic problems, the polynomials in a lex Gröbner basis can have awkwardly large coefficients and these can become problematic if standard numerical root-finding techniques are applied to generate approximations to the points in $\mathbf{V}(I)$. This is true especially in the higher dimensional analogs of the situation in Exercise 14 in Chapter 2, §7—systems that are often said to be in "shape lemma form." So other symbolic recipes for computing the solutions of these systems have been developed, all of which make heavy use of linear algebra in the quotient ring $\mathbb{C}[x_1, \ldots, x_n]/I$, a finite dimensional vector space over $\mathbb{C}$. Chapter 2 of COX, LITTLE and O'SHEA (2005), Chapter 2 of DICKENSTEIN and EMIRIS (2005), or Chapter 2 of COHEN et. al. (1999) present the background about multiplication matrices, eigenvalues, trace forms etc. that form the groundwork for these methods. Several different project topics could be generated from the material there. For example, one

project could simply be to learn how the linear algebra leads to the statement and proof of what is now often called Stickelberger's Theorem—the statement that the eigenvalues of the multiplication matrix for a polynomial f give the values of f at the points in $\mathbf{V}(I)$—Theorem (4.5) and Corollary (4.6) in Chapter 2, §4 of COX, LITTLE and O'SHEA (2005). Another project could deal with the application of these methods to *real* root counting and real root isolation for systems in several variables. Another very interesting project topic would be to investigate the idea of a *rational univariate representation* (RUR) for the solutions introduced in ROUILLIER (1999). An RUR expresses the coordinates of the points in $\mathbf{V}(I)$ as rational functions of the roots of an auxiliary polynomial equation where the variable is a so-called separating element—usually a linear combination of the coordinates—taking distinct values at the distinct points in $\mathbf{V}(I)$. The rational functions involved are typically significantly simpler than the polynomials in a lex Gröbner basis for the same ideal. Moreover the methods used to compute RUR's come from the same circle of ideas about multiplication matrices on $\mathbb{C}[x_1, \ldots, x_n]/I$, their traces, and so forth. The Maple `Groebner` package contains a `RationalUnivariateRepresentation` command that can be used to compute realistic examples.

3. **Gröbner Basis Conversion via FGLM**. In Chapter 10, we gave a version of Buchberger's algorithm that used Hilbert functions to convert a Gröbner basis with respect to one monomial order into a Gröbner basis for the same ideal with respect to another monomial order. We mentioned that there were other methods known for these Gröbner basis conversions, including the FGLM algorithm for zero-dimensional ideals. A number of different project topics could be developed in this area. The FGLM algorithm is also based on the vector space structure of the quotient $\mathbb{C}[x_1, \ldots, x_n]/I$ discussed in topic 2 above; the original source is FAUGÈRE, GIANNI, LAZARD, and MORA (1993). See also Chapter 2 of COX, LITTLE and O'SHEA (2005). This is now implemented in most of the computer algebra systems discussed in Appendix C but it is also a good programming exercise. Connections with the Buchberger-Möller algorithm for computing the vanishing ideal of a finite collection of points were developed in MARINARI, MÖLLER, and MORA (1993).

4. **Singular Points, Dual Curves, Evolutes, and other Geometric Applications**. The geometrical material on singular points of curves and envelopes of families of curves discussed in §4 of Chapter 3 could be extended in several different directions to give interesting project topics. A first topic might involve learning some of the theoretical tools needed for a more complete understanding of curve singularities: the Newton polygon, Puiseux expansions, resolution of singularities by quadratic transformations of the plane, etc. A good general reference for this is BRIESKORN and KNÖRRER (1986). Another beautiful and classical topic here would be to study the construction of the dual curve of a projective plane curve, finding the implicit equation of the dual by elimination, and perhaps discussing the Plücker formulas for curves with only nodal and ordinary cuspidal singularities; FISCHER (2001) and BRIESKORN and KNÖRRER (1986) are good sources for this. The envelope of the family of normal

lines to a plane curve is also known as the *evolute* of the curve, and these were studied intensively in classical differential and algebraic geometry, see for instance BRUCE and GIBLIN (1992). Evolutes also arise naturally in considering the critical points of the squared Euclidean distance function from a point to a curve. So they are closely connected to the question of finding the point on a curve closest to a given point—a typical constrained optimization problem. DRAISMA, HOROBEŢ, OTTAVIANI, STURMFELS, and THOMAS (2013) contains a beautiful discussion of the connection and introduces the *Euclidean distance degree* of a curve as a new invariant. That article also discusses farreaching generalizations to analogous higher-dimensional situations and discussions of many interesting applications of these ideas to areas such as geometric modeling, computer vision, and stability in control theory. Some of this is quite advanced, but this reference is a goldmine of interesting ideas.

5. **Implicitization via Resultants**. As mentioned in §6 of Chapter 3, resultants can be used for elimination of variables, and this means they are applicable to geometric problems such as implicitization. A nice project would be to report on the papers ANDERSON, GOLDMAN and SEDERBERG (1984a), ANDERSON, GOLDMAN, and SEDERBERG (1984b) and MANOCHA (1994).The resultants used in these papers differ from the resultants discussed in Chapter 3, where we defined the resultant of two polynomials. For implicitization, one needs the resultant of three or more polynomials, often called *multipolynomial resultants*. These resultants are discussed in COX, LITTLE and O'SHEA (2005). On a different but related note, GALLET, RAHKOOY and ZAFEIRAKOPOULOS (2013) discusses the general problem of the relation between a univariate generator of an elimination ideal and elements of the elimination ideal computed by means of resultants.

6. **The General Version of Wu's Method**. In our discussion of Wu's method in geometric theorem proving in Chapter 6, §4, we did not introduce the general algebraic techniques (characteristic sets, the Wu-Ritt decomposition algorithm) that are needed for a general theorem prover. This project would involve researching and presenting these methods, and possibly considering their relations with other methods for elimination of variables. Implementing them in a computer algebra system would also be a possibility. See WANG (2001) for a discussion of characteristic sets and CHOU (1988) and WU (2001) for complete presentations of the relations with geometric theorem-proving. The article WU (1983) gives a summary. Also, AUBRY, LAZARD and MORENO MAZA (1999) compares different theories of triangular sets of polynomial equations including characteristic sets and JIN, LI and WANG (2013) gives a new algorithmic scheme for computing characteristic sets.

7. **Molien's Theorem**. An interesting project could be built around Molien's theorem in invariant theory, which is mentioned in §3 of Chapter 7. This theorem gives an expression for the so-called Molien series of a finite matrix group G over $\mathbb{C}$ (that is, the generating function for the dimensions of the homogeneous components of the ring of invariants of G analogous to the Hilbert-Poincaré series studied in Chapter 10, §2):

$$\sum_{m=0}^{\infty} \dim(\mathbb{C}[x_1, \ldots, x_n]_m^G) t^m = \frac{1}{|G|} \sum_{g \in G} \frac{1}{\det(I - tg)}.$$

The algorithm given in STURMFELS (2008) can be used to find generators for $\mathbb{C}[x_1, \ldots, x_n]^G$. This can be used to find the invariants of some larger groups than those discussed in the text, such as the rotation group of the cube in $\mathbb{R}^3$. Molien's theorem is also discussed in Chapter 7 of BENSON and GROVE (1985) and Chapter 3 of DERKSEN and KEMPER (2002).

8. **Computer Graphics and Vision**. In §1 of Chapter 8, we used certain kinds of projections when we discussed how to draw a picture of a 3-dimensional object. These ideas are very important in computer graphics and computer vision. Simpler projects in this area could describe various projections that are commonly used in computer graphics and explain what they have to do with projective space. If you look at the formulas in Chapter 6 of FOLEY, VAN DAM, FEINER and HUGHES (1990), you will see certain 4×4 matrices. This is because points in $\mathbb{P}^3$ have four homogeneous coordinates. More extensive projects might also consider the *triangulation problem* in computer vision, which asks for a reconstruction of a 3-dimensional object from several 2-dimensional images produced by cameras viewing the object from different viewpoints. Techniques from algebraic geometry have been applied successfully to this question. The basic ideas are discussed in HEYDEN and ÅSTRÖM (1997) and the beautiful article AHOLT, STURMFELS and THOMAS (2013) studies the resulting multiview ideals and varieties using tools such as universal Gröbner bases, multigraded Hilbert functions and Hilbert schemes. Needless to say, much of this is beyond the scope of the topics discussed in this book but large parts of the article AHOLT, STURMFELS and THOMAS (2013) will be accessible because the presentation is very concrete and smaller special cases can be computed explicitly. There are also connections with the article DRAISMA, HOROBEȚ, OTTAVIANI, STURMFELS, and THOMAS (2013) mentioned in topic 4 above in the case that the images are "noisy" and no exact reconstruction exists. In that case, the problem is to determine a 3-dimensional structure that comes as close as possible to matching what is seen in the 2-dimensional images.

9. **Gröbner Fans, Universal Gröbner Bases, Gröbner Basis Conversion via the Gröbner walk**. How many different reduced Gröbner bases are there for any particular ideal? Is that collection finite or infinite? The so-called *Gröbner fan* of an ideal is a collection of polyhedral cones in $\mathbb{R}^n$ that provides a way to see that there are only finitely many different reduced Gröbner bases. Understanding the Gröbner fan also gives a way to produce *universal Gröbner bases* for ideals—finite collections of polynomials that are simultaneously Gröbner bases for all possible monomial orderings. All of this is discussed, for instance, in Chapter 8 of COX, LITTLE and O'SHEA (2005) and STURMFELS (1996). One project topic would be simply to understand how all this works and possibly to generate some examples. The software package gfan authored by A. Jensen is the current standard for these calculations and the Sage, Macaulay2, and Singular systems discussed in Appendix C incorporate interfaces to gfan.

The structure of the Gröbner fan also gives the background needed for the other Gröbner basis conversion method that we mentioned in passing in Chapter 10, the so-called Gröbner walk algorithm. The original source for the Gröbner walk is COLLART, KALKBRENER and MALL (1998) and this algorithm is discussed in Section 5 of Chapter 8 in COX, LITTLE and O'SHEA (2005); more efficient versions such as the so-called fractal walk have been developed as well. Versions of the walk have been implemented in several of the computer algebra systems mentioned in Appendix C, including Maple, *Mathematica*, Singular (hence Sage), and Magma.

10. **Gröbner Covers**. As we have seen in Chapter 6, many systems of polynomial equations that arise in applications naturally contain symbolic *parameters* appearing in their coefficients. Understanding how and whether specializing those parameters to particular constant values changes the form of a Gröbner basis for the corresponding ideal and affects the number and form of the solutions of the system is often extremely important. Weispfenning's theory of *comprehensive Gröbner bases* and his algorithm for computing them was the first major step here. WEISPFENNING (1992) is the original source for this; BECKER and WEISPFENNING (1993) gives a very brief description. More recently, the theory of *Gröbner covers* presented in MONTES and WIBMER (2010) has provided a way to find a simpler decomposition of the parameter space into segments on which the Gröbner bases of specializations have a constant "shape." Projects in this area could have a theoretical orientation, or could focus on implementation. As of July 2014 a package for the Singular computer algebra system is under development (see http://www-ma2.upc.edu/montes/). Another possibility for a larger project would be to include the article MONTES and RECIO (2014), which uses Gröbner covers to discover "missing hypotheses" for automatic *discovery* of theorems in elementary geometry, an extension of the automatic theorem proving considered in Chapter 6.

11. **Gröbner Bases for Modules and Applications**. The notion of an ideal $I \subseteq k[x_1, \ldots, x_n]$ can be generalized to a *submodule* $M \subseteq k[x_1, \ldots, x_n]^r$ and there is a natural way to define term orders (touched on briefly in Chapter 10, §4) and Gröbner bases for modules. The basic definitions can be found in ADAMS and LOUSTAUNAU (1994), BECKER and WEISPFENNING (1993), COX, LITTLE and O'SHEA (2005), KREUZER and ROBBIANO (2000), and EISENBUD (1999). Indeed, even the theory of Gröbner bases for ideals naturally involves modules such as the module of syzygies on a set of generators for an ideal or their leading terms, so KREUZER and ROBBIANO (2000) develops the theory for ideals and for modules simultaneously. One possible project here would be to understand how this all works and how Buchberger's algorithm generalizes to this setting. This is implemented, for example, in CoCoA, Macaulay2, Sage and Singular and it can be emulated via a standard trick in Maple [the idea is discussed in Exercise 6 of Chapter 2, Section 5 in COX, LITTLE and O'SHEA (2005)]. Another project topic building on this would be the application of modules to the construction of multivariate polynomial *splines*—piecewise polynomial functions of a given degree on a given polyhedral decomposition of a region in $\mathbb{R}^n$ with

a given degree of smoothness. This is discussed in Chapter 8 of COX, LITTLE and O'SHEA (2005) and the sources cited there. Other applications that might be considered in a project include methods for multivariate Padé approximation [see FARR and GAO (2006) for the latest work on this] and related decoding algorithms for certain error control codes [see Chapter 9 in COX, LITTLE and O'SHEA (2005)].

12. **Border Bases**. As we know from Chapter 5, if I is a zero-dimensional ideal, the monomials in the complement of $\langle \mathrm{LT}(I) \rangle$ form a vector space basis for $k[x_1, \ldots, x_n]/I$ and linear algebra in those quotient rings has appeared in several of the project topics listed above. From work of Stetter, Möller and Mourrain, it is known there are other ways to find good monomial bases for $k[x_1, \ldots, x_n]/I$ yielding normal forms modulo I and special sets of generators for I that are different from the corresponding information obtained from any Gröbner basis. Moreover, some of these alternatives yield representations of $k[x_1, \ldots, x_n]/I$ that make it easier to compute good numerical approximations for the points in $\mathbf{V}(I)$. There is now a well-developed algebraic theory of border bases for I that parallels Gröbner basis theory, but with some interesting twists. Several interesting project topics here might involve presenting this theory or implementing border division and normal forms in a computer algebra system. Chapter 4 of DICKENSTEIN and EMIRIS (2005)—by Kehrein, Kreuzer, and Robbiano—contains an excellent summary of this theory.

13. **Algebraic Statistics**. The rapidly developing field of algebraic statistics is based on the idea that many statistical models (i.e., families of probability distributions) for discrete data can be seen as algebraic varieties. Moreover the geometry of those varieties determines the behavior of parameter estimation and statistical inference procedures. A typical example is the family of binomial distributions. The probability that a binomial random variable X (based on n trials with success probability θ on each trial) takes value $k \in \{0, 1, \ldots, n\}$ is

$$p_k = P(X = k) = \binom{n}{k} \theta^k (1 - \theta)^{n-k}.$$

Viewing these as components of a curve parametrized by real θ satisfying $0 \leq \theta \leq 1$, we have a subset of the real points of a rescaled rational normal curve of degree n lying in the hyperplane defined by the equation $p_0 + \cdots + p_n = 1$. Given some number of observations we might want to estimate θ using maximum likelihood estimation, and this leads to a constrained optimization problem involving polynomial equations. A good introduction to the basics of model construction and experimental design can be found in PISTONE, RICCOMAGNO, and WYNN (2001). A discussion of algebraic techniques for maximum likelihood estimation appears in Chapter 2 of DRTON, STURMFELS, and SULLIVANT (2009). One of the main applications of these ideas so far has been in genomics. For students with the requisite background, the Jukes-Cantor models studied in Part I of PACHTER and STURMFELS (2005) could form the basis of a more extensive project. A different sort of application to design of

experiments can be found in Chapter 4 by Kehrein, Kreuzer, and Robbiano in
DICKENSTEIN and EMIRIS (2005). This draws on the material on border bases
discussed in the previous topic description.

14. **Graph Coloring Problems and Sudoku.** The final project idea we will pro-
pose involves the use of polynomials and varieties to study the solution of various graph coloring problems and related questions. The first discussion of this
connection that we know of appears in BAYER (1982), which uses polynomial
methods to solve the three-coloring problem for graphs. Section 2.7 of ADAMS
and LOUSTAUNAU (1994) contains a discussion of this as well. More recently,
a number of authors have presented applications of these ideas to the popu-
lar Sudoku and similar puzzles. ARNOLD, LUCAS and TAALMAN (2010) discusses different polynomial translations focusing on a 4×4 version of the usual
Sudoku. Chapter 3 of DECKER and PFISTER (2013) presents one particular
polynomial translation and gives Singular procedures for generating the relevant polynomial ideals and solving standard Sudoku puzzles. Several different
sorts of projects would be possible here from more theoretical discussions to
implementations of one or more approaches, comparisons between them, and
so on.

There are many other places where instructors can look for potential project topics for students, including the following:

- COX, LITTLE and O'SHEA (2005) includes material on local rings, additional
  topics in algebraic coding theory, and applications to combinatorial enumeration
  problems and integer programming that could serve as the basis for projects.

- ADAMS and LOUSTAUNAU (1994) contains sections on minimal polynomials
  of field extensions and integer programming. These could serve as the basis for
  interesting projects.

- EISENBUD (1999) has a list of seven projects in section 15.12. These are more
  sophisticated and require more background in commutative algebra, but they also
  introduce the student to some topics of current interest in algebraic geometry.

- KREUZER and ROBBIANO (2000) and KREUZER and ROBBIANO (2005) contain
  a large number of *tutorials* on various topics (usually at least one at the end of
  each section). These would be especially good for smaller-scale projects where
  the path to be followed by the student would be laid out in detail at the start.

If you find good student projects different from those listed above, we would be
interested in hearing about them. There are a *lot* of wonderful things one can do with
Gröbner bases and algebraic geometry, and the projects described in this appendix
barely scratch the surface.

References

J. Abbott, A. Bigatti, G. Lagorio, *CoCoA-5: A System for Doing Computations in Commutative Algebra* (2014), available at http://cocoa.dima.unige.it

W. Adams, P. Loustaunau, *An Introduction to Gröbner Bases*. Graduate Studies in Mathematics, vol. 3 (AMS, Providence, 1994)

C. Aholt, B. Sturmfels, R. Thomas, A Hilbert scheme in computer vision. Can. J. Math. **65**, 961–988 (2013)

D. Anderson, R. Goldman, T. Sederberg, Implicit representation of parametric *curves and surfaces*. Comput. Vis. Graph. Image Des. **28**, 72–84 (1984a)

D. Anderson, R. Goldman, T. Sederberg, Vector elimination: a technique for the implicitization, inversion and intersection of planar parametric rational polynomial curves. Comput. Aided Geom. Des. **1**, 327–356 (1984b)

E. Arnold, S. Lucas, L. Taalman, Gröbner basis representations of sudoku. Coll. Math. J. **41**, 101–112 (2010)

M. Aschenbrenner, Ideal membership in polynomial rings over the integers. J. Am. Math. Soc. **17**, 407–441 (2004)

M.F. Atiyah, I.G. MacDonald, *Introduction to Commutative Algebra* (Addison-Wesley, Reading, MA, 1969)

P. Aubry, D. Lazard, M. Moreno Maza, On the theories of triangular sets. J. Symb. Comput. **28**, 105–124 (1999)

J. Baillieul et al., Robotics. In: Proceedings of Symposia in Applied Mathematics, vol. 41 (American Mathematical Society, Providence, Rhode Island, 1990)

A.A. Ball, *The Parametric Representation of Curves and Surfaces Using Rational Polynomial Functions*, in *The Mathematics of Surfaces, II*, ed. by R.R. Martin (Clarendon Press, Oxford, 1987), pp. 39–61

D. Bayer, The division algorithm and the Hilbert scheme, Ph.D. thesis, Harvard University, 1982

D. Bayer, D. Mumford, *What Can Be Computed in Algebraic Geometry?*, in *Computational Algebraic Geometry and Commutative Algebra*, ed. by D. Eisenbud, L. Robbiano (Cambridge University Press, Cambridge, 1993), pp. 1–48

D. Bayer, M. Stillman, A criterion for detecting m-regularity. Invent. Math. **87**, 1–11 (1987a)

© Springer International Publishing Switzerland 2015
D.A. Cox et al., *Ideals, Varieties, and Algorithms*, Undergraduate Texts in Mathematics, DOI 10.1007/978-3-319-16721-3

D. Bayer, M. Stillman, A theorem on refining division orders by the reverse lexicographic order. Duke J. Math. **55**, 321–328 (1987b)

D. Bayer, M. Stillman, *On the Complexity of Computing Syzygies*, in *Computational Aspects of Commutative Algebra*, ed. by L. Robbiano (Academic Press, New York, 1988), pp. 1–13

T. Becker, V. Weispfenning, *Gröbner Bases* (Springer, New York-Berlin-Heidelberg, 1993)

C.T. Benson, L.C. Grove, *Finite Reflection Groups*, 2nd edn. (Springer, New York-Berlin-Heidelberg, 1985)

A. Bigatti, Computation of Hilbert-Poincaré series. J. Pure Appl. Algebra **119**, 237–253 (1997)

E. Brieskorn, H. Knörrer, *Plane Algebraic Curves* (Birkhäuser, Basel-Boston-Stuttgart, 1986)

J.W. Bruce, P.J. Giblin, *Curves and Singularities*, 2nd edn. (Cambridge University Press, Cambridge, 1992)

B. Buchberger, Ein algorithmus zum auffinden der basiselemente des restklassenrings nach einem nulldimensionalen polynomideal, Doctoral Thesis, Mathematical Institute, University of Innsbruck, 1965. English translation An algorithm for finding the basis elements of the residue class ring of a zero dimensional polynomial ideal by M.P. Abramson, J. Symb. Comput. **41**, 475–511 (2006)

B. Buchberger, *Groebner Bases: An Algorithmic Method in Polynomial Ideal Theory*, in *Multidimensional Systems Theory*, ed. by N.K. Bose (D. Reidel Publishing, Dordrecht, 1985), pp. 184–232

M. Caboara, J. Perry, Reducing the size and number of linear programs in a dynamic Gröbner basis algorithm. Appl. Algebra Eng. Comm. Comput. **25**, 99–117 (2014)

J. Canny, D. Manocha, Algorithm for implicitizing rational parametric surfaces. Comput. Aided Geom. Des. **9**, 25–50 (1992)

S.-C. Chou, *Mechanical Geometry Theorem Proving* (D. Reidel Publishing, Dordrecht, 1988)

H. Clemens, *A Scrapbook of Complex Curve Theory*, 2nd edn. (American Mathematical Society, Providence, Rhode Island, 2002)

A. Cohen, H. Cuypers, H. Sterk (eds.), *Some Tapas of Computer Algebra* (Springer, Berlin-Heidelberg-New York, 1999)

S. Collart, M. Kalkbrener, D. Mall, Converting bases with the Gröbner walk. J. Symb. Comput. **24**, 465–469 (1998)

D. Cox, J. Little, D. O'Shea, *Using Algebraic Geometry*, 2nd edn. (Springer, New York, 2005)

H.S.M. Coxeter, *Regular Polytopes*, 3rd edn. (Dover, New York, 1973)

J.H. Davenport, Y. Siret, E. Tournier, *Computer Algebra*, 2nd edn. (Academic, New York, 1993)

W. Decker, G.-M. Greuel, G. Pfister, H. Schönemann, *Singular 3-1-6—A* computer algebra system for polynomial computations (2012), available at `http://www.singular.uni-kl.de`

W. Decker, G. Pfister, *A First Course in Computational Algebraic Geometry*. AIMS Library Series (Cambridge University Press, Cambridge, 2013)

H. Derksen, G. Kemper, *Computational Invariant Theory* (Springer, Berlin-Heidelberg-New York, 2002)

A. Dickenstein, I. Emiris (eds.), *Solving Polynomial Equations* (Springer, Berlin-Heidelberg-New York, 2005)

J. Draisma, E. Horobeţ, G. Ottaviani, B. Sturmfels, R. Thomas, *The Euclidean distance degree of an algebraic variety* (2013). arXiv:1309.0049 [math.AG]

M. Drton, B. Sturmfels, S. Sullivant, *Lectures on Algebraic Statistics.* Oberwohlfach Mathematical Seminars, vol. 39 (Birkhäuser, Basel-Boston-Berlin, 2009)

T.W. Dubé, The structure of polynomial ideals and Gröbner bases. SIAM J. Comput. **19**, 750–775 (1990)

D. Dummit, R. Foote, *Abstract Algebra*, 3rd edn. (Wiley, New York, 2004)

C. Eder, J. Faugère, *A survey on signature-based Gröbner basis algorithms* (2014). arXiv:1404.1774 [math.AC]

D. Eisenbud, *Commutative Algebra with a View Toward Algebraic Geometry*, 3d corrected printing (Springer, New York-Berlin-Heidelberg, 1999)

D. Eisenbud, C. Huneke, W. Vasconcelos, Direct methods for primary decomposition. Invent. Math. **110**, 207–235 (1992)

J. Farr, S. Gao, Gröbner bases and generalized Padé approximation. Math. Comp. **75**, 461–473 (2006)

J. Faugère, A new efficient algorithm for computing Gröbner bases (F4). J. Pure Appl. Algebra **139**, 61–88 (1999)

J. Faugère, *Finding All the Solutions of Cyclic 9 Using Gröbner Basis Techniques*, in *Computer Mathematics (Matsuyama, 2001)*, Lecture Notes Ser. Comput., vol. 9 (World Scientific, River Edge, NJ, 2001), pp. 1–12

J. Faugère, A new efficient algorithm for computing Gröbner bases without reduction to zero F5. In: *Proceedings of ISSAC'02, Villeneuve d'Ascq, France*, July 2002, 15–82; revised version from http://www-polsys.lip6.fr/~jcf/Publications/index.html

J. Faugère, P. Gianni, D. Lazard, T. Mora, Efficient change of ordering for Gröbner bases of zero-dimensional ideals. J. Symb. Comput. **16**, 329–344 (1993)

G. Fischer, *Plane Algebraic Curves* (AMS, Providence, Rhode Island, 2001)

J. Foley, A. van Dam, S. Feiner, J. Hughes, *Computer Graphics: Principles and Practice*, 2nd edn. (Addison-Wesley, Reading, MA, 1990)

W. Fulton, *Algebraic Curves* (W. A. Benjamin, New York, 1969)

M. Gallet, H. Rahkooy, Z. Zafeirakopoulos, *On Computing the Elimination Ideal Using Resultants with Applications to Gröbner Bases* (2013). arXiv:1307.5330 [math.AC]

J. von zur Gathen, J. Gerhard, *Modern Computer Algebra*, 3rd edn. (Cambridge University Press, Cambridge, 2013)

C.F. Gauss, *Werke*, vol. III (Königlichen Gesellschaft der Wissenschaften zu Göttingen, Göttingen, 1876)

R. Gebauer, H.M. Möller, *On an Installation of Buchberger's Algorithm*, in *Computational Aspects of Commutative Algebra*, ed. by L. Robbiano (Academic Press, New York, 1988), pp. 141–152

I. Gelfand, M. Kapranov, A. Zelevinsky, *Discriminants, Resultants and Multidimensional Determinants* (Birkhäuser, Boston, 1994)

P. Gianni, B. Trager, G. Zacharias, *Gröbner bases and primary decomposition of polynomial ideals*, in *Computational Aspects of Commutative Algebra*, ed. by L. Robbiano (Academic Press, New York, 1988), pp. 15–33

A. Giovini, T. Mora, G. Niesi, L. Robbiano, C. Traverso, *"One sugar cube, please," or Selection Strategies in the Buchberger Algorithm, in ISSAC 1991, Proceedings of the 1991 International Symposium on Symbolic and Algebraic Computation*, ed. by S. Watt (ACM Press, New York, 1991), pp. 49–54

M. Giusti, J. Heintz, *La détermination des points isolés et de la dimension d'une variété algébrique peut se faire en temps polynomial*, in *Computational Algebraic Geometry and Commutative Algebra*, ed. by D. Eisenbud, L. Robbiano (Cambridge University Press, Cambridge, 1993), pp. 216–256

L. Glebsky, *A proof of Hilbert's Nullstellensatz Based on Groebner bases* (2012). arXiv:1204.3128 [math.AC]

R. Goldman, *Pyramid Algorithms: A Dynamic Programming Approach to Curves and Surfaces in Geometric Modeling* (Morgan Kaufman, Amsterdam, Boston, 2003)

D. Grayson, M. Stillman, *Macaulay2, a Software System for Research* (2013), version 1.6, available at http://www.math.uiuc.edu/Macaulay2/

G.-M. Greuel, G. Pfister, *A Singular Introduction to Commutative Algebra*, 2nd edn. (Springer, New York, 2008)

P. Griffiths, *Introduction to Algebraic Curves*. Translations of Mathematical Monographs, vol. 76 (AMS, Providence, 1989)

P. Gritzmann, B. Sturmfels, Minkowski addition of polytopes: computational complexity and applications to Gröbner bases. SIAM J. Discrete Math. **6**, 246–269 (1993)

J. Harris, *Algebraic Geometry, A First Course*, corrected edition (Springer, New York, 1995)

R. Hartshorne, *Algebraic Geometry* (Springer, New York, 1977)

G. Hermann, *Die Frage der endlich vielen schritte in der theorie der polynomideale*, Math. Ann. **95**, 736–788 (1926)

A. Heyden, K. Åström, Algebraic properties of multilinear constraints. Math. Methods Appl. Sci. **20**, 1135–1162 (1997)

D. Hilbert, *Über die Theorie der algebraischen Formen*, Math. Ann. **36**, 473–534 (1890). Reprinted in *Gesammelte Abhandlungen*, vol. II (Chelsea, New York, 1965)

D. Hilbert, *Theory of Algebraic Invariants* (Cambridge University Press, Cambridge, 1993)

J. Hilmar, C. Smyth, Euclid meets Bézout: intersecting algebraic plane curves with the Euclidean algorithm. Am. Math. Monthly **117**, 250–260 (2010)

H. Hironaka, Resolution of singularities of an algebraic variety over a field of characteristic zero I, II. Ann. Math. **79**, 109–203, 205–326 (1964)

W.V.D. Hodge, D. Pedoe, *Methods of Algebraic Geometry*, vol. I and II (Cambridge University Press, Cambridge, 1968)

M. Jin, X. Li, D. Wang, A new algorithmic scheme for computing characteristic sets. J. Symb. Comput. **50**, 431–449 (2013)

J. Jouanolou, Le formalisme du résultant. Adv. Math. **90**, 117–263 (1991)

M. Kalkbrener, *Implicitization by Using Gröbner Bases*, Technical Report RISC-Series 90-27 (University of Linz, Austria, 1990)

K. Kendig, *Elementary Algebraic Geometry*, 2nd edn. (Dover, New York, 2015)

F. Kirwan, *Complex Algebraic Curves*. London Mathematical Society Student Texts, vol. 23 (Cambridge University Press, Cambridge, 1992)

F. Klein, *Vorlesungen über das Ikosaeder und die Auflösung der Gleichungen vom Fünften Grade* (Teubner, Leipzig, 1884). English Translation, *Lectures on the Ikosahedron and the Solution of Equations of the Fifth Degree* (Trubner, London, 1888). Reprinted by Dover, New York (1956)

J. Koh, Ideals generated by quadrics exhibiting double exponential degrees. J. Algebra **200**, 225–245 (1998)

M. Kreuzer, L. Robbiano, *Computational Commutative Algebra*, vol. 1 (Springer, New York, 2000)

M. Kreuzer, L. Robbiano, *Computational Commutative Algebra*, vol. 2 (Springer, New York, 2005)

T. Krick, A. Logar, *An Algorithm for the Computation of the Radical of an Ideal in the Ring of Polynomials*, in *Applied Algebra, Algebraic Algorithms and Error-Correcting Codes*, ed. by H.F. Mattson, T. Mora, T.R.N. Rao. Lecture Notes in Computer Science, vol. 539 (Springer, Berlin, 1991), pp. 195–205

D. Lazard, *Gröbner Bases, Gaussian Elimination and Resolution of Systems of Algebraic Equations*, in *Computer Algebra: EUROCAL 83*, ed. by J.A. van Hulzen. Lecture Notes in Computer Science, vol. 162 (Springer, Berlin, 1983), pp. 146–156

D. Lazard, *Systems of Algebraic Equations (Algorithms and Complexity)*, in *Computational Algebraic Geometry and Commutative Algebra*, ed. by D. Eisenbud, L. Robbiano (Cambridge University Press, Cambridge, 1993), pp. 84–105

M. Lejeune-Jalabert, *Effectivité des calculs polynomiaux*, Cours de DEA 1984–85, Institut Fourier, Université de Grenoble I (1985)

F. Macaulay, On some formulæin elimination. Proc. Lond. Math. Soc. **3**, 3–27 (1902)

D. Manocha, Solving systems of polynomial equations. IEEE Comput. Graph. Appl. **14**, 46–55 (1994)

Maple 18, Maplesoft, a division of Waterloo Maple Inc., Waterloo, Ontario (2014). http://www.maplesoft.com

M. Marinari, H. Möller, T. Mora, Gröbner bases of ideals defined by functionals with an application to ideals of projective points. Appl. Algebra Eng. Comm. Comput. **4**, 103–145 (1993)

Mathematica 10, Wolfram Research, Inc., Champaign, Illinois (2014). http://www.wolfram.com/mathematica

H. Matsumura, *Commutative Ring Theory* (Cambridge University Press, Cambridge, 1989)

E. Mayr, A. Meyer, The complexity of the word problem for commutative semi-groups and polynomial ideals. Adv. Math. **46**, 305–329 (1982)

R. Mines, F. Richman, W. Ruitenburg, *A Course in Constructive Algebra* (Springer, New York-Berlin-Heidelberg, 1988)

B. Mishra, *Algorithmic Algebra*. Texts and Monographs in Computer Science (Springer, New York-Berlin-Heidelberg, 1993)

H.M. Möller, F. Mora, *Upper and Lower Bounds for the Degree of Groebner Bases*, in *EUROSAM 1984*, ed. by J. Fitch. Lecture Notes in Computer Science, vol. 174 (Springer, New York-Berlin-Heidelberg, 1984), pp. 172–183

A. Montes, T. Recio, Generalizing the Steiner–Lehmus theorem using the Gröbner cover. Math. Comput. Simul. **104**, 67–81 (2014)

A. Montes, M. Wibmer, Gröbner bases for polynomial systems with parameters. J. Symb. Comput. **45**, 1391–1425 (2010)

D. Mumford, *Algebraic Geometry I: Complex Projective Varieties*, cCorrected 2nd printing (Springer, New York-Berlin-Heidelberg, 1981)

L. Pachter, B. Sturmfels (eds.), *Algebraic Statistics for Computational Biology* (Cambridge University Press, Cambridge, 2005)

R. Paul, *Robot Manipulators: Mathematics, Programming and Control* (MIT Press, Cambridge, MA, 1981)

G. Pistone, E. Riccomagno, H. Wynn, *Algebraic Statistics: Computational Commutative Algebra in Statistics*. Monographs on Statistics and Applied Probability, vol. 89 (Chapman and Hall, Boca Raton, FL, 2001)

L. Robbiano, On the theory of graded structures. J. Symb. Comp. **2**, 139–170 (1986)

F. Rouillier, Solving zero-dimensional systems through the rational univariate representation. Appl. Algebra Eng. Comm. Comput. **5**, 433–461 (1999)

P. Schauenburg, A Gröbner-based treatment of elimination theory for affine varieties. J. Symb. Comput. **42**, 859–870 (2007)

A. Seidenberg, Constructions in algebra. Trans. Am. Math. Soc. **197**, 273–313 (1974)

A. Seidenberg, On the Lasker–Noether decomposition theorem. Am. J. Math. **106**, 611–638 (1984)

J.G. Semple, L. Roth, *Introduction to Algebraic Geometry* (Clarendon Press, Oxford, 1949)

I.R. Shafarevich, *Basic Algebraic Geometry 1, 2*, 3rd edn. (Springer, New York-Berlin-Heidelberg, 2013)

L. Smith, *Polynomial Invariants of Finite Groups* (A K Peters, Wellesley, MA, 1995)

W. Stein et al., *Sage Mathematics Software, version 6.3*. The Sage Development Team (2014), available at http://www.sagemath.org

B. Sturmfels, Computing final polynomials and final syzygies using Buchberger's Gröbner bases method. Results Math. **15**, 351–360 (1989)

B. Sturmfels, *Gröbner Bases and Convex Polytopes*. University Lecture Series, vol. 8 (American Mathematical Society, Providence, RI, 1996)

B. Sturmfels, *Algorithms in Invariant Theory*, 2nd edn. Texts and Monographs in Symbolic Computation (Springer, New York-Vienna, 2008)

I. Swanson, On the embedded primes of the Mayr-Meyer ideals. J. Algebra **275**, 143–190 (2004)

C. Traverso, Hilbert functions and the Buchberger algorithm. J. Symb. Comput. **22**, 355–376 (1997)

P. Ullrich, Closed-form formulas for projecting constructible sets in the theory of algebraically closed fields. ACM Commun. Comput. Algebra **40**, 45–48 (2006)

B. van der Waerden, *Moderne Algebra, Volume II* (Springer, Berlin, 1931). English translations, *Modern Algebra, Volume II* (F. Ungar Publishing, New York, 1950); *Algebra, Volume 2* (F. Ungar Publishing, New York, 1970); and *Algebra, Volume II* (Springer, New York-Berlin-Heidelberg, 1991). The chapter on Elimination Theory is included in the first three German editions and the 1950 English translation, but all later editions (German and English) omit this chapter

R. Walker, *Algebraic Curves* (Princeton University Press, Princeton, 1950). Reprinted by Dover, 1962

D. Wang, *Elimination Methods*, Texts and Monographs in Symbolic Computation (Springer, Vienna, 2001)

V. Weispfenning, Comprehensive Gröbner bases. J. Symb. Comput. **14**, 1–29 (1992)

F. Winkler, *On the complexity of the Gröbner bases algorithm over $K[x, y, z]$*, in *EUROSAM 1984*, ed. by J. Fitch. Lecture Notes in Computer Science, vol. 174 (Springer, New York-Berlin-Heidelberg, 1984), pp. 184–194

W.-T. Wu, *On the decision problem and the mechanization of theorem-proving in elementary geometry*, in *Automated Theorem Proving: After 25 Years*, ed. by W. Bledsoe, D. Loveland. Contemporary Mathematics, vol. 29 (American Mathematical Society, Providence, Rhode Island, 1983), pp. 213–234

W.-T. Wu, *Mathematics Mechanization: Mechanical Geometry Theorem-Proving, Mechanical Geometry Problem-Solving and Polynomial Equations-Solving* (Kluwer, Dordrecht, 2001)

Index

Abbott, J., 611, 627
Adams, W., 218, 624, 626, 627
admissible geometric theorem, 322
affine cone over a projective variety, *see* cone, affine
affine Dimension Theorem, *see* Theorem, Affine Dimension
affine Hilbert function, *see* Hilbert function, affine
affine Hilbert polynomial, *see* polynomial, affine Hilbert
affine space, *see* space, affine
affine transformation, *see* transformation, affine
affine variety, *see* variety, affine
Agnesi, M., 24
Aholt, C., 623, 627
algebra over a field, 277
  finitely generated, 277
  homomorphism, 278
  reduced, 278
algebraic statistics, 625–626
algebraically independent, 306, 327–329, 333, 337, 507–511, 513
algorithm
  algebra (subring) membership, 349, 369
  associated primes, 231
  Buchberger's, 80, 90–97, 109–119, 413, 539–543, 548, 557–560
  Closure Theorem, 225
  computation in $k[x_1, \ldots, x_n]/I$, 251
  ComputeM (F_4 SymbolicPreprocessing), 571
  consistency, 179
  degree by degree Buchberger (homogeneous ideals), 542

dimension (affine variety), 491
dimension (projective variety), 493
division in $k[x]$, 38–40, 54, 171, 241, 284, 288, 335
division in $k[x_1, \ldots, x_n]$, 33, 62–70, 248, 255, 280, 315, 335, 349, 355, 413, 569, 582, 583
  Euclidean, 42–44, 95, 96, 161, 170, 171, 187, 462
F_4, ix, 567–576, 590, 617
F_5, ix, 576, 580, 581, 585, 589
FGLM, 564, 590, 621
finiteness of solutions, 251, 606
Gaussian elimination (row reduction), 9, 54, 94, 166, 548, 549, 567, 568, 570, 572, 575
greatest common divisor, 41, 44, 187, 196
Gröbner walk, 564, 590, 608, 623
Hilbert driven Buchberger, ix, 550–567, 590
HPS (Hilbert-Poincaré series), 556–558
ideal equality, 94
ideal intersection, 193–195
ideal membership, 97, 151, 184
ideal quotient, 205
improved Buchberger, 109–119
irreducibility, 218
least common multiple, 196
Matrix F_5, 589
polynomial implicitization, 135
primality, 218
primary decomposition, 231
projective closure, 419
projective elimination, 429
pseudodivision, 335–336
radical generators, 184

© Springer International Publishing Switzerland 2015 635
D.A. Cox et al., *Ideals, Varieties, and Algorithms*, Undergraduate Texts in Mathematics, DOI 10.1007/978-3-319-16721-3

radical ideal, 184
radical membership, 184, 185, 325
rational implicitization, 139
regular s-reduction, 583, 591
resultant, 172
Ritt's decomposition, 337, 342, 622
saturation, 206
signature-based Gröbner basis, 576–591
tangent cone, 528
altitude, 322, 331
Anderson, D., 140, 622, 627
Arnold, E., 626, 627
ascending chain condition (ACC), 80, 82,
        92, 115, 202, 212, 409, 478, 572
Aschenbrenner, M., 620, 627
Åström, K., 623, 630
Atiyah, M.F., 230, 627
Aubry, P., 622, 627
automatic geometric theorem proving,
        319–343, 624
automorphism of a variety, 264

Bézier, P., 21
    cubic, see cubic, Bézier
Baillieul, J., 314, 627
Ball, A.A., 27, 627
Barrow, I., 24
basis
    minimal, 36, 74, 93
    minimal Gröbner, see Gröbner basis,
        minimal
    of an ideal, see ideal, basis for
    reduced Gröbner, see Gröbner basis,
        reduced
    standard, 78
Bayer, D., x, 76, 117, 128, 540, 620, 626,
        627
Becker, T., 60, 83, 84, 116, 194, 218, 309,
        543, 624, 628
Benson, C.T., 357, 623, 628
Bernoulli, J., 24
Bezout's Theorem, see Theorem, Bezout
Bigatti, A., 557, 611, 627, 628
bihomogeneous polynomial, see polyno-
        mial, bihomogeneous
birationally equivalent varieties, 273–276,
        302, 512, 514
blow-up, 536–538
Brieskorn, E., 464, 467, 621, 628
Bruce, J.W., 143, 148, 152, 622, 628
Buchberger's Criterion, 86–88, 91, 92,
        104–113, 568, 569, 573, 576, 584
Buchberger, B., vii, 79, 116, 248, 314, 628

Caboara, M., 117, 628
Canny, J., 140, 628
centroid, 331–333, 343
chain
    ascending, of ideals, 79, 80
    descending, of varieties, 82, 212, 226,
        409
characteristic of a field, see field
characteristic sets, 337, 342, 622
Chou, S.-C., 335, 338, 342, 343, 622, 628
circumcenter, 332
cissoid, see curve, cissoid of Diocles
classification of varieties, 238, 260, 275
Classification Theorem for Quadrics, see
        Theorem, Normal Form, for Quadrics
Clemens, H., 464, 628
closure
    projective, 417–422, 431, 432, 439, 502,
        505, 506, 536
    Zariski, 131, 199–205, 208, 209, 211,
        216, 219, 226, 282, 287, 531, 532, 537
Closure Theorem, 108, see Theorem,
        Closure
CoCoA, see computer algebra systems,
        CoCoA
coefficient, 2
cofactor, 171, 280, 549, 597
Cohen, A., 620, 628
Collart, S., 564, 624, 628
collinear, 321–323, 331, 332
colon ideal, see ideal, quotient
commutative ring, see ring, commutative
complexity, 116–119, 255, 620
comprehensive Gröbner basis, see Gröbner
        basis, comprehensive
computer aided geometric design (CAGD),
        20–22
computer algebra systems
    CoCoA, 184, 231, 495, 611–612, 624,
        627
    FGb package, 617
    gfan, 623
    Macaulay, 540
    Macaulay2, 184, 231, 495, 613–614, 623,
        624, 630
    Magma, 590, 617, 624
    Maple, 231, 495, 560, 590, 604–606, 621,
        624, 631
    Mathematica, 607–608, 624, 631
    MATLAB, 617
    REDUCE, 617
    Sage, 495, 608–610, 623, 624, 632
    Singular, 184, 231, 495, 590, 614–616,
        623, 624, 626, 628

cone
    affine, 405, 411, 412, 494, 498, 527, 536
    projectivized tangent, 537–538
    tangent, 520, 527–534
configuration space, *see* space, configuration
    (of a robot)
congruence (mod *I*), 240
conic section, 6, 27, 142, 437, 439
consistency question, 11, 46, 179, 185
constructible set, 131, 226–228, 309
control
    points, 21, 22, 28
    polygon, 21, 28
coordinate ring of a variety, *see* ring,
    coordinate (*k*[*V*])
coordinates
    homogeneous, 385, 388, 396
    Plücker, 445–447
coset, 382, 383
Cox, D., 127, 170, 254, 462, 564, 566, 581,
    620–625, 628
Coxeter, H.S.M., 357, 628
Cramer's Rule, 165, 166, 171, 428, 597
criterion
    rewriting, 585, 589
    sygygy, 585
cross ratio, 331
cube, 25–26, 357, 362–363, 372, 382, 623
cubic
    Bézier, 21, 27
cubic, twisted, *see* curve, twisted cubic
curve
    cissoid of Diocles, 25–26
    dual, 383, 464, 621
    family of, 147
    folium of Descartes, 142
    four-leaved rose, 12, 154
    rational normal, 420–421, 625
    strophoid, 24
    twisted cubic, 8, 19–20, 31, 33, 36, 69,
        88, 181, 207, 264, 266, 267, 400,
        402–404, 406, 415–418, 420, 490,
        505–507
cuspidal edge, 267
Cuypers, H., 620, 628

Davenport, J.H., 196, 628
Decker, W., 615, 626, 628
decomposition
    minimal primary, of an ideal, 229–230
    minimal, of a variety, 215–216, 454, 473
    minimal, of an ideal, 216
    primary, 229–231, 606, 610, 612, 614,
        616

degenerate case of a geometric configura-
    tion, 324, 326, 327, 329, 334, 340,
    466
degeneration, 467
degree
    of a pair, 540, 569
    of a projective variety, 505–506
    total, of a monomial, 2
    total, of a polynomial, 2
    transcendence of a field extension, 513
    transcendence, of a field extension, 514
    weighted, of a monomial, 433, 436, 566
dehomogenization
    of a polynomial, 399, 528, 543, 545, 590
derivative, formal, 47, 173, 247, 516, 523
Derksen, H., 369, 370, 623, 629
descending chain condition (DCC), 82, 212,
    226, 229, 409
desingularization, 538
determinant, 118, 163, 279, 312, 428, 445,
    525, 596–597
    trick, 280
    Vandermonde, 46
Dickenstein, A., 127, 140, 625, 626, 629
Dickson's Lemma, 72, 73, 75, 77
difference of varieties, 199, 425
dimension, ix, 3–11, 101, 117, 145, 249–
    252, 254–258, 260, 275, 287, 311,
    318, 469–473, 477–514, 518–523,
    533
    at a point, 520, 533
    question, 11, 486–525
Dimension Theorem, *see* Theorem,
    Dimension
discriminant, 173, 353
division algorithm, *see* algorithm, division
    in *k*[*x*] or *k*[*x*₁, . . . , *xₙ*]
dodecahedron, 363
dominating map, *see* mapping, dominating
Draisma, J., 622, 623, 629
Drton, M., 625, 629
dual
    curve, *see* curve, dual
    projective plane, 395, 436
    projective space, 405, 445
    variety, 383
duality
    of polyhedra, 363
    projective principle of, 387
Dubé, T., 116, 629
Dummit, D., 171, 173, 285, 352, 513,
    595–597, 629

echelon matrix, 51–52, 79, 95–96, 447, 450, 547–549, 568–570, 572, 573
Eder, C., 577, 585, 587, 589, 629
Eisenbud, D., 60, 184, 218, 624, 626, 629
elimination ideal, *see* ideal, elimination
elimination order, *see* monomial ordering, elimination
elimination step, 122
Elimination Theorem, *see* Theorem, Elimination
elimination theory, 17
   projective, 422
Emiris, I., 127, 140, 625, 626, 629
Enneper surface, *see* surface, Enneper
envelope, 147–152, 154–155, 621
equivalence
   birational, 273–276, 302, 512
   projective, 437–444, 449, 451
error control coding theory, 625
Euclidean distance degree, 622
Euler line, 332, 333, 343
Euler's formula, 405
evolute, 622
extension step, 122
Extension Theorem, 108, *see* Theorem, Extension

F_4 algorithm, *see* algorithm, F_4
F_5 algorithm, *see* algorithm, F_5
factorization of polynomials, 47, 83, 186, 188, 193, 195, 236, 239, 276, 453, 456, 458, 461, 463, 594, 595
family of curves, *see* curve, family of
Farr, J., 625, 629
Faugère, J.-C., ix, 564, 567, 574–577, 585, 587, 589, 621, 629
Feiner, S., 623, 629
Fermat's Last Theorem, *see* Theorem, Fermat's Last
FGLM algorithm, *see* algorithm, FGLM
fiber of a mapping, 238, 281–283, 286–289, 511
field, 1, 593
   algebraically closed, 5, 34, 132, 159–160, 164, 168–170, 176–184, 199, 202–204, 207, 210–211, 405, 410–412, 414, 419, 426–427, 431–434, 442–444, 450, 460, 491, 493, 498–503, 509–511, 524, 528
   finite, 1, 5, 36
   infinite, 3–4, 36, 134, 138, 208, 284, 287, 408, 409, 507, 533
   of characteristic zero, 188, 355, 364, 365, 368, 373, 377, 615

of finite (positive) characteristic, 188
of fractions, 268, 511, 614
of rational functions, 1, 15, 306, 336, 456, 612, 616
of rational functions on V ($k(V)$), 268–275, 413, 511–513
final remainder (in Wu's Method), 340
finite generation of invariants, 361, 368
finite morphism, 287, 289
finiteness question, 11, 251–255, 606
Finiteness Theorem, *see* Theorem, Finiteness
Fischer, G., 464, 621, 629
Foley, J., 623, 629
folium of Descartes, *see* curve, folium of Descartes
follows generically from, 328
follows strictly from, 325
Foote, R., 171, 173, 285, 352, 513, 595–597, 629
forward kinematic problem, *see* kinematic problem of robotics, forward
Fulton, W., 462, 464, 629
function
   algebraic, 129
   coordinate, 258
   defined by a polynomial, 3–4, 234–238, 257, 262–264, 507
   identity, xvi, 260, 270–273
   polynomial, 479, 487
   rational, 15, 136, 167, 268–275, 462
function field, *see* field, of rational functions on V ($k(V)$)
Fundamental Theorem of Algebra, *see* Theorem, Fundamental of Algebra
Fundamental Theorem of Symmetric Polynomials, *see* Theorem, Fundamental, of Symmetric Polynomials

Gallet, M., 174, 622, 629
Gao, S., 625, 629
von zur Gathen, J., 40, 43, 46, 117, 127, 595, 629
Gauss, C.F., 348, 629
Gaussian elimination, *see* algorithm, Gaussian elimination (row reduction)
Gebauer, R., 116, 629
Gelfand, I., 170, 630
genomics, 625
Geometric Extension Theorem, *see* Theorem, Geometric Extension
Gerhard, J., 40, 43, 46, 117, 127, 595, 629
gfan, *see* computer algebra systems, gfan
Gianni, P., 184, 218, 564, 621, 629, 630

Giblin, P.J., 143, 148, 152, 622, 628
Giovini, A., 116, 630
Giusti, M., 117, 630
$GL(n, k)$, *see* group, general linear
Glebsky, L., ix, 177, 630
Goldman, R., 27, 140, 622, 627, 630
graded lexicographic order, *see* monomial
    ordering
graded reverse lexicographic order, *see*
    monomial ordering
gradient, 10, 145–146, 149
graph
  coloring, 626
  of a function, 6, 134, 260, 264, 401, 434,
    444, 523
Grassmannian, 447
Grayson, D., 613, 630
greatest common divisor (gcd), 41–45,
    186–187, 196
Greuel, G.-M., 76, 184, 615, 628, 630
Griffiths, P., 464, 630
Gritzmann, P., 117, 630
Gröbner basis, 45, 78–128, 135–136,
    139–141, 149–160, 194–195, 205,
    237–238, 244, 248–255, 269, 280,
    305–310, 325–331, 342, 349–351,
    369–370, 375, 376, 407, 410–411,
    416–419, 430–433, 490, 491,
    527–529, 539–591, 604–617
  and linear algebra, 546–548, 567
  below signature M, 584
  comprehensive, 309, 624
  conversion, 551, 563, 590, 621, 624
  criteria for, 86–88, 111–113, 568, 584
  dehomogenized, 544
  homogeneous, 540–543, 590
  minimal, 92
  module, 624
  reduced, 93–94, 116, 179, 185, 221–225,
    407, 544, 551
  signature, 584
  specialization of, ix, 160, 220, 306–310,
    315–316, 624
  universal, 623
Gröbner cover, ix, 309–310, 331
Gröbner fan, 623
Gröbner, W., 79
group, 448, 595
  alternating, 363
  cyclic, 356
  finite matrix, 356
  general linear ($GL(n, k)$), 356
  generators for, 359–360

Klein four-, 360
  of rotations of cube, 362
  of rotations of tetrahedron, 363
  orbit of a point, 378
  orbit space, 378
  projective linear group ($PGL(n + 1, k)$),
    448
  symmetric, 352, 356–357
Grove, L.C., 357, 623, 628

Harris, J., 447, 630
Hartshorne, R., 211, 287, 289, 630
Heintz, J., 117, 630
Hermann, G., 184, 218, 630
Heyden, A., 623, 630
Hilbert Basis Theorem, *see* Theorem,
    Hilbert Basis
Hilbert driven Buchberger algorithm, *see*
    algorithm, Hilbert driven Buchberger
Hilbert function, 486–498, 552
  affine, 487–491, 496
Hilbert polynomial, *see* polynomial, Hilbert,
    610, 612, 614, 616
  affine, *see* polynomial, affine Hilbert
Hilbert, D., 370, 473, 630
Hilbert-Poincaré series, 552–564, 610, 612,
    614, 616
Hilmar, J., 462, 630
Hironaka, H., 79, 630
Hodge, W.V.D., 447, 630
homogeneous coordinates, *see* coordinates,
    homogeneous
homogeneous ideal, *see* ideal, homogeneous
homogeneous polynomial, *see* polynomial,
    homogeneous
homogenization
  of a polynomial, 181, 400, 495, 543, 562,
    590
  of an ideal, 415–419, 494
  $(x_0, \ldots, x_n)$- of an ideal, 431–432
homomorphism, *see* ring, homomorphism
Horobeţ, E., 622, 623, 629
Hughes, J., 623, 629
Huneke, C., 184, 218, 629
hyperboloid, *see* surface, hyperboloid of one
    sheet
hyperplane, 399, 405, 438
  at infinity, 397, 418
hypersurface, 399
  cubic, 399
  dimension of, 498
  nonsingular quadric, 442–444, 446
  quadric, 436–451
  quartic, 399

quintic, 399
    tangent cone of, 527

icosahedron, 363
ideal, 29
    basis for, 31, 35
    binomial, 257
    colon, *see* ideal, quotient
    complete intersection, 505
    determinantal, 118, 421, 549
    elimination, 122–127, 425–434, 610, 612, 616
    generated by a set of polynomials, 29
    Gröbner basis for, *see* Gröbner basis
    homogeneous, 407–408, 540–564
    homogenization, 415–419, 494
    homogenization of, 543
    in a ring, 244, 594
    intersection, 192, 413, 606, 612, 614, 616
    irreducible, 229
    maximal, 209
    Maximum Principle, *see* Maximum Principle for Ideals
    monomial, 70–74, 469–483, 487–493, 552–557
    of a variety ($\mathbf{I}(V)$), 32–35
    of leading terms ($\langle \mathrm{LT}(I) \rangle$), 76
    of relations, 373
    P-primary, 229
    primary, 228–231
    prime, 207–211, 216–218, 228–229, 373, 414
    principal, 41, 42, 82, 176, 245, 432
    product, 191, 413, 614
    projective elimination, 425–434
    proper, 209–211
    quotient, 200, 425, 606, 612, 614
    radical, 36, 182–184, 196, 216, 244, 253, 259
    radical of ($\sqrt{I}$), 182, 409, 614
    saturation, 202–205, 411, 425–429, 435, 606, 612, 614, 616
    sum of, 189, 413
    syzygy, 373
    weighted homogeneous, 433, 436, 566
ideal description question, 35, 49, 73, 77
ideal membership question, 35, 45, 49, 84, 97–98, 151, 184, 620
ideal–variety correspondence, 232
    affine, 183, 408
    on V, 259
    projective, 408
Implicit Function Theorem, *see* Theorem, Implicit Function

implicit representation, 16
implicitization, 17, 133–140, 551, 562–564
    via resultants, 170, 622
improved Buchberger algorithm, *see* algorithm, improved Buchberger
Inclusion-Exclusion Principle, 480, 484
index of regularity, 489
infinite descent, 276
inflection point, *see* point, of inflection
integer polynomial, *see* polynomial, integer
integral domain, *see* ring, integral domain
integral element over a subring, 280, 381
invariance under a group, 360
invariant polynomial, *see* polynomial, invariant
inverse kinematic problem, *see* kinematic problem of robotics, inverse
irreducibility question, 218
irreducible
    components of a variety, *see* variety, irreducible components of
    ideal, *see* ideal, irreducible
    polynomial, *see* polynomial, irreducible
    variety, *see* variety, irreducible
irredundant
    intersection of ideals, 216
    primary decomposition, 229
    union of varieties, 215
isomorphic
    rings, 243
    varieties, 238, 260–265, 509
Isomorphism Theorem, *see* Theorem, Isomorphism
isotropy subgroup, 382

Jacobian matrix, *see* matrix, Jacobian
Jensen, A., 623
Jin, M., 622, 631
joint space, *see* space, joint (of a robot)
joints (of robots)
    ball, 296
    helical ("screw"), 296
    prismatic, 292
    revolute, 292
    "spin", 304
Jounanolou, J., 170, 631
Jukes-Cantor models, 625

Kalkbrener, M., 139, 564, 624, 628, 631
Kapranov, M., 170, 630
Kehrein, A., 625, 626
Kemper, G., 369, 370, 623, 629
Kendig, K., 504, 522, 523, 631

kinematic problem of robotics
    forward, 294, 297–302
    inverse, 294, 304–314
kinematic redundancy, 318
kinematic singularity, 311–314
Kirwan, F., x, 462, 464, 631
Klein four-group, *see* group, Klein four-
Klein, F., 357, 631
Knörrer, H., 464, 467, 621, 628
Koh, J., 620, 631
Koszul syzygy, *see* syzygy, Koszul
Kreuzer, M., 60, 76, 543, 624–626, 631
Krick, T., 184, 631

Lagorio, G., 611, 627
Lagrange interpolation polynomial, *see*
    polynomial, Lagrange interpolation
Lagrange multipliers, 9–10, 99
Lasker-Noether Theorem, *see* Theorem,
    Lasker-Noether
Lazard, D., 117, 127, 546, 564, 621, 622,
    627, 629, 631
leading coefficient, 60
leading monomial, 60
leading term, 38, 60
    in a vector (signature), 578, 581
leading terms, ideal of, *see* ideal, of leading
    terms ($\langle \mathrm{LT}(I) \rangle$)
least common multiple (LCM), 84, 196
Lejeune-Jalabert, M., x, 631
lexicographic order, *see* monomial ordering
Li, X., 622, 631
line
    affine, 3, 389
    at infinity, 389, 453
    limit of, 530
    projective, 389, 399, 402, 436, 444, 445
    secant, 529–533
    tangent, 143, 145–146
Little, J., 127, 170, 254, 462, 564, 566, 581,
    620–625, 628
local property, 462, 515
locally constant, 462
Logar, A., 184, 631
Loustaunau, P., 218, 624, 626, 627
Lucas, S., 626, 627

Macaulay, F., 170, 631
Macaulay2, *see* computer algebra systems,
    Macaulay2
MacDonald, I.G., 230, 627
Magma, *see* computer algebra systems,
    Magma
Mall, D., 564, 624, 628

manifold, 523
Manocha, D., 127, 140, 622, 628, 631
Maple, *see* computer algebra systems,
    Maple
mapping
    dominating, 514
    identity, 260, 270–273
    polynomial, 234
    projection, 129, 234, 282, 286
    pullback, 262, 274
    rational, 269–274, 302
    regular, 234
    Segre, 421, 443, 444, 449
    stereographic projection, 23, 275
Marinari, M., 621, 631
Mathematica, *see* computer algebra systems,
    Mathematica
matrix
    echelon, *see* echelon matrix
    group, 355–357
    Jacobian, 310–312, 522
    permutation, 356
    row reduced echelon, *see* echelon matrix
    Sylvester, 162–163
Matsumura, H., 533, 631
Maximum Principle for Ideals, 223, 226
Mayr, E., 117, 632
Meyer, A., 117, 632
Mines, R., 184, 218, 595, 632
minimal basis, *see* basis, minimal
Mishra, B., 342, 632
mixed order, *see* monomial ordering
module, 566, 580, 589, 591, 624
Möller, H., 116, 621, 625, 629, 631,
    632
Molien's Theorem, *see* Theorem, Molien's
monic polynomial, *see* polynomial, monic
monomial, 2
monomial ideal, *see* ideal, monomial
monomial ordering, 54–60, 73–74, 407,
    604, 607, 609, 611, 613, 615
    elimination, 76, 123, 128, 281, 562
    graded, 416, 488, 489, 491, 541, 543,
        551, 574
    graded lexicographic (grlex), 58
    graded reverse lexicographic (grevlex),
        58, 117
    inverse lexicographic (invlex), 61
    lexicographic (lex), 56–58, 98–101,
        122–123, 135, 139, 156–160, 177,
        194, 205, 308, 347–348, 551, 574
    mixed, 75
    product, 75, 308, 309
    weight, 75–76

Montes, A., ix, 309, 331, 624, 632
Mora, F., 116, 632
Mora, T., 116, 564, 621, 629–631
Moreno Maza, M., 622, 627
Mourrain, B., 625
multidegree (multideg), 60
multinomial coefficient, 371
multiplicity
    of a root, 47, 144
    of a singular point, 153
    of intersection, 144–146, 458–462
Mumford, D., 117, 522, 536, 627, 632

Newton identities, 352, 354–355
Newton polygon, 621
Newton–Gregory interpolating polynomial,
        see polynomial, Newton–Gregory
        interpolating
Newton-Raphson root finding, 127
Niesi, G., 116, 630
nilpotent, 244, 278
Noether normalization, 284–287, 510
Noether Normalization Theorem, see
        Theorem, Noether Normalization
Noether's Theorem, see Theorem, Noether's
Noether, E., 366
nonsingular
    point, see point, nonsingular
    quadric, see quadric hypersurface,
        nonsingular
normal form, 83
Normal Form for Quadrics, see Theorem,
        Normal Form, for Quadrics
normal pair selection strategy, 116
Nullstellensatz, 34, 35, 108, 131, 184, 203,
        221, 252, 254, 259, 329, 419, 519, 529
    Hilbert's, 5, 179, 181, 183, 184, 200, 333
    in $k[V]$, 259
    Projective Strong, 411–412, 494
    Projective Weak, 410–411, 427, 454, 502
    Strong, 183, 325, 410, 412, 491
    Weak, ix, 176–180, 210, 211, 252, 260,
        410
numerical solutions, 127, 255

O'Shea, D., 127, 170, 254, 462, 564, 566,
        581, 620–625, 628
octahedron, 363
operational space (of a robot), see space,
        configuration (of a robot)
orbit
    G-, 378, 381
    of a point, 378
    space, 378, 381

order, see monomial ordering
order (of a group), 356
orthocenter, 331
Ottaviani, G., 622, 623, 629

Pachter, L., 625, 632
Pappus's Theorem, see Theorem, Pappus's
parametric representation, 14–17, 233
    polynomial, 16, 134, 563
    rational, 15, 138, 266
parametrization, 15
partial solution, 123, 125, 129
partition, 240, 309
path connected, 462, 465
Paul, R., 314, 632
Pedoe, D., 447, 630
pencil
    of curves, 406
    of hypersurfaces, 406
    of lines, 395
    of surfaces, 260
    of varieties, 260
permutation, 163, 346
    sign of, 163, 596, 597
Perry, J., 117, 628
perspective, 386, 390, 392, 395
Pfister, G., 76, 184, 615, 626, 628, 630
$PGL(n + 1, k)$, see group, projective linear
        group ($PGL(n + 1, k)$)
Pistone, G., 625, 632
plane
    affine, 3
    Euclidean, 319
    projective, 385–392, 451
Plücker coordinates, see coordinates,
        Plücker
point
    critical, 102, 622
    Fermat, of a triangle, 334
    nonsingular, 146, 152, 462, 519, 520,
        522–523, 525
    of inflection, 153
    singular, 8, 143–146, 151–153, 155, 402,
        449, 519–522, 525–526, 533
    vanishing, 386
polyhedron
    duality, 363
    regular, 357, 363
polynomial
    affine Hilbert, 489–491, 494–496, 505
    bihomogeneous, 433–434, 436
    elementary symmetric, 346–348
    Hilbert, 493–495, 606
    homogeneous, 351

homogeneous component of, 351
integer, 163
invariant, 358
irreducible, 185, 594–595
Lagrange interpolation, 97
linear part, 516
monic, 583
Newton–Gregory interpolating, 485
partially homogeneous, 424
reduced, 47
S-, 85–88, 90–92, 104, 109–116, 194,
    220, 257, 315, 413, 539–543,
    546–548, 558–564, 567, 569, 573,
    576–580, 606, 610
square-free, 47
symmetric, 346
weighted homogeneous, 433, 436, 566
polynomial map, see mapping, polynomial
PostScript, 22
power sums, 351–352, 366–368
primality question, 218
primary decomposition, see decomposition,
    primary
primary decomposition question, 231
primary ideal, see ideal, primary
prime ideal, see ideal, prime
principal ideal domain (PID), 41, 176, 245
product order, see monomial ordering
projection mapping, see mapping, projection
projective closure, see closure, projective
projective elimination ideal, see ideal,
    projective elimination
projective equivalence, see equivalence,
    projective
projective line, see line, projective
projective plane, see plane, projective
projective space, see space, projective
projective variety, see variety, projective
projectivized tangent cone, see cone,
    projectivized tangent
pseudocode, 38, 599–602
pseudodivision, see algorithm, pseudodivi-
    sion
    successive, 337
pseudoquotient, 336
pseudoremainder, 336
Puiseux expansions, 621
pullback mapping, see mapping, pullback
pyramid of rays, 390
Python (programming language), 609

quadric hypersurface, 399, 436–451
    nonsingular, 442–444, 446
    over $\mathbb{R}$, 442, 448

rank of, 440
    singular, 448
quotient
    field, see field, of fractions
    ideal, see ideal, quotient
    vector space, 486
quotients on division, 62–66, 83, 413, 605,
    607, 610

radical
    generators of, 184
    ideal, see ideal, radical
    membership, see algorithm, radical
        membership
    of an ideal, see ideal, radical of
Rakhooy, H., 174, 622, 629
rank
    deficient, 311
    maximal, 311
    of a matrix, 9, 311, 441
    of a quadric, 440, 441
rational
    function, see function, rational
    mapping, see mapping, rational
    univariate representation, 621
    variety, see variety, rational
real projective plane, 385, 388–390
Recio, T., 331, 624, 632
recursive, 225, 556
REDUCE, see computer algebra systems,
    REDUCE
reduced Gröbner basis, see Gröbner basis,
    reduced
regular sequence, 504, 589
regularity, index of, see index of regularity
Relative Finiteness Theorem, see Theorem,
    Relative Finiteness
remainder on division, 38, 62–68, 70, 83–84,
    86, 91–92, 94, 98, 104, 107, 109, 172,
    248–251, 269, 349–350, 369–370,
    376, 413, 541, 546, 558–564, 567,
    569, 577–580, 605, 607, 610, 612,
    613, 616
representation
    lcm, 107, 221
    standard, 104, 549, 569, 573, 576
resultant, 161–168, 451, 454–460, 462, 622
    multipolynomial, 140, 170, 622
Reynolds operator, 365–366
Riccomagno, E., 625, 632
Richman F., 184, 218, 595, 632
Riemann sphere, 397, 403
ring
    commutative, 3, 236, 242, 359, 593

coordinate ($k[V]$), 257–269, 277–278, 286, 287, 359, 377–378, 381, 507–509, 594
  finite over a subring, 279, 381
  finitely generated over a subring, 288
  homomorphism, 243, 374, 508, 544
  integral domain, 236–237, 258, 268, 511, 594
  isomorphism, 243, 244, 247, 258, 264–265, 270, 273–274, 285–287, 374, 376–378, 507, 509, 512
  of invariants ($k[x_1, \ldots, x_n]^G$), 358, 359
  polynomial ring ($k[x_1, \ldots, x_n]$), 3
  quotient ($k[x_1, \ldots, x_n]/I$), 240–247, 374, 383, 505, 594, 621, 625
Robbiano, L., 60, 76, 116, 543, 611, 624–626, 630–632
robotics, 10–11, 291–314
Roth, L., 447, 632
Rouillier, F., 621, 632
row reduced echelon matrix, *see* echelon matrix
Ruitenburg, W., 184, 218, 595, 632

Sage, *see* computer algebra systems, Sage
saturation, *see* ideal, saturation
Schönemann, H., 615, 628
Schauenburg, P., ix, 157, 224, 227, 632
secant line, *see* line, secant
Sederberg, T., 140, 622, 627
Segre map, *see* mapping, Segre
Segre variety, *see* variety, Segre
Seidenberg, A., 184, 218, 632
Semple, J.G., 447, 632
Shafarevich, I.R., 504, 521, 522, 632
sign of a permutation, *see* permutation, sign of
signature Gröbner basis, *see* Gröbner basis, signature
signature of a vector, 580, 581
Singular, *see* computer algebra systems, Singular
singular locus, 521, 524, 525
singular point, *see* point, singular
singular quadric, *see* quadric hypersurface, singular
Siret, Y., 196, 628
Smith, L., 370, 632
Smyth, C., 462, 630
solving polynomial equations, 49, 98, 255, 620
S-polynomial, *see* polynomial, S-
space
  affine, 3

configuration (of a robot), 294
joint (of a robot), 294
orbit, 378, 381
projective, 388–392, 396
quotient vector, 486
tangent, 516–521
specialization of Gröbner basis, *see* Gröbner basis, specialization of
s-reduction (signature reduction), 581–582
s-reduced vector, 583
stabilizer, 382
standard basis, *see* basis, standard
Stein, W., 608, 632
stereographic projection, *see* mapping, stereographic projection
Sterk, H., 620, 628
Stetter, H., 625
Stillman, M., 76, 117, 128, 540, 613, 620, 627, 630
Sturmfels, B., x, 117, 330, 369, 370, 372, 622, 623, 625, 627, 629, 630, 632
subalgebra (subring), 287, 349, 359, 369
subdeterminants, 421, 549
subgroup, 596
subvariety, 258
Sudoku, 626
sugar, 116
Sullivant, S., 625, 629
surface
  Enneper, 141
  hyperboloid of one sheet, 270
  quadric, 237, 421, 443, 448
  ruled, 103, 444
  tangent, 19–20, 101, 133, 135–136, 233–234, 535
  Veronese, 239, 421, 433
  Whitney umbrella, 141
S-vector, 583
Swanson, I., 620, 633
Sylvester matrix, *see* matrix, Sylvester
symmetric group, *see* group, symmetric
symmetric polynomial, *see* polynomial, symmetric
syzygy, 580
  Koszul, 581, 586
  on leading terms, 110

Taalman, L., 626, 627
tangent cone, *see* cone, tangent
tangent line to a curve, *see* line, tangent
tangent space to a variety, *see* space, tangent
Taylor series, 552
Taylor's formula, 517, 534
term in a vector of polynomials, 578

tetrahedron, 363
Theorem
    Affine Dimension, 491
    Bezout's, x, 451–467
    Circle, of Apollonius, 322, 330, 339–342
    Closure, 131–132, 140, 142, 199–200,
        219–228, 282, 289, 434, 509, 510
    Dimension, 493
    Elimination, 122–124, 126, 128,
        135–136, 150, 152, 194, 423, 430
    Extension, 125–127, 132, 136–170, 379,
        422
    Fermat's Last, 13
    Finiteness, 251, 278, 281, 283
    Fundamental, of Algebra, 4, 178
    Fundamental, of Symmetric Polynomials,
        347–351
    Geometric Extension, 130–131, 422–423,
        432
    Geometric Noether Normalization,
        286–287
    Geometric Relative Finiteness, 282–283
    Hilbert Basis, 31, 77, 80–82, 175, 176,
        216, 218, 245, 368, 371, 407, 431
    Implicit Function, 318, 523
    Intermediate Value, 462, 465
    Isomorphism, 247, 374, 524
    Lasker-Noether, 229–230
    Molien's, 368, 622
    Noether Normalization, 284–285, 287
    Noether's, 366, 371
    Normal Form, for Quadrics, 439–444
    Pappus's, 332, 333, 343, 393, 466
    Pascal's Mystic Hexagon, 463, 465, 466
    Polynomial Implicitization, 134
    Projective Extension, 426–429, 597
    Rational Implicitization, 138
    Relative Finiteness, 280–281, 283, 511
Thomas, R., 622, 623, 627, 629
Tournier, E., 196, 628
Trager, B., 184, 218, 630
transcendence degree, see degree,
    transcendence of a field extension
transformation
    affine, 303
    projective linear, 437
    quadratic, 621
transposition, 596
Traverso, C., 116, 550, 630, 633
triangular form (system of equations),
    337–339
triangulation problem (computer vision),
    623

twisted cubic
    curve, see curve, twisted cubic
    tangent surface of, see surface, tangent

Ullrich, P., 227, 633
unique factorization of polynomials, 595
uniqueness question in invariant theory, 361,
    373, 376
unirational variety, see variety, unirational

van Dam, A., 623, 629
van der Waerden, B., 170, 633
Vandermonde determinant, see determinant,
    Vandermonde
variety
    affine, 5
    irreducible, 206–218, 237, 239, 258, 267,
        268, 326, 327, 337, 377, 409, 413,
        418, 453, 503
    irreducible components of, 215–227,
        326–328, 409
    linear, 9, 399
    minimum principle, 228
    of an ideal ($\mathbf{V}(I)$), 81, 408
    projective, 398
    rational, 273
    reducible, 236
    Segre, 421–422
    subvariety of, 258
    unirational, 17
    zero-dimensional, 252
Vasconcelos, W., 184, 218, 629
Veronese surface, see surface, Veronese

Walker, R., 458, 462, 633
Wang, D., 342, 622, 631, 633
weight order, see monomial ordering
weighted homogeneous ideal, see ideal,
    weighted homogeneous
weighted homogeneous polynomial, see
    polynomial, weighted homogeneous
Weispfenning, V., 60, 83, 84, 116, 194, 218,
    309, 543, 624, 628, 633
well-ordering, 55–56, 66, 73, 347, 575, 590
Whitney umbrella, see surface, Whitney
    umbrella
Wibmer, M., ix, 309, 624, 632
Wiles, A., 13
Winkler, F., 117, 633
Wu's Method, 335–343, 622
Wu, W.-T., 343, 622, 633
Wynn, H., 625, 632

Zacharias, G., 184, 218, 630
Zafeirakopoulos, Z., 174, 622, 629
Zariski
    closure, *see* closure, Zariski

dense, 510
dense set, 200, 211, 216, 221, 510
Zelevinsky, A., 170, 630
zero divisor in a ring, 500, 504, 594